Lecture Notes in Networks and Systems

Volume 482

The series "Lecture Notes in Networks and Systems" publishes the latest developments in Networks and Systems—quickly, informally and with high quality. Original research reported in proceedings and post-proceedings represents the core of LNNS.

Volumes published in LNNS embrace all aspects and subfields of, as well as new challenges in, Networks and Systems.

The series contains proceedings and edited volumes in systems and networks, spanning the areas of Cyber-Physical Systems, Autonomous Systems, Sensor Networks, Control Systems, Energy Systems, Automotive Systems, Biological Systems, Vehicular Networking and Connected Vehicles, Aerospace Systems, Automation, Manufacturing, Smart Grids, Nonlinear Systems, Power Systems, Robotics, Social Systems, Economic Systems and other. Of particular value to both the contributors and the readership are the short publication timeframe and the world-wide distribution and exposure which enable both a wide and rapid dissemination of research output.

The series covers the theory, applications, and perspectives on the state of the art and future developments relevant to systems and networks, decision making, control, complex processes and related areas, as embedded in the fields of interdisciplinary and applied sciences, engineering, computer science, physics, economics, social, and life sciences, as well as the paradigms and methodologies behind them.

Indexed by SCOPUS, INSPEC, WTI Frankfurt eG, zbMATH, SCImago.

All books published in the series are submitted for consideration in Web of Science.

For proposals from Asia please contact Aninda Bose (aninda.bose@springer.com).

More information about this series at https://link.springer.com/bookseries/15179

Francesco Calabrò · Lucia Della Spina ·
María José Piñeira Mantiñán

Editors

New Metropolitan Perspectives

Post COVID Dynamics: Green and Digital
Transition, between Metropolitan and Return
to Villages Perspectives

Set 1

 Springer

Editors
Francesco Calabrò
Dipartimento PAU
Mediterranea University of Reggio Calabria
Reggio Calabria, Reggio Calabria, Italy

Lucia Della Spina
Mediterranea University of Reggio Calabria
Reggio Calabria, Italy

María José Piñeira Mantiñán
University of Santiago de Compostela
Santiago de Compostela, Spain

ISSN 2367-3370   ISSN 2367-3389   (electronic)
Lecture Notes in Networks and Systems
ISBN 978-3-031-06824-9   ISBN 978-3-031-06825-6   (eBook)
https://doi.org/10.1007/978-3-031-06825-6

This Springer imprint is published by the registered company Springer Nature Switzerland AG
The registered company address is: Gewerbestrasse 11, 6330 Cham, Switzerland

Preface

This volume contains the proceedings for the fifth International "NEW METROPOLITAN PERSPECTIVES. Post COVID Dynamics: Green and Digital Transition, between Metropolitan and Return to Villages' Perspectives", scheduled from May 25–27, 2022, in Reggio Calabria, Italy.

The symposium was promoted by LaborEst (Evaluation and Economic Appraisal Lab) of the PAU Department, Mediterranea University of Reggio Calabria, Italy, in partnership with a qualified international network of academic institution and scientific societies.

The fifth edition of "NEW METROPOLITAN PERSPECTIVES", like the previous ones, aimed to deepen those factors which contribute to increase cities and territories attractiveness, both with theoretical studies and tangible applications.

This fifth edition coincides with what is most likely the end of the COVID pandemic that began in 2020. The global health emergency, despite having been a phenomenon limited in time, has acted as an accelerator of some changes in behavior and in the organization of activities associated with the ever-increasing spread of ICT.

The phenomena are too recent and still ongoing to fully understand the implications they will have on settlement systems, but the conclusion reached at the previous edition of New Metropolitan Perspectives seems to be confirmed: from many of the works presented at the Symposium, a reduction in the relevance of the localization factor emerges with ever greater clarity, at least in the ways known so far from the times of the Industrial Revolution, bringing to light more and more a paradigm shift in the center-periphery dualism.

In fact, the phenomenon that in the past led to the birth of the modern city, the need to concentrate people and activities in small areas, seems to be decreasing: the progressive spread of smart working and the digital modality for the provision of services (just think, e.g., of the digital services of the Public Administration or online commerce) significantly reduces the gaps in terms of accessibility to goods and services between metropolitan cities and marginalized areas, such as inland areas.

But this edition of the symposium also coincides with the start of a new phase for European policies, guided toward the green and digital transition, for the period 2021-27, by the European Green Deal, especially through the tool of the Next Generation EU.

The links between new technologies and sustainability tend to focus on the role played and that can play the city at EU level in fighting climate change.

Many of the contributions collected in this volume address the issue of the green transition through multidisciplinary points of view, dealing with very different issues such as, for example: infrastructures and mobility systems, green buildings and energy communities, ecosystem services and the consumption of soil, providing interesting information on the main trends in progress.

The changes in individual behavior and social organization, associated with the digital transition, are illustrated by the contributions that have addressed the issue of rules and of social innovation practices that are prefiguring new forms of governance for the regeneration of settlement systems. In this context, the issues of the new declinations of the concept of citizenship were also addressed, also with reference to the need to create favorable contexts for individual initiative and entrepreneurship, especially for young people, as a possible response to the challenge of employability for the new generations.

In this context, territorial information systems take on a leading role, together with apps capable of making territories increasingly smart.

The substantial investments planned by the EU to support the green and digital transition in the coming years require multidimensional evaluation systems, capable of supporting decision makers in selecting the interventions most capable of pursuing the objectives. The financial resources used for the implementation of the policies are borrowed from future generations, to whom we will have the obligation to be accountable for our work.

Unfortunately, at the time of writing we must also register serious concerns for the future of humanity, stemming from the risks of the spread of the conflict between Russia and Ukraine. In addition to the obvious concerns about the suffering that wars always cause to civilian populations, this situation makes future scenarios even more uncertain: It is clear that the circulation of goods, people and ideas will be increasingly conditioned by future geopolitical balances.

The ethics of research, in the disciplinary sectors that the Symposium crosses, invites us to feed, with scientific rigor, policies and practices that make the territory more resilient and able to react effectively to catastrophic events such as the pandemic or the war: We hope to know the outcomes of these courses in the next editions of the New Metropolitan Perspectives symposium.

For this edition, meanwhile, the more than 300 articles received allowed us to develop 6 macro-topics, about "Post COVID Dynamics: Green and Digital Transition, between Metropolitan and Return to Villages' Perspectives" as follows:

1. Inner and marginalized areas local development to re-balance territorial inequalities

2. Knowledge and innovation ecosystem for urban regeneration and resilience
3. Metropolitan cities and territorial dynamics. Rules, governance, economy, society
4. Green buildings, post-carbon city and ecosystem services
5. Infrastructures and spatial information systems
6. Cultural heritage: conservation, enhancement and management.

And a Special Section, Rhegion United Nations 2020-2030, chaired by our colleague Stefano Aragona.

We are pleased that the International Symposium NMP, thanks to its interdisciplinary character, stimulated growing interests and approvals from the scientific community, at the national and international levels.

We would like to take this opportunity to thank all who have contributed to the success of the fifth International Symposium "NEW METROPOLITAN PERSPECTIVES. Post COVID Dynamics: Green and Digital Transition, between Metropolitan and Return to Villages' Perspectives": authors, keynote speakers, session chairs, referees, the scientific committee and the scientific partners, participants, student volunteers and those ones that with different roles have contributed to the dissemination and the success of the Symposium; a special thank goes to the "Associazione ASTRI", particularly to Giuseppina Cassalia and Angela Viglianisi, together with Immacolata Lorè, for technical and organizational support activities: without them the Symposium couldn't have place; and, obviously, we would like to thank the academic representatives of the University of Reggio Calabria too: the Rector Prof. Marcello Zimbone, the responsible of internationalization Prof. Francesco Morabito, the chief of PAU Department Prof. Tommaso Manfredi.

Thank you very much for your support.

Last but not least, we would like to thank Springer for the support in the conference proceedings publication.

Francesco Calabrò
Lucia Della Spina
Maria José Pineira Mantinan

Organization

Programme Chairs

Francesco Calabrò — Mediterranea University of Reggio Calabria, Italy
Lucia Della Spina — Mediterranea University of Reggio Calabria, Italy
María José Piñeira Mantiñán — University of Santiago de Compostela, Spain

Scientific Committee

Ibtisam Al Khafaji — Al-Esraa University College of Baghdad, Iraq
Shaymaa Fadhìl Jasim Al Kubasi — Koya University, Iraq
Pierre-Alexandre Balland — Universiteit Utrecht, Netherlands
Massimiliano Bencardino — Università di Salerno
Jozsef Benedek — RSABabes-Bolyai University, Romania
Christer Bengs — SLU/Uppsala Sweden and Aalto/Helsinki, Finland
Adriano Bisello — EURAC Research
Mario Bolognari — Università degli Studi di Messina
Nico Calavita — San Diego State University, USA
Roberto Camagni — Politecnico di Milano, Presidente Gremi
Sebastiano Carbonara — Università degli Studi "Gabriele d'Annunzio" Chieti-Pescara
Farida Cherbi — Institut d'Architecture de TiziOuzou, Algeria
Antonio Del Pozzo — Università degli Studi di MessinaUnime
Alan W. Dyer — Northeastern University of Boston, USA
Yakup Egercioglu — Izmir Katip Celebi University, Turkey
Khalid El Harrouni — Ecole Nationale d'Architecture, Rabat, Morocco
Gabriella Esposito De Vita — CNR/IRISS Ist. di Ric. su Innov. e Serv. per lo Sviluppo

Robert Triest   Northeastern University of Boston, USA
Claudia Trillo   University of Salford, UK
Gregory Wassall   Northeastern University of Boston, USA

Internal Scientific Board

Giuseppe Barbaro   Mediterranea University of Reggio Calabria
Concetta Fallanca   Mediterranea University of Reggio Calabria
Giuseppe Fera   Mediterranea University of Reggio Calabria
Massimiliano Ferrara   Mediterranea University of Reggio Calabria
Tommaso Isernia   Mediterranea University of Reggio Calabria
Giovanni Leonardi   Mediterranea University of Reggio Calabria
Tommaso Manfredi   Mediterranea University of Reggio Calabria
Domenico E. Massimo   Mediterranea University of Reggio Calabria
Marina Mistretta   Mediterranea University of Reggio Calabria
Carlo Morabito   Mediterranea University of Reggio Calabria
Domenico Nicolò   Mediterranea University of Reggio Calabria
Adolfo Santini   Mediterranea University of Reggio Calabria
Simonetta Valtieri   Mediterranea University of Reggio Calabria
Giuseppe Zimbalatti   Mediterranea University of Reggio Calabria
Santo Marcello Zimbone   Mediterranea University of Reggio Calabria

Scientific Partnership

SIEV - Società Italiana di Estimo e Valutazione, Rome, Italy
SIIV - Società Italiana Infrastrutture Viarie, Ancona, Italy
SIRD - Società Italiana di Ricerca Didattica, Salerno, Italy
SIU - Società Italiana degli Urbanisti, Milan, Italy
SGI, Società Geografica Italiana, Roma, Italy

Organizing Committee

ASTRI Associazione Scientifica Territorio e Ricerca Interdisciplinare
URBAN LAB S.r.l.
ICOMOS Italia, Rome, Italy

New Metropolitan Perspectives
5TH INTERNATIONAL SYMPOSIUM
REGGIO CALABRIA MEDITERRANEA UNIVERSITY
EDITION 2022

Contents

Contents

Green Buildings, Post Carbon City and Ecosystem Services

Inner and Marginalized Areas Local Development to Re-Balance Territorial Inequalities

The Valuation of Unused Public Buildings in Support of Culture-Led Inner Areas' Small Villages NRRP Strategies: An Application Model in Southern Italy

Giuseppina Cassalia$^{(\boxtimes)}$ ⓘ and Francesco Calabrò ⓘ

Mediterranea University of Reggio Calabria, Reggio Calabria, Italy
giuseppina.cassalia@unirc.it

Abstract. The present study introduces an experimental model of Economic Feasibility Project with the aim of enhancing unused public buildings in Sant'Alessio Siculo, Messina (Sicily, Italy). It has been used the SostEc model in order to highlight its effectiveness as an evaluative tool to be used throughout the decision-making process. This contribution is part of a broader research designed to provide decision support tools able to enhancing geographically isolated, peripheral areas at risk of depopulation and marginalization. The preliminary results of the study presented emphasize how the challenge of green and digital transition is today more linked than ever with the local capital and the territory's real needs. Under the general and current conditions of scarce financial resources, the evaluation approach presented subsequently support the public decision makers facing the choices between alternative programs and projects. The first results of the study lead to the identification of strategic areas of intervention, in accordance with the territorial needs expressed and the EGD goals: building capacity to facilitate information, guidance and educational modules on youth green entrepreneurship; supporting local policy makers in defining systemic territorial strategies toward green and digital transitions.

Keywords: Evaluation · Strategic programming · Economic feasibility · Cultural heritage · Inner areas

1 Introduction

The debate of recent years has highlighted how the territory becomes a crucial variable to explain the opportunities that are inherent in some regions and the roadblocks slowing down the development process of others [1, 2]. At the European and national levels, integrated local development policies are now considered indispensable to achieve smart, sustainable and inclusive growth.

The paper is the result of the joint work of the authors. Although scientific responsibility is equally attributable, the abstract and Sects. 2 and 3 were written by G. Cassalia while Sects. 1 and 4 were written by F. Calabrò.

Nowadays, this is a topic strongly linked with the growing need for consistent and systematic evaluation processes, capable of supporting the public decision makers facing the need to choose between alternative programs and projects, under the general-and current-conditions of scarce resources [3]. According to the orientation of the European Commission and the Organization for Economic Co-operation and Development, strategic evaluation is perceived as an extremely important tool for planning and designing policies and interventions and for the effectiveness and efficiency of their implementation, as well as for creating consensus among the actors involved in the decision-making process [4]. The challenge is to develop an effective methodological practice that enables public administrations to undertake multi-objective decisions, while at the same time ensuring active participation in the decision-making process of all stakeholders.

The paper contributes to this current issue, presenting the first stage of a research study applied to the territory of Sant'Alessio Siculo, Messina (southern Italy). The study aims at the implementation of local development models based on the endogenous resources of a marginal area. Presenting the results of the application of the SostEc model, the assessment defines the baseline needed to draw upon policy considerations inherent in the design of appropriate territorial enhancement policies, highlighting the importance of the evaluation approach to the decision-making process. In conclusion, this preliminary study's results provide a contribution to the scientific debate on the role of the Economic Feasibility for Valorization of Cultural Heritage in the dynamics of sustainable development, stressing the critical elements and opportunities for building an integrated valorization strategy.

2   Toward a Renovated Attractiveness of Small Villages: Culture and Tourism in Times of Green and Digital Transition

2.1   Post-Covid 19 Pandemic European and National Policy

The speed of technological progress in recent years has increased significantly. Technological innovation and the development of ICT have changed the rules of living physical and human spaces rewriting the parameters of work, social behavior and territorial governance, with a significant change of the setting patterns. Besides, the digital transition and a smarter and greener use of technologies is a key goal of the European Green Deal [5]. The challenge is to fostering green strategic technologies in order to trigger new business growth in EU urban/rural areas, creating digital services to enhance the competitiveness of enterprises and improving the quality life of people.

Furthermore, the COVID-19 pandemic - and the policies and practices put in place to contain the infection and the economic crisis - have accelerated the spread of ICT producing tangible effects on the settlement [6, 7]. The replacement of physical interactions with "virtual" contacts has used consolidated technologies but has accentuated their pervasiveness, generating impacts of different nature.

The complexity of the choices that governmental bodies at the national and local levels are facing requires an ability to implement complex decision-making processes. Despite the evolution of the territorial cohesion over the European programming cycles, marginal areas—in Italy, for instance—are still facing economic, social and environmental problems, resulting in unemployment, depopulation, marginalization, disengagement, or degradation of historic and architectural capital [8]. Will the green and digital transition claimed by European Green Deal (EGD) reverse the core–periphery dualism? Is the Covid-19 socio-economic impact affecting this transition at local scale?

Among the measures and tools providing significant financial support for reforms and investment in pandemic times, in Europe there is the NextGenerationEU. An economic recovery package with a more than €800 billion temporary instrument to help repair the immediate economic and social damage brought about by the coronavirus pandemic. NextGenerationEU also includes €50.6 billion for Recovery Assistance for Cohesion and the Territories of Europe (REACT-EU). It is a new initiative that continues and extends the crisis response and crisis repair measures delivered through the Coronavirus Response Investment Initiative and the Coronavirus Response Investment Initiative Plus. It is said it will contribute to a green, digital and resilient recovery of the economy. The funds will be made available to the European Regional Development Fund (ERDF), the European Social Fund (ESF), the European Fund for Aid to the Most Deprived (FEAD) [9].

On the other hand, Member States are working on their recovery and resilience plans to access the funds under the Recovery and Resilience Facility. As concern Italy, The European Union has set aside € 191.5 billion for the NRRP through grants and loans from the RRF (Recovery and Resilience Facility). Italy is supplementing the amount with € 30.6 billion through the Complementary Fund, financed directly by the State, for a total of 222.1 billion. The European Commission has granted €4.7 billion to Italy under REACT-EU to support the country's response to the coronavirus crisis and to contribute to a sustainable socio-economic recovery [10]. The Culture and Tourism Sector, the one taken into consideration by the present study, will benefit of 8 billion. The funds are intended for the revival of culture and tourism as strategic sectors for the growth and development of Italy. Among the objectives there is the increase of the Italian attractiveness by improving the tourism system and paying particular attention to the regeneration of the small villages (named "borghi") through the promotion of participation in culture, the relaunch of sustainable tourism and the protection and enhancement of parks. In addition, the private sector, citizens and communities will also be involved both in terms of promoting sponsorships and through multi-level governance. This approach is in line with the "Faro Convention" on the value of cultural heritage for society, and with the European Framework for Action for Cultural Heritage, which calls for the promotion of integrated and participatory approaches, to generate benefits in the four pillars of sustainable development: economy, cultural diversity, society and environment [11, 12].

By improving a model to facilitate stakeholders' approaches in Strategic Development Process (SDP), this contribution provides a participative approach as a process where individuals, groups, and organizations choose to take an active role in making decisions that affect them [13]. To this purpose, a case study is proposed in order to outline perspectives for a Culture-led strategic program of small cultural sites to be developed. The area is located in the northernmost part of the Sicily region, and it's identified as marginal area of the Southern Italy. In general, its geographical district is marked by the presence of the potentialities and criticalities that can be found in the entire South of Italy: rich of historical, cultural and natural heritage and unemployment, depopulation and social vulnerability.

2.2 Unused Built Public Investments in Times of Scarce Resources

The allocation of resources and investment strategies are crucial issues for the cultural sector. In recent decades, the idea of cultural built heritage as a weight bearing on the public sector has evolved into the concept of a strategic asset capable of increasing the attractiveness and competitiveness of some territories [14].

This important evolution in theoretical terms encounters, however, a strong limit from the increasingly strict constraints placed on public spending, which have constituted a real, generalized collapse of investments. Although the need to attract private resources and skills to the sector and to enhance the synergies between cultural heritage and tourist accommodation has long been highlighted, it is clear that the role of public investment remains crucial [15].

Moreover, given the low absorption rate of EU Structural Funds, the debate in Italy continues about the extent to which, after decades of subsidies, the European Regional Policy is effective. This explains the attention to the programming strategies of these resources and to the selection of the territories and themes to invest in, so as to arrive at the methodologies for measuring the impacts actually obtained [16, 17].

The fight against the progressive desertification of Inner Areas is one of the objectives of territorial policies on a European, national and regional scale. Achieving this goal requires an integrated approach, which let simultaneously face problems of a different nature: infrastructure, economic, social, etc. In the Inner Areas of Italy, in particular Southern Italy, there are a huge number of unused buildings, public or of public value whether they are equipped with special historical and architectural features. The conditions of abandonment of these buildings significantly contribute to the degradation of inhabited centres: recovering them can be a key factor in the improvement of the architectural-urban quality in order to raise attractiveness, which in turn is a strong element in the fight against depopulation. However, the recovery desire of the unused buildings deals often with management difficulties: frequently, the functions assumed for their rescue cause unsustainable management costs for local administrations [18, 19].

Two basic requirements arise from this:

- Identifying those functions coherent with the system of real needs of citizens and with the general objective system of local development;
- Identifying the optimal conditions for the involvement of private individuals in the investment and / or the management.

In order to make a conscious choice and reduce the risk of faults, the public decision-maker needs to know in advance the economic implications of both the possible destinations for the goods concerned and the possible forms of public-private partnership. The SostEc model applied to the aforementioned case study, allows to:

- deduce the intended use of real estate from an objectives' system that takes into account the objective conditions of the territories and the subjective indications of the stakeholders;
- check the feasibility / economic sustainability of the intended uses in relation to different management models which imply different forms of public-private partnership.

For brevity, please refer to Calabrò, Della Spina (2019) [20] the detailed illustration of the SostEc model; this article illustrates a case study concerning an unused public building in Sant'Alessio Siculo, a small village 8 miles far from the well-known Taormina in Sicily.

This contribution is part of a broader research designed to provide decision support tools able to tackle the risk of educational poverty in the area and aims at strengthening the link between the university and the world of school, families and the community, ensuring quality training and enhancing geographically isolated, peripheral areas at risk of depopulation and marginalization: places of strategic significance, educational communities with a high level of quality and social well-being, memory places cradle of historical, artistic, cultural and environmental heritage.

3 The Valuation Culture in Support of Culture-Led Inner Areas' Small Villages Strategies

3.1 A Specific Enhancement Project for the Territory: The SostEc Model

The SostEc model is the experimental model of the economic feasibility Project concerning the evaluation of unused public buildings; it can be, as well, considered as an effective tool in the whole decision-making process. Besides, people can adopt it to identify the intended uses corresponding to the territory needs, the local development policies and the inherent characteristics of the property to be enhanced. Actually, the first step of the model development consists of setting up those knowledges in an effort to identify the main problems and the vocations in the referring context and naturally, these lead to gain a greater awareness of the intended choices [21, 22].

3.2 The Case Study Area

The territory, with its landscape, represents the synthesis of all the physical, cultural, natural, settlement and immaterial elements and describes its identity and uniqueness. The analysis of such contest is therefore essential to understand the ongoing territorial dynamics and identify those endogenous resources from which new development dynamics may arise. The first step to create the cognitive framework is the statistical analysis collecting the most representative and exploitative data of the demographic, employment, social, economic and productive aspects, linked to the subsequent cognitive analysis of the cultural heritage - material and immaterial - featuring the reference area [23].

The Municipality of Sant'Alessio Siculo, object of the present study, is part of a geographical district located in the north-eastern part of the metropolitan area of Messina, Sicily. The district bounded upstream by the Peloritani mountains, it is crossed by four rivers: the D'Agrò, Savoca, Dinarini and Nisi streams, within whose valleys fall the 18 municipalities of the so-called Valli joniche dei Peloritani. In general, the Geographical District is marked by the presence of the potentialities and criticalities that can be found in the entire South of Italy. On the one hand, the resources offered by the rich historical, cultural and natural heritage in its tangible and intangible features. On the other hand, the worrying employment issue, about the youngest segments of the population. Also, can be outlined issues related to the hydrogeological structure of the territory and its widespread accessibility by the primary and secondary road network. From a morphological and infrastructural point of view, the area tends to be homogeneous, although some socio-economic and historical-cultural emergencies are still evident. The analysis aims at identifying these peculiarities, intending to enhance their value as a system, with particular reference to the valleys.

S'Alessio Siculo is an Italian town of 1 460 inhabitants (about 8000 in the summer season). It is part of the Valle d'Agrò area and overlooks the sea. S. Alessio Siculo has become a center with a tourist vocation, thanks to its stony beach and due to the tourist settlements and hotel structures that have been built in recent years. Villa Genovese, also known as "Villa del gallo", was indicated by the municipal administration as the structure on which to apply the SostEc model. It was built in the late 1800 s and then abandoned. The analysis of the local community point of view is essential to understand the ongoing territorial dynamics and identify those endogenous resources from which new development dynamics may arise. As far as resulted from questionnaires, with the acquisition of the Villa Genoese, the Municipality wants to provide its local community and tourists with a stable structure where to set up recreational and cultural activities. Therefore, the recovery and re-functionalization of Villa Genovese has a significant meaning: that of creating an important cultural center, enhancing the historical and cultural features of S. Alessio. Hence, the intake survey showed how remarkable the tourism sector is for the entire economic system of the Municipality. To find a better development strategy, it was therefore considered the cultural events promotion as appropriate in order to relaunch the economy of the entire Valle d'Agrò. The interviews submitted to the stakeholders (local administrators, entrepreneurs, professionals and associations) were important in defining the fact-finding survey that brought out an analysis of the points of view and perceptions of the different categories regarding the main problems and vocations of the area [24]. Each investment is intended to produce impacts that do not end in the production of assets or services capable of satisfying the need/s. For these reasons, each future stakeholder has a specific role and an interest in implementing that type of intervention. The project choices related to the hypothesis of reuse were synthetic; in fact, a functional program has been devised with adequate spaces for the different uses that has allowed to verify the consistency between the intrinsic characteristics of the building and the hypotheses of reuse. Compared to the strategies and objectives identified by the administration and the local community, hypotheses of reuse consistent with the planned actions have been developed [25].

3.3 The Management Model and the Feasibility Study

The need to address the public resources lack for the management of heritage-related activities leads to the consideration of partnership forms with private entities in order to meet the needs of the territory [26].

Even in this situation, the SostEc model is useful in identifying the possible form of involvement of privates and setting the economic conditions related to the basis of the partnership. If the private is just the manager of the intervention, its sustainability is verified, if he is also a promoter, it is necessary to establish whether there is a need for public co-financing of the investment and, in this case, the feasibility of the intervention is verified.

The identification of the form of private involvement, through the SostEc Model, takes place declaring the evaluation again, in order to identify feasible and / or sustainable solutions, as to dismiss those that do not possess these requirements.

For the management models of the Cultural Community Centre a profit model (P) was proposed as it pursues a profit-oriented purpose and must obtain adequate revenues to cover high fixed costs. With this type of management, there is a greater efficiency compared with public subjects and a flair flexibility characterizes those who are privately profitable. Even with management models already defined, they may vary over time. Moreover, it is necessary to check what a territory is able to express from the managers perspective. This verification of the financial economic balance of the management permits informed choices to be made, thus reducing the chances of error and giving greater transparency to the decision-making process.

From an economic point of view, the study preliminarily provides the evaluation of investment costs followed by an analysis of the economic dynamics in the management phase. The main purpose is to verify the economic feasibility/sustainability of the hypothesized form of public-private partnership: it verifies the existence of sufficient conditions of convenience for private subjects in the realization and/or management of a project in line with the public objectives from which it derives [27]. The Economic and Financial Plan is divided into four phases highlighted in Table 1.

Table 1. Phases of economic and financial plan.

Phase	Denomination	Contents
1	Stima Estimate of Investment Cost	Works; equipment and furnishings; marketing
2	Stima Estimate of Revenue	Identification of productions and/or services; estimate of their unit selling price; identification of the target; estimate of demand
3	Stima Estimate of Management Costs	Management model; human resources; management costs (consumables, services, personnel)
4	Project Income Statement	Cash flow analysis; feasibility/sustainability

* Source: Calabrò F., Cassalia G., Lorè I. (2021) The Economic Feasibility for Valorization of Cultural Heritage. The Restoration Project of the Reformed Fathers' Convent in Francavilla Angitola. In: NMP 2020. Smart Innovation, Systems and Technologies, vol 178. Springer, Cham. P 1111.

The estimate of the investment costs (see Table 2), being in a preliminary phase of defining the project choices (Phase 1), took place through synthetic-comparative procedures by the use of parametric (or mixed) estimates for functional elements or significant samples.

Table 2. Estimate of Investment Costs (VAT included).

Cost items	€
A - For auction-based works	
Building Works - Restoration and refunctionalization of spaces	€ 213.041,00
B - Supplies and services	
Equipment and furniture	€ 14.859,00
C – Marketing and promotion	
Communication plans, MKT and events	€ 1.000,00
TOTAL INVESTMENT	
VAT included	**€ 228.900,00**

The estimate of revenues (Phase 2) is articulated in the following actions: the identification of the productions and services to be provided; the estimate of their unit sales price; the identification of the target; the estimate of the demand and revenues [19, 20]. The letter is a function of the tourism potential of the area, the intended uses of the functional program, the attractiveness of the asset and the effectiveness of communication and marketing strategies. Revenues were estimated for each service according to the

Table 3. Estimate of costs and revenues.

Estimate of the Annual Revenue at Fully Operational

	Unit Price	Quantity	Revenue
Example Activity 1	10,00 €	6.920	€ 69.200,00
Example Activity 2	4,00 €	692	€ 2.768,00
Example Activity 3	20,00 €	5.000	€ 100.000,00
Tot. Annual Revenues			**€ 171.968,00**

Estimate of the Annual Costs at Fully Operational for Human Resources

	Unit Price	Quantity	Total Cost
Service Manager	€ 1.521,00	1	€ 18.504,00
Service Manager Asst.	€ 1.200,00	1	€ 14.400,00
Tot. Annual Costs for Human Resources			**€ 32.904,00**

Estimate of the Annual Management Costs for Services

	Unit Price	Quantity	Total Cost
Utilities	€ 3.100,00	1	€ 3.100
Cleaning	€ 14.400,00	1	€ 14.400
Other costs (e.g. mainten. service)	€ 3.200,00	1	€ 3.200
Marketing	€ 8.500,00	1	€ 8.500,00
Raw material	€ 12.500,00	1	€ 12.500,00
Tot. Annual Management Costs for Services			**€ 41.700**

tourist seasonality and to the increases from the conference activities. Table 3 shows the summary data of costs and revenues for each service.

The verification of the economic and financial sustainability concerned the considered management model with profit managers with investment entirely borne by the Municipality (public entity) in order to enhance the tourist attractiveness of the area. The check was carried out on the assumption that the provision of cultural services linked to the Cultural Community Centre is devolved to the voluntary activities of cultural associations, while the economic activities are carried out by a publishing company present on the area [4]. Tables 4 show the interventions sustainability during the management phase at fully operational in 3 years.

Table 4. Income statement project for the cash flow analysis.

Production Value (A)	
Revenue from sales and services	171.968,00 €
Production Costs (B)	
Raw, subsidiaries and consumables materials and goods	12.500,00 €
Services	26.100,00 €
Personnel	32.904,00 €
Amortization – Building Investments	64.950,00 €
Amortization – Supplies and services Investments	6.250,00 €
Risk fund and extraordinary maintenance	17.500,00 €
Difference between Value and Production Costs (A-B)	**11.764,00 €**

The last section is aimed at verifying the management sustainability of the project, through the verification of the balance sheet in the year at full capacity, through the Cash Flow Analysis. As shown in Table 4 the difference between Value and Production Costs allows the management body the building's maintenance and the provision of services and cultural activities for the local community and tourists.

3.4 Preliminary Results

The need of local administrations to proceed with a progressive and constant conversion of their building assets has inevitably led to an ever more solid integration between urban planning, financial instruments and public-private partnerships that aim for the pursuit of economic and social objectives from which derive benefits for all parties involved [28].

By improving a model to facilitate stakeholders' approaches in Strategic Development Process (SDP), the ongoing research provides a participative approach as a process where individuals, groups, and organizations choose to take an active role in making decisions that affect them. The preliminary results of the study demonstrate that the SostEc Model helps in:

- identifying use's destinations which meet the needs of the territory, local development policies and the intrinsic characteristics of the asset to be exploited;
- identifying the possible form of involvement of private entities: that is, if there are realistic conditions, given by sufficient profitability for their participation in the investment or if, rather, they can provide support only in the management phase;
- determining the economic conditions to be established for the partnership: whether there is a need for public co-financing of the investment or management or whether, on the other hand, the revenue generated by the project is sufficient to ensure its feasibility.

The identification of the form of involvement of private individuals, through the SostEc Model takes place by reiterating the assessment, in order to identify feasible and/or sustainable solutions and to exclude those that do not possess such requirements.

4   Conclusion

A local green and digital transition implies a drastic transformation of the territorial system and, hence, a lot of challenges. Transition calls for an integrative local approach. We find that theory has not yet developed sufficiently to address the practical and local challenges. Part of the problem is that what has been developed has too little connection with local practice. We conclude that the development of theoretical knowledge must be better attuned to the needs of the practitioners.

Climate change, environmental degradation and digital divide are a factual threat to Europe and the world. To overcome these challenges, Europe needs a new growth strategy that will transform the Union into a modern, resource-efficient and competitive economy.

European Green Deal (EGD) represents an innovation-driven development strategy for Europe towards a reorganization of priorities, making sustainable development the top strategic priority.

At the governance level, the new EGD strategy raises several crucial multilevel governance challenges. Actors who were not at the center of the European integration process like the regions, or absent, as cities and communities are now likely to play a crucial role. The links between new technologies and sustainability tend to focus on the role played and that can play the city at EU level in fighting climate change. By contrast, this role cannot be based exclusively on experiences that emerge mainly from prosperous cities or capitals, because these do not reflect Europe as a whole.

Culture and tourism represent a key resource for growth, jobs and social cohesion in rural and remote areas and small historic towns, characterized by a widespread cultural wealth and remarkable landscapes.

A culture-led Inner Areas' Small Villages Strategy, umbrella strategy of this research project, provides an opportunity to promote jointly the green and digital transition. A series of investments, in culture and sustainable tourism also favour the diffusion of more responsible environmental behaviour and induce citizens to shift their consumption habits, thus favouring the emergence of the circular economy.

Furthermore, access to cultural content fosters digitalisation and acquisition of digital competences, being prime fields of development and experimentation of emerging technologies, such as augmented and enriched reality, the Internet of Things and Artificial Intelligence. The development of tourism and culture in inner areas is indeed an important line of action for the NRRP-M1C3 in Italy. Interventions will be carried out to valorise the great heritage of history, art, culture, and traditions present in small Italian towns with a great natural, landscape and cultural potential.

Inner areas are often fragile demographic and social contexts with high environmental risks. Interventions are planned to enhance the great heritage of history, art, landscape, culture and traditions present in small Italian towns, favoring the rebirth of ancient agricultural structures and traditional crafts (e.g. handicrafts). It will support the activation of entrepreneurial and commercial initiatives, including new ways of receptivity such as diffuse hospitality and hotels, for the revitalization of the socio-economic texture of the places, countering the depopulation of the territories and favoring the preservation of the landscape and traditions and redevelopment of rural and historical buildings.

Supporting participated projects of cultural-based urban regeneration, focused on local communities, in which municipal administrations will be the protagonists, in partnership (co-design) with public and private actors, organized social institutions, third sector, cultural associations, and foundations.

Interventions are also planned for the redevelopment of public real estate degraded and/or unused, intended for social and cultural, educational services.

The research will continue by formalising the evaluation model, obviously of multicriteria nature, that is able to provide the public decision-maker with a stronger and more unbending tool for managing the needed information for a conscious choice.

References

1. United Nations General Assembly: Transforming our world: the 2030 Agenda for Sustainable Development (2015)
2. Todling, F., Tripl, M.: One size fits all? Towards a differentiated regional innovation policy approach. Res. Policy **34**, 1023–1209 (2005)
3. OECD: Sustainable development governance structures in the European Union. In: Institutionalising Sustainable Development, pp. 67–88. OECD Publishing, Paris (2007)
4. European Commission: Europe 2020: a strategy for smart, sustainable and inclusive growth. COM (2010) 2020 final (2010)
5. European Commission: Communication from the commission to the European Parliament, the European Council, the Council, the European economic and social Committee and the Committee of the regions. The European Green Deal, COM/2019/640 final (2019)
6. OECD: The territorial impact of Covid-19: managing the crisis across levels of government © OECD (2020) https://read.oecd-ilibrary.org/view/?ref=128_128287-5agkkojaaa&title=The-territorial-impact-of-covid-19-managing-the-crisis-across-levels-of-government
7. Chernick, H., Copeland, D., Reschovsky, A.: The fiscal effects of the covid-19 pandemic on cities: an initial assessment. National Tax J. **73**(3), 699–732 (2020) https://doi.org/10.17310/ntj.2020.3.04(2020)
8. De Mascarenhas, F.: Abandoned villages and related geographic and landscape context: guidelines to natural and cultural heritage conservation and multifunctional valorization. European Countryside **3**(1), 21–45 (2011)
9. European Commission: Communication from the commission to the European Parliament, the European Council, the European economic and social committee and the committee of the regions. Europe's moment: Repair and Prepare for the Next Generation. Brussels: European Commission (2020). https://eur-lex.europa.eu/legal-content/EN/TXT/PDF/?uri=CELEX:52020DC0456&from=EN
10. European Commission: Directorate-General for Budget, The EU's 2021–2027 long-term budget & NextGenerationEU : facts and figures, Publications Office (2021). https://data.europa.eu/doi/10.2761/808559
11. European Commission: Directorate-General for Education, Youth, Sport and Culture, European framework for action on cultural heritage, Publications Office (2019). https://data.europa.eu/doi/10.2766/949707
12. Council of Europe: Framework Convention on the Value of Cultural Heritage for Society, open for signature at Faro on 27 October 2005

13. Bombino, G., Calabrò, F., Cassalia, G., Errante, L., Vinci, V.: Inclusive strategic programming: methodological aspects of the case study of the Jonian Valleys of Peloritani (Sicily, Italy). In: Gervasi, O., et al. (eds.) ICCSA 2021. LNCS, vol. 12954, pp. 109–119. Springer, Cham (2021). https://doi.org/10.1007/978-3-030-86979-3_8
14. Calabrò, F., Iannone, L., Pellicanò, R.: The historical and environmental heritage for the attractiveness of cities. the case of the Umbertine Forts of Pentimele in Reggio Calabria, Italy. In: Bevilacqua, C., Calabrò, F., Della Spina, L. (eds.) NMP 2020. SIST, vol. 178, pp. 1990–2004. Springer, Cham (2021). https://doi.org/10.1007/978-3-030-48279-4_188
15. Rao, K.H., et al.: Public-private partnership and value addition: a two-pronged approach for sustainable dairy supply chain management. IUP J. Supply Chain Manag. 10(1), 15 (2013)
16. UVAL: L'Italia secondo i Conti Pubblici Territoriali. I flussi finanziari pubblici nel settore cultura e servizi ricreativi. Atti del convegno di presentazione, Collana Materiali UVAL n.3 (2014)
17. Tramontana, C., Calabrò, F., Cassalia, G., Rizzuto, M.C.: Economic sustainability in the management of archaeological sites: the case of Bova Marina (Reggio Calabria, Italy). In: Calabrò, F., Della Spina, L., Bevilacqua, C. (eds.) ISHT 2018. SIST, vol. 101, pp. 288–297. Springer, Cham (2019). https://doi.org/10.1007/978-3-319-92102-0_31
18. Tajani, F., Liddo, F.D., Guarini, M.R., Ranieri, R., Anelli, D.: An assessment methodology for the evaluation of the impacts of the COVID-19 pandemic on the Italian housing market demand. Buildings 11(12), 592 (2021)
19. Dolores, L., Macchiaroli, M., De Mare, G.: Sponsorship's financial sustainability for cultural conservation and enhancement strategies: an innovative model for sponsees and sponsors. Sustainability 13(6), 9070 (2021). https://doi.org/10.3390/su13169070
20. Calabrò, F., Della Spina, L.: La fattibilità economica dei progetti nella pianificazione strategica, nella progettazione integrata, nel cultural planning, nei piani di gestione. Un modello sperimentale per la valorizzazione di immobili pubblici in partenariato pubblico-privato Rivista LaborEst 16 Special Issue (2016). http://dx.medra.org/10.19254/LaborEst.16.IS
21. Campolo, D., Calabrò, F., Cassalia, G.: A Cultural route on the trail of Greek Monasticism in Calabria. In: Calabrò, F., Della Spina, L., Bevilacqua, C. (eds.) ISHT 2018. SIST, vol. 101, pp. 475–483. Springer, Cham (2019). https://doi.org/10.1007/978-3-319-92102-0_50
22. Calabrò, F., Cassalia, G., Lorè, I.: The economic feasibility for valorization of cultural heritage. the restoration project of the reformed fathers' convent in Francavilla Angitola: the Zibìb Territorial Wine Cellar. In: Bevilacqua, C., Calabrò, F., Della Spina, L. (eds.) NMP 2020. SIST, vol. 178, pp. 1105–1115. Springer, Cham (2021). https://doi.org/10.1007/978-3-030-48279-4_103
23. Bombino, G., Calabrò, F., Cassalia, G., Errante, L., Vinci, V.: The knowledge phase of the strategic programming: the case study of the Jonian Valleys of Peloritani (Sicily, Italy). In: Gervasi, O., et al. (eds.) ICCSA 2021. LNCS, vol. 12954, pp. 307–320. Springer, Cham (2021). https://doi.org/10.1007/978-3-030-86979-3_23
24. Torrieri, F., Oppio, A., Rossitti, M.: Cultural heritage social value and community mapping. In: Bevilacqua, C., Calabrò, F., Della Spina, L. (eds.) NMP 2020. SIST, vol. 178, pp. 1786–1795. Springer, Cham (2021). https://doi.org/10.1007/978-3-030-48279-4_169
25. Séraphin, H., Platania, M., Spencer, P., Modica, G.: Events and tourism development within a local community: the case of Winchester (UK). Sustainability 10(10), 3728 (2018). https://doi.org/10.3390/su10103728
26. Porter, M.E., Kramer, M.R.: Creating shared value. Harv. Bus. Rev. 1(2), 2–17 (2011)
27. Lichfield, N.: Economics in Urban Conservation. Cambridge University Press, Cambridge (1988)
28. Pearce, D.W., Mourato, S.: The Economics of Cultural Heritage. World Bank Report, CSERGE, University College London (1998)

Communities' Involvement for the Reuse of Historical Buildings. Experiences in Italian Marginal Areas

Caterina Valiante[✉] and Annunziata Maria Oteri

Department of Architecture and Urban Studies (DAStU), "Fragilità Territoriali" Research Project, Departments of Excellence Initiative 2018–2022, Politecnico di Milano, Milano, Italy
{caterina.valiante,annunziatamaria.oteri}@polimi.it

Abstract. Throughout the Covid-19 emergency phase in Italy, interesting practices of reuse of architectural heritage were fostered by the community impulse in small, depopulated towns in inner areas. This unprecedented situation led small communities of remote areas to find urgent solutions to entirely new problems. Some contexts activated initiatives for enhancing the health and social services, some others for strengthening the home care, others for readjusting existing spaces according to new needs. Through the case of the reuse of the Convento dei Padri Riformati into a hospital in Madonie and other similar initiatives promoted in other Italian inner areas, this contribution aims at highlighting the potentialities connected to the creation of services for the community by reusing the existing built heritage. In the experiences this contribution wants to illustrate, the role of "local community" was revealed to be essential during the pandemic since every practice of reuse is conceived, realized, or managed through the participation of the community and mainly destined to it. Concurrently, the objective is to introduce some reflections on the importance of the functions and uses introduced and their impacts on the territory: many times, the architectural heritage is intended for tourism-related facilities. However, in some cases, other functions connected to public services can profitably preserve and relaunch historical buildings, thanks to constant use and maintenance, assuring at the same time a broader involvement of the community at all levels.

Keywords: Architectural heritage · Inner areas · Community

1 Premise: Local Communities at the Test of the Pandemic

Recently, the interest in depopulated and marginal areas increased thanks to their partial repopulation by smart workers during the various lockdowns. A sort of enthusiasm invites one to hope that such provisional returns could become permanent in a short time. The optimism has to be downsized if one only reflects that the pandemic has been amplified rather than reducing fears and uncertainties in the local communities who cope with isolation, lack of essential services, and mistrust in central and local government. At the same time, debates and reflections on the value and resilience of local communities in inner areas have been rising at different levels to underline the contrast with the fragility of big towns at the test of the pandemic. Resilience, sense of nearness, and capability helped local communities to react to the pandemic and reorganise themselves, for example, concerning the issues of cure and solidarity. These "small retching of self-defence" [1][1] find fertile ground in those territories that have faced marginality and civil inequalities over time as they had to replace with creativity and talent what lacks. The unprecedented pandemic situation led small communities of remote areas to find urgent solutions to entirely new problems.

In this frame, traditions of territories and communities play an important role. In many cases, for example, the background of that particular area seems to suggest practices which have been characterised that area over time. As some of the following examples show, a sort of vocation, strictly related to the sense of "locality" [2] of the area inspires initiatives and experiences. The success of such an approach is strictly related to awareness of the communities on the material and immaterial values of the place where they live and produce. For this reason, the many spontaneous micro-projects developed during the pandemic revealed themselves more productive than programmed policies, which are mainly unaware of the sense of places. For those who study marginal territories and the related processes of transformation, in-depth analyses of such initiatives – and the cultural processes which inspired them – would be helpful to transform such experiences, developed with the emergency, into structured strategies for hindering depopulation despite the pandemic. Particularly in the preservation of architectural heritage, which is the focus of this paper, where programs and interventions are still mainly promoted through a top-down approach, the suggestions from communities could be illuminating.

Some contexts activated initiatives for enhancing health and social services, others for fostering home care, others for readjusting existing spaces according to new needs. In the experiences that this contribution wants to illustrate, the role of the "local community" was revealed to be essential during the pandemic. These practices activated during the last two years were conceived, realised, or managed through the participation of the community and mainly destined to it, and they could become a model also when the emergency will be over.

Before deepening these initiatives, it seems necessary to premise what we mean for "community", which is not intended for the natives or the residents. Still, it is perceived as a constantly evolving entity composed of different and multifaceted singles or groups of people that are somehow related to the territory. Community is a kaleidoscope defined by the existences of those who live in that place or context [3], and it is an incessant production in the past and the present [2]. Contrary to commonly believed, community means sharing spaces and activity without sharing sense and identity [4, 5]. A nostalgic and illusory approach still conceives the local community as a set of dense interpersonal relationships, shared values and defended boundaries as a counterpart of modernity. Community is also often related to a supposed notion of identity capable of guiding individuals or groups, which is a misleading and ideological interpretation of community

---

[1] From the presentation of G. Carrosio, "*Cittadinanza e aree interne in Italia*" at the conference "*Un'agenda di ricerca per le fragilità territoriali*", DAStU Dipartimento d'eccellenza Fragilità territoriali, Politecnico di Milano, March 26th, 2019.

[2, 6]. As the current scientific debate on Italian marginal areas demonstrates [7, 8], the concept of "community" has been widened, aiming at a more inclusive consideration of the various groups of people which can be involved in the reactivation process of an "inner area". In reaction to the Covid-19 pandemic, some initiatives had been promoted thanks to the community's readiness to face an emergency that sometimes implicates a transformation or a rearrangement of the community's composition and organization.

Fig. 1. A view of the town of Petralia Sottana from the *Convento dei Padri Riformati* (photo by C. Valiante, 2021).

1.1   The Case of the *Convento Dei Padri Riformati* in Petralia Sottana, Madonie

A particularly interesting case is represented by the reuse strategy activated by the municipality of Petralia Sottana, in Sicily[2] (see Fig. 1) for supporting the local health services. In 2020, due to the pandemic, part of the town's local hospital was converted into a covid unit. Therefore, some administrative and medical offices and a nursing home

---

[2] The Inner Area "Madonie", one of the pilot-areas selected within the Italian National Strategy for Inner Areas, is one of the case studies that the author is investigating in the framework of her PhD thesis - developed in the PhD program in Preservation of the Architectural Heritage at Politecnico di Milano, Department of Excellence on Territorial Fragilities. The thesis is mapping and studying practices of reuse of built heritage in wider local development initiatives in Italian inner areas.

were moved to the former *Convento dei Padri Riformati* (see Fig. 2, 4, 5). This complex, built between 1603 and 1663, is composed of a two-level central building and a church, which is now deconsecrated.

Fig. 2. Outside view (current entrance) of the *Convento dei Padri Riformati* in Petralia Sottana, Palermo (photo by C. Valiante, 2021).

The area to which Petralia Sottana belongs, the *Madonie*, is a mountain area located in the northwest part of Sicily, in Palermo province; it was historically characterised by an economy based on agricultural activities organized within a feudal system and by the significant presence of religious orders. Over the centuries, these processes shaped the territory, which still nowadays is characterized by various small historical centres, but also an extensive scattered rural heritage [9] and a widespread ecclesiastic heritage: in fact, this area holds one-third of the entire religious heritage of the Sicily Region [10–12]. Due to the economic shifts that occurred since the end of the Second World War and the migration flows towards the industrial sites of the coastal areas, the area had been affected by a severe depopulation trend (see Fig. 3) and a great amount of this built heritage is underused or abandoned.

The historical stratification of local dynamics and practices in the management of the territory and the quite strong cohesion among the municipalities of the area (joined in an administrative institution called *Unione di Comuni*) led, in the last decades, to the promotion of various strategies for the reuse of the local built heritage [15]. In particular, the *Convento dei Padri Riformati* in Petralia Sottana was firstly given to the

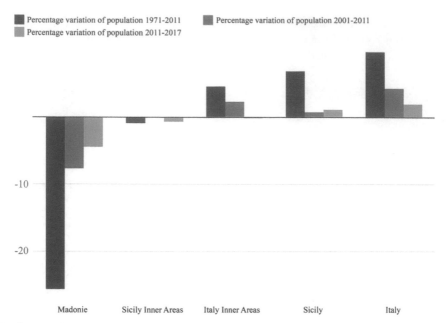

Fig. 3. Variation of population in the Madonie area [13, 14; Diagram created with Datawrapper]

municipality by the diocese of Messina in the XIX century [16]. After various decades of abandonment and different uses, such as military facilities or centres for war refugees in the XX century, the local administration promoted a complete restoration of the complex for cultural uses (conference centre, concert hall, etc.) in the 2000s. The interventions did not transform the volumetry but improved the internal and external conditions of the building. In 2015, the complex was included in the project "*Valore Paese – DIMORE*" promoted by the State Property Office and other institutional actors for the valorization and effective management of a public real estate with a particular historical value, located in tourism-attractive contexts for fostering local development processes [17]. In 2017, through a public consultation among local institutions, the municipality published a call for the valorization concession (Concessione di valorizzazione, Art. 3 bis L. 410/2001, comma 259, L. 296/2006) up to thirty years. The selected project planned to allocate functions very similar to the previously existing ones: congress and meeting centre with accommodations, concert hall, and spaces for events in general. Therefore, the local administration rented out the complex for several years, assuring relatively constant basic maintenance. In this phase, the wider community was not directly involved in the decision of the allocated functions, which were mainly destined to tourism or events.

However, in 2020, during the first pandemic emergency, as previously reported, the municipality, through a free loan agreement, made available the convent complex to the local health service, placing there some important health facilities, such as specialized clinics, an extended care unit, and other offices: thereby an everyday use, with more frequent monitoring, has been ensured, also allowing a wider part of the local community

Fig. 4. Cloister of the *Convento dei Padri Riformati* in Petralia Sottana, Palermo (photo by C. Valiante, 2021).

to make use of this heritage[3]. In October 2021, after more than one year of this new management, the municipality officially asked the Province Health Unit to leave the complex in order to set a new call for its governance and to re-establish the previous functions related to cultural activities [18]. The previous practice of reuse of this complex was conceived without the direct participation of the community, but for facing the emergency, this common good had been transformed into essential services for the entire population. This dynamic can be more in favour of social aspects than economic and touristic ones, but it allows more inclusive uses for a wider community. In addition, this everyday use is preferable for the historical buildings since it guarantees constant care and maintenance. For these reasons, unlike the current situation, this kind of reuse initiatives should be more likely fostered in the contexts of inner areas in order to become a permanent or long-term choice.

---

[3] Part of the information reported were collected thanks to the kind availability of Alessandro Ficile, president of the local development agency of Madonie and the major or Petralia Sottana, Leonardo Neglia.

Fig. 5. Deconsecrated church of the *Convento dei Padri Riformati* in Petralia Sottana, during the use of the local health service (photo by C. Valiante, 2021).

2 Beyond Tourism and Culture Related Functions for Reuse Projects

The described case of the convent represents one of the many examples of reuse of built heritage in the town of Petralia Sottana and the Madonie area in general. The significant number of complexes or buildings that are not used anymore because of the population shrinking and socio-economic shifts prompt the community to find ways to use the local built heritage to avoid the loss of these testimonies due to abandonment. Recently, in Petralia Sottana have been activated projects and strategies for the reuse of the former secondary school into a living lab, various abandoned residential buildings of the historical city centre into public housing, the former slaughterhouse into a conference hall, and other reuse projects of part of the historical centre were also promoted in the closer towns of Madonie, such as Petralia Soprana, Gangi, and San Mauro Castelverde. But this approach to reuse is not just a recent tendency. Throughout history, the vast presence of religious institutions in this area had always led the local community to find strategies and ways to take advantage of the legacies they left to the territory. For instance, in the last century, a very frequent case was the transformation of the large spaces of convents or monasteries into schools. Nowadays, the socio-economic and demographic situation is changed. However, the diffused inclination of the community to the reuse of architectural heritage is still there. It pushes to transform the existing buildings again according to the new needs: but rarely the projects aim at introducing public services for the citizenship, more often they are dedicated to creating touristic attractions or facilities, or cultural activities not directly addressed to the community and its present needs [19].

From the perspective of the local policies, an important local institution that supports the strategies activated in the territory is the Local Development Agency of Madonie (Agenzia di Sviluppo Locale Delle Madonie), which was born in 1997 for the management of the Territorial Pact of Madonie (Patto Territoriale Delle Madonie) and now handles the territorial promotion, the technical and administrative support to local actors and the interlocution between public and private stakeholders. This area is also the place of a highly relevant national public policy, the National Strategy for Inner Areas (SNAI), which selected the 21 municipalities of Madonie as one of the pilot-areas where apply a place-based development program [20]. In reuse strategies promoted at the national, regional or municipality level, the architectural heritage is more often associated with functions related to tourism and culture, which are often linked in national programs (see also the Italian Recovery Plan). But these are not the only sectors that can be related to the reuse of built heritage: also other functions, connected for instance to public services, can profitably preserve and relaunch historical buildings, thanks to constant use and maintenance, assuring a wider involvement of the community at all levels. Moreover, providing services for citizenship has a positive impact on the territory and the community in general, especially in marginal areas lacking in essential services (education, health, and mobility). Through the case presented by this contribution, but also through many other marginal or not-urban contexts, we can observe various and different dynamics of the relaunch of the territory without constantly recurring to the mass tourism or "commercial" culture exploitation. These experiences also demonstrate the feasibility of these kinds of interventions in similar context of marginalisation. According to Arturo Lanzani's studies, some Italian areas located within underdeveloped valleys, hills, and mountains (such as Appennino Modenese, Valli Nure e Trebbia, south-west Cuneo Province, Piedmont-Lombardy northern border area, Valbelluna and Friuli piedmont) present six different development perspectives [21]. They partially involved the reuse of the architectural heritage related to various sectors, such as innovative primary, post-manufacturing, new residency, slow tourism and social, education, administrative services. In other cases, the population's demands are fulfilled by the community cooperatives, such as the long-established experiences in the *Appennino Emiliano e Piacentino-Parmense,* which take care of the local heritage at different levels, providing different kinds of services (for instance, delivery and e-commerce services during the pandemic). In another context, the local association provides activities and playrooms for children, reusing existing buildings, and organizing events for spreading awareness of the local tangible and intangible heritage, such as the experience of *"Casa delle Agriculture Tullia e Gino"* in Salento. All these dynamics are strictly connected to some peculiar local resources or practices, which had been recently fostered and linked to a wider network to reactivate and strengthen the buildings and the territory in a broad sense. In this bottom-up perspective, the role of the community is crucial. Especially in inner areas, during the last decades, groups of citizens or local administrators tried to provide essential services through the promotion of projects of community welfare, regenerating part of the built fabric and managing the territorial resources [22, 23].

2.1 Conclusions

In the field of architectural preservation, the theme of the reuse of built heritage in marginal areas has been widely investigated [24], showing how this issue is central in the remote places' day-to-day life management, but the recent destabilising situation induces to new reflections about the possible future perspectives for this precious heritage. The reported case study of Petralia Sottana demonstrated the potentiality of the reuse of the historical built heritage as a tool for promoting communities' involvement and providing services addressed to the population. However, sometimes the communities' interests and the economic situation of the small municipalities struggle to converge in inner areas. In such contexts, affected by social and economic marginalisation, the feasibility represents an essential point. Since the lack of resources can hamper the realisation and durability of reuse initiatives from different points of view, national and local policies should support these bottom-up dynamics for gaining a more inclusive approach to inner areas. This contribution aims at illustrating the opportunities related to the creation of services for the community through the reuse of existing built heritage. The experiences mentioned above outlined alternative uses to the tourism-related facilities, meeting the local community's needs. The case of Petralia showed how the local actors reacted to an emergency, but it is necessary to properly sustain these initiatives in order to transform them into structural and long-term strategies. The viability of such reuse initiatives relies on economic sustainability assured by long-lasting activities promoted and addressed to the community and on the tendencies of the specific context and the local capabilities and expectations. The practices here presented show the achievability of such interventions under certain circumstances and should be investigated to understand potentialities suitable also for other marginalised territories and criticalities that could be hindered thanks to community, academic and institutional support.

References

1. Carrosio, G.: Cittadinanza e aree interne in Italia at the conference "Un'agenda di ricerca per le fragilità territoriali, DAStU Dipartimento d'eccellenza Fragilità territoriali, Politecnico di Milano, 26th March 2019
2. Torre, A.: Luoghi. Produzione di località in età moderna e contemporanea, Donzelli, Roma (2011)
3. Tantillo, F.: Comunità. In: Cersosimo, D., Donzelli, C. (eds.) Manifesto per riabitare l'Italia, pp. 91–95. Donzelli, Roma (2020)
4. Pasqui, G.: Città, popolazioni, politiche, Jaca Book, Milano (2008)
5. Pasqui, G.: La città, i saperi, le pratiche, Donzelli, Roma (2018)
6. Bettini, M.: Radici. Tradizione, identità, memoria, il Mulino, Bologna (2016)
7. De Rossi, A. (eds.) Riabitare l'Italia: le aree interne tra abbandoni e riconquiste, Donzelli, Roma (2018)
8. Barbera, F., De Rossi, A. (eds.) Metromontagna. Un Progetto per riabitare l'Italia, Donzelli, Roma (2021)
9. D'Amore, A.: The Madonie farms: from signs of neglect to potential. growth factors of an area of Western Sicily. In: Oteri, A.M., Scamardì, G. (eds.) One needs a town. Studies and perspectives for abandoned or depopulated small towns, International Conference Reggio Calabria, ArcHistoR EXTRA 7, a supplement of ArcHistoR n. 13, Mediterranea University, Reggio Calabria (2020)

10. Accordo di Programma Quadro, Regione Siciliana (2018). https://www.agenziacoesione.gov.it/wp-content/uploads/2020/10/Madonie.pdf. Accessed 28 Nov 2021
11. Minutella, A.G.: Il Progetto di Architettura per la qualificazione dell'Identità Territoriale. Spazio Pubblico e Paesaggio. Una ciclovia per il turismo sostenibile nella Città a rete Madonie-Termini, PhD thesis, Supervisor Prof. Francesco Cannone, co-supervisor Prof. Marcello Panzarella, University of Palermo (2015)
12. Prescia, R.: Abandoned small towns in Sicily. Strategies and proposals for recovering and enhancement. In: Oteri, A.M., Scamardì, G. (eds.) One Needs a Town. Studies and Perspectives for Abandoned or Depopulated Small Towns, International Conference Reggio Calabria, ArcHistoR EXTRA 7, Mediterranea University, Reggio Calabria (2020)
13. Istat, Censimento permanente della popolazione in Sicilia (2019). https://www.istat.it/it/archivio/253856. Accessed 28 Nov 2021
14. National Strategy for Inner Areas indicators, Sicily Region (2017). https://www.agenziacoesione.gov.it/strategia-nazionale-aree-interne/la-selezione-delle-aree/. Accessed 28 Nov 2021
15. Barca, F., Zabatino, A.: Le Madonie, il Mulino, 6/2017, 1006, (2017)
16. Amico, V.: Dizionario Topografico della Sicilia, Salvatore Di Marzo Editore, Palermo (1859)
17. Agenzia del Demanio, Progetto Valore Paese DIMORE, https://www.agenziademanio.it/opencms/it/progetti/valorepaesedimore/. Accessed 28 Nov 2021
18. Madonielive, online newspaper https://madonielive.com/2021/11/18/petralia-sottana-il-comune-chiede-allasp-di-liberare-lex-convento/. Accessed 28 Nov 2021
19. Oteri, A.M.: Historical Buildings in Fragile Areas. Problems and New Perspectives for the care of Architectural Heritage, ArcHistoR Anno VI, n. 11 (2019). https://doi.org/10.14633/AHR118
20. Barca, F., Casavola, P., Lucatelli, S. (eds.) Strategia nazionale per le aree interne: definizione, obiettivi, strumenti e governance, Materiali UVAL, n.31 (2014)
21. Lanzani, A.: Medio-metro-pede montagna. In: Barbera, F., De Rossi, A. (eds.) Metromontagna. Un Progetto per riabitare l'Italia, Donzelli, Roma (2021)
22. Carrosio, G., Luisi, D., Tantillo, F.: Aree interne e coronavirus: quali lezioni? In: Fenu, N. (eds.) Aree interne e covid, Letteraventidue, Siracusa (2020)
23. Sforzi, J., Teneggi, G.: Le imprese di comunità come strumento di welfare rurale. In: "SOCIOLOGIA URBANA E RURALE" 123/2020, pp. 29–45 (2020)
24. Oteri, A.M., Scamardì, G. (eds.) One needs a town. Studies and perspectives for abandoned or depopulated small towns, International Conference Reggio Calabria, ArcHistoR EXTRA 7, a supplement of ArcHistoR n. 13, Mediterranea University, Reggio Calabria (2020)

Action Research for the Conservation of Architectural Heritage in Inner Areas: Towards a Methodological Framework

Marco Rossitti(✉) (iD)

Politecnico di Milano, Via Bonardi 3, 20133 Milano, Italy
marco.rossitti@polimi.it

Abstract. Based on the acknowledgment of architectural heritage role for sustainable territorial development, the international scientific debate and the related official documents have stressed the importance of communities' engagement in its conservation in the last decades.

However, despite the wide recognition of the benefits from communities' active participation in conservation activities, their engagement is still marginal in practice and related to spontaneous experiences with no institutional support.

Thus, it is urgent to properly ground conservation decision processes on communities' engagement, especially in marginal areas, where the ongoing demographic shrinking phenomena call for local communities' centrality in planning strategies.

Based on these premises, the paper proposes a methodological framework to guide Action Research approaches to architectural heritage conservation in inner areas. It moves from investigating the importance of communities' engagement in heritage conservation and enhancement with a particular focus on inner areas' territorial dimension. Then, it analyzes the possible research approaches to conservation issues and identifies the most suitable one into the Action Research approach. Consequently, it delves into the theoretical groundings and main features of Action-Research processes to propose a methodological framework for supporting communities' engagement in conservation. Finally, the implications of adopting such a methodological framework are discussed, including the possible room for improvement and the critical aspects of its effective implementation.

Keywords: Action research · Architectural heritage · Inner areas

1 Introduction

Nowadays, the international policy scenario is focused on exploring ways to voice citizens in public affairs by setting innovative and participative processes [1]. These processes are also investing the cultural sector, even if this policy field still has a lot of room for improvement in promoting participative formats.

In this context, architectural heritage plays a fundamental role. Indeed, cultural heritage's potential for sustainable development is widely recognized [2], and local communities' active engagement in heritage affairs can enhance and trigger this potential.

F. Calabrò et al. (Eds.): NMP 2022, LNNS 482, pp. 26–36, 2022.
https://doi.org/10.1007/978-3-031-06825-6_3

This awareness of communities' role in cultural heritage conservation and management results from a long-lasting scientific debate, reflected in relevant international conventions and policy documents, starting in the second post-war with the *Convention for the Protection of Cultural Property in the Event of Armed Conflict* [3].

The milestones of this debate towards recognizing the community's role for architectural heritage conservation can be identified in the *Declaration of Amsterdam* (1975), the *Charter of Krakow* (2000), and the *Framework Convention on the Value of Cultural Heritage* for Society (2005).

The *Declaration of Amsterdam*, with the *European Charter of Architectural Heritage* (1975), introduces meaningful reflections on the relationship between local communities and their architectural heritage. More in detail, it highlights the importance of citizens' engagement by stating the necessity for public authorities to take decisions «in the public eye, using a clear and universally understood language, so that the local inhabitants may learn, discuss and assess the grounds for them» [4].

Some years later, the *Charter of Krakow* gives stronger emphasis to communities' role for cultural heritage recognition and management by stating that «each community, by means of its collective memory and consciousness of its past, is responsible for the identification as well as the management of its heritage» [5].

Then, in 2005 the European *Framework Convention on the Value of Cultural Heritage for Society* introduces the definition of *heritage community* as «community [...] of people who value specific aspects of cultural heritage, which they wish, within the framework of public action, to sustain and transmit to future generations». Furthermore, the Convention emphasizes the «shared responsibility for cultural heritage and public participation» [6].

This agreement on the importance of community engagement for architectural heritage conservation is constantly echoed also in the most recent international documents [7, 8]. It grounds itself on the several opportunities, both for society and heritage, stemming from a people-centered approach to conservation.

2 Architectural Heritage Conservation and Communities' Engagement

The opportunity of communities' engagement for architectural heritage, well expressed by international conventions and official documents, immediately finds its reasons in the gradual shift of interests in the conservation field from the mere preservation of the material culture to the relationship between people and places. Indeed, by looking at conservation as a social practice «continually recreating social networks and historical and cultural narratives that underpin these binding relations» [9], it becomes evident how architectural heritage conservation cannot prescind from a direct interplay with the communities living and experiencing that heritage.

Furthermore, the co-benefits from heritage conservation are widely recognized: heritage can play an active role in the lives of communities, and communities' support is beneficial to conservation [10]. Indeed, on the one hand, architectural heritage stands as a unifying feature for communities and a carrier of knowledge, in which people can discover both their common roots and some potential for creativity and innovation [11, 12]. On the other hand, people-centered approaches to heritage harness communities' capacity to improve conservation. Local communities, indeed, are endowed with skills and resources that outlast political or expert structures and can be complementary to professional knowledge [13].

Finally, this new emphasis on the social process of conservation, as defined by Erica Avrami [14], arises as an adequate answer to the need to deal with the tension between the static nature of the conservation theory and the ongoing wide cultural and social changes [15].

However, despite the acknowledged importance and benefits of communities' engagement in the scientific debate, the conservation practice remains in the expert domain [16]. In addition, most successful people-centered in conservation affairs suffer from the lack of support from public institutions, thus showing their incapacity to tap into the additional resources that local communities can bring [17].

In this context, it becomes urgent to properly ground conservation decision processes on communities' engagement by proposing structured frameworks that can also be used in the public policy domain without reducing participation to a simple 'label'.

3   Architectural Heritage Conservation in Inner Areas: Which Research Approach?

The investigation of the opportunities from community engagement in architectural heritage conservation makes it clear the need to frame decision processes in which local communities gain a primary role as much as experts/researchers in the conservation field.

Adopting such a perspective towards heritage conservation is even more necessary in territorial contexts, affected by marginalization dynamics, as in Italian inner areas. Here, indeed, the demographic shrinking phenomena led the Italian government to launch the National Strategy for Inner Areas in 2014 [18] by adopting a place-based and people-based approach to welfare recalibration and local development initiatives [19, 20]. However, even in this straightforward policy's implementation, the lack of structured methodological frameworks to guide conservation decisions is evident, thus limiting local communities' scope for action.

As previously mentioned, the definition of a methodological framework supporting decisions about architectural conservation in inner areas cannot prescind from the vibrant interaction between communities and researchers in the conservation field, whose knowledge and skills are essential and complementary.

Thus, the research domain needs to identify the most appropriate approach to take as a reference for defining a possible methodological framework. In this sense, given the importance of communities' engagement, it is helpful to compare the three leading research approaches developed in social sciences: *Conventional Research, Applied Research,* and *Action Research.*

Based on adapting the one made by Hilary Bradbury [21], this comparison highlights the main features of the three different research approaches and their strengths, weaknesses, and opportunities for architectural heritage conservation in inner areas (Table 1).

Table 1. Comparison among the three leading research approaches in social sciences (Author's adaptation to the architectural heritage conservation issue of the comparison proposed by Hilary Bradbury)

	Conventional Research	Applied Research	Action Research
Purpose	To understand	To improve	To understand and improve
Orientation	Researching 'on'	Researching 'for'	Researching 'with'
Researcher	External to the research context	Invited expert	Embedded with the research
Stakeholders	Subject of the research	Clients of the researchers	Co-researchers
Evidence	Qualitative and quantitative data	Qualitative and quantitative data	Experiential, partial, dialogic, intuitive, qualitative, and quantitative
Learning process	Knowledge is developed at a distance from the research context	Linear approach, based on understanding stakeholders' needs and attempting to deal with them	Learning is integrated within the research process through an iterative process
Strengths for conservation	Rigorous and deterministic approach to conservation, deploying a set of techniques to come to a positive outcome	Expert knowledge placed at the service of local needs	Definition of 'best solutions' in terms of conservation as a consequence of discussion and negotiation Possibility to define unique practices without replicating mainstream models
Weaknesses for conservation	Weak relationship with the research context	Risk of ignoring new learning opportunities and of uncritically applying general solutions to specific issues	Difficulty in summarizing and expressing different outcomes in quantitative terms Apparent lack of scientific rigor and concern for objectivity
Benefits for	Serves the academic community in improving its knowledge	Serves the clients of the research	The work belongs to people involved in the research process and allows them to develop new capacities and skills

(continued)

Table 1. (*continued*)

	Conventional Research	Applied Research	Action Research
Action outcomes	Communication and publication of the new knowledge and information produced	Short term win for the stakeholders through solving one specific issue	Action is conceived as a constant element in the research path

By reading through Table 1, *Conventional Research*, which has represented the leading approach to heritage conservation over the last century, stands out as incompatible with the need to engage communities effectively. *Applied Research* allows to place expert competencies at peoples' service to solve specific issues, but not to take full advantage of the opportunities of engaging communities. On the contrary, *Action Research*, considering communities' members as co-researchers and placing mutual and iterative learning at the core of the research process [22], perfectly fits the requirements of a people-centered approach to conservation issues.

For these reasons, it can be beneficial to develop a methodological framework to support decision processes for architectural heritage conservation in inner areas in light of an *Action Research* approach.

4   Action Research for the Conservation of Architectural Heritage in Inner Areas: Towards a Methodological Framework

The definition of an *Action Research* methodological framework for architectural heritage conservation requires delving into the theoretical groundings and main features of *Action Research* processes. Indeed, this preliminary understanding is fundamental to identifying the crucial steps for an *Action Research* process at the core of the methodological framework proposal.

4.1   Action Research: Theoretical Groundings and Main Features

Action-Research can be defined as: «a participatory process concerned with developing practical knowing in the pursuit of worthwhile human purposes. It seeks to bring together action and reflection, theory and practice, in the pursuit of practical solutions [...] and the flourishing of individual persons and their communities» [23].

This comprehensive definition highlights the *Action-Research* distance from the conventional way of researching conceiving relationships, knowledge, and the link to practice. *Action-Research*, indeed, is both about working towards practical outcomes to improve people's well-being and about producing new knowledge through the constant integration of action and reflection [24].

Such an innovative orientation to scientific research finds its grounding in some branches of contemporary philosophy, as *hermeneutic epistemology*, the *new science* coming from the Self-Organization Theory, and the *dialogic-communicative rationality*.

Indeed, all these different branches try to overcome the dominant dichotomies in Western thinking, as the one between knowledge and action, by interpreting the knowledge process as a continuous and iterative process, in which subject and object are mutually modified [25].

The two main features of an *Action-Research* approach, thus, can be identified in:

• The innovative way of conceiving the relationship between knowledge and action. Indeed, they are no longer considered as two consequential phases of a decision process but as two inseparable aspects of a unique *learning-by-doing* process;
• The demolition of the traditional and hierarchical distinction between the expert/researcher and the research object.

These detected main features nourish all the phases of effective *Action-Research* processes, that can be outlined as [26]:

1. Co-definition of the problems to be examined;
2. Co-generation of relevant knowledge about them;
3. Co-learning and co-design of actions;
4. Taking actions;
5. Interpreting results.

4.2 Action Research for the Conservation of Architectural Heritage in Inner Areas: A Possible Methodological Framework

The examination of the *Action-Research*'s main features and theoretical groundings makes it even more evident the opportunity of adopting this research approach for the conservation of architectural heritage in inner areas.

On the one hand, the innovative role of co-researchers and co-designers assigned to local stakeholders and communities completely meets the needs of inner areas. On the other hand, the dual purpose of *Action-Research* processes, working both on achieving practical outcomes and generating new helpful knowledge, perfectly blends with nowadays' conservation concerns. These concerns, indeed, are related to the necessity not to limit the architectural heritage conservation process to restoration works or reuse decisions but to develop long-lasting processes based on new significance recognition and innovative management forms by local communities [27].

Based on these premises, it can be promising to define a methodological framework, resting on the *Action-Research* approach, to support and guide architectural heritage conservation decisions.

This methodological framework can stem from linking to each of the five phases of effective *Action-Research* processes [26] a set of possible research techniques and methodologies to support it (see Fig. 1).

The proposed research techniques and methodologies, also characterizing the few existing experiences of *Action-Research* applied to heritage conservation [28–30], include both conventional and formal research techniques, and informal and innovative ones, qualitative and quantitative research methodologies. This kind of integration well expresses the *Action-Research* nature of a mixed-method research strategy, where the particular mix of methods must be contextually determined [31].

The necessity of contextually determining the methodologies to be selected raises a relevant point about the methodological framework application: it does not aspire and cannot be exhaustive, but, in light of the advocated place-based and people-based approach [32], it stands as a structured reference track to be integrated or revised according to the specificities of the inner area and the architectural heritage under study.

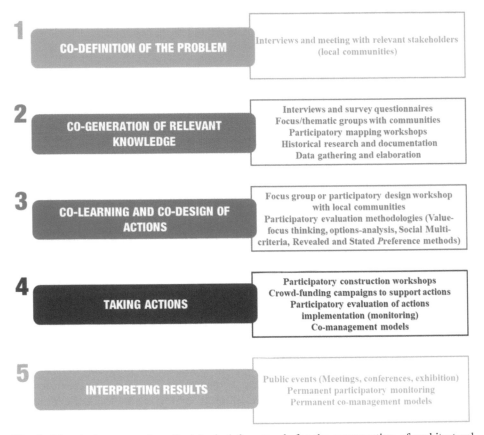

Fig. 1. The *Action-research* methodological framework for the conservation of architectural heritage in inner areas

Finally, by reading through the different phases and related tools in the methodological framework, it is evident that evaluation methodologies can play a crucial role by producing constant feedback to improve the process and productively nourish the iterative *learning-by-doing* effort [1].

5 Conclusions

The paper presents a methodological framework, conceived in light of the Action-Research approach, to support and guide architectural heritage conservation decisions in inner areas.

The adoption of such a framework, on the one hand, allows making communities' engagement effective in conservation issues. On the other hand, it provides decision-makers with support for implementing *Action-research* processes, thus addressing the lack of such experiences in the public institutions' domain.

The proposed methodological framework doesn't and cannot aspire to be exhaustive in displaying all the available or useful methodologies or techniques to implement an *Action-Research* approach for heritage conservation. It must be conceived as a structured track that can be integrated or revised to better fit the specificities of the territorial context and the architectural heritage under analysis and its local community.

Furthermore, by reading through the different phases and the related tools in the framework, it is evident that evaluation methodologies can play a crucial role in supporting the decision process implementation and feeding the constant *learning-by-doing* effort.

further research efforts will be devoted:

- to enrich and specify the array of methodologies and techniques proposed in the framework by performing a detailed analysis of the experiences of *Action-Research* applied to heritage conservation. Indeed, in scientific literature, it is possible to find different initiatives of *Action-Research* applied to conservation issues, and some refer to inner areas' territorial context [33–35]. Indeed, these experiences cover different phases of the *Action-Research* process by resorting to various tools. Thus, their cross-reading can provide a broad overview of the potentialities of such an approach for heritage conservation.
- to better investigate the role of evaluation in each phase of the methodological framework;
- to apply the framework to a specific territorial context within the Italian inner areas municipalities to address relevant heritage conservation issues.

Indeed, implementing such a methodological framework can bring relevant implications in the conservation field. It can provide a structured and methodologically robust guide to community engagement in heritage issues, thus overcoming the long-standing skepticism towards bottom-up approaches in conservation [16] and making community participation more than a simple label. In this sense, it becomes necessary for researchers to properly balance and choose the tools to be used and carefully tailor them in light of the communities' response to the process.

References

1. European Union: Participatory Governance of Cultural Heritage. Report of the OMC (Open Method of Coordination) working group of Member States' experts. Publications Office of the European Union, Luxembourg (2018) https://op.europa.eu/en/publication-detail/-/public ation/b8837a15-437c-11e8-a9f4-01aa75ed71a1, Accessed 23 Dec 2021
2. Baltà Portolés, J.: Cultural Heritage and Sustainable Cities. Key Themes and Examples in European Cities. UCLG Committee on Culture Reports 7 (2018) https://www.agenda21cult ure.net/sites/default/files/report_7_-_cultural_heritage_sustainable_development_-_eng.pdf Accessed 23 Dec 2021
3. Unesco: Convention for the Protection of Cultural Property in the Event of Armed Conflict with the Regulations for the execution of the Convention (The Hague, 14 May 1954). (1954) http://www.unesco.org/new/fileamin/MULTIMEDIA/HQ/CLT/pdf/1954_C onvention_EN_2020.pdf Accessed 23 Dec 2021
4. Congress on the European Architectural Heritage: The Declaration of Amsterdam (Amsterdam, October 1975) (1975) https://www.icomos.org/en/and/169-the-declaration-of-ams terdam, Accessed 23 Dec 2021
5. International Conference on Conservation "Krakow 2000": The Charter of Krakow 2000. Principles for Conservation and Restoration of Built Heritage (Krakow, October 2000). (2000) https://www.triestecontemporanea.it/pag5-e.htm Accessed 23 Dec 2021
6. Council of Europe: Council of Europe Framework Convention on the Value of Cultural Heritage for Society (Faro, 27 October 2005). European Treaty Series 199. (2005) https://rm. coe.int/1680083746 Accessed 23 Dec 2021
7. European Parliament: Report towards an integrated approach to cultural heritage for Europe (2014/2149(INI)). (2015) https://www.europarl.europa.eu/doceo/document/A-8-2015-0207_EN.pdf Accessed 23 Dec 2021
8. United Nations: Transforming Our World: The 2030 Agenda for Sustainable Development. (2016) https://sustainabledevelopment.un.org/content/documents/21252030%20A genda%20for%20Sustainable%20Development%20web.pdf Accessed 23 Dec 2021
9. Smith, L.: Discourses of heritage: implications for archeological community practice. Nuevo Mundo Mundo Nuevos. (2012). https://doi.org/10.4000/nuevomundo.64148
10. Galla, A. (ed.): World Heritage: Benefits Beyond Borders. Cambridge University Press, Cambridge (2012)
11. Jokilehto, J.: Engaging Conservation. Communities, place and capacity building. In: Chitty, G. (ed.) Heritage, Conservation and Communities. Engagement, Participation and Capacity Building, pp. 17–33. Routledge, Abingdon (2017)
12. Walter, N..: Everyone loves a good story. Narrative, tradition and public participation in conservation. In: Chitty, G. (ed.) Heritage, Conservation and Communities. Engagement, Participation and Capacity Building, pp. 50–64. Routledge, London (2017)
13. ICCROM: People-Centred Approaches to the Conservation of Cultural Heritage: Living Heritage. (2015) https://www.iccrom.org/sites/default/files/PCA_Annexe-2.pdf Accessed 27 Dec 2021
14. Avrami, E.: Heritage, values and sustainability. In: Richmond, A., Bracker, A.L.: (eds), Conservation: Principles, Dilemmas and Uncomfortable Truths, pp. 177–183. Elsevier/Butterworth-Heinemann, pp. 177–183 (2009)
15. Glendinning, M.: The Conservation Movement: A History of Architectural Preservation – Antiquity to Modernity. Routledge, London (2013)
16. Rashid, M.: Introduction. In: The Geometry of Urban Layouts, pp. 3–11. Springer, Cham (2017). https://doi.org/10.1007/978-3-319-30750-3_1

17. Wijesuriya, G., Thompson, J., Court, S.: People-centred approaches. Engaging communities and developing capacities for managing heritage. In: Chitty, G. (ed.) Heritage, Conservation and Communities. Engagement, participation and capacity building, pp. 34–49. Routledge, London (2017)
18. DPS: Strategia nazionale per le Aree interne: definizione, obiettivi, strumenti e governance. Documento tecnico collegato alla bozza di Accordo di Partenariato trasmessa alla CE il 9 dicembre 2013. (2013) https://www.agenziacoesione.gov.it/wp-content/uploads/2020/07/Strategia_nazionale_per_le_Aree_interne_definizione_obiettivi_strumenti_e_governance_2014.pdf, Accessed 27 Dec 2021
19. Carrosio, G.: A place-based perspective for welfare recalibration in the Italian inner peripheries: the case of the Italian strategy for inner areas. Sociologia e Politiche Sociali 19(3), 50–64 (2016). https://doi.org/10.3280/SP2016-003004
20. Saija, L., Pappalardo, G.: From enabling people to enabling institutions. a national policy suggestion for inner areas coming from an action-research experience. In: Bevilacqua, C., Calabrò, F., Della Spina, L. (eds.) NMP 2020. SIST, vol. 178, pp. 125–134. Springer, Cham (2021). https://doi.org/10.1007/978-3-030-48279-4_12
21. Bradbury, H.: Introduction: How to Situate and Define Action Research. In: Bradbury, H. (ed.) The SAGE Handbook of Action Research. 3rd edn. Sage Publications, London (2015)
22. Ponzoni, E.: Windows of understanding: broadening access to knowledge production through participatory action research. Qual. Res. 16(5), 557–574 (2015). https://doi.org/10.1177/1468794115602305
23. Reason, P., Bradbury, H. (eds.) Handbook of Action Research. Participative Inquiry and Practice. 1st edn. Sage Publications, London (2001)
24. Reason, P., Bradbury, H.: Introduction. In: Reason, P., Bradbury, H. (eds.) The SAGE Handbook of Action Research. Participative Inquiry and Practice. 2nd edn. Sage Publications, London (2008)
25. Saija, L.: Prospettive di ricerca-azione nella disciplina urbanistica. Infolio 19, 49–52 (2007)
26. Prodan, R., Fahringer, T.: Introduction. In: Grid Computing. LNCS, vol. 4340, pp. 1–11. Springer, Heidelberg (2007). https://doi.org/10.1007/978-3-540-69262-1_1
27. Chirikure, S., Manyanga, M., Ndoro, W., Pwiti, G.: Unfulfilled promises? Heritage management and community participation at some of Africa's cultural heritage sites. Int. J. Herit. Stud. 16(1–2), 30–44 (2010). https://doi.org/10.1080/13527250903441739
28. Falanga, R., Nunes, M.C.: Tackling urban disparities through participatory culture-led urban regeneration. Insights from Lisbon. Land Use Policy 108, 105478 (2021). https://doi.org/10.1016/j.landusepol.2021.105478
29. Gianfrate, V., Djalali, A., Turillazzi, B., Boulanger, S.O.M., Massari, M.: Research-action-research towards a circular Urban system for multi-level regeneration in historical cities: The case of Bologna. Int. J. Design Nature Ecodyn. 15, 5–11 (2020). https://doi.org/10.18280/ijdne.150102
30. Carmichael, B., et al.: A methodology for the assessment of climate change adaptation options for cultural heritage sites. Climate 8(8), 88 (2020). https://doi.org/10.3390/cli8080088
31. Greenwood, D.J., Levin, M.: Social science research techniques, work forms, and research strategies in action research. In: Greenwood, D.J., Levin, M. (eds.) Introduction to Action Research. 2nd edn. Sage Publications, London (2007)
32. Calvaresi, C.: Le Aree Interne, un problema di policy. Territorio 74, 87–90 (2015). https://doi.org/10.3280/TR2015-074015
33. Oppido, S., Ragozzino, S., Micheletti, S., Esposito De Vita, G.: Sharing responsibilities to regenerate publicness and cultural values of marginalized landscapes: Case of Alta Irpinia, Italy. Urbani Izziv 29, 125–142 (2018). https://doi.org/10.5379/urbani-izziv-en-2018-29-supplement-008

34. Salvatore, R., Chiodo, E., Fantini, A.: Tourism transition in peripheral rural areas: Theories, issues and strategies. Ann. Tour. Res. **68**, 41–51 (2018). https://doi.org/10.1016/j.annals.2017. 11.003
35. Pappalardo, G.: Community-based processes for revitalizing heritage: Questioning justice in the experimental practice of ecomuseums. Sustainability **12**(21), 9270 (2020). https://doi.org/ 10.3390/su12219270

Community-Driven Initiatives for Heritage Acknowledgement, Preservation and Enhancement in European Marginal Area
The case of Roşia Montană (Romania)

Oana Cristina Tiganea[✉] and Francesca Vigotti

Department of Architecture and Urban Studies, Politecnico di Milano, Milano, Italy
{oanacristina.tiganea,francesca.vigotti}@polimi.it

Abstract. By briefly reviewing community-driven approaches for the preservation of widespread heritage, the paper poses some reflections on the role of bottom-up approaches as trigger of top-down recognitions of properties set in marginal areas. In the Romanian context, mining activities have significantly contributed to designing the territories. When the extraction activities cease, the traces of production remain as parts of what might be considered a difficult but extraordinary legacy, which should be interpreted as a system. The material inheritance concerns not only the buildings but also the systems of relations with the environment and the memory of those who lived in these production places. The debate around recognizing these sites as "heritage" is long and strongly discussed at the international level. The preservation of the heritage and environment of these sites has been at the centre of civil mobilization in the last twenty years, which in time has raised national and international attention. Such propositions completely fit in the case of Roşia Montană Mining Landscape, the most recent recognized "cultural landscape" property in Romania by the World Heritage Centre (UNESCO). Based on literature review and research within the field of mining heritage, the paper is a first attempt to analyse the case of Roşia Montană Mining Landscape as part of a nation-wide territorial *constellation* of former gold mining sites, which in the last decade are going through deindustrialisation simultaneously with the acknowledgement and re-signification processes as heritage.

Keywords: Mining cultural landscape · Heritage · Bottom-up initiatives · Marginal territories · Romania

1 Built Heritage and Community-Driven Actions in European Marginal Areas: An Introduction

At the European level, the so-called "inner peripheries" [1] and "marginal territories" [2], which in many cases coincide with rural and mountainous areas, have been at the centre of debates in the last decade. National territorial cohesion policies have been put in place throughout Europe as experimental, participatory processes to reduce social, economic, and territorial inequalities. While focusing on accessibility to essential services, such policies have often indicated cultural heritage and rural capital as drivers

© The Author(s), under exclusive license to Springer Nature Switzerland AG 2022
F. Calabrò et al. (Eds.): NMP 2022, LNNS 482, pp. 37–46, 2022.
https://doi.org/10.1007/978-3-031-06825-6_4

to promote growth. The drafting of strategies for the development of marginal areas and their implementation has also raised the attention of academics and professionals from heterogeneous backgrounds on these territories. In this scenario, the theme of the preservation of built heritage in marginal and depopulated areas, which in many cases is abandoned, has been explored as a possible resource for development [3]. The COVID-19 pandemic has exacerbated territorial gaps, underlined structural differences, and fuelled critical aspects. In response, the European Commission has announced its Recovery Plan, Next Generation EU, in 2020. The Plan represents an unprecedented fund to *"help repair the immediate economic and social damage brought about by the coronavirus pandemic"* [4]. As of November 2021, almost all EU Countries have submitted their National Recovery and Resilience Plan: some States have included interventions on built heritage as part of the action lines dedicated to the cultural sector [5, 6].

Besides the top-down plans and policies drafted by the European Commission and national authorities, some bottom-up approaches have been developed in marginal areas by local communities and stakeholders to regenerate their territories and identities. These approaches often start with maintenance and reuse actions or follow a process of "reappropriation" of once productive, now abandoned, areas. Even though many of these initiatives have started before the pandemic, community-based actions for heritage conservation in "peripheral" areas might represent a unique testbed, in the current context, towards social integration and patrimonial acknowledgement. The engagement of local communities to protect their heritage might be elicited by specific events, interpreted as imminent threats, e.g., demolitions, relocation, induced abandonment, or as the result of long-term processes and dynamics. In some cases, the bottom-up approach of community mobilisation to safeguard heritage might be followed by top-down listings and actions.

While on the one hand, listings might raise attention towards the heritage protection, catalysing resources both in financial, legislative, and research terms, on the other, official recognitions might concur to unprecedented issues, e.g., exceed tourism carrying capacity limits. These aspects are particularly critical in sites set in marginal areas, which often face difficult access to essential services and must deal with the progressive loss of local communities, such as in the case of Roșia Montană (Alba County, Romania).

2   Roșia Montană as Part of a Broader Territorial *Constellation*

After a ten-year bureaucratic process, and after twenty years of community-driven initiatives, on July 27th, 2021, Roșia Montană (Alba County, Romania) (Fig. 1) was nominated on the World Heritage list as a mining landscape. Like many other mining sites listed as World Heritage Properties, *Roșia Montană Mining Landscape* is categorised as "cultural landscape", representing the only property of the State Party recognised as such, as well as its first site listed as "in danger" [7, 8]. Also, the revised report submitted in early 2020 for the candidacy of Roșia Montană as World Heritage property mentions the interpretation of the proposed core zone to include those areas and buildings "directly or indirectly connected to mining" [9].

Romania's architectural preservation practice is defined by a series of continuities and discontinuities throughout the 20th century, strongly influenced and interlinked with the

Fig. 1. Roșia Montană mining landscape. Source Google Earth, Maxar Technologies, 2021.

political setting following drastic consequences on the built environment and the overall cultural perception and acknowledgement of patrimonial values [10–13]. Considering this, the concept of "cultural landscape" has been introduced only recently, during the post-socialist years of (re)organisation and (re)definition of architectural practice and preservation legal framework. Thus, in 2000, the urban and territorial planning tools were issued, which included the acknowledgement, protection and safeguarding at the landscape level of the cultural heritage through the Law concerning the Planning of the National Territory (L.5/2000) [14]. Moreover, the newly introduced landscape concept was elaborated about recognising the patrimonial value of industrial legacies at the territorial level, e.g., industrial landscape. In consequence, in 2000, Roșia Montană, together with other gold mining settlements from Apuseni Mountains, was nominated as *protected areas*, e.g., Roman pits from Roșia Montană (Alba County), Roman Steps Shaft from Brad (Hunedoara County), and "Adamul Vechi" mine from Baia de Criș (Hunedoara County), in advanced of the issue of the Historic Monuments Act in 2001 [15]. These mining settlements are only a few production points, component elements of the broader industrial territory defined as "Gold Quadrilateral", shaped, and transformed continuously in more than two millennials of gold extraction activity and generating a "geo-mining system" as illustrated and described by the researcher Gabriela Pașcu [16]. In this scenario, entire regions were rapidly shaped by territorial interconnections and hierarchies developed based on the industrial flux principles, establishing a *constellation* of production places generated from isolated working communities to cities. Thus, new communities were established and set around these production places, in settlements designed and built to sustain a particular way of living centred around the concept of "industrial culture", typical for both western and eastern European industrial development.

Roșia Montană site presents some of the most complex and long-lasting testimonies of Roman gold extraction activity, as the candidacy dossier to become UNESCO property indicates it, formed by at least seven-kilometre-long underground galleries evaluated for their highly engineered works and waterwheels in four underground localities chosen for their high-grade ore. To this layer are added everything that came after the Roman period in mining exploitation, together with the intense industrialisation phase of the $18^{th} - 20^{th}$ century that drastically shaped the above-ground [17–19]. As architect Stefan Bâlici, manager of INP, describes this endurance translated into a "*systematic and profound interrelation between the natural setting and cultural phenomena from deep down into the mountain, all the way to the surface, from topography to fauna and flora and to the human communities of the area, which produced one of the richest and most spectacular cultural landscapes of Romania, and possibly of Europe*" [20]. Therefore, the cultural heritage from Roșia Montană can be defined starting from the underground and categorised in the many layers that overlapped in time, e.g., *underground mining system* consisting of galleries, extraction chambers, vertical workings, shafts, drainage channels; *above-ground*, the technical and technological network defined in various time periods for the ore processing, the railway and water networks (over 100 water reservoirs built on the slopes around the mine); *the Roman archaeological sites*; *the architectural and urban heritage* dating the $18^{th}–20^{th}$ century defined by multi-ethnic and multi-cultural testimonies, nonetheless united by the mono-industrial feature of the settlement; the "*small historic waste heaps and dumps*" formed in time, which have transformed the landscape giving its characteristic "undulating shapes"; and, nonetheless, the agropastoral landscape and way of living specifically for Apuseni Mountains [21, 22].

Looking at the long-lasting mining activity from Roșia Montană, the fluid transformation of the entire territory and landscape is considered interrupted with the 1948 socialist nationalisation and kick off the "forced industrialisation" [23]. The post-1948 period marks a shift at the national level in a matter of industrial approach, e.g., property status, technological upgrades, territorial systematisation, which shaped the landscape and communities furthermore and, most probably, with more evident environmental consequences in Roșia Montană such as the passage from underground to open-cast mining. The start of open-cast mining in 1970 and carried out until 2006 determined the physical disappearance of some previous pre-industrial and industrial testimonies. What is left from this last industrialisation phase – first under the socialist State during 1970–1994, then in a privatised manner during 1994–2006 – is perceived as "*a picture of abuse, social, environmental, cultural, and even economic*" [19].

But probably the high impact mining activity had on the defined geographical area of "Gold Quadrilateral", Roșia Montană included, can be comprehended when analysing the deindustrialisation process and its effects. Once the main mining activity ceased to function, the entire *constellation* suffered severely in a chain reaction presenting the signs of economic, social, and cultural shrinkage. Given these premises, the proposition is that the *continuing landscape* of Roșia Montană might be interpreted and approached within the broader territorial system: once at the centre of a complex production system, after 2006 forced ceasing of mining activities, the territory has started a transition as marginal area, only to gain back limelight as a heritage site.

3 Community-Driven Initiatives for (Re)Defining the *Cultural Identity*

Because of civic mobilisation, multiple community-driven initiatives have been developed in Romania during the last two decades. Focusing on the protection and management of scattered heritage, such approaches are also generated as a response to the lack of top-down planning tools and legislation by the State. Many of these initiatives occurred in similar territorial, economic, and sociocultural settings defined by that of the small mining towns. Such context was defined as the foundation of a homogenously distributed urban scheme, in the framework of the socialist urban theory [24, 25].

Roşia Montană, Baia Sprie, Brad, Anina, and Petrila (Fig. 2) result from long-term history, started in most cases centuries ago, which carry a diversified inheritance and a palimpsest of functions and communities. These small mining towns have been the centre of a diversified ecosystem of *bottom-up* initiatives, contributing to their re-signification as heritage sites and an increased sense of place attachment by communities. All developed after the 2006 general mining shutdown and in part still ongoing, the initiatives carried in the former mining sites were co-developed by communities and architects, archaeologists, artists, and heritage professionals. Starting from punctual actions, more complex and multidisciplinary research-in action projects were structured in all the sites [26].

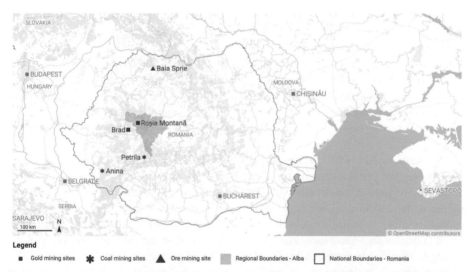

Fig. 2. The location of small-town mining sites of Anina, Petrila, Brad, Baia Sprie and Roşia Montană. Elaboration by the Authors via Datawrapper.

Among the case-studies mentioned above, Roșia Montană represents the "most famous" at the international level and probably the most extended community-driven initiative at the national level, triggered during early 2000s due to the mining plans with high environmental and cultural heritage impact. In 1999, in a period of economic and political transition, while the open-cast mining activities were still in operation, a foreign company which became Roșia Montană Gold Corporation (RMGC), with political support, started a plan aimed to enlarge the open-air mining activities on the site using dynamite and cyanide-gold extraction, with consequences in the demolition of most of the villages and the relocation of inhabitants. The mining project proposed by RMGC would have meant the relocation of *"910 households, displacement of about 2000 persons from 740 houses and 138 flats, destruction of four mountains and natural monuments, a lake of cyanide and toxic waste covering over 1800 hectares of land, demolished houses and buildings cultural patrimony such as the famous Roman Galleries and last but not least, un-burying their ancestors through the destruction of nice cemeteries and eight churches"* [27]. Such intentions lead, in time, to a crescent community-driven attempt to preserve and ensure use Roșia Montană mining landscape, where the civic right to environment and cultural heritage was raised and strongly interlinked for the first time in post-socialist Romania. Thus, civil society claimed the right to a just transition of the site, shifting from the mining activity to a renovated social and economic model based on long-term sustainable use of existing resources. Given this proposition, the landscape and scattered built heritage of Roșia Montană were considered fundamental capitals to foster the site's development. The movement, which grew in time at the national and international level, is known as *Save Roșia Montană* [28], and challenged publicly the political class and new democratic principles established in post-1989 Romania while pushing the top-down initiative directed towards its official acknowledgement as a World Heritage site.

Among other initiatives, a significant effort devoted to the knowledge, conservation and classification of architectural heritage has been carried out by *ARA – Architecture. Restoration. Archaeology*, a national cultural association. The organisation fosters actions to preserve the built environment by engaging the local population and volunteers in restoration workshops, disseminating the results of activities to the public and the scientific community. An example of such activities is the "Adopt a House" program. Through the initiative, the NGO provides specialistic knowledge and technical skills to assist the owners of houses in the maintenance operations of buildings in Roșia Montană [29, 30].

As mentioned in Sect. 2, the community-driven approaches to safeguard the scattered heritage of the site started years before the candidacy process of the site as World Heritage. The bottom-up mobilisation described above resulted from the issues affecting the conservation of Roșia Montană mining landscape to gain, at first, visibility of national heritage organisations. In the early 2000s, the local ICOMOS Romania chapter raised concerns regarding the threats to the site related to the development of open-air mining extractions as foreseen by the plans of Roșia Montană Gold Corporation [31]. To these appeals followed the interest by international bodies. In 2013, the site was selected to become part of the "7 most endangered" program by Europa Nostra [32] because of the candidacy file presented by Pro Patrimonio, an international non-profit, non-governmental organisation. The justification of the nomination recognises actions fostered by local organisations (ARA, Albunurs Maior, Pro Patrimonio Foundation) to safeguard the Roşia Montană heritage system. The path towards the recognition of Roşia Montană as World Heritage was based also on these initiatives, which have underlined the main critical aspects and threats, but also the potentialities and opportunities that the site might encounter in the nomination process as UNESCO property.

4 Possible Threats and Opportunities, from Past to Future: Some Remarks and Open Questions

In the last twenty years, Roşia Montană has passed through a series of status (re)definition from economic, environmental, cultural, and civic perspective as follows: from a centre of gold extraction became simultaneously the marginal deindustrialized and forced depopulated area as well as the centre of civic revival of Romanian post-socialist society. Furthermore, the UNESCO nomination contributes to transforming the site in a centre of culture with focus on the environmental and cultural heritage safeguarding and enhancement within an overall post-industrial strategy. However, the lack of such a strategy or overall top-down vision, tools, and mechanisms of the way this site can be redefined from an economic and cultural perspective, determined its simultaneously nomination as "World Heritage in Danger" [33].

The "in danger" status conferred by the World Heritage Centre raises diverse considerations on, on the one hand, the existing threats connected to the possible advancement of the open-cast mining in the area. On the other, the concomitant inclusion of the Property as World Heritage and as an endangered site embraces the concerns as exposed by the bottom-up initiatives in the area, strongly related to communities' perception of their context as a layered process still on-going, which should be preserved, cared of, and transmitted as a whole in time.

After reviewing the top-down and bottom-up initiatives, a main question remains open to further discussions: what does the "in danger" label mean to the different stakeholders involved in Roşia Montană? In the other sites briefly discussed (Anina, Petrila, Brad, Baia Sprie) which are the possible consequences of the lack of an official recognition of possible threats to these sites?

As the bottom-up movements and community-based actions have brought to the attention not only the evident, exogenous large-scale events that threat Roşia Montană, but also the endogenous, subtle, and small-scale ones, it seems clear that civic mobilizations have, at least partially, contributed to advancement in tackling risks and threats affecting the property from different perspectives. At the same time, the cases of other small mining towns discussed in the paper show how the partial, if not absent, recognition of immediate and longer threats might lead the sites to abandonment and, ultimately, to disappearance.

But probably what is still lacking in Romanian setting is an overall vision of the post-industrial *constellation* considered from its complex territorial interexchange and the role that Roşia Montană could play as a centre. To arrive to such a broader approach, administrative limits should be overpassed together with their political constraints, pushing towards a complex perception of the Apuseni Mountains *goldscape* [34].

This raises a different paradigm in reference to the possibility, or not, of coexistence of the industrial and cultural production on the same site. While one looks at the site from a perspective of its natural and environmental resources consumption, e.g., tangible disappearance and transformation of built and natural elements, the latter is based on the same tangible resources' maintenance and preservation in view of "cultural consumption". Moreover, it requires a shift in mentality that is based on the idea that the economic stability is linked with the sole industrial production, much appreciated and rooted in the Eastern European post-socialist countries [35].

But for this, the site needs its community to embrace, promote, and sustain the "cultural consumption". Roşia Montană, despite its well-deserved obtained visibility within the national and international scenes due to its tenacious civic initiatives, remains a contemporary case of induced depopulation area where the local, now World Heritage, remains stripped from its intangible manifestation able to maintain alive the spark of the "industrial culture and identity" as in case of Brad, Petrila, or Anina.

In this optic, if maintained the vision of the overall *goldscape* constellation in the Apuseni Mountains, Roşia Montană could gain a wider territorial community deeply shaped and rooted in those exact elements that define the authenticity and integrity of the mining landscape.

The theme of the article represents a common reflection of the two authors, while the direct contribution is divided as follows: Francesca Vigotti contributed with the first and third part, and Oana Cristina Tiganea contributed with the second and fourth part. The review and editing of the text were accomplished together.

References

1. ESPON, POLICY BRIEF - Inner peripheries in Europe Possible development strategies to overcome their marginalising effects, p. 3 (2018)
2. Moscarelli, R.: Marginality: from theory to practices. In: Pileri, P., Moscarelli, R. (eds.) Cycling & Walking for Regional Development – How Slowness Regenerates Marginal Areas, pp. 23–38. Springer, Cham (2020)
3. Oteri, A.M., Scamardi, G.: Un paese ci vuole. Studi e prospettive per i centri abbandonati e in via di spopolamento. ArcHistoR Extra no. 7 (2020)
4. Recovery plan for Europe. https://ec.europa.eu/info/strategy/recovery-plan-europe_en Accessed 27 Dec 2021
5. Culture Action Europe, Culture in the EU's National Recovery and Resilience Plans. The state of play one year after the launch of the campaign to earmark 2% in the post-pandemic strategies, November 2021
6. Italia Domani, Piano Nazionale di Ripresa e Resilienza, Attrattività dei Borghi, https://italia domani.gov.it/it/Interventi/investimenti/attrattivita-dei-borghi.html Accessed 27 Dec 2021
7. World Heritage Centre, Romania. https://whc.unesco.org/en/statesparties/ro Accessed 27 Dec 2021
8. World Heritage Centre, List of World Heritage in Danger. https://whc.unesco.org/en/danger/ Accessed 27 Dec 2021
9. Institutul Naţional al Patrimoniului, Roşia Montana Mining Landscape (Romania), Progress report on the measures required to ensure the protection and management of the potential OUV of the property, as identified by ICOMOS, p. 1, January 2020

10. Iuga, L.: Reshaping the Historic City under Socialism: State Preservation, Urban Planning and The Politics of Scarcity in Romania (1945 – 1977), PhD Dissertation CEU, Budapest (2016)
11. Iamandescu, I.: Patrimoniul industrial in Romania: despre stadiul inventarierii specializate. In: Arhitectura, vol. 4–5, pp. 68–69 (2017)
12. Grama, E.: Socialist Heritage. The Politics of Past and Place in Romania. Indiana University Press, Indiana (2019)
13. Tiganea, O.: The conservation of the industrial heritage: theoretical approaches and territorial experimentations in the case of Anina (Romania). In: ArcHistoR VII(13), 342–379 (2020)
14. Lege nr. 5 din 2000 privind aprobarea Planului de amenajare a teritoriului național – Secțiunea III – Zone protejate, MO. Nr. 152/12 aprilie 2000 (Ro); Law n. 5 from 2000 concerning the national territorial systematization plan, Section III – Protected Areas, published in the Official Bulletin n. 152/April 12th, 2000
15. Legea nr. 42 din 18 iulie 2001 privind protejarea monumentelor istorice, publicata in MO Partea I nr. 407/24 iulie 2001 (Ro); Law n. 42 issued on July 18th, 2001, concerning the protection of historic monuments, published in the Official Bulletin Part I, n. 407/July 24th, 2001
16. Pascu, G.: Patrimoine Industriel-Minier, Facteur de Développement Territorial. Complexité et enjeux en Roumanie, en comparaison avec la France et la Grande-Bretagne. PhD Thesis, FAU Timisoara & Universite Jean Monner de Saint-Etienne (2015)
17. Wollman, V.: Mineritul metalifer, extragerea sarii si carierele de piatra din Dacia Romana, Cluj-Napoca (1996)
18. Wollmann, V.: Patrimoniul preindustrial si industrial al Romaniei, (Vol. I), Honterus, Sibiu (2010)
19. Balici, S.: Rosia Montana. An overview on the question of cultural heritage. In: Caiete ARA 4, 205–228 (2013)
20. Ibidem
21. Lista monumentelor istorice 2015, Judetul Alba (Ro); Historic Monuments List 2015, Alba County (En). https://patrimoniu.gov.ro/images/lmi-2015/LMI-AB.pdf Accessed 29 Dec 2021
22. World Heritage Centre, Roșia Montană Mining Landscape. https://whc.unesco.org/en/list/1552/ Accessed 29 Dec 2021
23. Musteata, S., Cozma, E.: Community heritage: case of Rosia Montana mining landscape in Romania. In: Trskan, D., Bezjak. S. (eds.) Archaeological Heritage and Education. An international perspective on History Education, Slovenian National Commission for UNESCO, Ljubljiana (2020)
24. Cucu, S.: Orasele din R.S. Romania. Probleme de geografie economica (summary of the doctoral thesis), Alexandru Ioan Cuza University, Iasi (1977)
25. Ronnas, P.: Urbanization in Romania. A geography of social and economic change since independence. Ph.D. thesis, The Economic Research Institute – Stockholm School of Economics, Stockholm (1984)
26. Tiganea, O.: Taking action towards the enhancement of mining heritage in Romania. In: Bevilacqua, C., et al. (eds.): NMP 2020, SIST 178, pp. 1–13 (2021)
27. Velicu, I.: Demonizing the sensible and the 'revolution of our generation' in Rosia Montana. In: Globalizations 12(6), 846–858 (2015)
28. "Save Rosia Montana" Facebook page. https://www.facebook.com/saverosiamontanacanada Accessed 29 Dec 2021
29. Association ARA. http://www.simpara.ro/asociatia/ Accessed 29 Dec 2021
30. Adopt a house at Roșia Montană. https://www.adoptaocasa.ro/en/ Accessed 29 Dec 2021

31. ICOMOS Romania: Romania – Heritage at risk in Rosia Montana. In: Truscott, M., Petzet, M., Ziesemer, J. (eds.) ICOMOS, Heritage at Risk - Patrimoine en Peril/Patrimonio en Peligro. ICOMOS WORLD REPORT 2004/2005 ON MONUMENTS AND SITES IN DANGER, pp. 201–203, K.G. Saur, Munchen (2005)

32. Most Endangered, Roşia Montană Mining Landscape in Transylvania, Romania http://7moste ndangered.eu/sites/rosia-montana-mining-landscape-in-transylvania-romania/ Accessed 27 Dec 2021

33. Rosia Montana local administration is lacking the authorized urban tools, such as PUG – General Urban Masterplan and PUZ – Zonal Urban Masterplan of the UNESCO site, which should give all further indications and guidelines in matter of heritage and environmental safe-guarding, maintenance, economic and socio-cultural development in base of the Romanian legal framework https://whc.unesco.org/en/danger/ Accessed 30 Dec 2021

34. By *goldscape* is intended the entire territory of Apuseni Mountains known as "Gold Quadri-lateral", shaped, and transformed by the gold mining activity from pre-Roman times to contemporary means of production and extraction. The entire area presents homogenous layers of cultural heritage, with different pats during the deindustrialization process, and patrimonial acknowledgement as well

35. Velicu, I.: De-growing environmental justice: Reflections from anti-mining movements in Eastern Europe. In: Ecological Economics, vol. 159, pp. 271 – 278 (2019)

Hydrological Drivers and Effects of Wildfire in Mediterranean Rural and Forest Ecosystems: A Mini Review

Domina Delač[1], Bruno Gianmarco Carrà[2], Manuel Esteban Lucas-Borja[3], and Demetrio Antonio Zema[2(✉)]

[1] Faculty of Agriculture, University of Zagreb, Zagreb, Croatia
[2] AGRARIA Department, Mediterranean University of Reggio Calabria, Località Feo di Vito, 89122 Reggio Calabria, Italy
dzema@unirc.it
[3] Department of Agroforestry Technology, Science and Genetics, School of Advanced Agricultural and Forestry Engineering, Castilla La Mancha University, Campus Universitario s/n, 02071 Albacete, Spain

Abstract. Wildfires are key drivers of changes in hydrological processes in the Mediterranean environment. Hydrologic events in the post-fire period are important to understand, because these events can contaminate major water supplies, damage critical infrastructure, influence human safety and increase soil degradation. Several effects of wildfire influence soil hydrology, but the most important are water infiltration and soil water repellency (SWR). In this paper, we reviewed studies that have focused water infiltration, SWR and the resulting runoff and soil loss after wildfire from a selection carried out on bibliometric databases. We have found a great variability in the results of the various studies, with findings sometimes conflicting, due to the high complexity of post-fire hydrology and variability of experimental conditions. The complex relationships between soil properties that must be considered while attempting to reach a comprehensive conclusion suggest the need of further research, which should explore as much as possible the intrinsic variability of soil properties and wildfire characteristics. Understanding and capturing the mechanisms underlying soil processes following wildfire are of great importance for a successful ecosystem management under changing climatic conditions.

Keywords: Post-fire hydrology · Wildfire · Infiltration · Repellency · Surface runoff · Soil erosion

1 Introduction

Wildfire is a common phenomenon and part of the Mediterranean ecosystem. However, in the recent decades the number and intensity of wildfires over the entire Mediterranean area have increased [29]. How wildfire affects the environment depends on several factors, such as fire severity, vegetation characteristics (species, flammability, density, etc.), soil properties, topography, and meteorological conditions before and after the fire [30].

F. Calabrò et al. (Eds.): NMP 2022, LNNS 482, pp. 47–55, 2022.
https://doi.org/10.1007/978-3-031-06825-6_5

The hydrological response of burned soils to the post-fire rainstorms is a critical concern for land managers, these events increase runoff and erosion, contaminating the water bodies, damaging infrastructures, influencing humans safety and increasing soil degradation [36]. Wildfire impacts on soil hydrology by reducing or completely removing the vegetation cover. Other effects are the reduction in water infiltration, soil hydrophobicity (also known as "soil water repellency", SWR), and alterations in many other physico-chemical properties of the soil. For instance, the ash and charred residues, which are produced by fire, are the main reason of SWR. After fire, heavy changes in organic matter content are observed, and these changes affect the equilibrium of physico-chemical properties of the soil.

From this short analysis, it is evident that the number of factors influencing the post-fire hydrology is large, and therefore is crucial to collect knowledge about these factors driving the extremely complex post-fire hydrological processes. Low infiltration and hydrofobicity are two key issues in Mediterranean soils, which are dominated by the infiltration-excess runoff generation mechanism [25] and show natural repellency [6]. Wildfire reduces water infiltration and induces SWR, which lead to heavy changes in post-fire soil hydrology, resulting in increased runoff and soil erosion rates.

This mini-review analyses the main studies on the main factors that drive the post-fire hydrological response of rural and forest areas under Mediterranean conditions. Its aim is a better understanding of water infiltration and SWR as the main hydrological changes in areas affected by wildfires in the Mediterranean region as well as giving information about surface runoff and soil erosion as a response to wildfire disturbances under Mediterranean conditions.

2 Materials and Methods

A selection of the key studies (researches, reviews and short communications) has been carried out in the main bibliographic database (Scopus and Web of Science), covering the last 20–30 years. The papers have been extracted by these databases adopting the keywords of this paper, and a further selection has been operated based on the number of citations of the papers, and authors' experience and opinion. Figure 1 reports the methods and criteria adopted to identify, analyze and discuss the topics of the study.

3 Results and Discussion

3.1 Water Infiltration

Depending on soil temperature, a wildfire strongly modifies the physico-chemical properties of the soil surface, influencing the hydrological response to precipitation [9]. Soil organic matter is partially or totally burned depending on the heat released by the fire, and as a consequence the soil structure is modified. The most evident effect may be the changes in soil's hydraulic properties, and especially water infiltration. In the fire-prone Mediterranean environment, water infiltration has been widely studied as the most common soil property that is changed by the wildfire effects [4, 25]. After a fire, a supply of

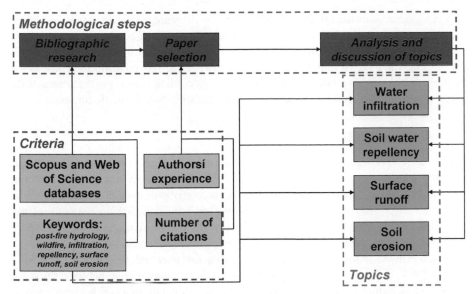

Fig. 1. Flow-chart of the methods and criteria adopted to identify, analyze and discuss the topics of the study.

highly erodible fine ash and charcoal is usually present on the soil surface. The infiltration capacity of the underlying soil is often reduced after a fire, which tends to increase surface runoff and thus the likelihood of erosion.

The disturbance in infiltration caused by fire is evidenced by its decrease in comparison to the unburned areas [35]. The main reasons are the loss of plant cover (with its beneficial effects on water infiltration due to the root systems) during combustion and the clogging of soil pores by ash residues [40]. However, the decrease in infiltration is also controlled by a repellent layer that forms during fire [12], and they are often putted in the concerns together [3].

[21] studied the infiltration rate in a semi-arid region of Northern Israelaffected by wildfire. In general, the infiltration values were highest, intermediate, and lowest in the heated, direct fire, and unburned soils, respectively, and these differences decreased with consecutive rainstorms. In a study by [38], in Central Portugal, the variables associated to infiltration (soil moisture and SWR) explained the runoff response under dry conditions (15%) slightly better than wet conditions (13%). Similar results were achieved for erosion, where 13% of soil loss variability was explained during wet periods and 36% during dry periods. The changes in water infiltration also depends on fire severity. In fires at low severity, much of the litter and ground remain, thus infiltration is usually higher than in areas affected by moderate and high-severity fires [35]. Furthermore, [11] observed that the infiltration rate in water repellent soils can be 25 times lower compared to similar soil without hydrophobicity. In the case of extreme SWR, infiltration may not occur at all during heavy rainstorm events [35]. According to [32], burned areas which are more exposed to insolation can record lower infiltration than areas that are more in the shadows.

On the other side, fire effects on soil infiltration does not always lead to reduction in infiltration. Many authors found higher infiltration rate in burned areas, however there effects are manly attributed that to ash layer effects immediately after fire [5], which absorbs rainwater. [20] found that moderate severity fire improved infiltration, due to the creation of many water-retaining pores in the soil. According to [15] infiltration capacity is not straightforward and often shows different response to wildfire disturbance.

3.2 Soil Water Repellency

Fire-induced SWR can be described as a relationship between heating temperature and soil properties at the time of contact with fire. Water repellency has a major effect in regulation of soil hydrology [13], since it is a characteristic of the soil which reduces its affinity for adsorption of water. The hydrophobic level of a soil after fire depends mainly on the intensity of fire [19]. Depending on its severity and persistence, SWR can reduce or entirely prevent soil wetting for periods ranging from seconds to months [14].

Since '1960s, many authors have found that fire may induce, enhance and even remove SWR, depending on burning conditions [2]. It is known that, between the combustion temperatures from 170 to 280 °C, the hydrophobicity in a soil horizon tends to be enhanced, while above 280 °C it is completely destroyed [2]. SWR is usually found on the soil surface or a few centimetres below and parallel to the mineral soil surface [14]. Furthermore, many authors state that SWR is more related to the composition of organic matter than on quantity [9, 41]. By comparing the results of different studies on SWR, caution is warranted owing to differences mainly in soil properties and burn severity. Soils that contain more than 3% organic matter always induce SWR. Other factors, such as the amount and type of litter consumed and soil moisture immediately before burning, can intensify or reduce the SWR.

The soil properties themselves can determine SWR during burning [1]. [27] found a pronounced SWR in calcareous soils than in acidic ones. The clay content and mineralogy play a crucial role in reducing the soil potential to become water repellent during heating [1, 26]. Concerning finest and coarser fractions, [10] reported that SWR is mostly pronounced on coarser fraction due to their smaller surface area [13]. On the other side, [27] found a greater persistence of SWR for the finest fraction of aggregates. Furthermore, [26] reported that some specific soil like "terra rossa", common in the Mediterranean environment show very low susceptibility to SWR both in burnt and unburned conditions.

3.3 Surface Runoff

The occurrence of wildfire produces a so-called "window-of-disturbance" [33], in which the hydrological response is increased. In this window, the soil is more exposed to runoff and erosion compared the unburned areas, and thereafter the pre-fire rates progressively restore throughout periods from few months to several years.

Fire reduces or totally removes (when at high severity) the vegetation cover and litter. Therefore, interception and evapo-transpiration decreases, and more precipitation turns to runoff. Furthermore, the bare soil due to vegetation burning is more exposed to splash erosion and particle detachment [42]. As previously outlined, this effect sums up to the changes in soil hydraulic properties (SWR and water infiltration). These two factors are those that mostly shape surface runoff. Both factors changes the conditions of soil surface regarding the water properties in the contact with fire [2]. The Mediterranean soils generate surface runoff when rainfall exceeds the soil's infiltration rate, and therefore a reduced infiltration rate and a connected disturbed hydraulic conductivity cause enhanced surface runoff [35]. In the literature established rainfall threshold can be found for generating surface runoff [28].

From this, it results that post-fire hydrology can be heavily altered by fires. However, the modifications in soil hydrology are different with the fire severity. Under low severity fires, soil the unburned areas. In contrast, heating is negligible, and surface runoff and soil erosion rates show limited increases compared to high severity wildfires. In contrast, during the wildfires the soil reaches very high temperatures, and strong water SWR and very low infiltration capacity can be observed [31].

According to [28], the increase in post-fire runoff is attributed to SWR and bare soil exposed. Furthermore, the remaining duff layer below the ash can also create water repellent patches enhancing surface runoff. This patchiness increases the spatial variability of the soil properties and adds complexity to understanding post-wildfire runoff and erosion responses. However, there are available research about low surface overland flow in areas where SWR is high, and high post-fire runoff has also been documented where SWR is absent. These observations indicate that SWR is not always necessary high for producing extreme floods [28]. An increase in bare soil result in an increase in the connectivity of water repellent soil patches. A threshold of 60–70% of bare soil seems to explain much of the post-wildfire surface runoff in a study by [23].

[7] in an eleven-year investigation concluded that surface runoff were not only influenced by vegetation regrowth on burned areas but also by type of vegetation and its effects on SWR. [21] observed that, when infiltration was highest, the runoff rate was the lowest, and in the opposite situation, when infiltration showed the lowest rate high surface runoff was measured. [16] found that soil moisture is negatively correlated with surface overland flow. Surface runoff was almost two time faster than in unburned plots. Cumulative runoff and runoff/rainfall ratio was also nearly twice higher in burned compared to unburned soils. [22] stated that surface runoff is related to many factors, such as the vegetation cover, rainfall intensity, soil properties, slope steepness and exposure, and fire intensity, but these authors demonstrated that rainfall intensity is dominant factor in runoff generation process. Moreover, surface runoff decreases by one order of magnitude over time from the wildfire date [21]. In a Mediterranean pine forest, [7] found that recovering vegetation three years after burning reduced overland flow by 80% compared to the runoff measured in bare soils six months after the fire. In Mediterranean shrubland areas, [17] measured at the plot scale during the first post-fire year total runoff and soil loss equal to 19.4 mm and 561 gm^{-2} in intense fire, and 14.7 mm and 326 gm^{-2} in the moderate fire, respectively, which contrasts with the very low runoff in control plots, 3.82 mm, and soil loss, 8.56 gm^{-2}. Surface runoff still remains high in burned areas comparatively with unburned area eight years after fire. [39] evaluated different hydrological regimes caused by wildfire disturbance, and showed that surface runoff altered by wildfires lead to increased annual water yields and peak flows for several post-fire years or even decades. Regarding the differences in fire severity, it has been shown less

surface runoff recorded following moderate or low fire severity than high severity fires [34]. On the other side, [18] reported relatively low differences between overland flow on high severity and unburned plot with 80% and 74%, respectively.

3.4 Soil Erosion

Erosion rates following wildfires appear to be remarkably variable, with reported increases ranging from negligible to extreme [36]. For example, [35] reviewed data from 25 studies of slope erosion and reported post-fire erosion rates ranging from 0.1 to $414\,t\,ha^{-1}\,year^{-1}$, a variation of more than three orders of magnitude [36]. Fire is prevalent across a wide range of landscapes with almost as much variability in erosion-related processes and characteristic as in landscapes that are not burned. The reasons for the wide variation among burned landscapes, such as differences in precipitation events, soil erodibility, vegetation cover, and slope justify why erosion rates exhibit a tremendous range of variability. The detachment of soil particles by rainsplash or overland runoff and their transfer downslope are very sensitive to the types of changes in land surface properties caused by fire [23]. In particular, a general reduction in vegetation cover - and especially ground vegetation and litter - makes the soil vulnerable to raindrops and reduces the rainfall storage so that erosive surface flow is more likely to occur. Many researchers consider these factors (increased runoff and rainsplash erosion) to be the most important factor leading to increased erosion following fires [34, 37, 38]. It is generally believed that these effects are related to the severity of the fire. There are a number of other factors that can influence the process of soil erosion by water after a fire, including slope gradient and aspect, soil thickness and its spatial variation, spatial distribution of soil thickness, climate (especially rainfall amount and intensity) [28]. For example, slope can affect the amount of plant matter burned and the rate of vegetation recovery along with the rate of soil drying, which in turn can affect its retention properties and the amount of soil lost. Attention has also been drawn to a number of other erosion processes such as landslides, debris flows and gravity sliding of dry particles [35].

Many research efforts are aimed at assessing the amounts and rates of erosion by water during this disturbance period. Under increasing erosion rates measured on field plots in the Mediterranean areas, [34] reported mean erosion rates after fire, respectively, of 0.39 for low, 3.28 for moderate, and 10.8 tons $ha^{-1}\,y^{-1}$ for high severity fire. In only five instances, these rates exceed $10\,t\,ha^{-1}$ in any single year of measurement. In half the field plot studies, annual post-wildfire rates do not exceed $1\,t\,ha^{-1}$ [34]. Other authors have noted relatively low erosion rates in the Mediterranean areas. There are actually reports of reduced rather than increased post-wildfire erosion compared with unburnt areas and all are based in the Mediterranean [21, 24]. [29] compiled a number of plot-based studies, mainly in the Iberian Peninsula, and found that most reported losses of $10\,t\,ha^{-1}$ in the first year after a fire. The statement that those erosion rates following wildfires are low is supported by a comparison of erosion data for all Europe by [8]. These authors pointed out the generally lower erosion rates in the Mediterranean areas compared to the other European regions, which they attribute in particular to the stony soils. [22] monitored catchment-scale post-wildfire erosion, and recorded a comparatively low sediment yield of $0.036\,t\,ha^{-1}\,y^{-1}$ in Mount Carmel, Israel.

The other two studies (Rubio et al. 1997; Mayor et al. 2007) reported rates that are similar to or actually higher than many recorded at the field plot scale, and are not dissimilar to the wide range of values reported for a small number of catchment studies worldwide [35]. Although, erosion is conventionally expected to reach its maximum during the first year after wildfire and decline thereafter, in some studies highest erosion is delayed until later post-fire years, in some cases in the third and even the fifth year after fire [28, 34]. This could be supported by the fact that high variability of rainfall amount and intensity in the Mediterranean ecosystem affect slow recovery of vegetation, so that soils may be without surface vegetation cover and increase the erosion rates in later than in initial years post-fire [34].

4 Conclusion

In rural and forest areas under Mediterranean semi-arid conditions, the impact of fire on hydrological response depends on many factors, and water infiltration and SWR are presumably the most critical. This mini-review has presented a short state-of-the-art of the studies focusing soil hydrology after wildfires as inter-relation between these key drivers and the hydrological response in term of runoff and erosion. However, the results of the various studies conducted over the years to investigate the effects of wildfire on soil hydrologic properties are sometimes contradictory and strictly depend on the complex relationships among soil properties that must be considered when attempting to reach a comprehensive conclusion. The contrasting effects of wildfire on water infiltration and SWR in Mediterranean burned areas suggests the need of further research, which should explore as much as possible the intrinsic variability of soil properties and wildfire characteristics, which are the main reasons of the variable response of surface runoff and soil erosion. Understanding and capturing the mechanisms underlying soil processes following wildfire are of great importance for a successful ecosystem management under changing climatic conditions. Due to climate change, the Mediterranean environment suffers from the changes and unpredicted occurrence of wildfires.

Overall, to the authors' honest opinion, research is still far from the comprehensive and unambiguous understanding of the complexity of fire impacts on the forest ecosystems. Moreover, intense discussions about the effectiveness if post-fire management actions are still open, since the optimal solution for the different ecosystems strictly depends on the intrinsic characteristics of the diverse impacted components (water, vegetation, soils) in fire-affected areas. On these perspectives, more experiences are needed to clarify the key drivers of the wildfire effects on the burned forests, and to identify the most effective post-fire management strategy, which should to be tailored to the site and wildfire characteristics.

References

1. Arcenegui, V., et al.: Factors controlling the water repellency induced by fire in calcareous Mediterranean forest soils. Eur. J. Soil Sci. **58**(6), 1254–1259 (2007)
2. Arcenegui, V., et al.: Soil water repellency. In: Pereira, P. et al. (eds.) Fire effects on soil properties, pp. 81–87. CSIRO Publishing (2019)

3. Beatty, S.M., Smith, J.E.: Dynamic soil water repellency and infiltration in post-wildfire soils. Geoderma **192**(1), 160–172 (2013)
4. Cerda, A.: Changes in overland flow and infiltration after a rangeland fire in a Mediterranean scrubland. Hydrol. Process. **12**(7), 1031–1042 (1998)
5. Cerdà, A.: Post-fire dynamics of erosional processes under Mediterranean climatic conditions. Zeitschrift fur Geomorphologie **42**(3), 373–398 (1998)
6. Cerdà, A., Doerr, S.H.: Soil wettability, runoff and erodibility of major dry-Mediterranean land use types on calcareous soils. Hydrological Processes: An International Journal. **21**(17), 2325–2336 (2007)
7. Cerdà, A., Doerr, S.H.: The influence of vegetation recovery on soil hydrology and erodibility following fire: an eleven-year investigation. Int. J. Wildland Fire **14**, 4 (2005)
8. Cerdan, O., et al.: Rates and spatial variations of soil erosion in Europe: a study based on erosion plot data. Geomorphology **122**(1–2), 167–177 (2010)
9. Certini, G.: Effects of fire on properties of forest soils: a review. Oecologia **143**(1), 1 (2005). https://doi.org/10.1007/s00442-004-1788-8
10. Crockford, H., et al.: Water repellency in a dry sclerophyll eucalypt forest — measurements and processes. Hydrol. Process. **5**(4), 405–420 (1991)
11. DeBano, L.F.: The effect of hydrophobic substances on water movement in soil during infiltration. Soil Sci. Soc. Am. J. **35**(2), 340–343 (1971)
12. DeBano, L.F.: The role of fire and soil heating on water repellency in wildland environments: a review. J. Hydrol. **231–232**, 195–206 (2000)
13. DeBano, L.F.: Water Repellent Soils: a state-of-the-art. General Technical Report - US Department of Agriculture, Forest Service (1981)
14. Doerr, S.H., et al.: Soil water repellency: Its causes, characteristics and hydro-geomorphological significance. Earth-Sci. Rev. **51**(1–4), 33–65 (2000)
15. Doerr, S.H., Moody, J.A.: Hydrological effects of soil water repellency: on spatial and temporal uncertainties. Hydrol. Process. **18**(4), 829–832 (2004)
16. Fernández, C., et al.: Runoff and soil erosion after rainfall simulations in burned soils. For. Ecol. Manage. **234**, S191 (2006)
17. Gimeno-García, E., et al.: Influence of vegetation recovery on water erosion at short and medium-term after experimental fires in a Mediterranean shrubland. CATENA **69**(2), 150–160 (2007)
18. González-Pelayo, O., et al.: Hydrological properties of a Mediterranean soil burned with different fire intensities. CATENA **68**(2–3), 186–193 (2006)
19. Granged, A.J.P., et al.: Short-term effects of experimental fire for a soil under eucalyptus forest (SE Australia). Geoderma **167–168**, 125–134 (2011)
20. Imeson, A.C., et al.: The effects of fire and water repellency on infiltration and runoff under Mediterranean type forest. CATENA **19**(3–4), 345–361 (1992)
21. Inbar, A., et al.: Forest fire effects on soil chemical and physicochemical properties, infiltration, runoff, and erosion in a semiarid Mediterranean region. Geoderma **221–222**, 131–138 (2014)
22. Inbar, M., et al.: Runoff and erosion processes after a forest fire in Mount Carmel, a Mediterranean area. Geomorphology **24**(1), 17–33 (1998)
23. Johansen, M.P., et al.: Post-fire runoff and erosion from rainfall simulation: contrasting forests with shrublands and grasslands. Hydrol. Process. **15**(15), 2953–2965 (2001)
24. Kutiel, P., Inbar, M.: Fire impacts on soil nutrients and soil erosion in a Mediterranean pine forest plantation. CATENA **20**(1–2), 129–139 (1993)
25. Lucas-Borja, M.E., et al.: Short-term changes in infiltration between straw mulched and non-mulched soils after wildfire in Mediterranean forest ecosystems. Ecol. Eng. **122**, 27–31 (2018)
26. Mataix-Solera, J., et al.: Can terra rossa become water repellent by burning? A Lab. Approach Geoderma **147**(3–4), 178–184 (2008)

27. Mataix-Solera, J., Doerr, S.H.: Hydrophobicity and aggregate stability in calcareous topsoils from fire-affected pine forests in southeastern Spain. Geoderma **118**(1–2), 77–88 (2004)
28. Moody, J.A., et al.: Current research issues related to post-wildfire runoff and erosion processes. Earth Sci. Rev. **122**, 10–37 (2013)
29. Pausas, J.G., et al.: Are wildfires a disaster in the Mediterranean basin? a review. Int. J. Wildland Fire **17**(6), 713–723 (2008)
30. Pereira, P., et al.: Environments affected by fire. Elsevier Ltd., Amsterdam, NL (2019)
31. Pereira, P., et al.: Post-fire soil management. Current Opinion Environ. Sci. Health **5**, 26–32 (2018)
32. Pierson, F.B., et al.: Impacts of wildfire on soil hydrological properties of steep sagebrush-steppe rangeland. Int. J. Wildland Fire **11**(2), 145–151 (2002)
33. Prosser, I.P., Williams, L.: The effect of wildfire on runoff and erosion in native Eucalyptus forest. Hydrol. Process. **12**(2), 251–265 (1998)
34. Shakesby, R.A.: Post-wildfire soil erosion in the Mediterranean: review and future research directions. Earth Sci. Rev. **105**(3–4), 71–100 (2011)
35. Shakesby, R.A., Doerr, S.H.: Wildfire as a hydrological and geomorphological agent. Earth Sci. Rev. **74**(3–4), 269–307 (2006)
36. Jahren, P., Sui, T.: Erosion. In: How Water Influences Our Lives, pp. 161–178. Springer, Singapore (2017). https://doi.org/10.1007/978-981-10-1938-8_8
37. Vieira, D.C.S., et al.: Does soil burn severity affect the post-fire runoff and interrill erosion response? a review based on meta-analysis of field rainfall simulation data. J. Hydrol. **523**, 452–464 (2015)
38. Vieira, D.C.S., et al.: Key factors controlling the post-fire hydrological and erosive response at micro-plot scale in a recently burned Mediterranean forest. Geomorphology **319**, 161–173 (2018)
39. Wagenbrenner, J.W., et al.: Post-wildfire hydrologic recovery in Mediterranean climates: A systematic review and case study to identify current knowledge and opportunities. J. Hydrology. **602**, 126772 (2021)
40. Woods, S.W., Balfour, V.N.: The effect of ash on runoff and erosion after a severe forest wildfire, Montana, USA. Int. J. Wildland Fire **17**(5), 535–548 (2008)
41. Zavala, L.M., et al.: How wildfires affect soil properties. a brief review. Cuadernos de Investigación Geográfica. **40**(2), 311 (2014)
42. Zema, D.A.: Postfire management impacts on soil hydrology. Current Opinion in Environ. Sci. Health **21**, 100252 (2021)

Agro-Food Wastewater Management in Calabria and Sicily (Southern Italy): A General Overview and Key Case Studies

Demetrio Antonio Zema[⊠]

AGRARIA Department, Mediterranean University of Reggio Calabria, Loc. Feo di Vito, 89122 Reggio Calabria, Italy
dzema@unirc.it

Abstract. The choice of the most suitable management system for wastewater of agro-food industries is a difficult task for firm owners, since the convenience of each system depends on the specific case. This overview discusses advantages and constraints of the main systems for wastewater of oil mills, citrus industries and wineries in Sicily and Calabria (Southern Italy). Intensive depuration, lagoons, constructed wetlands, energy valorization, reuse for agronomic purposes, valorization in biorefineries are the main focus of this discussion. Some examples of successful case studies in agro-food wastewater management on the real scale are proposed. It has been highlighted that, while the treatment in centralized activated sludge plants is viable for agro-food industries in large urban areas, depuration in lagoons or constructed wetlands are good options for firms in rural areas. Agronomic utilization is feasible in industries with enough land to irrigate, adopting proper application protocols to avoid soil degradation and crop damage. Anaerobic digestion is promising, but only when agro-food wastewater is blended with other substrates (agricultural or animal residues). The possible extraction of value-added compounds is potentially interesting, but the biorefinery processes are not technically mature and generally not convenient as far as now.

Keywords: Lagoons · Constructed wetlands · Agronomic reuse · Anaerobic digesters

1 Introduction

The management of wastewater produced by the agro-food industries (the so-called "agro-food wastewater", hereafter indicated as "AFW") is a severe concern for firm owners and public authorities, due to the economic and environmental implications. These problems are due to many factors, such as: (i) the large volumes of wastewater; (ii) the content of pollutants (e.g., organic matter, inhibiting compounds); (iii) the seasonality of production processes (concentrated in few months per year); and (iv) the small size and territorial dispersion of the industries [27]. The need to protect the health of soil and water bodies from pollution has required the issue of strict environmental rules, which pose additional technical and economic difficulties for wastewater management in local

F. Calabrò et al. (Eds.): NMP 2022, LNNS 482, pp. 56–65, 2022.
https://doi.org/10.1007/978-3-031-06825-6_6

agro-food industries. The environmental issues linked to the wastewater management are extremely important in some regions of Southern Italy, due to the specific climatic semi-arid conditions (scarcity of water resources and high degradation rates of soils (particularly for erosion, salinization and desertification [10].

The scientific literature proposes systems for wastewater management. The most suitable solution must be chosen case by case, accounting for many technical and economic factors (type, size and location of the agro-food industry, production processes and machineries, regional regulations, management procedures and so on). Since this choice is not easy, general guidelines are needed for industry owners and local authorities in view of reducing the cost of effluent disposal and minimize soil and water pollution.

This paper carries out an overview of AFW management in Sicily and Calabria (Southern Italy). In these regions, the management of wastewater produced by olive oil mills (OMW), citrus processing industries (CPW) and wineries (WW) is an important problem, since the main products of these industries (olive oil, citrus juice, and wine) are the backbone of the agro-food sector. The advantages and constraints of each management system are shortly discussed. Some examples of successful case studies in AFW management on the real scale are proposed.

2   Overview on Agro-Food Wastewater Characteristics

2.1   Production of Agro-Food Wastewater

The processing chain of agro-industrial fruits (olive, citrus, grape) mainly consists of fruit temporary storage, washing, grading and sorting, product extraction (olive oil, citrus juice, and wine), possible finishing, and packaging and storage. Generally, these industries produce solid (pomace, citrus peel or marc) and liquid (wastewater) residues, and the ratio solid/liquid is variable. The production of wastewater is in the range 60% (oil mills) - 100% (citrus industries) of the weight of the processed fruits [1]. The typical components of AFW, coming from the residual organic matter derived from fruit processing, are suspended solids, organic acids, sugars, inhibiting compounds (polyphenols in OMW, essential oils in CPW, and tannins in WW) as well as cleaning agents for machineries and facilities [2, 21].

2.2   Quali-Quantitative Characteristics of Agro-Food Wastewater

The quantity of fruits processed in agro-food industries is characterized by a noticeable inter-annual variability, which depends on the fluctuations of the fruit production year by year and affects wastewater production. AFW volumes show a seasonal qualitative and quantitative variability, due to the variable amount of fruit processed, water consumption per weight unit of processed fruit, and plant management. Moreover, AFW production is characterized by a large weekly and daily variability, due to plant inactivity in the night and weekend. Finally, the physico-chemical properties of AFW are largely variable industry by industry and over time, depending on the processing techniques and some of the factors above.

The main concerns linked to AFW treatment are the low pH, the high concentration of organic compounds, and the noticeable content of inhibiting compounds (polyphenols, essential oils or tannins in OMW, CPW and WW, respectively) [17]. The high acidity is an important problem if the wastewater must be reused for agronomic purposes. The high content of organic matter may consume the dissolved oxygen if the wastewater is disposed in surface water bodies, with consequent death of aquatic fauna. The significant presence of inhibiting compounds slowdowns or even blocks the biological processes in the depuration plants, due to its bacteriostatic action [5, 6, 27].

3 Management Systems of Agro-Food Wastewater

3.1 Intensive Treatment

AFW is usually treated in intensive biological plants prior to the disposal in the municipal sewer system or in water bodies. In general, activated sludge processes are mainly used. In the conventional plants, the organic matter and other polluting compounds are degraded by aerobic or facultative bacteria. However, this system is complex and expensive in AFW treatment, since the intensive plants are not efficient, and require skilled personnel and a large amount of electrical energy. The main drawbacks of intensive plants are: (i) the instability of the biochemical processes, because the AFW properties are variable over time; (ii) the frequent out-of-running of the plants, due to peaks in inhibiting compounds, low pH and scarcity of nutrients in AFW streams; (iii) high costs of personnel and energy required for plant operation; and (iv) complex start-up at the beginning of the processing season [13, 20].

However, AFW intensive depuration is practically a compulsory solution for agro-food industries that are in urban or industrial areas, which are served by large municipal plants. Here, the high-strength effluents are blended to municipal and/or industrial wastewater and thus diluted, reducing its effects on biological processes. The treatment in centralized plants is more effective than in on-site conventional plants, due to economies of scale [13].

Regarding the experiences of AFW treatment in intensive plants in Southern Italy, [8], treating CPW in two aerobic granular sludge sequencing batch reactors, reported an organic load removal efficiency of approximately 90%, which decreased to 75% at higher organic loads, due to the excessive acidic conditions. A promising alternative to conventional activated sludge plants for AFW are the membrane bioreactors (MBRs), which integrate flow permeation by membranes with the biological process for the final solid–liquid separation. However, the MBR fouling leads to frequent plant shutdowns and requires energy the membrane. On this regard, [11] proposed a pretreatment of CPW, applying a series of aerobic granular sludge technology and MBR operation in pilot plants, resulting in the almost total removal of the organic matter in wastewater. However, the higher resistance to filtration observed in the plant might have severely affected the membrane life.

Finally, AFW is rarely treated in plants based on physical and/or chemical processes (for instance, evaporation or flocculation), due to their high cost and reliability [27].

3.2 Lagoons

Lagoons (or stabilization ponds) are extensive natural systems to treat wastewater of urban or industrial origin. These systems are low-demanding and environmentally sustainable, since the construction, maintenance, and management are simple and cheap. However, lagoons must have a large storage capacity of wastewater, and therefore these systems require large areas of land and long storage times. Other problems are the emission of unpleasant odors, and thus the lagoons must be installed far from residential zones (in agricultural areas or in marginal lands).

When these constraints are overcome, the use of lagoons is one of the most suitable alternatives for AFW treatments. Generally, wastewater is stored for a long time (weeks or months) in one or more ponds (installed in parallel or in series), where aerobic or anaerobic micro-organisms degrade the organic matter and other polluting compounds in AFW. Thanks to the long storage, the variation of physico-chemical properties of AFW is very slow, and therefore the micro-organisms adapt to adverse environmental conditions (such as the low pH, and high organic loads and concentrations of inhibiting compounds).

The efficiency in removal of organic compounds is generally variable with maximum values up to 80–90%. This variability depends on the seasonal weather changes. Often the lagoons are anaerobic (that is, the wastewater is not supplied with oxygen). Anaerobic lagoons show an organic load removal of over 90% and do not require energy. However, the anaerobic processes are very long, and the unpleasant odors of anaerobic lagoons may be beyond the tolerance limits.

The aeration (using pneumatic or mechanical devices) reduces the odor emission; moreover, the aerated processes are much faster compared to the anaerobic lagoons. However, the most severe constraint of aerated lagoons is the high demand of energy, but this need is much lower compared to the activated sludge plants. Some researchers have proposed the use of aerobic-anaerobic lagoons to treat the CPW of a Sicilian industry. At the pilot scale, [25] tested some lagoon schemes with different air flows and times as well as the adaption of micro-organisms growing in the stored AFW to very high concentrations of essential oils. In another study by the same authors, the energy demand was reduced compared to the real-scale lagoons, by lowering the aeration power compared to the real-scale plants and limiting aeration time to the night (12 h only), when electricity cost is lower [24]. Again in pilot plants installed in Calabria, [4] analyzed the effects of different air flow rates and times, concentrations of polyphenols and organic matter:nitrogen ratios in OMW, and found removal efficiency of organic matter up to 90% and of polyphenols up to 64%.

An interesting case study for AFW treatment in lagoons is the system treating the effluents of a large citrus processing industry of Eastern Sicily. The lagoon (capacity of 10000 m^3 and depth of 7 m) consists of an earth pond with plastic waterproofing, and is aerated by 1, 2 or 3 floating aerators operating only during the ght hours and the weekend (Fig. 1). A study by [3] 2013) showed a low energy demand compared to the activated sludge plants, and a satisfactory efficiency in organic matter removal (on average 60% with peaks of 95%) also on effluent with very high concentrations of essential oils.

Fig. 1. The lagoon system treating the effluents of a large citrus processing industry of Sicily.

3.3 Constructed Wetlands

Various studies have demonstrated that the constructed wetlands (CWs) are efficient, at low cost and maintenance and energy-saving systems for depurating wastewater of different origin, especially urban wastewater [16, 23]. CWs exploit the combined actions of soil, plants and micro-organisms to remove the polluting compounds in wastewater [7]. The removal efficiencies are generally satisfactory but lower compared to the activated sludge plants, and depend on weather variability. However, the phyto-depuration must be adopted with caution systems for treating AFW, due to the specific physico-chemical characteristics of these effluents. The high acidity, the content of phyto-toxic compounds, the high organic matter:nitrogen ratio of the AFW can hamper vegetation growth and development, and even kill plants. Coupling pre-treatments (targeted at removing the suspended solids and aerating the treated wastewater) with CWs is a valid option to avoid these constraints, and to propose the phyto-depuration as an effective system for AFW treatment. Thanks to this solution, CWs become attractive for middle-sized agro-food industries in areas with warm climate conditions [14].

In Eastern Sicily, the use of a multistage constructed wetland system is emblematic for winery wastewater treatment with subsequent effluent reuse for irrigation (Fig. 2). The system depurates about 3 m^3/day of wastewater produced by a small winery. The CW system shows mean removal rates of 70–80% for organic matter and suspended solids, and 40–55% for total nitrogen and phosphates. These rates give an average quality of the CW effluent under the limits of the Italian regulations. Vegetation growth and development are regular, and phytotoxicity effects are not observed [19].

Fig. 2. Aerial view of the CW system treating the wastewater of a winery in Eastern Sicily (source: Google Map).

3.4 Energy Valorization

A promising solution to treat AFW is the use of anaerobic or aerobic reactors to convert the organic compounds of wastewater in biomethane (by anaerobic digestion) or ethanol or biohydrogen (by fermentation). Moreover, the by-products of these processes (e.g., the digestate that is the residue of the anaerobic reactors) can be spread on soil and valorized as organic conditioner and/or fertilizer. However, the anaerobic digestion of AFW is not a technically easy process, since the wastewater has a low content of dry matter (which requires the treatment of high wastewater volumes) and the concentration of inhibiting compounds is very high [18]. This constraint requires large anaerobic digesters or pre-treatments of raw AFW using physico-chemical processes.

In contrast, the anaerobic digestion is an effective treatment technique for high-strength AFW flows with a limited concentration of inhibiting compounds and an adequate pH. Blending the AFW with other agricultural (e.g., olive pomace, citrus peel, grape marc) or animal (manure) substrates is advisable, in order to limit the constraints of the anaerobic processes [22]. On this regard, [26] proposed an algorithm to identify the most energy-yielding blend for anaerobic digestion of agro-food by-products, and validated the method in agro-food industries of Calabria. Some real-scale anaerobic digesters fed by AFW blended to other substrates have been recently installed in both Sicily and Calabria. The case study of the "Fattoria della Piana" cheese factory in Calabria is emblematic, since its anaerobic digester (volume of 7500 m^3) is fed by a blend of manure, and liquid and solid residues of olive oil mills and citrus industries (Fig. 3). The electrical energy (over 8000 MWh/yr) and biomethane (yielding about 3300 MWh/yr of thermal energy and fuel for firm vehicles) produced by the digester feed the firm processes as well as a small village (3000 families) in the surrounding of the cheese factory; the digestate is used as fertilizer in the field crops.

Fig. 3. The anaerobic digester treating the wastewater of a cheese factory in Calabria (source: https://fattoriadellapiana.it/la-centrale-a-biogas).

The production of biohydrogen or bioethanol (since the ethanol is in high concentrations in WW, [21]) are other viable options for energy valorization of AFW. To the best knowledge of the authors no lab- or real-scale experiences have been found in Calabria and Sicily.

3.5 Agricultural Utilization

The most common option of AFW management in Southern Italy is its reuse for agronomic purposes. This option is beneficial, since: (i) provides an additional irrigation source for croplands in semi-arid areas; (ii) supplies soils with organic matter and other important compounds, improving fertility and physical characteristics; (iii) it is cheap and easy, thanks to the possibility to use the common irrigation plants (sprinkler and micro-irrigation with pre-filtration) or farm machineries. The main constraint is the need of marginal or cultivated lands in proximity to the agro-food industry, considering that the transport of AFW from the industry to the application lands may be expensive on long distances. Furthermore, land spreading of AFW must be practiced with caution, since the application of wastewater can damage some properties of the receiving soils or be in some cases (e.g., during germination of herbaceous crops) toxic to plants. Therefore, the soil's suitability for AFW application must also be evaluated with care in order to avoid the risks of soil degradation and groundwater pollution, and preserve plant health and yield. In the case of OMW, a review by Barbera et al. (2013) has analyzed the effects of spreading wastewater on soil properties and crop yields. These authors have confirmed that OMW is an important fertilizing agent for croplands, but, at the same time, have proposed some restrictions for OMW application based on soil characteristics.

The Italian regulation (Ministry Decree 185/2003) for wastewater irrigation in agriculture is very stringent, due to large number of water parameters to be monitored and the strict acceptance limits of many of these parameters. This dissuades farmers from adopting AFW land spreading on large scale [15]. Some Regions of Southern Italy (e.g., Calabria) have issued a regional plan of OMW agricultural utilization at 1:250000 scale together with technical rules for the land spreading of OMW. The overlay of the geographic distribution of the suitable soils for OMW land spreading on the map of olive production districts has shown that this practice is largely feasible for this region [9].

3.6 Biorefinery Valorization

The presence of added-values phytochemical and bioactive compounds (such as, for instance, flavonoids) in AFW suggests the possibility of its valorization in biorefineries for use in the food, cosmetic, and pharmaceutical industries. Polyphenols and flavonoids are antioxidant compounds, essential oils are substrates for perfume and cosmetic production, and flavonoids have many beneficial effects on human health. Many of these compounds have anti-inflammatory, antimicrobial, antiallergic, anticarcinogenic and antihypertensive activities [12]. However, this biorefinery approach has been limited by process complexity and dilution level of the added-value compounds in AFW. As far as now, these AFW valorization patterns requires sophisticated technologies (specific pretreatments with physical and biological agents followed by tailored recovery procedures [12]), whose adoption and operation increase processing costs. The number of companies trading in these components is increasing, but their recovery generally requires expensive purification or enrichment procedures and the management of the pretreated by-products and waste represent an environmental problem [12] As far as now, we think that these patterns are not currently mature for large-scale applications on AFW. AFW may be also used, as done for urban wastewater [28], to irrigate energy crops, such as Arundo donax, Phragmites australis or Typha latifolia, for renewable energy production.

4 Conclusions

The choice of the most suitable management system for AFW is not easy, since many factors of different nature influence the technical, economic and environmental effects of these systems. This overview has indicated that, while the treatment in centralized activated sludge plants is viable for agro-food industries in large urban areas, depuration in lagoons or constructed wetlands are good options for firms in rural areas. Agronomic utilization must be carried out in industries with enough land to irrigate, adopting proper application protocols to avoid soil degradation and crop damage. Anaerobic digestion is promising, but only when AFW is blended with other substrates (agricultural or animal residues). The possible extraction of value-added compounds is potentially interesting, but the biorefinery processes are not technically mature and generally not convenient as far as now.

The short analysis of the case studies has shown the good economic and environmental performance of relatively novel systems for AFW treatment. These examples are encouraging in view of a broader implementation of these systems, in order to ultimately convert AFW from a waste to a resource.

References

1. Algieri, A., et al.: The potential of agricultural residues for energy production in Calabria (Southern Italy). Renew. Sustain. Energy Rev. **104**, 1–14 (2019)
2. Amor, C., et al.: Application of advanced oxidation processes for the treatment of recalcitrant agro-industrial wastewater: a review. Water **11**(2), 205 (2019)
3. Andiloro, S., et al.: Aerated lagooning of agro-industrial wastewater: depuration performance and energy requirements. Journal of Agricultural Engineering. 7 (2013)
4. Andiloro, S., et al.: Depuration performance of aerated tanks simulating lagoons to treat olive oil mill wastewater under different airflow rates, and concentrations of polyphenols and nitrogen. Environments **8**(8), 70 (2021)
5. Andreottola, G., et al.: Biological treatment of winery wastewater: an overview. Water Sci. Technol. **60**(5), 1117–1125 (2009). https://doi.org/10.2166/wst.2009.551
6. Barbera, A.C., et al.: Effects of spreading olive mill wastewater on soil properties and crops, a review. Agric. Water Manag. **119**, 43–53 (2013)
7. Cirelli, G.L.: I trattamenti naturali delle acque reflue urbane. Sistemi editoriale. Arzano. (2003)
8. Corsino, S.F., et al.: Aerobic granular sludge treating high strength citrus wastewater: analysis of pH and organic loading rate effect on kinetics, performance and stability. J. Environ. Manage. **214**, 23–35 (2018)
9. Costantini, E.A.: Manual of methods for soil and land evaluation. CRC Press, USA (2009)
10. Costantini, E.A.C., Dazzi, C.: The Soils of Italy. (2013)
11. Di Trapani, D., et al.: Treatment of high strength industrial wastewater with membrane bioreactors for water reuse: effect of pre-treatment with aerobic granular sludge on system performance and fouling tendency. J. Water Process Eng. **31**, 100859 (2019)
12. Federici, F., et al.: Valorisation of agro-industrial by-products, effluents and waste: concept, opportunities and the case of olive mill wastewaters: valorisation of agro-industrial by-products, effluents and waste. J. Chem. Technol. Biotechnol. **84**(6), 895–900 (2009)
13. Kapellakis, I.E., et al.: Olive oil history, production and by-product management. Rev Environ Sci Biotechnol. **7**(1), 1–26 (2008). https://doi.org/10.1007/s11157-007-9120-9
14. Kataki, S., et al.: Constructed wetland, an eco-technology for wastewater treatment: A review on types of wastewater treated and components of the technology (macrophyte, biolfilm and substrate). J. Environ. Manage. **283**, 111986 (2021)
15. Licciardello, F., et al.: Wastewater tertiary treatment options to match reuse standards in agriculture. Agric. Water Manag. **210**, 232–242 (2018)
16. Masi, F., et al.: Wineries wastewater treatment by constructed wetlands: a review. Water Sci. Technol. **71**(8), 1113–1127 (2015). https://doi.org/10.2166/wst.2015.061
17. Mateo, J.J., Maicas, S.: Valorization of winery and oil mill wastes by microbial technologies. Food Res. Int. **73**, 13–25 (2015). https://doi.org/10.1016/j.foodres.2015.03.007
18. Messineo, A., et al.: Biomethane recovery from olive mill residues through anaerobic digestion: A review of the state of the art technology. Sci. Total Environ. **703**, 135508 (2020)
19. Milani, M., et al.: Treatment of Winery Wastewater with a Multistage Constructed Wetland System for Irrigation Reuse. Water. **12**(5), 1260 (2020). https://doi.org/10.3390/w12051260

20. Niaounakis, M., Halvadakis, C.P.: Olive processing waste management: literature review and patent survey. (2006)
21. Scoma, A., et al.: High impact biowastes from South European agro-industries as feedstock for second-generation biorefineries. Crit. Rev. Biotechnol. **36**(1), 175–189 (2016)
22. Selvaggi, R., et al.: Assessing land efficiency of biomethane industry: a case study of Sicily. Energy Policy **119**, 689–695 (2018). https://doi.org/10.1016/j.enpol.2018.04.039
23. Toscano, A., et al.: Comparison of removal efficiencies in Mediterranean pilot constructed wetlands vegetated with different plant species. Ecol. Eng. **75**, 155–160 (2015)
24. Zema, D.A., Andiloro, S., Bombino, G., Caridi, A., Sidari, R., Tamburino, V.: Comparing different schemes of agricultural wastewater lagooning: depuration performance and microbiological characteristics. Water Air Soil Pollut. **227**(12), 1–9 (2016). https://doi.org/10.1007/s11270-016-3132-4
25. Zema, D.A., et al.: Depuration in aerated ponds of citrus processing wastewater with a high concentration of essential oils. Environ. Technol. **33**(11), 1255–1260 (2012)
26. Zema, D.A.: Planning the optimal site, size, and feed of biogas plants in agricultural districts. Biofuels, Bioprod. Bioref. **11**(3), 454–471 (2017)
27. Zema, D.A., et al.: Wastewater management in citrus processing industries: an overview of advantages and limits. Water **11**(12), 2481 (2019)
28. Molari, G., et al.: Energy characterisation of herbaceous biomasses irrigated with marginal waters. Biomass and Bioenergy **70**, 392–399 (2014)

River Transport in Calabrian Rivers

Giandomenico Foti[1]([✉]) [iD], Giuseppe Barbaro[1] [iD], Giuseppe Bombino[2] [iD],
Giuseppina Chiara Barillà[1] [iD], Pierluigi Mancuso[3], and Pierfabrizio Puntorieri[1]

[1] DICEAM Department, Mediterranea University of Reggio Calabria (Italy),
Via Graziella, loc. Feo di Vito, 89122 Reggio Calabria, Italy
giandomenico.foti@unirc.it
[2] Agriculture Department, Mediterranea University of Reggio Calabria (Italy), Via Graziella,
loc. Feo di Vito, 89122 Reggio Calabria, Italy
[3] Public Works Department, loc. Germaneto, Calabria Region, 88100 Catanzaro, Italy

Abstract. The issue of river transport does not only concern river basins but also influences the equilibrium of the neighboring coasts. In fact, the eroded material within the river basin can reach the beaches near the river mouth. Therefore, high river transport can act as a natural nourishment while low river transport can cause shoreline erosions. The paper describes the methodology adopted to quantify river transport in Calabrian rivers. Calabria represents an interesting case study due to its geomorphological peculiarities. Indeed, most Calabrian rivers are characterized by a torrential and irregular hydrological regime, with extensive dry periods and frequent sudden flooding, caused by short and intense rainfalls. Also, many of these rivers have very wide beds with coarse grain size. This combination of hydrological and granulometric characteristics causes high solid transport, and the relative variations can alter the coastal dynamics and the shoreline evolution near the river mouths. The methodology described in this paper was applied in each river basins and is divided into four main phases, as follows: morphometric characterization, estimate of average yearly precipitation and temperature, estimate of Gavrilovic model coefficient, and estimate of river transport. Gavrilovic's model was used as it is particularly reliable for rivers such as the Calabrian ones.

Keywords: River transport · Gavrilovic model · Calabria · Fiumare

1 Introduction

River transport is a topic of interest in the field of planning and management of both river and coastal areas [1–5]. Indeed, this issue does not only concern river basins but also influences the equilibrium of the neighboring coasts as the eroded material within the river basin can reach the beaches near the river mouth [6–9]. Therefore, high river transport can act as a natural nourishment while low river transport can cause shoreline erosions and coastal and river dynamics should be analyzed together [10–16]. The equilibrium conditions of this system depend on both anthropogenic and natural factors [17–20]. The main anthropogenic factors are the construction of hydraulic structures such as dams and weirs [21], the withdrawal of river sediment [22], the expansion of inhabited centers [23–25] and the construction of ports and coastal defense works [26–29]. The main natural factors are the wave climate [30–32], the action of extreme events such as floods, sea storms or a combination of these [33–42].

Solid material is transported across rivers through two main mechanisms, bed load and suspended load [43], and can be quantified using many models and formulas [44–50]. On the other hand, at the basin scale an important mechanism is the soil erosion by water (WSE) [51]. This latter mechanism is very important in basins such as those of Calabria, characterized by an irregular hydrological regime and by high solid transport [52–54]. In this context, the Gavrilovic model [55] is particularly useful for quantifying the solid transport caused by WSEs.

The paper describes the methodology adopted to quantify river transport in Calabrian rivers. The methodology was developed by Foti et al. [56] in the Allaro river and, in this paper, it has been applied to all Calabrian basins. This methodology is divided into four main phases, as follows: morphometric characterization, estimate of average yearly precipitation and temperature, estimate of Gavrilovic model coefficient, and estimate of river transport.

2 Site Description

Calabria is a region in southern Italy, located at the tip of the typical Italian "boot" in the center of the Mediterranean Sea enclosed by two seas, the Tyrrhenian and the Ionian, by the Strait of Messina and by the Gulf of Taranto, each of them with different climatic characteristics and with different fetch extensions (Fig. 1). From the morphological point of view, Calabria is characterized by hills and mountains, with a percentage of less than 10% of flat lands. The main massifs are Pollino, Sila and Aspromonte, all with a maximum altitude of the order of 2000 m. The main coastal plains are that of Sibari, on the Ionian coast in the Gulf of Taranto, and those of Lamezia Terme and Gioia Tauro, both on the Tyrrhenian coast. Its narrow and elongated shape means that it has over 750 km of coastline, with an alternating mainly sandy and pebbly beaches, and high coasts, with the main headlands are those of Capo Rizzuto, on the Ionian coast, and of Capo Vaticano, on the Tyrrhenian coast.

Most Calabrian rivers (locally called *fiumare*) [57, 58], are characterized by a torrential and irregular hydrological regime, with extensive dry periods and frequent sudden flooding, caused by short and intense rainfalls. Also, many of these rivers have very wide beds with coarse grain size. This combination of hydrological and granulometric characteristics causes high solid transport, and the relative variations can alter the coastal dynamics and the shoreline evolution near the river mouths [59, 60].

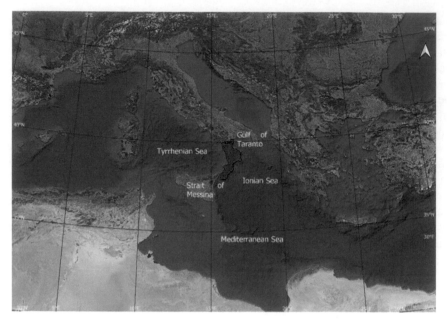

Fig. 1. The Calabrian region (shown with red dotted line), in the center of the Mediterranean Sea.

3 Methodology

The methodology was applied in each river basins and is divided into four main phases, as follows: morphometric characterization, estimate of average yearly precipitation and temperature, estimate of Gavrilovic model coefficient, and estimate of river transport using the Gavrilovic model [55] (Fig. 2).

The input data of the first phase are the cartographic data available in the Open Data section of the Calabrian Geoportal (http://geoportale.regione.calabria.it/opendata). These data are the Digital Terrain Model (DTM) with square mesh of 5 m and the shape-files of the Calabrian Rivers and have been processed on QGIS version 3.10.7 *"Coruna"*. The outputs of this phase are the area, the perimeter, the main stream length, the total stream length, the average heights, the average slope and the time of concentration, estimated by the formulas of Giandotti [61], Kirpich [62] and NRCS [63].

The input data of the second phase are the shapefile of the Calabrian gauges, available in the OpenData section of the Calabrian Geoportal, and the time series of precipitation and temperature records of each gauge, available in the Historical Data section of the Calabrian Multi-Risk Functional Center (http://www.cfd.calabria.it/). Before estimating the average precipitation and temperatures values of each basin, the gauges with a non-statistically significant time series were neglected. Subsequently, the influence area of each analyzed gauge was estimated using the Thiessen polygon method [64], corresponding on QGIS to the Voronoi polygons function. Finally, the average yearly precipitation and temperature values of each basin were calculated as a weighted average of the values recorded by each gauge with weight equal to the influence area.

About the third phase, the input data are the land cover data from the Corine Land Cover project fourth level relates to the years 2018 and freely available on the government agency website "*Istituto Superiore per la Protezione e la Ricerca Ambientale* (ISPRA)" (https://www.isprambiente.gov.it/it/attivita/suolo-e-territorio/copertura-del-suolo/corine-land-cover). A value of each Gavrilovic model coefficients was associated with each land use category of the Corine Land Cover. The average values of these coefficients of each basin were calculated as a weighted average of the coefficient values of each land use category, with weight equal to the area of each land use category.

In the last phase, river transport was estimated, in terms of the annual average volume of detached soil due to surface erosion, using the Gavrilovic model [55] which depends on: the potential annual volume of detached soil; a retention coefficient, proposed by Zemljic [65]; a temperature coefficient; the average yearly precipitation; a soil protection coefficient X, function of the type of vegetation cover; a erodibility coefficient Y, function of type of rock; an erosion typology coefficient; the average slope of the basin; the basin area; the basin perimeter; the average height of the basin; the main stream length and the average yearly temperature.

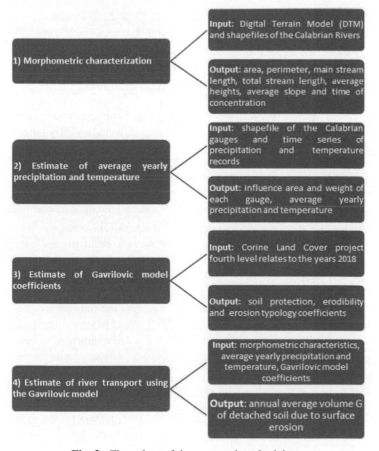

Fig. 2. Flow chart of the proposed methodology.

4 Results

The main results are reported in Table 1, which shows the 20 Calabrian basins with the greatest annual average volume of detached soil due to surface erosion in descending order. In addition, area and specific average volume of detached soil are shown for each basin.

The table highlights that the river with the highest annual average volume of detached soil is the Lao, over 150,000 m^3/year. Neto and Allaro also exceed 100,000 m^3/year and in total there are 16 rivers characterized by over 50,000 m^3/year of detached soil. Instead, in terms of specific average volume of detached soil, the river with the greatest value is Alaca, over 1000 m^3/year*km^2 followed by Allaro, almost 800 m^3/year*km^2, and Condojanni, over 750 m^3/year*km^2, and then Catona, Careri, Torbido, Bonamico and La Verde with between 500 and 600 m^3/year*km^2. All these rivers are located in the

Table 1. 20 Calabrian basins with the greatest annual average volume of detached soil due to surface erosion in descending order. Also, area, and specific average volume of detached soil for each basin.

River basin	Annual average volume of detached soil [m^3/year]	Area [km^2]	Specific average volume of detached soil [m^3/year*km^2]
Lao	152742	600.4	254
Neto	107877	1073.3	101
Allaro	102671	130.1	789
Torbido	88674	160.5	552
Savuto	83681	411.5	203
Petrace	80774	422.3	191
Bonamico	75361	136.4	552
Mesima	75053	815.2	92
Tacina	69373	426.9	162
Ancinale	65879	173.3	380
Amato	62990	443.8	142
Corace	61560	294.4	209
La Verde	58189	117	498
Trionto	58078	288.8	201
Careri	50949	92.1	553
Condojanni	50941	66.5	766
Amendolea	45683	150.4	304
Alaca	44247	41.1	1076
Crocchio	39889	129.7	308
Catona	39474	68.5	576

Ionian coast, with the exception of the Catona which is located in the Strait of Messina, and are *fiumare*. Instead, Lao and Neto and Savuto, Petrace, Mesima, Tacina and Amato are all rivers and have specific erosions between 100 and 250 m^3/year*km^2. These results confirm that the *fiumare* are characterized by a high river transport, as described in Sect. 2.

5 Conclusions

The paper describes the methodology adopted to quantify river transport in Calabrian rivers. This methodology was applied in each river basins and is divided into four main phases, as follows: morphometric characterization, estimate of average yearly precipitation and temperature, estimate of Gavrilovic model coefficient, and estimate of river transport. Gavrilovic's model was used as it is particularly reliable for rivers such as the Calabrian ones, characterized by a torrential and irregular hydrological regime. Indeed, the results obtained confirm that the *fiumare* are characterized by a high river transport, greater than rivers in terms of specific average volume of detached soil.

This issue is very important not only in the river area but also has consequences in the coastal area as the river transport variations can alter the coastal dynamics and the shoreline evolution near the river mouths. Therefore, river transport is a topic of interest in the field of planning and management of both river and coastal areas.

Funding. This work was funded by the Public Works Department of Calabria Region.

References

1. Sicilia, C.L., Foti, G., Campolo, A.: Protection and management of the Annunziata river mouth area (Italy). J. Air, Soil Water Res. **6**, 107–113 (2013)
2. Barbaro, G.: Master plan of solutions to mitigate the risk of coastal erosion in Calabria (Italy): a case study. Ocean Coastal Manag. **132**, 24–35 (2016)
3. Kantamaneni, K., Phillips, M., Thomas, T., Jenkins, R.: Assessing coastal vulnerability: Development of a combined physical and economic index. Ocean Coast. Manag. **158**, 164–175 (2018)
4. Viavattene, C., Jiménez, J.A., Ferreira, O., Priest, S., Owen, D., McCall, R.: Finding coastal hotspots of risk at the regional scale: the coastal risk assessment framework. Coast. Eng. **134**, 33–47 (2018)
5. Mucerino, L., et al.: Coastal exposure assessment on Bonassola bay. Ocean Coast. Manag. **167**, 20–31 (2019)
6. Short, A.D.: Handbook of Beach and Shoreface Morphodynamics. Wiley, New Jersey (2000)
7. Li, X., Zhou, Y., Zhang, L., Kuang, R.: Shoreline change of Chongming Dongtan and response to river sediment load: a remote sensing assessment. J. Hydrol. **511**, 432–442 (2014)
8. Dada, O.A., Qiao, L., Ding, D., Li, G., Ma, Y., Wang, L.: Evolutionary trends of the Niger Delta shoreline during the last 100 years: responses to rainfall and river discharge. Marine Geol. **367**, 202–211 (2015)
9. Dada, O.A., Li, G., Qiao, L., Asiwaju-Bello, Y.A., Anifowose, A.Y.B.: Recent niger delta shoreline response to Niger River Hydrology: conflicts between forces of nature and humans. J. Afr. Earth Sc. **139**, 222–231 (2018)

10. Barbaro, G., Foti, G., Mandaglio, G., Mandaglio, M., Sicilia, C.L.: Estimate of sediment transport capacity in the basin of the Fiumara Annunziata (RC). Rendiconti Online Società Geologica Italiana **21**(1), 696–697 (2012)
11. Barbaro, G., Foti, G., Sicilia, C.L.: Coastal erosion in the South of Italy. Disaster Adv. **7**, 37–42 (2014)
12. Barbaro, G., Foti, G., Sicilia, C.L., Malara, G.: A formula for the calculation of the longshore sediment transport including spectral effects. J. Coast. Res. **30**, 961–966 (2014)
13. Natesan, U., Parthasarathy, A., Vishnunath, R., Kumar, G.E.J., Ferrer, V.A.: Monitoring longterm shoreline changes along Tamil Nadu. India using geospatial techniques. Aquatic Proc. **4**, 325–332 (2015)
14. Yang, Z., Wang, T., Voisin, N., Copping, A.: Estuarine response to river flow and sea-level rise under future climate change and human development Estuarine. Coast. Shelf Sci. **156**, 19–30 (2015)
15. Acciarri, A., Bisci, C., Cantalamessa, G., Di Pancrazio, G.: Anthropogenic influence on recent evolution of shoreline between the Conero Mt. and the Tronto R. mouth (southern Marche, Central Italy). Catena **147**, 545–555 (2016)
16. Barbaro, G., Fiamma, V., Barrile, V., Foti, G., Ielo, G.: Analysis of the shoreline changes of Reggio Calabria (Italy) Int. J. Civil Engin. Tech. **8**(10), 1777–1791 (2017)
17. Addo, K.A.: Shoreline morphological changes and the human factor. Case study of Accra Ghana J. Coast. Conserv. **17**(1), 85–91 (2013)
18. Amrouni, O., Hzami, A., Heggy, E.: Photogrammetric assessment of shoreline retreat in North Africa: anthropogenic and natural drivers. ISPRS J. Photogramm. Remote. Sens. **157**, 73–92 (2019)
19. Ozpolat, E., Demir, T.: The spatiotemporal shoreline dynamics of a delta under natural and anthropogenic conditions from 1950 to 2018: A dramatic case from the Eastern Mediterranean. Ocean & Coastal Management **180**, 104910 (2019)
20. Zhang, R., Chen, L., Liu, S., Zhang, H., Gong, W., Lin, G.: Shoreline evolution in an embayed beach adjacent to tidal inlet: The impact of anthropogenic activities. Geomorphology **346**, 106856 (2019)
21. Zema, D.A., Bombino, G., Boix-Fayos, C., Tamburino, V., Zimbone, S.M., Fortugno, D.: Evaluation and modeling of scouring and sedimentation around check dams in a Mediterranean torrent in Calabria. Italy J. Soil Water Conserv. **69**(4), 316–329 (2014)
22. Foti, G., et al.: A methodology to evaluate the effects of river sediment withdrawal: the case study of the Amendolea River in Southern Italy. Aquat. Ecosyst. Health Manage. **23**(4), 465–473 (2021)
23. Manca, E., Pascucci, V., Deluca, M., Cossu, A., Andreucci, S.: Shoreline evolution related to coastal development of a managed beach in Alghero, Sardinia. Italy. Ocean Coast. Manag. **85**, 65–76 (2013)
24. Versaci, R., Minniti, F., Foti, G., Canale, C., Barillà, G.C.: River anthropization: case studies in Reggio Calabria. Italy. WIT Trans. Ecol. Environ. **217**, 903–912 (2018)
25. Foti, G., Barbaro, G., Barillà, G.C., Frega, F.: Effects of anthropogenic pressures on dune systems—case study: Calabria (Italy). J. Marine Sci. Eng. **10**(1), 10 (2022)
26. Barbaro, G.: Saline Joniche: a predicted disaster. Disaster Adv. **6**(7), 1–3 (2013)
27. Prumm, M., Iglesias, G.: Impacts of port development on estuarine morphodynamics: Ribadeo (Spain). Ocean Coast. Manag. **130**, 58–72 (2016)
28. Valsamidis, A., Reeve, D.E.: Modelling shoreline evolution in the vicinity of a groyne and a river. Cont. Shelf Res. **132**, 49–57 (2017)
29. Miduri, M., Foti, G., Puntorieri, P.: Impact generated by Marina of Badolato (Italy) on adjacent coast. In: Proc. 13th Int. Congr. Coastal and Marine Sciences, Engineering, Management and Conservation MEDCOAST, pp. 935–945 (2017)

30. Barbaro, G., Foti, G., Malara, G.: Set-up due to random waves: influence of the directional spectrum. Int. J. Maritime Engin. **155**, A105–A115 (2013)
31. Almar, R., et al.: Response of the Bight of Benin (Gulf of Guinea, West Africa) coastline to anthropogenic and natural forcing, Part 1: Wave climate variability and impacts on the longshore sediment transport. Continental Shelf Research **110**, 48–59 (2015)
32. Kroon, A., de Schipper, M.A., van Gelder, P.H.A.J.M., Aarninkhof, S.G.J.: Ranking uncertainty: Wave climate variability versus model uncertainty in probabilistic assessment of coastline change. Coastal Eng. **158**, 103673 (2020)
33. Fiori, E., et al.: Analysis and hindcast simulations of an extreme rainfall event in the Mediterranean area: the Genoa 2011 case. Atmosph. Res. **138**, 13–29 (2014)
34. Boudet, L., Sabatier, F., Radakovitch, O.: Modelling of sediment transport pattern in the mouth of the Rhone delta: Role of storm and flood events. Estuarine, Coast. Shelf Sci. **198**, 568–582 (2017)
35. Hagstrom, C.A., Leckie, D.A., Smith, M.G.: Point bar sedimentation and erosion produced by an extreme flood in a sand and gravel-bar meandering river. Sedim. Geol. **377**, 1–16 (2018)
36. Destro, E., et al.: Coupled prediction of flash flood response and debris flow occurrence: application on an alpine extreme flood event. J. Hydrol. **558**, 225–237 (2018)
37. Scionti, F., Miguez, M.G., Barbaro, G., De Sousa, M.M., Foti, G., Canale, C.: Integrated methodology for urban flood risk mitigation: the case study of Cittanova (Italy). J. Water Resour. Planning Manag. **144**(10), 05018013 (2018)
38. Barbaro, G., Petrucci, O., Canale, C., Foti, G., Mancuso, P., Puntorieri, P.: Contemporaneity of floods and storms. A case study of Metropolitan Area of Reggio Calabria in Southern Italy. In: Proc. New Metropolitan Perspectives (NMP, Reggio Calabria, Italy), Smart Innovation, Systems and Technologies vol. 101, pp. 614–620 (2019)
39. Zellou, B., Rahali, H.: Assessment of the joint impact of extreme rainfall and storm surge on the risk of flooding in a coastal area. J. Hydrol. **569**, 647–665 (2019)
40. Barbaro, G., Foti, G., Nucera, A., Barillà, G.C., Canale, C., Puntorieri, P., Minniti, F.: Risk mapping of coastal flooding areas. Case studies: Scilla and Monasterace (Italy). Int. J. Safety Security Engin. **10**(1), 59–67 (2020)
41. Canale, C., et al.: Analysis of floods and storms: concurrent conditions. Italian J. Eng. Geology Environ. **1**, 23–29 (2020)
42. Canale, C., Barbaro, G., Foti, G., Petrucci, O., Besio, G., Barillà, G.C.: Bruzzano river mouth damage to meteorological events. International Journal of River Basin Management, 1–17 (2021)
43. Yanshuang, Z., Yong, L., Xiaohua, Z., Hongxia, S.: Discussion on the mechanism of the differences of sediment transport capability of the different alluvial reaches in the Yellow River. Proc. Environ. Sci. **10**, 1425–1430 (2011)
44. Meyer-Peter, E., Müller, R.: Formulas for bed-load transport. In: Proc. 2nd Meeting Int. Association for Hydraulic Structures Research Delft (The Netherlands), pp. 39–64 (1948)
45. Schoklitsch, A.: Handbuch des Wasserbaues. Springer, Vienna (1962)
46. Wu, W., Vieira, D.A., Wang, S.S.Y.: 1D numerical model for nonuniform sediment transport under unsteady flows in channel networks. J. Hydraul. Eng. **130**(9), 914–923 (2004)
47. Zavattero, E., Du, M., Ma, Q., Delestre, O., Gourbesville, P.: 2D Sediment transport modelling in high energy river - application to Var River France. Procedia Eng. **154**, 536–543 (2016)
48. Török, G.T., Baranya, S., Rüther, N.: 3D CFD modeling of local scouring, bed armoring and sediment deposition. Water **9**, 56 (2017)
49. Sun, Z.L., Gao, Y., Xu, D., Hu, C.H., Fang, H.W., Xu, Y.P.: A new formula for the transport capacity of nonuniform suspended sediment in Estuaries. J. Coast. Res. **35**(3), 684–692 (2019)
50. Rahman, S.A., Chakrabarty, D.: Sediment transport modelling in an alluvial river with artificial neural network. J. Hydrol. **588**, 125056 (2020)

51. Terranova, O., Antronico, L., Coscarelli, R., Iaquinta, P.: Soil erosion risk scenarios in the Mediterranean environment using RUSLE and GIS: an application model for Calabria (southern Italy). Geomorphology **112**(3–4), 228–245 (2009)
52. van der Knijff, J.M., Jones, R.J.A., Montanarella, L.: Soil erosion risk assessment in Italy. European Soil Bureau Research Report, EUR 19044EN. Office for Official Publications of the European Communities, Luxembourg (1999)
53. Grimm, M., Jones, R.J.A., Montanarella, L.: Soil Erosion Risk in Europe. European Soil Bureau Research Report, EUR 19939 EN. Office for Official Publications of the European Communities, Luxembourg (2002)
54. Grimm, M., Jones, R.J.A., Montanarella, L.: Soil Erosion Risk in Italy: A Revised USLE Approach. European Soil Bureau Research Report No. 11, EUR 20677 EN. Office for Official Publications of the European Communities, Luxembourg. (2003)
55. Gavrilovic, S.: Méthode de la classification des bassins torrentiels et équations nouvelles pour le calcul des hautes eaux et du débit solide. Serbia, Vadoprivreda (1959) (in French)
56. Foti, G., Barbaro, G., Bombino G.: Application of remote sensing to estimate river sediment transport in the Calabrian basins (Italy). In: Proceedings of the 6th International Conference on Water Resource and Environment (WRE), 23–26 August 2020 (2020)
57. Sorriso-Valvo, M., Terranova, O.: The calabrian fiumara streams. Z. Geomorphol. **143**, 109–125 (2006)
58. Sabato, L., Tropeano, M.: Fiumara: a kind of high hazard river. Phys. Chem. Earth **29**, 707–715 (2014)
59. Barbaro, G., Bombino, G., Foti, G., Borrello, M.M., Puntorieri, P.: Shoreline evolution near river mouth: case study of Petrace River (Calabria, Italy). Regional Stud. Mar. Sci. **29**, 100619 (2019)
60. Foti, G., Barbaro, G., Bombino, G., Fiamma, V., Puntorieri, P., Minniti, F. Pezzimenti, C.: Shoreline changes near river mouth: case study of Sant'Agata River (Reggio Calabria, Italy). European J. Remote Sensing **52**(sup.4), 102–112 (2019)
61. Giandotti, M.: Previsione delle piene e delle magre dei corsi d'acqua. Memorie e studi idrografici. Servizio Idrografico Italiano, Italy (1934) (in Italian)
62. Kirpich, Z.P.: Time of concentration of small agricultural watersheds. Civil Engin. **10**(6), 362 (1940)
63. Natural Resources Conservation Service (NRCS): Pondsplanning, design construction. Agriculture handbook. United States Department of Agriculture (USDA), USA (1997)
64. Fiedler, F.R.: Simple, practical method for determining station weights using Thiessen polygons and isohyetal maps. J. Hydrol. Engin. **8**(4), 219–221 (2003)
65. Zemljic, M.: Calcul du débit solide – Evaluation de la végétation comme un des facteurs antiérosif, International Symposium Interpraevent, Villaco (1971) (in French)

The Impact of Spatial Location of Function Inside Building to Improve Distinguishing Architectural Forms as an Urban Landmark

Al Khafaji Ibtisam Abdulelah Mohammed[✉] and Maha Haki

Department of Architecture, Al-ESRAA University College, Baghdad, Iraq
alkhafajiibtisam@yahoo.com, maha@esraa.edu.iq

Abstract. Great buildings and places attract people and the close environment profited of this phenomenon, the beauty, which discovered part by part used to be much more attractive than visible in whole. This research discuss that the mental image of any place, is not a copy of reality, but rather as a simulation of that reality, it is an intellectual reconfiguration of the event or phenomenon, and the clarity of this image is related to the possibility of distinguishing forms, function and spaces as a spatial landmark. In all these processes, the important point is the attempt to involve man in the process of interpretation architectural environment. Due to the important roles of functions and location inside buildings, the objective aim of this research is to indicate how we can improve distinguishing architectural form as a landmark by choosing good location of functions inside buildings. In order to capture this aim, three basic concepts have been identified and framed into a conceptual structure, (functions as generative forces, distinguishing forms, spatial locations) then we used self-assessment practical tools to test hypothesis. Results explained that functions and location inside building create a spatial centers in urban environments and create a field of forces that attract people, central and peripheral location have the same importance in terms of improving distinguishing by attracting peoples, while distinguishing forms in widespread location depend on the possibilities of people to fined continuity of connections between these locations.

Keywords: Spatial forces · Functions location · Distinguishing landmark

1 Introduction: Intelligibility of Urban Environments

Since the beginning, the awareness of man and his reality based on interpretation, since this interpretation enabled him to control and to overcome this reality. This happen by three basic actions (perceptions & cognition, getting knowledge and mental images). Perception is the point of arrival to the last sensory station, Knowledge, refers generally to the action by which man can mentally control a particular subject in order to discover its distinctive characteristics, while the mental image indicates the meaning, essence, truth and identity.

© The Author(s), under exclusive license to Springer Nature Switzerland AG 2022
F. Calabrò et al. (Eds.): NMP 2022, LNNS 482, pp. 75–85, 2022.
https://doi.org/10.1007/978-3-031-06825-6_8

1.1 Mental Image-Definitions

Architectural studies have confirmed that the image of any building or street is not an exact copy of reality, but rather a simulation of that reality. It is an intellectual reconfiguration of the event or phenomenon and may be associated with some meanings interpreted by person, this depends on the ability to retrieve and receive of each person, on the environment in which he lives and on the recipient's culture [1]. Kevin Lynch defined the mental image as a representation of form and elements in the mind of the recipient; he divided this image into three components (structure, identity, meanings) and five elements (lines, edges, sectors, nodes, and landmarks) [2]. Environmental perceptions are the product of a two-sided process, the recipient and the environment. The environment provides stimuli and relationships, while the recipient organizes, gives and selects meaning for what he sees in the environment. Accordingly, a different mental image is formed from one recipient to another. The individual provides a structure for his self-knowledge, and the sum of this knowledge, values and meanings are organized according to certain rules that affect behavior patterns [3]. The mental image that individuals hold to the external physical world represents sensory impressions that individuals form about their surroundings and are mentally and psychologically linked to previous experiences centered in the subconscious. The most important physical and moral components of the mental image that the individual holds with those of the members of his society and integrate with them to form a general picture that expresses what individuals draw and remember from the elements of the environment. The clarity of the geometrical form and aesthetic characteristics of the space help people to absorb it [4]. Rapport defined the mental image as the internal representation by the individual. In addition, this affects his spatial behavior. Mental image is defined as a series of psychological variables that individuals acquire so that they encode, save and retrieve information about their space, environments, components, elements, locations, distances, directions, and the structure as a whole [5]. Lynch found some collective agreements between recipients of the same group, reflected in their mental image, this agreement is necessary to design an architectural and urban environment that can be used by many individuals [3].

1.2 Intelligibility of Architectural Image

The role of common knowledge among individuals is to increase the clarity of their mental image, as individuals formulate an objective world according to their common motives. It is possible to determine the characteristics of the system that provides objective knowledge because it will be realized in itself or to knowledge prior to the experience, is represented by ordinary relations (complete repetition, proportionality, perfect symmetry), and confirms its objectivity [6]. Lynch sees that the intelligibility of the city's structure is related to the visual quality, and legible architectural and urban environment can be defined as the environment whose visual elements (Paths, Edges, Landmark, Nodes, Visual districts) can be distinguished easily, and can be connected into a unified whole. The legibility of mental image, according to Lynch's definition, is also linked to the person's mental map, and the clear environment helps the person to form a clear picture of his urban environment and then interact with that environment effectively [7].

City whose landmarks, paths, centers, boundaries, and transition points can be distinguished clearly and easily, and in which these elements are interconnected and integrated into one clear formation, is the city that has a high ability to appear Image ability, then the fixed parts and activities of the users play role in the clarity of the mental image. There are many visual problems that affect the clarity of this image, as lack of integration of visual elements, mixing and overlapping of visual elements, weak boundaries, isolation of some elements, breaks in continuity, ambiguities, branching, lacks of character, lacks of differentiation [1]. Cullen presents the idea of–Art of Relationship–in the analysis of the spatial structure of urban environment. The distinction of relationships depends on visual perception, because it is related to memory of persons and experiences. Emotions represent strong forces to move the mind, then the environment stimulates production an emotional response, voluntary or unwilling and this depends on location and vision [8]. Mukhten defined the intelligibility of urban form as the way in which persons perceive and understand urban environment then give reactions towards that environment. This legibility is linked to:

1- The qualitative characteristics of the place through which the city acquires its identity.
2- Human movement and selection of movement axes within a specific urban environment.
3- The vitality of the urban environment, which is achieved due to the multi-functionality and mixed uses of the land.
4- The visual dimension of the city's scenery, which in turn is linked to the city's spaces and streets, in addition to facades, roofs, floors, architectural monuments and signposts, in addition to street furniture [7].

2 Research Problem–Aims–Hypothesis and Method

From all information mentioned above, we can specified research problem, aim, hypothesis and method Table 1.

Table 1. Research problem, aim, hypothesis and method (Source: by researcher)

Research problem	Reviewing previous theories, we found some knowledge gap related to the absence of a comprehensive theoretical indicators that can be used to explain how we can improve distinguishing architectural form as a landmark by choosing good locations for functions
Aim	To explain how we can improve distinguishing forms as a landmark by choosing good location for functions inside building
Hypothesis	The research suppose that: 1- good location of functions inside building acts as center of visual attractive forces. 2- Attractive forces created by functions and location increase the possibility of distinguishing building as a spatial landmark
Methods	: due to the imperial nature of our research, we will organize themes in conceptual structure then we will use self-assessment questioner to test variables

Table 2. Conceptual structure

Basic concepts	Sub-basic concepts	Values
Functions as centres of forces	Characters	Compatibles, local, Holistic, Generative, strong, salient, weak
	Generative forces of functions	Attract attention, attract movements, generate life and beauty, give felling of centrality
	Location	Central, peripheral, widespread
Distinguishing forms	Elements to be distinguish	Boundaries, centres, Movements. Physical forms and elements, Functions
	Techniques for improving distinguishing	Attracting, clarity, diversity, sense of order, understanding meaning, geometrical properties, strengthen relations, geometrical coherence
Spatial location of functions		Laws of Emergence, Theory of Location, Theory of Partitioning

3 Conceptual Structure

3.1 Functions as Centers of Spatial Forces

Spatial Centers (functional, social, and cultural) are a distinct physical system that occupies a certain volume in space and possesses the essence of compatibility. Table 2 The holistic urban environment includes multiple centers and it is difficult to draw boundaries that define and surround this whole [8]. Local center and living system has a specific activity and boundaries, both of them can be distinguished. These centers generate a feeling and mental impression regarding to the way the space works and the possibility of its occupancy. Spatial center is not an abstract point, but rather a field of organizational power in an architectural composition, which gives a feeling of centrality. These centers are compatible with each other and give a sense of life and beauty, this can be achieved now when the centers support and strengthen each other. These centers increase unity and intensify space structures. We found three types of relations connect them, which called "helping relations" (shape-element/element-space/shape-space relationship). These relationships work to increase values of life in the centers, as each center works to attribute and restore life to the other centers and make it more vital and effective [8]. There are three types of centers:

1- Strong centers: They are coherent and give the holistic image.
2- Salient centers, they are effective and distinct, appear in space, constitute totality and comprehensiveness, and are capable of appearing and being visible, forming every comprehensive, compatible and full of life.
3- Weak centers (latent): They are unconnected and invisible.

The importance of these centers does not depend only on their shape, but on their position within the holistic composition, their importance is related to the amount of their strength to attract attention and form vital centers. Each strong center includes many centers, and it is possible to feel the emergence and presence of different centers and at multiple levels, and this leads to increased interaction with the formation. Other trends are far away and that the difference for forces is what determines the characteristics of the visual and spatial field as a vital active center or not [9].

3.2 Distinguishing Architectural Form

Appley In his research entitled *Why buildings are known*, he focused on the concept of clarity of the world and focused on the characteristics of the physical form and considered that clear boundaries are one of the most important factors that distinguish the form from its surroundings and highlight it from its surroundings [10]. Jan Jacobs in her book The Death and life of great American cities, she indirectly defended the concept of diversity and distinguishing, where she asserted that the streets are the beating heart of the city and can be distinguished as streets full of life through the diversity of functions [11]. On the other hand, Lynch emphasizes the visual urban form, and in his research (Notes on city satisfactions), he emphasized the sensory and psychological effects of the physical form, which achieve a sense of happiness and contentment, among these effects are the sense of orientation, happiness in the presence of noticeable differences, rhythm, the sense of order [3]. Rapport, mentioned that places in the city have different meanings and affect different situations and different social identities, he emphasized the distinction between streets that are used for crossing only (functionally successful) and those that give a sense of life [5]. Alexander in his book, The Nature of Order, discussed deeply the existence of 14 geometric properties which strengthen form and improve distinguishing, like: Alternating Repetition–(Positive Space)–Level of scale–Thick Boundaries–Good Shape–Local Symmetrical–Deep Interlock and Ambiguity–Contrast–Gradient–Roughness–Echoes–The Void–Simplicity–Inner Calm and Non-Separateness [8]. Geometrical Coherence is another concept discussed by Salinger to distinguish a lively environment. Geometric compatibility is a qualitative characteristic through which the cooperation between the function of the parts takes place to reach the whole with a successful and distinctive function, Such as:

Coupling: There are no unconnected parts in the model.
Diversity: Emphasis on the connection of different elements.
Forces: Forces at the functional level are more influential than at the global level.
Boundaries: The different models are interconnected by their outer edges.
Hierarchy: Organize the elements in a hierarchical way.

Interdependence: The elements and models at their different levels do not depend on each other symmetrically, as the totality requires all other levels, and the opposite is not true [9]. There are different visual treatments that can be carried out to confirm the clarity of architectural forms and improve attracting people, such as Singularity, Continuity of the elements and boundaries, Clarity of connection, Changing Direction, Emphasis on Motion Awareness, Time sequence of the formation elements [12]. Al-Assaf and other explained that density and repteteness as necessary symptoms for any form to function as a work of art. They concluded that architectural work is aesthetically pleasing when formal design language is syntactically dense, its content is semantically dense, and the design process is replete [13].

G.E. Karepov, explained the process of improvement of the modern city includes a competent structural organization of the landscape and recreational areas. According to the laws of urban development, this factor stimulates the social activity that stimulates historical progress. Living sculpture symbolizes close connection of human and natural resources, promotes environmental consciousness, and brings attention to preserving current pieces of organic architecture [14].

3.3 Spatial Location of Functions

3.3.1 Laws of Emergence (Compactuses and Linearity)

There are many different laws that strengthen spatial system and others confuse it, and this affects the behavior and effectiveness of systems. The Paradox of Centrality indicates that there are internal and external relationships in the spatial system and the increase in the internal integration of the system means an increase in the external isolation. The circular shape can be considered the most appropriate form for this law as the integration is strong in the center and less powerful in the periphery. On the movement system, as the movement increases with short axes that pass through the center and decreases at the ends, the sequential distribution of the complementary forces from the center towards the edges decreases in linear forms, as the integration is stronger at the edges of the system and outside it. The Paradox of Visibility showed the difference between the metric (standard) properties and the optical properties of space, as the linear arrangement of the elements increases the space depth and decreases clarity. In addition, the increase in (metric isolation) increases the visual integration. Since the space system must interact with the external and internal world, the space system must have a balanced shape between the central and linear form. Visual clarity is linked with the clarity of the mental image, as the lines and surfaces are organized to reach a different state of integration and clarity [15].

3.3.2 Location

Location has a strong impact on increasing or decreasing degree of movement frequency, as well as the effectiveness of this movement, and determines whether this movement has a positive impact in terms of being communicative or direct. The urban form, which includes multiple functional activities with different locations, affects the person, his movement and his effectiveness. Within the same space [15]. The theory of attraction indicates that the buildings represent attractions through their location; there is a relationship between the movement of the user and the location of the buildings on the one hand, and the size of the surrounding spaces on the other hand. And to the building blocks with varying degrees of attraction according to the ability of the surrounding spaces to create movement [16]. Additions and locations of some buildings within an existing fabric has a clear impact on the feature of integration, the choice of the site at the local level has a great impact on the characteristic at the holistic level. There are several factors that affect the spatial characteristics, as the shape of the building, the added function, the size of the added function and the location of the function [15].

3.3.3 Theory of Partitioning

It is a set of laws that determine the nature of the integration property in the case of adding or subtracting some shapes from the space system. As the increase or decrease of the isolation property depends on the addition or elimination of some shapes. Deleting shapes from the space center increases isolation and reduces integration. Integration is reduced whenever added shapes are elongated. Repeated addition or deduction leads to an increase in the isolation property. The addition or deduction from the center increases the isolation more than the addition and deduction are from the corner or the ends [15].

4 Case Study/Practical Method–Results and Conclusion

Research aim is to show the impact of spatial location of functions inside building to improve distinguishing architectural forms as a landmark. We have used questionnaire (self–assessment practical tool), table [3] for testing variables mentioned in table [5]. This tool based on best practices and standards in business process architecture, design and quality management. In this practical tools, a series of questions are used to identify to what extent your architectural initiative is complete in comparison to the requirements set in standard. To facilitate answering the questions, there is a space in front of each question to enter a score on a scale of (1) to (5) we used six project (contemporary Islamic mosques and centers) to be applied due to the variety of form and functions that connect to them for their important location in urban public spaces, which attract people visually.

A score of (1) would mean that the answer is not clear at all, where a (5) would mean the answer is clear and defined. When the question is not applicable or you do not want to answer it, you can skip it without affecting your score and leave it empty. After you have responded to all the questions, Compute your average score for that section and get results [17] Table 3, 4 and 5.

Table 3. Self–assessment practical tool

1........	2.........	3.........	4.....	5......

Table 4. Dependent and independent variables

Independent variables	Techniques for improving distinguishing Attracting, clarity, diversity, sense of order, understanding meaning, geometrical properties, strengthen relations, geometrical coherence
Dependent variables	Distinguishing landmarks

Table 5. Results from case study

	Central location	Peripheral locations	Widespread location
1-attracting	55	15	15
2-singularity	10	10	15
3-clearity &simplicity	5	45	5
4-continiuity	5	5	35
5-geometrical properties	10	10	10
6-coherence	5	5	5
7-strengthen relations	5	5	5
8-understanding meaning	5	5	10

5 Discuss Results and Conclusions

Usually the great buildings are connected with great public spaces. This was the case Egyptian pyramids and the ceremonial procession road, as well as the procession road from the Ishtar Gate to the ziggurat in Babylon, Similar situation took place through ages, the temple was present in the city structure through the multiple vistas, usually well thought and composed. The same situation took place in the Renaissance when the role of public spaces became more important and along with the discovery of the perspective the vistas became longer and created with the awareness of the peculiarities of human sight. In Islamic architecture we see a strong connection to the mind as a result of impart a kind of spirituality to the place [18], from analyzing results we found that central location of the most important function improve distinguishing forms by: attraction people to a central vocal point of forces, singularity, clarity, understanding meaning. The peripheral locations increase clarity through uniqueness, simplicity of organization, continuity of shapes boundaries, clarity of connection points, and differentiation in directing movement. Through the diversity in the style of the formal and visual treatments adopted by the designer to deliver the desired message. As well as through the ease of vision from long and close distances and ease of perception in addition to excellence through the strength of the edges. As for the widespread location, the results showed that distinguishing form was achieved by focusing on the continuity of the boundaries, the clarity of the joint points of connections and the differentiation in directing of movement Fig. 1.

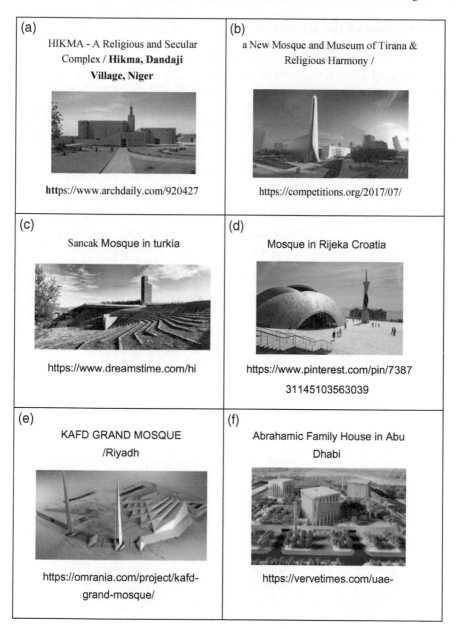

Fig. 1. a-f Projects to be applied–Islamic mosques and centers

6   Conclusion of the Research

The research concluded that the peripheral and central location have the same importance in terms of improving distinguishing by attracting peoples, while distinguishing
forms in widespread location depend on the possibilities of people to fbined continuity
of connections between these different locations. Therefore, enhancing the designer's
ability to deal with architectural functions as a center of forces by using contextual
data, and reconsidering the possibility of benefiting from locations data and exploring
its various aspects. The goal is to create an architecture that is consistent with the spatial
locations in which it is built first, and then the extent of its survival in the memory of the
users of the place Fig. 2.

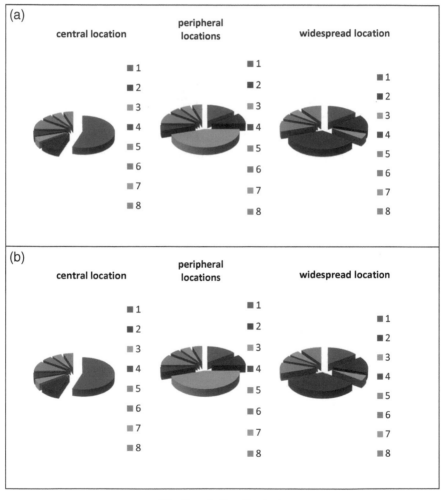

Fig. 2. a, b: Results charts

References

1. Lang, J.: Creating Architectural Theory: The Role of the Behavioral Sciences in Environmental Design, p. 136. Van Nostrand Reinhold, London (1987)
2. Kubasi, A., Fadel, S.: The inspired image in the urban contest, master thesis submitted to the university of technology, pp. 16–17 (2000)
3. Lynch, K.: The Images of Cities, pp. 6–46–288. London (1960)
4. Kirer, R.: Urban Space, p. 15. Rizzoli International Publications, ISBN 0847802337, 9780847802333 (1979)
5. Rapoport, A.: human aspects of urban form, pp. 20–120. Pergaman Press, London (1977)
6. Al-Jubouri, B.A.: The impact of syntethic changing to improve communicative media, nonpublished master thesis, p. 45. University of Technology, Iraq (1999)
7. Moughtin, C.: Urban design: Street and Square, Amstrdam, Third Edditions, pp. 218–220 (2003)
8. Christopher, A.: The Nature of Order, pp. 82–116. Taylor & Francis (2002). The Process of Creating Life, pp. 114–116 (2002), ISBN 0–9726529–0–6
9. Salingaros, N.A.: Complexity and urban coherence. J. Urban Des. **5**, 291–316 (2000)
10. Appleyard, D.: The use of streets. J. Urban Des. **9**(1), 3–22 (2004). https://doi.org/10.1080/1357480042000187686. p. 29
11. Gutman, R.: The Street Generation. In: Streets, On. (ed.) Sanderson) MIT Press, p. p48. Mass, Cambridge (1986)
12. Al din Yosef, M.S., Lynch, K.: The Image of the City, pp. 10–11. Online Publishing (1983)
13. Al-Assaf, N., Dahabreh, S.: The aesthetics symptoms of architectural form: The case of Barcelona Museum of Contemporary Art Conference: ARCHDESIGN: Current Trends in Architectural Design and Methodologies At: Turkey (2014)
14. Karepov, G.E., Lipovtseva1, S.R.: Small Architectural Forms In The Organization Of Urban Environment. Siberian Federal University Digital Repository, p. 681 (2018)
15. Hillier, B., Vaughan, L.: The City as One Thing, pp. 178–168–263–125–170–280–283. The Bartlett University College, London (2007)
16. Zupan, Puckered, B: Ursign ban space for Pedestrian, p. 15. MIT Press, Mass (1975)
17. A. Art of service, pp. 9–14 (2018)
18. Shafi'i, F.M.: Arabic Islamic Architecture Past–Present & Future, p. 50. King Saud University (1995)

Shaping Evolving Rural Landscapes by Recovering Human-Nature Harmony Under the Beautiful Countryside Construction in China

Yapeng Ou[✉]

Xi'an University of Architecture and Technology, Xi'an 710055, China
Ouyap721@163.com

Abstract. Rapid rural modernization in China since the 1980s has led to mutating rather than evolving rural landscapes. The on-going major state initiative of Beautiful Countryside Construction, aiming at economic, political, social, cultural and ecological construction of rural China, provides a precious opportunity for regenerating rural landscapes. This research is aimed to explore how recovering a harmonious human-nature relationship has contributed to the generation of evolving rural landscapes during the implementation of the initiative. To begin with, based on a literature review, it argues the relevance of classical Chinese philosophy on human-nature harmony to an organic evolution of rural landscapes. Then, it looks into the implementation of the initiative in Meixian County, Shaanxi Province. In particular, it analyzes the principles of the initiative and the role of heritage revitalization in recovering human-nature harmony. It is found that traditional knowledge and values have been integrated into heritage revitalization on the one hand and played a significant role in recovering a harmonious human-nature relationship and thereby making rural landscapes evolving on the other.

Keywords: Human-nature harmony · Beautiful Countryside Construction · Traditional knowledge and values · Heritage revitalization · Rural landscape

1 Introduction

Human beings are now living in the Anthropocene, a new geological epoch where they have transformed the earth's surface and caused environmental change to an unprecedented degree (Zalasiewicz et al. 2011). Under this broad context, the resilience and adaptability of rural communities to environmental uncertainties is one of the major areas that present key challenges for rural areas in the early 21st century (Woods 2012). Yet, rural uncertainties are not limited to environmental ones; instead, they come manifolds: socially, culturally and economically. Rural China, for instance, is currently going through drastic restructuring at spatial, economic and social levels (Long and Liu 2012). Increasing pressure from rapid urbanization and the accompanying land use changes are serious threats to the cultural landscapes in rural China (Yu et al. 2016). Consequently, a rural-urban tension has emerged. This tension in part manifests as a seemingly "peaceful" rural-urban integration, which has greatly improved rural living conditions and quality of life on the one hand. On the other hand, however, rural-urban integration has also raised the bad need "to protect and promote the rural landscape, geographical space, social, historical and cultural features in the new rural construction" (Zhang and Huang 2012).

In fact, the New Socialist Countryside Construction 社会主义新农村建设[1], as a national mission with historic significance, has in recent years attached unprecedented attention to the relevance of rural Heritage to socioeconomic development of rural China. To guide its implementation, the Beautiful Countryside Construction 美丽乡村建设, as a conceptual and methodological framework, was clearly proposed at the national level for the first time in the "No. 1 Central Document of 2013"[2], based on the concept of "Beautiful China 美丽中国"[3]. Ever since, the Beautiful Countryside Construction has become a consistent national policy, successively integrated into various government documents, such as the "No. 1 Central Document of 2014", the "National New Urbanization Plan (2014–2020)" and so on, which all require vigorously improving the rural living environment and building beautiful countrysides. The Beautiful Countryside Construction is aimed to coordinate the intertwining economic, political, cultural, social and ecological aspects of rural development. To meet the requirements of the New Socialist Countryside, the Beautiful Countryside Construction has systematically set up a series of goals ranging from production development, well-off living, civilized social ethos, clean and tidy village to democratic management. What underpins the coordinated aspects and goals of the Beautiful Countryside Construction is the necessity to restore and harmonize the human-nature relationship in rural China. On the one hand, during the Beautiful Countryside Construction, heritage, both tangible and intangible, is utilized or revitalized as a supportive tool for cultural display, which often implies "a project of 'improvement' and of building 'quality' among the 'backward' rural population (Oakes 2012: 381)". On the other hand, rural ecological progress has been prioritized to integrate poverty alleviation and environmental protection, and meanwhile restore rural ecology and improve residential quality.

Given the context above, this research seeks to investigate how human-nature harmony has been recovered through the implementation of the nationwide Beautiful Countryside Construction initiative. To begin with, based on a literature review of Chinese

---

[1] This historic task was first proposed on the occasion of the Fifth Plenary Session of the 16th Central Committee of the Communist Party of China in October 2005, putting forward such specific requirements as "production development, well-off living, civilized social ethos, clean and tidy village to democratic management" for rural construction.

[2] The "No.1 Central Document" is the first document issued each year by the Central Committee of the Communist Party of China and the State Council of the People's Republic of China. It is a cross-sectoral policy document at the national level focusing on such top priorities as agriculture, rural areas, and farmers' issues.

[3] The concept of "Beautiful China" was put forward in the report of the 18th National Congress of the Communist Party of China, which stated: "striving to build a Beautiful China and realizing the sustainable development of the Chinese nation". The report put forward the brand-new concept of "coordinated urban and rural development" to build a Beautiful China, emphasizing that the construction of ecological civilization should be placed in a prominent position and integrated into all aspects and the whole process of economic, political, cultural, and social development.

classics, it traces the philosophical root of the "beautiful countryside" concept and argues the relevance of classical Chinese philosophy on human-nature harmony to the organic evolution of rural landscapes. Then, it case studies the implementation of the Beautiful Countryside Construction initiative in Meixian County, Shaanxi Province. In particular, it analyzes how traditional knowledge and values based Heritage revitalization and reinterpretation contribute to the recovering of a harmonious human-nature relationship.

2 Human-Nature Harmony and Evolving Rural Landscapes

According to Chinese classic philosophy, such as Taoism, human and nature are barely separable. As Zhuangzi[4] remarks, "Heaven, Earth, and I were produced together, and all things and I are one 天地与我并生, 而万物与我为一"[5], which clearly shows his deep interest in "integrating all things of the world into 'One' rather than dividing them into 'Many'" (Ames 2009: 163). In fact, "the grand harmony of all things太和万物[6]", as the highest ideal of Zhuangzi, then, is "the core value that runs through Confucianism and Taoism (Ou 2012: 53)." "Harmony" in essence suggests a network of harmonious relationships, either the relationship between human and nature (Taoism) or the relationship between human and society (Confucianism). There is little doubt that this harmonious relationship between human and nature is critical to moderating rural transformations. On the one hand, it helps balance the need for rural socioeconomic development and ecological protection, which is the premise of rural sustainability. On the other hand, it can prevent the rural landscape from mutating due to excessive human activities that are beyond the environmental capacity. Therefore, the human-nature harmony can serve as a guiding principle of the Beautiful Countryside Construction, in that it can fuel socioeconomic development without compromising ecological progress.

However, in the real world, this vital human-nature harmony is often times put in jeopardy. Indeed, one of the fundamental features of modernity is the separation of the natural and the human (Woods 2004). Besides human-nature divide, spatially there is a rural-urban divide, and temporally a tradition-modernity divide. This deep-rooted binarist ideology has considerably affected theories and practices of development. Rural China, for example, has long blindly imitated cities while breaking away from traditions. As a result, since the 1990s, "the organic evolution of rural landscape has gradually degenerated at four levels of layout, fabric, pattern and form" (Wang and Qian 2015: 17). Given that culture and ecological progress are two priorities of the Beautiful Countryside Construction, it can serve as a precious opportunity to recover human-nature harmony. Now the question is, can an organic evolution of rural landscape be achieved by recovering human-nature harmony?

The answer is yes. When referring to the management of historic urban landscape (HUL) (2016), Taylor considers sustaining the evolving and dynamic nature of landscape,

---

[4] Zhuangzi (c. 369 BC - c. 286 BC) is credited with writing the book under the title *Zhuangzi*, which is one of the foundational texts of Taoism.

[5] From *Zhuangzi: Inner Chapters: The Adjustment of Controversies: Chapter 9* (《庄子内篇•齐物论: 9》), extracted from <http://ctext.org/>.

[6] For the complete chapter, see *Zhuangzi: Outer Chapters: The Revolution of Heaven: Chapter 3* (《庄子外篇•天运: 3》), extracted from <http://ctext.org/>.

morphological continuity and material-immaterial balance as three pivotal principles. Similar to urban landscape, rural landscape is a complex system characterized by totality, interrelations and multidimensional dynamic processes. Therefore, rural landscape is subject to dynamic forces in the economic, sociocultural and environmental spheres that have shaped and still keep shaping it (Winchell and Koster 2010; Tapiador 2007). In this sense, recovering human-nature harmony can contribute to an organic evolution of rural landscape in three ways. First and foremost, recovering human-nature harmony can help form a benign and moderate human-nature interaction, which is fundamental to the evolution of landscapes towards continuity rather than disruptive fragmentation that leads to landscape mutation. Second, by recovering human-nature harmony, there is very likely to be a shift away from the conventional binarist thinking towards a tradition-modernity linkage. This linkage can function as a "buffer zone" that is able to mitigate the negative externalities stemming from rural modernization. In so doing, it can protect the countryside from losing its *genius loci* and social ethos by preserving pathways for the transmission of traditional values and knowledge. Last but not the least, recovering human-nature harmony can contribute to sustainable rural development by creating a synergy between the cultural and the natural and between the tangible physical landscape and the intangible values. Such a synergy can stimulate endogenous heritage based spatial innovation in rural areas and therefore upgrade rural landscape while reshaping social fabric of the rural community.

3 Beautiful Countryside Construction and Human-Nature Harmony Recovering

3.1 Principles of the Beautiful Countryside Construction Initiative

The Beautiful Countryside Construction, aiming at building up a rural China that is economically and culturally dynamic and ecologically sound, has provided a precious opportunity for integrating a human-nature dimension into rural modernization. This integration promises the recovering of human-nature harmony and upgrading of the rural landscape, wherein rural Heritage, both tangible and intangible, has a significant role to play. Practically speaking, rural construction in China over the last two decades have demonstrated two characteristics. On the one hand, it is nothing new that in rural China, heritage preservation and display are popularly viewed by many as powerful tools of modernization and development (Oakes 2012). Recent years, an emerging "historic reconstructionism" has also been observed across rural China. Yet, displaying culture and Heritage and a retroist approach to restoring a historic landscape, both as a visual representation of "restored" temporospatial continuity and the *quality* of rural modernization and development has long proved problematic. On the other hand, while the cultural dimension has been increasingly integrated into the rural construction process, the human-nature dimension has remained mostly ignored. In recent years, it has been increasingly recognized that physical, sociocultural and eco-environmental dimensions of rural landscapes are of equal importance for the New Rural Construction and ultimately for rural sustainability. In response, the New Rural Construction should shift from a sheer focus on digital growth so as to help foster the concurrent "cultural and ethical civilization 精神文明" (Zhang 2006) and ecological civilization (Pan 2021; Schwartz and Cobb 2018).

To translate this very concept into reality, the Beautiful Countryside Construction must differentiate itself from previous cultural display practices in the form of physical beautification or "historical reconstructionism". Instead, it should not only protect built heritage and beautify the living environment, but more importantly revitalize both the built and the intangible heritage, including crafts, beliefs, memory, traditional values, traditional knowledge and so on. Equally important, it should create a synergy between culture and nature while revitalizing Heritage with a holistic approach. Therefore, it must, in the first place, avoid brutally complying with the far-reaching ideology of "discard the old and cultivate the new 破旧立新", and then integrate "old" rural traditions of the profound farming civilization and Confucian culture (Zhang 2006) as well as the embedded sociocultural values and traditional knowledge conducive to recovering human-nature harmony into the New Rural Construction process. In other words, only by being rooted in the territorial characteristics, traditions and Heritage and simultaneously "transmitting and discarding" them in a critical and creative way based on modern needs and contemporary values will the Beautiful Countryside Construction foster social innovation and achieve a vital balance between material affluence and ecological and cultural-ethical progress and between form and content. This will finally help both regenerate the rural landscape by fostering a *new* living and meaningful rural morphology and contemporizing its functionality and recover its temporospatial continuity.

3.2 Revitalization of Traditional Knowledge and Sociocultural Values

Traditional knowledge and sociocultural values, as integral elements of Heritage, are often considered critical to social innovation and development. In terms of the former, as a collective knowledge, traditional knowledge can make a significant contribution to sustainable rural development, serving as "a source of social progress and knowledge innovation" (Xue and Guo 2009: 141). Clarke (1990) maintains that traditional knowledge needs to be integrated into present development, because it is "environmentally sound" and highlights respect for the long-term requirements of nature, an intrinsic value of sustainable development. This means that traditional knowledge, when properly revitalized, is able to recover human-nature harmony and bridge the gap between the past and the present. As for traditional sociocultural values, rural landscapes are typical cultural landscapes resulting from harmonious human-nature interactions over time. As a result, rural landscapes are "imbued with value systems, traditional knowledge systems and abstract frameworks" (Taylor 2009: 7). These traditional values and knowledge are the cornerstone of community cohesion and source of positive social ethos, hence of great importance for fostering human-nature and human-society harmony.

As the examples from Meixian County in Shaanxi Province demonstrate, local heritage revitalization practices during the Beautiful Countryside Construction have followed a principal principle: recovering human-nature harmony while regenerating rural landscapes. To help recover human-nature harmony, traditional knowledge and sociocultural values have played a significant role. As immaterial heritage, traditional knowledge and sociocultural values have been revitalized and reinterpreted in a way that rural landscapes and rural spaces as a whole are reshaped and upgraded. A typical example is the revitalization of *laochi*, a traditional flood pond which used to be common in almost all villages in the Guanzhong region for rainwater collection and flood control. Since the 1980s, with rapid rural modernization, *laochi* was gradually abandoned as local people deemed it as "old-fashioned" and thus no longer valued it. During the Beautiful Countryside Construction, *laochi* has been revitalized to restore rural water ecology while beautifying rural landscapes. Revitalizing this kind of traditional knowledge is very beneficial to the Beautiful Countryside Construction, because first, it is very cost-efficient to address rural problems compared to the conventional exogenous and technocratic solutions which often times appear "scientific" but turn out to be incompatible with local contexts. Second, traditional knowledge revitalization, as the revitalized *laochi* demonstrates (Fig. 1), showcases how human-nature harmony can generate both socioeconomic and environmental benefits for local communities. Third, it can reshape in an innovative way and upgrade rural landscapes while meeting contemporary needs. For instance, the regenerated natural landscape of *laochi* serves as an empathetic space for the presentation and reinterpretation of folkloric culture and local identity, as well as for socialization. Therefore, *laochi* is transformed into a cultural landscape that meets local villagers' actual social and cultural needs (Ou 2017).

Fig. 1. Overall landscape of the laochi in Tongzhai Village, Qishan County. © *Yapeng* Ou (2017)

Besides traditional knowledge revitalization, the Beautiful Countryside Construction in Meixian County has also visually reinterpreted traditional values in the form of mural paintings. These mural paintings basically illustrate traditional values such as filial piety, diligence, thrift, mutual respect, moderation and so on. Meanwhile, they also convey core socialist values[7]. These traditional and contemporary values can help establish positive social ethos and ethics in rural communities, which are fundamental to fostering human-nature harmony in their patterns of production and ways of life. It is worth noting that, traditional values are also embedded in ecological restoration practices to highlight

---

[7] The core socialist values consist of 12 values at national, social and individual levels. The national values are prosperity, democracy, civility, and harmony 富强、民主、文明、和谐; the social values are freedom, equality, justice, and the rule of law 自由、平等、公正、法治; and the individual values are patriotism, dedication, integrity and friendship 爱国、敬业、诚信、友善.

the importance of human-nature harmony, such as the extended ecological restoration project in the southern floodplain of the Weihe River. The overall design concept of this project has incorporated major values related to a livable rural habitat, such as traditional landscaping, lotus culture, ecological knowledge and Confucian values, which altogether create a thriving landscape of harmonious human-nature coexistence (Fig. 2).

Fig. 2. Lotus garden in Hedi Village, Meixian County. © *Yapeng* Ou (2017)

4 Conclusion

Rural China is undergoing unprecedented transformations with deepening urbanization and rural modernization. At a time when rural transformations tend to manifest as landscape mutation, the Beautiful Countryside Construction initiative provides a precious opportunity to restore and harmonize the human-nature relationship in rural China, which is critical to moderating rural transformations. Recovering human-nature harmony can contribute to an organic evolution of rural landscape in terms of 1) helping form a benign and moderate human-nature interaction; 2) driving an ideological shift away from the conventional binarist thinking towards a tradition-modernity linkage; 3) and contributing to sustainable rural development by creating a synergy between the cultural and the natural and between the tangible physical landscape and the intangible values. As the examples of local Beautiful Countryside Construction practices from Meixian County, Shaanxi Province demonstrate, heritage revitalization has helped recover human-nature harmony. Traditional knowledge and values have fueled heritage revitalization on the one hand and played a significant role in recovering human-nature harmony on the other.

Due to space limitation, the present research did not investigate the relevance of recovering human-nature harmony to the transformation of rural economic landscape. Besides, it did not summarize the pros and cons of the examples. Future research may elaborate on these two aspects.

References

Ames, R.T.: Seeking Harmony Rather than Uniformity. Peking University Press, Beijing (2009)

Calafati, A.G.: 'Traditional knowledge' and local development trajectories. Eur. Plan. Stud. **14**(5), 621–639 (2007)

Clarke, W.C.: Learning from the past: Traditional knowledge and sustainable development. Contemp. Pac. **2**(2), 233–253 (1990)

Ilbery, B.: The Geography of Rural Change. Routledge, New York (2014)

Long, H., Liu, Y.: Rural restructuring in China. J. Rural. Stud. **47**, 387–391 (2012)

Oakes, T.: Heritage as improvement: Cultural display and contested governance in rural China. Modern China **39**(4), 380–407 (2012)

Ou, Y.: Traditional knowledge and sustainable rural development: On the revitalization of *Laochi* in Shaanxi Province, China. In: Proceedings of the Fifth Annual International Conference on Sustainable Development (ICSD) (2017), September 18–19, 2017, New York, USA

Ou, Y.: Intercultural conflict and dialogue from the perspective of multicultural discourse (Graduation Thesis). MA, Harbin Institute of Technology, Harbin (2012)

Pan, J.: China's Global Vision for Ecological Civilization: Theoretical Construction and Practical Research on Building Ecological Civilization. China Social Sciences Press, Beijing; Springer, Singapore (2021)

Schwartz, W.M.A., Cobb, J.B.: Putting Philosophy to Work: Toward an Ecological Civilization. Process Century Press, Anoka (2018)

Tapiador, F.J.: Rural Analysis and Management: An Earth Science Approach to Rural Science. Springer Science & Business Media, Berlin, Heidelberg (2007)

Taylor, K.: The historic urban landscape paradigm and cities as cultural landscapes. Challenging orthodoxy in urban conservation. Landscape Res. **41**(4), 471–480 (2016)

Taylor, K.: Cultural landscapes and asia: Reconciling international and southeast asian regional values. Landscape Res. **34**(1), 7–31 (2009)

Winchell, D.G. and Koster, R.: Introduction: The dynamics of rural change: A multinational approach. In: Winchell, D.G. et al. (eds.) Geographical Perspectives on Sustainable Rural Change, pp. 1–23. Brandon University, Rural Development Institute/Brandon University, Manitoba (2010)

Woods, M.: Rural Geography: Processes Responses and Experiences in Rural Restructuring. Sage, London (2005)

Woods, M.: New directions in rural studies? J. Rural. Stud. **28**, 1–4 (2012)

Yu, H., Verburg, P.H., Liu, L., Eitelberg, D.A.: Spatial analysis of cultural heritage landscapes in rural China: Land use change and its risks for conservation. Environ. Manage. **57**(6), 1304–1318 (2016). https://doi.org/10.1007/s00267-016-0683-5

Zalasiewicz, J., Williams, M., Haywood, A., Ellis, M.: The anthropocene: A new epoch of geological time?". Philos. Trans. Math. Phys. Eng. Sci. **369**(1938), 835–841 (2011)

Zhang, Z. and Huang, Y.: The Livability and characteristics of rural villages in the context of urban-rural integration development: A Case of Jiang Yi Village, Guangdong Province, China. In: Proceedings of the 6th International Association for China Planning Conference (IACP) (2012)

王竹 and 钱振澜: "乡村人居环境有机更新理念与策略". 西部人居环境学刊 **30**(2), pp. 15–19 (2015)

薛达元 and 郭泺. "论传统知识的概念与保护". 生物多样性 **17**(2), pp. 135–142 (2009)

张清奎: "传统知识、民间文艺及遗传资源保护模式初探". 知识产权 **16**(2), pp. 3–9 (2006)

Cultural Democracy, Cultural Ecosystems and Urban Development: Grassroot Initiatives at the Crossroads of Social and Cultural Goals in Bologna

Francesca Sabatini[(✉)] [iD]

Department of Architecture, University of Bologna, Bologna, Italy
f.sabatini@unibo.it

Abstract. The paper analyzes the grassroot cultural ecosystem of Bologna in the theoretical framework of cultural democracy and urban development. The first section sets the theoretical framework, grounded on cultural infrastructure, cultural ecologies, cultural democracy and their role in urban development. The second section presents the context of Bologna, the adopted methodology and the findings: it maps grassroot cultural activities with a societal orientation, observes their relevance as places for socialization and as a widespread cultural infrastructure, and analyzes the institutional actions and policy initiatives that sustain this ecosystem. The third section draws conclusions and advocates for a diversified cultural agenda for cities that support the indigenous, grassroot cultural ecology.

Keywords: Cultural democracy · Urban development · Cultural economy

1 The Cultural Infrastructure of the City

1.1 A Tale of Two 'Cities': Historical Prologue

The history of urban cultural development is marked by power statements. The position of cultural venues in the urban grid; the investments in creative segments and forms of art; the development of creative clusters, have all reflected for centuries the views of those who asserted power or contested it.

This is true for the city as a built environment as well as for the city as an economic engine and 'a living entity': Richard Sennett has recently made this distinction more explicit by positing the difference between the *ville*, i.e., the city in its physical, tangible articulation, and the *cité*, a more ungraspable, yet vivid concept, being "the character of life in a neighbourhood, the feelings people harboured about neighbours and strangers and attachments to place" (Sennett 2018: 15).

In the cultural domain, both the built environment and the city as an entity made of nodes and fluxes reflected (and still reflect) power structures – as Harvey (2006: 359) has put it, "in the space of power, power does not appear as such; it hides under the organization of space".

F. Calabrò et al. (Eds.): NMP 2022, LNNS 482, pp. 94–104, 2022.
https://doi.org/10.1007/978-3-031-06825-6_10

Since the XIX century, the attention of planners and decision makers has been placed mainly on a single dimension – the monumental and the centripetal, the nodes of the *ville* that were able to attract resources and capital, further reasserting social, economic and political power.

In the modern scenario of policymaking, this attitude remained unchanged for decades, reflecting views and patterns that belonged to the XIX century city and its cultural development. At the same time, two major shifts were occurring in the culture-led urban economy: the first is that cities became actors on a global scale, competing over scarce resources without the intermediation of the Nation-State. The exchange goods that cities began using for attracting investments, social capital and boost their economy was precisely cultural goods, cultural capital and the spillovers that cultural development is said to produce in urban development. In the post-industrial economy, then, culture is not simply the symbolic manifestation of capital in the urban space: it is capital itself, and of the most fruitful one.

As a consequence, cities have begun to develop governance models and policy tools that could lead to culture-led urban development. The first stage of this phase has led to some pioneering examples of culture-led regeneration which has made use of iconic architecture and large-scale projects to regenerate brownfields. This has happened mostly in a twofold way: the former is the construction from the scratch of iconic sites, as has happened with the Guggenheim in Bilbao and the Sidney Opera House at the beginning of the XXI century, or with the Elbphilharmonie in Hamburg in more recent times (Balke et al. 2018). The latter is the adaptive reuse of brownfields and factories (High 2017).

Most of the culture-led regeneration actions of this early stage, however, shared with the XIX century approach two controversial aspects: first, these interventions were not responsive to the nature of places, their cultural specificities and the development patterns they had followed up to that point. (Dickson 2017; Müller 2015). Second, they have for a long time shared the same, homogenous view of culture, which reflected the ideals and aesthetics of the dominating class.

1.2   From Centripetal Development to Scrambled Eggs: Cultural Democracy and Cultural Ecologies

New economic and political approaches have counteracted this dominant perspective in the past years. For what concerns the political approach, there has been a resurgence of the discourse around "cultural democracy". The term has a twofold connotation: the first refers to the democratization of art, i.e., to the bringing of "great art to everyone" (as the Arts Council titled its report on culture and creativity in 2013). This meaning, however, was still seen as patronizing, trying to impose the dominant culture over the widest number of cultural consumers (Gross, Wilson 2020). The second approach is that of cultural democracy in and of itself, and is an approach that widens the term "culture" to a broader array of creative expressions – going beyond the boundaries of "high" arts. This approach developed initially "from the bottom-up", with artists and collectives adopting a societal-oriented approach to their arts (which became known as community arts) (Jeffers, Moriarty 2017).

While for a long time the attention of policymakers has been placed almost exclusively on the *ville* – on the most tangible manifestations of heritage and creativity, the contemporary focus seems to be on mores, intangible heritage and lifestyles; this occurred with the relatively recent introduction of the notion of "intangible heritage", which was only provided an operational definition by UNESCO in 2001[1], alongside with a stream of research (and of policy actions) that has started to focus on the production and dynamic evolution of culture rather than on heritage as a static concept (Bertacchini et al. 2012). This epistemological twist, coinciding with a broader understanding of the notion of culture as an anthropological phenomenon, aims to foster a more inclusive cultural policy: in particular, drawing from the work of Sen (1999) on sustainable development, cultural democracy has been intended as a way to promote "cultural capabilities", a collective right by which people would be "free to 'shape the meanings and structures of their social existence together" (Smith 2010).

This shift, in turn, has coincided with another twist in the economic sphere, producing a new, ecological approach to culture and ecosystems.

The idea of city as a collectively construed entity and of creativity as a collective phenomenon is not new. Already in the 1940s Hayek understood knowledge as sparse through society. In more recent times, with the foundational research of Walter Santagata (2006), cultural districts as aggregators of cultural producers and creative practices were identified in their primary features: these included their density, knowledge spillovers, and common culture (be it tangible or intangible, tacit/implicit or explicit). This stream of research also led to the theorization of cultural commons, i.e. cultural resources produced and appropriated by a community, and identified by three dimensions: space, community and culture (Bertacchini et al. 2012). Regarding the territorial dimension, Cooke and Lazzeretti (2008) have placed an emphasis on the intersection between cultural clusters, creative cities and local economic development, while Borin and Donato (2015) have explored the potential of intellectual capital in cultural networks and advocated the implementation of cultural ecosystems for the competitive advantages of territories.

Cultural policy, though quite slowly, proved receptive with respect to economic research on cultural ecologies and on cultural democracy, as a new agenda is being put forward to promote a greater access to creative and cultural means of production, and for a widening of subsidized creative forms of expression (Gross, Wilson 2020). At the local level, many administrations have developed plans and strategies that favor the thriving of a creative economy (Grodach 2017). This approach to urban development was amply boosted by supra-national policies – an example being the birth of UNESCO's Creative Cities Network (2004), which attached the label of 'creative' to the city as a whole. These policies present an ambivalent tendency: on the one hand, they instigate competition among cities, which aim to use the label of "creative" as an advantage (Ponzini and Rossi, 2010) - as in the case of European Capitals of Culture. On the other, they foster cooperation and exchange, and produce networking effects (Rosi 2014).

Both the economic discourse and the political one, however, are missing two crucial aspects: first, the competition-based cultural development of cities has led to an understanding of culture as an economic engine, with cultural value for urban communities only coinciding with its market value (money). Second, while great attention has been placed on creative clusters and flagship projects, very little has been said about the contribution of the grassroot initiatives, associations and creative practices that not only set

---

[1] https://ich.unesco.org/en/events?meeting_id=00047.

the basis for a creative economy, but are at the core of communities' well being, actually putting in place a wider distribution of the cultural means of production (Comunian 2011).

Both at the research and at the operational level, therefore, what is missing is, first, an actual ecological understanding of cultural phenomena; it is necessary to exit the "particular tropes which dominate global understanding of cultural practice" (Perry, Symons 2019: 63), and, second, a strategy for the actual implementation of cultural democracy: embracing multiple, concentric cultural agendas and supporting grassroot, indigenous initiatives that allow for the rise of the creative sector in the first place.

The following section will present empirical insights from the city of Bologna, in Italy, concerning both cultural ecologies and cultural democracy: by illustrating the grassroot cultural ecosystem of the city, and the policy initiatives that have been put in place for its development, it will present an original contribution on the economic and political conjunction of cultural ecology and cultural democracy, open a potential stream of research on the topic with its further development and offer indications for the future of cultural urban policy.

2 Mapping the Grassroot and Analysing the Plicy

2.1 The Cultural Ecosystem in Bologna: The Context

The city of Bologna is a particularly lively environment, grounded on four main pillars: a thriving creative economy, a strong university, a long-standing tradition of collaborative governance and of active citizenship.

The creative economy of the city is historically rooted in two major grounds: one is the lively cultural milieu, embracing opera and illustration, folk music and craftsmanship; the other is the industrial one, as the Emilia-Romagna Region has been characterized by a strong manufacturing sector, ranging from fashion to engines, which was able to shift to a creative, design-intensive phase in the post-industrial turn of the century through innovation, design and a maker culture which has sustained the creative sectors (Bertacchini, Borrione 2013; Pfleger 2004).

Innovation and knowledge production have been developed thanks to, incubated, or been reflected by the university, which in Bologna plays a strong role in urban development. Attracting the second widest number of students from other Regions in Italy (ISTAT 2016), the Bologna University acts as a connecting tissue between several urban actors, with its students contributing to the social, economic and political life of the city.

For what concerns collaborative governance, Bologna comes from a strong sociopolitical tradition of shared governance, which in time has led to the creation of multiscalar governance forms and to the involvement of a multifarious set of stakeholders, ranging from entrepreneurs to civil society (Iaione 2017). This tradition has its root on civic action, which has both come in the form of dialogue with and struggle against the established municipal authority – social centres have aligned themselves with this latter scenario and have provided, over the years, a more or less compact political counterpart to institutional governance (Mudu 2012). Nevertheless the municipal administration has adopted a series of tools, which will be better described in Sect. 2.4, to facilitate collaboration and to enable shared governance.

Regarding active citizenship, it has taken multiple forms over time, ranging from the illegal and antagonistic to citizens-led organizations and activities. For what concerns the former, Bologna holds the primacy in Italy for the *centri sociali*, often illegally occupied structures which have become cultural, social and political nodes of the city, organizing activities for the community (from local markets to language courses). Bologna holds yet another primacy in this respect, being the city with the highest density of *centri sociali* in Italy for the number of its inhabitants (Mudu 2004). For what concerns the latter, citizens initiatives constellate the cultural and social infrastructure of the city, and it is on this segment that the present paper will focus.

2.2 Identifying Socially Engaged Cultural Grassroot Initiatives: The Method

The research uses grounded theory to observe and map socially engaged cultural grassroot activities in the city of Bologna. By moving between empirical findings and theory, the research will observe existing definitions on grassroot cultural initiatives to define the research field, and further implement the literature by providing theoretical insights from the mapping.

Within the diversified scenario of multifarious actors of Bologna, the present paper places its focus on grassroot cultural and social practices that lay the ground for the cultural infrastructure of the city. Drawing from complexity theory (Comunian 2011), the research is grounded on the fact that adhocracy, local specificities and the complexity of networks go beyond the simplifications of advocacy and the narrow focus of policy-making, and that the cultural ecosystem of a city needs to be understood in the light of all its components – grassroot cultural initiatives, in particular, are being increasingly regarded in the light of the role they play in *making* a city – in particular, they bring to life what Sennett has identified as the *cité*.

The types of organizations referred here, and mapped in the following section, cover different creative sectors, have different aims and focuses, and yet share some transversal characteristics that make them observable through a coherent lens: they are grassroot initiatives with both a cultural focus and a societal objective – i.e., they are born from citizen initiatives with the purpose of contributing to the well-being of their community or of the city as a whole, and develop cultural, creative and artistic practices and activities, "acting as socially obliged citizens who feel morally committed to helping shape alternatives to existing conditions, stressing social activism as essential in creating art…" (Kaddar et al. 2020: 1).

The mapping takes into account the whole city of Bologna. The criteria set for the research were societal goals and a cultural vocation; research was done through the triangulation of different sources: an initial mapping was made of the 123 Cultural, Creative and Recreational Associations registered in Bologna, which was then triangulated with the 4 main *centri sociali* and with the 67 *circoli Arci*, a network that supports the birth of neighbourhood associations and local social spaces. The organizations were then filtered and the research was refined by excluding the artistic organizations and the social cooperatives which did not have both vocations stated in their mission and/or their website. The contents of the websites were analysed to produce further inferences over the nature of the organisations and their activities. The organizations whose websites have been inactive for longer than a year have been excluded from the mapping.

The mapping has also deliberately left out the initiatives that have been implemented within the framework of the Bologna Regulation for the Care and Regeneration of the Urban Commons, which require specific attention (see Sect. 2.4). Only grassroot initiatives that have raised without institutional inference have been considered.

2.3 Grassroot Initiatives in Bologna: Places of Sociability and Cultural Experimentation as Urban Nodes

Spatial Organisation and Networks. 81 organisations were mapped that shared artistic and societal objectives, and that originated from the grassroot initiative of the Bologna citizens. The mapping showed an obvious densification in the city centre, and yet the scenario is lively across neighbourhoods, with a greater agglomeration of cultural services in historically lively and popular neighbourhoods, San Donato, Cirenaica and Bolognina (all adjacent to one another, in the North-East quarter of the city). Only *centri sociali*, are located in the centre or in its immediate surroundings (the city walls). *Circoli Arci* have historically represented a cultural and social infrastructure in Italy in general, and in Bologna specifically. While very diverse in nature and scope, they represent a transversal network and permeate the whole urban fabric, being among the most widespread associative forms in the city. Bologna is, additionally, the only city in Italy where three different cultural organisations (Laminarie, Ateliersi and Associazione Oltre) have joined the Trans Europe Halles network, a cultural network of adaptively reused buildings across the continent.

Balance of Artistic and Societal Objectives. The fact that organizations featured both artistic and societal objectives does not imply, anyways, that all organizations showed the same balance between the two dimensions. In the map (Fig. 1) this distinction has been made by indicating with a yellow mark the organisations that were more focused on artistic goals, and with a blue mark those in which the societal dimension prevailed; green marks identify *centri sociali*, while purple marks stand for *circoli Arci*. The preëminence of one vocation over the other, however, is by no means exclusive of the other dimension – quite the reverse, the two dimensions connote one another, meaning that artistic-oriented organisations elicit sociability to produce some form of collaborative, relational art, while the societal-oriented organisations develop and run artistic projects and activities to elicit proximity and sociability.

The artistic-oriented organisations primarily focus on theatre and the performing arts as a tool to develop a sense of community and promote well-being. There are, of course, exceptions, with Link being a multifunctional cultural space promoting emerging artists and collectives, or CACCA being an experimental multimedia research over food, health, and society. Whatever their creative domain, all the organisations link the idea of experimentation and discovery to that of togetherness, and see exchange with keen people as a necessary precondition for artistic and social innovation.

The societal-oriented organisations promote intercultural exchange through joint learning, horizontal exchanges and cooperative activities that orbit around culture – exhibitions, workshops and laboratories for learning, upskilling and the production of artefacts. In these organisations, culture appears as an intermediary practice for eliciting sociability and producing togetherness. The societal focus is also widely diversified: some organisations focus on lifelong learning and informal education, others on the integration of migrants, others on elders.

Organisations as Spatial Nodes. The grassroot cultural infrastructure in Bologna has emerged from neighbors' meeting places, from former recreating spaces (such as Dopolavoro, the workers' afterwork) or from the search of artists' collective for a place where to experiment and practice, which ended up reactivating misused, abandoned or neglected spaces and benefitting the community. The mapped spaces (especially the *centri sociali*, but also *circoli Arci* and some artistic spaces) share two characteristics: first, they are hybrid and multifunctional, often hosting more than one organisation and allowing for the development of different activities (either simultaneously or asynchronously). The space shared by CACCA and Panem et Circenses, as well as the multifunctionality stated by many Neighborhood centers listed in the Arci network, or the space of Café de la Paix, where fair trade markets and book presentations intertwine, testify for the relevance of place as a physical incubator of people and practices. Second, and in reason of their ability to aggregate functions and humans, they are urban nodes (Sennett 2011), which catalyze sociability and creativity, though at a scale which has for long been overlooked by cultural policymaking – a scale which, however, enables cultural democracy to exit rhetoric and enter the streets of a city in a capillary way.

Different Types of Political Expressions. As has been said, the city of Bologna is historically rooted in left-wing politics, and this is homogeneously reflected in the composition of its cultural environment: the *circoli Arci*, of the solidarity cafés, the *centri sociali* and all the different stratifications show a similar orientation, going against discriminations and fascism, promoting cooperation and the dialogue between cultures. There arc, as is natural, levels of political engagement, with the arts-driven organizations being more generally (if not rhetorically) focused on the unprejudicial exchange of ideas with other individuals, and the *centri sociali* being more radically focused on the intersectionality between cultural struggles, the recognition of the diversity of backgrounds, work precariousness, and the right to the city as a right to access its services, be them cultural, social, medical. In this respect, also the levels of engagement with the administration has been diverse, with some organizations engaging in open conflict (it is the case of *centri sociali*, again) and others being incorporated in local policy. This testifies for the heterogeneity of cultural experiences in the city, and to point out that not all actors of the cultural scenario have been at the forefront of cooperation in Bologna.

Fig. 1. The Bologna map of grassroot cultural organisations, comprising *circoli Arci*, *centri sociali*, and other cultural and social associations that are citizens-led. The purple spots identify *circoli Arci*, the green ones are *centri sociali*, yellow identifies arts-driven organizations and blue identify social-driven organizations.

2.4 Cultural Democracy in Bologna's Urban Cultural Policy

Neighbourhood Labs. FIU has engaged in an intensive dialogue with the neighbourhoods and their key actors (active citizens, associations) to understand the needs and wants of residents. Local cultural organizations play the role of mediators between the municipality and the community, and the labs are organized around a theme or topic of interest for the residents.

School of Collective Actions. A fund is allocated by FIU to projects, either organized by private citizens, civic organisations or cultural associations which aim to develop ideas that improve the life of communities and residents. The winning projects are followed by the staff of the Municipality to receive support and training for project management and development and capacity building.

Bologna Regulation for the Care and Regeneration of the Urban Commons.
Bologna has been the first city in Italy to adopt a regulation concerning the urban commons, abandoned urban assets to be reappropriated collectively through an agreement between the Municipality and the citizens. The regulation has led to the stipulation of more than 280 agreements with the municipality (Pais et al. 2017), and allows for a wide range of interventions, from temporary actions to the actual shared management of resources.

Cultura Bologna. The Municipality continuously publishes calls and tenders for cultural activities to be developed in the city, outsourcing cultural services to its citizens with the aim to produce local-made and tailored answers to the cultural needs of the city and of its neighborhoods. The calls vary in scale and scope, spanning from the summer activities in the University Area to new musical production, to the management of historically relevant sites.

While these actions show a receptive public administration, they shouldn't be mystified: the budget allocated to these actions is still dramatically limited and can, controversially, be read as a tool for consensus building, while the greatest part of the fund allocations are still attributed to large-scale projects which have historically caused unsustainable development patterns in city. Similarly, the dangers of participatory practices, such as that of only including the already active and motivated citizens, excluding marginal fringes (Fung 2004).

The analysis has shown, nonetheless, the presence of a historically rooted lively cultural ecosystem in the city, which has potentially resulted in the administration developing new tools and practices for participatory governance; the Municipal approach, in turn, sustains and supports this cultural ecosystem with funding and development tools, creating a dual movement between the top-down and the bottom-up, ultimately producing a collaborative ecosystem (Iaione 2015).

3   The Dual Movement of Cultural Democracy in Bologna – Lessons Learned and Conclusive Remarks

The research has presented bottom-up initiatives and top-down approaches in culture-making in the city of Bologna. Its original contribution to the literature consists in the combination of these two dimensions, the grassroot and the institutional, observed together: the mapping, unique in both its method and findings, has allowed for the observation of a dimension often overlooked in the cultural ecology of a city, the grassroot and citizen-led, which contributes to urban reactivation and to the creative economy in a widespread and sustainable way, making culture accessible to citizen on a local basis, and thus representing an actual stance of cultural democracy on a territory. The policy analysis, though limited in scope, allows for the observation of the municipal action for enabling cultural democracy, the potential result of a dual intention: on the one hand, that of enabling cultural opportunity (Gross, Wilson 2020) for the citizens, based on an already existing cultural ecology, and on the other, that of boosting the grassroot, indigenous cultural economy further. This allows for the identification of a dual movement: on the one hand, the presence of a strong cultural ecology in Bologna has caused the administration to rethink its relationship with the citizens and to develop tools that support the cultural system in more democratic ways, while these tools have further boosted the development of the cultural ecology in and of itself.

At the same time, however, the fact that Bologna has incorporated collaboration in its institutional action causes a twofold issue: the first is the narrative of cooperation, which is often used as a palliative to urban illnesses – cooperation is indeed a significant part of urban governance, but it is counterbalanced by practices of predatory urbanism and evictions of informal settlements such as *centri sociali*, which have been subject to violent actions from the part of the police by reason of their illegal occupation of vacant buildings and spaces; those spaces, nonetheless, produced a value for the city which has not been replaced after their eviction, thus resulting in a loss for the urban dwellers. The second controversy lies beyond the narrative, and precisely in the fact that the Municipality has been able to incorporate milder forms of contestation, thus re-absorbing political dissonance, while responding violently to other more radical configurations of urban politics. This complex dualism, which is not grasped by official narratives and which also implies further value judgements in the present mapping, has to be necessarily observed in order for the complex political scenario of the city to be fully grasped, and further elaborations will have to be made on these layers.

Future steps of the research will include a triangulation of the grassroot mapping with other, more "institutional" cultural infrastructures in the city (such as museums and theatres), a social network analysis and the search for patterns of causality and correlation between the policy level and the grassroot one, which have here been observed in parallel.

The research has the potential to also produce policy guidelines for other contexts, based on a similar analysis (which could be scalable and replicable), for the identification of the grassroot cultural ecosystem of a city, its local characteristics and specificities, and consequently for the development of tools and actions to sustain such ecosystem, thus enabling cultural democracy through policy design.

References

Arts Council England, Great Art for Everyone. Report (2013)

Balke, J., Reuber, P., Wood, G.: Iconic architecture and place-specific neoliberal governmentality: Insights from Hamburg's Elbe Philharmonic Hall. Urban Stud. **55**(5), 997–1012 (2018)

Bertacchini, E., et al: Cultural commons: A new perspective on the production and evolution of cultures. Cultural Commons. Edward Elgar Publishing (2012)

Bertacchini, E.E., Borrione, P.: The geography of the Italian creative economy: The special role of the design and craft-based industries. Reg. Stud. **47**(2), 135–147 (2013)

Borin, E., Donato, F.: Unlocking the potential of IC in Italian cultural ecosystems. J. Intellectual Capital(2015)

Comunian, R.: Rethinking the creative city: The role of complexity, networks and interactions in the urban creative economy. Urban Stud. **48**(6), 1157–1179 (2011)

Cooke, P., Lazzeretti, L. eds.: Creative cities, cultural clusters and local economic development. Edward Elgar Publishing (2008)

Dickson, A.: Is the Bilbao effect on urban renewal all it's cracked up to be? Financial Review 17 December 2017. https://www.afr.com/world/europe/is-the-bilbao-effect-on-urban-renewal-all-its-cracked-up-to-be-20171211-h02fru (2021). Accessed 18 Feb 2021

Fung, A.: Empowered Participation: Reinventing Urban Democracy. Princeton University Press, Princeton (2004)

Grodach, C.: Urban cultural policy and creative city making. Cities **68**(2017), 82–91 (2017)

Gross, J., Wilson, N.: Cultural democracy: an ecological and capabilities approach. Int. J. Cult. Policy **26**(3), 328–343 (2020)

Harvey, D.: Spaces of Global Capitalism. Verso, London (2006)

High, S.: Brownfield Public History. The Oxford Handbook of Public History, p. 423 (2017)

Iaione, C.: Algoritmo Bologna-Il Rapporto CO-Bologna sui primi tre anni di sperimentazione della collaborazione civica a Bologna (2017)

Iaione, C.: Governing the urban commons. Italian J. Pub. L. **7**, 170 (2015)

Jeffers, A., Moriarty, G. (eds.): Culture, Democracy and the Right to Make Art: The British Community Arts Movement. Bloomsbury, London (2017)

Kaddar, M., et al.: Artistic city-zenship: How artists perceive and practice political agency in their cities. J. Urban Aff., 1–19 (2020)

Mudu, P.: I Centri Sociali italiani: Verso tre decadi di occupazioni e di spazi autogestiti, pp. 69–92 (2012)

Mudu, P.: Resisting and challenging neoliberalism: The development of Italian social centers. Antipode **36**(5), 917–941 (2004)

Perry, J., Symons, J., eds.: Cultural Intermediaries Connecting Communities: Revisiting Approaches to Cultural Engagement. Policy Press (2019)

Pfleger, D.: Aspect of Emilia Romagna as An Economic Development Model (2004)

Müller, A.K.: From Urban Commons to Urban Planning – or Vice Versa? "Planning" the Contested Gleisdreieck Territory, pp. 410–459. In: M. Dellenbaugh et al. (eds.), Urban Commons. A World Beyond Market and State, pp. 9–25. Birkhauser, Basel (2015)

Pais, I., de Nictolis, E., Bolis, M.: "Evaluation of the Pacts of Collaboration Approved through the Regulation for Collaboration between Citizens and the Administration for the Care and Regeneration of the Urban Commons. Empirical Evidences from the Case of Bologna", 1 September 2017

Ponzini, D., Rossi, U.: Becoming a creative city: The entrepreneurial mayor, network politics and the promise of an urban renaissance. Urban Stud. **47**(5), 1037–1057 (2010)

Rosi, M.: Branding or sharing?: the dialectics of labeling and cooperation in the UNESCO creative cities network. City, Cult. Soc., **5**(2), 107–110 (2014)

Santagata, W.: Cultural districts and their role in developed and developing countries. Handb. Econ. Art Cult. **1**, 1101–1119 (2006)

Sen, A.: Development as Freedom. Oxford University Press, Oxford (1999)

Sennett, R.: Building and Dwelling: Ethics for the City. Farrar, Straus and Giroux, New York (2018)

Sennett, R.: The Foreigner. Notting Hill Editions, London (2011)

Smith, C.: What Is a Person? Rethinking Humanity, Social Life, and the Moral Good from the Person Up. University of Chicago Press, Chicago and London (2010)

Re-think Building Codes for Indoor Air Quality

Alberto De Capua[✉]

Department of Architecture and Territory, Mediterranea University of Reggio Calabria,
Reggio Calabria, Italy
adecapua@unirc.it

Abstract. Assuming the need to create safe and healthy environments as an essential priority, the text aims to promote the design and use of organic regulatory tools, able to promote conditions of well-being, in relation to the places where the building is located, the activities that are exercised in the interior of it and the construction methods. This is the need to "accompany" the designers in the evaluation and management of health risks, especially those related to indoor air pollution (IAQ), indicating effective protocols for prevention, control and correction. It is proposed, in a non-manual key but problematic, a reinterpretation of the Municipal Building Regulation; conceptually different, operationally alternative to those in progress. Bringing this regulatory instrument back to its original usefulness and meanings, the aim is to make it reassume a role of control and direction of the building activity, which expresses its effects at the moment of examining the requests for intervention and in the verification of the quality of the constructions. An objective that can be achieved only through collaborative actions and according to common and shared methodological and operational lines, between structures operating on the territory, with different technical-building and hygienic-sanitary competences.

Keywords: Sustainability · Energy · Living · Indoor air quality

1 The Challenge of Complexity

For some time now we have been witnessing a series of alarms that continually warn us and remind us that our well-being is leading to an exaggerated and ever increasing use of energy and resources, as well as the deep and irreversible transformation of natural systems, social inequality; in practice, it is the overall impact of the human species that continues to grow.

In truth, we had already been warned of this uncontrolled growth back in 1972 by the report [1] commissioned by the Club of Rome from the Massachusetts Institute of Technology and drawn up by a group of researchers led by Dennis Meadows. They argued forcefully that planetary ecological constraints on resource use and continuous polluting emissions would profoundly affect the future of the planet in the 21st century.

The hoped-for renewal of society, through technological, cultural and institutional transformations aimed at preventing the impact of mankind from exceeding the carrying capacity of the planet, unfortunately did not take place and in the 1992 edition [2], which

F. Calabrò et al. (Eds.): NMP 2022, LNNS 482, pp. 105–119, 2022.
https://doi.org/10.1007/978-3-031-06825-6_11

reviewed the 1972 report, the authors confirmed what they had written twenty years earlier, denouncing another dramatic result: in fact, humanity had already exceeded the limits of the earth's carrying capacity. The considerations that arise in the latest edition [3], that of 2004, confirm and weigh down the previous pessimism of the authors: despite the undoubted progress made in the scientific field, humanity is currently living in the field of unsustainable development.

The Nobel Prize for economics Joseph Stiglitz [4], arrives at such dramatic results: "(...) if we do not find a way to limit environmental damages, to save energy and conserve other natural resources, as well as to slow down the warming of the planet, we are destined to disaster (...)".

"Ecology" and "environment" have become the key words of the third millennium: a media bombardment that has helped overcome the insurmountable barrier of indifference and insensitivity. The growing universal alarmism on the subject has cracked some of those certainties behind which most of us have long hidden: that progress had inevitable costs without escape and that, above all, the consequences would be paid in several centuries. "The hypothesis that in four hundred years our descendants would have no gasoline did not trouble us much. However, the prospect that our children are already living on a collapsing planet, the idea that the future of the environment directly concerns us and that the bill is coming to us well in advance of our predictions, changes our outlook. As a result, the concept of sustainable development has now entered the vocabulary and agenda of politics, industry and economics" [5].

This situation has worsened considerably also due to the spread of urban systems that has inevitably increased the vulnerability of natural systems making it more difficult to implement concrete measures towards sustainability. The city is the most widespread habitat of the human species. Today in Europe at least 75% of the population lives in urban areas. In no other place as the city man has modified and altered the environment, determining the increase of material and energy flows and promoting environmental fragmentation. After decades of speculation not only in construction but also in energy, after having transformed three quarters of our housing stock in a thermal trap that devours oil and gas, even if there is a widespread regret, an effect of the growing environmental anxiety, we must unfortunately admit that the main problem of humanity for the next decades will be to reduce the consumption of fossil fuel because the survival of millions of people will depend on it. Because of the order of magnitude of the problem, one cannot help but demand greater accountability from those whose profession involves buildings.

The building sector is responsible for more than a third of global carbon dioxide (CO_2) emissions, and it accounts for nearly 40% of all the waste that humans produce, which contains enormous amounts of embodied energy. We live in cities connected to nature through markets and technology. All cities must import food, fuels, and materials, and all cities are markets. The environmental challenges posed to the city vary according to the level of economic activity.

The theme of the third edition of the World Conference on the Future of Science is The Energy Challenge. The challenge is: how to produce the energy necessary for the development of well-being, protecting our most precious asset, the health of man and the planet; and also how to make understand and accept the solutions that science proposes to a confused and frightened society. If in fact energy indicates something positive and vital, in recent years the term is increasingly associated with air pollution, ecological disasters, global warming, climate change, unsustainable costs.

There are many reasons, today, that make us say that the restoration of compatibility between transformation and environment, between artifact and nature, between production needs and global security needs, represent the real challenges for the architecture of the next decades. From many quarters we are invited to consider a new approach that, rather than making us speak in abstract terms of sustainable materials and technologies, suggests, to reintroduce expressions long out of use that refer to appropriate materials and technologies; making this term take on new and more complex meanings with the recovery of the needs of health and the concept of "material place" of the project [6]. Sustainable does not mean self-sufficient. To embark on a path toward sustainability, a country must improve the health and well-being of the collective, reduce environmental impacts, engage in recycling of materials, and use energy efficiently.

More and more frequently we talk about flexibility, which has become the key word that big cities, today the stage of the biggest environmental dysfunctions, must "adopt" to counteract the monoculture that kills any urban form.

The worst evil for a city is the inability to open up to the new and to adapt to changed circumstances. The city is bound to grow, but growth poses a problem of sustainability.

The ways in which sustainable development is encouraged and promoted vary according to the schools of thought and the history of each. Almost all of them, however, agree that the indispensable measures to be taken, in the construction sector, concern the adoption of verification criteria to be used by the designer, on the waste of resources, on harmful emissions or on any activity that weakens or threatens the availability of resources, on the man-made environment; on the health and well-being of humans.

But it is not that simple. The difficulties also stem from the fact that our legislative and regulatory apparatus, especially national, is incapable of defining adequate measures and methods of verification. Even if there are more and more attempts to identify precise standards for living comfort; to identify precise reference thresholds to be followed in the design, construction and production phases of materials and components. Even today, the question remains open.

The situation in which we find ourselves is actually more serious than the most pessimistic forecasts: in fact, the report of the Energy Watch Group, set up by some German parliamentarians with the participation of scientists and economists, contains a merciless comparison between the forecasts made by experts from the IEA, the International Energy Agency, where there is a glaring and worrying gap between what was predicted and reality.

And again the conclusions of the United Nations Report on Climate Change (IPCC) are ominous. It is predicted that by the end of this century, as a result of human activities, the global temperature will probably increase by 3° Celsius, to reach a level not reached by planet Earth since the Pliocene era, three million years ago. Scientists warn us that a climate change of this magnitude will endanger civilization itself and the future of the planet.

At the same time, international conferences that want to avert this dramatic prediction follow one another without respite.

Environment, ecosystem, greenhouse effect, harmful emissions, citizens' health, energy consumption, sustainable development. It is a list, which could be even longer, of terms that are part of our everyday life and that show how much the human "footprint" today is disproportionate, comparable, as Gianfranco Bologna says "to the geophysical forces that have profoundly shaped our planet, in billions of years".

A situation that must be faced and can no longer be postponed. The problem of improving our relationship with natural systems is increasingly becoming one of the great political emergencies of the immediate future. Scientific knowledge and the best available technology, today, provide us with theoretical and practical tools to seriously address these problems and try to reverse current trends.

If we do not learn to anticipate events and, in some way, to "govern" them, they will mark our destiny and, from this point of view, we could also suffer effects that could lead us to truly dramatic situations. The main path of true development, as stated by economist Georgescu Roegen, is to get more and more utility from fewer and fewer resources. Not that of the destructive and bold growth whose causes are: the increase in population, the increase in consumption per capita and the impactive technologies (Holdren equation).

The pressure exerted by social metabolisms on natural ones, even more increased by the overall increase in population with consumerist lifestyles, is now clearly unsustainable. If consumption trends continue to grow, as seems to be the case, without corrective action, the outlook cannot be rosy. The signs of suffering of the planet's ecosystems are very evident and our pressure on their regenerative and assimilative capacities is now considerable. The paths proposed by sustainability are feasible and practicable; however, it is necessary a real revolution of sustainability that is, fundamentally, a "cultural" revolution.

These are issues that, even if they inevitably concern the construction sector, the strategies and solutions to be adopted are not only its responsibility. It is necessary that large sectors of the scientific and political community define a line of demarcation between what must and must not be done with a logic not only of interdisciplinarity, but of transdisciplinarity, the only one capable of constructing a global thought capable of articulating the different areas of knowledge. (..) Understanding the interdependence of cultural systems and ideas is now more necessary than ever. This will help to change our way of thinking, giving us one more tool to escape the abyss towards which the planet seems destined (…).

2 Technology and Environment

The new scenario implies the need to relate decision-making processes for design to different categories of reference - political, social, economic - involving different orders of complexity in design practice. And yet, in this phase of great investments, our country is considerably behind on the issues that guide the innovation of building production. The reflection to make is how much this expansive and developmental phase has contributed to qualify the architectural project and govern, in an environmental sense, the transformations. In response to this, the signals that arrive are, in fact, very weak and with difficulty we read in the cities, in the suburbs a new season in the model of transformation of the territory, a new contribution to the quality of life that comes from a vision of the 2000s of the building products.

What does it mean today to realize a quality project? In recent times quality has often coincided with certain requirements considered, rightly, fundamental or perhaps necessary for a new way of building that is attentive to environmental issues, but above all to management costs.

It is well known that the architectural project is confronted, today, with an increased complexity, both theoretical and technical, which requires broader and more articulated skills. A new complexity that presupposes, as Nicola Sinopoli [7] states:

"(...) a formation in the field of the project of architecture, aware of its formal, physical, economic aspects, and of constructability, of the complex contexts (city, client, production...) in which today such project is realized, of the evolution of the demands, of the technological resources of the technologies of the questions related to the energy and to the sustainability, of the aspects related to the industrial culture, to the production and to the innovation in the picture, obviously, of the actual internationalization of the markets of the materials, of the components, of the business performances and of the professional services (...)".

The generalized objective of saving resources and using renewable energy sources has now become a necessary strategy that fully involves architecture, since the environmental weight of buildings in industrialized countries has grown considerably in recent years.

The fact that the building industry is the main responsible for the consumption of all energy produced and huge amounts of CO_2 emissions, must make people realize that the project is now at the center of a revolutionary change in the way buildings are designed and built. A change so radical in its importance and scope that it is a real opportunity that will finally "force" us to face the double challenge of energy waste and impact on people's health [13; 15].

Technology is configured, for its current and past history, as a discipline, able to undertake new design paths. Thanks to the latest technological advances, it is now possible to design and construct "healthy" buildings that create their own needs from local renewable energy sources, configuring themselves "(...) to serve as both power plants and habitats (...) [8].

Although the issue of environmental sustainability is one that must be addressed globally, through coordinated policies, the ability to activate strategies depends on the local scale.

In Italy, many public administrations have begun to introduce environmental sustainability criteria in their building regulations, assigning volume "premiums" or incentives to sustainable construction.

Unfortunately, to date there are still few that address the issue in an integral manner, taking into account all the different parameters that contribute to a rigorous environmental assessment. The lowest common denominator present in all regulations that have adopted environmental sustainability guidelines is the promotion of energy saving through the increase of the level of thermal insulation of the building envelope. In fact, since energy consumption accounts for 70–80% of the environmental impacts generated by the building in its life cycle, energy saving is given a leading role in environmental protection.

In the recent past, the Italian Government took an important step, facilitating and regulating the energy optimization of buildings through the application of advanced heating and cooling systems. The most important aspect is the promotion of a fiscal policy in the field of housing linked above all to the performance of buildings by encouraging the implementation of energy-environmental renovation programs of the public heritage. Not thinking, however, that houses increasingly sealed to ensure energy savings have repercussions on another issue: the lack of ventilation and the stagnation of stale and polluted air. The most recent regulations on energy saving - made more attractive by tax breaks - push towards situations of this kind.

In truth, our country is slow to address issues that other countries have long since resolved. In spite of the fact that from many sides, it is stressed how serious the situation is now and the repeated alarms of environmentalists and scientists, there is a considerable delay in complying with agreements that would lead to an extraordinary improvement of our development. Consequently, there is an extreme difficulty in spreading a culture capable of correctly combining the issues of economic development with strategic environmental budgets for the preservation of natural resources and systems over time.

And the European Union, whose technical policy member states must conform to, has included environmental protection among its strategic objectives for some years now. In the construction sector, this should be translated into a drastic limitation of the consumption of resources, seeking a general improvement of environmental comfort and increasing the conditions of livability and urban identity. The environmental parameters define new requirements that affect the project, transforming the established structures. The current debate on sustainability presents evident contradictions. If, on the one hand, we are in the presence of a metabolization of its general principles, on the other hand, we have not yet reached a stage in which the methods of implementation or the results of concrete eco-friendly architecture are widespread and controllable.

In response to these needs, tools have been developed, some time ago, to support environmental design and evaluation of the building through, design criteria oriented to sustainability (energy saving, water saving and recovery, recycling of materials, measures to ensure a healthy indoor air quality). Starting from these "lists" of requirements, multi-criteria evaluation tools have been developed, called "score systems", which associate these criteria with a merit score, based on the degree of satisfaction of the requirement verified through indicators [9].

Especially at the international level, the development of such systems was born at the urging of the builders, who expressed the need to "certify" the realization of buildings with high energy performance and low environmental impact, based on consolidated references and with the endorsement of reliable reference structures: BREEAM and LEED, which are the only real environmental certification systems for buildings, have found considerable success on the market. Thanks also to the acquired awareness on the part of end users and large real estate investors, who have expressed the need for tools to guarantee the quality of the buildings they purchase.

In fact, sustainable skyscrapers are emerging, especially in the United States, certified by LEED, whose influence is not marginal for the definition of the commercial value of the construction to which it is attributed. In the past, building a "green" building involved an increase in costs of 20%, currently this increase is estimated at around 5% and is amply rewarded by the increase in profits that a building with good sustainability characteristics can generate. Energy cost containment can reduce operating costs by up to 70%. According to the Indoor Environment Department at Lawrence Berkley National Laboratory, this provides colossal indirect savings by reducing illness and disease to which office and home occupants are subjected. This becomes an incentive for constructing buildings that meet the LEED standard, which can access funding and substantial tax reductions.

However, scoring systems have several critical issues. First of all, their approach is apparently performance-based, in fact because it is articulated in so many indicators it ends up being prescriptive. In addition, they do not have a "systemic" approach because the approach is aimed at "optimizing" individual elements of the project without an overall verification of results: it is taken for granted that the sum of performances corresponds to the final overall performance of the building, but this does not always happen, also because often the optimal design solution to meet a certain requirement is to the detriment of other requirements (the project is always a compromise of often conflicting needs). Finally, there is a total lack of a life cycle approach, especially in the criteria related to the choice of materials and building components: even the indicators related to the verification of energy consumption calculate separately the energy incorporated in materials and the energy in use, without a unitary balance of the entire life cycle.

Sustainable development is certainly one of the major challenges of this century: the priority objective of a new sustainable design should be to properly select materials and components of the building in order to reduce, in the first instance, especially its most significant energy consumption attributable to the operating phase. The design for sustainability must base its proposals on the comparative assessment of the environmental implications of different solutions technically, economically and socially acceptable, and must result in the creation of materials, products, components and services designed taking into account their entire life cycle. Knowing the environmental impact of the life cycle of materials is therefore essential to make the right choice: the basis of any choice is a comparison between different options that allow to perform the same function.

3   Reducing Emissions and Saving Energy

Energy saving and environmental sustainability are considered fundamental objectives in the realization of new buildings and in the requalification of the built environment, so much so that they occupy prominent positions both in the debate on contemporary architectural design and in the processes of definition of future regulatory frameworks. Unfortunately, most of the buildings are constructed without paying attention to the environment and energy, and although there is a significant increase of interest from technicians and administrations, qualified solutions remain a niche.

Moreover, user behavior, due to this situation, is hardly oriented to the control of energy consumption. Being able to have a comfortable indoor microclimate all year round, given that the buildings we live in, in most cases are not able to guarantee it, often makes us resort to the help of technology, which over time has fatally improved, but has also contributed to aggravate a global situation.

Energy blackouts, climate change, polluting emissions are the most obvious aspects of the need to rethink our way of doing construction. This is due to the increasingly alarming inadequacy of the current way of producing energy, in meeting the demands of consumption, but also for the impossibility of thinking about an increase in production and consumption of energy without an immediate impact on the quality of life.

The awareness campaigns carried out by the Ministry of Environment, the new European and national regulations, the tax relief for interventions that increase the level of energy efficiency, as well as the continuous increase in fuel costs have made the reduction of energy needs of buildings become an intent shared by most of the community.

Today's times require that as many people as possible learn to think in an environmentally friendly way. In small and big daily choices.

There are countless technical solutions that show a greater interest in environmental issues, but too often outside our country. If Germany and Great Britain have reduced polluting emissions by 3%, thanks to the use of renewable energy instead of coal, Japan is the world leader in photovoltaic roofs and ISO 14001 certified products (more than ten thousand).

The analysis of the Italian building production of the years two thousand, through the filters of innovation, provides a great discomfort: we are faced, for the most part, to monofunctional buildings, with outdated energy solutions and technological components in the building product for nothing different from those of the seventies, no trace of environmental sustainability in the projects, no attention to energy saving.

In the field of sustainability, combined with energy saving, pollution reduction, bio-technology, the life cycle of the building product, on which the European Union is pushing, we can only record some timid experimental experience, some innovative but new building regulations and little else and even less if we use the filter of technological innovation and information communication technology, apart from the use of information technology in the design studies, there is very little engineering in construction.

It is therefore evident that the energy issue is central to the construction sector, and in turn central to the concrete implementation of emission reduction. And it is equally clear that it is a cultural issue.

The consumption and the impacts of the building sector are distributed on two fronts: on the one hand the production, the installation and the disposal of building components, on the other hand the management of the building system during its use.

But it is not believed that the energy problem can be solved by intervening on individual components. There are several reasons for this. The first is that the environmental quality of the building does not derive from the sum of components and materials, but from their interaction within a unitary organic system, which is precisely the building organism. The second is that the building is closely related to the way of living in places and therefore to the culture, the technical capacity of individuals and communities that live there.

These observations open the field to a way of acting that is not limited to identifying innovative solutions of materials, components and techniques, but that places them in the context of a more complex technological, environmental and social treatment. The design aimed at energy saving must be able to control and act on three different levels: environmental, typological and technical construction.

4 A New Building Hygiene

The problem of air quality in non-industrial confined environments, since a few decades, has assumed considerable importance certainly for the importance of the technical implications - design but especially for the problems of impact that air quality has on people's health. The scientific community, due to the occurrence of specific pathologies and situations of discomfort in the occupants of homes and offices, has reiterated the need to investigate the sources and agents responsible, to quantify the impact on health of exposure to particular substances or materials and, of course, to indicate possible remedies and appropriate technical solutions. Terms such as *Sick building syndrome* and *Building related illness* are now treated with insistence. Homes, schools, hospitals that until a few years ago escaped any control of air quality - indoor air quality - and healthiness of the environment are now the subject of specific studies; however, there is still an extreme difficulty of communication between the different fields involved in research.

These issues are of growing interest in Europe, whose European Community has always given priority to research into the well-being and health of its citizens.

"We have to worry about dusts, gases, microorganisms, odors" (...) it should be noted that in total, in this very important field of air purity, which has worried hygienists for so long, we are actually very poorly armed to establish the needs.

The actuality of this consideration by Blachere, from 1971, confirms that our living is characterized by a condition of discomfort that is increasingly caused by the houses in which we live, the factories and offices in which we work, and the cities in which we live.

The attention paid to the building as a "problem within the problems", more and more disturbing that invest the future of our species, seems justified on the basis of at least three considerations:

- it is always opportune that reflection on the ecological question starts from the everyday, from the well-being and health of the individual;
- there is, especially in our country, still an immense cultural and behavioral gap around these issues;
- it is necessary to address the same scientific and technological research towards the development of knowledge capable of placing the quality of life as a central theme of study.

For some years, in the most developed countries, it has been noted the emergence of a new issue of primary interest for public health: the air we breathe inside the non-industrial buildings is more polluted than the outside air and this is particularly serious because most of the population spends most of the time of their lives in such places. The house that after decades of building policy oriented towards objectives of "quantity" had long since lost that aura of sacredness that has historically distinguished it, has now become a dangerous place exposed to radiation of natural and artificial origin.

Houses, hotels, offices, schools present concentrations of pollutants dangerous to health higher than those in outside. The causes of this situation are attributable to several factors, such as the introduction of new products in the building cycle, the lack of attention to design and technical solutions, the decrease in ventilation due to the increased sealing of buildings.

The problem is aggravated by the fact that today we are far too distracted by problems of energy efficiency, energy conservation and energy saving. And such indications are absolutely antithetical to strategies for solving indoor problems that prefer more ventilated and less sealed environments.

But indoor pollution is not a "new chapter" of risk conditions linked to particular alterations in air quality, it is added to those already known in the history of public hygiene and that have concerned the relationship between environmental conditions, the built environment and the individual: requirements for healthiness, risks from air pollution and safety in the workplace. The role of the impact of outdoor climatic conditions on the indoor microclimate as well as the relationships between indoor and outdoor pollution are obvious and need not be repeated here.

But the possibility that the home itself may pose a threat is particularly unpalatable, though not a new idea. Whenever large masses of the population have become urbanized, problems of danger related to the built environment have arisen, such as epidemics and diseases due to the unhealthiness of living and working places. The building hygiene of the last century had focused its attention on the pathologies resulting to people from the permanence in humid and unhealthy living places. Their spread was facilitated by overcrowding, promiscuity and lack of hygienic conditions. Pathologies such as rickets, tuberculosis, various forms of rheumatism and pulmonary diseases found ideal conditions for development in cold, humid environments where sunlight rarely entered.

From the observation of the relationship between the built environment and disease was born the concept of public health that led, centuries later, to the modern urban planning and hygiene-building regulations, with the requirements of habitability and well-being.

Retracing its history, we recall: the ordinances of the Roman era on the maximum height of buildings, the proposals of Vitruvius and Alberti [10] to make the building healthier and more durable, up to the nineteenth-century legislative measures aimed at improving housing conditions, Broggi's proposals on hygiene and housing decorum, but also the rationalist ideology of air, light and green.

Lavoisier had the intuition that to "spoil" the air contributed the by-products of organic combustion of tissues expelled through the respiratory tract, consisting mainly of carbon dioxide, rather than the reduction of oxygen content. This hypothesis, however, began to be questioned since the middle of the last century, as it was noted that, under conditions of normal occupation of the environments, the rate of CO_2 in the confined air never reached the values for which it could be considered harmful ($\approx 5\%$), as per experimental tests, and that in the case of considerable crowding the sense of discomfort and disturbances affecting the occupants were felt long before the percentage of CO_2 had reached these harmful values. It was therefore very likely that the air pollution reached an unbearable level and was highlighted by discomfort and intolerance not due to the production of CO_2, but of other substances. The latter were identified in some volatile organic products, whose entity is linked to the presence of man in the environment and therefore were called by the hygienists who studied them, Brown-Sequard and D'Arsonval, anthropotoxins. Little was known about the nature of these substances; in general they were gases eliminated by the lungs, products of skin perspiration and sweating and were difficult to measure.

Since the content of CO_2 in the air could, however, be measured exactly, it was possible to assume for the sought-after index of tolerance the value of the percentage of CO_2 above which the first discomfort could occur, even if their cause was not attributable to the CO_2 content mentioned above. Thus, the criterion was established that in order to maintain the air in an occupied closed environment in healthy conditions, the percentage of CO_2 [9] should not exceed one per thousand (Pettenkofer's anthracometric index). Subsequently, with the widespread development of industrial production processes, the attention of hygienists has shifted to the study of working environments - where production processes can introduce into the environment high concentrations of pollutants - identifying the risks of exposure to which people were subjected and defining the technical and regulatory precautions to be taken.

Pollution thus begins to take over not only industrial cities but also ancient cities, sociologists and scholars of hygiene and public health, even if they do not yet speak of ecology, operate among several difficulties the first proposals for intervention, through technical measures and health laws, to try to stop the growing environmental contamination.

The first examples of environmental intervention, however, were carried out right inside those industries that were the main causes of pollution, and this was done not so much for philanthropic reasons or for workers' claims, but to make work more productive by controlling the thermal conditions, humidity and ventilation of the environments. The fact that this circumstance then contributed to a decrease in the amount of infections and illnesses contracted at work is in fact a coincidental episode, since much more will have to pass before this requirement becomes essential, even if respect for it is still often achieved in terms of regulations rather than reality.

The introduction of industrial processes therefore determines a new kind of environment and a new way of life and requires an increasing use of scientific instruments and increasingly sophisticated techniques that consequently trigger further changes.

In the last twenty years or so, air quality has once again become a major health concern, given that many people spend most of their time indoors.

The increasing complexity of the production process has made it difficult to understand the interrelationships between different technical acts. The advent of industrialization and, within this, the contribution of chemistry, has introduced in the construction field hundreds of new materials that should have perfected the living and building, but that, in reality, have made it more difficult to control the quality of the built and the control of the correspondence of this to the needs of users, more and more distant from the process for the production of housing.

The search for solutions that would allow the exploitation of valuable areas have made designers forget the beneficial influence, both psychological and physical, of air and natural light. The haste to build and the cost of labor has imposed the use of materials that are easier to use, but often little tested.

The spread of "do it yourself" has put in the hands of the common man highly dangerous products, if used without due caution.

All these actions, but especially the combination of them, has had the effect of worsening the quality of indoor air.

The discovery of indoor pollution is therefore, an aspect directly linked to a new way of thinking about environmental and hygiene requirements, which have been neglected for too long in the construction design process.

The current building situation unable to respond to the changes in demand and the emergence of needs that have occurred in recent decades is closely linked to the qualitative aspects of indoor air.

"It is necessary, as Guido Nardi [11] says, to deepen the role of design both in determining some causes of the current situation of discomfort and in allowing ways out of what seems to be an irretrievably compromised situation. From this point of view, a clarification is necessary: the causes of the widespread discomfort with the built environment are to be found not only in the increasingly precarious competence of designers (architects or others), but above all in the inability to culturally interpret contemporary reality, both in terms of materials and techniques, and in terms of the increasingly rapid transformations of behavior that affect society.

The prevention of indoor pollution can be carried out by acting simultaneously on different levels: that of information, that of control and that of normative or regulation.

Designing and constructing a healthy building, as recommended by the various studies that have been dealing with the problem for years, means:

- That the different actors in the building process must work synergistically;
- that the different responsibilities of those who have to ensure that the users' demand for healthy buildings is observed are well understood;
- that there is a correct approach to the initial phase of the project in the given environmental context;
- that there is real scientific knowledge on the part of planners and designers, and builders must ensure that new and old knowledge is always applied;
- define laws, regulations, and standards so that the roles and responsibilities of the final product are clear;

The designer's task is to provide a healthy building [14], in which all risks of indoor pollution are reduced as much as possible by acting, in particular, on:

- the control of factors that worsen air conditions;
- the reduction of polluting sources (choice of products);
- the confinement of polluting sources (design of plant location);
- dilution of concentrations (ventilation design);
- expulsion of pollutants (evacuation devices).

5 Innovation in Local Building Regulations

Today, the overall quality of living is affirmed in order to face and solve the new social and market conditions in a way connected to the real evolution of society, overcoming the concept of healthiness, not in order to demolish it but to include and enhance it. In this regard, the trend that we try to affirm here takes note of the profound transformation that has affected in recent decades the overall sector of building regulation, that of the project and that of construction techniques, and proposes a completely different approach than in the past.

The perspective is to bring back every regulatory instrument to its original meaning and purpose. In this sense, the Building Regulations must be a real instrument of control and direction of the building activity, which has its effects at the time of examining requests for interventions and carrying out checks on the quality of buildings, with the intention of developing a whole series of collaborative aspects - according to common and shared methodological and operational lines - between structures that operate with technical-building and hygienic-sanitary competences.

This work wants to identify the means and methodologies through which the two sectors of construction and sanitation can be structured with progressive cohesion, starting from the point of view and the reason that the separation has increased the amount of constraints, contradictions, discretion, possible interpretations, not always and perhaps not more to the benefit of the welfare and real safety of end users.

Among other things, the procedures and methods of control are not homogeneous in different realities and not very consistent within them, creating uncertainties of interpretation, overlaps and contradictions that result in objective difficulties for operators and citizens.

The product of the research is aimed at renewing the entire apparatus of guidance and control of the activities of protection, use and construction of the territory. A regulatory intervention that wants to renew the regulatory apparatus building - municipal, specifying the objective of driving to the achievement of safe and healthy environments. It introduces new indexes in the current regulation, new performance aimed at achieving conditions of well-being in relation to the places where the building is located and the activities that are exercised in the internal environments of it. It is proposed, that is, a tool that leads to unity the complex problem of environmental hygiene and safety, taking into account the risks that may arise.

Through the study of the regulations in force in complementary sectors and disciplines, procedures, design criteria, requirements and controls are established, to be exercised by looking at the global aspects of the problem; thus overcoming the limits of a sectoral regulation.

The fundamental principles of the proposed tool are:

- The adoption of normative prescriptions of a demanding-performance type and, of a set of quality levels, clear, measurable and released from specific constructive references, resulting from a series of basic needs (safety and well-being) and controllable in a systematic way.
- The attribution to the designer and the builder of the responsibility to ensure the compliance of projects and works to the requirements and the municipality to be able to carry out no more paper checks, but substantial on the work carried out.
- The explication of all performance characteristics (mandatory and recommended requirements) of the real estate unit, as well as its metric and dimensional data, through the establishment of a special descriptive data sheet.

The proposed legislation has as a consequence a substantial revision of competences and controls in the building process. In fact, once all requirements and reference parameters have been precisely defined and made explicit in advance, there is no need for opinions which, by their nature, are justified only where there is room for evaluations not covered by precise regulatory provisions. In the hypothesis outlined by the new norms, all the technical investigations pertaining to the local authority become, instead, substantially, controls of correspondence to what is disposed in the Building Regulations systematizing and reunifying in this way the framework of competences.

The renewed municipal building regulation must be the fundamental seat of this new philosophy and can be so under two conditions:

1. the first is that it be supported by a regional (or state) law that allows the repeal of certain technical disciplines, contained in primary regulatory sources, replacing them with provisions of a demanding-performance nature;
2. the second is that the choice of the exigency-performance approach be made fully and coherently, replacing a system based on the prior regulation of every aspect with another new system, based on the relative freedom of forms and methods and on the rigorous control of results.

In the product envisaged, there are undoubtedly greater difficulties of control and verification compared to traditional instruments, especially in terms of preparation of operators, availability and suitable equipment and, perhaps, definitions of further knowledge and specific methodologies. The problem is certainly open, but the solution can only be evolutionary. From this point of view, it is essential to be convinced that there can be no seriously managed regulation, but above all no quality policy, without adequate experimentation, without control of real quality in actual operating conditions and without adequate technical support information. With the support of the exigency and performance methodology it is possible to overcome that dispersed exercise of single competences that often results in a general de-responsibilization on the results of the control process; this implies the realization of an interdisciplinary and multidisciplinary approach, codified in a unitary normative instrument, according to the hypothesis just schematized, defined independently from who will then operate the single disciplinary interventions of standardization and control. The standard regulation is expressed through a series of minimum performances that must be possessed by buildings and housing. The satisfaction of these performances constitutes an effective threshold, below which non-compliance corresponds to a judgment of impossibility to a healthy, safe and comfortable use. The proposed legislation has as a consequence a substantial revision of the responsibilities in the building process and control that will have to develop according to:

– *Design Verifications.* Many of the currently proposed calculation procedures are unreliable and discriminating, and should be improved in light of the latest disciplinary findings. They are the most important form of error prevention and should be conceived as self-directed checks.
– *On-site audits.* Performed by a third party, they function as an external control to supervise the self-directed control, but in this sense they must be comparable with

the results of the verifications on the project and allow, when necessary, to start a diagnosis of the causes of non-compliance.

– *Third Party Control.* It allows to identify "technical structures" able to accumulate a historical memory of data on pathologies, errors and defects, through self-directed control. It is a learning and correction tool available to the design structures, but in this perspective the identification of responsibilities and competencies must be clear.

From this framework emerges a direct empowerment of design and operators in general, which reasonably reduces multiple and overlapping controls and simply and clearly identifies who is responsible and for what. Redefining the roles and competences of the public administration, the highly simplified procedural control can be overlapped by a substantial control on the real quality of the interventions; this role, freed from margins that today are too often discretionary, should be supported by appropriate equipment and professionalism for control and verification.

References

1. Meadows, D.H., Meadows, D.L., Randers, J.: The Limit to Growth. Universe book, New York (1972)
2. Meadows, D.H., Meadows, D.L., Randers J.: Beyond the limits. Chelsea Green Publishing Company (1992)
3. Meadows, D.H., Meadows, D.L., Randers J.: Limits to Growth. The 30-Year Update. Chelsea Green Publishing Company (2004)
4. Stiglitz, J.: La globalizzazione che funziona. Einaudi, Torino (2007)
5. Veronesi, U.: Le guide di Repubblica: Energia & Scienza, 7 Settembre (2007)
6. De Capua, A.: Tecnologie per una nuova igiene del costruire. Gangemi Editore, Roma (2008)
7. Sinopoli, N., Tatano, V.: Sulle Tracce dell'Innovazione–Tra Tecniche e Achitettura. Franco Angeli ed., Milano (2002)
8. Rifkin, J.: Economia all'Idrogeno. Oscar Mondadori Milano (2002)
9. De Capua, A.: Il Miglioramento Della Qualità Ambientale Indoor Negli Interventi Di Riqualificazione Edilizia. Legislazione Tecnica, Roma (2019)
10. Alberti, L.B.: De re aedificatoria, Milan, Il Profilo, 1966, book I, vol. 2, I, p. 50 (1845)
11. Nardi, G.: "Costruire distrattamente. Incongruenze e contraddizioni nell'impiego odierno delle tecniche progettuali e costruttive" in Baglioni A. / Piardi S. *Costruzioni e salute.* F. Angeli, Milano (1990)
12. Blachere, G.: Saper costruire. Hoepli, Milano (1971)
13. della Salute, M.: Direzione generale della comunicazione e dei rapporti europei e internazionali (2015). www.salute.gov.it
14. Zannoni, G., et al.: Gas Radon, tecniche di mitigazione. Edicom, Monfalcone (2006)
15. Özdamar, M., Umaroğullari, F.: Thermal comfort and indoor air quality. Int. J. Sci. Res. Innovative Technol. 5(3) (March 2018)

Design Opportunities Towards the Ecological Transition of Villages, Cities, Buildings and Dwellings

Lidia Errante[✉]

Department of Architecture and Territory, 'Mediterranea' University of Reggio Calabria, Via dell'Università 25, Reggio Calabria, Italy
lidia.errante@unirc.it

Abstract. The environmental quality of cities is a central issue in the contemporary debate, due to multiple factors related to the technological and energy obsolescence of buildings, spaces and urban infrastructures, the dynamics of resource consumption and the production of harmful emissions and waste. Principles of environmental quality can be found in national and supranational strategies, defining the directions and perspectives towards which orient solutions and actions for transforming the built environment. This contribution analyses the wide range of design solutions - formal, technological, energetic - that can be adopted to improve the environmental quality of cities and residential districts, acting on the quality of life of communities and the quality of socio-spatial dynamics. The different contexts — the village, the city, the neighbourhood, the building and the dwelling — will be investigated considering the vulnerability of the functional arrangement of the indoor and outdoor spaces of everyday life and according to the new modes of use imposed by the circumstances of the pandemic. Will be eventually considered the goals and target for the most recent and relevant national and international strategies in the matter of sustainability: the UN Agenda 2030 and the Sustainable Development Goals, the New European Green Deal and the Recovery and Resilience Plan.

Keywords: Ecological transition · Environmental quality · Design technology

1 The (Urban) World We Live in a Post-pandemic Perspective

The urban environment in which we move during our everyday life activities is framed by different dimensions, spaces, and functions. In the contemporary, the global focus on sustainability has increased the awareness of public opinion about the unecological dynamics through which these activities are carried out and performed. Consumption, energy, waste production, clean air, green spaces, social cohesion, are just a few among the main concerns of contemporary times. These topics can be addressed from different perspectives and in various areas of production: energy, industry, food, farming, cultivation, waste management, commercial, building, services. Among these, the urban dimension is the one that holds the majority of them in an infrastructural way.

© The Author(s), under exclusive license to Springer Nature Switzerland AG 2022
F. Calabrò et al. (Eds.): NMP 2022, LNNS 482, pp. 120–135, 2022.
https://doi.org/10.1007/978-3-031-06825-6_12

The flows of goods and people, the use of energy and resources, the production and the management of waste, the demand and the supply chain, are all dynamics that occur within the urban world and enable human activities. According to that, the city — the built and the urban environment — is where the ecological consequences of progress and modernity can be displayed and hopefully resolved.

The unsustainable dynamics of the urban scale can be distinguished into two spheres. The infrastructural one, related to flows of energy and resources for the primary systems of the city, and mobility of goods and people and the related issues of consumption of energy from non-renewable sources and production of harmful emissions. On the other hand, the construction sector also implies the use of a significant amount of energy and the production of different kinds of waste and emissions during both the construction and operation phases of a building. According to the 2020 Global Status Report for Buildings and Construction of the UN Environment Programme, the global share of energy consumed by the residential sector is 35% (mostly by residential buildings, with 22%), more than industries (32%) and transport (28%). This proportion remains similar for emissions, with 38% for the construction sector compared to 32% for industry and 23% for transport. In the share of emissions, the construction sector is split differently from energy consumption. The main emission sources are residential buildings, accounting for 17% of direct and indirect emissions, but the construction industry accounts for 10%. In this case, the gap between emissions from buildings and the construction industry is less wide (UN, 2020). Also, buildings in an operational state alone account for 55% of global electricity consumption.

In terms of emissions, consumption of resources and production of construction waste, new construction sites have a huge impact on the environment, particularly for the use of traditional wet construction technologies. As well as generating greater use of resources - i.e. the quarrying of aggregates or the use of water to moisten mortar - the buildings constructed in wet technologies cannot be reused at the end of their life cycle. For this reason, in a perspective of urban sustainability, this contribution reflects on the recovery of existing buildings and their constant maintenance, as opposed to demolition and reconstruction, and the use of prefabricated, ecological, and reusable materials.

The post-pandemic perspective has led the discussion on sustainability to a deeper reflection on sustainable mobility. During the first lockdown, the decrease in industrial production and fossil fuel transport use has drastically improved global air quality (WMO, 2021). Aerosols from human activities have the most dramatic effect on human health and air quality, and it is the most impact factor to the amount of PM2.5 in densely populated areas. In 2020, during the early stages of the global pandemic, the unprecedented reduction of urban mobility, transport and aviation, was also determined by the economic downturn associated with the COVID-19 (WMO, 2021).

Economic and ecological awareness has also determined different mobility choices. Individual and more sustainable means of transportation, such as bicycles, e-bikes or e-scooters became very popular.

This desirable transition towards more ecological solutions is often registered where the road infrastructure permits this choice. In many urban, peri-urban and country areas, the mobility grid, the morphological conditions or, in some cases, the demographic, do not allow the spread of such sustainable means. Many European governments have allocated funding and launched policies for transforming road layouts and integrating technologies and equipment to encourage individual and public sustainable means of transport. A fast response was also made possible for the strategic and funding guidelines proposed by the European Union were already in line with these processes. Sustainable mobility was already a priority and a necessity for the EU and their members, and the circumstances of the pandemics have accelerated the speed of change (Lozzi et al., 2020). On the other hand, the contribution reflects on design and technology alternatives that may adapt to different contexts, to address global sustainable and ecological concerns in particular areas with specific issues.

2 Strategies for the Ecological Transition

The efforts put on the perspective of the ecological transition are not so recent. "Ecological transition" is the name of the paradigm by which we define the many national and supranational strategies towards a more sustainable world. In the last 50 years many approaches, policies and actions have been mobilized for that aim, through the dedication of political and social activists and movements.

Through the decades, civil society has absorbed these concerns, increasing awareness on the topic of sustainability and ecology, leading to a reversal approach in the daily-life patterns of use and consumption. On the individual scale, the credit for this progressive ecological transition goes, in general, to how the sustainable development guidelines of the most recent agendas and policies are disseminated. Next to the media, the community awareness actions carried out by the United Nations and the European Union through community planning have undoubtedly made it possible to spread the principles of sustainability and ecology widely. They have also provided an operational and factual understanding of the benefits of sustainable actions and behaviours, as well as their feasibility, by relating the change to the real possibilities of those being asked. They have also provided or developed in situ, specific tools for the sustainable transformation of places and human activities in consultation with local communities, adopting flexible and replicable methods of analysis and transformation. This is the case in UN-Habitat programmes around the world.

The dynamics of the Covid-19 pandemic have played a central role in this, prompting us to review our consumption patterns and give greater impetus to the ecological transition of our habits but also a better use of living spaces. In the last year, the real estate sector has also seen an increased focus by buyers on housing solutions with outdoor spaces, balconies and gardens, good views, bright and well-ventilated rooms and at least one extra room for study or indoor leisure activities. These considerations have to be integrated with the whole discourse of sustainability because they reflect the changes imposed by contemporary times and needs. Although, it is important to understand how this global-scale event — from supranational strategies and policies to unexpected socio-economic or health-related issues — can impact the city form, the urban services and the quality of life in a sustainable and ecological perspective.

In the following paragraphs, the contribution will address three strategies that, to many extents, address sustainable development and ecological transition, finding for each of them the implication on the built environment.

2.1 The 2030 Agenda and the Sustainable Development Goals[1]

Since its publication in 2015, the 2030 Agenda has dominated the debate on sustainable development for its ambition to lead the transformation of our world, as its title claims.

The Agenda is a "programme of action for people, planet and prosperity" with a strong emphasis on the proactive nature of the proposed strategies. Sustainable development is connected to the eradication of poverty and the ecology approach, as essential components of social and environmental sustainability. The 17 Goals and 169 Targets define a Universal Agenda that the "areas of crucial importance for humanity and the planet". Keywords as People, Planet, Prosperity, Peace, Collaboration, are used to indicate the mutual effect of physical and social transformations of the world and the effort required by institutions and communities to impact the quality of life, the environment and democracy.

The 2030 Agenda emphasizes poverty, which end is a precondition for sustainable development. It is clear that to break out of the vicious circle of the dominant capitalist socio-economic model and thus to adopt more ecological and sustainable behaviour, individuals and communities must be put in a position economically and culturally to make conscious choices, not conditioned by the market or by limited access to products and services considered more environmentally friendly.

According to this vision, the actions and programmes related to the Agenda aimed at developing countries, where poverty is accompanied by a general condition of scarcity, hunger, disease, and lack of physical, mental and social well-being. Such goals were conceived in times of great challenges to sustainable development, by imbalances and inequalities, threats to ecosystems and global health. For that, the SDGs are still robust in times of pandemics or energy crises.

The 17 SDGs express the complexity of the contemporary world with principles that can be applied to developed and under-developed countries with different scales and priorities. The urban and built environments play a central role in the reaching of SDGs under three main aspects: the socio-physical dimensions of housing, public spaces and the opportunity to carry out diverse outdoor, accessible and collective activities. It is also important, since the more recent energy crisis, that the Agenda promotes the production of green, sustainable and accessible energy by renewable sources.

Capability and opportunities are concepts that can be applied to the city environment in a broader sense to promote and achieve well-being according to the right design strategies. In this sense, the design approach may involve the transformation of the urban, built or social environment, promoting sustainable habits.

The 2030 Agenda recognises the sustainable management of cities as an asset for the growth and development of urban settlements "to promote cohesion among communities, personal security and stimulate innovation and employment". For the first time, some specific public and urban spaces and activities are given greater weight in the overall strategy: this is the case of public space and sport in SGDs 11.

Also, the monitoring of the actions promoted for the achievement of SDGs and targets consider the impossibility to quantify the results of some targets for which there

---

[1] The information in the paragraph on the United Nations 2030 Agenda for Sustainable Development, where not specifically mentioned, refers to the document: United Nations (2015).

are no clear numerical targets, encouraging individual states to start experimenting and providing data in this regard.

The Ecological Transition of the City in the UN 2030 Agenda.
The issue of the quality of urban space is relevant to the Agenda and we can find elements of it in several targets of the 17 SDGs. In Goal 3, Target 3.6 declares, by 2020, to halve the global number of deaths and injuries from road accidents, implying a clear reorganisation of the hierarchy of urban routes to allow more space and safety for pedestrians and cyclists, to the detriment of the presence of cars. This target is included in the same way as access to health services.

Target 6 refers to the accessibility and sustainable management of water, also regarding the infiltration from landfills and pollutants that may contaminate groundwater.

Target 7 is particularly interesting in terms of ecological transition, as it refers to universal access to clean energy, provided by cheap, reliable, sustainable and modern systems. Innovation in energy services and production from renewable sources requires large investments in research and technology, which Target 7a advocates.

Target 9, for resilient infrastructure and industrial innovation, certainly also addresses the construction industry and its impact on natural ecosystems.

But it is, in particular, Target 11 that refers, explicitly, to cities and communities. Its targets are related to social and physical accessibility, safety and healthy, physical quality of housing and basic services and their affordability, sustainable and safe public transportations, inclusive urbanization. Special attention is paid to the needs of the most vulnerable and their right to participate in the sustainable transformation of the city. Three specific targets are directed to the resilience of urban settlements in case of natural disasters. The emphasis is on safeguarding the world's cultural and natural heritage and the protection and prevention of human and economic losses, as well as on the quality of air and waste management.

These objectives and targets have clear repercussions on the management of the city as an urban ecosystem, necessitating process and design changes that will be better addressed in the following paragraphs. These changes, somewhat accelerated by the circumstances of the global pandemic, concern both the infrastructure of public space, i.e. services, mobility and essential urban activities and the stock of buildings, with particular reference to the residential stock. This mainly concerns the reduction of polluting emissions, the use of non-renewable resources, the production of energy from renewable sources, the containment of consumption and waste, physical and social accessibility, air quality (indoor and outdoor) and the containment of the effects of global warming in the built environment.

2.2   The New European Green Deal

The European Green Deal is intended as a strategy to fight climate change and environmental degradation and its effects on the ecosystem. The declared ambition is to transform the EU into a modern, efficient, sustainable and competitive economy.

In particular, by 2050 the European Green Deal sets specific targets to stop generating net greenhouse gas emissions, to achieve a resource-intensive economy and to fight to ensure that no person or place is neglected. As in the case of the 2030 Agenda - which was also ratified at the EU level for the management of funds, resources, strategies and projects of member states - the environmental, economic and social dimensions of sustainability are clearly expressed. In the specific case of the European Green Deal for 2050, the theme of ecological transition is more clearly defined and gives a clearer and more decisive orientation to the strategies that member states will have to implement to achieve the desired results.

The European Green Deal is also being developed given the negative effects of the Covid-19 pandemic: one-third of the €1,800 billion investment in the Next Generation EU recovery plan and the EU's seven-year budget will finance the European Green Deal. The EGD is devoted to the quality of life, health, and well-being of the people and the urban and built environment. Great attention is paid to the natural environment, biodiversity and the main resources — air, water, soil. Energy efficiency is promoted in the renovation of buildings and public transportation, supported by the use and production of renewable sources. Technology innovation plays a central role, also in the design of products that can be upcycled, repaired, or reused. The industrial sectors are mostly affected by this transformation and new skills and professions may arise from this opportunity.

More in detail, the EU intends to be climate neutral before 2050, protecting humans, animals and plants by reducing pollution, helping industries to become a world leader in clean technologies and products, and contributing to a just and inclusive transition. According to these aims and goals, the actions of the EGD address climate, environment and oceans, energy, transportation, agriculture, regional investments and development, industry, research and innovation. In this perspective, the city is the ground of major experimentation to incentivise the decarbonisation of the energy sector and reduce emissions by sustainable renovating the existing building stock and introducing cleaner means of public transportation.

For each objective, the EGD have foreseen several programmes and policy tools. Climate neutrality is advocated through the European Climate Act, the European Climate Pact, the Adaptation Strategy, as well as through Climate Diplomacy and participation in various climate change conferences such as COP 26. The Pact is based on an open and inclusive approach that multiplies the impact of actions through the dissemination and sharing of good practices. The focus areas activated with priority in the initial phase are green areas, green transport, green buildings and green skills. These will be increased over time to cover all EGD actions. The European Climate Act is the actual regulatory instrument of the European Green Deal and defines intermediate targets of at least 55% GHG reduction by 2030. The law also includes measures to monitor progress and provides for the adjustment of actions accordingly. The EU Adaptation Strategy, adopted on 24 February 2021, defines how the EU can adapt to climate change and become climate resilient by 2050. The strategy aims to make adaptation smarter and systemic, using a robust data set and risk assessment tools available for each scale and level of interested people. The Climate-ADAPT 10 Case Studies (2018), gather urban design, architectural and technological interventions across Europe. Among the many, the green corridor of Passeig de Sant Joan in Barcelona (ES), the Green roofs in Basel (CH) where are combined mitigation and adaptation measures, or the management of heavy rains and storm-water in Copenhagen (DK) according to urban design principles.

In line with the previous 2030 Agenda, the three main principles of this action are: ensuring secure and affordable energy supply for the EU developing a fully integrated, interconnected and digitized energy market priorities energy efficiency, improving the energy performance of our buildings and developing an energy sector based largely on renewable sources. To this end, the Green Deal Europe objectives for the transition to clean and sustainable energy are: build interconnected energy systems and better-integrated networks to support renewable energy sources; promote innovative technologies and modern energy infrastructure; Increase energy efficiency and promote eco-design of products; Decarbonizing the gas sector and promoting smart integration between sectors; Empower consumers and help the Member States tackle energy poverty; Promoting EU energy standards and technologies worldwide; Develop the full potential of EU offshore wind energy.

2.3 The Italian National Recovery and Resilience Plan (PNRR)[2]

From a post-pandemic perspective, the EU and its member countries have agreed on recovery strategies and policies that become an opportunity to implement green and digital transition-oriented actions. Italy, in the framework of Next Generation EU and Next Generation Italy, has elaborated the National Recovery and Resilience Plan (PNRR). The economic recovery of the country, hit by the health crisis, passes through social and environmental recovery, as well as that of the productive sector. Particular attention is given to fragile population groups, such as women and young people looking for work, and to places with high hydrogeological vulnerability and climate change, such as coastal and inland areas. A key aspect is also linked to the digital transition to seize all the opportunities for renewal and development of the productive sector. The six missions of the plan are Digitalization, Innovation, Competitiveness; Culture and Tourism; Green Revolution and Ecological Transition; Infrastructure for Sustainable Mobility; Education and Research; Inclusion and Cohesion; Health. According to Mario Draghi, the plan is intended to be an instrument for the country's innovation and modernization, capable of tackling contemporary challenges through "planning capacity and concreteness".

The ecological transition is the core of the new Italian and European development model and implies actions to reduce polluting emissions, prevent and fight land degradation, and minimize the impact of production activities on the environment, to improve the quality of life and environmental safety, as well as to leave a greener country and a more sustainable economy to future generations. The ecological transition can also be a pivotal factor in increasing the competitiveness of our production system, encouraging the start-up of new, high added-value business activities and encouraging the creation of stable employment.

The green revolution and the ecological transition are part of Mission 2 in which, among the various possible actions, there is a clear reference to the energy efficiency of public and private real estate and the promotion of sustainable mobility, with related urban infrastructure.

---

[2] The information in the paragraph on the PNRR, where not specifically mentioned, refers to the document: Consiglio dei Ministri (2021).

From the perspective of the ecological transition of urban mobility, Italy has numbers with a clear imbalance between the use of cars (with 663 cars per 1,000 inhabitants in 2019) and the extent of the rail network (only 28 km per 100,000 inhabitants, compared to 47 km / 100,000 inhabitants in Germany) with the oldest fleet of vehicles in Europe. It, therefore, remains a priority to encourage the use of environmentally friendly individual and public means of transport, including through a review of traffic hierarchies, primarily by providing the infrastructural opportunity for their safe use. To this end, the PNRR allocates almost 60% of its resources to Mission 2, of which about 24% to energy transition and sustainable mobility and about 15% to energy efficiency and upgrading of buildings. The actions under Mission 2, Green Revolution and Ecological Transition which can be applied to the urban and built environment refer in particular to sub-actions M2C2 (Renewable Energy and Sustainable Mobility) and M2C3 (Energy Efficiency and Renovation of Buildings).

Through the latter, the aim is to strengthen energy efficiency by increasing the level of efficiency of buildings, one of the most virtuous levers for reducing emissions in a country like ours, which suffers from a stock of buildings with more than 60% of the stock over 45 years old, both in public buildings (e.g. schools, judicial citadels) and private buildings, as already initiated by the current 'Super bonus' measure. An innovative element that concerns the culture of the project is that relating to almost zero energy buildings (nZEB) is in line with the energy strategy for the national building stock. This objective is pursued through a general simplification and acceleration of procedures for the implementation of energy efficiency measures.

Indeed, Mission 5, Inclusion and Cohesion, incentivizes the creation of quality urban and residential services, as sub-Mission M5C2, Social Infrastructure, Families, Communities and the Third Sector, with investments in Urban Regeneration and Social Housing to reduce social exclusion and degradation, Integrated Urban Plans and the Innovative Housing Quality Programme.

The investment in urban regeneration projects are direct to municipalities with a population of more than 15,000 inhabitants and provides contributions for urban regeneration, reduction of territorial marginalisation and social decay. The Innovative Housing Quality Programme of PNRR considers the new construction of housing structures and the recovery of the existing ones to reduce housing difficulties in particular degraded areas, focusing mainly on green innovation and sustainability. From a design point of view, it is interesting that the financing proposals will be selected assessing the environmental, social, cultural, urban-territorial, economic-financial and technological-processual impact of the projects.

3   Design Opportunities for Environmental Quality

The strategic, political and regulatory instruments above mentioned defining an action framework in which each role — academic, governmental, public and local — is defined. In this sense, design and technological research have a central position in experimenting with solutions that, at different scales, are capable to guide the path towards ecological transition tangibly and understandably. Also, the dissemination means used by the academic research involve a process of constant review and benchmark of the results.

The three national and supranational strategies described in the previous paragraphs represent the logic succession and the progressive definition of theoretical and design paradigms linked to the ecological transition and the specific technologies required to implement it across several areas of intervention. The design opportunities connected to the ecological transition of the built and urban environment, its infrastructures and spaces, accelerate the sustainable development of villages, cities, buildings and homes by applying principles of environmental quality and innovation.

The criteria of ecological design should be oriented to:

- the correct planning of the intervention, account its impact on the social, economic and environmental spheres; the flexibility of the interventions, which means planning according to scenario steps and reuse existing buildings (Bergevoet and van Tuijl, 2016);
- aim at circularity in the production, execution and management stages (De Capua and Errante, 2021);
- the improvement of the urban and built environment quality — in housing, public spaces and mobility — according to the right green technology;
- the regeneration of the social commons (Manzini, 2018).

To achieve the ambitious objectives promoted by the most recent national and supranational strategies, the design of space - urban, architectural, technological - must once again play a central role as a mediator between the environment, society, the market and the state, and no longer the opposite. The sensitive nature of the project must once again be put at the service of the specific characteristics of territories, places, communities, individuals, subjects, skills and opportunities. For this reason, it seems appropriate to conclude this contribution with a brief examination of the design opportunities for the ecological transition that can be adopted for the urban and built environment. The village is considered with that of the city, because of their great morphological and infrastructural differences such as to determine, for the same objective, methods and actions that are quite opposite. This sphere also includes urban space, whether public, semi-public, semi-private, private or residential. Opportunities for the renovation of residential buildings and housing are then described, which have a decisive weight in the ecological transition strategy with particular reference to the Italian case and the PNRR.

3.1 Villages and Cities

Although they seem very different, villages and cities — or parts of them — share some characteristics. In particular, the contribution refers to the Modern city[3], developed during the 20th century due to the need to expand the tertiary sector and production, and then evolving a place of consumption and affirmation of the dominant life model.

– The Modern city, by its very constitution, can be considered as a sum of small towns that are functionally self-sufficient;
– Small towns and modern residential neighbourhoods present similar scales, dimensions and proportions;
– Modern residential districts are characterised by housing densities similar to those of a village;
– Modern housing estates and small towns often have similar scales, sizes and proportions;
– The urban structure has often remained unchanged, limiting the full potential of entire communities and contributing to poor quality of life.

The circumstances of the pandemic have further highlighted that quality of urban life is not an abstract concept and it is connected to specific spatial issues as;

– lack of outdoor social, recreational, sport, communal and collective activities;
– lack of healthy, green, safe and accessible public spaces;
– lack of opportunities and possible interaction;
– physical and social degradation;
– lack of activities and services accessible only by reaching other districts/villages.

The pandemic circumstances have suggested looking to the village model, which recalls in the collective imagination an ideal of well-being and quality of life. Small towns are well known for the quality of life that they offered in the past as well in the present times, for their well-balanced position between the natural environment, social relations, cultural heritage and local economies. All these features lead to physical configurations that allow communities to feel rooted and connected with that place, creating a sense of belonging which is often missing in the Modern residential district.

In this sense, villages and contemporary cities have the same need to reconstitute an adequate physical and social infrastructure, made up of a mobility network and aggregation spaces that can also be configured as environmental protection devices, off-grid energy production, containment of the heat island effect and absorption of harmful emissions. In particular, design and technological opportunities in the built environment can revolve around public space infrastructure and urban ecosystem services (Errante, 2021; De Capua, Errante, Palco, 2021; 2019; De Capua and Errante, 2019). For example:

---

[3] The concept of the modern city refers to the urban development of large and medium-sized European cities in the period between the two wars, characterised by large-scale urban plans and the management of cities according to homogeneous zones, according to a rationalist vision and aesthetic (cf. Samonà, 1978, Benevolo, 1980).

- definition of green areas with the planting of tree species able to absorb harmful substances and emissions from urban mobility;
- increasing the permeability of the soil through new areas of technological green areas able to interact positively with the sewerage network infrastructure;
- integrate, in urban regeneration and public space transformation projects, areas dedicated to the production of energy from renewable sources such as photovoltaic or kinetic surfaces, which can be used to supply public lighting or charging stations;
- in places affected by flooding from heavy rain, identify suitable places for temporary water harvesting following the model of water squares;
- redefining the hierarchy of urban mobility routes by providing adequate space for sustainable individual transport and related service infrastructure (e.g. e-bikes, e-scooters, etc.);
- integrating appropriate strategies for improving urban accessibility into mobility redefinition strategies, with a special focus on access to homes and public utility buildings, and the definition of guided routes to essential services;
- transforming boulevards into green corridors.

3.2 Residential Building and Dwellings

Villages and cities, with particular reference to suburbs, are subject to equal and opposite conditions of degradation. While both suffer from a general and widespread structural, technological and energetic degradation of the building stock, the suburb suffers from depopulation, while the city fears to overcrowd predicted by the increase of the urban population foreseen by the UN by 2050. Concerning this problem, it should be pointed out that the housing stock in the village was built at a time and with construction technologies very different from those used during the construction of the modern city, where we can also encounter prefabrication systems in addition to traditional construction technologies. This certainly invites the adoption of appropriate technologies for the requalification of buildings.

The design implications related to urban living quality are particularly interesting in the framework of public housing districts, where to value two complementary aspects: the technological obsolescence of the public housing stock and the widespread lack of variety of the residential public outdoor spaces. According to those concerns, public residential estates are widely recognised by contemporary social and academic debates. The pandemic has exacerbated the perception of these issues.

The reflections put forward here are the result of broader post-doctoral research in Sustainable Building developed at the "Mediterranea" University of Reggio Calabria. The research refers in particular to the relations between the quality of life and quality of space in the public residential estates. According to different urban and typological forms, can be found several patterns of use and consumption performed by different groups of the same community. Analysing these socio-spatial dynamics, the research wanted to define a socio-technical framework, highlighting the opportunities and peculiarities of the social, environmental and economic context. These aspects, as analytical categories, can give an account of the housing condition from a qualitative and quantitative point of view and about needs (expressed and potential) and their satisfaction.

The Italian housing stock is one of the most scarce and obsolescent in Europe. The public housing stock in Italy represents only 4% of the total compared to much higher percentages in other countries (France 16,5%, United Kingdom 18%, Denmark 21%, The Netherlands 30%) (Pittini, 2019). Of the entire residential stock, both public and private, more than 16 million dwellings (51% of the national stock) were built before 1970, according to speculative models of low building quality, in the absence of anti-seismic regulations and 90% of cases having an energy class measurable between F and G (Camera dei Deputati, 2020). A rapid transformation of housing or buildings to adapt to the circumstances of the pandemic can happen only in synergy with actions of ordinary and extraordinary maintenance.

In this perspective, the performance requirements for a better quality of living can be considered:

- energy efficiency;
- air and ventilation;
- spatial functionality.

These three variables establish new causal connections with the quality of designed space at the domestic scale. From a methodological point of view, it would be useful to systematically observe the habits of use and consumption of the residents according to their environment and opportunities. The correspondence of that environment to the needs of the occupants and vice-versa determines the degree of actions that may require new configurations of the space. The assumption is that the characteristics of the domestic space can stimulate or inhibit individual or collective behaviour. For instance, the use of artificial lighting during daylight hours or the use of heating or cooling systems depending on the right — or bad — exposure and orientation of the building and the construction technology adopted.

These considerations alone are not enough to evaluate the quality of space as a performance requirement at the neighbourhood scale. The environmental quality of the residential space (Zucchi, 2011) can be pursued, in an eco-systemic logic, according to sustainability principles that refer to the control of the environmental performance and impact of the transformation process. These principles translate into possible interventions — applicable to both new construction and existing buildings — with results on morphology and use.

Design Criteria

According to international strategies and policies on the improvement of public and private buildings, the main problem to overcome is energy efficiency. Due to climate change, our buildings are no longer able to respond effectively to the requirements of indoor environmental comfort, making numerous design measures necessary (De Capua, 2019; 2002).

The parts of the building that are of most interest from an energy point of view are undoubtedly the building envelope - façade, roof, windows and doors - and the central heating systems. Correct design and choice of appropriate construction technologies can generally limit the energy demand for the building's operation. In particular, bio-climatic architecture principles can be adopted, which tend to satisfy thermal comfort requirements without the aid of air conditioning systems and to control the internal microclimate by exploiting the correct exposure and orientation of the building. Such design strategies, defined as passive, can exploit the sun, wind, water, soil and vegetation to optimise thermal exchange between the building and its surroundings. According to several authors (Ben Bonham, 2020; Los, 2013; Gallo, 2011; Sasso, 2011) the key criteria of bioclimatic design are:

- location;
- orientation;
- solar radiation on the envelope (i.e. shading);
- natural ventilation inside the building;
- thermal insulation according to the accumulation of solar energy avoiding dispersion (in winter);
- promoting thermal inertia with shading and natural ventilation (in summer).

According to Mac Lean and Silver (2021) and Los (2013) many environmental and bioclimatic design actions can be pursued to mediate between shapes and functions, technological tools and natural resources. Among the many:

- The correct thermal insulation of the envelope drastically reduces the energy needs of buildings related to heating, cooling and ventilation systems. The so-called "thermal insulation" is the driving action of the most recent funding for the energy recovery of residential buildings and can be carried out inside or outside the building envelope, using sustainable and ecological plant-derived materials. This must be combined with the insulation of windows and doors to avoid draughts or thermal bridges.
- The solar chimney improves the ventilation of the building by exploiting the principle of rising warm air. This action can be combined, in the appropriate contexts, with heating systems with a radiant floor or wall panels that use geothermal energy.
- The green roof plays an insulating role for the whole building with important energy-saving advantages and, in some cases, self-production. From an environmental point of view, the green roof absorbs heat, reduces the use of electrical appliances and therefore harmful emissions, filters the air by eliminating or capturing polluting particles and contributes to the microclimate of the surrounding area by reducing the heat island effect.
- The butterfly roof, also known as an inverted roof, has a V shape that allows it to integrate systems to improve the energy performance of the building, such as photovoltaic, solar thermal or rainwater collection and storage. The choice of this shape also has an impact on the building organism, typology and functional spaces. Appropriately oriented, it allows a considerable increase in the height of the walls facing east and west with an increase in the brightness of the rooms. The opening of appropriate light wells helps natural ventilation and increases the overall efficiency of a traditional photovoltaic roof.

The applications of bioclimatic design and environmental design have various effects: on architectural language; on space and functions, both public and private; on the management of urban public services, such as mobility; on the development of new materials; on the complementarity of technological and digital tools for the control of energy consumption; on the optimisation of waste and dispersion; on the ability to dialogue with the natural environment more harmoniously.

This also implies that traditional or regional building techniques can be the starting point for innovative strategies in the building sector. This type of research and development falls within the objectives of the regulations and national and supranational policies already described, concerning the need to innovate the industrial production processes involved in the construction sector, with benefits on the use of non-renewable resources, the production of clean and accessible energy, the reduction of waste and harmful emissions.

4   Conclusions

The contribution highlighted, in a descriptive form, the most recent, urgent and concrete design — and applied research — opportunities in the ecological transition perspective. Several issues have been addressed in defining an overview of the topic. First of all, the necessary input provided by supranational and national strategies which, unlike in the past, identify precise lines of action, funding and evaluation tools, and therefore precise selection criteria to calibrate appropriate and targeted interventions in the various spheres of green innovation. It was highlighted goals, targets and funding programmes mainly directed to the urban ecosystem, its physical and social infrastructure, the built environment, its buildings and their occupants. The transversality of the strategies here described — the 2030 Agenda for Sustainable Development, the new European Green Deal, the Italian National Recovery and Resilience Plan — shows that the governments possess the tools to control the complexity of the challenges posed by the contemporary tomes, but they require there a further incentive in terms of design. Issues as sustainability, the protection of ecosystems, the efficient use of resources, the response to climate and pandemic crises, and the physical, social and psychological wellbeing of individuals and communities are the objectives that innovative design must pursue.

The architectural project finds a new unity in these reflections. The urban, architectural and technological dimensions of the architectural project are intensified both in their disciplinary independence and in their necessary combination, to make the project flexible and adaptable to the new needs and performances that human settlements must satisfy.

The ecological transition requires a functional, spatial and cultural reorganisation of living spaces, on which this contribution has briefly commented, while bearing in mind the implications on other sectors, such as industry and the production of energy from renewable sources. The transport sector and the transformation of housing are those most affected by this change and are already undergoing a radical functional and organisational implementation. In this sense, it is worth pointing out that a fundamental innovation in the direction of ecological transition also relates to service design, which represents a process innovation capable of combining strategy and action and increasing the chances of positive impact. Generally speaking, the methodology of service design proposes an approach capable of applying general criteria to a particular case, a necessary process in the pursuit of smart adaptation actions. Strategies and design actions for transformation that can be pursued at the urban scale will not necessarily be applicable in the context of hamlets and villages, just as there will be differences in approach between a coastal and an inland settlement. This is for different reasons related to the climatic, geographical, morphological, cultural, infrastructural, architectural context, or more generally to different priorities, resources and opportunities.

Several actions can be pursued, including those identified in the previous paragraphs, and declined in many variations depending on the context of the application. The design innovation, in this sense, is given by the ability to combine functions, technologies and materials most properly. For this reason, in the transformation of space - indoor or outdoor - or of service in light of ecological transition, one cannot disregard the analysis of the social context, the habits of use of communities, individuals and users.

According to the most recent strategies, policies and funding instruments, we can observe a progressive shift from actions and design opportunities aimed at the urban environment — mobility, public spaces and more generally at urban regeneration, which are still fundamental and present — towards actions direct at the renewal of the built environment. The energy and structural upgrading of buildings, the replacement or addition of parts of the building organism, the adoption of bio-architecture and passive building criteria, the choice of eco-compatible materials, with a green or eco-label, produced from renewable energy sources, are unequivocally recognised by the strategies and made compulsory both by the sector's regulations and for access to funding. In addition, the fact that there is more emphasis on the rehabilitation of existing buildings than on new construction and that strategies are increasingly oriented towards less land consumption makes us confident that the design and process innovations pioneered by applied research are being progressively absorbed by policy instruments.

References

Ben Bonham, M.: Bioclimatic double-skin façades. Routledge, New York (2020)

Bergevoet, T., van Tuijl, M.: The Flexible City. Sustainable solutions for a Europe in Transition. Nai10 publishers, Rotterdam (2016)

Camera dei Deputati: Servizio Studi XVIII Legislatura. Risparmio ed efficienza energetica. (2020)

Climate ADAPT — 10 case studies. Publications Office of the European Union, Luxembourg (2018)

Consiglio dei Ministri: Piano Nazionale di Ripresa e Resilienza #Nextgenerationitalia (2021)

De Capua, A.: La Qualità Ambientale Indoor nella Riqualificazione Edilizia. Legislazione Tecnica (2019)

De Capua, A.: Nuovi Paradigmi per il Progetto Sostenibile. Contestualità, Adattabilità, Durata, Dismissione. Gangemi Editore (2002)

De Capua, A., Errante, L.: Interpretare lo spazio pubblico come medium dell'abitare urbano. Agathón – International Journal of Architecture, Art and Design (6), 59–72 (2019)

De Capua, A., Errante, L., Palco, V.: Co-existing with COVID-19. Design for resistance. Urban studies and public administration. 3(2), 69-86 (2020). Scholink Inc.

De Capua, A., Errante, L., Palco, V.: Methods and techniques for sustainbale urban living: between seismic vulnerability and urban sustainability. In: Bevilacqua, C., Calabrò, F., Dalla Spina, L. (eds.) New Metropolitan Perspective, pp. 592–605. Springer Nature Switzerland AG (2021)

Errante, L.: Hybrid communities and resilient places. Sustainability in a post-pandemic perspective. In: Sposito, C. (ed.) Possible and preferable scenarios of a sustainable future, pp. 32–45 (2021)

European Environment Agency. Climate-ADAPT 10 case studies. How Europe is adapting to climate change (2018)

Gallo, P.: Recupero bioclimatico edilizio e urbano: Strumenti, tecniche e casi studio. Sistemi Editoriali (2011)

Los, S.: Geografia dell'architettura. Progettazione bioclimatica e disegno architettonico. Padova: Il Poligrafo (2013)

Lozzi, G., Rodrigues, M., Marcucci, E., Teoh, T., Gatta, V., Pacelli, V.: Research for TRAN Committee – Structural and Cohesion Policies. Brussels (2020)

Mac Lean, W., Silver, P.: Environmental Design Sourcebook. RIBA, London (2021)

Manzini, E.: Politiche del quotidiano. Progetti di vita che salvano il mondo. Edizioni di comunità (2018)

Pittini, A.: The State of Housing in the EU 2019. Housing Europe (2019)

Sasso, U.: Il nuovo manuale europeo di bioarchitettura. Mancosu Editopre (2011)

United Nations: Transforming our world: the 2030 Agenda for Sustainable Development. A/RES/70/1 (2015)

World Metereological Organization: State of Global Climate 2021 WMO Provisional Report (2021)

United Nations Environment Programme: 2020 Global Status Report for Buildings and Construction: Towards a Zero-emission, Efficient and Resilient Buildings and Construction Sector. Nairobi (2020)

Zucchi, V.: La qualità ambientale dello spazio residenziale. Franco Angeli Editore (2011)

Green Transition Towards Sustainability. Design, Architecture, Production

Alberto De Capua and Lidia Errante[✉]

Department of Architecture and Territory, Mediterranea University of
Reggio Calabria, Reggio Calabria, Italy
{adecapua,lidia.errante}@unirc.it

Abstract. The scientific knowledge and technologies available provide us with theoretical and practical tools to face environmental challenges, reminding us of responsibility to review the relationship between natural systems and production, consumption and our behaviors. This task can no longer be postponed.

These considerations underpin the contemporary debate on sustainable development and ecological transition and guide national and trans-national policy agendas. On the other hand, civil society is also becoming increasingly attentive to these issues in an attempt to reverse current trends. Every field of knowledge is involved in the formulation of possible solutions, individual or system, aimed at the analysis and containment of the climate emergency, so that dramatic circumstances can be prevented in terms of loss of life and biodiversity. The field of architecture and design are also involved in this reflection, because of their close correlation with the industrial sector on the one hand and with the social partners on the other. In both cases design plays a role of mediation between economic and production needs and social demands concerning the right to the city and urban democracy.

Keywords: Green transition · Sustainability · Technology design

1 Introduction

The semantic field of sustainable development includes terms such as environment, ecosystem, greenhouse effect, harmful emissions, citizens' health, urban health, resource consumption, clean energy, pollution, poverty. These terms, which are part of our daily lives, are also the objectives that humanity as a whole has set itself to achieve through sustainability policies to reduce the ecological footprint and intervene in the health of the planet.

The scientific knowledge and the best available technology provide us with theoretical and practical tools to face these challenges, reminding us of responsibility to review the relationship between production systems, consumption habits, and the natural ecosystem. This task can no longer be postponed.

These considerations underpin the contemporary debate on sustainable development and ecological transition and guide national and trans-national policy agendas. On the

F. Calabrò et al. (Eds.): NMP 2022, LNNS 482, pp. 136–145, 2022.
https://doi.org/10.1007/978-3-031-06825-6_13

other hand, civil society is also becoming increasingly attentive to these issues in an attempt to reverse current trends. Every field of knowledge is involved in the formulation of possible solutions, individual or system, aimed at the analysis and containment of the climate emergency so that dramatic circumstances can be prevented in terms of loss of life and biodiversity. The field of architecture and design are also involved in this reflection, because of their close correlation with the industrial sector on the one hand and with the social partners on the other. In both cases, design plays a mediation role between economic and production needs and social demands concerning the right to the city and urban democracy. This mediation relationship, however, includes a fourth part, consisting of the environmental ecosystem in which the project is located, the living society and the economic and productive activities that operate. The role of the designer is characterized by a strong ethical connotation that is inevitably transferred to the results of his project, at any scale, depending on his spatial, technological, material choices and that relates to the surrounding environment. Such choices can improve the quality of the environment, can worsen a condition of degradation, can cooperate in its functioning and sustainability and, finally, can exploit the environment as a primary resource without harming the ecosystem balance.

In this respect, some practical and theoretical perspectives can be distinguished and will be briefly explored in the subsequent paragraphs of the contribution.

1. The relationship between industrial production and architecture in the field of bio-construction and bio-components, regarding the use of the waste of agricultural production for the production of materials for building use;
2. The relationship between the design and regeneration of urban public spaces for the improvement of the environmental quality of cities, regarding green design, urban ecosystem services and the agricultural micro-production of urban gardens;

The principles of the circular economy that guide the ecological transition of sustainable architectural and urban design through the use of recycled materials.

2 Sustainability. A Complex Concept

The environmental emergency we are experiencing is reminded daily by the media, which regularly recount the risks to the health of the planet, the dramatic natural disasters that are taking place in every corner of the world. Every year the news reports that the planet has exceeded the estimate of available resources for the current year in a day now known as "Overshoot Day". For the rest of the year, the planet will consume resources from the following year in a vicious circle. In 2020 Overshoot Day was reached on August 22, 5 months ahead of the end of the year.

The need for a paradigm shift to combat phenomena such as uncontrolled urbanisation, intensive crops and livestock farming, waste generation, pollution of the seas, oceans and rivers is palpable. A similar concept had already been expressed in 1968 by Garret Hardin with his theory of the "Tragedy of the Commons". Hardin [1968] says that the increase in population and the consequent increase in resource consumption will have serious repercussions on the environmental ecosystem due to its intensive exploitation. In this sense, the commons are considered goods such as the atmosphere, oceans, rivers and fish, and all those goods that escape public or private property, but that bring a benefit to the community. The latter governs its use through unwritten rules, social structures, traditions, or in some cases laws. Nobel Laureate Elinor Ostrom [1990] further expanded this concept in the 1990s by stating that commons can be preserved through their collective management, corroborating this perspective through numerous field experiences.

The concept of sustainability as we know it is introduced at the first UN conference on the environment. The risks to the stability of the environmental ecosystem had been suggested as early as 1972 by the "Limits of Development" report commissioned by the Club of Rome at the Massachusetts Institute of Technology. They argued that the uncontrolled use of natural resources and continuous polluting emissions would profoundly affect the future of the planet in the 21st century and called for a renewal of society that should limit man's impacts on the planet. Finally, in the Brundtland report in [1987], sustainability is defined as a condition of development capable of ensuring that the needs (environmental, economic and social) of present generations are met without preventing future generations from realizing their own. From 1992, with the Rio Conference, the environment, development and sustainability will be linked to the present day. On this occasion, however, it was found that humanity had already exceeded the limits of the Earth's sustenance. The considerations of the last edition of 2004 confirmed and aggravated the previous pessimism of the authors: despite scientific advances, the planet has entered a phase of unsustainability.

The growing awareness on the issues of sustainability and resource consumption is demonstrated by the policy statements on development, issued by international bodies and governments throughout Europe and the world. Nevertheless, for us to hope for an improvement, it will change our lifestyles and our models of well-being, on which our ecological footprint depends. The required change invests and involves all fields of knowledge, being and production in a systemic way.

The 2030 Agenda [2015] captures the multi-scalar scope of this change, managing to orient the 17 Sustainable Development Goals towards a holistic dimension of environmental and, economic and social sustainability. Sustainable Development Goals (SDGs) and their targets cover a complex range of transversal and overlapping fields of action aimed at improving the living conditions of society and the planet. In particular, for the topics covered by the contribution, it is interesting to mention Objectives 11, 12, 13, 14, and 15, namely:

- Goal 11. Sustainable Cities and Communities. This objective aims at the construction of inclusive, safe, sustainable and sustainable urban settlements for the benefit of its inhabitants, ensuring housing, basic services, public spaces and quality modes of transport. None other things, the aim is to reduce the environmental impact and resource consumption of cities, while also containing the risk of environmental disaster.
- Goal 12. Responsible Consumption and Production. SDG 12 aims to ensure sustainable development and production models with a particular focus on the management and efficient use of natural resources, the reduction of waste from food production,

chemicals management and waste in general. Particular emphasis is placed on the dissemination of information useful for changing production and management lifestyles and models, for individuals, private companies and public bodies.

- Goal 13. Fight against Climate Change. The objective aims to adopt urgent measures to combat the consequences of climate change, strengthening resilience and adaptability to climate-related risks and environmental and natural disasters. In this respect, it is suggested that both concrete law enforcement and awareness-raising measures be integrated into member states' policies and strategies.
- Goal 14. Life Underwater. To conserve and use marine resources sustainably by combating and reducing pollution, protecting and restoring ecosystems, combating destructive forms of fishing.
- Goal 15. Life on Earth. To protect, restore and promote the sustainable use of the earth's economy, forests and biodiversity through proper forest management, the fight against desertification and soil degradation. The objective is aimed at all-natural habitats, both wet and dry, the protection of which should be the primary objective of the planning tools.

This premise, together with the focus on the United Nations Sustainable Development Goals, is necessary to highlight two elements that will be used in the following paragraphs. The first concerns the general perspective in which to frame sustainable development, for which all systems are connected and the change made on one has a reflection, albeit imperceptible, on the others. The second concerns the complex, multi-scaled and multidimensional spheres of sustainable development, which require the adoption of holistic, rather than sectoral, approaches.

3   Fields of Action and Dimensions of Sustainability

The complexity expressed by the concept of sustainable development leaves ample room for experimentation, both theoretical and practical, for the formulation of approaches and methodologies able to optimize production processes. Specifically, the authors operate in the academic field in the field of Architecture Technology with particular attention to the environmental quality of indoor and outdoor spaces, residential buildings and public spaces. In these areas, the fallout from some sustainable practices integrated into the processes of new construction, redevelopment and architectural and urban regeneration will be deepened.

3.1   Sustainability, Architecture and Production

The world of industrial production and that of construction and architectural production are closely linked. In this sense, we can talk about two systems that are partially overlapping and mutually driving, being also areas of the macro-economic and market system. This causal link is not secondary to the issue of sustainability. The construction production process generates a demand to which the industrial system responds drawing on scientific and technological progress and producing materials, components, building elements. The field of architecture is, therefore, able to "generate" production, consumption and services that are not always ecological or sustainable, which in turn produce toxic waste, pollutants and substances harmful to man and the environment.

The latest regulations introduce sustainability principles in the building process, oriented to regulate the relationship between the building and the natural system, the optimization of material and technical-constructive characteristics to protection needs, the use of passive building operating systems for lighting and natural air conditioning. The construction sector has been the main instrument of building speculation and over-consumption of energy resources. To date, the contemporary city's housing heritage is a thermal trap, highly energetic and dependent on non-renewable energy resources such as oil and gas.

The theme of energy consumption in architecture concerns not only the heating and cooling systems of buildings but also considers the flows of energy and materials for the production of the building. Tackling energy efficiency in use alone can lead to neglect of the wholesomeness and comfort of indoor environments, maintainability and durability. In this sense, when energy-saving measures do not respond to a system strategy, I end up having the opposite effect, negatively affecting other aspects, such as indoor air quality.

The contribution of research in this direction is making enormous progress, through the formulation of interpretative frameworks and proposals to address the problem of quality in environmental construction. Nevertheless, there is a difficulty in producing project control tools that can cover the different thematic areas and disciplines involved. The results obtained are often strictly thematic and fail to address the problem in its complexity.

Another problem relates to the slowness with which the ecological transition in the construction sector is taking place. This delay can be due to the leading role of the design phase and the role of the professional who is often unable to properly combine the issues of economic development with strategic environmental budgets for the conservation over time of natural resources and systems. On the other hand, the pressures exerted by the numerous economic crises have led to the material and technological choices made for new construction being driven more by the production sector than by creative genius or sustainability requirements.

In the 1970s, in connection with the energy crisis linked to the Crisis in the Middle East and its impact on the oil market the increase in the cost of oil used for heating systems suggested that consumption should be reduced as much as possible through better insulation of buildings, by reducing ventilation and recirculating indoor air in larger building complexes. On this occasion, at least 60,000 insulating substances and laying techniques consisting of substances of chemical origin whose harmfulness had not been verified were introduced. The removal of drafts and the blocking of window openings have led to a reduction in the minimum standards of indoor air quality. This episode demonstrates how a wrong design choice can have serious repercussions on the environmental quality of the planet and the quality of life of individuals. The market offers solutions that address the problem according to logic and dynamics that do not always meet the needs of ecological transition. Instead, the project must drive change. The greatest constraint on the widespread diffusion of sustainable technologies for energy saving lies in the fact that the basis of scientific knowledge for their application takes place in a language that is not that of designers but that of physicists or systems engineers.

The green or sustainable value that many professionals attribute to their projects lies, unfortunately, in the integration of "eco-gadgets" (solar collectors, wind generators, photovoltaic panels, etc.) or vegetative inserts. In some cases, alternative technologies and materials with eco-labels are adopted, but the design process has not changed. Even when technology provides us with application software to monitor the energy consumption level of buildings, they fail to question the process, since the resulting design outcome concerns the increase in the performance of a single technological element to rebalance the algorithm and fall within the predetermined values. These operations are no longer sufficient: it is not a question of adding a requirement or a particular technology, but of integrating the environmental theme into all choices. We are paying the price of having built, for too long, in the complete indifference of conditions to the context, without paying attention to the waste of energy or using materials but at the same time denying its nature.

The approach of bio-construction or bio-architecture, on the other hand, has the merit of questioning the principles of the design process, with a focus on renewable energy sources and the integration of natural materials with construction techniques with low energy impact. Eco-compatibility is sought at all stages of the life cycle of the building, both newly built and in the case of redevelopment. In addition to the technological aspects, the principles of bio-construction provide for an in-depth study on the environment concerning orientation, exposure, soil composition, arrangement of surrounding buildings, the number of trees, maximum and minimum temperatures, humidity and everything that can be considered according to the optimization of the building process and the performance of the artefact. A bio-architecture should allow natural ventilation and maximize 'natural' heating and cooling through a prevalent exposure to the south of the environments and effective material, technological and plant choices. The presence of large green areas in the vicinity of buildings has not only an aesthetic or recreational value but contributes to the overall cooling of the external environment and the shading of the interiors. Another fundamental factor is that of water-saving, providing technologies capable of conveying rainwater or white water for irrigation of gardens or toilet drains. The most recent evolution of the concept of bio-construction comes from the experimentation of Biotope, a low-carbon house entirely made with the waste of agricultural processing. The prototype was made by Danish company Een Til Een and is located in Middelfart, DK. Woods, weeds and waste from agricultural production are used, impregnated in an organic resin that increases its strength and durability. From this conglomerate, a construction panel is made.

Such searches are becoming more and more widespread. In 2017 the design company ARUP published the report "The Urban-Bio Loop" which illustrates how organic food waste is recycled as building materials and bio compounds. These elements can be used, in building an organism, such as internal partitions, sound-absorbing panels, flooring, furnishings, heat-insulating panels, façade coverings. Many natural fibres can be used: flax and hemp, cellulose, corn, sunflower, sugar cane, peanuts, bananas, potatoes, rice and wheat, pineapple. Each of these fibres has different characteristics that meet performance requirements that can completely replace traditional and polluting building components (Arup 2017).

3.2 Sustainability in the Urban Environment

The issue of environmental sustainability in an urban context can be addressed starting from the very recent debate on the pandemic crisis. Although the two topics may seem different and distant, there are numerous features between the Sustainable Development Goals, air quality, urban space and the containment of pandemics. The post-pandemic perspective follows and summarizes what are the fundamental themes of urban environmental sustainability, in the close correlation between the quality of spaces and the quality of life (De Capua et al. 2020).

Liveable, walkable cities with parks, public spaces for sport and outdoor recreation are a fundamental factor in countering two phenomena that put the health of the environment and individuals at risk. The presence of green spaces, with mixed vegetation, high and low, allows to purify of the air and cool the temperature in the case of areas affected by strong volumes of vehicular traffic. There is a correlation between prolonged exposure to pollutants present in the air, such as nitrogen dioxide (NO2), and covid-19 mortality cases (Ogen, 2020). Nitrogen dioxide is a toxic component that enters the atmosphere because of the burning of fossil fuels and it is associated with numerous risk factors such as hypertension, cardiovascular diseases, obstructions and pulmonary and respiratory dysfunction in children and adults, with serious effects on diabetes and the immune system (Ibid.). According to the WHO (World Health Organization, 2003), the health risks produced by NO2 and its derivatives are real and the world's population should be protected from exposure to these pollutants. Mapping the presence of these pollutants also showed that there is a spatial coincidence with the outbreaks of contagion (Ogen, 2020).

Similarly, prolonged exposure to particulate matter also produces numerous risks. In early April 2020, the Department of Biostatistics at Harvard, Boston, had tracked an increase in mortality rates concerning the presence of particulate matter (Wu et al. 2020). This data is confirmed by the Italian Society of Environmental Medicine (SIMA) that have found traces of RNA of the coronavirus on particulate matter, which would then be able to transport it.

Air quality in the urban environment has always raised the concerns of policy, environmentalists and citizens about the consequences for public health. The present context invites the reflection on urban ecology and indicates drastic measures to fight air pollution, encouraging the identification of "clean air" areas within residential areas, to minimize car travel and implement green grass surfaces – able to trap particulate matter – along the busiest roads, and to encourage a healthier lifestyle, also in carrying out sports activities daily. Local authorities are making great efforts to equip cities with greener, environmentally friendly and low-emission means of transport and want to encourage the use of sustainable individual means of transport as bicycles. From a design point of view, this implies the implementation of a safer cycle-pedestrian network that allows citizens to reach the main places of interest without the use of the car as the sole option.

The first lockdown in March 2020 demonstrated how the forced isolation of the majority of the population, associated with the closure of numerous production activities, has significantly reduced emissions of pollutants into the atmosphere (De Capua et al. 2020).

A further issue that emerged during the pandemic crisis is the growing interest of citizens in public spaces belonging to residential buildings, the care, use and maintenance of open and green areas. The need to spend a lot of time inside the apartment has allowed citizens to rearrange the functions and the importance of the home space, or the enjoyment of the free time as for the view from their window, but also their ecosystem and productive potential.

Increasing the presence of urban green areas can bring numerous environmental benefits. Firstly, it helps to reduce the heat island effect by lowering temperatures and rebalancing the thermal gap between urban, peri-urban and rural areas. A greater presence of vegetation implies an increase in permeable surfaces and relative absorption capacity of meteoric waters. Lawn-treated plant surfaces can catch particulate matter. The combination of parks and urban gardens, through the introduction of different crops and species, can increase biodiversity. In addition, these activities play a role as a social catalyst, having a positive impact on the cultural aspects of sustainability. "Investing in the culture of the garden means engaging in a widespread action of food education to raise awareness of the seasonality of productions, respect for the environment and the real characteristics of the food that is brought to the table", claim Carmelo Troccoli, the Director of the campaign "Campagna Amica" by Coldiretti, emphasises that "it is a need for knowledge increasingly felt in the cities".

In general, the construction of a green infrastructure helps to provide a wide range of ecosystem services capable of producing social, ecological and economic benefits, off-balancing the presence of "grey" infrastructure (roads, railways, buildings, etc.).

3.3 Circular Economy

The economist Herman Daly (2002) — one of the founders of the ecological economy — ironically defines sustainability as an annoying "buzz" in which he points out all that is good and suggests some principles for a finally sustainable economy:

- Exploiting renewable resources at a rate that does not exceed the regeneration capacity of the ecosystem;
- Limit the use of resources, to produce less waste that can be absorbed by the ecosystem;
- To exploit non-renewable resources at a rate which, as far as possible, does not exceed the rate of introduction of renewable substitutes.

The above takes up the principles of the circular economy applied to construction, in which in addition to limiting the use and exploitation of resources, a strategy of selective demolition, recycling and reuse of building components is implemented, treated to be re-entered into the production circuit. According to the Ellen MacArthur Foundation, the circular economy is designed to be able to regenerate itself: by emulating biological flows, capable of being reintegrated into the biosphere, so that technical flows are destined to be revalued without entering the biosphere.

In this sense, the circular economy sees the city as a mine of materials, according to the principle of urban mining, in which the building is understood as a material bank from which to draw elements and materials to be reused or recycled as secondary raw material. As described in the previous paragraphs, this principle can be applied to different urban scales in terms of mining, and to different levels of production in terms of recycling. The most futuristic aspect is the re-entry of the waste of demolition into the production process of a new product or material. Although the process looks very similar to that for bio-construction, not all demolition waste can be recycled.

For this to be possible, the building must possess particular material, physical or technical characteristics. Wet construction technologies can be partially recovered as long and costly processes as a result of the elimination of materials that could compromise their reuse, such as plant plastic components, polyurethane foams, waterproofing sheaths, mortars and adhesive glues. Dry construction technologies lend themselves best to the principle of disassembly (or Design for Disassembling) which are a source of post-consumer materials. On the other hand, those which are recovered from the waste from the processing of industrial products not necessarily intended for construction are called pre-consumption. Among the latter, recycled plastic fabrics are made from the waste of the tetra pack to make the mooring poles of the Venice lagoon (GBCI, 2019). Similarly, wood and glass fibres are used to produce different types of materials and insulating panels, as well as from heterogeneous plastics it is possible to produce, by stamping, flooring coating panels.

These materials are not ecological in themselves, in the sense that they are highly part of the traditional industrial production system, with the effects in terms of emissions that these entail. On the other hand, the principles of the circular economy make it possible to call into question the production process and the building process at least in terms of reducing processing waste.

4 Conclusions

Despite the limitations of the production process and the construction process towards the ecological transition, it is necessary to underline the important goals achieved in all the priority sectors. Measures to make transport systems greener; improving the energy efficiency of buildings; revisions to eco-friendly design directives and eco-label; progress in research and the channels of finance for industrial innovation. However, negative trends persist, including the still growing demand for natural resources and the increase in energy consumption in transport.

The fundamental objectives of the international community are clear, to pursue sustainable development containing climate warming, the path to follow is not yet clearly outlined. The private market invests funds to ecologically improve production processes and use more renewable energy sources.

What is the complex system of principles that animate the path to sustainability today? On the one hand, the ones aimed at the protection of the health of people and environment. On the other, social and economic concerns, that affect the cultural dimension of use, consumption and production patterns. In any case, these perspectives imply a change necessary for the protection of future generations. Unfortunately, these principles are often overstated by market interests. For example, the Dow Jones Sustainability

Index (DJSI), one of the world most prestigious indexes, is a global reference for sustainability that has recently rewarded companies such as Terna, Finmeccanica, Fiat, Enel and Pirelli. It is somehow interesting that the industrial sector, without the support of a strong regulatory and control policy, can produce a realistic auto-evaluation of their progress in sustainability. Of course, the market can and could drive a change, but not if that is increasingly distant from the real demands of sustainability in the broadest sense.

Certainly, the ecological and social transition to sustainability must move from a renewed concept of a resource that calls into question every area of consumption. Starting from the foundations of the building process is an excellent beginning, for it is responsible for 50% of neat energy consumption. On the other hand, the transition to a new paradigm implies a multi-dimensional perspective in solving such complex problems. Also specifically in the construction sector, the rethinking of conventional and standardised solutions requires the adoption of an integrated strategy, capable of optimizing the exploitation of natural resources and systems, the protection of health and that of the environment. Without questioning our development models, we will continue to find temporary, and therefore unsustainable, solutions to lasting and increasingly intricate problems.

References

Arup: The Urban Bio-Loop- Growing, making and regenerating (2017)

Brundtland, G.: (s.d.). Report of the World Commissions on Environemnt and Development: Our Common Future. United Nation General Assembly Documenti a/42/427. (1987)

De Capua, A., Errante, L., Palco, V.: Co-existing with COVID-19. Design for resistance. Uurban Studies and Public Administration 3(2), 69–86. (2020)

Daly, H.: Beyond Growth. Einaudi (2002)

General Assembly of the United Nations: Transforming our world: the 2030 Agenda for Sustainable Development (2015)

Green Building Council Italy: Circular economy in construction (2019)

Hardin, G.: The tragedy of the commons. Science, New Series **162**(3859), 1243–1248 (1968)

Ogen, Y.: Assessing nitrogen dioxide (NO2) levels as a contributing factor to coronavirus (COVID-19) fatality. Science of the Total Environment, 1–5 (2020)

Ostrom, E.: Governing the Commons: The Evolution of Institutions for Collective Action. Cambridge University Press (1990)

World Health Organization: Health Aspects of Air Pollution With Particulate Matter, Ozone and Nitrogen Dioxide: Report on a WHO Working Group. Bonn, Germany (2003)

Wu, X., Nethery, R.C., Sabath, B.M., Braun, D., Dominici, F.: Exposure to air pollution and COVID-19 mortality in the United States: A nationwide cross-sectional study. medRxiv. (2020)

Take Care. The Body of Architecture, the City and the Landscape

Francesca Schepis[✉] [iD]

Architecture and Territory Department, University Mediterranea of Reggio Calabria,
Via dell'Università 25, 89124 Reggio Calabria, Italy
francesca.schepis@unirc.it

Abstract. The considerations presented are intended to offer a point of view on an aspect proposed by the session: *Cultural sustainability, urban commons and care of places*. And to do so, in particular, with respect to the relationship of scale of architecture, of the city, of the landscape framed in the broad theme of urban design in Italy. Starting from some theoretical reflections and from some evaluations of exemplary design experiments, possible intervention strategies applicable to the regional territory in compliance with the principles of sustainability are outlined, focused on the enhancement of the landscape and economic heritage of the city in which they are inserted and vaults the improvement of the quality of life, accessibility and mobility, functional *mixité* and the promotion of social relations.

Keywords: Landscape · Transcalarity · Sustainability

1 Foreword

"The City is like a Big House, and the House Itself is a Small City"
Leon Battista Alberti, De Re Aedificatoria, 1452.

"Only if We Have the Ability to Dwell Can We Build"
Martin Heidegger, Building, Dwelling, Thinking, 1951.

In 1452 Leon Battista Alberti presented to Pope Nicholas V what is fully considered the first treatise on modern architecture, the *De Re Aedificatoria*. In opening the ten books that make up the text, Valeria Giontella, Italian curator of the latest version published for the types of Bollati Boringhieri, invites not only the historian but also the architect to read Alberti's pages. *The art of building (L'arte di costruire)*, this is the translation chosen in relation to the more widespread *The Architecture (L'architettura)*, introduces a lively world in which theoretical and scientific knowledge are combined with the construction technique and attendance at the construction site [1]. The professional figure that Alberti outlines has traits that can still be recognized as valid, rich in disciplinary skills but also open to the ability to administer the various components that affect the definition of a good architectural project. The first of the books has a defining character, "the lineaments", as if the author first wanted to establish a common vocabulary of terms to provide readers with a shared basic knowledge. In listing the six parts that make up the architecture – the region, the territory, the distribution of the rooms, the walls, the roof and the openings – he also intends to bring the complexity of knowledge and construction back to a few simple recognizable principles. The communicative power and expressive audacity of this thought can only be compared to the enunciation of the well-known five points of Le Corbusier. The region is mainly intended as a portion of the territory in which the project is to be established and includes, together with the morphological aspects, also considerations relating to the right orientation and correct exposure, the ease of drawing on primary resources and the quality of the environmental components.

Through a series of examples Leon Battista Alberti shows how the choice of an adequate building site represents the primary condition not only for the construction of good architecture but, certainly, also for guaranteeing a better quality of life for future inhabitants. Before going on to illustrate the aspects close to the scale of the building, to which the other five parts are dedicated, he intends to offer a reflection on a larger scale, trying to find a mediation between the urban and environmental aspects and those inside the house and housing. The centuries that separate us from these considerations, even if projected into a totally different context, do not in any way reduce their speculative value and disciplinary scope.

The fruitful season of urban design in Italy which for about fifty years, from the post-war reconstruction to the beginning of the new millennium, has seen the succession of multiple theoretical positions together with prolific experimental achievements, has been the testing ground on which both architects and urban planners have measured themselves by trying to find more points of contact than to highlight scientific differences [2]. Yet the rapid growth of cities linked more to political, economic and social reasons than to precise planned planning choices has, in fact, excluded architects and urban planners from urban formation processes, and from the almost uniform scenario that unfolds today in our eyes appears more like a homogeneous and indefinite expanse of isolated, autonomous, identical buildings, extraneous to the context in which they are located. This situation knows no regional boundaries, to the point that the Italian territory – just to limit our reflection to the national scale – can be assimilated to a single large periphery where historical and monumental centres and the few pieces of landscape not yet built can be found distinct [3].

2   The Measure of the Contemporary City. The Case of Calabria

In the transition to the new millennium, the trust in the urban project, as an elective scale to measure the passage from the house to the city and as an operational tool for the mediation between theory and construction, has gradually diminished and the positions of the critical-disciplinary debate have lost their meaning, their decision-making power, limiting themselves to recording the phenomena in progress. Without a definable design dimension and also suspended by the temporal scansion, the transformations of the man-made space, flattened in global conformity, concern individual buildings that incorporate the idea of the city in the bigness of the magnified self-referential object; or they abandon themselves to the contemplation of the landscape in a rewriting of the zero degree of architecture; they still become ephemeral manifestations of fashion, aware of having renounced the idea of lasting forever [4].

The contemporaneity has accustomed us, therefore, to sudden jumps between the geographical dimension and that of the accommodation, forgetting the relationship between the morphology of the territory and the urban fabric which also seems to want to regain a central theoretical and technical position when we are called to confront ourselves in an unavoidable way with the existing city [5]. The idea of economic, social and energy sustainability passes from the ability to grasp the limits imposed by the lack of unlimited resources as fixed points around which to build a renewed thought in terms of the relationship between home, city and landscape.

Like the entire Italian peninsula, the state of the architecture and urban aggregates of Calabria does not present a better condition, further exasperated, at times, by an endemic delay, typical of the southern regions, to the increasingly rapid response times required of the political, social, technical action to intercept the useful economic resources to invest in the building heritage so as to keep up with rapid social transformations.

The Calabrian territory offers very different landscape scenarios, a coastal development close to 800 km, a mountain scenery with elevations even higher than 2,000 m above sea level, a fragmentation into valleys determined by the scanning of the river beds, three large-scale productions that they divide into as many great stretches from north to south. The remarkable territorial heterogeneity is superimposed on a dusty diffusion of distinct inhabited centres in over 400 municipal administrations, which can include the complex size of metropolitan cities up to the fraction of a few dozen families. All these aspects highlight the difficulty of being able to offer a single and unambiguous mode of intervention as regards the issues of territorial planning, urban definition, resource management. The social phenomena of migration affecting the entire Mediterranean basin, counterbalanced by a continuous exodus of young people who leave Calabria for better study and work opportunities, as well as by the permanence of cultural and linguistic minorities, make the regional picture even more varied that resist even in a time of total global crushing.

It is therefore logical to believe that the remarkable differentiation of places and inhabitants corresponds to a plurality of contemporary ways of dwelling.

3   Developments and Expectations

The study proposed here intends to think about the relationship between architecture, city and landscape, defining it starting from the Italian debate on the broad theme of urban design. The term of *unity* must be understood, as easily deducible, in the meaning given by Giuseppe Samonà in the well-known collection of essays *L'unità architettura urbanistica* [6]. The reasons for this authoritative reference lie in the conviction that the current condition of disciplinary fragmentation – imposed exclusively by the needs and provisions of the teaching system – has, in fact, divided not only the competences but also the general vision of architecture. As if the individual parts that make up a project can from time to time be treated from a specialist sector point of view. The direct relationship between housing and the architecture of the city is considered by Samonà – and widely shared within the animated debate on the urban project in Italy – a logical consequence based on a rigorous organic vision of the space in which human life takes place. He reminds us repeatedly that the predisposition of places for dwelling – not the distinct reasons of an economic, political, functional, social, etc. – is the main objective of the architectural project.

Unity recognized not in the homologation of a common language but given by the quality of theoretical research, design practice and construction technique at the same time. The constant reference between the different art forms – with generous exchanges with the visual, literary, photographic, cinematographic ones, etc. – laid solid foundations for the development of Italian architecture. After the years of reconstruction, there was a need to take stock of the knowledge and collective sense of what had been produced up to then and of what still had to be done.

The current state of the Italian and southern city forces us today to rediscover a unitary thought for the project, a non-partialized vision on how to imagine the future that is close to us. The key to sustainability, understood in its broadest sense, offers us the opportunity to read the processes of political, economic, social transformation and to borrow them into an idea of space. Necessarily open, plural, metamorphic, a sort of operable urban matrix.

From a comparative study between the state of ERP in Italy and the more contemporary management of Social Housing, substantial differences emerge and some data on which it is appropriate to dwell. The total building capital built in Italy with public funds amounts to approximately 1,000,000 housing units, 70% built in the early 1980s. It is immediate to note that a substantial part of these requires urgent interventions which, starting from extraordinary maintenance, may involve rethinking interventions, including typological and functional ones, or even the total revision of the plant component up to the seismic safety of the structures. This figure joins the shared goal of recovering existing assets and reducing land consumption.

On the other hand, the resources allocated by the "Fondo Investimenti per l'Abitare" (AIF) managed by "Cassa Depositi e Prestiti", which adheres to it together with the "Ministero delle Infrastrutture e dei Trasporti" and private individuals (banking and insurance groups and private pension funds), envisage both the construction of new residential structures and the recovery of existing ones. Here it is only worth remembering that the distribution of the Fund on the national territory does not take place in an equitable manner and that only 7% of the resources are destined for the South, against 68% for the North and 25% for the centre, by reason of what was explained by "Cassa Depositi e Prestiti" about "the different availability of subjects present on the territory able to present projects that are coherent in terms of implementation and at the same time profitable" [7].

A recent survey conducted by Tortuga – a think-tank made up of students, researchers and professionals dealing with economics and social sciences and offering professional support to institutions, associations and companies – highlights the different distribution of new residential interventions in Italy, framing it in a broader study on the relationship between public and private resources. The controlled opening to private investment funds would represent for the Italian panorama, like the best European experiences, a not negligible economic driver. ANCE - National Association of Building Constructors and ENEA - National Agency for New Technologies, Energy and Sustainable Economic

Development also come to similar considerations, which converge in the belief that the development of the country passes from the production scientific research, design experimentation on the typology, on innovative materials, on the technologies to be used and on the renewable energies on which to invest, on the natural resources to preserve.

A decisive role in this is played by "Federcasa" – borrowed in 1996 by the "Associazione Nazionale Istituti Autonomi per le Case Popolari" (ANIACAP) of 1950 – and "Federcostruzioni" as well as the Social Housing Foundation, promoter of the FIA among other things. The Social Housing Foundation, which works on the national territory, is responsible for promoting overall intervention strategies – from the planning phase to the social participation in the proposal phase, from the finding of financing funds, including private ones, to the management of public tenders for the realization of works – in the field of social residence, it has identified and categorized different types of housing needs. From traditional accommodation to collective residence for the elderly; from the temporary home for a single individual to the student residence, the requests for co-working and co-dwelling accelerate the promiscuity of forms of dwelling in a single building that can no longer be defined by residential typologies based on standardized families.

A fundamental aspect of the question is added to the rectification of the construction systems and plant elements, represented by the typological redefinition of the buildings and interior spaces of the house – up to the mutation into hybrid architectures, given by the typological combination of aggregations in line, in rows, in tower, with a balcony also all co-present within a single settlement – and from the reconfiguration of the external spaces of relationship common to the inhabitants of the same building – there is a frequent tendency to convert courtyards into productive cultivated gardens that open in the form of public gardens for all the citizens of the neighbourhood, in a rediscovered vision of the scale of neighbourhood unity [8].

Finally, the quality of regional sustainable dwelling, independent of the size of the intervention, can be supported by some constants to be considered in the design phase on the possible ways of interweaving relationships with the context. Only to limit ourselves to exemplifying three spatial conditions, which can then assume the most disparate configurations and plastic articulations and material definitions, and which define the thresholds of interior, intermediate and exterior.

These spaces represent the sensitive part of the architecture as they stabilize relationships and references with the surrounding context, vehicles of communication and an opportunity for new social aggregation. The spaces intended for common life can thus be declined on at least three levels:

- spaces inside the architectural organism, such as multipurpose common rooms, also intended for co-working, study rooms, kindergartens or meeting centres for the elderly of the community;
- intermediate spaces of relationship between the building and the outside, such as common loggias placed on the different floors of the buildings and the landscape that are placed in relation with the city, or even on the architectural ground floor of the architectural organism that opens onto the external green spaces and relate to the mobility and accessibility of the area (Fig. 1);

– outdoor spaces, which manage the direct relationship with the city at different degrees of pedestrian or vehicular accessibility. These spaces are also those dedicated to the implementation of participatory actions such as the management of urban gardens or games for children, but also for the management of energy resources (waste, materials, supply of alternative plants,…).

Fig. 1. Le Corbusier, *Immeubles-villas*, not located, 1922. Extract from Le Corbusier and pierre jeanneret, *Œuvre complète*, volume 1, 1910–1929. © FLC/ADAGP.

The idea that the common spaces are taken care of by all the inhabitants of the building is a cardinal principle. This is true for common rooms and common loggias, but it is fundamental for outdoor spaces. Each resident will be able to have a part of the land destined for a vegetable garden and cultivate it. Shared management will root a sense of belonging and respect for the place. Furthermore, as already expressed above, a more intense relationship between the inhabitants of the place and the spaces for common life, whether they are internal or external to architecture, expressing the guidelines that are being implemented in the innovative paradigmatic projects in terms of renewed sustainable sensitivity (Fig. 2).

Fig. 2. Le Corbusier, *Immeubles-villas*, not located, 1922. Extract from Le Corbusier and pierre jeanneret, *Œuvre complète*, volume 1, 1910–1929. © FLC/ADAGP.

4   Conclusions. *Take Care*

On the consideration of the heterogeneity of the regional territory, the casuistry expands if, to the already existing collective housing complexes, all the state-owned and religious structures in decommissioning or entire small communities no longer inhabited are also added, just to identify the two extreme conditions.

The functional reconversion of military or conventual artifacts inside the consolidated city represents a great potential for the national panorama. These types of architecture are partly in the process of being decommissioned or abandoned to the texture of the thin lines of the fabric, in evidence as a scale in relation to the thinness of the residences. The architecture of the second wave of the industrial revolution erected between the end of the nineteenth century and the beginning of the twentieth century can also be assimilated to military or religious complexes due to the size and position in the territory.

In recent years we are witnessing the proliferation of small-scale experiences of considerable cultural interest and of high symbolic value, such as the recovery of abandoned historic villages by public and private investors, implemented through a political-cultural project. total production. The smaller towns and inland areas that never cease to emanate an ancestral sense of dwelling and of the constructive knowledge handed down, are, like the large designer residential blocks, counted among the fields of investigation and of greatest interest [9].

Redevelopment and recovery, architectural and urban regeneration, retrofitting up to demolition and reconstruction methods are the operational strategies that can be considered to apply in the field of urban transformation starting from the thick expanse that characterizes cities. Just a mention on the method of demolition and reconstruction that could lead to significant results, certainly, in terms of architectural and urban quality, but also in terms of energy performance and seismic safety. The expenditure of means necessary for the plant-technological and seismic overhaul for adaptation to the most stringent current legislation, for example being limited to those architectures, residential complexes or urban parts to be preserved for their recognized architectural-compositional quality by intervening in a more decisive way where recovery is not strategically convenient. Reference is not made here to the historical heritage, to fine architecture, to historic centres, even monumental ones that deserve a different consideration, but to the restoration of the modern. Term that concerns the careful work on most of the urban fabric and buildings that make up our cities today. Not a conservation in a historical-philological sense but a reformulation downstream of the changed needs of the society that inhabits them.

References

1. Giontella, V.: Introduzione. In: Alberti, L.B. (ed.) De re aedificatoria (1452), I Libro, cura ed. it. Giontella, V.: L'arte di costruire. Bollati Boringhieri, Torino (2010)
2. Ferrari, M.: Il progetto urbano in Italia. 1940–1990. Alinea editrice, Firenze (2005)
3. Neri, G.: Due modernità, un progetto. In: Tornatora, M., Schepis, F. (a cura di) Rigenerare. Strumenti e strategie di progetto per un abitare sostenibile. Progetti per Gioia Tauro. Librìa, Melfi (2013)
4. Gregotti, V.: Tre forme di architettura mancata. Einaudi, Torino (2010)
5. Corbellini, G.: Housing is back in town. Breve guida all'abitazione collettiva. LetteraVentidue Edizioni, Siracusa (2012)
6. Samonà, G.: L'unità architettura urbanistica. Scritti e progetti (1929–1973). Franco Angeli, Milano (1978)
7. Tortuga, Edilizia pubblica: a che punto siamo, cosa possiamo fare, Econopoly. Numeri idee progetti per il futuro, blog de "IlSole24ore". https://www.econopoly.ilsole24ore.com/2019/06/24/edilizia-pubblica-cosa-fare. last access 24 June 2019
8. Schepis, F.: Un milione di case. Modelli dell'abitare sociale. In: Calzolaretti, M., Mandolesi, D. (a cura di) Rigenerare Tor Bella Monaca. Quodlibet, Macerata, pp. 231–233 (2014)
9. Lauria, M. (a cura di): Che fine hanno fatto i centri storici minori? Atti del Seminario di studi. Centro stampa d'Ateneo, Reggio Calabria (2009)

Service Innovation: A Literature Review of Conceptual Perspectives

Giulia Freni[✉]

Mediterranean University of Reggio Calabria, Reggio Calabria, Italy
giulia.freni@unirc.it

Abstract. In recent decades, theoretical and empirical contributions on the topic of service innovation have increased significantly, creating an established and growing body of research. In fact, the debate has been enriched over time, feeding on the socio-economic changes happened on recent times, opening up new interesting frontiers of theorization and a starting point for future ones.

Despite the amount of research on the subject, the debate has evolved along lines not always shared by all the scholars who, frequently, propose different conceptualizations, definitions and theories regarding service innovation; there is no common core concept at the moment. This paper examines the different lines of theorization related to service innovation research which has occurred over the decades, explaining the perspectives and the ideas that generated them, and highlighting how each of those drove the discussion about service innovation. The aim of the paper is to propose a framework of the evolution of theories on service innovation, analyzing the scientific literature produced on the subject and distinguishing it into homogeneous categories, which share the same approach and objectives.

Keywords: Smart city · Services · Service innovation

1 Introduction

The theme of the smart city and all its related declinations – e.g. smart villages, smart lands – has been introduced over the last few decades and has been the subject of a debate that is actually at its most intense point.

Although the construction of the concept *smart city* is still a work in progress, the definition indicates a place where, through use of digital solution and ICT infrastructures, it is possible to optimise and make more efficient infrastructures and services for citizens.

The emergence of this new paradigm is not unexpected; it is the result of a 'historical-evolutionary' process of thinking that has changed for adapting to the changing times.

The progressive change in the way we live and the instruments at our disposal for city planning has generated theorizations on what should be the best way to shape the city.

F. Calabrò et al. (Eds.): NMP 2022, LNNS 482, pp. 154–163, 2022.
https://doi.org/10.1007/978-3-031-06825-6_15

These theories responded to the socio-cultural needs and potentials of the time in which they were generated. In recent years, the concept of the *digital city*, which originated from the synergy between the technological boom and its potential for urban development, has been replaced with a new urban ideal: the *smart city*.

This one means going a step further: the smart city is in fact a digital *and* intelligent city, which is not entirely driven by latest available technologies, but makes efforts to use them in optimised applications.

In the smart city, technology is not the main driver of the urban reality, but it's conceived as a useful tool to serve citizens (and not other way around).

Currently, the scientific community has focused on different perspectives of the smart city: technological, symbolic and, above all, collaborative. The principle of collaboration between all the players in the urban ecosystem is not sufficient, but it is necessary for an urban reality to be classified under the label of smart city.

In accordance of these conceptualisations, the concept itself of the smart city seems to be ascribable to the macro category of services, in line with the most recent theories developed in that field.

In their most recent conceptualisation, services are defined as "the application of specialized competences through deeds, processes, and performances for the benefit of another entity or the entity itself" [1].

It is evident how this definition allows a re-conceptualization of the concept of smart city: this, in fact, involves the use of ICT infrastructures ("specialized competences through deeds, processes, and performance") to obtain a synergy between a better quality of life for citizens ("for the benefit of another entity") and an efficient, sustainable, effective and interactive place ("or the entity it-self").

The smart city is also characterized by a collaborative logic, where citizens (like customers in services) interact with the public in order to co-create value and share knowledge for achieving a specific aim.

This conceptual framework can offer the possibility of having new tools available in the smart city study agenda; if the latter can be assimilated into the category of services, some of the studies that have been done on services, which are more substantial because they originated previously, can be a theoretical cauldron from which to draw.

Given that the debate on the smart city often focuses on the concept of innovation, this paper presents the evolution of studies on service innovation, highlighting the different perspectives that emerge from different contributions, both theorical and/or empirical. The analysis of these successional perspectives allows to explore the concepts and theorizations underlying them and to highlight the scientific point of breakthrough in which the service innovation research topic has become an autonomous one, emancipating itself from the earlier manufacturing based one.

2   Service Innovation: A Literature Review of Conceptual Perspectives

The term "service innovation" refers to innovation and to related practices that may be implemented in the diversified context of service provision [2], that may involve changes in various dimensions of an existing service [3]. Originating from the discipline of marketing, studies on service innovation and on the impact it may have are a recent branch of the scientific literature, which tends to become increasingly interdisciplinary and highly contributed [3–5].

Although innovation has traditionally been associated with the manufacturing sector and technology, shifting its focus analysing sectors that have always been considered disconnected from it is the objective towards which the scientific community has been moving for some decades now.

Over time, increased investments in manufacturing-related innovation have led to the emergence of several conceptualizations, analysis and measurement tools with a product, tangible and technological-based logic. The aim of these was to look for an innovative component strongly oriented towards technological progress and, consequently, towards increasing the value of the product output.

The socio-economic changes of the last decades of the twentieth century, the strengthening of the role of services in the contemporary economies and the theoretical rehabilitation of the concept of service itself have stimulated the desire to further explore the line of study regarding innovation related to them. It was hypothesised that the perception of services as non-innovative was a consequence of the approach with which they had always been analysed, based on the application of the same concepts and tools used to measure innovation in the manufacturing industry.

Services in fact have their own different configuration, which can hardly be standardised and categorised, because they are highly variable; in relation to this nature, innovation can take on a multidimensionality yet to be defined in its complexity [6].

The need to understand the nature of service innovation is thus seen as a priority field of study. Its importance is linked in direct proportion to the growing importance of services in modern societies, both independently and integrated into the offerings of manufacturing firms, according to a phenomenon so impressive that scholars have coined a neologism to define it: *servitisation* [7].

The debate, enriched by the cross-fertilization between different disciplines, has evolved along lines not always shared by all scholars who frequently propose different conceptualizations and definitions of innovation in services.

For ease of understanding, therefore, the studies that have been carried out over the years on the subject of service innovation are often grouped together in homogeneous groups according to a classification proposed by Coombs and Miles in 2000. They divided studies into three different categories, based on as many study perspectives. These vary in their basic assumptions, their different views on service innovation and their different objectives: assimilation, demarcation and synthesis [8]. Although Coomb and Miles were the first to introduce in the scientific literature such a successful nomenclature for the three different approaches to the study of innovation in services, revived in several review studies [3, 9–11], a few years earlier Gallouj and Weinstein distinguished the presence of two different approaches to the subject, one adopting a predominantly *technological* perspective, the other *non-technological* [12].

The first approach that can be outlined, defined as *assimilation*, is based on the analytical process of assimilating innovation in services to that of manufacturing, judging it as equal. This egalitarian approach leads to adopt in the analysis of innovation in services the same conceptualizations, tools and instruments for measuring innovation used in the more familiar – and well-known – manufacturing sector, focusing mainly on technological development. The prevalent attention when measuring innovation performance according to this approach is mainly focused on researching and analysing a purely visible type of innovation, whose key element is the introduction of systems, technological components and the development of ICT applications, evident in the output or process phase. This technological perspective has led that, in some studies, this approach (in its entirety, or in its most radical positions) is directly defined as *technologist* [10, 12], to underline its scientific focus. A work often associated with the technologist perspective is the Reverse Product Cycle (RPC) model of Barras, based on the theory that information and communication technologies have an enabling role in relation to service innovation, and focused on analysing the effects of the adoption of such technologies (which always have an external origin) [13].

Many studies associated with the perspective of assimilation were not born, however, with the specific intent of analysing services, but the use of the "technology" factor as a founding principle for the taxonomic schemes they theorised had, as a natural consequence, the subordination of those economic sectors in which innovative performance is more interdisciplinary and also includes non-technological factors, defining services as the "dark side" [14] of the economy. For this reason, the subsequent literature has almost unanimously indicated that these studies belong to the category of assimilation, thus formed in retrospect as a corollary of theoretical studies of originally broader scope and often different focus.

Pavitt, the author of one of the theories nowadays counted in the assimilation approach, has the objective, for example, of proposing a classification of manufacturing firms based on the different trajectories of the prevailing innovative activities, closely linked to the capacity – and competence – of these sectors in technological change [15].

Using a strongly inductive methodology, Pavitt proposes a taxonomy that identifies four major sectoral categories: *science-based firms*, *specialised suppliers*, *scale intensive producers*, *supplier dominated firms*. The first two categories are identified by Pavitt as strong active players in the production of development and technological change, while the last is indicated as dependent – in this sense – on the others.

Although this classification proposal was intended to be universally valid, applying a technology-based taxonomy to all economic activities does not allow to detect innovation trajectories in those sectors that do not rely primarily on technological change. Consequently, in Pavitt's taxonomy, all services would generally tend to be included among *supplier-dominated firms*, which are last in the classification as autonomous innovativeness.

In a study published in 2001, Miozzo and Soete offer a reinterpretation of Pavitt's classification in order to apply it to services. The aim is to propose a technological taxonomy of services, based on their technological links with other economic sectors [16]. Services are thus divided into three categories: *supplier dominated*, *scale-intensive physical and information networks* and *specialised suppliers/science-based*. While the first category is not particularly active in contributing to technological advancement because it relies mainly on other factors (such as the professional skills of the service provider), the other two are particularly sensitive to the development and application of technologies that have a particular impact on the growth of services.

Soete and Miozzo, therefore, are the first to question Pavitt's principle that all services belong to the *supplier-dominated firms* sector, isolating and categorizing separately the technology-intensive service sectors.

Although the study does not openly state what is meant by innovation (the objective, in any case, is not to categorize services on the basis of the level of innovation, but on the basis of technological characteristics), it is evident that the term *innovation* is used in the text as a synonym for *technology*; given the frequency with which these words are used interchangeably, it appears that the basic idea is a conceptual assimilation between the two, and this has led the subsequent scientific literature to classify this contribution as part of the assimilation approach.

The taxonomy proposed by Evangelista [17], which embraces and supports the pre-existing studies of Pavitt and Miozzo and Soete (from whom it takes up the conceptual scheme of innovation), proposes a taxonomy that divides services into four categories, in increasing order of innovation intensity: *technology users services* (services that are exclusively users of technologies developed externally by other sectors); *interactive and IT services* (services that develop innovative practices through interaction with clients); *science and technological based services* (services that generate innovation and new technological skills, which they can then export to other sectors); *technical consultancy services* (services that produce innovative activities and technological change designed specifically to meet clients' needs).

Even in this case, therefore, it is possible to find a conceptual assimilation between the term *innovation* and *technology*; the evidence is that the *technology users services*, close to the Pavittian archetype of *supplier-dominated services* that do not develop technologies autonomously, are defined by Evangelista as the least innovative group. They are, however, the most representative sample, being the most widespread group of services at the time of publication of the article in Italy and the one with the highest percentage of employees.

The conclusion of this analysis is the idea that, in terms of innovation, the service and the manufacturing sectors share more similarities than differences, reiterating the need to develop a common theory of innovation analysis, in view of the detected capacity of services to innovate by generating and/or adopting new technologies (especially thanks to the development and diffusion of ICT).

Is thus evident the analogy of this study with the leading theory of the assimilation perspective, which combines services and manufacturing with an approach focused primarily on technological trajectories and indicators and, for this reason, often judged as too limited to accurately describe innovation in services [9].

The *demarcation* approach is characterised by the aim of highlighting the specific characteristics of services, demarcating them from the manufacturing sector and researching specific theories. This study perspective appears, therefore, as totally antithetical to the assimilation perspective, advancing the intention to develop an individual field of study for services, creating from anew models built on those specificities of services perceived as distinctive.

A study by Djellal and Gallouj, published in 2000, is directed towards this objective of contributing to the development of an autonomous concept of innovation in services [18].

Their hypotheses are tested on the basis of data collected from a survey carried out in France in 1997, and confirm that in service companies the set of information, skills and knowledge that generate innovation often finds its origin in customers (this was the case for 76% of the survey sample) and in personal contacts, rather than in technological consultancy. In addition to this result, it is confirmed that innovation is not necessarily linked to the presence of a specialised department such as R&D, but rather to a flexible organisation and the collaboration of the customers themselves.

The relatively minor influence of the R&D department on the management of innovation in services was also noted in a study by Sundbo a few years earlier [19]; an empirical analysis had in fact allowed him to note that innovation - in the service sectors he analysed - does not rely on a specific department and scientific results. It is difficult, therefore, to indicate a unique management model of the innovation process shared by all sectoral typologies, since there is more than one; in the case, however, of medium-large service companies the organisation of the innovation process is described by Sundbo according to a model that divides it into four sequential phases (*idea generation*; *transformation into an innovative project*; *development*; *implementation*).

However, this schematic framing of the innovation process is not synonymous with a linear development. The trajectories of innovation are often confused, not easily programmable; however, it appears necessary to plan and manage - especially the more rational phases, such as development - in order to increase the chances of successfully carrying out the innovative process.

Services, thus, innovate; and technology is not always the only determining factor in this process, but frequently only a medium. The dimensions through which innovation can take place, in order to achieve new (or renewed) service functions, may be various, combined (or not) with each other: according to den Hertog's multidimensional model, for example, they may consist of a *new service concept*, a *new interaction with the customer*, a *new system of values or business partners*, a *new revenue model*, a *new technological or organisational service delivery system* [20].

The studies categorized in the demarcation approach often share some common aspects in terms of scientific methodology and objectives: they start from empirical research (analysing the theme of innovation in services according to a deductive methodology) in order to develop theoretical models on the basis of them and are typically aimed at identifying and proposing non-traditional types of innovation, specifically focused on specific service sectors.

Drejer, in particular, raises concerns about the demarcation approach, especially in relation to the risks that a perspective that focuses too much on the specificity of services may involve; the one of judging as exclusive to services theories that would also apply well to the manufacturing sector, and of proposing a concept of innovation that is too far away from the original economic conceptualizations [9]. However, it is frequent to find in such studies a mention of the chance that the results obtained may also be adoptable in the manufacturing context [18–20] (although it often starts from a thinking exclusive to services), in the view that, in the future, innovative practices will tend to converge increasingly [21].

The most recent synthesis approach - also referred to as integrative approach - consists in fading away the debate on the exclusivity of services and service-related innovation, in order to explore practices and theories with a completely integrated approach between services and manufacturing, in order to unify the two sectors, not by subordinating one to the other (as was usual in the assimilation approach), but by adopting a broader view that includes both. In this perspective, a number of studies that were carried out in the perspective of assimilation have been revised in order to update them to the synthesis approach to service innovation [23].

The starting point of these studies, however, is generally focused - empirically or theoretically - on the world of services and their characteristic aspects. Hipp and Grupp, for example, list six, most of which are not entirely new to the debate [22]. *Intangibility* (analysed from a practical point of view: intangibility makes it difficult to protect innovative service activities by patents and intellectual property, which may inhibit innovative initiatives); *the human factor and integration with customers* (which, more than anything else, require innovation that is often linked to non-technological factors); *the organisation of innovative processes* (innovation in services is often not systematic because it is not linked to an R&D department); *the types of innovation output; the structure of the service sector* (often smaller companies); *and regulatory issues.*

A pioneering paper on this integrative perspective is that of Gallouj and Weinster, published in 1997, who were among the first to propose a universally applicable model of innovation management and analysis [12]. In order to achieve this, the first step was to adopt an approach that overcame the traditional dichotomy between goods and services, proposing a product formalization that would be good for both sectors, on which to construct the factors generating innovation.

For this, they adopted the characteristic-based strand: a perspective that was generated by Lancaster's theory [24], according to which any type of good can be defined as a set of objective characteristics. A good may have several characteristics at the same time (in fact, this is the norm), which may, at times, be similar to those generated by another good, and may also vary over time and through combinations with other goods.

These characteristics become fundamental because the consumer does not buy goods as such, but precisely because of the characteristics they possess which, in his eyes, take on utility value.

Saviotti and Metcalfc [25] proposed to divide these characteristics into three fundamental typologies: the final characteristics of the good or service (the characteristics of the product seen from the consumer's point of view); the internal technical characteristics; the process characteristics (linked to the methodology by which a good or service is produced).

Gallouj and Weinstein start from this approach in order to extend it to services: the process characteristics are no longer considered as an autonomous typology (but are incorporated by the internal technical characteristics), and a new typology of characteristics is theorised, the competence characteristics (both internal, of the supplier, and external, of the client; they can originate from experience, interaction, updates, etc.). This new typology of characteristics is added to the original framework of Saviotti and Metcalfe in order to solve the particularly critical lack of service relationship elements.

The product, therefore, is no longer identified as a physical good or an intangible service, but becomes the result of the action of these vectors of characteristics; any change (whether planned or not) interacting with this system, influencing one (or more) terms of the vectors of characteristics produces innovation. The different ways in which such a change may modify the starting set-up of the system is reflected in different modes of innovation.

Although some studies make an almost identical subdivision of the theoretical perspectives on the service innovation topic, but under different headings and nomenclatures (and eventually highlighting some subcategories), the categorisation of existing researche proposed by Coombs and Miles, analysed here, is sufficiently comprehensive and describes the different lines of research concerning the relationship between services and innovation. These, even more synthetically, move along two main lines. The first one assumes the postulate of innovation as technical update and technological change, taking into consideration, above all, the technological component; the other one, which over the years tends to be increasingly free from the innovation/technology binomial, taking it into consideration but also accompanying it with further reflections on other indicators which can also generate innovation.

As noted, the trajectory of the scientific debate on innovation in services has followed a coherent development, comparable to a natural life cycle, whose maturation steps are represented by the three approaches described above [26], ideally placed in a chronological and propaedeutic order. The first scholars, naturally adopting a technologist and materialist perspective that better suited a society rooted in manufacturing, began to include the service sector in their studies, even if with a subordinate perspective, perhaps unconsciously paving the way for this field to be, sooner or later, analysed independently. This first "evolutionary" phase is slowly giving way to other perspectives that have a decidedly more service-oriented approach; the integration perspective, in particular, having passed the emerging phase, is in full expansion.

The chronological demarcation between the three approaches does not always seem to be so clear-cut. Although they have a different specific weight depending on the time, they are often coexistent and do not seem to have been preparatory to one another. On the contrary, the different perspectives continue to exist in parallel, because each study uses the one that fits best with the focus it intends to adopt [27]. There is no doubt, however, that research on the subject of innovation in services – regardless of the perspective adopted by individual scholars – has matured over time following an evolutionary trend. In fact, Carlborg distinguishes three different historical phases which, starting from the initial phase of formation (1986–2000), have driven research towards a phase of maturity (2001–2005) and, finally, of multidimensionality (2006–2010) [3]. As of today, the debate is far from over: the most recent researchs on the topic, however, tend to be less about proposing universal models and seem to be more about empirical analyses [28], which develop the theme of innovation in relation to specific contexts. It has been pointed out, however, that these types of research often do not dialogue with those of a purely theoretical nature [29], a synthesis that would be necessary, however, to implement knowledge on the gaps that exist today. The most recent literature reviews [3, 28, 30–32] have thus highlighted the need to continue investigating this issue [31, 32] and the challenges associated with it [33], in light of the advances that have been made [34].

References

1. Vargo, S., Lush, R.: Why "service"? J. Mark. Sci. **XXXVI**, 25–38 (2008)
2. Durst, S., Mention, A., Poutanen, P.: Service innovation and its impact: What do we know about? Investigaciones Europeas de Dirección y Economía de la Empresa **XXI**, 65–72 (2015)
3. Carlborg, P., Kindström, D., Kowalkowski, C.: The evolution of service innovation research: a critical review and synthesis. Serv. Ind. J. **XXXIV**(5), 373–398 (2014)
4. Ordanini, A., Parasuraman, A.: Service innovation viewed through a service-dominant logic lens: a conceptual framework and empirical analysis. J. Serv. Res. **XIV**(1), 3–23 (2010)
5. Toivonen, M., Tuominen, T.: Emergence of innovations in services. Serv. Ind. J. **XXIX**(7), 887–902 (2009)
6. Ostrom, A.L., Paraduraman, A., Bowen, D.E., Patrìcio, L., Voss, C.A.: Service research priorities in a rapidly changing context. J. Serv. Res. **XVIII**(2), 127–159 (2015)
7. Vandermerwe, S., Rada, J.: Servitization of business: adding value by adding services. Eur. Manag. J. **VI**(4), 314–324 (1988)
8. Coombs, R., Miles, I.: Innovation, Measurement and Services: the new problematique. In: Metcalfe, J.S., Miles, I. (eds.), Innovation Systems in the Service Economy, pp. 83–102 (2000)
9. Drejer, I.: Identifying innovation in surveys of services: a Schumpeterian perspective. Res. Policy **XXXIII**(3), 551–562 (2004)
10. Droege, H., Hildebrand, D., Forcada, M.A.H.: Innovation in services: present findings, and future pathways. J. Serv. Manag. **XX**(2), 131–155 (2009)
11. Vence, X., Trigo, A.: Diversity of innovation patterns in services. Serv. Ind. J. **XXIX**(12), 1635–1657 (2009)
12. Gallouj, F., Weinstein, O.: Innovation in services. Res. Policy **XXVI**(4–5), 537–556 (1997)
13. Barras, R.: Towards a theory of innovation in services. Res. Policy **XIX**, 215–237 (1990)
14. Gallouj, F.: Innovation in services and the attendant old and new mith. J. Socio-Econ. **XXXI**, 137–154 (2002)
15. Pavitt, K.: Sectoral pattern of technical change: toward a taxonomy and a theory. Res. Policy **XIII**(6), 343–373 (1984)
16. Miozzo, M., Soete, L.: Internalization of services: a technological perspective. Technol. Forecast. Soc. Change **LXVII**, 159–185 (2001)
17. Evangelista, R.: Sectoral patterns of technological change in services. Econ. Innov. New Technol. **IX**, 183–221 (2000)
18. Djellal, F., Gallouj, F.: Patterns of innovation organisation in service firms: postal survey results and theoretical models. Sci. Public Policy **XXVIII**(1), 57–67 (2001)
19. Sundbo, J.: Management of innovation in services. Serv. Ind. J. **XVII**(3), 432–455 (1997)
20. den Hertog, P., van der Aa, W., de Jong, M.W.: Capabilities for managing service innovation: towards a conceptual framework. J. Serv. Manag. **XXI**(4), 490–514 (2000)
21. Sundbo, J., Gallouj, F.: Innovation as a loosely coupled system in services. Int. J. Serv. Technol. Manag. **I**(1), 15–36 (2000)
22. Hipp, C., Grupp, H.: Innovation in the service sector: the demand for service-specific innovation measurement concepts and typologies. Res. Policy **XXXIV**(4), 519–521 (2005)
23. Chen, M., Chang, Y.: Service Regime and Sectoral Patterns of Innovation in Services: a Reinvestigation using Miozzo & Soete Taxonomy. Paper on Innovation, Strategy and Structure, Druid, Denmark (2011)
24. Lancaster, K.J.: A new approach to consumer theory. J. Political Econ. **XIV**, 133–156 (1966)
25. Saviotti, P.P., Metcalfe, J.S.: A theoretical approach to the construction of technological output indicators. Res. Policy **XIII**, 141–151 (1984)
26. Gallouj, F., Savona, M.: Innovation in services: A review of the debate and a research agenda. J. Evol. Econ. **XIX**(2), 149–172 (2009)

27. Kowalkowski, C., Witell, L.: Typologies and frameworks in service innovation. In: Bridges, E., Fowler, K. (eds.) The Routledge Handbook of service research insights and ideas, pp. 109–130. Routledge, New York. (2020)
28. Antons, D., Breidbach, C.F.: Big data, big insights? advancing service innovation and design with machine learning. J. Serv. Res. **XXI**(1), 17–39 (2018)
29. Gustafsson, A., Witell, H., Witell, L.: Service innovation: a new conceptualization and path forward. J. Serv. Res. **XXIII**(2), 111–115 (2020)
30. Snyder, H., Witell, L., Gustafsson, A., Fombelle, P., Kristensson, P.: Identifying categories of service innovation: a review and synthesis of the literature. J. Bus. Res. **LXIX**(7), 2401–2408 (2016)
31. Witell, L., Snyder, H., Gustafsson, A., Fombelle, P., Kristensson, P.: Defining service innovation: A review and synthesis. J. Bus. Res. **LXIX**(8), 2863–2872 (2016)
32. Furrer, O., Kerguignas, J.Y., Delcourt, C., Gremler, D.D.: Twenty-seven years of service research: a literature review and research agenda. J. Serv. Market. **XXXIV**(3), 299–316 (2020)
33. Martin, B.: Twenty challenges for innovation studies. Sci. Public Policy **XLIII**(3), 432–450 (2016)
34. Djellal, F., Gallouj, F.: Fifteen advances in service innovation studies. In: Scupola, A., Fuglsang, L. (eds.) Integrated Crossroads of Service, Innovation and Experience Research-Emerging and Established Trends, pp. 39–65 (2018)

Garden Cities 2.0 and Revitalization of Depopulated Rural Communities. A Net Positive Resource Production Approach

Ruggero Todesco[✉]

Via San Domenico Savio is. 255/B n°96, 98122 Messina, Italy
ruggero.todesco@libero.it

Abstract. The decades long ongoing demographic trend suggests that huge mass of population will continue migrating away from small rural town towards the bigger urban centers adding to the already overstressed areas ulterior demographic pressure, resulting in a set of harsh unprecedented issues. In order to overcome the challenges which the built environment is due to face – both currently and in the near future- it is mandatory to highlight and enact a whole set of policies aimed towards the increase in attractiveness of small towns and in the decentralization of urban functions to the small belt of settlement in the outskirts of the bigger urban centers. The scarcely populated small centers, which are now suffering from degradation resulting from progressive abandonment and lack of maintenance throughout the years, offer an interesting opportunity to rethink the liminal rural spaces and enact governance policies through an integrated architectural, technological, and economical approach in order to achieve an increase in the livability of the territories outside the urban centers.

Keywords: Off-grid · Rural communities · Zero impact

1 Introduction

The XIX-XX centuries can be identified as the time frame where the basis for the modern way of life as we see it today was laid.

Among all the major changes that have taken place since then, we can certainly identify the phenomenon of the intensive development of the city (understood as a large urban agglomeration) based on the presence of a productive centre that provides for the sustenance of the area to which it belongs.

Today, the world in which we live has a dominant city-centric orientation: a large concentration of population is concentrated in large cities that have become powerful financial, economic and productive centres. Following the current trend, this phenomenon will become more and more marked: according to the estimates of the 2018 United Nations report, the percentage of the urban population currently exceeds that of the rural population, and by the year 2050 68% of the global population will live in cities [1].

© The Author(s), under exclusive license to Springer Nature Switzerland AG 2022
F. Calabrò et al. (Eds.): NMP 2022, LNNS 482, pp. 164–172, 2022.
https://doi.org/10.1007/978-3-031-06825-6_16

Although with different motivations [2], the global trend sees an uncontested shift of population towards an urban reality.

The variability of the causes and phenomena linked to this exodus, which has been going on for about two centuries, is so vast that behind the term *urban sprawl*, often used to define it, there is no precise definition, nor could there be.

This evolution, increasingly focused on the settlement of a centralised city, has stimulated a debate on the general unsustainability of this settlement practice, highlighting the different problems that large urban agglomerations have. For some time, these issues have been addressed through regulations, imposed with the aim of maintaining healthy, providing a (more) sustainable living environments and protecting inhabitants and built-up areas as much as possible. With the increase in world population - and, in turn, the increase of the area and density of the urban environment and the resources needed to maintain it - the scale of the problem has now become serious and urgent. For this reason, regulatory intervention alone no longer seems to be sufficient; instead, it is necessary to reflect on a paradigm shift that takes into account new settlement patterns that can be culturally, technologically and economically acceptable and sustainable.

It is not the first time that the scientific community has tackled this issue, proposing solutions to resolve it; however, today the debate on the depressurisation of urban realities is more lively than ever, motivated by the need for an increasingly urban world that requires a human redistribution, since the degree of global density is characterised by a profound imbalance.

In this sense, the focus is often placed on small towns, which today are often characterised by problems of abandonment and depopulation. Starting from them it could be possible to imagine and model a new settlement structure, better distributed, increasing the liveability of such places with strategies for a smart and sustainable settlement [3].

This paper comprises two parts. The first adopts an historical view in order to highlight the facts and the reasons of the evolution that has led to the actual global urban configuration. The second and latter part focuses on the possible strategies of revitalization of the depopulated rural areas, proposing some key concepts to focus on in order to boost the appeal and liveability of them.

2   From Villages to Modern Cities: A Retrospective on the Historical Context and on the Past Garden City Solution

Over the centuries, mankind's typical settlement development has been marked by a strategic territorial logic, characterized by the emergence and growth of settlements in territories that offered an advantage over others. This advantage was the discriminating factor that allowed the choice of the location in city would be founded: it could be economic, productive or social, such as the proximity to a particular resource, easy defensibility or the presence of waterways. Over time, these settlements took shape in a variety of realities; the evolution of each of them, in fact, did not proceed linearly and uniformly. Generally, however, their development took the form of a series of larger centers, in which the most important social, governmental and economic functions were centralized, and a constellation of smaller settlement centers, developed following the same logic of advantage, serving the most important settlements. The functions carried out by these small centers were essentially of productive character, at the service of the hypertrophic mass of population of the vast settlement area, which were located outside the walls, in the belt of the immediate vicinity of the major centers, or connective character, developed as settlements along the lines of connection of these major centers, so as to be easily accessible to each other on horseback or on the back of a donkey in a day's journey, thus ensuring greater safety in travel (Fig. 1).

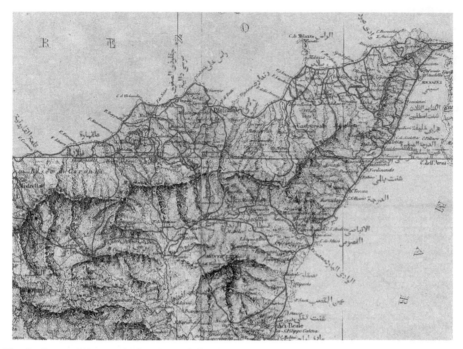

Fig. 1. Comparative map of ancient Sicily (M. Amari- A. H. Dufour, 1849) with the outline of the main path of connection between major urban centers.

The dissolution of such a settlement logic has been determined by the industrial revolutions and has occurred in conjunction with the sudden upheaval that, since the nineteenth century, has occurred in the social and governmental structure of European states. Moreover, a parallel aspect affecting the rapid shift in the city layout organization, deriving from the industrial revolution, was the development of the railway system whose main effect was to attract huge masses of population away from inland centers towards the coastline based settlements, radically altering the overall regional geography and landscapes.

Over time, the centralization of production and settlement in industrial cities led to the emergence of problems having opposite effects but caused by the same phenomenon, in both categories of historically stratified settlements. In the smaller centers, the mechanization of production has caused the rapid decline of productive activities, the eventual assimilation, often with a merely residential function, within the peripheries of the larger centers; in the cases of centers located along the connecting routes, the lack of the need to move on horseback has caused their depopulation until reaching, in the most extreme cases, total abandonment. In the most populous centers, the migratory phenomenon directed according to the countryside-city paths has determined an overpopulation of the latter, progressively deteriorating the quality of life inside and causing a series of additional and collateral problems caused by the exclusive hypertrophy of the city housing fabric and the resulting social pathology. Currently, in fact, the extension of today's metropolises has become such that the outermost ramifications of each are confused with those of neighboring metropolises, thus re-proposing the problem of sustainability and supply of resources on a larger scale.

While still in the XIX century the idea of the centralized industrial city stimulated collective ideals as the place of progress and was identified as the guideline for modernity, already in the XX century several scholars (such as Robert Owen, Charles Fourier or the economist Alfred Marshall) highlighted several problems related to this urban conception. They proposed utopian settlement strategies alternative to the centralized industrial city.

Among these contributions, some today sound more relevant than ever, and can become starting points for a contemporary reasoning, such as the concept of Garden City proposed by Howard [4], which was an alternative answer, free from political, moral or sociological considerations, to the problem of the management of overpopulated cities and the relationship between them and the countryside that provided for the sustenance of their citizens. In his essay *A Peaceful Path to Real Reform* [5], the urban planner Ebenezer Howard theorizes the model of a utopian city, the Garden City, a hypothesis of a standard settlement model, fully replicable in any place on earth that was able to

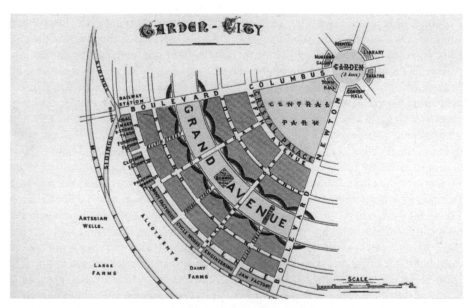

Fig. 2. Typological layout of a sixths-divided neighborhood in Howard's Garden cities.

rationalize land consumption, the movement of citizens and the relationship between urban agglomerations and the countryside (Fig. 2).

The garden city develops in a radial way, has a perfectly circular shape and is divided into districts of equal width, 60° each, by the six main arterial routes of the settlement. At the intersection of the three main arteries there is the main square bordered by the most representative buildings of city life: the City Hall, the theater, the library, the hospital, the museum and the concert hall and, moving from the center to the outer perimeter is divided into sectors with gradually decreasing population density, interspersed with wide avenues and belts of green city and agricultural green.

The contemporaneity of Howard's proposal consisted in the realization that the demographic pressure and congestion of the urban environment cannot be solved by the construction of new urban areas, but by a complete project of decentralization.

Although Howard's garden city experiments saw little application in the following years, they contain *in nuce* some exhortatory proposals towards the attention to those problems of environmental sustainability that persist in the current urban settlements and for which it has not yet been possible to find a univocal solution. They therefore show an intuition of the lines of action to counteract the ways in which man impacts on the surrounding environment.

3   Key Element for Revival Small Communities

Howard's concept of the garden city provides us with a good starting point for reasoning about the key aspects necessary for an optimal configuration of the reality of small urban centers. The circular form of the city and its internal subdivision, together with the six lines connecting the different urban centers, are hypothesized as solutions to the problem of minimizing the time and cost of transport and travel, as well as the idea that the city is also a garden that produces on site the food that will be consumed by the population.

The solutions put in place to mitigate the phenomena of the urban overpopulation, always look only to a category of settlements, or large extension or small size without considering the problem on a global scale, focusing almost entirely on the development of technologies and practices of governance able to alleviate the demographic pressure on large centers and conceiving the small centers exclusively in view of the decentralization of the functions of secondary importance of the major settlement nuclei. Those solutions frequently resort to the development of technological methods aimed at the mitigation of the effects caused by the high density of population, such as the decrease of air quality index, pollution, urban heat islands and noise pollution and delocalization of residential areas in the periphery of the cities and dislocation of trade, administrative and social life nodes in the epicentric areas of major settlements. Such practices become a measure of contrast with limited effects when confronted with settlements of massive scale, and it is evident that the solution to this problem must be sought not so much in the development of technological solutions that can improve efficiency, but in a radical change of paradigm. The current economic system postulates a substantial deregulation, at various levels and entities, of the market, which becomes free and capable of self-regulation through the natural competition of individuals within itself. This, however, determines a system that decreases the overall efficiency of the production output since, lacking a line of direction that specializes and sectorializes the territories, there is an increased competitiveness in some sectors, whose actors mutually limit the overall earning capacity, and a substantial monopoly in others. On the other hand, the strategies implemented for the redevelopment of villages in depopulation are often based on the concession of residences at advantageous prices, if not symbolic, without finding any strategy to increase the attractiveness of the village in the long term, which therefore, not being able to ensure a structured community life within itself is destined, in the long term, to become a mere residential center or a new progressive depopulation.

It is clear that today's development, except in sporadic virtuous cases, is the victim of a lack of adequate and coherently structured governance practices that can provide guidelines for the overall well-being of the population: to a centralization of functions and productive districts in limited spaces, it is necessary to oppose a practice of delocalized development that aims to determine a uniform increase of the entire territory, albeit each with the specializations deemed most appropriate, so that the agglomerations of small entities can gain attractiveness again not only in relation to services that can benefit the most important neighboring population centers. These practices are already present in embryonic stages on city territories, which are often organized in districts (industrial, agricultural, commercial, residential) following the instructions dictated by the planning instruments. What is desirable is the strengthening of these planning tools on a wide territorial scale, supra-regional and, when possible, supranational: the experiences gained so far involving geographical areas of inter-regional extension, although already run in for several decades, often concern only the planning of connections of the main urban centers and the management of common risks, resulting, however, deficient regarding the shared planning of development. As proof of this, it is possible to note how the planning of waste management in the Italian national context has substantially a "predatory" character in which the location of economically predominant activities follows the logic of the "race for profit" of the actors involved, since they are not efficiently located throughout the national territory, and often display marked regional differences. The desirable line of governance must tend to operate a constant and progressive analysis of the territories, which monitors the strengths and weaknesses and their evolution over time, so as to be sufficiently flexible and incisive to adapt to the sudden changes of a rapidly evolving society and that responds to the objective of environmental protection now becoming a compulsion that translates into a categorical imperative through appropriate regulatory tools.

The strategies that can be implemented to increase the perceived attractiveness of rural villages cannot disregard an economic and technological component that leads them to develop the competitive advantage necessary to make them more attractive than the overpopulated cities. Unlike major urban centers, in rural settlements the wide availability of space lends itself to an adequate dislocation of technological facilities that can constitute small communities based on a substantial self-sufficiency through the production of a surplus of resources necessary to ensure the livelihood of the community. By means of an integrated approach on the design of interventions aimed at the production of energy, food, water and thermal resources necessary to ensure quality standards and housing, it is possible to give new attractiveness to depopulated villages, defining them as centers of autonomous production of resources, totally disconnected from the centralized supply.

The great availability of space typical of rural centers allows the development of an economy based on the collection and extensive reuse of rainwater as a founding element of the settlement. The meteoric waters, especially in little anthropized environments as the rural territories have high qualitative chemical, physical and chemical parameters that make them, if captured, usable for irrigation purposes, if not drinkable in some cases, without the aid of purification treatments or through the use of low cost treatments, both in economic and energy terms that time. The collection of rainwater, by means of impermeable surfaces such as tanks or roofs of buildings is a practice known for thousands of years that, however, has been gradually abandoned in the last two centuries in favor of practices of centralized supply not always advantageous (Fig. 3).

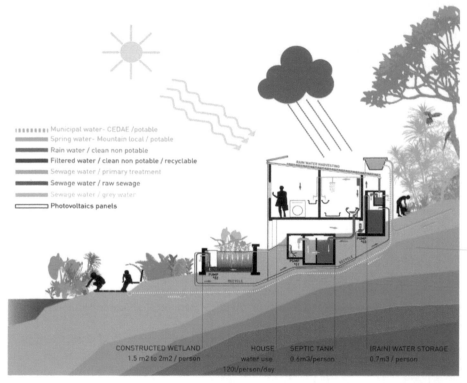

Fig. 3. Schematic of an off-grid water harvesting - wastewater treatment - constructed wetland system replicable in rural context [11].

Similar considerations can be made for the production of electricity, which mostly follows a logic that provides for the generation in large quantities of the resource in a few points and then distribute it through an extensive network that requires a constant and expensive maintenance and is still subject to numerous load losses during distribution. The production or capture in situ of these resources, in addition to ensuring greater efficiency and sustainability in their management, promotes the overall development of the urban fabric towards a view of conscious consumption, making users responsible who, not depending on a network, will be required not to consume resources when not necessary. From such surplus, energy and water, it is possible to ensure a productive layout based on the revitalization of rural economies, feeding the typical local crops by means of fertigation obtained from rainwater and, suitably treated by integrated systems of degreasing, anaerobic digesters ABR and constructed wetlands, from graywater coming from domestic users.

4 Conclusions

The aim of this paper was to analyze the highlight the conditions under whom it could be possible to revitalize depopulated rural centers.

In the paper is stressed how the implementation of a low cost technological infrastructure could serve as a carrier intervention for the revitalization of the same in productive agricultural, energy and water. Providing their own self-sufficiency, the attraction exercised on the economic level and raising the quality of life from the possibility of living without having to depend on the centralized distribution of resources, with the resulting expenditure, would contribute to the recovery of housing themselves, as well as the de-congestion of overpopulated urban centers. Far from wishing to provide services to urban centers, and without wishing to generate a serialized delocalization of productive units, this methodological proposal instead provides the basic minimum infrastructure through which to imagine a long-term development for each village, each of which can be declined according to its own specificity, intrinsic characteristics and ambitions. The added implementation of shared and coordinated services between neighboring settlements, such as shared services of soft mobility for connections, community management of resources and issues for land management practices and definition of common development strategies, is therefore the next step that from individual disconnected cellular infrastructures constitutes a structured and coherent network able to sustain itself and grow over time.

References

1. United Nations, Department of Economic and Social Affairs. Population Division: World Urbanization Prospects: The 2018 Revision (ST/ESA/SER.A/420). United Nations, New York, NY, USA (2019)
2. Bueno-Suárez, C., Coq-Huelva, D.: Sustaining what is unsustainable: a review of urban sprawl and urban socio-environmental policies in North America and Western Europe. Sustainability 12, 4445 (2020)
3. Liaros, S.: A Network of Circular Economy Villages: Design Guidelines for 21st century Garden Cities. Built Environ. Project Asset Manag. 12(3), 349–364 (2021). https://doi.org/10.1108/BEPAM-01-2021-0004
4. Howard, E.: Garden Cities of To-Morrow. The M.I.T. Press, Cambridge (1965)
5. Howard, E.: To-morrow: A Peaceful Path to Real Reform. Swan Sonnenschein & Co. (1898)
6. Venter, O., et al.: Sixteen years of change in the global terrestrial human footprint and implications for biodiversity conservation. Nat. Commun. 7, 12558 (2016)

7. Fenner, A.E., et al.: The carbon footprint of buildings: a review of methodologies and applications. Renew. Sustain. Energy Rev. **94**, 1142–1152 (2018)
8. Al-Chalabi, M.: Vertical farming: skyscraper sustainability? Sustain. Cities Soc. **18**, 74–77 (2015)
9. Shamshiri, R.R., et al.: Advances in greenhouse automation and controlled environment agriculture: a transition to plant factories and urban agriculture. Int. J. Agric. Biol. Eng. **11**(1), 1–22 (2018). https://doi.org/10.25165/j.ijabe.20181101.3210
10. McClements, D.J., et al.: Building a resilient, sustainable, and healthier food supply through innovation and technology. Annu. Rev. Food Sci. Technol. **12**(1), 1–28 (2021). https://doi.org/10.1146/annurev-food-092220-030824
11. https://www.aguacarioca.org/what-are-constructed-wetlands

(Un)earth Vulnerable Chile

Carlotta Olivari[1]([✉]) and Margherita Pasquali[2]

[1] Architectural Association, 36 Bedford Square, London WC1B 3ES, UK
carlotta.olivari@aaschool.ac.uk
[2] University of Trento, Mesiano 77, 38123 Trento, TN, Italy
margherita.pasquali@unitn.it

Abstract. Vulnerable topographies, geographies of conflicts and morphologies are reservoirs of resilience to react to the social, economic, and environmental neoliberal crisis. The aim of this contribution is to investigate the current state of vulnerability in relation to climate, politic, and economic crisis in extreme territories such as Chile. These territories are leftovers of the capitalist neoliberal mode of space occupation, yet they can represent reservoirs of resilience for developing different interactions with the landscape and its ecology. The first question that the paper aims to investigate is the relation between land and the informal issue in the extreme Chilean territory. The analytical reading of the Chilean *terre* that we want to propose is based on social, cultural, and natural geographical conditions. In order to spatialize the Chilean context, it becomes necessary to talk about *Espace* to introduce spatial categories into social criticism. Thus, an open process based on mapping is set to identify a methodology to identify the risk and resources of these extreme areas. Finally, the Gran Valparaíso region case is set to visualize these spaces as geographies of possibilities.

Keywords: Vulnerability · Informality · Landscape ecology

1 Introduction

1.1 The Geography of Crisis in Chile

What the crisis of the neo-liberal model, exhaustion of natural resources, political disorders, and natural hazards have in common? The crises we are living are interconnected, the prevalent political/economic models have been shaping the planet and the relations between humans and the earth. The land and its exploitation are core to understanding the current crises and the spatial and social transformations of the planet. The investigation is based on the potentials emergent on these crises, their interconnection, and how the question of land is woven through all of them. Especially in the so-called Global South (Garland Mahler 2017) it is possible to unveil processes of exploitation due to unequal socio-spatial context, in which value creation practices occur through the production of continuous fractures. In these contexts, extractive socio-spatial phenomena are strongly manifested. The term 'extractivism' indicates, at the same time a socio-economic system and a model of territorial appropriation, in which processes of subtraction of resources,

© The Author(s), under exclusive license to Springer Nature Switzerland AG 2022
F. Calabrò et al. (Eds.): NMP 2022, LNNS 482, pp. 173–181, 2022.
https://doi.org/10.1007/978-3-031-06825-6_17

material (raw materials, agricultural products), and intangible (cultures, knowledge, traditions local), reorganize spaces, and ecologies (Gudynas, 2010). In the Latin America context, extraction is almost always coupled with neo-colonial discourses and policies in which the state tries to shape its inhabitants (Svampa 2019) How to critically re-read these ongoing phenomena? Which strategies can be defined starting from the Global South?

The Chilean case is an emblematic example of the thrust end of a development model unable to face and imagine the social, democratic, and ecological complexity of the current global system. One of the most evident materializations of this context of crisis is the phenomenon of informality. The slums, *"campamentos"* in the Chilean case, spontaneous and precarious settlements, are the result of toma de la terra (literally "taking of land") as described by *"Catastro Nacional de Campamentos"* (MINVU 2020). Thus, the land becomes a key factor to analyze the Chilean *"campamentos"*: the extreme and symbiotic relationship between informal development and the morphological conformation of the Chilean natural context.

From the report *"Ministero de Desarrollo Social y Familia"* (Ministry of Social Development and Family 2020) it is possible to observe that the phenomenon of informality in the Chilean territory is not decreasing on the contrary: with the coming of the Covid-19 pandemic, with the ongoing climate crisis, with the process of extraction and depletion of resources, the phenomenon has intensified. Since 2019, the geography of crisis and conflict has opened a dialogue with the Chilean government. From the first insurrections in October 2019 to today's elections of Gabriel Boric, Member of the Chamber of Deputies of Chile, Chile has proven to be in a state of crisis that is constantly evolving and changing in the hope of future improvement.

The socio-political and geographical position that this contribution wants to take, must deal with a highly vulnerable and dynamic context. This is the set of factors that have led Chile to be an unstable territory, economically weak, and socially fragile. However, along with the socio-political instability, the extreme complexity of Chilean territory is also due to natural phenomena of an endogenous nature and has such a conformation that the topography itself becomes an urban fact: morphologically fragile (natural risks related to geography and topography), environmentally (unsustainable overexploitation of natural resources) and socially in crisis.

In this contribution Chile is considered as the most extreme representation of the tension between nature and built space. In particular, the Chilean *"campamentos"* are taken as an expression of this interaction: in an area of social disparity, spontaneous settlements prove to be highly adaptive to the morphology of the territory, by creating a new rural and social dimension.

2   Chilean Informal Territory: How to Design a New *"Terre"*

2.1   Extreme *Terre*: Risk and Resource in Chile

Today, it is no longer possible to talk about urban or rural. From the soil emerges rather the possibility of "returning to the land", in the sense of returning to the tangible effects of the phenomena of change to which our lands/soils are subject. In this contribution, we want to pay particular attention to how the *terre*, has been modified in the Chilean context. The French term *terre* includes in the same etymology land and soil (ground, soil, earth) and brings the politicized meaning of territory to its etymological root: from the Latin terra, from the Greek, Sanskrit root *ters*.

The current state of perennial unbalance has to be investigated by looking at the morphology of the Chilean "*terre*": extreme, vulnerable, risky and potential. This Extreme *Terres* are located in the utmost uncertainty: mankind has subjected them to a condition of risk that threatens their stability, their identity, their resource and has placed them at the extreme limit that they are able to bear.

In these places, man exploits their "land", which is the primary resource, choosing it as raw material for production processes (Lefebvre 1979). The resource extracted, such as mineral or water, has been exploited for years without giving it a value in the production process; or even more simply, man has never bothered to give back what he received from the "*terre*". Thus, these informal *terres* in the Chilean context, which contain precious minerals, water sources, fertile soils, forests, and many other resources; are put at greater risk every day. The risk they are subjected to (hydrogeological, seismic, fire, desertification) increases as their resources become more precious.

The morphological conformation of the Chilean territory has always been considered a "risk" and a "resource" for its highly significant value in the South American territorial context. Here, the risk and the resource are both closely linked to the conformation of the territory: its orography, the hydrographic system, the climatic consequences, and soil deterioration, due to overexploitation of water resources, mining, forestry, and agriculture, influence more and increase the incidence of hydrogeological and seismic risk. The set of natural and social resources and the conformation of the territory overlapped with the risk.

The informal settlements are chosen as a place of investigation because here the relationship between the risk, natural and anthropic, and between the resource is maximal. They sit at the edge of the urban infrastructural system so that they can access densely populated areas and services, fleeing desert and remote areas. It is precisely the relationship between the informal settlements, risk, resource, and proximity to services that is illustrated in the project "*Yuxtaposición Extrema*" (Olivari and Pasquali 2019), the extreme juxtaposition between risk and resource. Therefore, the experimental methodology developed in "*Yuxtaposición Extrema*" (Olivari and Pasquali 2019) is taken up. Further it is essential to explain why we investigate marginal territories: where there are informal settlements, more precisely on the margins of urban services, there are both raw materials for production systems and a high risk of seismicity and fires.

The aim of the contribution is to explain, investigate and spatialize the "*Yuxtaposición Extrema*" in the Chilean context. Geography as discipline and mapping as tool are used for representing the space and for understanding the development of the reference context: the effort of this research is to include the human scale in the Chilean sociopolitical *terre*. The research methodology becomes the design process in response to the endemic problems detected in different places: the research model is structured on the social, cultural, and natural conditions of the Chilean territory. "*Yuxtaposición Extrema*" represents the desire to create a methodological approach to estimate the vulnerable informal settlements, through a multi-scale and multi-level approach looking at fragmented territory. The aim is to underline the critical issues, potentiality, and sustainability of the informal space with respect to the morphological conformation of the territory.

2.2 Tool and Evaluation Method

The goal is not to propose certain and predefined solutions to the informal issue; rather to re-imagine a fragmented and extreme territory. Thus, the mapping agency becomes a key element in the proposed methodology: mapping is critically intended as an active and projective tool. The need is to critically read and visualize the Chilean territory which generates, influences, and composes its Extreme *Terres*. By identifying the mapping as an operative tool, the cartographic survey is considered as a potential tool in the transformation of the territory. It is, therefore, an element with not only a descriptive but above all a performative and political character. To critically evaluate the informal space, a methodology structured in phase is proposed.

First Phase. The proposed method of investigation needs to create parameters to evaluate the social, cultural, and natural conditions of the Chilean territory. Starting from an in-depth study of Lefebvre's concept of *espace* (Lefebvre 1979) and looking at the extreme *terre* of Chile, three parameters are proposed: Informality (I), Nature (N) and Urban (U): I-N-U (Olivari and Pasquali 2019).

Informality. I-Informality factors are related to the development of *campamentos* in the Chilean territory. The phenomenon of *toma silenciosa* (land taken silently to build precarious settlements) takes place where the conformation of the *terre* is in a continuous state of risk, isolated from the Urban space and its basic services (i.e. access to paved roads, energy, potable water, sewage service). The number of families living in informal settlements is constantly increasing (TECHO Chile 2020). Families are moving from poor areas of the city to an even more precarious, peripheral, and isolated ones. The "higher rents" implements a mechanism of exclusion of the poor population and, consequently, increasing the migration to the *campamentos*.

Nature. N-Nature parameter analyzes the endogenous Chilean territory risks. This territory is a natural island, isolated to the north, the Atacama Desert from Perú, surrounded by the Pacific Ocean to the south and west; the topography itself becomes an urban fact: social, spatial, and cultural relations are deeply linked to this territory. Chile is characterized by an extreme and endogenous nature: deserts, glaciers, earthquakes, tsunamis, and fires. In areas where the risk and natural vulnerability is greater, the *campamentos* appears. Then, Nature parameter is also considered as a resource.

Urban. Chile's urban distribution reflects social and class distinctions: *campamentos* are located on the outskirts of Chilean cities. One of the main drivers is the land access and cost: around urban settlements the land cost is higher than 1.2 UF[1]/m2 (53 US \$/m^2), the limit permitted cost to build *vivienda sociales* (social housing program). Because of that, the affordable housing system is detached from the city and its services, making them

---

[1] La Unidad de Fomento, UF, is a resettable unit of account in accordance with inflation, used in Chile.

less desirable. On the other hand, informal settlements take the free land, to have better connectivity to the services and works within the city. The U-Urban factor considers the level of (dis-)connectivity, access to services between informal settlements and the city.

Second Phase. The complex Chilean context has been redefined according to the I.N.U. parameters. These parameters become fundamental to establish a methodology based on mapping. Each of the parameters "Informal", "Nature" and "Urban" is categorized into specific related criteria [based on 2015–2019 GIS open data sources]. Taking the case study of the Gran Valparaíso Region, the most emblematic Chilean region for informal settlements, the selected criteria can be analyzed more specifically.

In the Gran Valparaíso Region, the *campamentos* are mainly located in the main coastal cities (Viña del Mar and Valparaíso), attracted by the economic thrust of Viña del Mar - Valparaíso conurbation. Therefore, the number of *campamentos* and population density are the first Informality based criteria selected. The *campamentos* are located in risk areas (presence of ravines, aridity, steep slopes): thus, the frequency of natural hazard such as fires, landslides or flash floods become the Nature criteria related, along with the soil composition and soil fertility. Each selected criterion is re-classified according to a normalized scale (1 = low, 10 = high) for risk and resource as shown in the exemplative Table 1.

Table 1. Data reclassification: GIS assisted multi-criteria evaluation.

Data reference	Criteria	Unit	Risk / Resource rating	
IDE Chile	**Campamentos density**	People/m2	*1_low risk*	*10_high risk*
FAO-UNESCO	**Soil composition**	Fertility rate	*1_low resource*	*10_high resource*

For example, the Informal risk refers the recent trends witnessed in Chile of an increase in the number of families living in informal settlements. The CIS Techo-Chile study evidences a process of urban displacement by which families are moving from poor areas of a city to an even more precarious, peripheral, and isolated one, often located at the margin of Chile's urban conglomerations. This phenomenon shows a significant geographical concentration of low-income citizens and their movements. In fact, the recent economic logic advocated by tenants is to maintain "high rents" to implement a mechanism of exclusion of the poor population and, consequently, to increase the migration to the *campamento* (TECHO Chile 2020). The Urban risk is considered through the infrastructural dis-connection of *campamentos* with respect to services and densely populated settlement centers, has become a crucial risk factor. As reported by the MINVU land registry study, the high level of precariousness of the *asentamientos* is characterized by the lack of basic services (hospitals, emergency rooms, schools, work). This disconnection increases the dependence of the *campamentos* from the near city which is dictated by the conformation of the land. In fact, most informal settlements develop along the perimeter of cities, within ravines or areas where the land is not consolidated and the natural hazard is very high, which is why despite the UF index being on average high, the land is not adequate for the increase in buildings in recent years (Fig. 1).

On the other hand, the set of natural and social resources are summarized in the reclassification of the criteria according to the resource gradient. Finally, the aim is to be able to visualize and compare tangible factors like the natural context in the informal development process described in the Matrix I.N.U. Fig. 2.

Matrix I. N. U.

Sémiologie Graphique	Code	Name	I-N-U		Code	Critical Maps	Final Map
	0.1	Chilean campamentos	I		I.0.1		
	0.2	Built cities	U		N.0.5		
Symbol	0.3	Viviendas sociales	U		I.0.6		
	0.4	Mines	U		N.0.9	m1. Riesgo	
	0.5	Natural hazards	N		U.1.1	Extreme relationship between campamentos density and risk density.	
	0.6	Families campamentos	I		N.1.8		
Symbol Gradient	0.7	Population cities	U		N.1.9		
	0.8	Population viviendas	U				
	0.9	Topography	N				
Lines Gradient	1.0	Grey Infrastructure	U		I.0.1		Yuxtaposición Extrema
	1.1	Political Borders	U		U.0.2		Extreme relationship
					U.0.3		between campamentos density and natural resources.
	1.2	Ecoregions	N		U.0.4		
	1.3	Flora	N		I.0.6	m2. Recurso	
	1.4	Geomorphologic units	N		U.0.7	Extreme relationship between campamentos density and natural resources.	
Fill Cathegorized	1.5	Soil Composition	N		U.0.8		
	1.6	Land use	N		N.0.9		
	1.7	Climate	N		U1.0		
					N.1.2		
Lines + Fill Cathegorized	1.8	Hydrography	N		N.1.4		
					N.1.7		
	1.9	Desertification	N		N.2.0		
Fill Gradient	2.0	Soil Potentiality	N				

Fig. 1. I.N.U. Matrix: GIS assisted multi-criteria evaluation method for the allocation of risk and resource areas to suit a specific objective based on a variety of attributes that the selected areas possess. Graphic elaboration Olivari and Pasquali, 2021

Third Phase. Having established the levels of interest, expressed in the I.N.U. matrix, two critical maps (Fig. 2) are developed at the regional scale by cross correlating the reclassified data of risk and resource. These maps make it possible to highlight the relationship between *campamentos* and Nature and Urban, underlining their risk and potential levels. Thus, the re-classified data take place through the map tool: more precisely, through the Geographic Information System (GIS) tool. The identified local vulnerabilities (floods, lack of a collective water system, lack of sewage, lack of drainage system, aridity, soil degradation, inequality) and resources (soil fertility, connectivity to services, endogenous ecology) are mapped. In particular, the use of Multi-Criteria Analysis (MCA) allows the overlapping of the two critical readings of the context: Risk and Resource make evident in the mapping process the extreme condition of this territory.

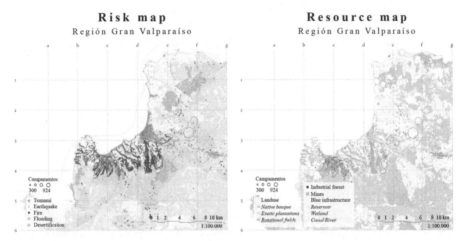

Fig. 2. Assessed risk and resource maps: mapping of risk and resource areas based on reclassified criteria. Graphic elaboration and mapping Olivari and Pasquali, 2021

Fourth Phase. *"Juxtaposition" is the action and effect of juxtaposing (putting something together or immediately to something else). The concept is formed by the Latin words iuxta ("next to") and positio ("position")*[2].

The Yuxtaposición aims to put things that are not similar next to each other. In particular, we refer to the *Yuxtaposición Extrema* between mapped risk and resource. The Chilean *campamentos* are an expression of this interaction: in an area of high resource and even higher risk, spontaneous settlements prove to be highly adaptive to the morphology of the territory, by creating a new rural and social dimension. *Campamentos* are the representation of the extreme juxtaposition: where the symbiosis/tension/relation/resistance between risk and resource is set.

A critical taxonomy (Fig. 3) is developed at the regional scale of the Chilean territory, based on the *Yuxtaposición Extrema.* The aim is to underline the critical issues, potentiality, and sustainability of the informal space with respect to the morphological conformation of the territory. The natural context becomes an essential dynamic resource in the informal development process. As stated above, studying the issue of informality means considering an apparatus of themes, conflicts, subjects, and politics that have been overlapped and accumulated over time without erasing/ negating themselves: the Chilean *campamentos* are its archives. It is necessary to deal here with an endogenous and dynamic logic in continuous growth.

---

[2] Definition taken from Encyclopedia Britannica.

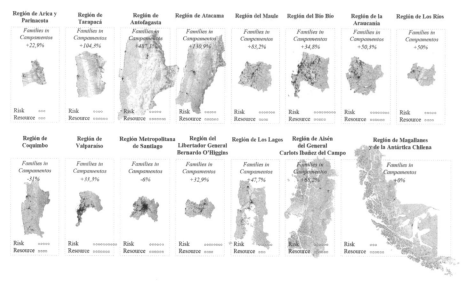

Fig. 3. Extreme Juxtaposition of Chilean Regions: Multi-Criteria analysis for the allocation of extreme areas according to risk and resource gradients. Graphic elaboration and mapping Olivari and Pasquali, 2021

3   Expected Results

The processual methodology is proposed to put these frameworks into a systemic approach that considers the feasibility and eventual fallibility of the developed process.

Moreover, the proposed methodology is considered as a process that cannot be replicated in actual results, but repeatable are the conditions for establishing, each time, a new complex and endogenous thought.

The aim of the research is to demonstrate how by working through critical analyses, the multi-scalar mapping can be adapted to specific needs: the informal element must deal with its instability, as it is settled down in risky areas. As the native indigenous populations in this area lived precariously and in continuous movement depending on the fickleness of nature. Rather than focusing our attention on large population centers, we investigate urbanization from the point of view of its putative "others", the areas that are commonly represented as opaque, minority, external.

Mapping is chosen as an operational project tool that has "the capacity to reformulate what already exists". (Corner 1999) A fundamental role is played by GIS software for its ability to return mappings consisting not simply of physical and geographical data but also semi-quantitative sociological, spatial economic data, for their processing and definition through a graphic code. By identifying the map as an active tool, the cartographic survey is considered as a selective process and as a potential and central tool in the transformation of the territory. It is, therefore, an element with a not only descriptive but above all performative and design character. "The geographical representation guides the action, it too produces territory", Dematteis affirms, "not only reflects the social relations and therefore the relations of society with the material environment, but it helps to determine. It acts on the things it wants to transform and records the response". (Dematteis 1985).

Attribution. The contribution is the object of research by the two authors: the contribution has been edited jointly by the two.

References

Dematteis, G.: Le metafore della terra: la geografia umana tra mito e scienza. In: Campi del sapere, 1a ed. Feltrinelli (Campi del sapere. Economia e società), Milano (1985)

Garland Mahler, A.: What/Where is the Global South? Global South Studies [Preprint] (2017). https://doi.org/10.1093/OBO/9780190221911-0055

Gudynas, E.: The new extractivism of the 21st century: ten urgent theses about extractivism in relation to current South American progressivism. Am. Program Rep. (2010). Center for International Policy, Washington, DC

Svampa, M.: Neo-extractivism in Latin America: Socio-environmental Conflicts, the Territorial Turn, and New Political Narratives. Cambridge University Press, Cambridge (2019). https://doi.org/10.1017/9781108752589

Lefebvre, H.: La produzione dello spazio. Pgreco. https://books.google.it/books/about/La_produzione_dello_spazio.html?id=wufRoAEACAAJ&source=kp_book_description&redir_esc=y (1979). Accessed 4 Jan 2022

Ministero de Desarrollo Social y Familia: COVID-19. Socioeconomic Survey of Chile. Summary of Main Results from the First wave (2020)

MINVU: Catastro de Campamentos, Ministerio de Vivienda y Urbanismo. https://www.minvu.gob.cl/catastro-de-campamentos/ (2020). Accessed 3 Jan 2022

Olivari, C., Pasquali, M.: Yuxtaposición Extrema. Politecnica. Maggioli Editore (0-n). https://www.maggiolieditore.it/yuxtaposicion-extrema.html (2019)

Corner, J.: The agency of mapping: speculation, critique and invention. In: Cosgrove, D.E. (eds.) Mappings. Reaktion. London (Critical views.), p. 311 (1999)

TECHO Chile: Monitor de Campamentos - Centro de Investigación Social. https://chile.techo.org/cis/monitor/monitor.php (2020). Accessed 3 Jan 2022

Exploring the Resilience of Inner Areas: A Cross-Dimensional Approach to Bring Out Territorial Potentials

Diana Rolando[1] ⓘ, Manuela Rebaudengo[2] ⓘ, and Alice Barreca[1](✉) ⓘ

[1] Architecture and Design Department, Politecnico di Torino, Viale Mattioli 39, 10125 Turin, Italy
alice.barreca@polito.it

[2] Interuniversity Department of Regional and Urban Studies and Planning, Politecnico di Torino, Viale Mattioli 39, 10125 Turin, Italy

Abstract. Italian Inner Areas are fragile territories often lacking in essential services and thus characterized by depopulation and degrade. In recent years, an innovative national policy, the National Strategy for Inner Areas (SNAI), funded 72 pilot cases to enhance their natural and cultural resources, counteracting marginalization and demographic decline. With the aim of bringing out territorial potentials, the paper started from an in-depth literature review, highlighting that resilience is widely dealt with at urban scale, neglecting -above all- remote and marginal territories. In order to bridging this gap, authors propose a 6-phases methodological approach to study different contexts through a quantitative analysis based on specific indicators. Shifting from the mono-dimensional approach (focused on vulnerabilities), the use of cross-dimensional indexes is suggested to explore territorial vulnerability and vibrancy (intended as local trigger to development and renewal). The selected case study is "Branding4Resilience" research project, where 4 Italian fragile areas were explored to define new resilient scenarios for local development. The results showed that the proposed approach effectively support the territory exploration and lay the ground for future evaluations useful also to better address specific local policies.

Keywords: Inner areas · Territorial capital · Territorial resilience · Territorial vulnerability · Territorial vibrancy · Local development

1 Introduction

The Italian territory is a complex system of cities, municipalities and rural towns, where larger urban areas, offering facilities, services and activities, act as attractors for the population. The most remote areas, often offering only essential services and thus characterized by depopulation and degrade, are defined "Inner Areas": they are fragile territories, far from the big cities, covering 60% of the national surface and hosting nearly 13.5 million people (22% of its population) [1].

F. Calabrò et al. (Eds.): NMP 2022, LNNS 482, pp. 182–190, 2022.
https://doi.org/10.1007/978-3-031-06825-6_18

The National Strategy for Inner Areas (SNAI) is an innovative national policy to counteract marginalization and demographic decline within Italian Inner Areas, by enhancing their natural and cultural resources, creating new employment circuits and new opportunities of local development through territorial cohesion (https://www.agenzi acoesione.gov.it/strategia-nazionale-aree-interne). In 2019, the SNAI selected 72 project areas (about one thousand municipalities, representing 2 million inhabitants) and funded them with a total amount of more than € 591 million, in addition to the allocations of ESI Funds from Operational Programme and other public/private funds. The impact of this policy is now being evaluated in order to measure the results of the public investments and to start again with the next funding cycle.

Considering that the opportunity of new economic activities and job creation is closely related to the supply of essential services (education, healthcare and mobility), which is therefore an absolute precondition for reversing the decline trend, in recent years many studies [2–7] have been undertaken for policies, development projects and widespread strategies for fragile areas. "B4R Branding4Resilience. Tourist infrastructure as a tool to enhance small villages by drawing resilient communities and new open habitats" is a research project of national interest (PRIN 2017 – Young Line) funded by the Ministry of Education, University and Research (MIUR), started in 2020 [8]. The project is coordinated by the Università Politecnica delle Marche and it involves as partners the Università degli Studi di Palermo, the Università degli Studi di Trento and the Politecnico di Torino. B4R investigates the potential of branding in Italian small villages, proposing the transformation of minimal tourist infrastructures as an engine for the development of more structural and resilient territories and local communities. (https://www.branding4resilience.it/).

In this framework of both research and tangible national strategies for the development of actions aimed at revitalizing fragile areas, it is important to be able to bring out the potentials of territories in a multi-dimensional perspective, characterized as far as possible by quantitative data and indicators that can allow comparisons between similar territories located in very different geographical areas.

In this research a study strongly based on a first quantitative exploration of already existing data was conducted, in contrast to the common approaches that reduced or neglected the quantitative approach to a more qualitative, faster, and more flexible evaluation, which allows a more dynamic and generic vision. The contribution proposes an analytical framework to be applied in the exploration of fragile areas, which allows a certain flexibility and adaptability to different contexts and not limited to the single area of observation. A methodological approach (based on a multidimensional diagram structured in different sub-dimensions of analysis) and tools (indicators and cross-dimensional indexes) are proposed to study the resilience of marginal territories.

All this, according to a new significant assumption, designed and applied in B4R project (of which the authors are part), requires a shift from a "vulnerability approach" to a "vibrancy approach": no longer focusing on the weaknesses of the areas, but exploring their dynamism, i.e., the presence of actions/initiatives that give rise, or can contribute to the areas revitalization.

After an in-depth examination of the most recent scientific literature, which highlights an important gap, the paper describes the proposed methodological approach (which should be interpreted in a scalar way) and the multidimensional evaluation approach which, through cross-dimensional indexes, overcomes "sector-specific" visions and provides a broad vision of the territorial potentials.

2 Background

With the aim of highlighting the methodological perspectives recently adopted by scholars to investigate the territorial resilience, vibrancy and vulnerability by means of indicators, a synthetic literature review is here presented. The research was focused only on scientific works indexed on the most common databases and published in the last years. Geographically, no limits have been set, even if, at different scales, mainly European and Italian works have been taken into consideration.

To capitalize the researches already conducted by other authors on the basis of their results, appropriate contributions were thus included in our analysis by inputting in Scopus database (https://www.scopus.com, last accessed on July 22nd 2021) three different search scripts such as: "ALL (inner areas AND resilience OR vibrancy OR vulnerability)" which reports 150 results; "ALL (resilience AND vulnerability AND vibrancy) that reports 22 results and "ALL (inner areas AND development index) that reports 11 results.

These initial results were then analysed and reduced, and some research gaps emerged. Many scholars tackled the vulnerability and vibrancy topic in urban areas while only few others faced the problem to bridging the gap between urban/central areas and inner territorial areas [9–16]. Generally, the analysed papers can be divided in two main groups in relation to the geographical scale that they consider in the research: a regional or supra-municipal scale and an urban or sub-urban scale. For the present research we considered indicators belonging to both levels but that can be adapted to a village, not-urban, and inner area scale. At regional or national level there are EU official reports and very interesting development and resilience indicators [11, 17–24] and composite indexes [25, 26], suitable also for vulnerability/vibrancy research but not very adaptable for analyses at municipal or sub-municipal scale. The biggest and more specific part of the literature is related to cities and urban areas, with resilience, development, vulnerability, and vibrancy indicators [16, 27–31], but, also in this case, due to the different nature of the territory, most of these indicators are not suitable for inner areas or villages. In fact, some indicators are designed to have variability in urban sub-areas and so, when calculated at the territorial scale, do not identify differences between neighboring municipalities (e.g. Difference between male and female employment rate at the age of 15–64). Furthermore, there are some indicators referring to issues that do not exist in non-urban areas (e.g. urban transformation index based on the approval/adoption/drafting of urban plans) and, on the contrary, other key themes are missing. For example, the farm density and their innovation level represent indicators not relevant in urban contexts, but fundamental in studying inner areas. Finally, it is commonly recognized the criticality of data availability: in urban contexts POIs data are often available, while in suburban areas data are usually aggregated at municipal level and therefore less precise (e.g. New residential buildings and related dwellings by period of construction).

In Italy there are two main initiatives that are building and applying indicators to study and monitor the national territory, including inner areas: the National Strategy for Inner Areas (SNAI), that is trying to assess the resilience of inner areas/rural areas and villages [32], and the ICity Rank indicators built up to monitor and rank the Italian cities. The latter measures, every year, cities capacity to adaptation (smartness) on the way to get more dynamic cities, more functional, more ecological, more liveable, more manageable, more innovative and more capable of promoting sustainable development by reacting to ongoing socio-economical changes through the use of new technologies [33]. It is worth mentioning also the Regional Digital Economy and Society Index (DESI), which tracks the progress of EU countries regarding their digital competitiveness, but it is also developed at urban level in the Piedmont Region in Italy to build indexes on sustainability and marginality. The Regional DESI Sustainability Index is an experimental multidimensional index, finalized to analyse local potentials for sustainable development. The assessment methodology was based on the work carried out by Forum PA "ICity Rank – 2019 annual report", partially borrowing its methodology and applying on all Piedmont municipalities, which were evaluated within groups of "similar" municipalities, based on the number of residents (small municipalities under 1000 inhabitants, medium-small 1000–5000, medium large 5000–25000 and large over 25000) and the area reference altimetric (plain, hill, mountain) [34].

By analysing this literature, some research gaps emerged. First, the concept of "territorial vibrancy" is very low investigated in the literature, while those of "territorial resilience" and "territorial vulnerability" are now almost developed. In general, the indicators presented in the literature refer to small scales (regions and territories) but with rather updated data, while if the scale is enlarged and therefore granular and not aggregated information is desired, the only multidimensional indicators found are related to urban areas. Fortunately, the lack of timely updated data in recent years is going to be reduced, thanks to the commitment of the various European nations to make as much data as possible available in open data format. In Italy, however, the situation is still very different from region to region, and the data availability mainly depends on the topics to be analysed: in fact, obtaining data that are even indirectly related to the national privacy law is still very difficult.

The abovementioned SNAI, given its purposes, is carrying out the analysis of the territories with a very effective series of indicators which, however, have the limit of being generally one-dimensional.

3   Methodological Approach

In order to investigate the resilience of inner areas we developed a methodological approach based on the calculation of indicators and cross-dimensional indexes aimed at exploring vulnerability and vibrancy of fragile territories. Figure 1 shows its six phases.

Firstly, it is necessary to define the main objectives of the analysis, which can be almost general but specifically related to the territorial context that needs to be explored and evaluated (Phase 1). In fact, the analysed fragile area can be related to just a municipality or to a group of few or several municipalities.

Then, an extended selection of indicators has to be gathered on the basis of the objectives and the territorial scale of the analysis in order to create a suitable knowledge

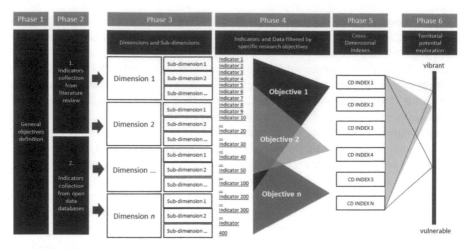

Fig. 1. The six phases of the methodological approach (Source: Authors' elaboration)

base, that is fundamental to set up the future development of the territory exploration (Phase 2). This phase must be supported by the literature, as well as by the analysis of the existing databases and data available for the areas to be explored.

The Phase 3 of the proposed approach is crucial to manage the numerous indicators previously identified. This important knowledge base has to be structured and organized by classifying each indicator in specific dimensions and sub-dimensions, which can change to represent the main issues of the analysis.

Therefore, more specific research objectives have to be outlined to more deeply analyse key aspects of the territory (Phase 4) and, according to them, a set of indicators can be selected to create cross-dimensional indexes (Phase 5). The previously identified dimensions and sub-dimensions can guide this selection even if it is highly suggested not to consider a single dimension/sub-dimension. It is important to highlight that the extended selection of indicators gathered during Phase 2 are not reduced, but just properly combined for the established purposes. The resulting cross-dimensional indexes can finally be used to further explore the territorial potentials (Phase 6). On this basis, different areas of the analysed territory can emerge as more or less vulnerable or vibrant, so that specific policies and actions can be strategically addressed.

4   Results

The proposed methodological approach was applied to explore the resilience of fragile territories. In particular, the general objective of the analysis (Phase 1) is to investigate the territorial potential of Italian small villages and inner areas in order to reactivate habitats and develop minimal tourist infrastructures as an engine for the development of more structural and resilient territories and local communities. Repopulation, local development and economic enhancement represent common challenges for several Italian territories: in this paper the four focus areas analysed in the context of the "Branding4Resilience" research project (https://www.branding4resilience.it/) were assumed as references. Although they are located in four Italian regions (Marche, Sicily, Trentino, Piedmont) and reflect different place identities, a common methodological approach can be applied to explore their potential and enhance their territories and communities.

To this aim a quantitative analysis based on specific indicators was set up: assuming the results achieved in other studies [9, 12, 15, 18, 21, 25, 26, 28], an extended selection of indicators was gathered in order to create a suitable knowledge base for the territory exploration (Phase 2). More than 400 indicators were identified and the main existing databases in the Italian context were analysed to check data availability for the considered areas. For example, the following data sources were examined: the Italian National Institute of Statistics (ISTAT), the National Strategy for Internal Areas (SNAI) Open Kit, the Open Coesione platform, Urban Index indicators, the Open Bilanci platform and different Geoportals at regional and municipal level.

Therefore, all the indicators were structured and organized in specific dimensions (Phase 3). Each dimension was then detailed in several sub-dimensions, able to gather and classify specific indicators (Fig. 2).

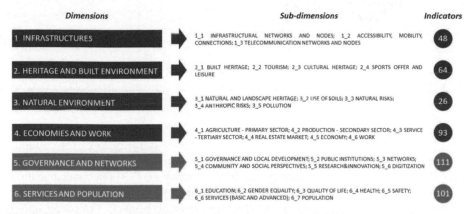

Fig. 2. Knowledge base dimensions, sub-dimensions and related indicators (Source: Authors' elaboration)

These dimensions and sub-dimensions supported not only the structuring and creation of a general knowledge base, but also the analysis of a series of specific issues of the territory able to highlight its vulnerability and vibrancy.

Therefore, it was necessary to outline more specific research objectives to address deeper analyses (Phase 4). For example, several key aspects were investigated such as the tourist offer, the commercial desertification, the presence of basic services, the capability of public administrations to attract funds, the presence of digital divide, the offer of sport and leisure activities, the presence of innovative companies. Figure 3 shows, for example, smooth and fast traffic networks, places of cultural or naturalistic interest and cultural and sports activities, to represent the entire cultural and leisure offer of the analysed area.

On this basis, a set of indicators from different sub-dimensions was selected and combined to create cross-dimensional indexes (Phase 5) and to further explore the territory vulnerability or vibrancy and bring out the territorial potentials (Phase 6). For

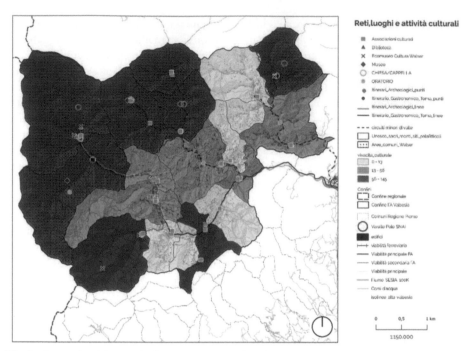

Fig. 3. Example of data and indicators representation: Cultural and leisure offer (Source: Authors'
elaboration)

example, according to the specific objective of creating infrastructures for cultural-
tourist fruition, several indicators from different sub-dimensions - 1.1 (Infrastructural
networks and nodes), 1.2 (Accessibility, mobility, connections), 1.3 (Telecommunication
networks and nodes), 2.2 (Tourism), 2.3 (Cultural heritage) 2.4 (Sports offer and leisure),
3.1 (Natural and landscape heritage), 4.3 (Service - tertiary sector), 5.3 (Networks), 5.5
(Research&Innovation) - were aggregated as follows:

$$CDI = f\left(Ind_{h,d};\ Ind_{k,d};\ Ind_{j,d};\ \ldots;\ Ind_{z,d}\right)$$

where CDI is the cross-dimensional index expressed by the different indicators ($Ind_{h,d}$,
etc.) belonging to the analysed different sub-dimensions.

5   Conclusion

The proposed methodological approach shows how quantitative analyses can support
the territorial resilience exploration of fragile areas, to which the current literature has
not yet provided an organic and comprehensive solution. The division into 6 subsequent phases allows to progressively build an important knowledge base scalable to
the specific territorial level of the research. The structure, organized in dimensions and
sub-dimensions, provides a more detailed overview context, enabling to select an appropriate set of dimensional indicators, depending on the survey specific objectives. The

construction of cross-dimensional indexes gives a cross-sectional reading of the territories, useful for targeted development policies based on specific territorial vulnerability and vibrancy issues. Therefore, the proposed approach can effectively support the territory exploration and lay the ground for future evaluations based on a possible integrated use of data, simple indicators and cross-dimensional indexes.

Nevertheless, this study highlighted some limitations which have to be considered in applying the proposed methodological approach to specific case studies. For example, the problem of data availability at municipal scale is still to be addressed, in striking contrast to the current era of big data. Also other critical aspects must be taken into account, such as: scaling constraints; problems with updating information; lack of availability of (open) source data, both punctual and aggregated; the need to set up continuous series of data to describe trends over time.

Starting from this first exploration, the selected case study (B4R research project) is currently addressing the challenge of planning at municipal level, in order to define specific and/or widespread projects that – starting from existing networks – can be the trigger for virtuous local actions.

Acknowledgements. B4R Branding4Resilience. Tourism infrastructure as a tool for the enhancement of small villages through resilient communities and new open habitats (Project number: 201735N7HP) is a research project of relevant national interest (PRIN 2017 - Young Line) funded by the Ministry of Education, University and Research (MIUR) (Italy) for the three - year period 2020 – 2023

References

1. Barca, S., Casavola, F., Lucatelli, P.: A strategy for inner areas in Italy: definition, objectives, tools and governance. Materiali Uval Series, Issue 31 (2014)
2. De Rossi, A.: Riabitare l'Italia: le aree interne tra abbandoni e riconquiste. Donzelli editore (2019)
3. Jachia, E., Osti, G.: AttivAree: un disegno di rinascita delle aree interne. Il Mulino (2020)
4. Brandano, M.G., Faggian, A., Urso, G.: Oltre le crisi. Rinnovamento, ricostruzione e sviluppo dei territori,le. Franco Angeli, Milano (2020)
5. Carrosio, G.: I margini al centro: l'Italia delle aree interne tra fragilità e innovazione. Donzelli editore, Roma (2019)
6. Gómez-Ullate, M.., Rieutort, L., Kamara, A., Santos, A.S., Pirra, A., Solís, M.G.: Demographic challenges in rural Europe and cases of resilience based on cultural heritage management. A comparative analysis in Mediterranean countries inner regions. Eur. Countrys. **12**, 408–431 (2020)
7. Martinelli, L.: L'Italia è bella dentro. Storie di resilienza, innovazione e ritorno nelle aree interne. Altreconomia, Milano (2020)
8. Ferretti, M., Favargiotti, S., Lino, B., Rolando, D.: B4R Branding4Resilience. Tourist infrastructures as a tool to enhance small villages by drawing resilient communities and new open habitats. In: Proceeding of the XXIII National Conference SIU DOWNSCALING, RIGHTSIZING. Contrazione demografica e riorganizzazione spaziale, pp. 346–354 (2021)
9. Mastronardi, L., Giagnacovo, M., Romagnoli, L.: Bridging regional gaps: community-based cooperatives as a tool for Italian inner areas resilience. Land Use Policy **99**, 104979 (2020)

10. Assumma, V., Bottero, M., De Angelis, E., Lourenço, J.M., Monaco, R., Soares, A.J.: A decision support system for territorial resilience assessment and planning: an application to the Douro Valley (Portugal). Sci. Total Environ. **756**, 143806 (2021)
11. Pontarollo, N., Serpieri, C.: JRC Technical Report. A composite policy tool to measure territorial resilience capacity (2018)
12. Pontarollo, N., Serpieri, C.: A composite policy tool to measure territorial resilience capacity. Socio-Econ. Plann. Sci. **70**, 100609 (2020)
13. Brunetta, G., et al.: Territorial resilience: toward a proactive meaning for spatial planning. Sustainability (Switzerland) **11**, 2286 (2019)
14. Pilone, E., Demichela, M., Baldissone, G.: The multi-risk assessment approach as a basis for the territorial resilience. Sustainability (Switzerland) **11**, 2612 (2019)
15. Hidalgo, D.M., Nunn, P.D., Beazley, H.: Uncovering multilayered vulnerability and resilience in rural villages in the pacific: a case study of Ono island, Fiji. Ecol. Society **26**, 26 (2021)
16. Olar, A., Jitea, I.-M.: Assessing the quality of local development strategies in Romania, evidence from 2014–2020 programming period. Manage. Econ. Eng. Agri. Rural Dev. **20**(2) (2020)
17. Fratesi, U., Perucca, G.: Territorial capital and the resilience of European regions. Ann. Reg. Sci. **60**, 241–264 (2018)
18. Di Pietro, F., Lecca, P., Salotti, S.: Regional economic resilience in the European Union: a numerical general equilibrium analysis. Spatial Econ. Anal. **18**, 287–312 (2021)
19. Giannakis, E., Bruggeman, A.: Regional disparities in economic resilience in the European Union across the urban–rural divide. Reg. Stud. **54**, 1200–1213 (2020)
20. Odei, S.A., Stejskal, J., Prokop, V.: Understanding territorial innovations in European regions: insights from radical and incremental innovative firms. Reg. Sci. Policy Pract. **13** (2021)
21. Graziano, P., Rizzi, P.: Resilience and vulnerability in European regions. Sci. Reg. **19**, 91–118 (2020)
22. Benczúr, P., Joossens, E., Manca, A.R., Menyhért, B., Zec, S.: JRC technical report. How resilient are the European regions? Evidence from the societal response to the 2008 financial crisis (2020)
23. Pontarollo, N., Serpieri, C.: Challenges and opportunities to regional renewal in the European Union. Int. Reg. Sci. Rev. **44**(1), 142–169 (2021)
24. Barbier-Gauchard, A., et al.: Towards a more resilient European Union after the COVID-19 crisis. Eurasian Econ. Rev. **11**, 321–348 (2021)
25. Booysen, F.: An overview and evaluation of composite indices of development. Soc. Indic. Res. **59**, 115–151 (2002)
26. Staníčková, M.: An overview and evaluation of methods for deriving composite indices of regional development in socio economic issues (2018)
27. Da Silva, J., Moench, M.: City Resilience Framework (2014)
28. Martín, C., et al.: Institutionalizing urban resilience: a midterm monitoring and evaluation report of 100 resilient cities. The Urban Institute (2018)
29. Barreca, A., Curto, R., Rolando, D.: Urban vibrancy: an emerging factor that spatially influences the real estate market. Sustainability (Switzerland) **12**, 1–23 (2020)
30. Barreca, A., Curto, R., Rolando, D.: Assessing social and territorial vulnerability on real estate submarkets. Buildings **7**(4), 94 (2017)
31. Barreca, A., Curto, R., Rolando, D.: Is the real estate market of new housing stock influenced by urban vibrancy? Complexity **2020**, 22 (2020)
32. Adam-Hernández, A., Harteisen, U.: A proposed framework for rural resilience – How can peripheral village communities in Europe shape change? Ager. Rev. de Estud. sobre Despoblación y Desarrollo Rural (28), 7–42 (2020)
33. Dominici, G., Fichera, D., Musacchio, C.: ICity Rank Annual Report (2019)
34. Piemonte, O.: Indice di sostenibilità. La classifica dei comuni Piemontesi per classi (2019)

Fostering Resilience in Inner Areas. The Sicani Case Study in Sicily

Barbara Lino[✉] and Annalisa Contato

Department of Architecture, University of Palermo, Palermo, Italy
{barbara.lino,annalisa.contato}@unipa.it

Abstract. In recent years, the theme of the regeneration of small centres in fragile and marginal areas has returned to the centre of the Italian debate and public policies, as well as resilience practices in which culturally-based regeneration projects, community cooperatives and resettlement processes are based on a spatial dimension capable of playing an active and unprecedented role.

The COVID-19 pandemic we are currently experiencing has reignited the debate on fragile areas, reinforcing the perception of those territorial inequalities between metropolitan and inner areas exacerbated by the crisis. In fact, the pandemic has further highlighted the serious shortcomings in the provision of services – first and foremost health and education services – and in the infrastructure of these territories, both material and digital, and has revealed the limits of incessant urbanisation and the concentration of settlement and infrastructure policies in large conurbations.

Starting from the framework of the research project of national interest "B4R Brand-ing4Resilience" and continuing in the wake opened by the SNAI but also overcoming it, Unipa Unit Research aims to define the need for a more inclusive settlement model in the Sicani Territory in Sicily (Italy) in order to rebalance the existing asymmetries by recharging the peripheral areas with a new centrality.

Keywords: Co-creative communities · Inner areas · Local development

1 Re-inhabiting Inner Areas: How Is This Possible?

1.1 The Vulnerability Factors of Inner Areas and Their Potentialities

Finding perspectives for inner areas is a highly relevant contemporary issue in Italy, whose territory is made up of a dense and differentiated network of urban-rural poles, which are disconnected from each other and present different levels of spatial and relational peripherality (Copus and Noguera 2016).

After the process that since the 50's has led to the gradual abandonment of inner areas and despite the constant presence of the metropolitan dimension in public policies, the territories on the margin have returned to become visible through public and collective actions in the last years (De Rossi 2018): especially the National Strategy for Inner Areas[1] (SNAI) (DPS 2013) has overturned the debate on the paradigms of the past, going beyond the opposition city/countryside and accompanying a process that guarantees access to basic services, but that also allows to promote medium/long-term policies attentive to the specificities of contexts and with a place-based approach.

These areas are characterised by objective difficulties that hinder their liveability and development, such as: lack of access to services and other opportunities such as work and social interaction; insufficient infrastructural interconnection, with repercussions also on productive development, both for the difficult supply of raw materials and for the connection with logistical nodes to access outlet markets (a fabric of diffuse micro-economy in decline characterises these territories); low quality of life of citizens in terms of social unease due to difficulties in accessing basic services such as education, health and digital connectivity. Furthermore, the experience of the COVID-19 pandemic has highlighted in all its rawness territorial differences, in terms of medical assistance and digital divide: rural communities need more and better digital connectivity to compensate for their remoteness.

SNAI – recently included among the 55 actions defined by the Italian Government for the post-COVID-19 economic recovery plan within the Next Generation – provides important lessons, in terms of the need to give prominence to the territories and space for co-design, the reconstruction of a new identity and renewed confidence in marginal areas, and the importance of reintroducing the territory into the policy horizon.

The wake opened up by the SNAI should be pursued and even surpassed, for example starting from the spatialization of policies that reactivate marginal spaces, not necessarily coinciding with the inland areas alone, and that place themselves from an observation point "not from the 'centres' towards the 'margins', but from the 'margins' themselves" (De Rossi and Mascino 2020: 51).

In order to return to inhabit inner areas of the Country, new infrastructures and basic services are needed, but also new perspectives are needed that solicit tools, policies and projects capable of radically changes of models of production and consumption and, to make these territories re-habitable by triggering new trajectories of regeneration. As widely testified also by the recent book "Aree Interne e Covid" (Fenu 2020), the country must commit itself to overcoming territorial disparities that are no longer acceptable, giving marginal areas a new centrality through policies and projects that look at the extreme variety of environmental, cultural and economic resources and, as stated in the recent "Manifesto per riabitare l'Italia" (Cersosimo and Donzelli 2020), territorial policies in which marginal areas find a renewed centrality.

In the face of obvious social, infrastructural and economic disadvantages, in several cases some communities of the inner areas are already trying to "diversify themselves locally" leveraging on relevant enabling context conditions, such as: a strong local identity in terms of quality of life, architectural and natural quality, a wide range

---

[1] The Italian national strategy named "Strategia Nazionale Aree Interne" (SNAI) represents an existing and emerging governance approach looking at official authorities but also at informal governance groups to overcome the classic opposition between rural and urban. SNAI has fuelled the attention of Italian policymakers towards the need for improving socio-economic conditions of people living in inner areas.

of exploitation of social and territorial capital in terms of both physical and human resources, accessibility of the real estate market and low cost of buildings, in front of a high architectural quality and typical characters to recover (Carta et al. 2018).

Metropolitan cities and inner territories may be 'opposites' in terms of living and working conditions, but they should cooperate and become interdependent, while preserving their different identities (Lino 2020). These areas "must be thought of and planned, on the one hand, as recipients of collective goods and services, and, on the other, as areas capable of producing and offering collective goods, which respond to needs expressed by the whole of society, and which take the form of services capable of strengthening new links between inland areas and cities" (Meloni 2020: 146). The small centres of the inner areas can become an extended cooperative system of centres linked to productive territories and metropolitan cities, new rural-urban archipelagos in which each municipality shares dwellings, public spaces and facilities, and contributes to balancing rural, urban and territorial development (Carta 2017). Therefore, collaboration between large urban nodes and rural areas must be promoted, moving towards incremental integration and interconnection, in terms of scale and territorial differences: from the network of small urban centres and rur-urban nodes to the development of their specialised function of porous hinge with the large poles/metropolitan cities in order to promote the creation of a large polycentric sub-regional system networked with others in the same region (Contato 2019).

1.2 The Objectives of the Research Branding for Resilience and the Case Study Sicani in Sicily

1.2.1 The Research Branding for Resilience

The research project "B4R Branding4Resilience. Tourist infrastructure as a tool to enhance small towns by drawing resilient communities and new open habitats"[2], aims to change the current imagery and make room for positions and areas of interest that can trigger new development dynamics for inner areas by addressing them as drivers of innovation and test fields for new development dynamics, looking at the potentials and resources specifically related to space, settlements and landscapes (Schröder et al. 2018). The project aims to develop exploratory scenarios (Ferretti and Schröder 2018) and relational models that can open up design-driven knowledge production for broader spatial strategies that support administrations in formulating development policies.

The B4R research project is structured in five phases, corresponding to as many work packages. Two of these phases are continuous during the entire project (Coordination, Communication, and Dissemination), while the other three, the main operative phases, used to carry out the research project. The first phase is 'Exploration', an analytic and mapping phase with the aim to analyse in depth the territorial context being researched.

---

[2] Branding4Resilience is a research project of national interest (PRIN 2017 Young Line) funded by the Ministry of Education, University and Research (MIUR) with a three years duration (2020–2023). The project is coordinated by the Università Politecnica delle Marche (national coordinator Maddalena Ferretti) and it involves as partners the Università degli Studi di Palermo (local coordinator Barbara Lino), the Università di Trento (local coordinator Sara Favargiotti) and the Politecnico di Torino (local coordinator Diana Rolando).

After the analysis and the interpretation of the data and dynamics of the territorial context, each Research Unit select a Focus Area (FA) for intervention towards regeneration. This phase's aim is to bring out networks, multi-scalar interactions, and intersections between territories and material and immaterial resources. For this phase, qualitative and quantitative tools have been defined with the aim to analyse each territory not only in its spatial dimensions, but also in its immaterial dimensions, such as formal and informal relationships, community identities, traditions, people, and the creativity of bottom-up initiatives. Within the framework of the project, the Research Unit of the University of Palermo[3] works in Sicani territory, in Southern Sicily, focusing on the topic of "Co-creative communities". The second operative phase is the 'Co-Design' with communities, with the goal of proposing useful transformations of small infrastructures in selected villages. The UniPA Research Unit explored the Sicani territory in November 2021, deepening and promoting a participatory co-design workshop in Santo Stefano Quisquina with the community. The third phase, that will be developed in next year, is the "Co-Visioning" processes in collaboration with local actors.

1.2.2   The Case Study Sicani in Sicily

The Focus Area (FA from now on) is made up of 18 municipalities and is situated in an intermediate position between the cities of Palermo and Agrigento, from North to South, and between the cities of Trapani and Caltanissetta, from West to East. It is part of a large portion of the regional territory located on the fringes of the polarisation of the large coastal urban areas (Palermo, Catania and Messina) (Fig. 1).

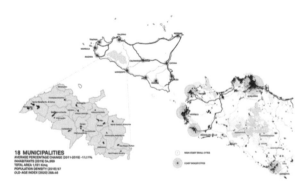

Fig. 1.   The Sicani focus area in southern Sicily. Source: ©Branding4Resilience, 2020–2023. Data processing and graphics: Barbara Lino and Annalisa Contato.

The 18 municipalities of the FA are non-coastal, hilly or mountainous centres whose territorial surface is spatially contiguous, and with a settled population per municipality of less than 7,000 inhabitants. The settlement structure is characterised by an archipelago of small urban centres (with a discontinuous and sparse pattern) and rural villages.

---

[3] The Unit Research of the University of Palermo is composed by: Barbara Lino (local coordinator), Annalisa Contato, Mauro Ferrante, Giovanni Frazzica, Luciana Macaluso, Francesca Sabatini.

From the point of view of accessibility, the FA shows evident criticalities. The area is connected to the nearest urban poles (Agrigento, Palermo and Castelvetrano) through two State Roads that cross the territory from North to South and from East to West. The A29 motorway does not serve any municipality, with the result that the average travel time to reach Palermo, the Sicily's capital, by private vehicle is about 1:50 h (with peaks of 2:40 h for the municipalities closest to Agrigento) and about 1:20 h to reach Agrigento (with peaks of 2 h for municipalities closest to Castelvetrano). The two airports, Birgi and Falcone-Borsellino, are on average 2 to 3 h away from the FA. In addition, critical issues also concern rail transport: a vast network of narrow-gauge railways that ran through the territory has now been completely decommissioned, with the result that no municipality in the FA has an active railway stop. Public transport connections are also poorly served and with a low frequency to the urban centres.

As regards the resources present in the FA, the territory is characterised by an uncontaminated rural landscape, archaeological sites far from the traditional tourist circuits and a precious historical heritage.

From a naturalistic and landscape point of view, there are important rivers. The richness of water is also characterised by the presence of lakes and thermal springs that express potential also from the point of view of a tourist-receptive use linked to health and well-being. There are also several nature reserves and numerous Sites of Community Importance (SIC) and Special Protection Area (ZPS) (Fig. 2).

Fig. 2. The Sicani focus area in southern Sicily: Sicani landscape. Source: ©Branding4Resilience, 2020–2023. Photo by Sandro Scalia for B4R.

Regarding the cultural-identitary resources, in the area there is an historical centre of ancient origin, 7 historical centres of medieval origin and 10 historical centres of new foundation. There are also 159 properties scattered among military, religious, residential and productive architectural buildings, and 30 museums.

Cultural events play an important role in keeping the territory alive in terms of culture and traditions and have the main purpose of highlighting the identity characteristics of the places, products of excellence, natural beauty and promoting art. This territory is also rich in typical products of excellence, quality craftsmanship and there are two Slow food presidia. These excellences of productions and the strong link between agri-food and the territory of origin is demonstrated by the presence of important DOP, IGP and IGT marks (products entered in the Register of Protected Designations of Origin, Protected Geographical Indication and, Typical Geographical Identification).

The richness of the villages in these territories has also received national recognition, such as the title of "Borgo dei borghi", awarded to Sambuca di Sicilia in 2016, and the title of "Borgo autentico d'Italia" awarded to Bisacquino.

Despite the evidence of marginality, such as low density, aging population, increasing emigration and socio-economic weaknesses, the Sicani area has several experiences that are generating an innovative social dimension: new eco-creative communities and neo-rural practices are emerging, such as repopulation processes resulting from public policies and private initiatives and relational tourism experiences.

2 Strategies, Policies and Co-creative Communities: Tracks of Vibrancy in Sicani Territory

2.1 Planning Dynamism

The Sicani territory have matured in recent years an intense vivacity of the system of local governance that is generating a significant proliferation of programs, coalitions and networks. There are the following local development policies: an area included in the SNAI, 2 LAGs (Local Action Groups), 4 Unions of Municipalities, 1 Generalist Territorial Pact, 1 Integrated Territorial Plan, 1 Integrated Territorial Development Plan, 1 Wine Route, 2 Tourist Districts. In addition, there are: membership of five national associations for the promotion of the territory (Borghi Autentici d'Italia, Città delle Ciliegie, Città del Bio, Città del vino, Comuni termali) and membership of a "Cultural Route of the Council of Europe" certified in 2009 (Iter Vitis). This shows how this territory is very active from the point of view of local development planning, as well as the existence of aggregative networks between municipalities with variable geography, that is relevant in terms of capacity building and opportunity in future development.

Among the various territorial coalitions resulting from local development programs and community initiatives that have been activated in the territory, the Sicani Inner Area stands out for its relevance and, the "GAL Sicani" (LAG), which has distinguished itself for a lively planning activity developed in the last two programming cycles.

The general objective of the Participatory Local Development Strategy (SSLTP) of the LAG is the strengthening and further development of the Sicani Rural Quality District[4] (DRQ Sicani) established in the previous programming 2007/2013 (PSL Gal Sicani 2007–2013). An important recent initiative of the LAG is the creation, in the 29 municipalities of the District, of the DRQ Sicani Functional Municipal Units, living labs in municipally owned assets that will aim to promote the territory and provide services to the communities. In addition, actions are planned to network municipalities and associations in the area, along with more than 30 interventions on municipal real estate now in disuse, investments also for outdoor activities, the strengthening of the Magna Via

---

[4] DRQ Sicani is a network of agricultural activities and production of goods and services that through the initiatives and funding of the PSL 2007–2013 has a collective brand in order to certify and enhance local products, natural resources and crafts and tourism and business activities.

Francigena and Via Francigena Mazzarense, in the sections where the paths intercept the municipalities belonging to the "GAL Sicani", the creation of new tourist routes or the improvement of paths within natural areas or historical centres to be enhanced. All these initiatives will be supported by the call of the PSR Sicily 2014–2020, measure 19 of the Local Development Strategy LEADER", Action of the PAL "Living and travelling in the Sicani DRQ".

On the whole, the different programs manifest the will to enhance the existing resources (cultural, landscape, natural, productive and tourist), promoting the territory, aiming at the strengthening of the job offer and the agricultural local production system, the reduction of the migration process in place and, therefore, the aging of the population, in addition to intervene in the provision of basic services. The programmatic policies, return at the same time a complex and fragmented picture, and the coexistence of different control rooms.

2.2 Co-creative Communities

In marginal territories is possible to recognise development trajectories that can be traced back to practices rooted in the territories and innovations produced locally. Some communities are already trying to "diversify locally" by leveraging relevant enabling context conditions (Carta et al. 2018).

In some cases, these practices are as fragile as the places on which they insist, but they must be observed carefully. There are communities that are trying to implement actions and strategics to encourage repopulation, through economic and fiscal incentives, or the valorisation of abandoned residential building heritage or, again, support for artistic production. In other cases, small communities offer land on a free loan to reactivate agriculture or – as in the case of community cooperatives – experimental forms of collaborative economics aimed at responding to new social needs, creating networks and community ties and, at the same time, proposing innovative and shared hybrid services, attentive to sustainability and environmental protection. In still other cases, different forms of tourism are promoted in which travellers no longer passively experience the territory, but become protagonists because they bring and exchange skills and values with the territory, its resources and inhabitants (Lino 2021).

In all these cases, resilience manifests itself as a cohesive and driving force. Working with an adaptive capacity and through innovative forms of local self-organisation, the practices in place draw energy from the characteristics of the spatial and social context, use (with new meanings) local identity resources (spaces, social capital, landscape and cultural heritage) and through a symbolic mediation operation, create shared value (economic, social, cultural), stimulate active community collaboration, modify spaces, attract new population and retain younger generations.

Despite the marginality of the Sicani area, is possible to recognize in it is resilience capability, is regenerative force thanks to its communities that from below give life to different experiences that are generating an new social dimension, co-creative communities and, neo-rural practices, such as repopulation processes resulting from public policies and private initiatives, creative activities and relational tourism experiences.

In Santo Stefano Quisquina, a shepherd poet has transformed his farm into an event venue and open-air gallery, the Andromeda Theatre, which attracts many international visitors. In addition, the community, through the strength of the associations present, is promoting the valorisation of the water resource and other actions aimed at territory promotion and reusing disused buildings and spaces in culturally active forms open to the community. This municipality was the place chosen to carry out (in November 2021) the first co-design workshop of the B4R project in the Focus Areas[5]. The co-design workshop in Santo Stefano Quisquina "Traces of Water, Community Imagery and Creativity" was conceived as a moment of approach to the territory in which, through design explorations, some of the central questions posed by the research theme were verified and explored: possible operational actions of territorial branding were explored, hinging on the theme of the water resource as a cultural paradigm and source of collective imagination, resilience and creativity for the its community. The metaphor of water has been declined in three design visions, three forms of water traces, and three project areas have been identified to test some possible transformation scenarios. The aim was to make water not only a resource to be guarded and protected, but the element on which to reconstruct the collective territorial imaginary, starting from the enhancement of the various water traces present and urban regeneration, the rethinking of forms of tourist reception, the unconventional use of the landscape and the valorisation of the human capital of artists and craftsmen present: the co-creative communities.

In Cianciana, in recent years, new inhabitants have settled from northern Europe and the United States in search of new models of living and working. In Sant'Angelo Muxaro, the community is exploring forms of relational tourism thanks to a local cooperative that promotes the area and is in dialogue with the LAG to set up the "Rete dei borghi Sicani" (Sicani Villages Network); the municipality is also promoting the "Re-Generation Project", which involves artists and the local community in regenerating the neighbourhoods of the small centre with street art interventions.

Another interesting ongoing process is that of Sambuca di Sicilia. In the context of the "Borghi italiani" – carried out by Airbnb under the curatorship of Federica Sala, in collaboration with MiBACT and under the patronage of Anci – with the aim of promoting the enhancement of 40 Italian villages, the project to enhance the value of "Casa Panitteri" was completed in 2018. An artist's residence located inside the Palazzo of the same name, which also houses the Archaeological Museum, and which can accommodate travellers and boast the visibility offered for a year on the homepage of the Airbnb platform, which has donated the works of art and its commission to the local administration. Owned by the municipality, the proceeds from the accommodation business are used to finance cultural projects aimed at the local community and initiatives to enhance tourism. Also, in 2019 the "Houses for 1 euro" initiative for the redevelopment buildings in the historic centre and in the Arab quarter of the seven alleyways has been a great boost, stimulating the private property market and reactivating other abandoned buildings (Lino 2021). The aim was to facilitate a process that was already under way and which in recent years has seen around 20 families of different nationalities buy a house and move in as residents (both

---

[5] Each Unit Research have to organize a co-design workshop to explore a part of the Focus Areas according to the paradigmatic nature of the issues it raises, with the aim of identifying possible operational branding actions which, imagined for the territory covered by the workshop, can represent scalable and adaptable responses for the entire FA, as well as potentially significant for the other FAs in the project.

permanent and temporary). With this initiative, the municipality triggered benefits not only from a tourist point of view but also from an entrepreneurial one, through a virtuous mechanism that creates work and development and stops the process of progressive depopulation. In July 2021, the municipal administration decided to repeat the initiative with the issuance of the "Announcement of alienation of municipal real estate houses at 2 euros", which provides for the auction of 18 properties, in need of renovation works.

3 Conclusions

The marginal territories of Sicily, reached through the lens of a new vision of development ask to set in motion an economic and social model that is able to reactivate latent resources, retain and attract the population and propose competitive models of life compared to those offered by large urban areas (Carta et al. 2017).

The research investigates the potentials of Sicani area and of territorial branding in drawing a resilient development, making rural-urban contexts attractive to new residents and users and transforming them into resilient models for communities. The analyses carried out have shown a stratified system of natural and cultural resources; tourism activities which may represent a potential for development and experiential tourism model experimentation; an important agricultural vocation with typical products of excellence; triggering of policies and spontaneous trends aimed at supporting forms of re-housing and regeneration trajectories; a collective brand identifying the Sicani's quality productions and tourist attractions as the "Sicani Rural Quality District"; the overlapping in the territories of an increasing number of control rooms, programs and projects of European and national origin that can be representative of an interesting form of local planning dynamism but also that asks for a not fragmented vision of development.

Therefore, the small centres of the Sicani area can be understood as a premise for: a greater cooperation between the coastal area close to the cities of Agrigento (to the South) and Palermo (to the North), transforming this "peripheral" territory into an innovative model of living and working; the opportunities linked to the built, natural and human capital can be more effectively linked to forms of relational tourism, through networking, the implementation of minimum infrastructures, but also to agriculture and entrepreneurial innovation, to creative districts, thanks to urban policies favouring spin-offs and spin-offs; the particular constellation of people already present – new settlers, temporary citizens and travellers – suggests that in the Sicani area the tourism of the future will be able to merge with different work/life models based on multiplace living.

The research perspectives, therefore, will move towards exploring how strategies, policies and projects can contribute to foster resilience, to define new alliances with coastal cities, to provide a new active settlement habitat for existing communities and potential new inhabitants, and to enabling innovation processes that, although still fragile, can be discerned in the signs of resilience and transformative trends detected. In addition, it will be necessary to recompose in a choral and coherent vision the ambitions of this community, whose resilience will have to strengthen the ability to "diversify locally" (Lino 2017a), drawing on the characteristics of the context, proposing innovation processes that are not limited to modifying the product but that forge new meanings and visions to enhance local identity resources, create shared value and new economies, stimulate active cooperation and attract new population (Lino 2017b).

References

Carta, M.: Planning for the rur-urban anthropocene. In: Schröder, J., Carta, M., Ferretti, M., Lino, B. (eds.) Territories. Rural-Urban strategies, pp. 36–53. jovis Verlag GmbH, Berlin, German, (2017)

Carta, M., Contato, A., Orlando, M. (eds.): Pianificare l'innovazione locale. Strategie e progetti per lo sviluppo locale creativo: l'esperienza del SicaniLab. FrancoAngeli, Milano, Italy (2017)

Carta, M., Lino, B., Orlando, M.: Innovazione sociale e creatività. Nuovi scenari di sviluppo per il territorio sicano. ASUR **123**, 140–162 (2018)

Cersosimo, D., Donzelli, C. (eds.): Manifesto per riabitare l'Italia. Donzelli Editore, Roma, Italy (2020)

Contato, A.: Policentrismo reticolare. Teorie, approcci e modelli per lo sviluppo territoriale. Franco Angeli, Milano, Italy (2019)

Copus, A., Noguera, J.: Le "periferie interne". Che cosa sono e di quali politiche necessitano?. Agriregionieuropa **12**(45), 10–13 (2016)

De Rossi, A. (ed.): Riabitare l'Italia. Comunità e territori tra abbandoni e riconquiste. Donzelli, Roma, Italy (2018)

De Rossi, A., Mascino, L.: Sull'importanza di spazio e territorio nel progetto delle Aree Interne. In: Fenu, N. (ed.) Aree Interne e Covid, pp. 48–54. LetteraVentidue, Siracusa, Italy (2020)

DPS-Dipartimento per lo Sviluppo e la Coesione Economica.: Strategia Nazionale per le Aree Interne: definizione, obiettivi, strumenti e governante. Accordo di partenariato 2014–2020. Dipartimento per lo Sviluppo e la Coesione Economica, Roma, Italy (2013)

Fenu, N. (ed.): Aree Interne e Covid. LetteraVentidue, Siracusa, Italy (2020)

Ferretti, M., Schröder, J.: Scenarios and Patterns for Regiobranding. jovis Verlag GmbH, Berlin, German (2018)

Lino, B.: A New Rur-Urban Utopia? Social innovation and the case of the Sicani area in Sicily. In: Schröder, J., Carta, M., Ferretti, M., Lino, B. (eds.) Territories. Rural-Urban strategies, pp. 110–117. jovis Verlag GmbH, Berlin (2017a)

Lino, B.: Innovazione sociale e triplice dimensione della connettività come asset strategici per i Sicani. In: Carta, M., Contato, A., Orlando, M. (eds.) Pianificare l'innovazione locale. Strategie e progetti per lo sviluppo locale creativo: l'esperienza del SicaniLab, pp. 41–47. FrancoAngeli, Milano, Italy (2017b)

Lino, B.: Branding as a lever for resilient transformation. In Transforming Peripheries Magazine. Topos. The international review of landscape architecture and urban design European landscape magazine, Special Issues (2020)

Lino, B.: Ri-generazioni a Sambuca di Sicilia. In: Ferreri, F. (ed.) Case a 1 euro nei borghi d'Italia. Sambuca di Sicilia: un esempio di successo nel governo del territorio, pp. 60–82. Dario Flaccovio Editore, Palermo, Italy (2021)

Meloni, B.: Aree Interne, multifunzionalità e rapporto con le città medie. In: Fenu, N. (ed.) Aree Interne e Covid, pp. 142–175. LetteraVentidue, Siracusa, Italy (2020)

Schröder, J., Carta, M., Ferretti, M., Lino, B. (eds.): Dynamics of Periphery Atlas for Emerging Creative Resilient Habitats. Jovis Verlag GmbH, Berlin, Germany (2018)

"Fit to 55": Financial Impacts of Italian Incentive Measures for the Efficiency of the Building Stock and the Revitalization of Fragile Areas

Manuela Rebaudengo[1] (ID), Umberto Mecca[2](✉) (ID), and Alessia Gotta[2]

[1] Interuniversity Department of Regional and Urban Studies and Planning, Politecnico di Torino, Viale Mattioli 39, 10125 Turin, Italy
[2] Architecture and Design Department, Politecnico di Torino, Viale Mattioli 39, 10125 Turin, Italy
`umberto.mecca@polito.it`

Abstract. Starting from the strategic initiative presented in 2020 by the European Commission and entitled "A wave of renovations for Europe: greening buildings, creating jobs and improving lives", Member States are now committed to stepping up efforts to renovate their building stock, with the precise goal of achieving climate neutrality in 2050. With the aim of exploring the potential appeal of incentive measures, the paper started with some considerations on the use and ownership of the building stock in Italy, frequently with a lack of maintenance. The proposed method, set up in three successive steps, verifies (for a hypothetical owner) the procedural feasibility and the financial sustainability of a new recent measure of energy efficiency, the *Superbonus 110%*. The selected case study is a complex of three buildings, built between the sixties and seventies in response to the strong demand for second homes and located in a small mountain municipality in the North-West of Italy. The results showed that the intervention allows, on the one hand, energy efficiency and, on the other hand, the initial and long-term convenience for the owner. However, a common strategy is needed for such policies, which -if left to the actions of individuals- may counteract the current potential for territorial resilience that seems to stimulate, especially in fragile territories.

Keywords: Energy efficiency · Financial sustainability · Territorial resilience · Decarbonisation · Real estate appraisal · Maintenance

1 Introduction

Extreme weather events, rising sea levels, ocean acidification and loss of biodiversity are tangible effects of the climate change affecting our planet. In order to cope with them and prevent even more catastrophic scenarios, it is now clear that we need to take decisive action on human activities, regulating them so as to limit carbon dioxide emissions into the atmosphere as much as possible. This is why, in 2019, the European Commission,

F. Calabrò et al. (Eds.): NMP 2022, LNNS 482, pp. 201–210, 2022.
https://doi.org/10.1007/978-3-031-06825-6_20

with the European Green Deal (or European Green Pact), proposed a series of policy initiatives to achieve climate neutrality in Europe by 2050.

To achieve this ambitious goal, a balance must be reached between CO_2 emissions and their absorption. However, this is not easy because in nature, the systems capable of absorbing carbon dioxide (also known as sinks), i.e. soils, forests and oceans, are not currently able to cope with the quantities emitted, which are estimated to exceed their absorption capacity by a factor of three [1].

Therefore, the EU, in order to reach its goal to become the first continent to remove as much carbon dioxide as it produces by 2050, must act on several fronts and in several sectors. Certainly a key role is played by the construction sector. In fact, buildings are the main consumer of energy (around 40% of the energy consumed in the EU) and are responsible for around 36% of energy-related CO_2 emissions due to their operation, in particular heating, cooling, and hot water [2].

This is why, in autumn 2020, a strategic initiative was presented by the Commission with the aim of stepping up efforts to renovate Europe's building stock. The initiative, entitled "A wave of renovations for Europe: greening buildings, creating jobs and improving lives" was conceived with the precise goal of achieving climate neutrality in 2050; however, it brings an additional objective: to contribute to the economic recovery of the EU from the COVID-19 pandemic [3].

To follow up on this strategy and make it a concrete legislative action, in July 2021 the Commission issued the "Fit for 55" package (recently revised: 15.12.2021), which sets out a series of initiatives that aim to make Europe's building stock more efficient and less dependent on fossil fuels, in two words: more resilient [4]. From 2027 all new public buildings are to be zero-emission and, from 2030, the same rule will apply to (new) private buildings. With regard to buildings, Member States are required to upgrade at least 15% of the buildings with the worst energy performance in order to improve their energy class from G to F. From the point of view of timing, this last requirement must be implemented within two different timescales depending on the use of the building: 2027 for non-residential buildings and 2030 for residential buildings [5].

The implementation of these actions will not only contribute to achieving climate neutrality, but also to alleviating energy poverty. It is estimated that there are more than 30 million extremely energy-hungry buildings in Europe, consuming two and a half times more than the average, and encouraging them to become more energy efficient will reduce their energy needs and thus their energy bills [5].

2 Background

At European level, a large number of buildings were constructed before 1970, i.e. before the first energy laws came into force [6]. it is easy to see, therefore, that action in this area is absolutely necessary. Obviously, a major limiting factor is the spending capacity of the owners of the buildings to be upgraded; therefore, significant and concrete economic support is needed to ensure that the measures can be implemented. Therefore, rather than through taxation, perhaps it would be useful to follow the example of some European countries (e.g. Belgium, France, Portugal, Italy, etc.) that have adopted tax incentives in various sectors, from residential to commercial or even public administration, precisely to stimulate intervention on buildings.

At the Italian level, a number of incentive measures have been introduced for some years, but the most promising one, due to the rates envisaged, was introduced in mid-2020: the 110% Superbonus (SB110). This new measure, which will be described later, is potentially available for all residential buildings that need to be made more efficient (provided that certain basic requirements are met) and therefore, the expectations related to the efficiency of the Italian building stock are certainly very high. Analyses carried out by the "Agenzia delle Entrate" (that is the Italian Inland Revenue Service) show that the number of residential buildings available, i.e. those not rented out and not inhabited continuously, exceeds the 20% threshold in 3,340 Italian municipalities [7].

If we exclude those with a strong tourist connotation, such as those located on the islands, or in the Ligurian coastal area, or even in mountain municipalities that, especially in winter, become popular ski destinations, it is easy to see that the remainder are inland municipalities (Fig. 1), i.e. those portions of the territory that over the years have seen a progressive depopulation, a general impoverishment and a consequent loss of value of the properties caused by the lack of demand on the market (also known as fragile areas or inner areas [8, 9]).

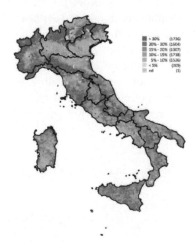

Fig. 1. Municipal distribution of the percentage of available housing (Source: Agenzia Entrate [7])

Therefore, in these places, the SB110 measure, together with the implementation of other services (e.g.: improvement of mobility services, upgrading of existing connections to the web, etc.), could become an important boost for their revitalisation. Thanks to this measure (and not only), buildings could be thoroughly renovated and made more efficient, better adapted to current needs and improved in terms of aesthetics and management [10].

The new incentive measure of SB 110, aimed at stimulating energy efficiency interventions in buildings for residential use, was created with a twofold purpose: to reduce energy consumption related to the use of housing units, reducing carbon dioxide emissions, and to stimulate the economic recovery of a sector that has been stagnant for years.

The measure, introduced by the Relaunch Decree in 2020, was subject to numerous clarifications and amendments, especially with regard to the time limits that make it applicable. For example, it was recently refinanced through the new Budget Law (approved at the end of 2021) and extended again, with different deadlines depending on the type of property unit (condominium, single property etc.) [11].

The measure provides for three different facilitation mechanisms: direct deduction (DD), invoice discount (ID) and credit transfer (CT) [12].

The direct deduction consists in the possibility of deducting 110% of the amount of the eligible works in 5 equal annual tranches up to the annual tax sum resulting from the income tax return. On the other hand, with the discount on the invoice, the beneficiaries will be able to ask their suppliers (of goods and services) for an advance contribution not exceeding the cost of the work itself. In practice, the company carrying out the works or the supply of materials or the provision of services will advance the contractually agreed amount from its own pocket and may subsequently transfer the accrued credit to banks or financial institutions. Lastly, the credit transfer mechanism provides that the beneficiaries can transfer the portion of the deduction due to them to a credit institution or financial intermediary, to companies and bodies, to other private individuals, to self-employed persons or companies, or to suppliers of goods and services [11].

Moreover, in order to be granted, the measure requires certain conditions to be verified in advance, the strictest of which concerns the actual state of the building. In particular, the condition must coincide with what was actually authorised by the competent authority at the time of construction (or subsequent restructuring).

There are essentially two types of eligible interventions: the main (or driving) ones and the secondary (or driven) ones, to be carried out together with at least one of the main interventions (Table 1).

Table 1. Summary description of the main and secondary interventions described in the Relaunch Decree

Main (or driving) interventions	Secondary (or driven) interventions
- thermal insulation of envelopes - replacement of winter air-conditioning systems in common areas - replacement of winter air-conditioning systems on single-family buildings or on building units of functionally independent multi-family buildings* - anti-seismic interventions	- energy efficiency interventions - installation of solar photovoltaic systems - infrastructure for recharging electric vehicles - interventions to eliminate architectural barriers

* a building unit is functionally independent if it is equipped with at least three of the following installations or objects of exclusive ownership: water supply systems; gas installations; electricity installations; winter air conditioning system [13]

In any case, in order to receive the contribution, it must be demonstrated that the interventions carried out have led to an improvement in the energy class (at least a double jump in class, e.g. from class G to class E) between the pre-intervention situation and the post-intervention situation. Furthermore, the maximum expenditure ceilings established by the Relaunch Decree must be respected, both as regards the total expenditure of each intervention and their individual parametric costs.

3 Materials and Method

Given the complexity of the measure introduced and briefly illustrated above, with this research work we have tried to define a working method to be followed in the case in which it is intended to request the facilitation of SB110 for an intervention of energy efficiency of a building (Fig. 2). The workflow to be followed starts from the analysis of the situation and the identification of the type of property, the type of ownership and the preliminary energy analysis of the property. Then, through the energy studies, we move to the design phase during which the mix of interventions must be chosen, and it must be verified that these determine at least a double jump in the energy class of the building. Finally, to understand which mechanism to choose among those provided by the measure (DD, ID, CT) we pass to a careful analysis of the economic convenience through the discounted cash flow analysis (DCFA).

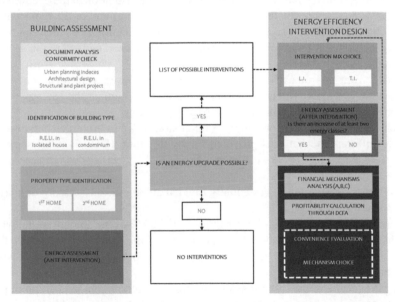

Fig. 2. Graphic schematization of the work method carried out (Source: Authors' content)

To test the workflow designed and specially to highlight how the measurement of SB110 can be impacted, from an economic point of view, on the available homes located in areas with low tourist connotation, it was then chosen a theoretical case study, a complex consisting of three buildings located in a small mountain municipality in the province of Cuneo, in the North-West of Italy. These buildings were built between the sixties and seventies in response to the strong demand for second homes that had occurred in that period and currently, due to the progressive depopulation of the place and the change in the tourist habits of the owners (and more generally of Italians), are mostly uninhabited.

Specifically, to achieve the objective set (the verification of economic benefits), it was considered sufficient to focus on a single building of the complex. This is developed along the East-West axis and has a reinforced concrete structure with masonry curtain walls with a variable thickness between 30 and 40 cm, except for the under-windows that have reduced thickness to accommodate the terminals of the heating system. The building, in addition to the presence of a basement and an attic unheated, stands for 7 floors above ground and houses within it a total of 41 housing units, each with independent generator to produce hot water. The heating system instead is centralized and in 2015 it was changed the generator that feeds it. As for the windows and doors, it is noted that these are made of wood and are equipped with double glazing, while the boxes for the housing of the shutters do not have any insulation. Finally, an analysis of the preliminary energy performance shows that the building currently falls into class D, i.e., in a medium-low class. Lastly, about the accessibility of the building there are significant architectural barriers, while about the general state of preservation of the facade system it can be said that this appears to be in good condition since it was recently painted (2018).

The intervention of planned efficiency consists in the insulation of the opaque surface dispersing, which represents the main intervention (driving) and in the replacement of windows, in the insulation of the shutter boxes, in the addition of shielding systems and in the elimination of architectural barriers, which instead constitute the secondary interventions (driven) [14].

Starting from this mix of interventions, energy analyses were then conducted to determine the new energy class of the building and to estimate the savings resulting from the lower consumption of primary energy for heating.

To calculate the different conveniences obtainable by a single entitled person according to the different mechanisms provided for by the norm (DD, ID, CT), some basic choices were made.

- the cost of the intervention was equally divided among the 41 building units (not in millesimal form);
- the cost relating to design has been estimated to be 8% of the amount of the works.
- an additional cost was considered, equal to 1.5% of the amount of the works, for the expenses to be paid to other professional figures (accountant) who will deal with the financial part;
- among the revenues, the lost costs deriving from the future non-acquisition of part of the energy required to meet the current energy needs have been added; in fact, it is estimated that following the application of the intervention, this will be significantly reduced;
- a time horizon of 5 years has been considered;
- in the case of the transfer of credit, it has been estimated that the amount actually reimbursed to the user is around 93% (due to the expenses to be paid to the credit institutions);

– in the case of the discount on the invoice, it has been hypothesised that this can only be stipulated with suppliers for 50% of the work carried out, while for the remaining part of the amount the individual entitled parties will have to request a direct deduction.

Finally, two different scenarios have been analysed, the first in which the amount of the work is entirely borne by the individual owners using equity, while the second envisages that this amount is entirely in debt and therefore the individual owners will have to pay interest on the capital loaned to the banks.

Based on the analyses carried out and the assumptions made, the profitability indicators (NPV and IRR) were then calculated, which make it possible to order the three mechanisms envisaged by the regulations according to their achievable economic advantage.

4 Results

We report below the results of the analysis carried out, in Table 2 are reported the data relating to the state of the property (pre-intervention), in Table 3 the main data obtainable with the development of planned interventions (post-intervention), finally, in Table 4 and Table 5 are reported the results relating to the verification of economic benefits for a single owner.

Table 2. Data concerning the pre-intervention building situation

Dispersant surface	Theoretical annual consumption (methane)	Annual heating cost (VAT included)	Energy class
$[m^2]$	$[m^3]$	$[€]$	$[-]$
1,883.87	69,032.2	59,020.87 €	D

Table 3. Data regarding the situation of the building post-intervention

Theoretical annual consumption (methane)	Annual heating cost (with VAT)	Annual heating saving	Attainable energy class	Cost of intervention (with VAT)	Deductible cost with SB110 (with VAT)
$[m^3]$	$[€]$	$[\%]$	–	$[€]$	$[€]$
49,758.4	42,542.24	28%	B	322,943.25	322,943.25

Table 4. Data highlighting the economic advantages achievable according to the mechanism chosen if the individual owner decides to self-finance (Equity)

Mechanism	VAN	TIR
DD	2,409.37 €	15,29%
CT	1,185.86 €	85,91%
ID	2,107.80 €	23,85%

Table 5. Data highlighting the economic benefits achievable according to the mechanism chosen in the event that the individual owner decides to borrow 100% of the capital necessary to implement the intervention (Debt)

Mechanism	VAN	TIR
DD	1,106.15 €	7,25%
CT	1,030.72 €	73,81%
ID	1,399.66 €	15,37%

5   Conclusion

Analyzing the results obtained, it is immediately clear that the intervention designed for the property considered can bring great benefits in terms of lower consumption of primary energy for heating and in terms of operating costs for individual owners. Also, because the combustion of a cubic meter of methane produces about 1.97 kg of CO_2 [15], if the intervention were implemented, it would save each year the emission into the atmosphere of about 38 tons of carbon dioxide.

In addition, by carrying out interventions of this type, not only there is an efficiency improvement of the building from the energy side, but if these are carefully designed, respecting the canons of the building on which they are carried out and the surrounding environment, they can lead to a general improvement in the state of preservation of the building. The latter, moreover, will also be directly reflected in the market value of the building itself and the real estate units it contains [16]. Clearly, the acquisition of value will also be proportional to the location of the building in question; in fact, it is clear that in areas such as the one considered, where over the years there has been a strong depopulation, simple interventions on buildings will not be enough to revitalize them. However, if these were carried out in coordination with structural interventions aimed at improving the services present on site, probably we could see a repopulation of these areas. Obviously, this situation would be desirable because on the one hand it would limit the consumption of land in other areas already densely populated and on the other hand there would be an economic recovery of these places now almost totally abandoned. Therefore, perhaps it could be useful in the coming years to reshape the rates of SB110, granting the maximum rate (110%) only in cases where it is decided to upgrade buildings located in fragile contexts (as inland areas) and at risk of being completely abandoned, to host permanent residents. As seen from the case study analyzed, these interventions would be in line with the provisions of the "Fit for 55" package and therefore would allow the Italian State to align itself with the policies for achieving climate neutrality by 2050 dictated by the European Commission.

References

1. European Parliament: What is carbon neutrality and how can it be achieved by 2050? https://www.europarl.europa.eu/news/en/headlines/society/20190926STO62270/what-is-carbon-neutrality-and-how-can-it-be-achieved-by-2050 (2021). Accessed on 05 Jan 2022
2. United Nations Environment Programme: 2021 Global status report for buildings and construction: towards a zero-emission, efficient and resilient buildings and construction sector. https://globalabc.org/sites/default/files/2021-10/GABC_Buildings-GSR-2021_BOOK.pdf (2021). Accessed on 05 Jan 2022
3. Council of the European Union: Preparation of the Council meeting (Transport, Telecommunications and Energy) on 11 June 2021 Proposal for a Regulation of the European Parliament and of the Council on guidelines for trans-European energy infrastructure and repealing Regulation (EU) No 347/2013 – General approach (2021). https://data.consilium.europa.eu/doc/document/ST-8923-2021-INIT/en/pdf. Accessed 05 Jan 2022
4. European Commission: Directive of the European Parliament and of the Council on the energy performance of buildings (recast) (2021). https://ec.europa.eu/energy/sites/default/files/proposal-recast-energy-performance-buildings-directive.pdf. Accessed 05 Jan 2022
5. European Commission: European green deal: commission proposes to boost renovation and decarbonisation of buildings (2021). https://ec.europa.eu/commission/presscorner/detail/en/ip_21_6683. Accessed 05 Jan 2022
6. Mecca, U., Moglia, G., Piantanida, P., Prizzon, F., Rebaudengo, M., Vottari, A.: How energy retrofit maintenance affects residential buildings market value. Sustainability 12(12), 5213 (2020). https://doi.org/10.3390/su12125213
7. Agenzia Entrate: Lo stock immobiliare in Italia: analisi degli utilizzi. GLI Immobili In Italia – 2019, (2019). https://www.agenziaentrate.gov.it/portale/documents/20143/2239117/1.+Lo+stock+immobiliare in+Italia+analisi+degli+utilizzi.pdf/138b6e74-f5a5-f574-c16c-7d6bee248b06. Accessed 05 Jan 2022
8. Italy's National Strategy for "Inner Areas" (SNAI) (2019). https://www.agenziacoesione.gov.it/strategia-nazionale-aree-interne/?lang=en
9. Modica, M., Urso, G., Faggian, A.: Do «inner areas» matter? conceptualization, trends and strategies for their future development path. Scienze Regionale 20, 237–265 (2021). https://doi.org/10.14650/99816
10. Mecca, B., Lami, I.M.: The appraisal challenge in cultural urban regeneration: an evaluation proposal. In: Lami, I.M. (ed.) Abandoned Buildings in Contemporary Cities: Smart Conditions for Actions. SIST, vol. 168, pp. 49–70. Springer, Cham (2020). https://doi.org/10.1007/978-3-030-35550-0_5
11. Agenzia Entrate: Superbonus 110. https://www.agenziaentrate.gov.it/portale/web/guest/superbonus-110%25 (2021). Accessed 05 Jan 2022
12. Mecca, U., Moglia, G., Prizzon, F., Rebaugengo, M.: Strategies for buildings energy efficiency in Italy: financial impact of Superbonus 2020. In: 20th International Multidisciplinary Scientific GeoConference SGEM 2020, Nano, Bio and green – Technologies for a Sustainable Future (2020). Available at: https://www.sgem.org/index.php/elibrary?view=publication&task=show&id=7722. Accessed 05 Jan 2022

13. Agenzia Entrate: Superbonus - interventi su unità immobiliare funzionalmente indipendente - ambito applicativo - Articolo 119 del decreto legge 19 maggio 2020, n. 34 (decreto Rilancio). https://www.agenziaentrate.gov.it/portale/documents/20143/4002792/Risposta_810_15.12.2021.pdf/0971f79b-c5c3-6164-e4f3-b733dc51be27 (2021). Accessed 05 Jan 2022
14. Gotta, A.: Riqualificazione di un complesso anni '60 a sampeyre come nuovo modello di residenza - Studio di applicabilità Superbonus 110% = Rehabilitation of a 1960s complex in sampeyre as a new model of residence - Study of Superbonus 110%, Bachelor's Thesis, Politecnico di Torino. https://webthesis.biblio.polito.it/16507/ (2020). Accessed 05 Jan 2022
15. Ministero della Transizione Ecologica: Tabella dei parametri standard nazionali per il monitoraggio e la comunicazione dei gas ad effetto serra ai sensi del decreto legislativo n.30 del 2013. Available at: https://www.mite.gov.it/sites/default/files/archivio/allegati/emission_trading/tabella_coefficienti_standard_nazionali_11022019.pdf (2019). Accessed 05 Jan 2022
16. Barreca, A., Fregonara, E., Rolando, D.: Epc labels and building features: spatial implications over housing prices. Sustainability **13**(5), 2838 (2021). https://doi.org/10.3390/su13052838

Human/Urban-Scapes and the City Prospects.
An Axiological Approach

Cheren Cappello[1](✉) ⓘD, Salvatore Giuffrida[2] ⓘD, Ludovica Nasca[1] ⓘD,
Francesca Salvo[3] ⓘD, and Maria Rosa Trovato[2] ⓘD

[1] Department of Architecture Design and Urban Planning, University of Sassari,
Piazza Duomo, 6, 07041 Alghero (Sassari), Italy
c.cappello@studenti.uniss.it
[2] Department of Civil Engineering and Architecture, University of Catania, Santa Sofia, 54,
95125 Catania, Italy
{salvatore.giuffrida,mariarosa.trovato}@unict.it
[3] Department of Environment Engineering, University of Calabria, Via P. Bucci, Cubo, 4487036
Rende (CS), Italy
francesca.salvo@unical.it

Abstract. This contribution deals with the issue of the representation in terms of value attributes of the municipalities characterized by complex evolutive processes involving their landscape, urban and architectural characteristics. With reference to the case study of the municipality of Syracuse (Italy), an assessment pattern has been drawn with references to the two dimensions of the urban and human capital, and based on the official dataset by the Italian National Institute of Statistics (ISTAT), on the Census Sections scale. Many indices have been coordinated within an observation-interpretation pattern involving the three main aspects of the evolutionary approaches concerning the natural eco-systems. A GIS-based representation of the numeric results coming from a Hierarchic Multidimensional approach allowed us to identify some of the main typical urban/human profiles referable to the areas more significantly characterized by landscape-urban and socioeconomic values, as well as the ones affected by decay processes.

Keywords: Urban eco-systems · MultiCriteria analysis · Urban-human capital

1 Introduction

1.1 Disciplinary Premises

The relationship between present and future in the analysis, evaluation and project processes is one of the main original contributions that Economic appraisal provides to the assessment aimed at complex social systems such as the territorial-urban ones. Aspects of priority interest for the science of evaluations emerge particularly on the urban scale, for three reasons: 1. the city is an omnipresent form of spatial organization along the evolution of the human species, and among all, the most fruitful one as for the ways of adapting by the settled communities to environment [1]; 2. the effective ability to turn

© The Author(s), under exclusive license to Springer Nature Switzerland AG 2022
F. Calabrò et al. (Eds.): NMP 2022, LNNS 482, pp. 211–220, 2022.
https://doi.org/10.1007/978-3-031-06825-6_21

constraints and opportunities generated by the synergy of human being and nature into resistant and recognizable languages and communication forms; 3. the greater or lesser longevity that these organizations present/promise, due to the coherent and consequential stratification of values and criticalities, and as a consequence of the intensity of the relationship between omnipresence and adaptation of the urban eco-socio-system.

The disciplinary evolution of Appraisal towards the urban issues involves the territorial-economic fairness [2, 3]; the latter – usually addressed in monetary terms, involving market (current or latent – therefore to the tangible and intangible components) – includes the complex cognitive and operational space of sustainability. The convergence of the urban dimension and the economic category of capital assumed here to represent the density of urban and human [4] value of the settled communities, selects the connotations of sustainability considering the components of the saliences and urgencies through which the volume and value of social capital [5, 6] can be represented [7].

1.2 Topics and Objectives

Omnipresence, adaptivity and resilience of the eco-socio-urban system involve the science of evaluations by enabling those cognitive functions that on the one hand represent the design generative potential, on the other border the creative impulse within the borders of evaluation; the latter supports accountability and care, therefore, sustainability and inclusion. Evaluation implicitly refers to rules and as such associates the creative processes with the conventions of social communication and collective intelligence [8]. These two dimensions converge as the "energy" aimed at the accumulation of capital, and the "shape" of the urban eco-socio-system evolution [9].

The topic of the proposed application is the representation in terms of value of the territorial-urban context of the Municipality of Syracuse (Italy), by means of a structured set of value attributes associated to different spatial and social units, from the perspectives of multiple criteria, referable to the two spheres of urban and human capital.

The paper is divided into five sections. The second outlines a synthetic territorial profile of the context identifying the areas more sensitive to convergence of development prospects and instances of preservation; the third summarizes the articulation of the multi-criteria analysis model; the fourth provides the results of the GIS mapping and valuation of territory; the fifth, discusses the results highlighting some operational limits and outlining the perspectives of the model as well as the main consistencies between the consolidated knowledge and the valuative added value of this experience.

2 Materials. Syracuse as a Multilayer, Multifaceted and Multi-prospects City

Syracuse is the capital of the homonymous province, one of the nine in Sicily, significantly representing the the main aspects of the region's landscape, urban, architectural, artistic and cultural value.

The municipality of Syracuse is home to 117,053 inhabitants and consists of a building stock of 10,596 buildings for a total surface of around 7.5 million square meters of surface area, with an average availability of 64 square meters per inhabitant.

The city is structured in five main areas: the ancient core, the islet of Ortigia, the original settlement dating back to the eighth century. b.C., and characterized by a considerable density of artistic, architectural and urban values that have made it one of the main tourist destinations all over the world; as a result, in the last twenty years it has become the target of a massive gentrification process and huge real estate investments; the Umbertine area built at the end of the nineteenth century, as a consequence of the demolition of the defensive city walls that surrounded Ortigia, and characterized by a settlement pattern based on a grid of regular blocks, and by large residential buildings of good architectural quality; the "Borgata di Santa Lucia", built between the end of the 19th and the beginning of the 20th century in the prospects of reducing the pressure of overcrowding in Ortigia and providing hygienic and functional conditions suitable for the middle-class; the northern part, developed mainly from the 1960s, which is divided into the Grottasanta, Akradina, Tiche, Neapolis, Epipoli, Canalicchio, Targia, Belvedere neighbourhoods; finally, to the west, Canalicchio and Carrozziere neighbourhoods, and to the south, the hamlet of Cassibile and the coastal settlement of Fontane Bianche (see Fig. 1).

To the multi-thousand-year history of Syracuse and its recent and decisive socio-economic transformation – due to the settlement of an impressive petrochemical area along a 27 km coastal stretch lying north the city of Augusta – correspond multiple criticalities and potentialities, which raise the interest of the science of evaluations nowadays strongly involved in the new researches concerning the ecological transition in progress.

The areas along the coast are of great interest in terms of the many perspectives of mainly intangible development, and in particular:

– the area of the island of Ortigia and the two ports, where a significant part of the overall value of the entire Municipality is concentrated, characterized, at the same time by the typical criticalities of the art cities, as well as: the rapid and decisive modification of the socio- economic; intense phenomena of filtering up, reduction of residents, massive real estate investments from the outside; distortion of the map of real estate values;
– the northern area, on the border of the aforementioned industrial settlement, is more directly affected by the effects of pollution, but medium/long-term development prospects are outlined in the industrial archeology sector in view of the expected decommissioning process of the plants;
– the area of the Maddalena Peninsula, rich in its landscape and naturalistic values in some way threatened by the increase in real estate interests;
– the hamlet of Cassibile and, on the water front, the settlement of Fontane Bianche, whose expansion compromises the landscape value of the coast.

Many other aspects, including critical issues and opportunities dotting this complex territorial area, cannot all be listed. This contribution summarizes the implementation/verification process of an analysis and evaluation model aimed at unifying and articulating the knowledge of this city-territory unit in the light of criteria and contents pertaining the relationship between value and evaluations.

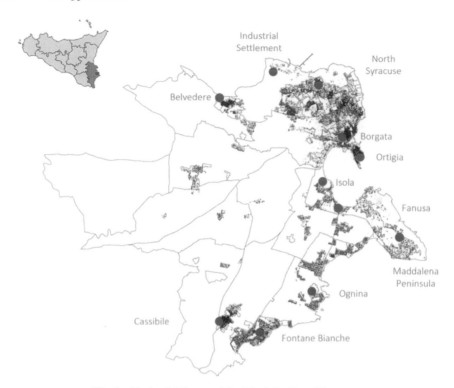

Fig. 1. Territorial frame of the Municipality of Syracuse.

3   Methods. An Urban Multiple Criteria Valuation Model

Urban and Human Capital were described with reference to the official data sources available in the various sections of the ISTAT (National Institute of Statistics) portal, which provide a numeric representation of the main factual terms and convey the most general expressions of the value judgment, its semantics and its metrics [10–14].

The entire database is made up of elementary information units grouped into the two dimensions of the aforementioned capital. Each form of data aggregation required first the construction of value functions, then the assignment of polarity and finally the attribution of hierarchical weighting systems in accordance with the dendogram that represents the organization of the information framework (see Fig. 2) [15–25].

The set of data comes from the Statistical Atlas of Municipalities (ASC), from the ISTAT database and from the ISTAT section 8000Census specifically aimed at representing aspects of vulnerability. The study units consist of the Census Sections (CS) whose variable size is inversely proportional to the population in them. At this scale, the statistical surveys mainly concern aspects relating to the extension and articulation of the Population, the presence and integration of Foreigners, the Housing, the structure of Families and the Extent and Condition of the Building stock.

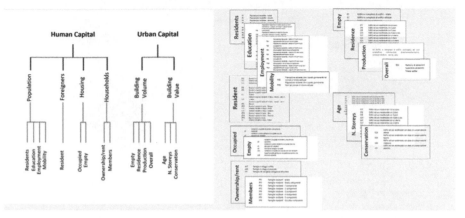

Fig. 2. The dendrogram with the main three valuation levels and a sample of the sheets containing the official data implemented.

4 Application Results

The mapping of the different levels of information is provided in the detail of the CS as a dashboard for synoptic consultation of the various observations and elaborations, connected to a territorial information system [26]; such information platform allows to compare the synthetic and generalized knowledge of the municipal context to the results provided by the evaluation and interpretation obtained by means of the model, progressively aggregating and turning into value judgements the elementary information units.

The description includes quantitative and qualitative aspects relating to some conditions of potential distress. Consequently, it is not surprising that the areas of the densest building fabrics, i.e. Ortigia and the Borgata, report average values, which reflect the typical condition of urban areas manifestly instable, albeit for quite different reasons.

A criticality affecting the clarity of mapping concerns the consistency of the referent (e.g. the population for human capital, the number of buildings for the urban capital) compared to the extension of the Census Section. It is frequent the case of a few people living in a large CS, mapped as a very extensive condition, which it is not. Similarly, a small number of buildings in a poor state of conservation within a large peripheral area would seem to indicate an extensive condition of decay.

4.1 Human Capital

The above interpretative process provided an early assessment of Human Capital through the aggregate variables concerning Population, Residents, Housings and Households (see Fig. 3). The resident population is described with 46 elementary information units, the reduction of which is represented in top left the map.

As for the gaps of representation of the social and material condition on the CS scale is it can be noticed the lack of economic data, present instead at the municipal scale. Consequently, the analysis of the socio-professional and income status was entrusted to

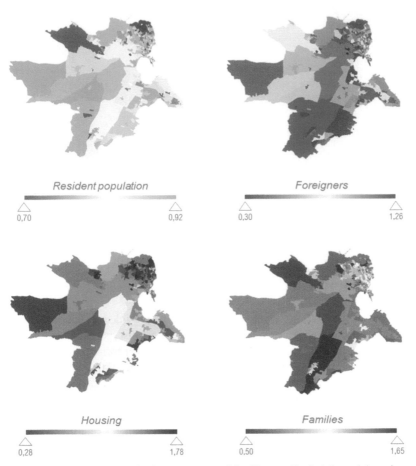

Fig. 3. Map of the main axiological components of the Human Capital (our elaboration of the Istat dataset)

indirect indicators such as, in this case, the degree of education of the resident population. Again, the data referring to the resident population are not able to represent the degree of education in the settlements where second homes prevail, such as Fontane Bianche or Ortigia, and therefore an index such as the degree of education is not supported by adequate semantics. This also applies to mobility, which is more intense due to the movements of students and therefore should have the same intensity both in the coastal area as Cassibile and in the area between Isola and La Fanusa.

Further singular aspects of this interpretative interaction between graphic and numerical representations concern the integration of the foreigners whose location is due to the presence of numerous farms in the central strip of the municipal territory near Cassibile and in different areas of the Borgata and of the North area of the city.

The results of the combined representations of human and urban capital must be interpreted with the same attention, for example with regard to the number of occupied dwellings and the asset status, indicated by the ratio between rented and owned housing, especially if related to the settlement quality of areas in which ownership prevails over rent and vice versa. Some examples: the Pizzuta district reveals the convergence between the localization of high-income families and a high degree of settlement quality; in Ortigia rented families prevail but this circumstance does not denote a low level of architectural quality; on the contrary, the prevalence of speculative uses of inactive real estate assets, denote an exemplary convergence of inefficiency and inequity; the Borgata, on the other hand, shows a significant mix between the two conditions, outlining future opportunities for a growth in the quality of settlement and care for real estate. An observation without any remarkable territorial connotations concerns the number and composition of families, poorly connected to the location and settlement quality. An empirical datum undergoing further investigation is the almost final abandonment of Ortigia by households due to the typical accessibility issues, and the progressive movement of the larger families from the Borgata to the northern area due to the better building quality in strictly functional terms guaranteed by buildings conceived according to contemporary living standards.

4.2   Urban Capital

As well as Human Capital, Urban Capital was observed and interpreted on the basis of the aggregations and normalizations of the various indices with reference to the consistency of the buildings in terms quantitative and qualitative terms (see Fig. 4). The observations concerned the presence of residential and productive intended uses and the total volumes built on the CS scale [27, 28].

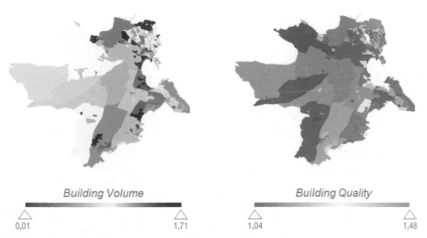

Fig. 4. Map of the main axiological components of the Urban Capital (our elaboration of the Istat dataset)

An important value term of the urban landscape is the widespread presence of historic buildings which was represented by combining the percentage of buildings based on number of floors and the number of dwells [29, 30].

Excluding the extra-urban areas, the maps have highlighted the difference between the landscape quality of the urban-historical fabrics, Ortigia and the Borgata, and the recent buildings in the northern area. Despite the effects of a representation in this case influenced by measurements on different scales, it is noticed that the denser sections of the buildings, i.e. those of the densest urban fabrics, more easily report average valuations decisively influenced by the broad spectrum of states of maintenance of the buildings dating back to the 60–80s of the last century.

5   Discussion and Conclusions

Complexity and contradictions of this territorial context are summarised in terms of the complementary and convergent categories of human and urban capital (see Fig. 5).

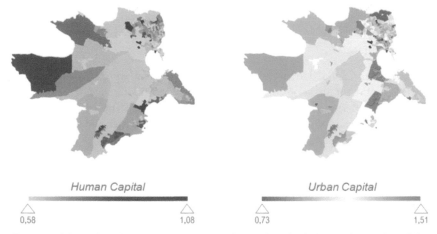

Fig. 5. Map of the main axiological components of the Urban Capital (our elaboration of the Istat dataset)

Such an abstraction is inspired by the prospects for enhancement implicit in the "economic-estimative narrative". The latter is due to the decline of the Petrochemical Settlement and the emergence of the well-known cultural industry of Syracuse in the last decades, unfortunately not free from the typical risks coming from the creation of huge flows of wealth in search productive and speculative uses. Effects of this trend are also revealed in the extra-urban area, such as in the Maddalena Peninsula and the southern coast in the area of Cassibile and Fontane Bianche. Here the naturalistic and landscape quality has become the privileged target of new real estate pressures due to the dragging effect of real estate prices in Ortigia and in the Umbertina area of the historic center of Syracuse. These potentials are matched by a set of concerns that concern not so much and not only the irreversible impacts of activities that interpret the territory as an opportunity for economic-entrepreneurial and real estate success, but rather the risk that the lack of coordination of these energies may increase the gap between the reasons for efficiency and the instances of fairness.

References

1. United Nations. The 2030 Agenda for Sustainable Development. http://sdps.un.org/goals (2015). Accessed 02 April 2021
2. Camagni, R., Capello, R.: Regional competitiveness and territorial capital: a conceptual approach and empirical evidence from the European union. Reg. Stud. **47**, 1383–1402 (2013)
3. Camagni, R.: Principi di Economia Urbana e Territoriale. Carocci, Rome, Italy (2011)
4. Coleman, J.S.: Social capital in the creation of human capital. In: Coleman, J.S. (ed.) Networks in the Knowledge Economy. Oxford University Press (2003). https://doi.org/10.1093/oso/978 0195159509.003.0007
5. de Hart, J., Dekker, P.: A tale of two cities: local patterns of social capital. In: Hooghe, M., Stolle, D. (eds.) Generating Social Capital: Civil Society and Institutions in Comparative Perspective, pp. 153–170. Palgrave, New York, NY, USA (2003)
6. Glaeser, E.L., Charles Redlick, C.: Social Capital and Urban Growth; Working Paper 14374. National Bureau of Economic Research, Cambridge, MA, USA, (2008). http://www.nber.org/papers/w14374. Accessed 10 Oct 2021
7. Giuffrida, S.: The true value. on understanding something. In: Stanghellini, S., Morano, P., Bottero, M., Oppio, A. (eds.) Appraisal: From Theory to Practice. GET, pp. 1–14. Springer, Cham (2017). https://doi.org/10.1007/978-3-319-49676-4_1
8. Trovato, M.R., Giuffrida, S.: The monetary measurement of flood damage and the valuation of the proactive policies in sicily. Geosciences **8**(4), 141 (2018)
9. Belsky, E.S., DuBroff, N., McCue, D., Harris, C., McCartney, S., Molinsky, J.: Advancing Inclusive and Sustainable Urban Development: Correcting Planning Failures and Connecting Communities to Capital. Joint Center for Housing Studies of Harvard University, Cambridge, MA, USA (2013)
10. ISTAT: Index of Social and Material Vulnerability. http://ottomilacensus.istat.it/fileadmin/download/Indice_di_vulnerabilit%C3%A0_sociale_e_materiale.pdf (2011). Accessed 01 Aug 2021
11. Associazione Nazionale Centri Storico Artistici (ANCSA): Centro Ricerche Economiche e Sociali del Mercato dell'Edilizia (Cresme). Centri Storici e Futuro del Paese. Indagine Nazionale sulla Situazione dei Centri Storici, ANCSA, Cresme, Italy (2017)
12. ISTAT: Rapporto Annuale 2020—La Situazione del Paese. https://www.istat.it/it/archivio/244848 (2020). Accessed 01 Aug 2021
13. ISTAT: Data Set 8mila Census. http://ottomilacensus.istat.it/. Accessed 01 Aug 2021
14. ISTAT: Territorial bases and census variables. https://www.istat.it/it/archivio/104317. Accessed 01 Aug 2021
15. Goodwin, N.R.: Five Kinds of Capital: Useful Concepts for Sustainable Development. http://ageconsearch.umn.edu/bitstream/15595/1/wp030007.pdf (2003). Accessed 10 Oct 2021
16. Barreca, A., Curto, R., Rolando, D.: Assessing social and territorial vulnerability on real estate submarkets. Buildings **7**(4), 94 (2017)
17. Trovato, M.R., Clienti, C., Giuffrida, S.: People and the city: Urban fragility and the real estate-scape in a neighborhood of Catania. Italy. Sustainability **12**(13), 5409 (2020)
18. Steiniger, S., et al.: Localising urban sustainability indicators: the CEDEUS indicator set, and lessons from an expert-driven process. Cities **101**, 102683 (2020)
19. Trovato, M.R.: A multi-criteria approach to support the retraining plan of the biancavilla's old town. In: Calabrò, F., Della Spina, L., Bevilacqua, C. (eds.) ISHT 2018. SIST, vol. 101, pp. 434–441. Springer, Cham (2019). https://doi.org/10.1007/978-3-319-92102-0_46
20. Tanguay, G.A., Rajaonson, J., Lefebvre, J.-F., Lanoie, P.: Measuring the sustainability of cities: an analysis of the use of local indicators. Ecol. Indic. **10**, 407–418 (2010)

21. Giuffrida, S., Ventura, V., Nocera, F., Trovato, M.R., Gagliano, F.: Technological, axiological and praxeological coordination in the energy-environmental equalization of the strategic old town renovation programs. In: Mondini, G., Oppio, A., Stanghellini, S., Bottero, M., Abastante, F. (eds.) Values and Functions for Future Cities. GET, pp. 425–446. Springer, Cham (2020). https://doi.org/10.1007/978-3-030-23786-8_24

22. Frigerio, I., Carnelli, F., Cabinio, M., De Amicis, M.: Spatiotemporal pattern of social vulnerability in Italy. Int. J. Disaster Risk Sci. **9**(2), 249–262 (2018)

23. Giuffrida S., Trovato, M.R.: A semiotic approach to the landscape accounting and assessment. an application to the urban-coastal areas. In: Salampasis, M., et al. (Eds). 8th International Conference, HAICTA 2017; Chania, Crete Island; Greece; 21–24 Sept 2017, CEUR Workshop Proceedings, pp. 696–708. Aachen, Germany (2017). ISSN: 16130073

24. Colavitti, A.M., Usai, N., Bonfiglioli, S.: Urban planning in italy: the future of urban general plan and governance. Eur. Plan. Stud. **21**, 167–186 (2013)

25. Trovato, M.R., Giuffrida, S.: The protection of territory from the perspective of the intergenerational equity. In: Mondini, G., Fattinnanzi, E., Oppio, A., Bottero, M., Stanghellini, S. (eds.) SIEV 2016. GET, pp. 469–485. Springer, Cham (2018). https://doi.org/10.1007/978-3-319-78271-3_37

26. Trovato, M.R.: Human capital approach in the economic assessment of interventions for the reduction of seismic vulnerability in historic centre. Sustainability **12**(19), 8059 (2020)

27. Napoli, G., Giuffrida, S., Valenti, A.: Forms and functions of the real estate market of palermo (Italy). Science and knowledge in the cluster analysis approach. In: Stanghellini, S., Morano, P., Bottero, M., Oppio, A. (eds.) Appraisal: From Theory to Practice. GET, pp. 191–202. Springer, Cham (2017). https://doi.org/10.1007/978-3-319-49676-4_14

28. Giannelli, A., Giuffrida, S., Trovato, M.R.: The beautiful city and the rent from information. Monetary axiology of the shape surplus [La città bella e la rendita d'informazione. Assiologia monetaria dell'eccedenza di forma]. Valori e Valutazioni **27**, 53–66 (2020)

29. Champion, T.: Urbanization, suburbanization, counterurbanization and reurbanization. In: Paddison, R. (ed.) Handbook of Urban Studies, pp. 143–161. SAGE Publications Ltd, 1 Oliver's Yard, 55 City Road, London EC1Y 1SP United Kingdom (2001). https://doi.org/10.4135/9781848608375.n9

30. Nasca, L., Giuffrida, S., Trovato, M.R.: Value and quality in the dialectics between human and urban capital of the city networks on the land district scale. Land **11**(1), 34 (2022)

Development Processes in European Marginal Areas: An Investigation in the UNESCO Gastronomic Creative City of Östersund in Sweden

A. Julia Grisafi$^{(\boxtimes)}$

Agriculture Department, Mediterranean University of Reggio Calabria, Reggio Calabria, Italy
julia.grisafi98@gmail.com

Abstract. Gastronomy has received increasing attention in recent years thanks to its ability to create not only economic development but also to protect the values of cultural heritage by directly linking people to traditions and landscapes. This means that local and regional natural and cultural resources can be used in the search for vibrant and sustainable regional development. An alternative to attract investors and enhance the area is to use this products and gastronomy as a resource. UNESCO's network of creative cities can help in the process of revitalizing places and their products. The purpose of this article is to explore the role of UNESCO in the development of remote and rural areas and to assess, after some time, how this development can influence the inhabitants and society of these places and small businesses working on a small scale. In particular, the inhabitants of a city that is part of UNESCO's network of creative cities judge how being part of this network for almost 10 years affects the economic and social development of the city, whether it actually reduces the economic gap between less-developed and central regions and whether there are points for improvement in the work of UNESCO in the opinion of the inhabitants of a city. This study is based on semi-structured and face-to-face interviews with people involved in Network related work in the city of Östersund in Sweden, which is part of the UNESCO Network of Creative Cities in the Gastronomy section since 2010. The results indicate that although indirectly, UNESCO has helped during this years in the development of the city but that there are many things that should be improved in the future to make its work more efficient. About the small-scale producers, they too have been indirectly helped by the fact that Östersund is part of the UNESCO Network of Creative Cities.

Keywords: UNESCO creative city · Gastronomy · Rural development

1 Introduction

Gastronomy has received increasing attention in recent years thanks to its ability to create not only economic development but also to protect the values of cultural heritage by directly linking people to traditions and landscapes.

F. Calabrò et al. (Eds.): NMP 2022, LNNS 482, pp. 221–233, 2022.
https://doi.org/10.1007/978-3-031-06825-6_22

This means that local and regional natural and cultural resources can be used in the search for vibrant and sustainable regional development. An alternative to attract investors and enhance the area is to use this products and gastronomy as a resource. UNESCO's network of creative cities can help in the process of revitalizing places and their products.

The purpose of this article is to explore the role of UNESCO in the development of remote and rural areas and to assess, after some time, how this development can influence the inhabitants and society of these places and small businesses working on a small scale. In particular, the inhabitants of a city that is part of UNESCO's network of creative cities judge how being part of this network for almost 10 years affects the economic and social development of the city, whether it actually reduces the economic gap between less-developed and central regions and whether there are points for improvement in the work of UNESCO in the opinion of the inhabitants of a city.

This study is based on semi-structured and face-to-face interviews with people involved in Network related work in the city of Östersund in Sweden, which is part of the UNESCO Network of Creative Cities in the Gastronomy section since 2010.

The paper is organised into five main sections. The next section reviews the literature on the role of networks and in particular the UNESCO network of creative cities related to Gastronomy. Section 3 briefly outlines the methodology whilst Sect. 4 and 5 highlight and discuss the results and conclusions respectively.

2   Literature Review

2.1   The Creative Cities

In today's globalised economy not only companies but also territories are in competition with each other. Cities and regions, in fact, are competing with each other on the international market for goods and factors of production, and this means that territories that are economically weak and less competitive are lagging behind and risk of being isolated and in decline. Knowledge factors and cultural, creative and gastronomic elements are becoming more and more important in order to reuse in an innovative way the resources present in the territory which, as a consequence, require increasing investments and this leads cities to always try to attract new investors and tourists and to expand international markets (Camagni 2002). In this context, the coming of the *experience economy* has given rise to the creative class, principal customer of all those enterprises that produce this type of good and services (Pine and Gilmore 1998; Florida 2012).

The main feature of these companies is that they create *cultural goods* that, as consumer goods, have a type of production that needs constant updating and input from workers. For this reason, *clusters* of small companies are created that work on projects with teams based on dynamic collaborations that change rapidly and frequently (O'Connor 2010). In this context, regional clusters and networks between cultural industries at regional level have assumed an important role as they are directly linked to the economic development of the place and the urbanization itself.

Cities thus became the main economic and social organizational unit of this creative period, and creativity in turn became the main incentive that allowed cities to develop economically and urbanistically and become attractive for the creative class (Florida

2012). *Creative class people*, however, do not move to creative places for services that are offered or potentially offered. These people are moved by the desire to live experiences in places where the environment allows new experiences and activities and also allows these people to express their creative identity and talent. The most important thing in these cities is that they become places for meeting and exchanging culture and ideas such as bars, museums, art galleries, restaurants, etc. Other influential and important factors are the high level of innovation and the bureaucracy which is leaner and therefore very attractive for entrepreneurs (Florida 2012; Landry 2011). The Creative class, infact, is not only the principal costumer of the creative industries but they work within them and the don't move to a place just because there is work availability, they choose a pleasant and attractive place where to live and there they create their own work instead (Skoglund and Laven 2019).

In recent decades networks have become increasingly important concepts. There are two elements on which to base the concept of network:

1- There must be cooperation both at organizational and behavioural level, which implies collaborations at the level of technological progress and innovation.
2- Cities must be recognised as economic actors that therefore have their own development strategy and can compete with each other, taking into account the fact that individual economic actors in cities can find interest in cooperation (Camagni and Capello 2005).

The great advantage of networks is that they can absorb the costs of maintaining a team or department for innovation research and development on the market.

Among the different networks of creative cities, the most famous network is the UNESCO Network of Creative Cities (Namyslak 2014; Pratt 2010). In essence, the UNESCO network promotes many cultural projects globally and provides new platforms for collaboration between partnerships and cities, as well as between the cities themselves. The main objective is to focus on creativity and to invest in it as a key element for the social and economic development of all cities in the network. These objectives are achieved through innovative activities, good practices, exchange of information, knowhow and skills (Namyslak 2014). It is a network of excellence that includes cities with similar specialisation profiles and includes 7 areas. These are literature, cinema, music, crafts and folk art, design, gastronomy and media arts. Today 180 cities participate in this network (en.unesco.org/creativecities).

One of the biggest advantages that a city receives from participating in a network of famous creative cities like UNESCO's, in addition to sharing information, knowledge and expertise within the network between the different partners, is certainly the fact that a brand like UNESCO's is synonymous with "quality" and "high level". For this reason, networks are often used for branding city purposes (Rosi 2014) especially in the already mentioned case of the prestigious UNESCO brand. This brand helps to improve the image of the city that owns it and, consequently, to attract investors and tourists and also the citizens of the city, who become more proud and content to live there by increasing the share capital (Pearson and Pearson 2015).

Gastronomy is one of the seven fields covered by the UNESCO Network of Creative Cities (Pearson and Pearson 2015). To be part of the network of UNESCO's creative gastronomic cities, you must, of course, reach a minimum standard required. UNESCO provides a list of guiding features to help cities that want to be part of the network reach this standard. These are very different from each other and may require quite obvious things such as owning a well-developed and characteristic gastronomy of the area, the availability of ingredients from the local environment used in traditional cuisine and owning local know-how, traditional cooking practices and methods, remained intact and untouched by technological and industrial progress; but also things such as respect for the environment and the promotion of local and sustainable products and the promotion of nutrition in educational institutions and the inclusion of educational programs for the conservation of biodiversity in cooking schools.

Cities that decide to aspire to be part of the network are also encouraged to speak and confront with cities that are already members of the network. Each year, applications are reviewed and examined by a jury of experts.

Although often not yet considered a creative industry, gastronomy tourism has recently become very popular among creative class tourists (Richards 2014). The reason why it is chosen is precisely because gastronomy creates a strong experience that involves almost all the senses (smell, taste, sight, touch) and is often characteristic of the place and therefore very much linked to the territory and the inhabitants of the place and their consumption (Laganà 2017). At the same time, gastronomy cannot be separated from culture, because the consumption of drinks and food automatically implies participation in the culture of the place or the area (Santich 2004). There are many ways in which gastronomy and tourism are linked together, some of these are given by the fact that food connects the local culture with the tourist, supports the local culture and encourages the production of distinctive food, and can develop new branding and marketing activities (Richards 2014).

Gastronomic tourism certainly offers many advantages and as Florida (2012) explains, cities that can offer more work and entertainment for creative people are destined to attract more people and are therefore destined to increase population. Consequently, all cities that can offer these services are creative cities and centres of development and could also be rural cities, which can be developed through tourism itself. Richards (2011) argues that tourists themselves can bring innovation but at the same time strengthen the link with traditions as that is the real reason that attracts tourists to rural areas. It is always they who stimulate the country to grow by encouraging the locals to stay and not to emigrate because that country can now offer jobs and services.

Food products, gastronomy, traditional dishes are what differentiates and characterizes one place from another, increasing its worldwide fame and its link with the local population and traditions and the place itself (Laganà 2017). Rural areas are strongly suited for this type of tourism because, as we have already explained, they are those that offer the most authentic experiences and are able to create the closest relationships with the surrounding nature, the locals and all traditions. It is for this reason that they are the ones that tourists, especially those looking for niche products and particular places, are launching and that are therefore potential subjects for development.

Despite these positive aspects that the creative industry is able to bring, there are some negative effects that must be taken in account. Firstly, it is important to ensure that the experience of the industry does not lead to gentrification. The fact that people who actually belong there are forced to move because of excessive property prices means that the link between the place and the local population will be lost. This will lead to a loss of identity for the inhabitants and villages who, overwhelmed by strong economic development, will force the real inhabitants of that place to relocate. In addition, there is often the danger that traditionality will become nothing more than a standardised industry and that it will once again lose its link with the land and population and not be renewed. In this case, the territory could represent traditions that perhaps no longer reflect the place itself, which in reality will be "false" for the tourist (Richard and Wilson 2007; Skoglund and Jonsson 2012). The development of the rural territory, in order to prevent places and people from degrading, must be sustainable. To ensure the sustainability of a place that lives on gastronomic tourism and guarantee its development, it is important to address the growth of the place and the relationship that people have with it, because tourists and businesses attracted to that place could put pressure on it (Rinaldi 2017). Development models must therefore be adapted to creative businesses by creating separate organisations to meet the needs of the different sectors involved in business (Skoglund and Jonsson 2012).

3 Methodology

A case study research method was followed and in this case the city of Östersund and the UNESCO network of Creative Cities were chosen (Yin 2003). Semi-structured qualitative interviews were used for data collection (Bryman 2014; Corbetta, 2015; Musolino et al. 2018; Musolino et al. 2020), and secondary sources were analyzed among which the candidacy document that the city provided to the UNESCO Network of Creative Cities.

In this case a targeted sampling approach was adopted (Bryman 2016; Corbetta 2015; Musolino et al. 2018; Musolino et al. 2020). Twelve candidates were selected either because they worked in sectors that involved companies operating on a small scale and therefore were in contact with them or because they were people carrying out studies on this topic and staff of the UNESCO office operating in the city of Östersund. The two people from the Mid Sweden University are privileged observers as they worked on the proposal of the candidacy of the Unesco City. Also the two interviewees who are part of the UNESCO office are privileged observers as they are part of the administrative. The remaining part of the can be considered stakeholders as they are involved in the development processes of the area or are producers and entrepreneurs in the area. In particular the people that have been interviewed are shown in Table 1.

The research questions refer to UNESCO, through the Network of Creative Cities, has acted on the territory. In particular, most of the questions refer to the economic and social development that this has possibly led to the gastronomy sector and local farmers. Other questions try to assess, to a certain extent, the work that has been done by UNESCO on the territory according to the opinions of the interviewees and for which aspects of the work could be hoped to improve. Some of the questions are quite specific and require a good knowledge of UNESCO's actions and the sector in which they operate.

The interviews were all conducted face-to-face in previously agreed places by correspondence via e-mail and only in one case was Skype used to proceed with the interview.

All interview where audio recorded with the consent of the interviewees. All the interviews took place in and around the city of Östersund because the city of Östersund was presented as a representative of the Swedish region of Jämtland within the UNESCO network of creative gastronomic cities.

The interviews covered the following fields:

1- General opinions and development of the city: understood as what the interviewees think of the city and if they perceive, to some extent, a development of the city since 2011 (i.e. since the city joined the UNESCO Network).

2- Collaboration between the parties and the role of UNESCO: respondents are asked whether the parties and all possible actors within the area of interest (the city and all the surrounding areas of the region) have collaborated with each other. Alongside this, they were asked whether UNESCO and the very fact that the city was part of this network contributed to the economic development of the city and the region.

3- Objectives and threats in the application for membership: the extent to which the objectives presented in the application for membership of the Network and the threats described were respectively achieved and solved according to the opinion of the respondents is examined.

4- What needs to be changed and/or improved: in this case we ask for feedback on the work of the Network and whether it has points to improve or focus on in the future.

5- "Would you recommend other cities to join this network? This last field is not a real interview area. It focuses on a particular question in the interview. The question is whether the advantages of being part of this network outweigh the disadvantages.

4   The UNESCO Creative Gastronomic City of Östersund

The management of the Östersund food sector is shared between the city of Östersund and the Jämtland region. The city is the main market for food products, and rural areas provide a working space for producers, thus strengthening urban-rural links.

The region's creative sector, which includes architecture, design, interactive software, film and television, music, publishing and the performing arts, is growing by 5–10% on an annual basis. Altogether there are about 2,000 companies and 4,500 employees in the creative industries of the region, most of which are based in Östersund but still, tourism is an important part of the local economy.

Characteristics and Strengths. Jämtland's cuisine is part of Nordic gastronomy, is very much linked to tradition and is characterised by a unique taste and quality products that cannot be obtained through industrial food processing. The ancient gastronomic tradition is characterised by specific food preservation techniques, while the ingredients used come partly from hunting activities, such as elk meat (symbol of the town of Östersund), reindeer, Arctic trout and occasionally bear and beaver. Other ingredients come from the harvesting of berries, such as blueberries and blackberries, and mushrooms, birch sap. In addition, agricultural and farming industries produce handicraft products such as cheese and dairy products, bread, strawberries and other products such as eggs and poultry, potatoes and vegetables. The region's natural and cultural resources have given the county a unique gastronomic profile within the Nordic cuisine. The production of food products is based on traditional processes and includes hundreds of producers involved in the processing of all types of alimony. The best- known transformation process is probably the processing of cheese, often made from goat's milk or Swedish mountain cattle. The goat's cheese matured in the cellar has always been produced in the same way in Jämtland County for thousands of years.

Next to these farmers and craftsmen and their productions there are some authorities at local, regional and national level that provide them with support through the offer of various services. Worthy of mention is certainly *Eldrimner*, the Centre Swedish national small scale production. Eldrimner has established a centre in Ås (near Östersund) in order to to give producers the opportunity to start their activities in the best possible way, and has been supporting entrepreneurs for 10 years with guides, workshops, study trips, product development, inspiration and product change with the aim of making flourish small-scale artisanal food production. The vouchers results achieved in Jämtland have made Eldrimner the national centre Swedish small-scale food in 2005. The goal of this center is to to assist the creation of new companies, to help them develop and expand offering its experience, support and assistance. The method used by Eldrimner is to provide skills tailored to the needs of the producers. Potential entrepreneurs and experts can participate together in a range of activities that contribute to a learning and development process. The practical and theoretical courses are linked to each other and to the participant's learning and development process. Eldrimner has adopted the producer's point of view and perspective and supports the creation of regional resource centres in other regions of Sweden. The centre has established contacts with experts in all areas of crafts and food production. In addition to the team of consultants, company consultants and specialists are employed at the centre by some of the leading European experts in various fields, such as curing techniques baked meat, baking methods, fruit preservation techniques and vegetable.

Another organization operating in the territory is *The Federation of Swedish Farmers – LRF –* that's an interest and business organisation for the green industry with approximately 140 000 individual members. Together they represent some 70 000 enterprises, which makes LRF the largest organisation for small enterprises in Sweden. Almost all cooperatives within Swedish agriculture and forestry are also members. LRF seeks to create the appropriate conditions for sustainable and competitive companies and to develop a favourable base for social life and enterprise in rural areas (LRF).

The public sphere consists of the Östersund Municipality, the County Council, Jämtland County Council and the other seven municipalities in the region, the University of Midsweden and other educational institutions. In particular the Mid Sweden University participated in a consistent way to the creation of the application form that let Östersund join into the UNESCO Network. The private sphere consists of a network of small food producers, distributors and retailers, tourism and event companies linked to the Jämtland Härjedalen Turism association, producers linked to local agriculture, The Food Academy and all cultural associations of workers linked to cultural and design institutions, and other entrepreneurs active and interested in the topic of gastronomy.

Local Weaknesses or Threats. Östersund and Jämtland is, in our globalized world, exposed to different threats coming from different fronts (economic, environmental and climate front) and influencing different factors. Locally produced food is not produced enough to meet local and regional demand and should therefore be aimed at increasing production, expanding the market and improving the level of knowledge of producers on issues such as food, health and the environment as it has superior nutritional and health qualities. Distances from the market for many of the companies is another problem. The rural structure of the region entails the risk of increasing costs in the long term, jeopardising competitiveness on the market for the products. A further problem is the social and economic perspective of the country, which shows a low and declining population and the exodus of young people and citizens migrating from this region to the urban regions of southern Sweden. Finally, there is the environmental challenge. The Jämtland region has been and still is exposed to environmental threats. Despite the fact that agriculture in the region is the cleanest in Europe in terms of the amount of pesticides and fertilisers used, there is the problem of deforestation by industry and the decline in biodiversity which is directly linked to it, in combination with livestock farming. Climate change, however, is certainly one of the external threats to the region and will, in a shorter-term perspective, cause warmer winters which in turn will affect winter tourism in Östersund and Jämtland.

5   The Results of the Interview

Different answers emerged from the interviews depending on the interview areas, whose are treated separately for convenience.

Development and General Opinions About the City. One of the first questions that were asked in the interview was what the respondent thought about Östersund being part of the Network. The people who were interviewed, despite being from different backgrounds, were very positive, proud and happy that Östersund was part of this worldwide network. However, it is important to note that one of the respondents (L) was not aware of this fact. In particular, the respondent in this case came from a peripheral part of the region. This is an important fact because it highlights the clearly deficient communication in the peripheral parts of the region. To motivate their positivity, respondents generally made it clear that they saw the participation of the city in this network as an important growth factor. Apart from the UNESCO network, all respondents found that the city has changed a lot since 2010 and has evolved in a positive direction. According to D, the inhabitants of the region have partly abandoned some of the typical behaviour of marginal areas by opening their minds, and this has further contributed positively to the development of the territory. Over the years, many events, festivals and meetings have been organized, comparing different countries and knowledge, and new jobs have been created. The way of seeing food is also changing and the importance and demand for products of a certain type is growing, and in food we also look for the taste and intrinsic quality of the product itself, such as the fact that it is locally-sourced or organic.

Collaborations Between Organisations and Institutions and the Role of UNESCO in the Development of the City. To the question concerning the correlation between UNESCO and the development of the city the answers become a little more discordant.

Most of the respondents admit that the fact that Östersund joined the network certainly influenced, even indirectly, the development of the city and the region. In particular, the fact that the network of gastronomic cities greatly encouraged restaurateurs to buy products from the Jämtland region. This has implicitly increased the local economy and the growth of the small-scale economy. Despite this, both A, B and E stated during the interview that they had no help or any link or collaboration with UNESCO and their representatives in Östersund during their projects. E argues that there have been some collaborations with Eldrimmer over the years but only with that body and never with UNESCO and the network of creative gastronomic cities despite the fact that the projects they worked on were, in some cases, related to the gastronomy business and small-scale productions of the Jämtland region. J, on the other hand, declared that he had no connection with UNESCO, the LAG or the LRF. Despite this obvious isolation, Eldrimmer is once again the only link.

Objectives Set Out in the Application. Some of the questions were put in relation to the application for membership. The questions referred to the objectives, problems, weaknesses of the region, and the negative or positive effects of joining the Network.

According to G, many of the projects that were carried out allowed the city of Östersund to achieve its objectives to a certain extent. We are therefore on the right track. According to C, the achievement of the objectives is still quite far away but in any case everything is moving in the right direction. The initiatives have certainly highlighted the region, the city and the gastronomic qualities it has to offer, this has led to a marked improvement in the tourism sector and the number of tourists per year has increased and above all is more diversified. The connections between the city and the rural part have also improved a lot and Östersund, together with the whole Jämtland region, has succeeded, thanks to the participation within the network of creative gastronomic cities, in improving its situation and becoming known, recognized and respected both nationally and internationally, improving its weaknesses and moving in the direction of achieving the goals it had set itself in its application to join this large and important project.

What Should be Changed and/or Improved in the Work of UNESCO on the Territory. In the penultimate section, questions focused on flaws, mismanagement, and things that respondents felt needed to be changed. According to C, everything should be valued more. In particular, the work of the UNESCO Office in festivals and initiatives using the UNESCO logo, which in fact, despite being synonymous with quality and reliability, is used very little in advertising.

According to D another important problem is the lack of funds for the organization of these events and if politicians were to be made aware of the importance and influence that participation in this network can have on the economic level, they would be more willing to provide funds that would increase both in number and size. In addition to this, the connections between retailers and restaurants with manufacturers and craftsmen should be further improved.

H thinks there is a need for greater collaboration between the parties, organisations and also with the UNESCO central office at the global level. In addition to this it is also important to raise awareness among politicians about the additional benefits that the city and the region could receive from funds to organize more events and festivals. Östersund within the Network is working to bring the rural part of the city closer to the market but still has transport problems and it would be nice, according to H, to be able to involve politicians as well and make them active participants in these issues. Finally, the logo has not been used well enough.

G believe the points that should be improved most are those concerning communication and tourism. As far as communication between parties and at different levels is concerned, the problems are the same as those experienced by H and it is easy to understand why as they both work at the UNESCO office in Östersund. As far as tourism is concerned, G argues that there is a need to increase proposals and tourist packages for all visitors.

"Would You Advise Other Cities to Join the UNESCO Network of Creative Cities?"
The latter part focuses on one question in particular: "Would you advise other cities to try to join the UNESCO network of creative cities? Why?".

The answers to this question have again been very homogeneous. H says that he would advise cities to try to join the network if they are fully aware of what they are facing. H points out that joining and remaining in the network has cost and will continue to cost many hours of work and an entire work office to ensure that this happens.

According to G, it's also a good idea for all cities to try to join the network. The advantages are many. The city becomes more important, it has the opportunity to get involved, to grow and improve. It 'also nice collaboration and many new experiences to which you can submit the city and its inhabitants. It is always very important to remember that behind all this there is constant and important work.

Also according to C it is considerably advantageous for a city to be able to enter a network of this level. This allows a great publicity to the city in question and helps a lot the economic growth of the city and the neighboring places. There are also many opportunities that arise, several international meetings that allow the exchange and comparison of ideas. Nevertheless, C also highlights the fact that behind the implementation and maintenance of this plan there is a whole team of people who are absolutely necessary and who must work constantly and for many hours.

6 Discussion and Conclusions

In the context of the inevitable shift towards a 'new economy' (Pine and Gilmore 1998), cultural and creative industries are constantly growing and find expression in the multiple attempts of cities to establish themselves as creative places (Pratt 2010; Rosi 2014; Scott 2006). To be classified as a creative city, the main economic driver should be creativity (Pratt 2010) and the place should attract creative entrepreneurs by exposing a rich and diverse culture that includes the organization of cultural events and a large number of restaurants, art galleries, museums, etc. (Landry 2011).

These characteristics can be clearly observed by looking at the member cities of the UNESCO network of creative cities. They all state that creativity is at the heart of their urban development strategy and host events such as food festivals and gastronomic congresses.

Gastronomy, in particular, has achieved a much better visibility in recent years thanks to its ability to strongly link the population to the cultural heritage. Local agriculture on a small scale and the artisanal food sector can therefore be important for local development and local culinary experiences and can also be used for tourism initiatives (Sjölander-Lindqvist et al. 2016).

The criteria to be met as a prerequisite for membership of the UNESCO Network of Creative Cities cover to a large extent the characteristics of a creative city, so as to ensure that only cities where creativity plays an important role can access the network.

According to Rosi (2014), membership of a network can help to brand a city, thus helping to increase the number of tourists and investors and increase the satisfaction and pride of residents (Pearson and Pearson 2015). A factor that accompanies the brand of the city as a gastronomic destination is the most frequent representation of the city by the media, both locally, nationally and internationally. Finally, residents also benefit from their city's membership and brand as they feel proud and satisfied to live there and appreciate the local gastronomy more by raising awareness among members. Other topics that were highlighted during the interviews concern the economic and social development that the city has undergone due to the fact that it is part of the network of creative cities and the role that the different bodies of the city have within the network, defects and problems that were found along the path of the city within the network.

As reported by the interviewees, the connections between the different bodies in the city and in the region are rather scarce, if not totally non-existent. The possibility of improving communication between agencies and the UNESCO office in Östersund should therefore be considered in order to establish collaborations and synergies as some projects are in common with all agencies. With regard to the development and improvement of the city, all respondents agree that the city has certainly evolved and improved and that it is most likely also thanks to UNESCO, albeit indirectly and in fact, being part of the UNESCO network of creative cities has encouraged and to some extent obliged all retailers and restaurateurs to buy products from the region and this has led to an increase in the demand for small-scale products within the region, thereby supporting and stimulating this economy.

The objectives presented in the application for admission to the network have not been fully achieved or have only been partially achieved but despite this, the respondents again agreed that they are moving in the right direction. Many things have been done, the city has become more visible, the culinary culture has improved considerably and the region is evolving and modernising and even with regard to the weak points of the region some, such as the links between the rural part and the city, are improving and even if not everything is already solved, we are moving in the right direction.

The respondents' answers to the question of what should be changed or improved were not very homogeneous however, it was in general pointed out that there is a need for greater communication, comparison and cooperation between the parties both within the region and globally with the central office of UNESCO. Another important thing is that the UNESCO logo has not been used properly for advertising purposes by the city and this is a pity because the logo is synonymous with quality and reliability and using it more means to reassure possible customers and thus attract more customers. It would also be useful to involve and raise awareness among politicians about the benefits that being part of this network brings to the city and its residents and to be able to allocate more funds to the events and festivals that are organized to promote the city, the region and its gastronomic qualities.

Another point that has come to light is that we should increase services for tourists to entice them more, this can be done through the organization of tourist packages that make it easy for customers to book.

Finally, all respondents agree that being part of the network brings significant benefits, in fact both from an economic and a social point of view there are considerable improvements. This is due to the tourism that is encouraged and the local producers that are supported, but also thanks to the comparison and exchange of information with totally different and distant countries but it is, however, important to remember that behind these enormous advantages there is also a huge amount of work carried out by skilled and competent people. Future research will be conducted on several case studies, with the aim to improve policy and business development in the tourism and CCI intersections in primarily rural contexts, in an international perspective.

It is therefore clear that the presence and actions of UNESCO in this area has helped and supported the territory and local business and that, even if there are still many improvements to be made, they are moving in the right direction. This kind of intervention is also desirable for other rural areas and marginal territories that could grow thanks to the intervention and the visibility that being part of an important organization such as UNESCO can give.

Acknowledgments. The research activities of this study have been carried out during an Erasmus Trainesheep in Mid Sweden University as part of the field work for my thesis in Agricultural Sciences and Technologies in the Mediterranean University of Reggio Calabria, Italy. The publication of the study was supported financially by the Research Project "Ruralscapes" (2020/2024) of the Mid Sweden University.

References

Boix, R., Capone, F., De Propris, L., Lazzeretti, L., Sanchez, D.: Comparing creative industries in Europe. Eur. Urb. Reg. Stud. **23**, 935–940 (2014)

Bryman, A.: Social Research Methods, 5th edn. Oxford University Press, Oxford (2016)

Camagni, R.: On the concept of territorial competitiveness: sound or misleading? Urb. Stud. **39**, 2395–2411 (2002)

Camagni, R., Capello, R.: The city network paradigm: theory and empirical evidence. In: Urban Dynamics and Growth – Advances in Urban Economic (2005)

Corbetta, P.: La ricerca sociale: Metodologia e Tecniche, vol. 3: Le Tecniche Qualitative; Il Mulino: Bologna, Italy (2015)

De Jong, A., et al.: Gastronomy tourism: an interdisciplinary literature review of research areas, discipline, and dynamics. J. Gastron. Tour. **3**, 131–146 (2018)

Florida, R.: The Rise of the Creative Class. Basic Books, New York (2012)

Laganà, V., Nicolosi, A., Skoglund, W., Marcianò, C.: Gastronomic culture and landscape imaginary: the costumer's opinion in regions of north and south Europe (2017)

Landry, C.: The creative city index. City Cult. Soc. **2**, 173–176 (2011)

Laven, D.,Skoglund, W.: Valuatin and evaluating creativity for sustainable regional development. Mid Sweden University, Mittuniversitet (2016). LRF. https://www.lrf.se/om-lrf/in-english/. Accessed Jan 2020

Moore, I.: Cultural and creative industries concept–a historical perspective. Procedia Soc. Behav. Sci. **110**, 738–746 (2014)

Musolino, D., Crea, V., Marcianò, C.: Being excellent entrepreneurs in highly marginal areas: the case of the agri-food sector in the province of Reggio Calabria. Eur. Countrys. **10**(1), 38–57 (2018)

Musolino, D., Distasio, A., Marcianò, C.: The role of social farming in the socio- economic development of highly marginal regions: an investigation in Calabria. Sustainability **12**(13), 1–20 (2020). MDPI, Open Access Journal

Namyślak, B.: Cooperation and forming networks of creative cities: Polish experiences. Eur. Plan. Stud. **22**(11), 2411–2427 (2014)

O'Connor, J.: The Cultural and Creative Industries: A Literature Review. Creativity, Culture and Education Series (2010)

Pearson, D., Pearson, T.: Branding food culture: UNESCO creative cities of gastronomy. J. Food Prod. Mark. **23**(3), 342–355 (2015)

Pine, B.J., Gilmore, J.H.: Welcome to the experience economy. Harward Bus. Rev. (July–August 1998)

Pratt, A.C.: Creative cities: tensions within and between social, cultural and economic development: a critical reading of the UK experience. City Cult. Soc. **1**, 13–20 (2010)

Richard, G.: Creativity and Tourism. The State of the Art. Tilburg University, TheNetherlands (2011)

Richard, G.: The role of gastronomy in tourism development. Tilburg University, The Netherlands(2014)

Richards, G., Wilson, J.: Tourism, Creativity and Development. Routledge (2007)

Rosi, M.: Branding or sharing? The dialectics of labeling and cooperation in the UNESCO creative cities network. City Cult. Soc. **5**, 107–110 (2014)

Santich, B.: The study of gastronomy and its relevance to hospitality education and training. Int. J. Hospitality Manage. **23**(1), 15–24 (2004)

Scott, A.J.: Creative cities: conceptual issues and policy questions. J. Urban Aff. **28**(1), 1–17 (2006)

Skoglund, W., Jonsson, G.: The potential of cultural and creative industries in remote areas. Nordisk kulturpolitisk tidsskrift **15**(2), 181–191 (2013)

Skoglund, W., Laven, D.: Utilizing culture and creativity for sustainable development: reflections on the city of Östersund's membership in the UNESCO creative cities network. In: Calabrò, F., Della Spina, L., Bevilacqua, C. (eds.) ISHT 2018. SIST, vol. 101, pp. 398–405. Springer, Cham (2019). https://doi.org/10.1007/978-3-319-92102-0_42

Sjölander-Lindqvist, A., Skoglund, W., Laven, D.: Regional foodscape and gastronomy development – the way forward to rural development in the Swedish periphery? In: Valuing and Evaluating Creativity for Sustainable Regional Development, Östersund, 11–14 September 2016 (2016)

Yin, R.K.: Applications of Case Study Research, 3rd edn. SAGE Publications, Thousand Oaks (2012)

Yin, R.K.: Case Study Research – Design and Methods, 5th edn. SAGEPublications, Thousand Oaks (2014)

UNESCO. https://en.unesco.org/. Accessed Aug 2019

UNESCO Creative Cities A. https://en.unesco.org/creative-cities/home. Accessed Sept 2019

UNESCO. Östersund Kommun. Application. http://www.unesco.org/new/fileadmin/MULTIMEDIA/HQ/CLT/pdf/CNN_Ostersund_Application_Gastronomy_EN.pdf. Accessed Sept 2019

Lifestyle Migration and Rural Development: The Experience of Kaxås in the Periphery of Sweden

Matilda Meijerborg, Fanny Sandström, and Daniel Laven^(✉)

Department of Economics, Geography, Law and Tourism/ETOUR Research Centre,
Mid Sweden University, Östersund, Sweden
Daniel.Laven@miun.se

Abstract. This paper reports on an innovative, bottom-up effort (Projekt Kaxås) to attract lifestyle migrants to the rural Swedish village of Kaxås. This study is situated against the broader backdrop of population and out-migration from many of Sweden's rural communities. The experience of Kaxås is unique in that, at the time of this writing, the population of the village is growing. Study findings suggest that social dimensions such as *solidarity* and *hospitality* were critical factors in shaping the decision for families to relocate to Kaxås. Findings from this study can inform similar efforts that are currently underway in rural communities in Sweden and throughout Europe.

Keywords: Lifestyle migration · Place development · Experienced sense of place

1 Introduction

Although there is growing interest in the countryside as a place to live in Sweden [1], the most sparsely populated areas, primarily in the north, are still experiencing a continued population decline [2]. The ability to attract people to rural areas is critical to local, regional and national growth, which is why focusing on issues related to development and attractiveness of rural areas has become a priority for political decision-makers, and regional strategists [3, 4]. Although such initiatives tend to be initiated and coordinated at the regional level, it is also becoming more common for local communities to get involved in these types of issues, particularly in more peripheral rural areas [5, 6]. Project Kaxås is one such example of this type of effort. The project began in 2019 and is aimed at reversing negative population trends and attracting young families to the village, with an immediate focus of keeping the village school alive [7]. By the end of 2021, the village more than doubled its population.

The purpose of this study is to examine why families, many of whom have moved for lifestyle reasons, decided to relocate specifically to Kaxås and in what ways Project Kaxås helped create the right conditions for their relocation.

A substantial portion of the existing research focuses on how planners and decision makers try to attract new residents from other countries [8, 9] and whether international

F. Calabrò et al. (Eds.): NMP 2022, LNNS 482, pp. 234–245, 2022.
https://doi.org/10.1007/978-3-031-06825-6_23

(im)migrants contribute to demographic and workforce-related changes in Sweden's rural areas [10, 11]. This study is a complement to earlier research because it focuses on those who move to rural areas in Sweden from other places within the country. Furthermore, in this context, the perspectives of lifestyle migrants have not previously been researched, yet these migrants have important experiences that can be used to gain a greater understanding of such grassroots initiatives and the reasons for their success.

2 Theoretical Context

The evolution of human geography has led to an established view of places as social constructions that are continuously re-created through the interplay between both human interactions and physical and symbolic attributes [12–15]. The concept of place, however, is frequently divided into three components: location, locale and sense of place [16]. Location is best described as the position of the place and the geographic point related to other places. Locale relates to the physical and social contexts that unfold in the place. Finally, sense of place refers to the perception of and feeling for a place, as well as the individual and collective significance attributed to the place. The purpose of this study is to examine why families, many of whom have moved for lifestyle reasons, decided to relocate specifically to Kaxås and in what ways Project Kaxås helped create the right conditions for their relocation.

According to Stenbacka [17, p. 141], a rural living environment is considered 'an arena where a form of identity is developed, and a certain lifestyle is adopted [...]'. A related concept is also 'the rural idyll' which describes the continuously (re)created idea of the countryside as a problem-free environment. The rural idyll is also characterized by living 'the good life', which often means closeness to nature but within a strong, social community [18]. The rural idyll is also associated with agricultural living that differs from post-industrial society by emphasizing self-sufficiency, small-scale and local community involvement [17]. Kåks and Westholm (1993, see [17] p. 43) report that migration into rural areas is often stimulated by notions of self-realization and that such relocation not only represents a change of physical spaces (housing and landscapes), but also mental spaces represented by perceptions and dreams.

Hjort [19] emphasizes that demographic and socio-economic factors as well as previous experiences influence the choice of living environment. This is similar to what Stenbacka [17] refers to as the physical, psychological and social forces that drive human mobility and relocation into specific places. Niedomysl [20] suggests that immaterial values play an increasingly larger role in the choice of where to live as people's economic circumstances improve. Thus, people often relocate to places they associate with possibilities to live a certain lifestyle and where they can picture a greater quality of life [21]. In this way, the place to which people move says a great deal about the kind of life they want to live [18]. Furthermore, Niedomysl [22] describes how people have an almost instinctive perception of what they find (un)attractive about a place, and that place attractiveness, from a migration perspective, is a very complex process as the individual compares different alternatives before finally deciding where to relocate [22]. To conceptualize place attractiveness in a migration context, Niedomysl [22] presented a theoretic framework (Fig. 1).

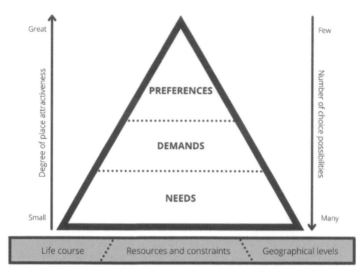

Fig. 1. Place attractiveness as conceptualized by and adapted from Niedomysl [22].

The foundation at the base of the pyramid consists of three concepts that have shown to be important in people's decision-making process. The first concept relates to the individual's current phase in their life course, which affects their needs, demands and preferences. The second concept is based on the resources and constraints they experience which may be economic or social factors. The geographic levels between which the relocation takes place are also significant as the number of factors that are considered usually increase in line with the geographic distance between the destinations of the move. The attractiveness of a place and the number of destination choices depends on the extent to which people's needs, demands and preferences are fulfilled [22]. How a place is perceived, created and portrayed thus plays a critical role in an individual's decision to relocate [3, 23].

Niedomysl and Amcoff [1] suggest that in order to attract people of working age to rural areas, jobs and good commuting possibilities are important parameters for the areas to be able to compete against towns. In contrast, Hjort and Malmberg [24] emphasize that attractive living environments are the crucial factor and suggest that the actual home as well as the concept of the 'rural idyll' may be more important than job opportunities. At the same time, difficulties related to getting a mortgage have also been identified as an obstacle to migration as in some cases the market value is lower than the cost of building [1].

Activities aimed at attracting citizens to rural areas often involve several actors and take place at different levels, from the most local level to national political agendas and global strategies [25, 26]. In Sweden, these issues are typically addressed within municipal administrations and their associated monopolies on planning decisions. However, Syssner [27] highlights that rural municipalities in Sweden largely lack defined and proper strategies for managing depopulation and attracting new residents, and that the measures taken are more reactive than proactive in nature.

Syssner and Meijer [6] describe that during recent years many researchers have called attention to the fact that different forms of strategies are needed to plan and develop areas of growth or decline, and that some researchers have even implied that a kind of paradigm shift is underway. The traditional view of place development is that formal actors create plans for buildings and physical infrastructure that are financed by public funds [6]. Place and rural development have thus mainly been linked to municipalities, regions and other authorities while the current paradigm increasingly includes both 'formal' and more 'informal' processes [28]. Rather than acting as 'passive producers' following development plans which traditionally have been formulated from above in a kind of community hierarchy, Boonstra [29] describes this in terms of residents increasingly acting as 'creative and active producers'. Li et al. [30] suggests that there is a risk traditional 'top-down' planning overlooking the real needs of both the local population and the potential immigrants, and that 'bottom-up' initiatives can capture these more effectively [31, 32]. Development initiatives that originate as bottom-up movements have been shown to be more common in rural communities than in towns and regions of growth [6]. Moseley [33] describes that revitalization of rural areas can only succeed if the development processes are founded on local, bottom-up factors and driven at a very local level. He suggests that it is only in this way that all human and material resources can be utilized.

3 Methods

A case study design was adopted for this exploratory research project focusing on the village of Kaxås (Sweden). Furthermore, this research should be considered as a study of an atypical case [34] given that unlike in many other rural communities with similar characteristics, Kaxås is currently experiencing positive population growth. Our hope was to gain a deeper understanding of the experiences of those lifestyle migrants as well as of the planning perspective behind the project. The empirical materials were gathered through seven qualitative, semi-structured interviews with families who recently relocated to the village as well as through continuous contact with the people associated with the project. The interviews were conducted in Swedish and lasted for approximately one hour. Four of the interviews took place at the local community center, one using Zoom, one over the phone and one in the family's new home in Kaxås. The study was carried out during April–June 2021. The collected material was then examined in order to identify subjective thoughts, feelings or experiences that were shared by several individuals in the form of an explanatory pattern [35]. The strength of the study is that it focuses on families moving to the same place which means that differences in the result cannot relate to the place as an 'object' but rather from the subjective experiences of place.

Study Context
Kaxås is a small village located in the centre of Sweden in the county of Jämtland. This is one of the largest counties in Sweden in terms of area, but it only contains 1.5% of the country's population [36]. Nature is ever-present and tangible in the area and in addition to the dramatic, steeply sloping landscape, there are also views of a lake and mountains as well as the surrounding peaks and forests (Fig. 2). Just as with other rural communities, the number of residents in the village has gradually decreased. At the same

time, the area has plenty of desirable qualities that should enable it to attract new people, not least its geographic location between Östersund, the largest city in the county, and Åre, the largest ski resort in Scandinavia.

Fig. 2. Kaxås. This photo is from the personal collection of Åsa Bornström-Högström and sourced from http://www.kaxasbygden.se/ with permission.

Kaxås offers basic services, a school, a pre-school and a handful of companies in different sectors, particularly agriculture and forestry. In addition, the ski area of Almåsa Alpin is located just a couple of kilometers from the village. When the popular village school was threatened with closure in summer 2019, Dan Olofsson, a wealthy IT entrepreneur with roots in the village of Kaxås, initiated Project Kaxås. The main purpose of the project is to attract young families to save the village school. Below is the first paragraph from the project's website:

> *If you are looking for quality of life close to nature, a safe place for your family and still want access to a functioning employment market, Kaxås located 60 km north-west of Östersund is an interesting new alternative, far away from the congestion of the city. Project Kaxås invests in young, responsible families seeking a unique place to live close to nature with an abundance of leisure activities. We customize solutions for families that want to give their children a safe, active and good upbringing and still want access to a meaningful professional career. We have access to tools that help enable specific solutions for different families [7].*

In addition to actively marketing Kaxås (mostly through social media), extensive efforts are underway to create housing options. Firstly, formerly desolate buildings have been identified and mapped and secondly, new homes designed by architects are being built in the area. They are known as Ekobyn (eco-village) Ladriket due to the holistic, sustainable approach taken in terms of both climate-smart and energy-efficient homes (Fig. 3). As part of the local community association, residents also have access to illuminated ski trails and a ski-waxing hut, a communal bake house, a paddle court, fishing cabin, boat, two snow-mobiles and for those with horses, access to stables, a paddock and pastures at cost-price [7].

Fig. 3. New housing in Kaxås. This photo is from the personal collection of Åsa Bornström-Högström and sourced from https://projektkaxas.se/har-hittar-du-boende-i-kaxasbygden with permission.

The work of the project, however, is not limited to housing opportunities. As it can be difficult for young families to obtain mortgages for homes in more sparsely populated rural areas, the project customizes solutions for financing if needed. The project also arranges weekends to which they invite people who are interested in seeing what Kaxås has to offer. There are also ambassadors involved in the project whose primary role is to provide support to families that are interested, or those that have already decided, to move to the area. As part of Project Kaxås, the Uppstart Kaxås foundation also focuses on creating more jobs in the area, and the project also helps with contacts and networks [7]. Next to the nearby Almåsa Alpin ski area, there are now several small cabins that can be used by visitors as well as by families who visit the area to find out more about what it's like to live there. There are also plans for a future co-working solution next to the ski area (Bångman, personal communication, 7 April 2021). Even though not all of the families have relocated to the village yet, Project Kaxås [7] states that it has attracted 104 new residents, 40 of which are children.

4   Key Findings and Discussion

If a rural place is to attract new residents, special attributes or circumstances are not sufficient. Some form of packaging of the place and its contents is also required. Both Eimermann [23] and Niedomysl [3] highlight the significance of how place(s) are created and portrayed within people's relocation decisions. Project Kaxås has been a key factor for the increasing number of residents in the village, which otherwise most likely would have remained unknown to these families. The result clearly indicates that Kaxås has certain qualities related to its geographical location as well as the physical landscape. All of the study participants mentioned this in terms of their dreams of a different lifestyle where they can live closer to nature and have the opportunity to pursue various activities such as hiking, kayaking, fishing and hunting.

The primary activity associated with Project Kaxås involves building good relationships with families who are interested in finding out more about the village and the lifestyle it offers. During the interviews, several families reported that once they discovered Kaxås and communicated their interest via the project website, a representative from Project Kaxås contacted them either the same day or the day after. The contact took place by phone, e-mail and through in-person meetings in the village. Another significant element that all of families (study participants) emphasized was the 'personal' and 'informal' contact as well how 'welcome' they had been made to feel by the project promoters along with the local community. In addition to helping find accommodation during their visit, some of the families also spent a day with either the project promoters or a project ambassador. These visits also included a tour of the village, the school, the plots for the new eco-village, and a dinner invitation at the home of a Kaxås resident. In several instances, it was particularly evident that the visit to Kaxås and the welcome received during that time was a determining factor in the decision to move. This pattern echoes Benson and O'Reilly [18] as well as Stjernström [37] observations about the ways in which tourism mobilities can inspire people to relocate by allowing people to picture how everyday life could look like in that place. A specific word that almost every family expressed was 'safety', and how 'feeling safe' characterized their entire process. Interestingly, both Burgess [38] and Hepple et al. [39] described good hosting in terms of a package of several components that are all targeted at creating a feeling of safety.

Place Attractiveness Within Lifestyle Migration
Through his theoretical framework, Niedomysl [22] described how the degree of place attractiveness as well as the number of potential destinations are affected by how well an individual's needs, demands and preferences are fulfilled. He also suggested that people's migration processes often involve comparing different options before making a final decision. However, the results in this study indicate an additional and less rational type of relocation process. Even though several families explained that they had thought about moving for quite a while, few of them had actually planned their relocation and fewer had looked at different options. Some families described that they hadn't even thought about relocating before they heard about Kaxås and they expressed that after some time turned into a question of 'Kaxås or nothing'. Several families explained that they decided very quickly to move to Kaxås once they had visited the village, in some cases within just a couple of days. Therefore, our study suggests an extension of the theoretical framework originally presented by Niedomysl [22] (Fig. 4).

This revised model illustrates how the families' stories relate to ideas about place attractiveness and relocation processes. However, due to the small scope of this study, the purpose of this extended model (Fig. 4) is to explore this existing theoretical framework to the families' experiences and in that way open doors to new perspectives on place attractiveness within lifestyle migration.

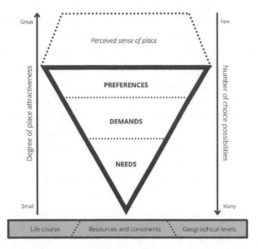

Fig. 4. Revised place attraction model.

First, the pyramid has been turned upside-down because of the families' desire for a different lifestyle rather than their needs and demands that formed the main motive to move to Kaxås. In addition to what can be traced to the families' needs, such as housing, school and work, several study participants also mentioned aspects that they neither had been looking for nor been aware of would influence their perception of place. We decided to call this the 'experienced sense of place'. Not only has this experienced sense of place influenced the degree of (place) attractiveness, but also, the number of destination choices considered. Since Kaxås fulfilled and, above all, exceeded the families' needs, demands and preferences thereby creating a high degree of positive sense of place, the need to look for other options decreased or completely disappeared. Interestingly, this result also indicates that their experienced sense of place is almost entirely due to the work of Project Kaxås. This will be described further in the following sections.

Clearly Defined Target Group and Removal of Barriers
The framework is based on the three concepts of life course, resources and constraints and geographical distance, which still in some ways have influenced the families' needs, demands and preferences. Many of the factors that can complicate relocation to rural areas have been addressed throughout the Project Kaxås concept. First, the project focuses on a limited and specific target group, i.e., 'young, responsible families' [7], who in general have similar needs, demands and preferences. Further, it has been shown that it is mostly families with children that move to rural areas [24, 40]. The families highlighted various social aspects that contributed to their decision making of relocating. Something that clearly has affected the decision is the fact that many other families are moving to the village during the same period of time. The families experienced the process of entering and getting the feeling of belonging to the new society as easy, especially considering that many of them lack social networks in the area. One of the study participants highlighted the importance of both a secure social setting since they all 'have been pulled up by the roots' and feel a bit lost together with being open minded

and searching for new social contacts. Another study participant expressed this in terms of 'getting a free pass straight into the local community'.

How resources and constraints affect people's living preferences is highly individual, but one potential constraint may be difficulties in finding work, which the project has helped to address through, for example, providing local contacts. Even though work has not been the main driver for relocation, availability of work is important for being able to live in a place. Another potential constraint for rural relocation may be difficulties in obtaining a mortgage [1], as experienced by some of the families. Since Project Kaxås has provided special financing opportunities it has been possible for some of the families to buy a house in Kaxås before solving their work situations. Further, the houses in Ekobyn Ladriket have been constructed to be climate smart and cost-efficient, which is another financial aspect. Buying a house in the eco-village automatically includes membership in the local community association. As previously mentioned, this includes certain sharing possibilities, which may reduce the need to own things individually. There is also great flexibility in terms of the configuration and choices regarding the new houses, which several families noted had impressed or surprised them. This resembles what Moseley [33] describes about grass-roots-based development and how they allow more imaginative, innovative and 'outside-the-box' ideas. This indicates that Project Kaxås has lowered the barriers that may arise based on the three aspects upon which the framework rests.

The Power of Grass-Roots Initiatives
An additional success factor of Project Kaxås can be said to stem from the local commitment, and how the local community and its residents increasingly act as 'creative and active producers' as expressed by Boonstra [29]. Li et al. [30] point out that traditional 'top-down' planning risks overlooking the real needs of current as well as potential future residents. Elshof and Bailey [41] and Moseley [33] emphasized that 'bottom up' planning should be considered a prerequisite to successful rural development efforts. Syssner [4] describes this in terms of potential cost-savings and that it enables local engagement. For example, the project promotors explained that much of their own personal engagement in the project derives from the fact that they live in the area themselves. They also described how they want to create the right conditions so new residents stay in the village for a long time and for them to be happy since they will become neighbors (Bångman, personal communication, 7 April 2021). The development in Kaxås is largely being driven at a grass-roots level and is based on local and collective ideas. The importance of local engagement was expressed by the project promoters who mentioned that the people of Kaxås know the area best and that their ideas, desires and needs are thus paramount to the growth of the village. Interestingly, study participants also mentioned this during the interviews and several of them said that they noticed the strong solidarity and engagement in the village, even during their first visit to Kaxås.

5 Conclusions

Even though Kaxås has certain geographic and financial resources few other rural villages have, the results nonetheless highlight lessons that can inspire other places and rural communities in order to attract new residents. This study shows that while practical and geographic circumstances made relocation possible, it was the social dimensions such as solidarity and hospitality that were the critical factors for relocation. For several families, the opportunity to 'try out' living in Kaxås was a decisive factor as it gave them an appreciation of how it would be to live in the village. If rural communities can offer people a chance to visit the place in an organized way, it may give potential lifestyle migrants a better sense of the place.

Furthermore, the results of the study indicate that strategies for hosting, which have traditionally been associated with the hospitality and tourism sectors, also is of relevance for development of rural areas when it comes to attracting new residents. Per-haps it is no longer solely about marketing place 'from a distance' with visual material but instead about inviting potential lifestyle migrants to see, feel and experience the place. Based on this study, it appears to be important for potential lifestyle migrants to get a feeling for the place as well as the possible lifestyle and social atmosphere. Thus, the personal dimension must not be under-estimated as relatively simple gestures were shown to be very important to the study participants. Their stories clearly describe that this helped them in terms of feeling more secure.

At the end of the day, it is not simply about creating an attractive place but about getting people to take the plunge and deciding to relocate. From a planning perspective, it may thus be important not to overlook the fact that people make place(s) and make the most of the human resources. In the case of Kaxås this has been done in forms of good hosting, ambassadors and welcoming activities which all have been important to the new residents. This grass-roots initiative has strong local commitment, not just from the project promoters but also from the rest of the local community. Such the personal engagement has clearly influenced the development of Kaxås.

Acknowledgements. 1. Matilda Meijerborg and Fanny Sandström equally share primary authorship for this paper.

2. This paper emerged from the thesis work of Matilda Meijerborg and Fanny Sandström. The original work is published in Swedish and available from the Mid Sweden University Library: http://www.diva-portal.org/smash/record.jsf?pid=diva2%3A1574190&dswid=-9899. Daniel Laven served as the academic supervisor for this thesis work.

3. Support for this paper was provided in part by the Ruralscapes project. Ruralscapes is a collaborative effort between Mid Sweden University and Universita di Meditteranea de Reggio Calabria and funded by the EU Leader Programme through the LAGs Skog, Sjö och Fjäll and Locridee.

4. The authors thank and acknowledge Lisa Cockette (http://www.anythingenglish.se/) for translation from Swedish to English.

References

1. Niedomysl, T., Amcoff, J.: Is there hidden potential for rural population growth in Sweden? Rural Sociol. **76**(2), 257–279 (2011)
2. Statistics Sweden Homepage. https://www.scb.se/hitta-statistik/artiklar/2018/befolkningen-okarsvagt-palandsbygden/. Accessed 1 June 2022
3. Niedomysl, T.: Migration and place attractiveness [doctoral thesis]. Uppsala University, Sweden (2006)

4. Syssner, J.: Mindre många – Om anpassning och utveckling i krympande kommuner. Dokument Press, Sweden (2018)
5. Glesbygdsverket: Kartläggning av strategier för att öka befolkningen i kommuner och regioner, Sweden (2007)
6. Syssner, J., Meijer, M.: Informal planning in depopulated rural areas. In: Hospers, G.H., Syssner, J. (eds.) Dealing with Urban and Rural Shrinkage: Formal and Informal Strategies, pp. 100–110. LIT Verlag GmbH & Co., Berlin (2018)
7. Projekt Kaxas Homepage. https://projektkaxas.se/. Accessed 1 June 2021
8. Eimermann, M.: Promoting Swedish countryside in the Netherlands: international rural place marketing to attract new residents. Eur. Urban Reg. Stud. **22**(4), 398–415 (2015)
9. Hedberg, C., Haandrikman, K.: Repopulation of the Swedish countryside: globalisation by international migration. J. Rural. Stud. **34**, 128–138 (2014)
10. Eliasson, K., Westlund, H., Johansson, M.: Determinants of net migration to rural areas, and the impacts of migration on rural labour markets and self employment in rural Sweden. Eur. Plan. Stud. **23**(4), 693–709 (2013)
11. Hedlund, M., Carson, D., Eimermann, M., Lundmark, L.: Repopulating and revitalising rural Sweden? Re-examining immigration as a solution to rural decline. Geogr. J. **183**(4), 400–413 (2017)
12. Relph, E.: Place and Placelessness. Pion, London (1976)
13. Tuan, Y.-F.: Space and place: humanistic perspective. Prog. Hum. Geogr. **6**, 211–252 (1974)
14. Philips, D., Robinson, D.: Reflections on migration, community and place. Popul. Space Place **21**(5), 409–420 (2015)
15. Gren, M., Hallin, P.-O.: Kulturgeografi – en ämnesteoretisk introduktion. Liber AB, Sweden (2003)
16. Cresswell, T.: Place – A Short Introduction. Blackwell Publishing Ltd., USA (2004)
17. Stenbacka, S.: Landsbygdsboende i inflyttarnas perspektiv – Intention och handling i lokalsamhället [doctoral thesis]. Uppsala University, Sweden (2001)
18. Benson, M., O'Reilly, K.: Migration and the search for a better way of life: critical exploration of lifestyle migration. Sociol. Rev. **57**(4), 608–625 (2009)
19. Hjort, S.: Socio-economic differentiation and selective migration in rural and urban Sweden [doctoral thesis]. Umeå University, Sweden (2009)
20. Niedomysl, T.: Tourism and interregional migration in Sweden: an explorative approach. Popul. Space Place **11**(3), 187–204 (2005)
21. Walford, N., Stockdale, A.: Lifestyle and internal migration. In: Smith, D.P., Finney, N., Halfacree, K., Walford, N. (eds.) Internal Migration: Geographical Perspectives and Processes, pp. 99–111. Ashgate Publishing Ltd., UK (2015)
22. Niedomysl, T.: Towards a conceptual framework of place attractiveness: a migration perspective. Geogr. Ann. **92**(1), 97–109 (2010)
23. Eimermann, M., Hayes, M., Korpela, M.: Conference proceedings from the lifestyle migration hub meeting [Konferenspublikation]. Umeå University, Sweden, 28–29 November 2019
24. Hjort, S., Malmberg, G.: The attraction of the rural: characteristics of rural migrants in Sweden. Scott. Geogr. J. **122**(1), 55–75 (2006)
25. Ikonen, R., Knobblock, E.: An overview of rural development in Sweden. In: Copus, A.K. (ed.) Continuity or Transformation? Perspectives on Rural Development in the Nordic Countries, pp. 90–110. Nordregio Report 2007:4, Sweden (2007)
26. Nilsson, B.: Meanings of 'The Local' in a Swedish rural development organization: all Sweden shall live! J. Rural Commun. Devel. **13**(2), 39–56 (2018)
27. Syssner, J.: Planning for shrinkage? Policy implications of demographic decline in Swedish municipalities. J. Depopul. Rural Dev. Stud. **20**, 7–31 (2015)

28. Syssner, J., Meijer, M.: Innovative planning in rural, depopulating areas: conditions, capacities and goals. In: Hagen, A., Higdem, U. (eds.) Innovation in Public Planning, pp. 151–169. Palgrave Macmillian, UK (2020)

29. Boonstra, B.: Planning strategies in an age of active citizenship: a post-structuralist agenda for self-organization in spatial planning [doctoral thesis]. Gent University [InPlanning series], The Netherlands (2015)

30. Li, Y., Westlund, H., Liu, Y.: Why some rural areas decline while some others not: an overview of rural evolution in the world. J. Rural. Stud. **68**, 135–143 (2019)

31. Hague, C., Jenkins, P. (eds.): Place Identity, Participation and Planning. Routledge Taylor & Francis Group, UK (2005)

32. Meijer, M.: Community-led and government-fed: comparing informal planning practices in depopulating regions across Europe. J. Rural Commun. Dev. **14**(4), 1–26 (2019)

33. Moseley, M.J.: Rural Development – Principles and Practice. SAGE Publications Ltd., USA (2003)

34. Denscombe, M.: Forskningshandboken: För småskaliga forskningsprojekt inom samhällsvetenskaperna (3 uppl.). Studentlitteratur AB, Sweden (2016)

35. Marshall, C., Rossman, G.B.: Designing Qualitative Research, 3rd edn. Sage Publications, USA (1999)

36. Regionfakta Homepage. https://www.regionfakta.com/jamtlands-lan/. Accessed 20 Apr 2021

37. Stjernström, O.: Flytta nära långt borta - Om sociala relationers betydelse förlångväga flyttningar [doctoral thesis]. Umeå University, Sweden (1998)

38. Burgess, J.: Perspectives on gift exchange and hospitable behavior. Int. J. Hosp. Manag. **1**(1), 49–57 (1982)

39. Hepple, J., Kipps, M., Thomson, J.: The concept of hospitality and an evaluation of its applicability to the experience of hospital patients. Int. J. Hosp. Manag. **9**(4), 305–318 (1990)

40. Amcoff, J.: Samtida bosättning på svensk landsbygd [doctoral thesis]. Uppsala University, Sweden (2000)

41. Elshof, H., Bailey, A.: The role of responses to experiences of rural population decline in the social capital of families. J. Rural Commun. Dev. **10**(1), 72–93 (2015)

Craft Breweries and the Corona Crisis–Exploring the Scandinavian Context

Wilhelm Skoglund[1]($\boxtimes$) and Øystein Rennemo[2]

[1] Mid Sweden University, Östersund, Sweden
wilhelm.skoglund@miun.se
[2] Nord University, Levanger, Norway
oystein.rennemo@nord.no

Abstract. This paper addresses the way the Corona pandemic has impacted the rapidly growing craft beer sector. More specifically, it does so through a qualitative study of two bordering regions in Sweden and Norway. Interviews with a total of 20 breweries highlight the pandemics major impact on the sector. The patterns that were identified in the two regions in our study in many ways point towards similar effects. Firstly, the study showed that the pandemic has hit craft breweries rather hard, but most have managed to survive and some have even increased production and sales. These breweries have done so through creative adaptability, such as finding alternative income sources, cutting production and reducing costs. Other dimensions that have been crucial for the breweries in this crisis have been a location with high population density, large numbers of domestic tourists, or a greater emphasis on selling to retailers instead of bars and restaurants. As the pandemic and its effects continues, taking on new shapes and patterns, the findings of this study could have relevance for the support systems of craft breweries and the creative economy. It would also be beneficial to perform follow-up studies with varying methodological approaches during later stages, or after the pandemic.

Keywords: Entrepreneurship · Craft beer · Corona · Regional development

1 Introduction

The evolution of the craft beer sector has successively changed the beer and brewing landscape across the world, and the large-scale, industrial breweries are now complemented by a craft brewing sector that has been growing steadily. Many previous studies point to the U.S. as the geographical origin of this evolution, which has since spread to most corners of the world. The characteristics of a craft brewery include a small size, independence from large scale brewers, and the use of traditional yet innovative brewing methods [1]. As many of these breweries are not only small but also relatively new in comparison to the rest of the brewing industry, they are also often in a more vulnerable position. With the outbreak of the corona pandemic in 2020, small-scale businesses

---

W. Skoglund and Ø. Rennemo—Contributed equally.

faced monumental challenges. In this study, the authors shed light on a number of small-scale brewers in Scandinavia in order to examine how the sector has managed during this crisis. The concept of creative adaptability, meaning the capacity to quickly react to stressful situations [2], has been central in our findings. In the context of this article, this includes finding new ways to cut costs or finding new and entrepreneurial ways to maintain sales in these challenging times. This study aims to contribute to the research front of the sector. It also aims to contribute to policy making in the creative industries, a sector often highlighted within sustainable regional and rural development, which here includes craft food and drinks.

2 Literature Overview

Studies in the craft beer sector have integrated a variety of dimensions. These dimensions encompass the background of the craft beer evolution as described by Garaviglia and Swinnen [3], whereas other studies [4, 5] have looked at the dimensions of entrepreneurial and small business motivation. Craft beer research has also been undertaken on branding aspects by Eberts [6] and Sjölander-Lindqvist et al. [7]; consumption by Carbone and Quici [8]; tourism opportunities by Murray and Kline [9], Fletchall [10], and Duarte and Sakellarios [11]; financing by Cabras and Higgins [12], Reid and Gatrell [13], and Skoglund [14]; place and spatial development by Esposti et al. [15] and Gatrell et al. [16]; policy by Malone and Lusk [17] and Williams [18]; and sustainability trends in the regional craft beer industry by Hoalst-Pullen, Patterson, Mattord and Vest [19]. For this study, it is also important to note that several scholars point to craft brewery agglomerations as relevant for local and regional economic development, for example, through place attractiveness and place revitalization [20, 21].

Several attempts to map out the field of study have been done [22], claiming that there are three categories of studies: place, space, and identity; production, markets, and consumer culture; and tourism perspectives and sustainability. Baianos' [23] study of 367 articles led to a division of research within the sector into seven fields: definitions, fiscal policy, innovation, safety, health, consumer knowledge, and sustainability. With respect to research needs, Smith et al. [24] claim that consumer behaviour, tourism, commercial development, and the local impact of craft breweries are relevant dimensions for further studies. Most of these studies come from the U.S, the U.K., Italy, and other parts of Europe, but there is an ever-growing number of studies from countries all over the world. Apart from discussions within forums related to the brewing industry, challenges breweries face due to the Covid-19 pandemic are naturally, due to the recent emergence of the virus, an under-studied area of research. The case study by Bivona and Cruz [25] is one of the few published studies on Corona crisis management in a craft brewing environment. The study highlights the need for business models that rely on the agile use of alternate revenue streams, existing resources and networks as a foundation to extract new knowledge and innovation. Pitts and Witrick [26] found that the pandemic significantly disrupted the macro and micro brewing business sectors in terms of sales of draft beer, since restaurants and bars and places of consumption were closed down. What remained was the packaged beer business, which meant that breweries with operations focused on off-site production and breweries that were able to adapt by increasing canning and bottling were those that were least impacted by the pandemic. Clarke et al. [27] looked at the Scottish craft brewing sector, noting that there was a successful response to the new challenges presented by the pandemic and that business activities were reorganized accordingly.

With a rich qualitative data set from Scandinavia, this paper will contribute to this small but important and growing field. It provides relevant information on the pandemic's impact on craft breweries in general, and it also contributes to knowledge about the rebound processes for the sector. The input on the development of the sector also aims to contribute to the policy making processes, as the recovery of this creative sector could have relevance for the recovery of the economy as whole, particularly the rural economy, which we study here. Lastly, the study adds to the rather thin body of research into the craft beer sector in Scandinavia, but also to a broader research perspective of the sector in contexts beyond the geographical location of this study.

3 Methodology

The geographical context for this study is the middle of Scandinavia, where the regions of Jämtland/Härjedalen (pop 132,000) and Trøndelag (pop 469,000) share a border that constitutes a large part of the over 1,600 km Swedish-Norwegian border. The two regions share a long cultural history with a strong and continuous cultural exchange. Trondheim is the main city in Trøndelag with 192,000 inhabitants, whereas Östersund is the only major city in Jämtland/Härjedalen with a population of 64,000. Both of the regions can be described as largely rural (particularly Jämtland/Härjedalen), rural here meaning areas comprised of open land and settlements with fewer than 3,000 residents [28] The two regions are normally characterized by a perforated border, with a high degree of cross-border activities in most sectors of society. During the pandemic, the border has been closed on several occasions, giving rise to debate and difficulties for businesses that rely on border shopping or tourism.

This study is built on a project undertaken in cooperation between researchers in Norwegian Trøndelag and Swedish Jämtland/Härjedalen. The research in the project used a mixed methods approach, whereas this particular study is more explorative and built on qualitative interviews with 9 and 11 craft brewers in Trøndelag and Jämtland/Härjedalen, respectively. The breweries studied in the two regions are categorized as craft breweries, but they also differ in size. For the purposes of this paper, the authors refer to breweries that produce less than 20,000 L a year as small, breweries that produce between 20,000–100,000 L a year as medium-sized, and breweries that produce more than 100,000 L a year as large. The interviews consisted of questions that varied in character, including question related to marketing, financing, policy, and the entrepreneurial activities behind these breweries. The interviews also included questions on how the breweries have been impacted and responded to the pandemic. The questions relating to the pandemic focused on the respondents short term experiences, long term expected challenges, possibilities for governmental support, as well as an open discussion question on the effects of the pandemic. The interviews were digitally carried out in the winter of 2020–2021, which was a period well into the global pandemic. This type of method can be characterized as a case study of a rather explorative type, with the selected Scandinavian region representing the case and multiple informants from each side of the border. The transcripts from the interviews were organized into nationally specific accounts, which were then analysed and coded. The nationally coded data was thereafter compared and coded into a number of mutual thematic effects from the pandemic.

4 Results

Findings from Jämtland/Härjedalen, Sweden
Jämtland/Härjedalen is a highly rural region, stretching from the middle of Sweden almost 500 km north. The region has long been characterized by forestry and trade, and low levels of industrialization. Gastronomically, the National Center for Artisan Food, Eldrimner, which is located outside Östersund, has been vital in re-energizing the region's small scale food and drink sector since its start in the 1990s. The region's brewery sector was heavily rationalized, which left it without a single brewery in the 1990s. Since then, the craft beer sector has grown strongly, and there are now 18 breweries located around Jämtland-Härjedalen. These breweries all have between 1–10 employees. Most of the brewers are men who have other jobs, and around half of them rely on the sale of beer to tourists in restaurants and bars in the region's tourism destinations. The other breweries rely more heavily on retail (to grocery stores, but primarily to the government-owned alcohol monopoly, Systembolaget), but they all share a tight cooperative network where logistics, festivals, mutual purchases, and even recipes are integrated. Due to the pandemic, many of the region's breweries experienced major difficulties due to the lack of consumption at restaurants and bars and the absence of beer festivals. These difficulties have been particularly visible in rural areas, which are dependent on international (including the dominating Norwegian) tourism, due to border closures or difficulties crossing borders.

We have lost between SEK 1.5 and 2 million here in the restaurant. It has been a disaster with the closed border between Sweden and Norway, with no Norwegian guests coming! (Small rural brewery no 1 (in tourism location)).

Meanwhile, other breweries have benefitted from the pandemic, seeing more tourist traffic than usual, even some breweries in rural locations. The background for this is an increase in domestic travel, which has also been caused by travel restrictions. Hence, the small-scale breweries located in domestic tourist hot spots were able to profit from increasing numbers of visitors, even though restaurants and bars faced partial restrictions.

People went to the Swedish mountains for vacation because that was the only vacation alternative. (Medium-sized rural brewery no 2).

The pandemic has also caused many people to want to support small and local businesses in order to keep them from going under. This has meant that local residents, as well as tourists to some extent, modified their beer consumption patterns.

People that have come here (summer mountain vacationers) have asked for local products, they come in and shop to support one of the small producers. (Medium-sized rural brewery no 6 (in tourism location)).

The changing tourism patterns have also brought new types of visitors to the region, tourists that seem to spend and consume more and thereby support the breweries that have attracted this group.

The pandemic has also led to another type of tourist here, who normally don't travel to the mountains, and who consume more. (Medium-sized rural brewery no 6 (in tourism location)).

A common trait among the brewers is that they are often so called "combinateurs", meaning that they have more than one source of income. Some of the "combinateurs" in the regions dependent on international (including Norwegian) tourism have also stressed that they encountered difficulties in their other sources of income.

Corona hasn't meant so much for our (small) brewery, but it has led to difficulties for our other (tourism) activities. (Small rural brewery no 8).

Some of the "combinateurs" have been employed outside of tourism, and they have managed to decrease their levels of production, hence lowering costs while waiting for demand to increase again. These breweries have thus managed to avoid going out of business.

We have decreased production, and instead we've been running more test batches and experimenting. (Small rural brewery no 1 (in tourism location)).

For the larger breweries, this has been more difficult to accomplish as they have permanent employees that need to be paid regularly, which means a need for continuous sales and income. However, the larger breweries more often sell to retailers, such as grocery stores, and most of all to Systembolaget. Systembolaget is the government owned alcohol retailer in Sweden (900 stores in total), and the breweries that sold their beer through these stores during the pandemic seem to have done rather well.

Sales at bars and pubs became rather dead, but sales to Systembolaget have increased. (Medium-sized rural brewery no 10).

Sales at Systembolaget and grocery stores like ICA and Coop have been good and meant that we have been alright.... whereas sales to restaurants have decreased... and events and sampling events have just been cancelled. (Large rural brewery no 11).

This leaves the breweries that have historically depended on sales to bars, pubs, and restaurants. They have struggled severely, and for the brewers that have not had access to alternative sources of income, the situation has been critical.

The pandemic has pretty much killed off all of our sales to restaurants, it is almost at zero. (Small rural brewery no 3).

For some of the breweries, government support has been crucial for survival, particularly the larger breweries with employees that have been furloughed. Some of the breweries have not needed to rely on this support and have instead downsized or turned to alternative income sources.

Findings from Trøndelag, Norway

The region of Trøndelag is located in the middle of Norway, has a long coastline along the Atlantic, and a long interior border with the Swedish region of Jämtland. Due to the topography created by the North Sea to the west, the large fiord (Trondheimsfjorden) in the middle, with small cities, farmland, forests and mountains in the east, Trøndelag is characterized by heterogeneity regarding its economic and industrial base. The region is home to a strong and growing craft beer sector. Today, Trøndelag hosts almost 40 craft breweries, most of which have started during the last 5–10 years and almost all by men. Like other regions, the brewers here have been heavily impacted by the pandemic, but the impact has varied significantly between brewers.

In rural areas, the breweries have survived because of the increase in domestic tourism. In both countries, domestic tourists have significant disposable income and therefore compensate for the absence of external tourism. This also connects to the fact that rural breweries have a product mix that often includes the sale of beer in local bars/restaurants combined with hospitality or other tourist activities.

> *July 2020 was a very good month for our brewery, because of the Norwegian summer holidays. Principally, the visitors were only Norwegians, or foreigners residing in Norway. Those people also spend more money than visitors from abroad. Therefore, our sales increased by 50% compared with the year before.* (Small rural brewery, no 1).

Another brewer indicates a similar experience:

> *We solved the problems. In March (2020) we added take away food and followed this track for some months. … Additionally, we have a storehouse where we can serve up to 20 persons, small parties. When the pandemic hit us, we expanded our serving space with this building.* (Small rural brewery no 2).

A similar experience has been reported by another small rural brewery with sales to restaurants and bars nearby:

> *The problems started in March (2020), and the decrease in sales to restaurants and bars started immediately. The situation appeared quite dramatic, and the future felt quite insecure, since our margins are tight. Then the summer came, a Norwegian summer holiday for all in Norway. What happened was that we doubled our sales in our farm store, which until then, was pretty insignificant. This almost compensated for the decrease in sales to bars and restaurants and saved us through the year.* (Small rural brewery no 3).

Medium-sized rural breweries, which are dependent on sales to bars and restaurants, seem to have faced other, more severe problems, and what saved their operations seems to have been a change in sales channels.

Well, the pandemic took away our face to face customers. The summer months have been the most important for us, meeting customers at festivals, pubs, events where people can taste and buy. And we can talk about our beer, new sorts of beer, and so on. Due to legislation in Norway, this is our main marketing channel, we are, for example, not allowed to promote our beer on Facebook. We lost that channel, but another one saved us; retailers and the government-owned liquor stores. Since Norwegian tourists spent much more money in Norway wherever they travelled, they also had to buy beer in the stores, where they also found our products. (Medium-sized rural brewery no 4).

Still, the same brewer confirms that breweries without the capacity to change their sales channels encountered significant problems:

Breweries that only focused on restaurants, bars, and night life have faced big, big problems; they have really suffered. We survived owing to the fact that we found a new marketplace (in retail). (Medium-sized rural brewery no 4).

In Trøndelag, the few, bigger craft breweries are located in urban areas. Before Covid-19, they focused on volume and sales to nation-wide chains of distributors, but also the licensed trade (to bars and restaurants). When beer consumption suddenly became a home activity within the family, licensed trade disappeared. Still, these urban breweries seem to have found a way to compensate through retail chains and new partnerships with other industrial oriented actors in the market.

We were not equipped to deal with the corona situation. Our owners had to find a strategic change of direction. They cancelled a cooperation (production, market- ing, and distribution) with X (a nationwide industrial brewery) and established a partnership with Y (a finance house) in our city. In the middle of this chaos, the 20th of April, I as a brewer got the message that all our business systems would cease to exist the first of June and was given six weeks to find a new accountant and establish new systems for everything. Now, in two weeks, we will start transporting our products on the basis of new negotiated conditions. When the pandemic hit, 60% of our sales volume was based on licensed trade (to bars and restaurants). Our margins in the cooperation with X were quite tight. When the licensed trade option closed, our chances for survival were dramatically reduced. But, because of the break with X, we took over the whole distribution chain, changed it to an in-house activity. This marked a turning point in our economy. The licensed trade is still absent, but sales to retail stores have gone up, and we now sell with far better margins. When the licensed trade hopefully starts again, we have much more solid ground to stand on. In that way, the corona situation has helped us. (Large urban brewery no 5).

Though we do not see the same dramatic reorganization, it is still possible to rec- ognize similar experiences among smaller and medium sized urban breweries. One of these breweries had no former cooperation with industrial breweries, such as brewer no 5.

You would expect a decrease in sales in 2020 due to the pandemic. In fact, we did well. Even though licensed trade disappeared, we could keep up with demand on the other markets (commodity stores and government owned liquor stores). In fact, we had problems keeping up with deliveries, we were constantly empty. Everything we produced at maximum capacity was sold immediately. 2020 was our best year ever! (Medium-sized urban brewery no 6).

5   Discussion and Implications

The patterns identified in the two regions in our study in many ways reflect similar effects from the pandemic. Below, these patterns have been grouped and merged into a number of thematic effects connected to the pandemic. Firstly, we find that the pandemic has hit craft beer breweries hard, which is in line with the findings in the study by Pitts and Witrick [26], which particularly elevated the loss of sales to bars and restaurants as an effect of Corona. However, in the present study, none of the breweries needed to close due to the pandemic. In Norway, one of the small rural breweries was forced to close, but this was due to the loss of key competence (the brewer). The rest of the 19 breweries have managed to survive so far, and some have even increased production and sales due to creative adaptability, meaning agility and quick reactions to the new situation [2]. Secondly, we find different patterns of survival, depending on location and population density. These patterns are particularly visible in Norway, with several of the breweries being located in Trondheim or other cities and having an advantage in their proximity to retailers and the government owned alcohol stores. In Jämtland-Härjedalen, the two urban breweries also fared well, benefitting from the same advantages as the urban Norwegian breweries, whereas the rural breweries have had more difficulties reaching retailers, such as Systembolaget (the nationally owned alcohol chain). Hence, leaning heavily on retail has been a key to survival, even success, so far in the pandemic, as consumers continue to purchase beer, though they consume it more at home and less in bars and restaurants. This connects once again to the study by Pitts and Witrick [26], pointing towards the importance of packaged beer during the pandemic, and the capacity to increase sales to retailers. This is linked to yet another effect: the breweries located in rural regions with high levels of domestic tourism have performed well without selling to retailers, since domestic tourism has increased dramatically in the two regions since 2020. It also seems that these tourists have been more interested in supporting and consuming local products than before the pandemic. This contrasts with rural breweries that are dependent on international tourism, which have been heavily impacted by the restrictions on the Swedish-Norwegian border and decrease in cross border tourism. The consumption and tourism patterns add a pandemic-dimension to what Smith et al. [24] have stated regarding further studies on the sector.

Another emerging thematic effect is that, even though small and rural breweries have been hit hard, they have still managed to survive due to their "combinateur" character, by offering alternative products or services, or by relying on other jobs. The flexibility of the brewers is often also characterized by their small size, which means that they do not have employees that need to be laid off. This connects to the creative adaptability concept, which appears to be highly relevant in crisis times when businesses need to find new solutions to keep sales up and costs down. Thus, the adoption of new business models and quickly finding alternative revenue streams without increased costs [25] – in other words, being creative and entrepreneurial – is the key to survival in a crisis such as the corona pandemic.

Although the findings of the present study come from the somewhat differing conditions of Norway and Sweden, the findings point in the same direction, which partly can be explained by their similar, to a large extent rural character. Another possible explanation could be connected to the cultural proximity of the studied regions, which in "normal" times has an open border with lots of roads and also a rail road crossing over. The somewhat similar legislative restrictions on selling alcohol, including beer, can also be a part of the explanation.

On the value of this study, there has not been much written on this topic globally, either from a beer or a craft beer perspective. Furthermore, the Scandinavian context is even more lacking in studies that look at the effects of the corona pandemic on the small business sector, particularly the craft beer sector. Hence, this study provides a relevant contribution to craft beer studies and small business studies in Scandinavia. The study also provides relevant input into feasible policy support for this sector and other small businesses, particularly in the cultural and creative economy with a focus on small scale food, drinks, and gastronomy. In this way, the study contributes practical perspectives and continues to build on earlier scholarly work by Nilsson et al. [20] and Reid [21], both of which highlighted the relevance of craft breweries for local and regional economic development.

6   Limitations and Future Studies

This study was based on a qualitative approach, mostly relying on a number of explorative, qualitative interviews in the Mid-north of Sweden and Norway. The information taken from the study has value due to its transferability across national borders and cultures. However, combining this with more quantitative data would enable statistical generalizations, which this study is not capable of delivering. Also, as the pandemic and its effects continue at the time of publication, follow up studies on the long-term effects on this sector are relevant for the continued development of the craft beer sector.

Acknowledgements. The data collection for this article was undertaken by a team of researchers from Nord University and Mid Sweden University. We would like to express our appreciation to all of our colleagues for this collaboration and the possibility to use the data for this text.

References

1. American Brewers Association (n.a.). www.brewersassociation.org/statistics-and-data/craft-brewer-definition/. Accessed 30 Dec 2021
2. Orkibi, H.: Creative adaptability: conceptual framework, measurement, and outcomes in times of crisis. Front. Psychol. **11**, 588172 (2021)

3. Garaviglia, C., Swinnen, J.F.: Economics of the craft beer revolution: a comparative international perspective. In: Garaviglia, C., Swinnen, J.F. (eds.) Economic Perspectives of Craft Beer: A Revolution in the Global Beer Industry, pp. 3–51. Palgrave Macmillan, Cham (2018)
4. Danson, M., Galloway, L., Cabras, I., Beatty, T.: The origins, development and integration of real ale breweries in the UK. Entrepreneurship Innov. **16**(2), 135–144 (2015)
5. Cappellano, F., Spisto, A.: Innovative milieu in Southern California: the case of the San Diego craft breweries. In: Calabrò, F., Spina, L.D., Bevilacqua, C. (eds.) ISHT 2018. SIST, vol. 100, pp. 314–321. Springer, Cham (2019). https://doi.org/10.1007/978-3-319-92099-3_37
6. Eberts, D.: Neolocalism and the branding and marketing of place by Canadian microbreweries. In: Patterson, M., Hoalst-Pullen, N. (eds.) The Geography of Beer: Regions, Environment, and Society, pp. 189–200. Springer, Dordrecht (2014)
7. Sjölander-Lindqvist, A., Skoglund, W., Laven, D.: Craft beer – building social terroir through connecting people, place, and business. J. Place Manag. Dev. **13**(2), 149–162 (2019)
8. Carbone, A., Quici, L.: Craft beer Mon Amour: an exploration of Italian craft consumers. Br. Food J. **122**(8), 2671–2687 (2020)
9. Murray, A., Kline, C.: Rural tourism and the craft beer experience: factors influencing brand loyalty in Rural North Carolina, USA. J. Sustain. Tour. **23**, 1198–1216 (2015)
10. Fletchall, A.M.: Place-making through beer drinking: a case study of Montana's craft breweries. Geogr. Rev. **106**(4), 539–566 (2016)
11. Duarte, A., Sakellarios, N.: The potential for craft brewing tourism development in the United States: a stakeholder view. Tour. Recreat. Res. **42**(1), 96–107 (2017)
12. Cabras, I., Higgins, D.: Beer, brewing, and business history. Bus. Hist. **58**(5), 609–624 (2016)
13. Reid, N., Gatrell, J.D.: Craft breweries and economic development: local geographies of beer. Polymath **7**(2), 90–110 (2017)
14. Skoglund, W.: Microbreweries and finance in the rural north of Sweden – a case study of funding and bootstrapping in the craft beer sector. Res. Hosp. Manage. **9**(1), 43–48 (2019)
15. Esposti, R., Fastigi, M., Viganò, E.: Italian craft beer revolution: do spatial factors matter? J. Small Bus. Enterp. Dev. **24**(3), 503–527 (2017)
16. Gatrell, J., Neil, R., Steiger, T.L.: Branding spaces: place, region, sustainability and the American craft beer industry. Appl. Geogr. **90**, 360–370 (2018)
17. Malone, T., Lusk, L.L.: Brewing up entrepreneurship: government intervention in beer. J. Entrep. Pub. Policy **5**(3), 325–342 (2016)
18. Williams, A.: Exploring the impact of legislation on the development of craft beer. Beverages **3**(18), 1–16 (2017)
19. Hoalst-Pullen, N., Patterson, M.W., Mattord, R.A., Vest, M.D.: Sustainability trends in the regional craft beer industry. In: Patterson, M., Hoalst-Pullen, N. (eds.) The Geography of Beer, pp. 109–116. Springer, Dordrecht (2014). https://doi.org/10.1007/978-94-007-7787-3_11
20. Nilsson, I., Reid, N., Lehnert, M.: Geographical patterns of craft breweries at the intraurban scale. Prof. Geogr. **70**(1), 114–125 (2018)
21. Reid, N.: Craft breweries, adaptive reuse, and neighbourhood revitalisation. Urban Dev. Issues **57**(1), 5–14 (2018)
22. Withers, E.T.: The impact and implications of craft beer research: an interdisciplinary literature review. In: Kline, C., Slocum, S.L., Cavaliere, C.T. (eds.) Craft Beverages and Tourism, Volume 1, pp. 11–24. Springer, Cham (2017). https://doi.org/10.1007/978-3-319-49852-2_2
23. Baiano, A.: Craft Beer: An Overview. Comprehensive Reviews in Food Science and Food Safety. Wiley (2020)
24. Smith, S., Farrish, J., McCarroll, M., Huseman, E.: Examining the craft brew industry: identifying research needs. Int. J. Hosp. Beverage Manage. **1**(1), 1–15 (2017)
25. Bivona, E., Cruz, M.: Can business model innovation help SMEs in the food and beverage industry to respond to crises? Findings from a Swiss brewery during COVID-19. Br. Food J. **123**(11), 3638–3660 (2021)

26. Pitts, E.R., Witrick, K.: Brewery packaging in a Post-COVID economy within the United States. Beverages **7**(1), 14 (2021)
27. Clarke, D., Bowden, J., Dinnie, K.: Illuminating craft brewers' experiences of dealing with Covid-19 and making fresh sense of what Covid-19 can do to/for craft beer: an intègraphic approach. In: Clarke, D., Ellis, V., Patrick-Thomson, H., Weir, D. (eds.) Researching Craft Beer: Understanding Production, Community and Culture in An Evolving Sector, pp. 49–72. Emerald Publishing Limited, Bingley (2021)
28. Glesbygdsverket. Fakta. Om Sveriges gles-och landsbygder. 2007. Glesbygdsverket, Stockholm (2007)

Cultural Heritage Digitalisation Policy as a Co-creation of Public Value. Evaluation of the Participatory Digital Public Service of Uffizi Galleries in Italy During the COVID-19

Maria Stella Righettini[(✉)] [iD] and Monica Ibba

Department of Political Science, Law and International Relations, University of Padua,
Via del Santo 28, 35123 Padua, Italy
mariastella.righettini@unipd.it

Abstract. The paper evaluates the public value created through the cultural heritage digitalisation policy adopted by the Uffizi Galleries by opening a Facebook page during the Covid-19 pandemic. Using quality-quantity techniques based on text mining, Uffizi's policy outcomes will be assessed by identifying citizens' digital behaviors and underlying generative causes. Evidence shows that citizens-users do not act as mere passive beneficiaries but as implementers of the digital public service through continuous interactions allowed by social networks. Consequently, they contribute to the co-production of shared value and a transformation of the relationship between the museum and its citizens-users in managing cultural heritage. Finally, the paper defines generative causes for citizen users' digital behavior. Thus, it allows a theory for the possible future successful transfer of cultural heritage digitalisation programs.

Keywords: Cultural heritage · Public value · Digital behavior · Sentiment analysis · Text-mining

1 Introduction

This work investigates an innovative cultural heritage managing policy adopted by the Uffizi Galleries during the outbreak of the Covid-19 pandemic and the resulting lockdown measures to identify the transformation of involved actors' actions. As we'll clarify, the Uffizi Galleries' strategy is an exciting case study to review the analytical frames of quality and accessibility of digital services and understand the citizens' digital behavior.

Before the Covid-19 pandemic, the digitization of cultural heritage was pursued by European and international agendas as an alternative strategy to the 'real' one for the achievement of better accessibility, conservation, and promotion of cultural heritage [1–3]. According to the Thematic Indicators for Culture in the 2030 Agenda[1], the increase

---

[1] A document developed in 2019 by UNESCO to establish a methodology that demonstrate the contribution of culture in the implementation of UN Sustainable Development Goals of the Agenda 2030 for Sustainable Development.

in digital cultural services is a means to consolidate a universal right to access cultural heritage, providing better cognitive and economic accessibilities [4]. The Council of Europe's Faro Convention also encourages the use of digital technologies as instruments to increase the accessibility of cultural heritage [5].

The outbreak and spread of the SARS-CoV-2[2] pandemic highlighted a need to ensure full accessibility and continuity of cultural services. Therefore, a need to accelerate the digitalisation process has emerged so far experienced a non-homogeneous relevance within Italy's political and institutional agendas.

Many museums and cultural institutes have increased their online presence during the lockdown measures [3, 6]. The Uffizi Galleries have opened a new Facebook page that has soon achieved resounding success, going from 0 to more than 120 thousand followers in just over one year. This paper will attempt to uncover the digital behaviors adopted by citizens-users accessing the museum's digital cultural heritage and their generative causes to measure the public value in terms of outcomes produced by Uffizi's digitalisation policy. Research results clearly show that digital behaviors differ from 'real' behaviors not only in accessing digital cultural heritage but also in their outcomes. Citizens-users adopt digital behaviors as policy beneficiaries and implementers at the same time. In so doing, they cause the production of public value that results in modifying the relationship between museum institutions and citizens-users in managing cultural heritage. Furthermore, the definition of underlying generative causes or causal mechanisms [7–9], will allow us to develop a causal theory on citizens-users digital behaviors, from which to derive generalizations on the links among context conditions, elements of policy design and outcomes achieved [10], to facilitate the possible favorable transfer of the program's success [11] and to stress the most beneficial responses to the program [9].

The paper is structured in three sections. The first section will briefly illustrate some key concepts and the research methodological approach. The second section will carry out an empirical analysis to identify the digital behaviors and their underlying generative mechanisms by analyzing trends and citizens-users' digital discourses and opinions using quality-quantity techniques based on text mining [12, 13]. Specifically, quantitative data concerning access to digital content on the Uffizi Galleries' Facebook page will be analysed. Later, a sentiment [12] and a content analysis [13] will be performed on 41,225 comments in Italian issued by citizens-users below digital contents published by Uffizi Galleries in their Facebook page between March 10, 2020 (the opening of the social page), and June 30, 2021[3]. Part of this section will be executed using the Iramuteq software[4]. Finally, the last section will present concluding considerations.

---

[2] The new coronavirus strain that has been associated with the Covid-19 pandemic.

[3] The data and comments were collected on 1st July and on 18th August 2021.

[4] Iramuteq is an open software (GNU GPL license), based on R software and Python language, and allows to perform statistical analysis on textual corpora. See http://www.iramuteq.org/.

2   Methods. The Cultural Heritage Digitalisation Through Social Networks

The analysis proposed in this paper focuses on three aspects of Uffizi's digitalisation process: the role of social networks in promoting accessibility to digital public services, thanks to a high ability to reach a wide range of people and opinions [12]; the contribution in policy innovation favored in the medium term by the emergency solution; the production of public value resulting from digital behaviors adopted by citizens-users.

Suppose digitization of cultural heritage is the conversion of cultural heritage's tangible and intangible assets from a physical to a digital form [14]. Similarly, the digitalisation of cultural heritage stems from a process of transformation of social life «around digital communication and media infrastructures» [14, p. 4], which means transforming social interactions into digital social interactions [15] through multiple digital media, including social networks.

The present analysis of Uffizi's digital social strategy adopts a definition of social networks as social spaces [16], which allow continuous synchronous and asynchronous interactions among citizens-users and citizens-users and museum institutions. Citizens-users – as individuals, entities, and organizations – are integrated within a highly interactive system. Thus, they co-produce the digitalisation policy at the implementation level [17] through the adoption of digital behaviors.

Social networks are digital «modern agoras» that generate big data. Therefore, it is possible to derive masses of information and opinions equivalent to the resulting of a survey [12, pp. 12–13]. Therefore, they are suitable for quality-quantity automatic analysis methods [16][5]. For this reason, the searching for digital behaviors and their generative mechanisms will be implemented in this study by the analysis of trends in digital access to cultural heritage and through sentiment[6] and a content analysis of comments made by citizens-users below digital contents published by the Uffizi Galleries on their Facebook page between 10 March 2020 and 30 June 2021.

3   Enhancing the Digital User-Experience as the Creation of Public Value

The Uffizi Galleries are a non-profit institution endowed with scientific, administrative, financial and accounting autonomy. They include the Uffizi Gallery, the Vasari Corridor, the Boboli Gardens and the Pitti Palace and they are one of the most important museums in the world, confirmed in 2019 as the most popular Italian museum with 4,391,895 visitors[7]. Like many other museums during the Covid-19 pandemic [6], the Uffizi have

---

[5] However, such analyses could be subject to a problem of representativeness, as over-representation of young people [16] or of other groups – given the increase in the average age of social networks' users in recent years (https://vincos.it/2018/08/16/facebook-in-italia-31-milioni-di-utenti-giovani-50/) – are possible. In addition, it is also possible that digital access to cultural heritage persists in excluding the same groups who were also excluded from physical use, not necessarily contributing to an increase in accessibility.

[6] «The analysis of the 'feeling' contained in a text» [12].

[7] See https://www.uffizi.it/news/uffizi-numeri-2019.

also increased their online presence. The 10 March 2020 they opened a social account on Facebook platform, which has reached more than 120 thousand of followers in just over one year. In the period between March 10, 2020, and June 30, 2021, the Uffizi Galleries shared on their Facebook page 558 digital contents, including profile updates, photos, videos and live recordings. As of August 18, 2021,[8] the same digital contents had 720,455 likes, 64,489 comments, and 9,132,681 views[9].

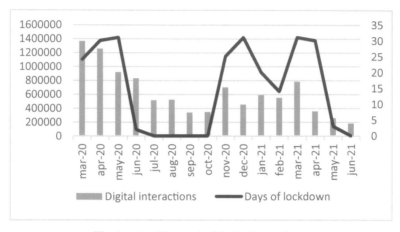

Fig. 1. Monthly trend of digital interactions.

As shown in Fig. 1, the monthly trend of the total amount of considered digital interactions is irregular and probably correlated to the days of lockdown due to the Covid-19 pandemic. Moreover, the museum presented the new social account connected to the pandemic context. As stated by the museum's director Eike Schmidt,[10] the Galleries' Facebook page has in part been created to offer a contribution to the community during the weeks of national closures. However, despite the initial emergency context, Fig. 1 clearly shows that the absolute number of digital interactions remained high during the whole analysed period. Therefore, it shows that the digitalisation policy has produced essential outcomes in the long term. In addition, an automatic content analysis of comments with Iramuteq software recognized that 'thanks'[11], 'beautiful'[12], 'interesting'[13], 'good morning'[14] and 'congratulations'[15] are the most frequently used words[16] in the comments, suggesting that the high number of interactions on the Uffizi's Facebook page

---

[8] Data collection date.

[9] Views refer only to videos and live recordings.

[10] Interview with the director of Uffizi Galleries Eike Schmidt at the Corriere della Sera of 1 April 2020, see https://www.facebook.com/corrieredellasera/videos/981605565574486.

[11] 18,428 frequencies.

[12] 3,381 frequencies.

[13] 3,168 frequencies.

[14] 3,100 frequencies.

[15] 2,947 frequencies.

[16] Also named occurrences.

is accompanied by the prevalence of positive sentiment around digital contents. To further confirm this hypothesis, sentiment analysis of almost all citizens-users' comments in Italian issued below the museum's digital Facebook contents published between 10 March 2020 and 30 June 2021 was undertaken, for 41,225 comments analysed.[17] The social network sentiment analysis was performed without the use of the software, considering the medium-sized [13] of the corpus[18], characterized by a substantial variability of its inner lexical words, consisting of personal uses of vocabulary or wide employment of expressions whose polarity is strongly influenced by citizens' individualities. Since our aim is the analysis of citizens users' digital behaviors, we considered document-level as the most appropriate level of analysis to perform sentiment analysis. Thus, each comment belongs to a diverse citizen-user and must be analysed as a whole, separately from the other comments and with the attribution of a single polarity [18, 19]. We first divided individual comments into personal and objectives [19]. If they contain emotions or opinions, they are considered personal. If not, they are considered objectives. Then, we attributed to subjective comments a positive polarity (if they express positive opinions or emotions), or negative polarity (if they say negative views or feelings), and to objective comments a neutral polarity [19, 20]. Figure 2 shows the analysis results

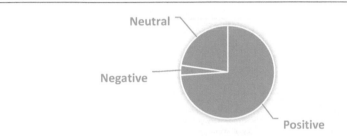

Positive
Congratulations for the skill and competence. I like the details. (Comment #26490, content of November 17, 2020).
Congratulations. Always exciting and very interesting. (Comment #6938, content of April 13, 2021).
Negative
There are problems not working the audio to anyone. (Comment #9995, content of March 23, 2021).
Interesting the whole series but tiring to follow despite the subtitles in Italian for the speed of the story. (Comment #30851, content of August 27, 2020).
Neutral
Is it true that married women wore their hair collected while unmarried women wore it loose? (Comment #8750, content of March 30, 2021).
Elena. (Comment #6349, content of April 16, 2021).

Fig. 2. The categories of comments' sentiment polarity.

---

[17] Compared to the total amount of comments on the museum's Facebook page for the same period as of August 18, 2021, equal to 64,489, those that did not contain text, were written in a diverse language from Italian, written by the museum, written in response to other comments were not considered in the analysis.

[18] Set of texts (comments in this specific case study) analysed.

and some examples of statements for the three categories, translated by the authors into English.

The negative comments category covers audio and framing issues, criticism, or suggestions for content choices. The neutral comments category includes comments that only contain other people's tags, proper or place names, numbers, and exclamations, questions, or remarks on the subject matter under discussion in the specific digital content that couldn't be attributed to different categories. The positive comments category contains favorable judgments or thoughts about digital content or art and cultural heritage.

Sentiment analysis results disclose a mainly positive feeling around museum-digital contents and highlight how access to digital cultural heritage implies a continuous interaction between citizens-users and museum institutions. Citizens-users express their opinions and emotions, chatting around digital content, thus establishing a relationship with the museum institution that translates into the learning by the museum of continuous and numerous feedbacks related to cultural heritage digital access.

But what are the reasons underlying digital behaviors related to cultural heritage digital access? To answer the question, we explored the topics resulting from citizens users' digital experience in more detail. First, we implemented an automatic content analysis

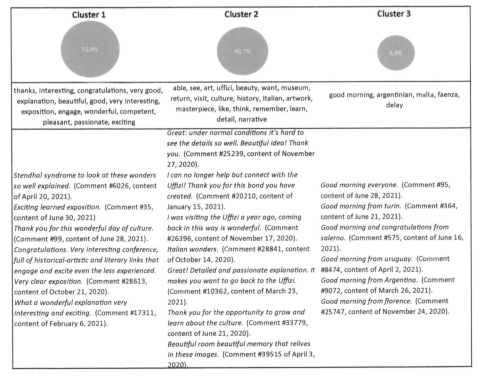

Cluster 1	Cluster 2	Cluster 3
52,4%	40,7%	6,9%
thanks, interesting, congratulations, very good, explanation, beautiful, good, very interesting, exposition, engage, wonderful, competent, pleasant, passionate, exciting	able, see, art, uffizi, beauty, want, museum, return, visit, culture, history, Italian, artwork, masterpiece, like, think, remember, learn, detail, narrative	good morning, argentinian, malta, faenza, delay
Stendhal syndrome to look at these wonders so well explained. (Comment #6026, content of April 20, 2021). *Exciting learned exposition.* (Comment #35, content of June 30, 2021) *Thank you for this wonderful day of culture.* (Comment #99, content of June 28, 2021). *Congratulations. Very interesting conference, full of historical-artistic and literary links that engage and excite even the less experienced. Very clear exposition.* (Comment #28613, content of October 21, 2020). *What a wonderful explanation very interesting and exciting.* (Comment #17311, content of February 6, 2021).	*Great: under normal conditions it's hard to see the details so well. Beautiful idea! Thank you.* (Comment #25239, content of November 27, 2020). *I can no longer help but connect with the Uffizi! Thank you for this bond you have created.* (Comment #20210, content of January 15, 2021). *I was visiting the Uffizi a year ago, coming back in this way is wonderful.* (Comment #26396, content of November 17, 2020). *Italian wonders.* (Comment #28841, content of October 14, 2020). *Great! Detailed and passionate explanation. It makes you want to go back to the Uffizi.* (Comment #10362, content of March 23, 2021). *Thank you for the opportunity to grow and learn about the culture.* (Comment #33779, content of June 21, 2020). *Beautiful room beautiful memory that relives in these images.* (Comment #39515 of April 3, 2020).	*Good morning everyone.* (Comment #95, content of June 28, 2021). *Good morning from turin.* (Comment #364, content of June 21, 2021). *Good morning and congratulations from salerno.* (Comment #575, content of June 16, 2021). *Good morning from uruguay.* (Comment #8474, content of April 2, 2021). *Good morning from Argentina.* (Comment #9072, content of March 26, 2021). *Good morning from florence.* (Comment #25747, content of November 24, 2020).

Fig. 3. The clustering of citizens-users' comments.

of comments using Iramuteq software, which resulted in the clustering of three semantic classes. Figure 3 displays, for each cluster, some significant words and comments' examples that were also in this case translated by the authors into English.

The first cluster is the largest one, comprising 52.4% of total analysed comments. It concerns predominantly emotional judgments and impressions about digital content and the digital cultural experience.

The second cluster is, on average large, comprising 40.7% of total analysed comments. It concerns the depiction of cultural heritage digital access' experience and is much less homogeneous than clusters 1 and 3. As can be seen in Fig. 4, which presents the primary forms and their networks within cluster 2, the digital access of the contents shared by the museum is a cultural experience that generates new pieces of knowledge, routines, leisure, and bonds, thanks also to the ability of the Uffizi's digital contents to

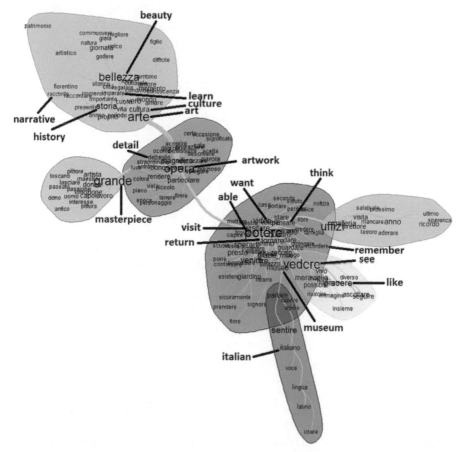

Fig. 4. Main words and their networks within cluster 3. Some significant terms also given in Fig. 3 have been provided with English translations.

create memories, emotions and the desire to see the museum. The one with Uffizi's digital offer is a daily appointment during which citizens-users actively immerse themselves in artworks, culture, and history's details.

The third cluster is the most concise, including 6.9% of total analysed comments. It concerns the moment of encounter among actors of digital social interactions, who welcome each other and recognise themselves as a community, expressing forms of greetings. An analysis of the χ^2 – that is, of the statistical dependence – among the different forms assumed by digital contents (live recording, photos, profile updates, or videos) and the three distinct clusters provides additional insights on digital fruition. The live recording is the digital content's form that encourages the most this manifestation of social relationships. See in this regard Fig. 5, in which it is undeniable the achievement of very high χ^2 values by the live recording bar within cluster 3.

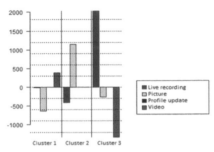

Fig. 5. The statistical dependence between the different forms assumed by digital contents and the three different clusters.

3.1 Causal Mechanisms

The Facebook social digitalisation policy adopted by the Uffizi Galleries at the outbreak of the Covid-19 pandemic takes the shape of sharing digital contents by the museum institution and the adoption of digital behaviors around the same contents by citizens-users. If the institution shares digital contents on a daily and unrestricted basis, in the same way, the citizens-users access, enjoy and interact with such contents, generating public value through the expression of opinions and emotions and the release of continuous and numerous feedbacks related to cultural heritage digital fruition.

The digital contents are transmitted in various forms by the museum, have as their object both works and stories of popularity and less known others, and concern explanations made mainly by the Galleries' staff or sometimes by external experts. The same digital contents are judged by citizens-users of high quality, resulting from a strong commitment on the part of the museum, which gives new institutional credibility to an institution that is already highly appreciated by the public. Furthermore, the accuracy and commitment placed in the creation of high-quality digital contents are highly appreciated by citizens-users because they allow them to engage with the contents, actively immersing themselves in artworks and dissertations, thus causing emotions, entertainment, social ties, and bringing to light details that would have been difficult to discover without the guidance of an expert and the possibility of a close view. We define digital commitment as the first generative mechanism based on these premises.

Furthermore, citizens-users are primarily satisfied with the digital fruition of contents shared by the museum. They express this satisfaction by releasing judgments that assume positive polarity in most analysed comments. Exposure to high-quality digital content, therefore, generates on citizens-users a feeling of pleasure that marks the importance, to create successful digitalisation policies, of creating content that understands the expectations of citizens-users. For this reason, we define satisfaction as the second generative mechanism.

Finally, these high-quality and satisfactory digital contents, shared by the museums daily and mainly in Italian, encourage new routines and a barely international digital community. Thus, just about 5% of total analysed comments are made in languages other than Italian, mainly English, Spanish and French. Moreover, digital contents arouse citizens-users a sense of attachment that takes the shape of an increase in both physical and virtual accessing intention. Comments prove the strong desire of citizens-users to continue to enjoy digital content or to visit the museum. This is partly a consequence of the Covid-19 pandemic, expression of a more generic wish to 'return to normal', and part of the fruition of digital contents. This finding is even more significant if we consider loyalty as a property that generally is not associated with physical museum fruition. There is no evidence of a causal link between satisfaction and re-visiting intentions [21]. In conclusion, we define loyalty as the third generative mechanism.

4 Closing Considerations: So, What?

In concomitance with the closures due to the Covid-19 pandemic that has affected the entire national territory and cultural institutions, the Uffizi Galleries have adopted a cultural heritage digitalisation strategy sharing on Uffizi's Facebook page of digital contents made by the museum itself. The evaluation of Uffizi's digitalisation policy made in this study has highlighted the ability of digital cultural content to generate public value by eliciting digital commitment, satisfaction, and loyalty. Driven by these emotional incentives, citizens-users access digital cultural heritage by establishing a continuous interaction with the museum, commenting on digital content, and telling the digital experience. Therefore, the effectiveness of digitalisation policy in creating public value concerns the adoption by citizens-users of digital behaviors, not only as mere passive beneficiaries but as implementers of the digital public service, which brings to the constant acquisition of feedback by the museum institution.

The paper has shown the importance of digitalisation policies in creating public value in cultural heritage management if combined with 'real' policies. But the adoption of digitalisation policies entails new costs for cultural institutions, which could be unsustainable for smaller institutions. A museum, or a museum network, could therefore look at the resources of the National Recovery and Resilience Plan (PNRR)[19], which allocates 6.68 billion euros to investments in tourism and culture, to make cultural heritage

---

[19] The investment and reform package adopted by Italy due to the Covid-19 pandemic and the NextGenerationEU (NGEU), the temporary European emergency response instrument.

accessible in ways not yet considered, by encouraging better and broader accessibility of their services and by transforming the relationship with citizens-users.

References

1. COM: 711, Commission recommendation of 27 October 2011 on the digitisation and online accessibility of cultural material and digital preservation (2011)
2. Adane, A., Chekole, A., Gedamu, G.: Cultural heritage digitization: challenges and opportunities. Int. J. Comput. Appl. **178**(33), 1–5 (2019)
3. Dominguez-Jiménez, J.S., Castilla-Agredano, B., González-Nieto, M., Moreno-Alcaide, M., Monterroso-Checa, A.: Preparing rural heritage for another kind of Covid pandemic: heritage digitalisation strategies in the Alto Gaudiato Valley and Subbetica of Cordoba, Spain. Sci. Res. Inf. Technol. **11**(1), 195–208 (2021)
4. United Nations Educational, Scientific and Cultural Organization, UNESCO: Culture 2030 indicators. Thematic Indicators for Culture in the 2030 Agenda (2019)
5. Council of Europe: Council of Europe Framework Convention on the Value of Cultural Heritage for Society, Council of Europe Treaty Series, vol. 199 (2005)
6. Vayanou, M., Katifori, A., Chrysanthi, A., Antoniou, A.: Cultural heritage and social experiences in the times of COVID 19. In: Proceedings of AVI 2CH 2020: Workshop on Advanced Visual Interfaces and Interactions in Cultural Heritage, AVI 2CH 2020. ACM, New York (2020)
7. Barzelay, M.: Learning from second-hand experience: methodology for extrapolation-oriented case research. Gov. Int. J. Policy Adm. Inst. **20**(3), 521–543 (2007)
8. Mayntz, R.: Mechanisms in the analysis of social macro-phenomena. Philos. Soc. Sci. **34**(2), 237–259 (2004)
9. Pawson, R.: Middle-range realism. Eur. J. Sociol. **41**(2), 283–325 (2000)
10. Melloni, E.: Dieci anni di Impact Assessment della Commissione Europea. Come funziona, a cosa serve e a chi serve. Rivista Italiana di Politiche Pubbliche **7**(3), 419–449 (2012)
11. Busetti, S., Dente, B.: Designing multi-actor implementation: a mechanism-based approach. Pub. Policy Adm. **33**(1), 46–65 (2018)
12. Ceron A., Curini L., Iacus S. M.: Social Media e Sentiment Analysis. L'evoluzione dei fenomeni sociali attraverso la Rete. Sxi – Springer per l'Innovazione, Milano (2014). https://doi.org/10.1007/978-88-470-5532-2
13. Tuzzi, A.: L'analisi del contenuto. Introduzione ai metodi e alle tecniche di ricerca. Carocci editore, Roma (2003)
14. Ginzarly, M., Srour, J.: Cultural heritage through the lens of COVID-19. Poetics (2021). https://doi.org/10.1016/j.poetic.2021.101622. Accessed 23 Dec 2021
15. Bloomberg, J.: Digitization, Digitalisation, and Digital Transformation: Confuse Them At Your Peril. Forbes. https://www.forbes.com/sites/jasonbloomberg/2018/04/29/digitization-digitalization-and-digital-transformation-confuse-them-at-your-peril/?sh=7a2102e82f2c. Accessed 23 Dec 2021
16. Smyrnaios, N., Ratinaud, P.: Comment articuler analyse des réseaux et des discours sur Twitter. tic&société **7**(2), 120–147 (2013)
17. Cataldi, L.: Coproduzione: uno strumento di riforma in tempi di austerity? Rivista Italiana di Politiche Pubbliche **10**(1), 59–86 (2015)
18. Vanaya, S., Belwal, M.: Aspect-level sentiment analysis on e-commerce data. In: Proceedings of the International Conference on Inventive Research in Computing Applications, ICIRCA, pp. 1275–1279. IEEE Xplore (2018)

19. Kaur, H., Mangat, V., Nidhi: A survey of sentiment analysis techniques. In: International Conference on I-SMAC (IoT in Social, Mobile, Analytics and Cloud), I-SMAC, pp. 921–925. IEEE Xplore (2017)
20. Wilson, T., Wiebe, J., Hoffmann, P.: Recognizing contextual polarity in phrase-level sentiment analysis. In: Proceedings of Human Language Technology Conference and Conference on Empirical Methods in Natural Language Processing (HLT/EMNLP), pp. 347–354. Association for Computational Linguistics, Vancouver (2005)
21. Hume, M.: How do we keep them coming?: examining museum experiences using a services marketing paradigm. J. Nonprofit Pub. Sect. Mark. **23**(1), 71–94 (2011)

The Necessary Digital Update of the Camino de Santiago

Rubén C. Lois-González[1](✉) (iD) and Xosé Somoza-Medina[2] (iD)

[1] University of Santiago de Compostela, 15705 A Coruña, Spain
rubencamilo.lois@usc.es
[2] University of León, 24071 León, Spain
somoza@unileon.es

Abstract. The Camino de Santiago (Way of Saint James) was considered in 1987 by the Council of Europe as the first European cultural route. Since then, hundreds of thousands of pilgrims from all over the world have walked through the traditional cultural landscapes of this corner of Europe seeking their own personal journey, an experience of introspection that, at least in theory, demands disconnecting from the world. The pandemic caused by COVID-19 has meant the greatest crisis in tourism in contemporary times, especially in mass destinations, revitalizing by contrast other forms and tourist places, such as cultural routes. From the political objective of economic and social recovery, public administrations are promoting the digitalization and use of new technologies in the dissemination and management of the Camino de Santiago as a cultural and tourist product. According to all the plans and projects planned, in the next years not only the mobile device will be a tool to help carry out the Way, it will become essential to achieve a completely satisfactory experience.

Keywords: Camino de Santiago · Tourism digitalization · Smart tourism · Apps · Digital technologies

1 Introduction

To a certain extent, walking is to move from one place to another and intensely enjoy this experience, but since the Middle Ages the plan of walking was not always fulfilled. As we have pointed out on certain occasions, along with the long and costly pilgrimages of several months, other pilgrims made pilgrimages by substitution (sending others to make the route) or made costly donations [1]. Leaving aside these historical evidences, today the act of realizing the Camino de Santiago is usually focused on walking slowly the route, but it involves an important effort of prior preparation and, once completed, it generates abundant narrative and graphic material that circulates through the networks. Therefore, the most emblematic of cultural itineraries is at the same time an intense experience of travel with effort and a set of previous activities of preparation and subsequent celebrations, many of which can only be carried out by using the Internet.

Consequently, the fact that public administrations and numerous private entities use the network to inform, communicate or sell on the Camino is an undoubted reality that has been tried to characterize [2, 3]. For two reasons, this reality has become particularly relevant in recent times. The first is the widespread use of big data and the recourse to digitalisation of all types of information on the route, which is now evident. The second, the impact of the COVID pandemic, has led to an intensification of blogs, information and forums on the situation. A route that is analyzed with concern, also with hope for the future and that is rethought in intense virtual debates that we have begun to study [3].

2   Tourism and Digitalization in PostCOVID Era

As noted in numerous studies and articles on the subject, tourism activity has been completely transformed into the digital era [4–6]. On the one hand, the network has made it possible to multiply the volume of information available about the whole world and, consequently, the destinations to which to travel. On the other hand, it has radically innovated the ways in which tourism products are traded. Third, it facilitates the change in the relationships between actors linked to the activity. Tourism related companies are less and less interested in closed holiday packages; they are obliged to diversify their offers. At the same time, the rigidity in the relations between citizens-consumers and companies-producers of the service cease to be linear and repeated. Today they can express themselves in many different ways [7, 8]. Tourists can directly connect and negotiate the book of a room, a guide to a destination or a ticket to an event. Companies can respect a high degree of self-organisation of the client and offer only very specific services in the whole tourist experience, among many possibilities.

2.1   Tourism in the XXI Century

In the present century the tourist activity has maintained a series of continuities with the immediately preceding stage, although progressive changes are detected that need to be commented. Among the permanencies, undoubtedly, the massive nature of the trips to rest and replace for a period, far from the place of habitual residence and work. Climate motivations continue to be very important to justify tourism practices, although with some novelties. Thus, the holiday period has become fragmented and has escaped monotony. People who goes to beach destinations, try to complete their experience with nature excursions, visits to historic centres or complementary practices guided by sustainability [9, 10]. Normally, the Sun and Beach destination is no longer the only one chosen for vacations throughout the year. Its hegemony has been tempered by emerging practices such as visiting scenically attractive cities or places, such as those situated in cultural locations, which proliferate throughout Europe. Tourism linked to these routes is perceived as less crowded, healthier and more environmentally friendly, as it is usual to resort to cycling or hiking to move around [3]. The cultural itineraries complement a climate tourism of Sun and beach, which is still the majority, but no longer covers all the expectations of the traveller.

As has been stressed, cultural itineraries are associated with a number of highly valued attributes of current tourism. First, the value of landscape, the encounter with the place, while enjoying a slow mobility [11]. Second, a cultural content that, although not predominant in many cases, should permeate the traveller's experience [12]. Third, the existence of such valued referents as healthy individual exercise, contact with nature and the availability of time for oneself [10]. Fourth, a new way of knowing different localities and regions from a linear route [13]. A sketch that becomes a programmed route from the intensive management of ICTs and social networks.

2.2 Cultural Routes as Resource for the Enhancement and Conservation of Cultural Heritage: Territorial Resilience and Digital Technologies

Contemporary times have taught us that reality shows two inseparable facets: tangible reality and the set of images it generates. This is especially appropriate for cultural routes, marked by monuments (more or less preserved and restored), sets of symbolic or landscape value, and all kinds of attributes that enrich them. At the same time, these monuments, ensembles and elements are photographed, digitized, reproduced and retouched to the fullest [14, 15]. In the case of cultural routes, this intensive use of ICTs is manifested in a constant publicity, well presented, to encourage visits and tourism. Monuments, landscapes and other elements of culture are especially praised. It is encouraged to travel them from messages very worked repeated monotonously: a millennial path, the origin of Europe, time to disconnect, etc. It is evident that this digital advertising chooses some promotional attributes, while despising others. However, as later the travel experience allows stopping in the territory, the subsequent emission of messages by tourists and pilgrim through social networks it is possible to reconstruct the original message with those new elements that visitors of a country or carriers of a predetermined belief are highlighting after their own experience in the cultural route [16].

2.3 Digitalization and New Technologies in Cultural Landscapes

The whole process of codifying and promoting itineraries underlines the importance of the landscape. A landscape always humanized, whether in a traditional way by taking care of the fields and forests, the building of historic cities and villages of vernacular architecture, or in recent times, where landscaping has highlighted certain perspectives, Urban rehabilitation reinterprets the built patrimony and the roads invade everything. As originally indicated by C. Sauer, the landscape is cultural or is not landscape [17, 18]. For tourism, especially that which places slow mobility at the centre of the experience, the landscape is one of the main attractions of the visit. "The tourist gaze" in the centre [19]. The gaze of a tourist who travels a historical itinerary full of monuments as central axis of the experience. We look at everything, from the intermediate scale that the landscape requires and from more concrete points of view such as those generated by an urban scene. Even more tangible experiences, such as gastronomy or the purchase of handicraft, are mediated by the evocation of productive agrarian landscapes or images of people making objects. The landscape envelops the traveller, the notion of natural route and the memory of the journey. All the digital elements of reproduction of this experience proliferate as a result of it, but also as attractive images for those who want to visit the routes.

3   The Necessary Digital Update of the Camino de Santiago

In recent years, the Camino de Santiago has undergone a process of continuous growth until the crisis caused by the COVID-19 pandemic. As can be seen in Fig. 1, the Way of Saint James was pilgrim in the second half of the 20th century by just a few dozen people annually, who were encouraged when the pilgrimage was made in a jubilee year. The jubilee year in Santiago corresponds to when the feast of Santiago (July 25th) coincides on Sunday, which happens in repeated series with alternations every 5, 6 or 11 years. The 1993 Compostelan Holy Year was used by public administrations and pro-Camino associations to promote the Jacobean route, with major investment in new accommodation infrastructure and international promotion campaigns. The route had been declared the first European cultural route by the Council of Europe in 1987 and was included in the UNESCO World Heritage List in December 1993 [20–22].

Starting in 1993, the continued work on the Camino de Santiago was fruitful and every year more pilgrims were able to return to their homes and count the excellent personal sensations that the Camino had given them. The following Holy Years (1999, 2004 and 2010) signified the confirmation of the expectations created through the territorial and strategic planning of a product of cultural and religious tourism that was still expanding [23].

The second decade of the 21st century marks a new shift in the observed trend (Fig. 1), until then the Holy Years marked peaks of inflection. However, between 2011 and 2019 there was a continuous increase, year after year, of about 8%, taking into account that 2016 was a Compostelan Holy Year. This fact came to confirm that the Way had become a tourist-cultural route fully contemporary, beyond its original meaning strictly religious [24].

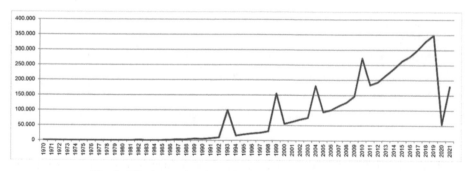

Fig. 1. Pilgrims to Santiago de Compostela 1970–2021 [25].

The year 2020 was the year of the pandemic crisis. In the months of April and May no person was registered on pilgrimage. In total in 2020, with the subsequent easing of the confinement measures, 54,144 people made a pilgrimage to Santiago, compared to 347,578 in 2019. The Camino de Santiago was to be reinvented anew, adapting to the new conditions. The different administrations and entities related to the Camino then bet, among other measures, to promote the digitalization and the use of the new technologies to continue developing the Camino de Santiago in the new post-COVID era.

3.1 The Role of the Spanish National Administration

In June 2020, the Archbishop of Santiago asked Pope Francis to extend the Jacobean year of 2021, due to the restrictions imposed by the pandemic on the pilgrimage. During the solemn opening of the Holy Door on 31 December 2020, the Apostolic Nuncio read the decree extending the Jubilee Year throughout 2022. Since 1122, for the first time in history, the Compostelan Holy Year will last 24 months.

The proposals to reactivate the Camino have come from different structures and entities. The above point is an example of how far the Christian Church has been involved in relaunching the Way. The Spanish government has also carried out different initiatives in this regard, among which the National Tourism Plan Xacobeo 2021–2022 should be highlighted. Announced in June 2021, it has a budget of 121 million euros, financed by the European funds Next Generation EU. Within the axes of action (Table 1), the fourth will allow the financing of five projects in the Camino de Santiago that will be framed within the Spanish Network of Intelligent Tourist Destinations (DTI), with concrete actions of development of digital solutions as apps or tourist guides, open technological platforms that provide information in real time, analysis and visualization tools of data, or installation of totems and interactive kiosks.

Similar calls at state level develop public funding tools for digitisation projects and new technologies, always under the Recovery, Transformation and Resilience Plan (PRTR), with four repeated objectives: green transition, digital transition, energy efficiency and improved competitiveness.

Table 1. Xacobeo 2021–2022 Tourism National Plan [26].

Axis	Descriptor	Mills €	%
1	Sustainable projects of maintenance and rehabilitation of the historical heritage in the Camino de Santiago	65,3	54%
2	National plan of tourism sustainability in Xacobeo destination	45	37%
3	Tourism Xacobeo product development program	5	4%
4	Smart tourist destinations program	5	4%
5	Xacobeo 2021–2022 international promotion program	1,1	1%

The Spanish DTI Network was created in 2013 and since then, different cities, towns and regions have adhered to this proposal put forward by the Secretariat of Tourism that proposes a future strategy based on governance, sustainability, accessibility, innovation and technology as backbone. Within this framework are included different programmes and calls that at the regional level are proposed as Cohesion Actions in Destination, to homogenize conditions and levels of quality of service in tourist districts, and at local level are specified in Sustainability Plans in Destination, where it is the different conditions of each municipality that structure the main objectives of the planning [27].

Among the digital solutions proposed at the state level, two projects already consolidated can be highlighted. The first of them is the app "Cno. de Santiago", developed by the National Geographic Institute (IGN) together with the Spanish Federation of Associations of Friends of the Way, which launched its first version in October 2015. This application has the excellent cartographic bases of the IGN, permits the download on line and off line of the different stages of the Roads, associate updated meteorological information, data on the artistic heritage or services offered in each place. The second digital proposal was promoted by the Cathedral of Santiago in January 2021 and is the app "Pilgrim Digital Credential", a tool that aims through stamps with QR technology to complement the traditional credential paper, where the stamps that accredit the pilgrimage are placed. Little by little, the QR stamps for the application are spreading among the hostels, parishes and local associations, to improve this digital proposal.

3.2 Different Strategies in the Regional Level

When the government of Madrid sent the draft of the National Tourism Plan Xacobeo 2021–2022 to different entities and institutions, one of the recipients that raised more allegations was the government of the Galician autonomous administration, because within the political dialectic (belong to different parties) understood that the central government had not heeded any of its considerations; Galicia, where all the Ways converged, was underrepresented and some digitization projects collided with their own projects.

In fact, different regions of northern Spain have based part of their tourism promotion strategies on the Camino de Santiago [28] and despite the fact that there are several inter-administrative bodies to coordinate policies on the Camino, it is common for there to be frequent lack of coordination in setting specific objectives, strategies or programmes. An example of this lack of coordination has to do with the apps about jacobean routes promoted by different autonomous communities, which only include information from their respective territories and when the person making the pilgrimage changes region must also change app, when entering a space with a different regional government.

The Camino de Santiago de Galicia App offers information on all Jacobean routes from the moment the pilgrim enters this region, such as the Portuguese Camino, which as it is originated in the neighbouring country is the reason for the main criticism of users of the IGN App who do not have any information about this international route. The App promoted by the Galician government has real-time information on events, offers, services, meteorology, or incidents on the routes. The augmented reality allows through the camera of the mobile to visualize points of interest and information about those places and has a log book where to store photos and texts that can be shared in social networks. On the other hand, "Camino Assist" is the application of the Camino de Santiago in Asturias that in addition to the above-mentioned functionalities incorporates an agreement with the insurance company Europe Assistance, which offers different types of insurance to the pilgrim.

Galicia approved in 2015 the Master and Strategic Plan of the Camino de Santiago (Table 2) that defined eight strategic lines and a total of 27 action plans associated to each line, with an estimated budget of 56.1 million euros [29]. Precisely the last of the action plans was defined as Impulse to Smart Camiño. A measure which, at the time of drafting the Strategic Plan, was last in place, and which, however, over the years and in the circumstances brought about by the pandemic, have occupied a pre-eminent

Table 2. Camino de Santiago 2015–2021 Master and Strategic Plan [29].

Line	Descriptor	Action plans
1	Coordination and administrative organization	6
2	Preservation and enhancement of the heritage values of the Camino	2
3	Conservation and maintenance of the Camino	2
4	Environmentally sustainable Camino	1
5	Strengthening of the city of Santiago as Goal of the Jacobean route	3
6	Enhancement of all pilgrimage routes to Santiago	2
7	Specialization and quality in the service of assistance to the pilgrim	6
8	Research, communication and dissemination of the Way and Jacobean culture	5

place in Galician regional policy on the Camino, with an investment of 10 million euros. From this concrete plan, the Galician region has driven the creation of the platform caminodesantiago.gal where people who want to pilgrimage can prepare their Jacobean experience with 360° panoramas, plan each day, connect with other pilgrims during the Camino and expose their experiences once the route is completed. In addition, under the Smart Camiño Plan, the provision of wifi in the public hostels of the Galician network, the installation of totems to charge the batteries of phones and tablets or the creation of the app for mobiles mentioned above has been generalized.

The Aragonese Tourism Strategy Plan 2021–2024 [30] promotes the digitisation of the sector to enhance communication before, during and after the trip, achieving an intensive and intelligent use of a digital technology ecosystem, including: the internet of things; augmented reality and virtual reality; artificial intelligence and robotization; geolocalisation and TIG; and big data, small data and open data.

The Euskadi region has also stood out for the digitisation strategy in the promotion and organisation of the Caminos de Santiago that run through its territory, introducing QR codes in all printed materials, promoting the GeoEuskadi App, with more detailed tracks than other applications and cybersecurity. Asturias, Castile and Leon, Navarre and Cantabria also promote the digitalisation of tourism in their regions with specific projects on the Caminos de Santiago.

3.3 Some Local Examples on Digitalization and Use of New Technologies Around the Camino

Finally, at the local level of the administration, a good number of municipalities that are crossed by the Camino de Santiago have carried out in recent years actions for the digitization of the tourism sector and the use of new technologies. These are cities and small tourist localities that have chosen to be included in the Spanish Network of Smart Tourist Destinations, created in 2013, which include, for example: Jaca, Logroño, Burgos, León, Ponferrada, Avilés, Gijón, Donostia, Vitoria, Noja, Santander, Badajoz, Salamanca, or Murcia.

As already mentioned, the Recovery, Transformation and Resilience Plan (PRTR) is the Spanish collective project that serves as an operational framework to mitigate the impact of the health crisis and adapt to the new post-pandemic scenario in which tourist demand has changed. Under this framework programme, Spain will obtain up to €140 billion from the Next Generation EU Funds and one of the four cross-cutting axes on which it is based is the digital transition. According to this PRTR, the digital transition is the bridge between innovation, productivity and sustainability, as well as being a backbone of territorial and social cohesion. The Strategy for Tourism Sustainability in Destinations [27], within the PRTR, was presented in July 2021 by the Ministry of Tourism, with an estimated budget of 1,904 million euros for the period 2021–2023.

Within this initiative, axis 3 sets out the actions that can be financed in the field of the digital transition of the tourism model, grouped under five headings: communication and access of resources and services to the destination tourist; management of the impact of tourism on the destination; development of destination tourism intelligence platforms and systems; destination tourism companies and providers; and public governance and their access to technology.

The management instrument for these measures is the Territorial Plan for Tourism Sustainability in Destination, which is configured as a mechanism for cooperative action between the national, regional, local and private sectors. Between ordinary and extraordinary calls will be approved in the next few years several dozens of territorial plans of this type, which will facilitate the progressive digitalization of the tourism sector in Spain. In the first regular call, in December 2020, 25 plans were approved and in the second, in July 2021, there were 23. In the first extraordinary, with Next Generation EU Funds, in December 2021, more than 300 proposals have been submitted.

4 Final Remarks

The digitalisation of the Camino de Santiago is a process that began about a decade ago, through pioneering projects and the personal application of new technologies to all our daily experiences. This digitalisation has accelerated as a result of the restrictions imposed by the COVID-19 pandemic, which has made it necessary to maintain security distances, to have real-time information or to carry out more procedures and actions in digital format.

The cultural routes, such as the Caminos de Santiago, are a healthy, sustainable, environmentally responsible, social-friendly and territorial cohesion-generating type of tourism, which can increase its socioeconomic impact in the coming years with its definitive digital transition. The channelling of public investment towards this objective should be seen as a positive attempt to transform the tourism model, a step forward in line with the changes suggested by the scientific and professional community of tourism.

The Spanish strategy of tourism sustainability in destination is a comprehensive, structured and ambitious initiative. The allocation of different instruments at each administrative level should seek to remedy the errors of previous planning calls, where political interests and political dialectics were colliding according to administrative territorial boundaries. If the major objectives set out in the PRTR are to overcome traditional inertia and translate into concrete action on all fronts in the coming years, the expectations of recovery of the Camino de Santiago as a resource fort the enhancement and conservation of cultural heritage will be unquestionable.

References

1. Lopez, L., Lois-González, R.C.: The voices of female pilgrims in medieval wills. The Jacobean devotion in Apulia (Italy). Gend. Place Cult. **24**(4), 482–498 (2017)
2. Santomil, D.: A imaxe exterior de Galiza no século XXI. University of Santiago de Compostela. A Coruña (2011)
3. Lopez, L., Lois-González, R.C.: New tourism dynamics along the wayn of St. Jams. Undertourism and overtourism to the post-COVID-19 era. In: Pons, G.X., Blanco, A., Navalón, R., Troitiño, L., Blázquez, M. (eds.) Sostenibilidad Turística: overtourism vs undertourism, pp. 541–553. Societat d'Historia Natural de les Balears. Palma (Illes Balears) (2021)
4. Cohen, E.: The changing faces of contemporary tourism. Society **45**(4), 330–333 (2008)
5. Hosteltur: Guía rápida de la nueva economía del turismo. Hosteltur, Madrid (2016)
6. Blanco, A., et al.: Diccionario de Turismo. Cátedra, Madrid (2021)
7. Pencarelli, T.: The digital revolution in the travel and tourism industry. Inf. Technol. Tour. **22**(3), 455–476 (2019)
8. Sorooshian, S.: Implementation of an expanded decision-making technique to comment on Sweden readiness for digital tourism. Systems **9**, 50 (2021)
9. Urry, J.: Sociology Beyond Societies: Mobilities for the Twenty-First Century. Routledge, London (2000)
10. Creswell, T., Merriman, P.: Geographies of Mobilities: Practices, Spaces, Subjects. Ashgate, Farham (2011)
11. Cosgrove, D.: Social Formation and Symbolic Landscape. The University of Wisconsin, Madison (1998)
12. Coleman, S., Eade, J., (eds.): Reframing Pilgrimage. Cultures in Motion. Routledge, London (2004)
13. Pileri, P., Moscarelli, R. (eds.): Cicling & Walking for Regional Development. How Slowness Regenerates Marginal Areas. Springer y Fondazione Politecnico di Milano, Cham (2021). https://doi.org/10.1007/978-3-030-44003-9
14. Stefanou, J.: The contribution of the analysis of the image of a place to the formulation of tourism policy. In: Briassoulis, H., van der Straaten, J. (eds.) Tourism and the Environment: Regional, Economic, Cultural and Policy Issues, pp. 229–238. Kluwer Academic, Dordrecht (2000)
15. McWha, M., Frost, W., Laing, J.: Travel writers and the nature of self: essentialism, transformation and (online) construction. Ann. Tour. Res. **70**, 14–24 (2018)
16. Lois-González, R.C., Lopez, L.: Liminality wanted. Liminal landscapes and literary spaces: the way of St. James. Tour. Geogr. **22**(2), 433–453 (2020)
17. Roger, A.: Court traité dy paysage. Gallimard, Paris (1997)
18. Bercque, A.: La pensée paysagière. Archibooks & Sautereau Ed., Paris (2009)
19. Urry, J., Larsen, J.: The Tourist Gaze 3.0. Sage, London (2011)
20. Frey, N.: Pilgrim Stories: On and Off the Road to Santiago. University of California Press, Berkeley, CA (1998)
21. Santos Solla, X.M.: Mitos y realidades del Xacobeo. Boletín de la A.G.E. **28**, 103–119 (1999)
22. Lois-González, R.C., Somoza Medina, J.: Cultural tourism and urban management in northwestern Spain: the pilgrimage to Santiago de Compostela. Tour. Geogr. **5**(4), 446–461 (2003)

23. Somoza Medina, X., Lois-González, R.C.: Improving the walkability of the Camino. In: Hall, C.M., Ram, G., Shoval, N. (eds.) The Routledge International Handbook of Walking, pp. 390–402. Routledge, Abingdon (2018)
24. López, L., Nicosia, E., Lois-González, R.C.: Sustainable tourism: a hidden theory of the cinematic image? A theoretical and visual analysis of the way of St. James. Sustainability **10**(10), 3649 (2018)
25. Pilgrim Office. Santaigo de Compostela Pilgrimage Statistics. Archicofradía del Apóstol, Santiago de Compostela (1970–2021)
26. Gobierno de España: Plan Turístico Nacional Xacobeo 2021–2022. Madrid (2021)
27. Gobierno de España: Estrategia de sostenibilidad turística en destinos. PRTR. Madrid (2021)
28. Somoza Medina, X., Lois-González, R.C.: Ordenación del Territorio y estrategias de planificación en los Caminos de Santiago Patrimonio Mundial. Investigaciones Geográficas **68**, 47–63 (2017)
29. Xunta de Galicia: Plan director y estratégico del Camino de Santiago en Galicia 2015–2021. Xunta de Galicia, Santiago de Compostela (2015)
30. Gobierno de Aragón.: Plan Aragonés de Estrategia Turística 2021–2024. Zaragoza (2021)

A Project of Enhancement and Integrated Management: The Cultural Heritage Agency of Locride

Francesco Calabrò, Giuseppina Cassalia, and Immacolata Lorè[✉]

Department PAU, Mediterranea University of Reggio Calabria, 89124 Reggio Calabria, Italy
immacolata.lore@unirc.it

Abstract. The paper is part of an applied research study on the Locride Area of the Metropolitan City of Reggio Calabria for the implementation of management and sustainable development models based on endogenous resources. The context covers an area of 1.355 Km^2, for a total of 42 municipalities; for each town the study has identified the relevant assets (tangible and intangible) as key elements to structure the demand for cultural tourism and a potential new offering and management: the Cultural Heritage Agency of Locride. The data collection lays the groundwork for the guidelines processing of the cultural tourism's growth of the area, according to the Next Generation EU instrument with its Recovery and Resilience Facility (the Italian PNRR). This paper represents the first step of a research study that employs an effective context analysis as a scientific basis for a systemic local development proposal based on a new management model of cultural heritage in a highly fragile environment. The analysis and surveys point out some strengths and weaknesses of the cultural heritage management, in terms of quantitative consistency, reputation on tourist market and current attractiveness highlighting the differences between intermediate and peripheral areas.

Keywords: Management model · Cultural heritage · Cultural tourism · Inner areas · Territorial analysis

1 Introduction

The study is the cognitive phase of an applied research activity in the Locride area of the Metropolitan City of Reggio Calabria, aimed at the implementation of management and sustainable development models based on endogenous resources [1]. The applicative element of the next phases is the economic construction of a *Cultural Heritage Agency* of a wide area; the aim is to design management models on an enhancement strategy (*heritage-led innovation*) in line with national and EU policies (M1C3: TOURISM AND CULTURE 4.0 - PNRR - NGEU) able to network the cultural resources. The Cultural

The paper is the result of the joint work of the authors. Although scientific responsibility is equally attributable, the abstract and Sects. 1, 4 were written by F. Calabrò; Sects. 2.1, 2.2 were written by G. Cassalia , and Sects. 2.3, 3 were written by I. Lorè.

Heritage increases competitiveness and attractiveness of a country, as highlighted by the Next Generation EU instrument that underlines the decisive role of culture in achieving goals, as a cornerstone and instrument of resilience, sustainability and social development. In this context the cultural heritage, once integrated and enhanced with other main assets of the area, it could act as driver of local development strategies, recognising the economic potential of cultural heritage in addition to the values in use [2, 3]. The investment on heritage increases value production lines (direct/indirect), as well as being part of the logic of sustainability of green transition [4]; this increases the variety of assets that heritage managers will have to take care of, expanding the social, economic and environmental skills and opportunities, as well as the variety and number of threats to heritage sites [5].

The most important challenge for the post COVID Cultural Heritage is to react starting from the potential of the cultural system, digital opportunities and innovative models in which the transition from physical protection to a multilevel management consider social, economic and environmental issues and takes a main role to communities (*heritage communities*) [6]. The increased importance to the relationship between heritage sites and the "context" involved a conceptual turning point; the physical boundaries no longer coincide with the site ones but in a series of stratifications that produce complex phenomena and new management challenges in which protection (no longer just defensive) has planning and proposing comparisons ability with contemporary society [7, 8]. The enhancement of the widespread cultural heritage contributes to collective well-being not only as protection action, but also as a sustainable response to the demand for work and as an urban renewal and revitalisation strategy of social contexts such as inner areas. In a theoretical view that leads to rethinking management models in areas characterized by considerable marginality, the paper copes with the weakness/opportunities of lagging areas of Calabria and the vision of the territory as a system of social, cultural and economic features [9]. Nowadays the management of cultural heritage is characterized by several objectives, and this entails the evaluation of a wide range of contexts, knowledge and values (for present and future generations). As part of this scenario the research study focuses on the Locride area, a fragile context whose heritage potential is not yet expressed due to management problems that do not ensure the economic sustainability and accessibility; it is necessary to recognise its cultural influence in terms of attractiveness and ability to develop innovative models. The aim is to analyse the cultural system complexity of the area in relation to the management models through a review and guidelines framework aimed at adopting the necessary measures to ensure its effectiveness as empowerment of the subjects involved [10, 11].

This study is part of the scientific debate on the economic evaluation of management models for cultural heritage to be implemented through innovative tools in order to verify the economic feasibility and the sustainability of projects and new partnership forms between public and private subjects. In the face of national and EU policies and strategies on cultural heritage, the search by public subjects for forms of partnership PPP that meet the needs of local communities is increasingly frequent; in this perspective, the article 111 of the Italian *Codice dei Beni Culturali e del Paesaggio* (Code of Cultural Heritage and Landscape) considers the action of private subjects as socially useful recognising its purpose of social solidarity. In the context of PPP forms the Article 6 precisely provides the participation of private subjects in this type of activity, according to the procedures of the following articles [12]. It is necessary to reaffirm the commitment to build new partnership models by a greater integration of culture and tourism, reducing barriers and facilitating effective partnerships between government, private and community organizations in these sectors, and using tourism and culture as a fundamental tool in development cooperation for recovering of the areas affected by crisis [13].

2 State of the Art

2.1 The Territorial Context

The study takes into consideration a 1,366.60 Km^2 area and includes 42 municipalities of the Locride area of the Metropolitan City of Reggio Calabria; for each municipalities the main cultural assets with organised management (monuments, urban agglomerations, archaeological areas) have been identified (see Fig. 1). The area is characterized by the wealth of the natural heritage (13/42 municipalities fall within the Aspromonte National Park) and by over 90 km of coastline (Costa dei Gelsomini) linked to the Greek and Roman archaeological sites [14]. The mobility system is structured into a regional level of longitudinal (A3, SS 106) and transversal (SGC Jonio-Tirreno) lines that connects the Locride area with the Piana di Gioia Tauro, and into a district level of transversal roads (SS 281-501-111) connected to the inner system with physical limits in the strengthening of the networks due to the geomorphological and hydrographic characteristics. The area is served by the FS Railway Line Taranto - Reggio C. (with stations in: Monasterace, Siderno, Gioiosa J., Locri) and connected to the tourist-fishing port of Roccella J., to the national airport of Reggio Calabria and to the international airport of Lamezia Terme (average distance: 115 km).

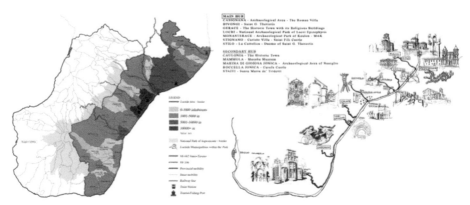

Fig. 1. The territorial context of the Locride Area of the Metropolitan City of Reggio Calabria, IT (top left). The *Atlas of Cultural Heritage of Locride* (top right). Image source: elaborations by authors.

Characteristic of the area is its location that makes it possible to transfer from the coast to mountain areas in a few kilometers; this affects the settlement fabric, on the one hand the inner historic centers, on the other hand the new agglomerations on the coast that constitute the most dynamic and economically relevant area affecting the migratory dynamics [15].

The case-study area has a specific historic and cultural identity, whose potential is not yet expressed due to management problems that do not allow the economic sustainability and accessibility to heritage and the offer of services for use. The main obstacle is the ownership and management that affer to different subjects and institutions, the economic resources, and the lack of knowledge and awareness of the communities that are now called to engage in the protection of heritage. Overall, the cultural sector of these territories has growth potential if supported by policies able to solve some of the main problems (fruition/services, planning, promotion, monitoring). As highlighted by Principle 2 of the revision of the ICOMOS International Charter for Cultural Tourism (1999), it is necessary to manage tourism at cultural heritage places through management plans informed by carrying capacity, planning instruments and monitoring of impacts on the natural and cultural values of the place as well as on the social, economic and cultural well-being of the host community [16, 17].

2.2 Methodological Considerations

The Locride area is characterized by territorial environments with a specific identity, whose cultural dimension is not easy to identify and measure; however, it is possible to base the evaluation on an accurate identification of key elements, able to describe by comparisons the state and evolution of the heritage wealth and its typological implication.

The objective of the study is to identify innovative management models (based on feasibility, sustainability and effectiveness criteria measured in the environmental, social and economic dimension), evaluating their application to different contexts through decision-making tools looking at the welfare economics and intangible aspects by the use of quantitative and qualitative parameters [18].

One of the main problems in the analysis of cultural heritage is the concept of "quality" and "comparability" towards a quantitative approach as the data collection (tangible and intangible) presents different analysis methods. Culture has quantifiable dimensions, as the "collective" one and of the related units of measurement, that allow to acquire and disseminate information as part of much broader processes [19].

The methodological approach involves the collection and investigation of direct surveys and statistical data aimed at defining the context, through the quantitative and qualitative analysis of the main variables of the cultural heritage system of Locride area with the support of graphic representations, summary tables and diagrams.

The research study is structured in the following phases:

- Phase 1 - Cognitive Frame (subject of this paper)
- Phase 2 - Definition of Strategy (Good Practices - Objectives - Interventions - Coherence with National and EU policies)
- Phase 3 - Construction of the Integrated Management Model
- Phase 4 - Evaluation of the Economic, Financial and Management Feasibility
- Phase 5 - Evaluation of Impacts (Indicators and their measurement).

In the first phase, after the analyses of the context and mapping of the main resources of the area an *Atlas of Cultural Heritage* was elaborated to identify the assets to be put into the strategic management system and evaluate the effectiveness of a management model for its development or updating [20, 21].

The applicative and experimental phase will identify pilot areas by defining intervention programs and actions coherent with the macro-objectives. In the field of evaluation by the economic and management construction of a *Cultural Heritage Agency*, different scenarios will be verified on economic and management feasibility highlighting the conveniences for the subjects involved in the valorisation process [22, 23].

It is possible to distinguish private subjects between profit and no-profit ones but in this context the most relevant distinction refers to the nature of activities; there are, in fact, many cases of no-profit subjects that carry out activities of economic relevance by varying only the tax regime (together with the no distributing profits) but not the cost structure as compared to profit ones. Particularly, the study will develop experimental tools and models aims at verifying if the economic conditions are satisfied, and which ones, if any, are appealing for the private involvement within the realisation and/or management of collective utility interventions [24].

The methodology is based on the definition of evaluative tools enabling the identification of effective and coherent concepts for understanding, planning and enhancing the territory in the cultural tourism sector. This preliminary stage of the project aims to verify the preconditions for the establishment of a local, attractive management system of cultural heritage sites of the Locride area, combining socioeconomic analysis, identification of thematic resources, and the study of financial territorial allocation. The paper's results contribute to the scientific debate on the role of the "knowledge phase" in the dynamics of sustainable development, highlighting critical elements and opportunities for implementing an integrated valorisation strategy of the case-study area.

2.3   Cultural Heritage Sector and Data Collection

After the analysis on the main cultural sites and the connected municipalities, in order to give an intermediate point of view between a general area and a more detailed one, it was carried out a classification in main and secondary hubs identifying the pole centers characterized by administrative, social and economic forms of integration. The territory under examination presents a multi-centrality model that can be delineated as follows: services to businesses (Siderno - Locri), services to citizen, businesses and accommodation facilities at a local level (Locri-Siderno-Gerace triangle, centers of Gioiosa J. and Marina di Gioiosa J., the coastal linear system of Ardore-Bovalino-Bianco).

Specifically, the first phase of the study involved the construction of the *Atlas of Cultural Heritage of Locride* (see Fig. 1) with the aim of surveying and analysing the cultural spaces of the area, detecting the fruition level (distances/times, routes, costs/tickets, opening hours, services) through the configuration of data forms and the visitor flow analysis [25]. In the economy of the construction of an integrated management model of the Locride, the main assets of historic and architectural relevance (archaeological areas, historic centers, architectures, museums, rock sites) have been surveyed and classified into main and secondary hubs on the basis of the cultural, historic and dimensional relevance; the analysis could be extended in a second phase to the intangible heritage and landscape that have similar but at the same time very different problems.

Specifically, the study takes as reference two of the most important UNESCO Conventions [26–28]. As highlighted in the declarations of the UNWTO/UNESCO World Conference on Tourism and Culture it is necessary to promote responsible and sustainable tourism management of cultural heritage by protecting and safeguarding its tangible and intangible values, ensuring the engagement of local communities and financing through investments and self-financing models, and promoting sustainable tourism management on the principles of resource efficiency and the quality of the visitors' experience [29].

3 The Tangible Heritage with Organised Management of the Locride Area – The Results of Survey

The analysis and surveys on tangible heritage with organised management allowed to identify some strengths and weaknesses of the cultural system of Locride area, in terms of consistency, notoriety on the tourist market and attractiveness; the cultural sites thus identified are conceived as new ordering centers of the territorial assets to be safeguarded and enhanced in all their components. The functions assigned to the sites must be extend to the broadest meaning of the integrated management, together with the study of the other potential sites of the widen cultural system able to promptly produce a series of coordinated and shared projects [30].

The summary of the main survey results shows that the tangible heritage with the most significant concentration is, as verification of the vocation of the area, the historic and architectural assets with 58% (defensive architecture 25%, historic architecture 18%, historic centers 15%), followed by the archaeological areas with 13% due to the important presence of 3 parks in Locri, Monasterace and Casignana (municipalities with a tourist vocation located along the coast); they are followed by museums with 11% and rock and religious sites with 9% (see Fig. 2).

The endowment of historic and architectural heritage consists of twenty-four architectures and monuments and eight historic centers with valuable characteristics. This indicator, which tends to be stable over time (except for the excavations expansions in archaeological context), takes into account – besides the well-known heritage located in the municipalities of Stilo (La Cattolica), Locri - Monasterace - Casignana (Archaeological Areas) and Gerace (The Historic Center) - a capillary presence (even in the inner areas) of culture sites in which the interpenetration of landscape and heritage is realised as a distinctive features of the territories identity and as an important asset in economic dynamics.

Almost all the assets/sites are public property and management (70%–59%) followed by the private (30%–35%) and mixed ones (6%). The correlation between the equipment of services for use and the management model was immediately identified; specifically, the public assets/sites with a private management system (mixed) have a greater services offer (opening hours, guide service, online ticket, updated communication).

With regard to services, only 2% of the sites/assets have a transport service while the 16% have a parking area for arrivals with private vehicles; 23% have an info-ticket office, 12% allows the online purchase and confirmation and 21% offers a guide service

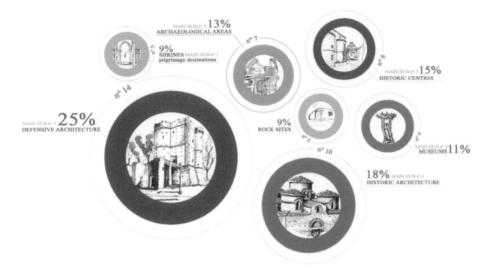

Fig. 2. Consistency and characteristics of the tangible heritage of the Locride Area of the Metropolitan City of Reggio Calabria (IT). Data processing by directly documented sources.

in the main European languages (audio guide 8%). Among the complementary services, 8% of the cultural sites have a refreshment area and 10% a bookshop (see Fig. 3).

On the effectiveness of communication, 30% of the sites/assets have an official website and 21% use social media for promotion (Facebook, Instagram) with a medium-low level of updating; 37% has paper information material (80% not updated) and only 12% use digital tools.

In reference to the state of conservation[1] most of the tangible heritage of the area (64%) has a medium level of conservation followed by the high and low (21%–15%).

A further study was conducted on the visitors flows to the sites/assets (2015–2019* pre Covid-19) that record the most significant turnout in the 2016 (see Fig. 4) in the summer season, with an annual positive peak in the first two weeks of August due to the tourist presences (mostly return tourism that repopulates the Locride area) in the centers along the coast (mainly linked to the offer of accommodation facilities and complementary tourist services); it was also registered that most of sites/assets (n ° 22) have free admission and the remainder (n ° 10) for a fee with an average ticket price of around € 4.00.

The identification and analysis of the consistency and characteristics of the cultural heritage of the Locride area highlighted the richness of territorial resources, but it also defined a cultural geography of the area confirming management problems on the organization of a cultural system of the area and on the offer of tourist services, with the exception of individual emergencies.

---

[1] High Level = The site/asset does not require maintenance interventions. Medium Level = The site/asset requires minor maintenance interventions. Low Level = The site/asset requires extraordinary maintenance interventions.

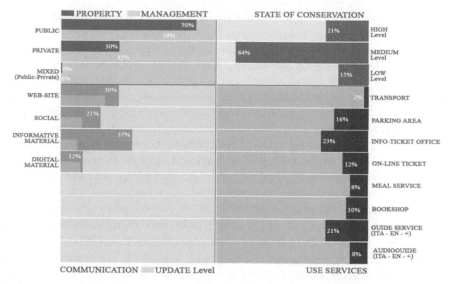

Fig. 3. Analysis of the services for use of the cultural sites/assets of the Locride Area of the Metropolitan City of Reggio Calabria (IT). Data processing by directly documented sources.

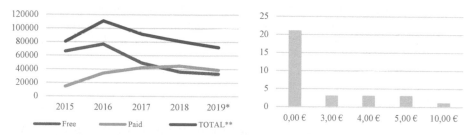

Fig. 4. Visitors flows (top left) and ticket prices trend (top right) relating to the cultural sites/assets of the Locride Area of the Metropolitan City of Reggio Calabria (IT). Data sources: MiC - Statistical Office. Homepage, http://www.statistica.beniculturali.it, last accessed 2021/11/21.

4 Concluding Remarks

Cultural Tourism is increasing, bringing with it challenges to the sites from wear-and-tear and the impacts that large numbers of visitors can have on their 'spirit of place'. In turn, this is forcing a search for methods of balancing conservation with the livelihood needs of local people and the right of tourists to enjoy such places. Cultural sites can benefit from tourism in several ways, in that additional funds for conservation can be generated from tourism, and the profile of the site can be raised, both of which help to generate greater government support. Furthermore, where local people experience economic benefits from these sites, they are also more likely to become aware of the importance of conservation in line with the Faro Convention [31, 32].

The results of survey show a cognitive frame of the area to which further investigations on territorial analysis and economic effects (not deriving from previous actions) must be added in order to study the impacts and the real needs to orient the following choices [33]. However, already in this phase, it is clear how is indispensable to work in a system strategy that rationalises and makes the Locride area competitive. This strategy will constitute the convergence point of local policies on the enhancement of territorial resources, based on a shared governance system, able to use the attractive and competitive potential of the cultural offer of the area. The study represents the starting point for the construction of an integrated model of management and enhancement for the cultural heritage of Locride; it allows a differentiated articulation of the interventions at a territorial level according to the main assets, needs and strengthness and weakness points, taking into account spatial, social, economic and cultural characteristics that determined the current structure. The proposal of a *Cultural Heritage Agency of Locride* has the potential as a driving for growth and development by the construction a shared project of a wide area with the aim of creating a network of cultural resources through heritage management and promotion of stakeholders collaboration. The research represents an opportunity to study a context in which they are present the effects of the ongoing changes and the need for a strategic direction. The context of the COVID-19 pandemic is revolutionising every sector and work activity, including the tourism and culture one. As ecological awareness increases and the need to be authors of culture is affirmed, the earth is reconquered not only in productive terms, but in mental and creative ones. By restoring the right of citizenship to their places, creating new forms of sharing, southern society discovers that it is possible to produce not only for the market, but for a local circular economy [34].

References

1. Mollica, E.: Valorizzazione delle risorse architettoniche, storiche e ambientali in area vasta della Calabria. De Franco, Reggio Calabria (2006)
2. MEF: Piano Nazionale di Ripresa e Resilienza (PNRR) - Next Generation EU (NGEU), Recovery and Resilience Facility (RRF) (2021)
3. EU: Next Generation EU – NGEU (2020)
4. European Commission: European Green Deal – EGD (2019)
5. Piccioli, C.: La valorizzazione del patrimonio culturale una opportunità per il superamento della crisi economica. In: I Conference Diagnosis for the Conservation and Valorization of Cultural Heritage, 9–10 December (2010)
6. COE: Faro Convention. Portugal (2005)
7. Volpe, G.: Per una innovazione radicale nelle politiche della tutela e della valorizzazione. In: De Tutela. Idee a confronto per la salvaguardia del patrimonio culturale e paesaggistico, pp. 109–115. Edizioni ETS, Pisa (2014)
8. Pallottino, E.: Cultura del patrimonio e progetti di valorizzazione contestuale. In: Ricerche di Storia dell'Arte, vol. 108, pp. 27–31 (2012)
9. Cassalia, G., Tramontana, C., Calabrò, F.: Evaluation approach to the integrated valorization of territorial resources: the case study of the Tyrrhenian area of the metropolitan city of Reggio Calabria. In: Calabrò, F., Della Spina, L., Bevilacqua, C. (eds.) ISHT 2018. SIST, vol. 101, pp. 3–12. Springer, Cham (2019). https://doi.org/10.1007/978-3-319-92102-0_1

10. Aas, C., Ladkin, A., Fletcher, J.: Stakeholder collaboration and heritage management. Ann. Tour. Res. **32**(1), 28–48 (2005)
11. Dolores, L., Macchiaroli, M., De Mare, G.: Sponsorship's financial sustainability for cultural conservation and enhancement strategies: an innovative model for sponsees and sponsors. Sustainability **13**(6), 9070 (2021). https://doi.org/10.3390/su13169070
12. D.Lgs. 22.11.2004, n. 42 and s.a.: Codice dei beni culturali e del paesaggio
13. UNWTO/UNESCO: Siem Reap declaration on tourism and culture – building a new partnership model. In: UNWTO/UNESCO World Conference on Tourism and Culture – Building a New Partnership, Cambodia (2015)
14. Cuthbert Hare, A.J. (edited by Maria Rosaria Costantino): Città della Calabria. Edizioni Monteleone, Vibo Valentia (2005)
15. Lacquaniti, L.: Morfologia ed evoluzione dei centri abitati della Calabria: considerazioni ed esempi. In: Boll. Soc. Geogr. It., pp. 31–37 (1946)
16. ICOMOS: International Cultural Tourism Charter: Reinforcing cultural heritage protection and community resilience through responsible and sustainable tourism management. ICOMOS ADCOMSC and ADCOM, 27th October and 3rd November 2021. To be adopted at the ICOMOS General Assembly 2022 (2021)
17. Tajani, F., Liddo, F.D., Guarini, M.R., Ranieri, R., Anelli, D.: An assessment methodology for the evaluation of the impacts of the COVID-19 pandemic on the Italian housing market demand. Buildings **11**(12), 592 (2021)
18. Cassalia, G., Lorè, I., Tramontana, C., Zavaglia, C.: Statistiche culturali - Il censimento del patrimonio culturale nell'area tirrenica della città metropolitana di Reggio Calabria. In: LaborEst, Reggio Calabria, vol. 13 (2016). http://dx.medra.org/10.19254/LaborEst.13.02
19. Pigou, A.C.: The Economics of Welfare. Macmillan and Co., London (1932)
20. Coleman, J.S.: Individual Interests and Collective Action: Studies in Rationality and Social Change. Cambridge University Press (1986)
21. Calabrò, F., Iannone, L., Pellicanò, R.: The historical and environmental heritage for the attractiveness of cities. The case of the Umbertine Forts of Pentimele in Reggio Calabria, Italy. In: Bevilacqua, C., Calabrò, F., Della Spina, L. (eds.) NMP 2020. SIST, vol. 178, pp. 1990–2004. Springer, Cham (2021). https://doi.org/10.1007/978-3-030-48279-4_188
22. Nesticò, A., Moffa, R.: Economic analysis and Operational Research tools for estimating productivity levels in off-site construction [Analisi economiche e strumenti di Ricerca Operativa per la stima dei livelli di produttività nell'edilizia off-site]. In: Valori e Valutazioni, vol. 20, pp. 107–128. DEI Tipografia del Genio Civile, Roma (2018). ISSN 2036-2404
23. Calabrò, F., Cassalia, G., Lorè, I.: The economic feasibility for valorization of cultural heritage. The restoration project of the reformed fathers' convent in Francavilla Angitola: the Zibìb Territorial Wine Cellar. In: Bevilacqua, C., Calabrò, F., Della Spina, L. (eds.) NMP 2020. SIST, vol. 178, pp. 1105–1115. Springer, Cham (2021). https://doi.org/10.1007/978-3-030-48279-4_103
24. Barile, S., Saviano, M.: Dalla Gestione del Patrimonio di Beni Culturali al Governo del Sistema dei Beni Culturali. In: Golinelli, G.M. (ed.) Patrimonio culturale e creazione di valore, Verso nuovi percorsi, pp. 97–148. Cedam, Padova (2012)
25. Calabrò, F., Della Spina, L.: La fattibilità economica dei progetti nella pianificazione strategica, nella progettazione integrata, nel cultural planning, nei piani di gestione. In: LaborEst, Inserto speciale n. 16. Università Mediterranea, Reggio Calabria (2018). http://dx.medra.org/10.19254/LaborEst.16.IS
26. Bellisario, M.G.: Dossier studi: strumenti per il sud e la presentazione dello "Studio ed il rilevamento dei dati sull'offerta relativa a musei, aree archeologiche e monumenti non statali delle regioni obiettivo 1". MIBAC-MISE- DPS. Edizioni Artemide (2006)
27. UNESCO: Convention Concerning the Protection of the World Cultural and Natural Heritage, Paris (1972)

28. UNESCO: Convention for the Safeguarding of the Intangible Cultural Heritage, Paris (2003)
29. ICCROM, ICOMOS, IUCN, UNESCO WH Centre: Gestire il Patrimonio Mondiale Culturale. Manuale delle Risorse. In: Managing Cultural World Heritage, ICCROM (2019)
30. UNWTO/UNESCO: Muscat declaration on tourism and culture: fostering sustainable development. In: 2nd UNWTO/UNESCO World Conference on Tourism and Culture, Muscat (2017)
31. Porfyriou, H., Yu, B. (eds.): China and Italy: Routes of Culture, Valorization and Management. Cnr Edizioni, Roma (2018)
32. Cochrane, J., Tapper, R.: Tourism's contribution to world heritage site management. In: Managing World Heritage Sites, pp. 97–109. Routledge, London, New York (2006)
33. Massimo, D.E.: Green building: characteristics, energy implications and environmental impacts. Case study in Reggio Calabria, Italy. In: Coleman-Sanders, M. (ed.) Green Building and Phase Change Materials: Characteristics, Energy Implications and Environmental Impacts, vol. 1, pp. 71–101. Nova Science Publishers, New York (2015)
34. Consiglio, S., Riitano, A.: Sud Innovation - Patrimonio culturale, innovazione sociale e nuova cittadinanza. In: Pubblico, professioni e luoghi della cultura. Edizioni Franco Angeli (2015)

New Technologies for Accessibility and Enhancement of Cultural Heritage Sites. The Archaeological Areas of Locride

Francesco Calabrò[1], Giuseppina Cassalia[1], Paolo Fragomeni[2],
and Immacolata Lorè[1(✉)]

[1] Department of PAU, Mediterranea University of Reggio Calabria, 89124 Reggio Calabria, Italy
immacolata.lore@unirc.it
[2] University of Ferrara, 44121 Ferrara, Italy

Abstract. The paper is the second phase of an applied research study in the framework of evaluation of impacts in the enhancement and promotion of cultural sites by the use of new technologies. The study investigates the existing characteristics of use and the level of accessibility in the archaeological sites of the Locride Area of the Metropolitan City of Reggio Calabria. The main aim is to define the advantages of adoption of digital tools evaluating their impacts in a management system and in the offer of services to the use; these tools can play important roles in experiencing (physical accessibility) and comprehending (perceptual accessibility) culture and heritage defining a satisfactory level of accessibility in order to encourage new promotion and management practices. New technologies and innovation are important vehicles for disseminating cultural values; they attract new visitors and enhance their knowledge about heritage. Recognising their enabling role as a driver for cultural sustainability, this study aims to investigate the use of digital tools as a very common and widespread medium of accessing heritage and supporting cultural tourism.

Keywords: Cultural heritage and tourism · Archaeological sites of Calabria · Accessibility · Innovation · New technologies · Visitor experiences · Monitoring

1 Introduction - Cultural Heritage, Digital World and Legislation

The study concerns the assessment of impacts in terms of accessibility, promotion, increase of visitors and knowledge of the archaeological sites of Calabria deriving from the use of digital tools; the first phase of the study focused on the technical tools of virtual reality by an immersive VR application with the reconstruction on scientific basis of the Temple of Punta Stilo in the Archaeological Park of the Ancient Kaulon (Monasterace Marina, RC) in the Locride Area of the Metropolitan City of Reggio Calabria. The VR

The paper is the result of the joint work of the authors. Although scientific responsibility is equally attributable, the abstract and Sects. 1, 5 were written by F. Calabrò; Sect. 2.1 was written by P. Fragomeni; Sects. 2.2, 2.3, 3 were written by I. Lorè, and Sect. 4 was written by G. Cassalia.

F. Calabrò et al. (Eds.): NMP 2022, LNNS 482, pp. 289–300, 2022.
https://doi.org/10.1007/978-3-031-06825-6_28

experimentation made it possible to return in the three dimensions (fully perceiving spatiality and scale) the archaeological asset (not characterized by elevated structures), including the setting and the design of a CAVE for the connected museum (MAK) in the field of "edutaintment" and "learning by consuming" (see Fig. 1) [1].

Fig. 1. The CAVE (Cave Automatic Virtual Environment) and VR Viewer adopted to the restitution of the Doric Temple of Punta Stilo, Monasterace Marina (RC). Image source: elaborations by P. Fragomeni.

After the Covid-19 pandemic and the development of storytelling approaches, the use of new technologies and innovation (immersive applications, virtual reality, serious games) in the Cultural Heritage and Archeology sector have acquired an ever increasing importance in the use, readability and understanding keeping alive the dialogue with heritage that it is extremely important in these historical period [2]. In the recent time public administration has in fact recognised ICT as important vehicles for accessibility and dissemination of cultural values and as a driver for cultural sustainability [3]; they are effective tools for increasing the inhabitants' awareness and so contribute to preserve cultural sites and for attracting new visitors and enhancing their knowledge about heritage [4]. It is necessary to enhance visitor experience through sensitive interpretation and presentation of the interconnections of tangible and intangible values of cultural heritage by the use of new technologies and innovation [5]. The article 6 of the italian Codice dei Beni Culturali e del Paesaggio (Code of Cultural Heritage and Landscape) defines the enhancement as the discipline of activities aimed at promoting knowledge of cultural heritage and ensuring the best conditions of use, including digital innovation, which is the subject of a specific plan (Piano Triennale per la Digitalizzazione e l'Innovazione dei Musei). Among the main objectives are the offer of accessibility tools and the improvement of services to the public by digital solutions (VR-AR-gaming experiences), customer satisfaction and monitoring actions on quality of services [6, 7]. As highlighted by Principle 3 of the revision of the ICOMOS International Charter for Cultural Tourism (1999), the authenticity, values and significance of places are often complex, contested and multifaceted, and every effort should be inclusive of appropriate, stimulating and contemporary forms of education and training, using opportunities of technology, including augmented reality and virtual reconstructions based on scientific research. Heritage practitioners and professionals, site managers and communities share the responsibility of interpretation and presentation cultural heritage that should be accessible to all, including people with disabilities [8]. Digital opportunities for heritage sites and managing subjects must not be a simple addition to the involving activity of public but must be included in a innovation-based strategy on the assumption that it is not the public that must change, but the system of use, accessibility and communication that must evolve, adapting to new audiences and requests and thus expanding the user base and the educational reach of culture sites and generating new income flow [9].

Recent research carried out by the Observatory for Digital Innovation on the level of digitization of italian cultural institutions confirms the trend towards the reformulation of their management and communication models [10, 11]. If at the beginning of the pandemic it was necessary to give an urgent response to the problems of the inevitable closure of the cultural sites, after a few months the management subjects realised the importance of adapting their communication models to the rapid transformations of society [12, 13]. These analyses highlighted the emergence of new business models based on the centrality of digital tools - one of the main points of the italian PNRR (National Recovery and Resilience Plan - Mission 1-Cultural heritage for the next generation) - that guarantee the public access in total safety but also at a distance [14]. The challenge is to continue to support the creation of innovative products made exclusively for digital use without forgetting the essential experience of direct contact with cultural heritage [15]. From the discussion on the Digital and Cultural Heritage union emerges the convergence towards a need for experimentation and mature applications forms that converge in a fascinating suggestion of environments made up of scenarios with a high cultural content (archaeological and monumental reconstructions) in which to develop social interactions that also have educational components [16]. Accessibility should be considered an important feature of these cultural sites' identity, meaning, and significance and derived from the intrinsic values of the site together with the internal sensibility of the people for the place – both bearers of its significance. Despite these cogitations we witness a puzzling cultural decline today in the cultural sites of Calabria that are not equally accessible. Archaeological remains are often considered uninteresting, dull, or even worse dangerous; this attitude results in unfair experiments not just with the past but an experiment with the future.

This research study aims to investigate the advantages of adoption of digital tools evaluating their impacts in a management system and in the offer of services to the use for archaeological areas of Calabria that showed the necessity of substantial investment in terms of both effort and resources, to bring about cultural tourism-based territorial promotion; one of the main challenges is to combine the need for scientific rigor to public needs by multisciplinarity and qualitative content analysis methodology. The research is structured in the following phases:

- Phase 1 - Digital World for Cultural Heritage - Technical application of VR [1];
- Phase 2 - Research Context - State of the Art (subject of this paper);
- Phase 3 - Data Survey and Colletion - Case Study (The Archaeological Sites of Calabria - The Locride Area);
- Phase 4 - Evaluation of the Economic, Financial and Management Feasibility [17];
- Phase 5 - Evaluation of Impacts (Indicators and their measurement).

2   The Research Context

2.1   The Archaeological Sites of Calabria - The Locride Area

The physical conformation of Calabria presents very diversified and often contrasting territorial areas which influenced the anthropization and commercial asset, even the productive and agricultural ones [18], in addiction to endogenous and exogenous factors that have determined the loss of a large part of the historic architectural heritage as the series of earthquakes that have irreversibly devastated the urban centers. Another important aspect is the position of Calabria in the center of the Mediterranean Sea wich influenced the settlement of the first Greek colonies in the 8th century BC along its coasts [19]. Although the location was favorable to commercial traffic, it has determined over the years an extreme condition of vulnerability, materialized with the continuous destruction of the ancient cities as the settlements of the Locride area of the Metropolitan City of Reggio Calabria; from Casignana to Locri, up to Kaulonìa, they not preserving elevated structures destroyed by earthquakes, demolition by enemies or dismantled to be used as reuse materials in new constructions; an emblematic case is the Norman Cathedral of Gerace built in the 11th century AD., where the columns and capitals were recovered from the ancient site of Locri and reused in the colonnade that divides its internal naves, or the four Byzantine columns of the Cattolica of Stilo, probably recovered from ancient Roman buildings [20]. The case study concerns the archaeological area of the *Kaulonìa polis*, that according to recent studies has been born under the brightest city of *Kroton*, but it never had an influential political and strategic role. In 388 BC. It was conquered and annexed to the territory of Locri by *Dionysus I*, tyrant of Syracuse, and in 356 BC by *Brettii*. The latest news on *Kaulonìa* concerns the second Punic War around 280 BC and its destructions that determined the end of the polis as an urban center leaving a few evidence of elevated structures [21]. On the other hand, there are important mosaic works, as the "Hall of dragons and dolphins", dated around the 3rd century BC., the largest Hellenistic mosaic in the South Italy, or the 3rd century mosaic room of the "House of dragon" and the base of the Doric temple of Punta Stilo (see Fig. 2).

It is clear that in such a context of events that have undermined the architectural integrity and created a space-time short circuit with the lack of tangible evidence, a capacity for real reading of the past becomes more necessary, no longer as an historicized present, but as an intelligible reality from which draw an innovative interpretation of the past; this entails not only the recovery of the historical memory of the events that characterizing these places, but also the magnificence of the architectural structures, civil or religious, in their settlement complexity.

Fig. 2. The Doric Temple of Punta Stilo (Monasterace Marina - RC) with no evidence of elevated structures due to destructions of the 2nd Punic War (280 BC.) that determined the end of the polis as an urban center. Image source: elaborations by authors.

2.2 State of the Art

The total and sudden closure of almost all cultural sites in the world due to the Covid-19 pandemic, an unprecedented dramatic event, has prompted immediate and unanimous responses from the scientific community with monitoring studies on the expected impacts (cultural, economic, social, logistic) from pandemic and conditioned reopening and on the relevance through the digital collection of opinions from institutions and individuals. The surveys collected by UNESCO, NEMO (Network of European Museum Organizations), and by the italian Direzione Generale Musei – MiC report the expected impacts on specific topics such as use and accessibility, safety of people and things, human resources, communication (digital use and social media), relations with territory, users (involvement, education, enhancement) and stakeholders.

The main impact area resulting from the investigations, is the communication system with digital access as the case of italian cultural sites. As reported by the latest Istat surveys, at the end of 2018 only half of the italian cultural sites used new technologies and digital tools (44.7%), had an official website (51.1%) and accounts on the most important social networks (53.4%), and only one site in ten (9.9%) offered a virtual experience [22, 23].

The recent surveys of the Observatory for digital innovation reported that today 95% of italian cultural sites have a website (>10% compared to 2020), 83% own at least one official social channel (>7% compared to 2020) and about 80% offer digital content (VR, App, VideoGames) in the visitor experience (from 40% in 2020) [10].

2.3 Case Study - The Archaeological Park of the Ancient Kaulon

In the case study of the Archaeological Park of the Ancient Kaulon, the combined demand of provide promotion and comprehension to the cultural site ensuring the traditional use has required a complex approach to the accessibility that followed the people's abilities and was inspired by the site's significant features in order to provide for a richer experience for visitors [24].

In the archaeological context of Calabria, the adopting of immersive and multimedia solutions (VR-AR) that are currently available on the market, may work toward filling the gap of these sites which are unable to offer adequate experiences for visitors, providing accessible solutions to cultural institutions.

The archaeological sites of Locride Area nothing has changed since before the Covid-19 pandemic; in 2021 only 30% of the cultural sites of the area have an official website and 21% use social media for promotion (Facebook, Instagram) with a medium-low level of updating; 37% has paper information material (80% not updated) and only 12% use digital tools (see Fig. 3).

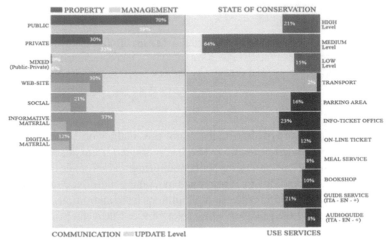

Fig. 3. Analysis of the services for use of the cultural sites in the Locride Area of the Metropolitan City of Reggio Calabria (IT). 2021 Data processing by directly documented sources. Image source: elaborations by authors.

Looking at the visitor flows of the case study, it was observed that the maximum peaks (2012, 2013, 2014, 2018) (see Fig. 4) correspond to the extraordinary openings to the public of the thermal complexes and mosaics (not regularly open for conservative reasons).

Similar sets of persistent problems have emerged at the all archaeological sites of Locride area open to the public. The digital tools are intended to lessen the likelihood that such problems develop and become irremediable at archaeological sites opened to the public. To prevent the park and the museum falling into oblivion during restoration periods, digital tools can intervene by supporting the relaunch of heritage; by exploiting the opportunities of digital in a correct and profitable way and by appropriate communication strategies, the managing subjects can intrigue and attract using an understandable and scientifically correct language, able to excite the public. Among the good practices "A Night in the Forum", an environmental narrative game on the Forum of Augustus in Rome co-produced in 2019 for the Sony Playstation VR by CNR in collaboration with Museo dei Fori Imperiali and the VRTRON private company (H2020 Reveal2 - European Commission project) [25]. The aim of the project was to use the Sony gaming platform of

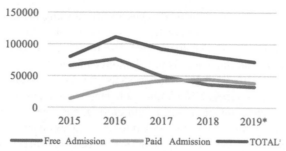

Fig. 4. Visitors flows relating to the archaelogical site of the Ancient Kaulon in the Locride Area of the Metropolitan City of Reggio Calabria (IT). Data sources: MiC - Statistical Office. Homepage, http://www.statistica.benicultu-rali.it, last accessed 2021/11/21. Elaborations by authors.

4.2 million users, as a tool for knowledge and promotion of the archaeological heritage in the field of tourism, expanding the presence of cultural sites (as Assassin's Creed in Monteriggioni - Tuscany and Father and Son - MANN) and creating a new economic model with direct (revenue) and/or indirect (presence and satellite activities) benefits for institutions [26].

3   Semantic Technologies as an Environment for Interdisciplinary Research Applied to Valorisation of Cultural Heritage

Concerns about economic and human resources (skills) and issues of use prevail in the expected changes; the new centrality of the local dimension is clearly recognised, both as a wealth of heritage and as a source of local users, the transformation of public demand (including those with special needs or conditions of fragility and marginality) and the need to rethink the cultural offer are felt, taking into account the new restrictions and potential of digital tools, to strengthen the cooperation and sharing of structures and activities.

In VR applied to Cultural Heritage, the passage from the decomposition of reality to digital memory is observed as an alternative and contingent transmission tool of knowledge able to increase the classical idea of a cultural asset as a "container" and extend it to a "catalyst" of internal resources (intangible aspect). In the field of virtual archeology an application must make methods, sources and documentation accessible and transparent as highlighted by the London Charter (Principles 3–4) and the Seville Charter (Principles 5–7) that affirm their importance to ensure the scientific integrity and results evaluating with respect to the contexts and purposes [27, 28].

Today the cultural heritage is internationally recognised as a powerful driver for economic development within the framework of cultural tourism as one of the leading tendencies that emphasises the unique cultures of a region as a tool for distinction in present-day context of globalisation and strong market competition [29]. In this context to reveal the cultural potential with its qualities and significance it is first necessary to provide accessibility to heritage as accessibility plays a key role when choosing both the adequate opportunity for sustainable use (activities familiar) and the approach to contemporary expression (interpretation and presentation) [30]. While the initial efforts were focused on improving the physical accessibility (leading the cultural sites to the loss of knowledge and alienation) today there is a growing international emphasis on the aspects of cognition and usage. The challenges arise from the development of our understanding of the cultural heritage and from the change of our expectations related to the contemporary standards for its accessibility because it is the individual who determines the dependence "value – accessibility". Only a balanced interaction of all accessibility types can guarantee the complete contact with cultural heritage: physical accessibility (associated with human senses and not limited to physical approach), intellectual accessibility (mental abilities on cognitive information) and economic accessibility. Economic accessibility in a broader sense refers to vitality of sites and the appropriate integration of constant or cyclic function that could meet the needs of society; the lack of use unavoidably leads to obsolescence and therefrom to alienation, deterioration, and even loss of heritage sites. Human needs and abilities (including people with disabilities) are considered to be the basis for seeking alternative and challenging solutions that could provide the balance between "sustainable use with equal access for all" and "preservation and contemporary expression of the heritage values" [31]. It is not about mere tourism industry but about rational and balanced economic activity (not excluding supplements and change) that enables the vitality of the heritage site moving beyond the passive approach towards resilience in order to guarantee the wholesome balance between the cultural and economic considerations [32].

4   Impacts, Relevance and Priorities: Towards Responses to an Overall Vision

How to raise citisens' awareness of the importance of cultural heritage in community life and the need for its protection and enhancement? The case study to the the Archaeological Park of the Ancient Kaulon deepens within the scientific debate the economic and development potential of heritage for territories, according to an interdisciplinary and technological innovation perspective, by the evaluation of impacts (accessibility, increase in the number of visitors and knowledge). A monitoring plan of a cultural site should specify the technologies, protocols, instruments, indicators, and standards that should monitor: the condition of cultural resources; visitor numbers, circulation, and satisfaction; community satisfaction; and the condition of facilities and infrastructure. As highlighted in the *Salalah Recommendation on Archaeological Parks and Sites* of ICOMOS, management involves establishing capacities for monitoring together with management planning and implementation that should be linked not only for the site, but also for the immediate surroundings and region where development is planned that is related to visitation [33, 34].

The paper focuses on the study and analysis of the impacts of multimedia applications and products on users of the archaeological sites of Calabria not sufficiently adequate to meet the interests and needs of visitors. The impacts include the increase in the level of satisfaction and knowledge of users, the growth in the number of visitors together with the increase in the ones returning to the site and the distribution of multimedia tools within a cultural network useful to improve accessibility and use. To meet the use needs of a cultural site it is necessary to conduct a careful analysis of user requirements (accessibility, quality of experiences, usability of tools) and profiling that identifies the audiences that the examined cultural site already reaches, but also the users that could reach it from the use of digital tools and new communication strategies. Each technology underlies a universe of specificities, opportunities and limitations; the Virtual Reality tool of the case study has different features useful for achieving high levels of accessibility with an active involvement of the user thanks to a sensorial and educational experience. In this context it is important the evaluation of user experience, a design dimension that focuses on the characteristics and needs of users, deriving from psychological and emotional manifestations of the interaction with technology and from technical characteristics and context [35]. The user experience is subjective and this aspect (expectation of use) clearly separates it from the quality of the digital product that can be evaluated by different and hedonic criteria of technology (ergonomics, design, usability); its evaluation still finds obstacles in the structuring of a methodology often referred to complex projects that involve a multidisciplinary approach and specific subjects on context.

To date many of the evaluation proposals are based on usability tests (technical evaluation scales) and customer satisfaction tests focused on quantitative data (mass survey on synthetic answers) that give a general trends of satisfaction [36]. They are highly useful for analysing the product-user relationship but not suitable for discovering the benefits deriving from the use of the (real or virtual) cultural asset by the user. For this reason, an evaluation methodology is needed that takes into consideration all these factors and involves quantitative and qualitative researches.

Among the tested methodologies there is the "tripartite analysis" developed by the Virtual Reality Laboratory of the CNR ISPC in Rome, that uses three evaluation tools (1. Observation - 2. Questionnaire - 3. Guided scenario) for quantitative (2) and qualitative (1. - 3.) analyses [37]. Direct Observation provides the first considerations on users and technologies, the Questionnaire is built to analise the user's knowledge (professional profile and education), needs and expectations [38] and to evaluate the project (usability and customer satisfaction) by statistical rules (percentage calculation and cross comparison) that return a general image of the trends on these topics; the Guided Scenario is the direct communication between operator and user (impressions and comments) during the use of the digital application [39]. The evaluation is considered in three different moments: before the installation of technological solutions (to study the exhibition environment and how visitors approach it), during the visit (to analyze the communicative effectiveness of the digital tools), after the experience (user feedback). As part of these research activities, between 2011 and 2018, 15 virtual installations were studied in laboratory and in site (museums, archaeological areas) analysing different topics (context, accessibility, usability of digital tools, public appreciation, quality of experience, expectations of curators and visitors). The users' feedback has shown that, to date, the practice of digital storytelling is not yet fully exploited within cultural sites: the assets are presented through factual information lists, captions and panels with no attention to the history and identity of the context, preventing an active involvement of visitors.

5   Conclusions

Virtual reconstruction is a great tool for promoting and enhancing cultural heritage as it improves accessibility to the often complex historical and archaeological data to anyone by a more engaging user involvement [40]. However, the attention towards new technologies in cultural sites is not lasting but is a function of the support ability of cultural contents and of the ease and naturalness of interaction. For this reason it is necessary to consider new technologies at a deeper level of use, working on a cognitive and emotional level that is connected to visitors' motivation of use of multimedia applications [41]. Immediate action and a medium to long-term strategy are needed for the functioning of the archaeological sites of Locride area need to consider the intrinsic and symbolic value of cultural sites for restoration, economic recovery and improve responsiveness. Identifying effects for the sector and proposing visions and prospects for intervention for the future management of these cultural institutions, offering the possibility of following the evolution of points of view and the emergence of impacts and perspectives especially after Covid-19 pandemic. Each site has its individuality and therefore requires an independent approach and in the recent years there is an increase of the need of establishment of guidelines for planning the accessibility to cultural heritage. On one hand, this is related to the fact that our democratic society is increasingly recognising the diversity of human needs and respectively the opportunities of using the cultural heritage. On the other hand, the emerging of new technical solutions in the information technologies era provides probably countless alternatives for contemporary expression and interpretation. The accessibility to cultural heritage today means the possibility for every human to visit, appraise, experience, and use the heritage site in a personal unique manner. Archaeological sites challenge even more a complex accessibility because of their ruined and fragmented character but a new (and much more real) perception of an archaeological sites could be provided for the visitor by using new technologies an communication strategies with unexpectedly good results.

References

1. Fragomeni, P., Lorè, I.: VR as (In)Tangible representation of cultural heritage. scientific visualization and virtual reality of the Doric temple of punta stilo: interference ancient-modern. In: Bevilacqua, C., Calabrò, F., Della Spina, L. (eds.) NMP 2020. SIST, vol. 178, pp. 1851–1861. Springer, Cham (2021). https://doi.org/10.1007/978-3-030-48279-4_175
2. ICOMOS: ICOMOS Venice Charter (1964)
3. Calabrò, F., Della Spina, L.: Prove tecniche di futuro: Atlantide è scomparsa di nuovo. In: LaborEst n°20, Mediterranea University of Reggio Calabria (2020)
4. Lopez, L., de los Ángeles Piñero Antelo, M., Gusman, I.: Risorse online e gestione patrimoniale. Il caso del Cammino portoghese. In: LaborEst n°18, Mediterranea University of Reggio Calabria (2019)
5. UNWTO/UNESCO: Muscat Declaration on Tourism and Culture: Fostering Sustainable Development. Second UNWTO/UNESCO World Conference on Tourism and Culture, Muscat (2017)
6. Lgs., D.: 42/2004 - Codice dei Beni Culturali e del Paesaggio (2004)
7. Mic: Piano Triennale per la Digitalizzazione e l'Innovazione dei Musei (2019)

8. ICOMOS: International Cultural Tourism Charter: Reinforcing cultural heritage protection and community resilience through responsible and sustainable tourism management. In: ICOMOS ADCOMSC and ADCOM, 27th October and 3rd November 2021. To be adopted at the ICOMOS General Assembly 2022 (2021)
9. Angelino Giorzet, G., Vellar, A.: Audience development and social media: Turin's museums between fans and anti-fans. In: "Sociologia della Comunicazione" 56/2018, pp. 141–164 (2018)
10. Nesticò, A., Maselli, G.: Declining discount rate estimate in the long-term economic evaluation of environmental projects. J. Environ. Account. Manage. **8**(1), 93–110 (2020). https://doi.org/10.5890/JEAM.2020.03.007
11. Manganelli, B., Tajani, F.: Optimised management for the development of extraordinary public properties. J. Property Investment Finan. **32**(2), 187–201 (2014)
12. Osservatorio per l'innovazione digitale: Extended Exsperience: la sfida per l'ecosistema culturale. Politecnico di Milano (2020)
13. Benintendi, R., De Mare, G.: Upgrade the ALARP model as a holistic approach to project risk and decision management. Hydrocarb. Process. **9**, 75–82 (2017)
14. MEF: Piano Nazionale di Ripresa e Resilienza (PNRR) - Next Generation EU (NGEU), Recovery and Resilience Facility (RRF) (2021)
15. Cicerchia, A., Miedico, C.: Rapporto Finale – Musei In_visibili. Visioni di future per I musei italiani per il dopo emergenza Covid-19. Fondazione Scuola dei beni e delle attività culturali (2021)
16. Palombini, A.: The rights of reproducing Cultural Heritage in the digital Era. An Italian perspective. ExNovo J. Archaeol. **2**, 49–62 (2017)
17. Calabrò, F.: Integrated programming for the enhancement of minor historical centres. The SOSTEC model for the verification of the economic feasibility for the enhancement of unused public buildings | La programmazione integrata per la valorizzazione dei centri storici minori. Il Modello SOSTEC per la verifica della fattibilità economica per la valorizzazione degli immobili pubblici inutilizzati. ArcHistoR **13**(7), 1509–1523 (2020)
18. Mollo, F.: Guida archeologica della Calabria. Rubettino, Soveria Mannelli (2018)
19. Cuteri, F.A.: Percorsi nella Calabria antica. Itinerari archeologici nelle province calabresi. Koinè Nuove Edizioni (2004)
20. Panaia, C.: Caulonia, storia di una polis. Città del Sole, Reggio Calabria (2018)
21. Mertens, D.: Città e monumenti dei Greci d'Occidente, Dalla colonizzazione alla crisi di fine V secolo a.C. L'erma di Bretschneider, Roma (2006)
22. NEMO: Survey on the impact of the COVID-19 situation on museums in Europe (2020)
23. UNESCO-ICOM: Covid-19: UNESCO and ICOM concerned about the situation faced by the world's museums (2020)
24. Campolo, D., Calabrò, F., Cassalia, G.: A cultural route on the trail of greek monasticism in Calabria. In: Calabrò, F., Della Spina, L., Bevilacqua, C. (eds.) ISHT 2018. SIST, vol. 101, pp. 475–483. Springer, Cham (2019). https://doi.org/10.1007/978-3-319-92102-0_50
25. Pescarin, S.: Videogames, Ricerca, Patrimonio Culturale. In: Educazione al patrimonio culturale e formazione dei saperi. Franco Angeli, Milano (2020)
26. Ferguson, C., van Oostendorp, H., van den Broek, E.L.: The development and evaluation of the storyline scaffolding tool. In: 11th International Conference on Virtual Worlds and Games for Serious Applications (VS-Games), pp. 1–8. IEEE (2019)
27. London charter for the use of 3D visualisation in the research and communication of cultural heritage. In: International Forum of Virtual Archaeology (2006)
28. Charter, S.: International charter for virtual archaeology. In: International Forum of Virtual Archaeology (2010)
29. World Tourism Organization: Indicators of Sustainable Development for Tourism Destinations: a Guidebook, Madrid (2004)

30. ICOMOS: The ICOMOS Charter for the interpretation and presentation of Cultural Heritage Sites. In: 16th General Assembly of ICOMOS, Québec (Canada) (2008)
31. Georgieva, D.: IT'S not binary
    or economic accessibility to cultural heritage. In: Annual of the University of Architecture, Civil Engineering and Geodesy Sofia, vol. 51 (2018)
32. Tramontana, C., Calabrò, F., Cassalia, G., Rizzuto, M.C.: Economic sustainability in the management of archaeological sites: the case of Bova Marina (Reggio Calabria, Italy). In: Calabrò, F., Della Spina, L., Bevilacqua, C. (eds.) ISHT 2018. SIST, vol. 101, pp. 288–297. Springer, Cham (2019). https://doi.org/10.1007/978-3-319-92102-0_31
33. ICOMOS: Salalah Guidelines for the Management of Public Archaeological Sites. 19th ICOMOS General Assembly, New Delhi (2017)
34. Massimo, D.E., Musolino, M., Fragomeni, C., Malerba, A.: A green district to save the planet. In: Mondini, G., Fattinnanzi, E., Oppio, A., Bottero, M., Stanghellini, S. (eds.) SIEV 2016. GET, pp. 255–269. Springer, Cham (2018). https://doi.org/10.1007/978-3-319-78271-3_21
35. Solima, L., Minguzzi, A.: Relazioni virtuose tra Patrimonio culturale, turismo e industrie creative a supporto dei processi di sviluppo territoriale. In: Proceedings of the XXIV Convegno Annuale di Sinergie, Il Territorio come Giacimento di Vitalità per l'Impresa, Lecce (2012)
36. Barrile, V., Malerba, A., Fotia, A., Calabrò, F., Bernardo, C., Musarella, C.: Quarries renaturation by planting cork oaks and survey with UAV. In: Bevilacqua, C., Calabrò, F., Della Spina, L. (eds.) NMP 2020. SIST, vol. 178, pp. 1310–1320. Springer, Cham (2021). https://doi.org/10.1007/978-3-030-48279-4_122
37. Pagano, A., Pietroni, E., Poli, C.: An integrated methodological approach to evaluate virtual museums in real museum contexts. In: Proceedings of the 9th Annual International Conference of Education, Research and Innovation (ICERI), Siviglia, Spain (2016)
38. Garrett, J.: Elements of User Experience: User-Centered Design for the Web. New Riders Press, Thosand Oaks (2002)
39. Rubin, J., Chisnell, D.: Handbook of Usability Testing. Wiley Publishing, New York (2008)
40. Pujol, L.: Archeology, museums and virtual reality. In: Digit HVM Revista Digital d'Humanitats, vol. 6, pp. 1–9, Barcelona (2011)
41. Goleman, D.: Emotional Intelligence: Why it can Matter More than IQ. Bloomsbury, London (1995)

Accessible Culture: Guidelines to a Cultural Accessibility Strategic Plan (C.A.S.P.) for MArRC Museum

Giuseppina Cassalia[1]([✉]) [iD], Claudia Ventura[2], Francesco Bagnato[1] [iD],
Francesco Calabrò[1] [iD], and Carmelo Malacrino[2]

[1] Mediterranea University of Reggio Calabria,
Reggio Calabria, Italy
giuseppina.cassalia@unirc.it

[2] Ministry of Culture, Museo Archeologico Nazionale di Reggio Calabria, Reggio Calabria,
Italy

Abstract. In recent years, the function of museums has changed, proposing a series of new methods to promote and display its collections and contribute towards collective cultural and social development. In this new vision, accessibility gains a central role, as scratching the surface of the universal design. This paper is focused on outlining barriers to cultural accessibility, solutions, and possible next steps to ensuring universal accessibility. There is still limited empirical investigation on the expectations, satisfaction, and challenges of museum's visitors. In setting the context of Cultural Accessibility Strategic Plan discussion, the article first outlines what accessibility involves, accessibility standards and guidelines, and presents the case study of the National Archaeological Museum of Reggio Calabria, Italy. In this vein, the study draws on a qualitative research approach employing data from experts and visitors to understand those individuals' perceptions, needs, and challenges. The guidelines are designed based on the European convention and practical experiences, as well as the national binding standards. In conclusion, the study offers several practical implications to help museum managers enhance the experience of multi-features visitors.

Keywords: Cultural heritage · Accessibility · Museum · Strategic planning

1 Introduction

Today there is a strong demand for innovation in the cultural offer, in particular in the Museums, which no longer neglects the value of the term accessibility. It comes in varying degrees and forms and for some time now is associated with the idea of inclusiveness, because the visit must be lived without barriers and differences, allowing everyone to access the available contents and information.

The paper is the result of the joint work of the authors. Although scientific responsibility is equally attributable, the Sects. 1 and 4 were written by F. Bagnato, Sect. 2 by F. Calabrò, Sect. 3.1. by C. Ventura, Sect 3.2 by G. Cassalia, while Sect. 3.3 was written by C. Malacrino.

F. Calabrò et al. (Eds.): NMP 2022, LNNS 482, pp. 301–312, 2022.
https://doi.org/10.1007/978-3-031-06825-6_29

In this sense, Museums may be seen a powerful resource for local development. They can inspire creativity, promote cultural diversity, help retrain local economies, attract visitors and generate revenue. It is also becoming increasingly clear that they can contribute to social cohesion, civic engagement, health and well-being [1, 2]. For several decades now, cities and regions have been drawing on these resources to implement heritage initiatives as part of their broader economic development strategies. National governments and municipal and regional administrations, the museum community and other stakeholders are increasingly interested in these issues. New ways are being sought to demonstrate the impact of culture and museums on local development in order to effectively channel public and private funding [3].

Museums, in recent years, have grown numerically and the widespread effort is to make them more accessible and more modern. The museum is increasingly asked to be places of social inclusion, territorial garrison to strengthen the cohesion and citizenship of the population, to become real attractions of the city, drawing the tourist offer of the territorial contexts with different scales [4, 5].

In the scientific literature, accessibility has seen complex contexts (cities, airports, hospitals). Few studies have considered the difficulties that a museum can imply. Over time, the museums have moved towards a complex system of relationships, with an increasing and articulated presence of services to promote social and cultural inclusion, participation, interaction and more active fruition, with consequent problems of overcrowding [6].

In this framework, the contribution addresses the complexity of cultural accessibility in the specific area of museums by outlining a design methodology and identifying possible selection criteria which are not alternative but inclusive to current strategies of universal design. The general focus concerns the imperative attention to be given to cultural accessibility strategies in museum spaces in respect of their protection and with a view to a renewed relationship with visitors; specific aims of the research deal with the identification and validation of a methodology and technical solutions aimed at making appropriate choices to the specificities of the places with the ambition of making the model replicable for other museum experiences and more generally tourist/cultural sites. In the cognitive phase, numerous case studies have been studied in which this logic, with different levels of intervention and design responses, has favoured an interaction with the visitor. The project is still in its preliminary phase, and the first results obtained are illustrated as conclusive remarks, opening up prospects for the real transferability of the Cultural Accessibility Strategic Plan (C.A.S.P.) for Museums and tourist/cultural sites.

2 Cultural Accessibility and Museums

The characterization of space, if designed to stimulate the senses during the search for a goal within a building, is fully part of the accessibility that deals not only with communication systems (signage) but also with the organization and connotation of space, as expressed by the concept of wayfinding [7]. Accordingly, the Orientation can be defined as the ability to determine and control one's own and another's position and/or displacement within a conceptual framework of spatial reference, as well as a willingness to deal with both known and unknown environments and people.

Following this reasoning, an environment can be considered accessible if any person, even with reduced or impaired motor, sensory or psycho-cognitive abilities, can access and move safely and independently. Making an environment "accessible" means, therefore, making it safe, comfortable and qualitatively better for all potential users. Accessibility should therefore be understood in a broad way as the set of spatial, distributive and organizational-managerial features that can ensure a real use of the places and equipment by anyone [8].

Orientation, in fact, is not only determined by the ability to accept the explicit instructions of the signs, but in a more complete way is characterized by the ease of reacting to sensations and stresses produced by surfaces, sounds, colors, smells that contribute to create a mental image of a space, the paths to enjoy it, its most relevant elements. Along a path, the user constantly tries to make the sensory inputs coincide with his own 'cognitive map', that is, the storage of spatial information that allow its use [9, 10]. Wayfinding is an additional parameter, in addition to many others, to improve the quality of a project that, once realized and used for some time, can be evaluated by users with ex post methodologies in which the perceptive and sensory qualities deriving from environmental factors objectively constitute a significant reference; the effectiveness of the design choices related to the ways of use of the spaces can therefore be derived directly from the experience of the users. Many studies on wayfinding have shown that the first environmental information depends on the direct contact of the senses with the environment [11]; a subject, in the act of exploring space, puts into operation all the senses and therefore it becomes effective to stimulate them in a synesthetic way. In orienting themselves people do not read the space in a systematic way, but they visualize the relevant elements, the more consistent pedestrian flows and then finally the orientation maps and the signs in general. In addition to the view, it is also effective to use the tactile sense with which you can perceive warmth, softness, roughness. Smell (powerful activator of emotions and memory) and hearing for the sounds produced by the acoustic contrast of the different materials of footfall can also contribute [12].

Therefore, acting on an effective wayfinding strategy based on environmental deductible information using materials in a communicative way can limit the loss of spatial cognition by visitors, positively affecting internal flows and therefore the optimal use of space. On a national scale, the attention to wayfinding in museums, although still limited to signage, is documented by its introduction among the factors that determine the quality of the structures[1].

In the European context, in many evaluation procedures, the communication system is marked as an accreditation parameter in the certification bodies. In the Accreditation Scheme (UK)[2] one of the parameters required is a strategy for comprehensible guidance; in Spain, accreditation, delegated to local authorities, places among the minimum requirements the provision of signage and information tools. The Access plan required in the Accreditation Scheme for Museums and Galleries should outline priorities for action

---

[1] Circolare n. 80/2016 MIBACT. http://musei.beniculturali.it/wp-content/uploads/2016/12/Rac comandazioni-in-merito-allaccessibilit%C3%A0-a-musei-monumenti-aree-e-parchi-archeo logici-Circolare-80_2016.pdf.

[2] Accreditation Scheme for Museums and Galleries in the UK: Accreditation Standard. http://www.artscouncil.org.uk/supporting-museums/accreditation-scheme-0.

based on museum access policy or statement and recommendations from the museum access assessment. The Herity program[3], spread internationally, examines four aspects to determine quality: the perception of cultural value; the state of maintenance; communication and information for the visitor; the quality of hospitality. It is clear, though not yet explicitly, that the universal accessibility requirement has a positive impact on the overall assessment [13].

In this context, the research addresses the complexity of universal accessibility in the specific area of museums by outlining a design methodology and identifying possible selection criteria not alternative but inclusive and additives to current strategies of cultural accessibility [14, 15]. The proposal within a general objective concerning the important attention to be given to cultural accessibility strategies in museum spaces in respect of their protection and with a view to a renewed relationship with visitors, specific objectives such as the identification and validation of a methodology of approach to the project and technical solutions aimed at making appropriate choices to the specificities of the sites with the ambition of making the model replicable for other museum experiences and more generally tourist/ cultural sites. In the cognitive phase, numerous case studies have been studied in which this logic, with different levels of intervention and design responses, has favored an interaction with the visitor. Projects in which wayfinding has also contributed to the revival of the image: Maryland Institute College of Art (MICA), Baltimore; the Museum of Contemporary Art (MOCA) in LA; the Ashmolean Museum, Oxford; the National Maritime Museum of the Royal Museums Greenwich, the World Soil Museum in Wageningen, the Netherlands; the Children's Museum of Pittsburgh, Pennsylvania; the Vitra Design Museum in Germany. The analysis has also extended to other uses (e.g. shopping centres and hospitals) to transfer their experiences and define the determining factors to be included in the project inputs.

3   Case Study

3.1   The National Archaeological Museum of Reggio Calabria

The National Archaeological Museum of Reggio Calabria was recognized as among the most prestigious archaeological museums in Italy when it was made autonomous by the MiBACT 2014 reform.

The building that houses it is one of the first in Italy to be designed with the exclusive purpose of a museum. Marcello Piacentini, one of the leading exponents of the early twentieth century, conceived of the museum to have a modern style of exhibition, after he visited the main museums in Europe. Located in the heart of the city, the museum is an important element of the landscape and the life of all Calabria. Located next to De Nava Square in the center of the city, the southern façade faces the shore, with its splendid views of the Strait of Messina [16].

The National Archaeological Museum was born from the merger of the State Museum with the Museum of Reggio Calabria. The latter was inaugurated on June 18th, 1882, to guard the numerous archaeological remains of the area. Its headquarters were initially in the facilities of the Municipal Library, but with the increase of the collections,

---

[3] International body for the quality management of cultural heritage http://www.herity.it/.

between 1887 and 1889, it was moved to a building located next to the Roman baths, recently discovered at that time. During the 1908 earthquake, the Museum building was severely damaged. This accelerated the process of establishing a national archaeological museum, strongly supported by Paolo Orsi, among others, who in 1907 was named the first superintendent of the excavation in Calabria. On May 22nd, 1948, an agreement was signed between the City Council of Reggio Calabria and the General Directorate of Antiquities of the then Ministry of Education. After acquiring these collections from the civic museum, the latter was closed.

The new permanent exhibition, which originated from the remodeling of the building that started in 2009, features 220 displays and is divided into four levels, which tell the story of the human population in prehistoric times until Roman Calabria, in chronological order. The visit begins on the second floor (level A – Prehistory and Protohistory, the Age of Metals), continues on the first floor (level B – Cities and sanctuaries of Ancient Greece), on the mezzanine (level C – Necropolis and daily life of Ancient Greece: Sybaris and Croton, Hipponion, Kaulon, Ciro Laos; and ends on the ground floor (level D – Reggio), where it is placed the room of Riace and Porticello Bronzes, in a setting equipped with an air-conditioning and an anti-seismic system, which can be accessed after a brief stop at the grape room and at the antipollution filter area. The floor of the basement (level E) is reserved to temporary exhibitions.

In recent years, the MArRC Museum is working hard in improving the accessibility to its collections and facilities. Although the term 'accessibility' usually tends to be thought of in connection to physical disability, in the particular cases of culture and heritage sites it also crucially involves the ability to perceive ideas and understand meanings and cultural values.

As previously stated, the term 'accessibility' is now customary to be defined as a generic concept and interpreted in relation to its double articulation as physical and perceptive accessibility. In general, accessibility means the ability of everyone to use a service or a cultural product, regardless of specificity (disability, age) or context of use. In other words, accessibility is, by definition, a matter of usability.

Hence, the main aim of the ongoing research project is to investigate the existing level of accessibility to tourism facilities and cultural potential, in order to define elements that play important roles in experiencing (physical accessibility) and comprehending (perceptual accessibility) culture and archaeological heritage. The intent of involving such important cultural Institute, is to present good practices in a specific case study aimed at defining a satisfactory level of physical and perceptual accessibility to culture and heritage, in order to encourage new promotion and management practices, as well as redefine the new paradigms of heritage interpretation.

3.2 Toward the Definition of a Cultural Accessibility Strategic Plan for (C.A.S.P.)

To remain relevant to changing user needs and expectations a Cultural Accessibility Strategic Plan for (C.A.S.P.) is needed to improve the physical, sensory and intellectual access to the collections, information about the collections and access to the building – identifying priorities, timescales and resources [17, 18]. The C.A.S.P. research project outlines priorities for action based on the Museum access policy and recommendations from the National Authorities[4]. As showed in Fig. 1 the access assessment will include: Access audit or access checklist; Work with focus groups, support; agencies, advocacy groups, charities, experts; Facilities checklist; Review of interpretation and collections use.

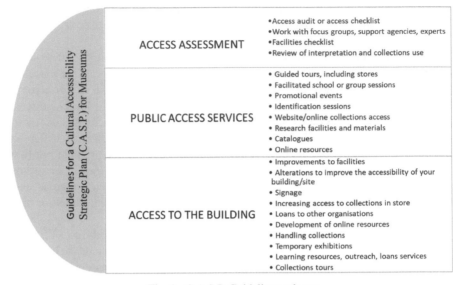

Fig. 1. C.A.S.P. Guidelines scheme

The assessment can be carried out in-house or with external expertise. It should be carried out within the last five years, or more recently if there's been a significant change to the buildings or displays. The Strategic Plan is also meant to outline the procedures followed for the Public access, including how information is provided on what is and is not available for public access; and how access to collections is provided both on display and in store and to information on the collections (Fig. 2).

A series of preliminary talks with different multidisciplinary skills were conducted (superintendencies, museum directors, architects, officials, visitors) in order to acquire information to build a database and propose project guidelines later [19].

In the MArRC case study, an operational sequence of studies and preliminary checks was carried out, including:

---

[4] Legge 3 marzo 2009, n. 18. Ratifica ed esecuzione della Convenzione delle Nazioni Unite sui diritti delle persone con disabilità, con Protocollo opzionale, fatta a New York il 13 dicembre 2006 e istituzione dell'Osservatorio nazionale sulla condizione delle persone con disabilità. Gazzetta Ufficiale Serie Generale n. 61, 14.03.2009. MiBACT, 2008. Linee guida per il superamento delle barriere architettoniche nei luoghi di interesse culturale. Decreto 28 marzo 2008. Gazzetta Ufficiale Serie Generale n. 114, 16.05.2008, Suppl. ordinario n. 127. MiBACT, 2016. Raccomandazioni in merito all'accessibilità a musei, monumenti, aree e parchi archeologici. Gruppo di lavoro istituito con D.D. 1 dicembre 2015 (rep. 7363). Circolare MiBACT 80/2016. Unwto, 2016. Highlights of the 1st UNWTO Conference on Accessible Tourism in Europe. San Marino 19–20 November 2014. World Tourism Organization (UNWTO), 32 pp.

- an in-depth analysis of the history and evolution of the museum, a survey of the context, number and type of visitors;
- subdivision of the museum into functional areas and related environmental units;
- analysis of flows and routes, identification of criticalities through configurational analysis;
- identification of the units on which to intervene to increase the communicativeness of the spaces;
- analysis of the environmental system, with particular attention to the requirements of usability, visual integration and transformability;
- analysis of existing technical solutions.

To support appropriate options to improve the visit's use "for all", the study proposes some criteria to realize a repertoire of effective technical solutions to facilitate the cognitive and behavioral processes of the user during the experience of visit to the museum [20, 21].

Fig. 2. C.A.S.P. - Reasoning by activities

The study has led to the assumption that there are no ideal solutions for everyone: any environment or product will always present difficulties of use or use for some specific users, as well as there will always be special situations that will require customized solutions. Think of the different forms of disability and the various related issues that make what is an obstacle for some individuals can be a fundamental element for others (the physical barriers for people in wheelchairs or for the visually impaired). It was therefore found the impossibility to design specifically for each disability and the awareness that there is no perfect solution "for everyone", leading to meet the needs of as many people as possible, setting aside the logic of standard and ordinary solutions.

Further research concerns the inclusion of the accessible museum in a virtuous tourist circuit. The path undertaken with this project will allow to develop a brand that distinguishes museums and tourist attractions that will conform to the accessibility standards defined during the years of development and experimentation. Extending the agreed standards to other museums and cultural institutions can become a stimulus to change that involves and reflects on the entire organization of cities, because an accessible city is of benefit to all.

3.3 Lessons Learned and Future Perspective

The issue of accessibility is undoubtedly one of the most crucial from the point of view of the livability of the built spaces and is therefore an essential quality feature of the property and its equipment. All this is all the more true for buildings of cultural interest, as they generally represent places of memory or "precious spa" for the community, to be used for activities and events that must still be accessible and "welcoming" for all, thus transforming constraints into opportunities for participation. In this perspective, the issues of accessibility "must be considered as normal design elements, such as safety, structural solidity, thermo-hygrometric comfort, building and urban regulations, economic availability, the same guiding principles of the restoration: distinguishability, reversibility, compatibility physique-chemistry, expressive authenticity. To remain relevant to changing user needs and expectations a Cultural Accessibility Strategic Plan (C.A.S.P.) is needed to improve the physical, sensory and intellectual access to the collections, information about the collections and access to the building – identifying priorities, timescales and resources [22]. On the basis of a study on accessible cultural tourism, new potential users of the museum were estimated.

In the main filed of "universal accessibility" the very first and major output of this preliminary research's phase regards the orientation's sector. It is necessary for the environment to provide as much useful information as possible to determine with reasonable accuracy its position with respect to the environment and to identify the most effective route to reach the desired destination. Within the places of cultural interest, in order to allow and encourage orientation, various strategies can be used, among which the main ones identified by the project are the identification of points and lines of reference, the design of appropriate signage and the use of maps that effectively represent the environment in which visitors find their selves [23].

- *Reference points*

The Museum should be able to positively stimulate the perception capacity of space and should be interesting, that is, rich in external sensory stimuli. The person should be able to easily recognize the connotation of the environment; for example, the type, the hierarchy of entrances, the location of the interior, their function, the subdivision and the identifiability of private and/or public spaces. In the case of MArRC, the suggestions concern the use of shapes, materials, colors and symbolic elements, through which to characterize the space giving a complete sense to what is present in the environment.

The reference points are visual, tactile, acoustic, olfactory, kinesthetic discrete information, easy to perceive and always found in the environment, which people can use to recognize precise places.

The guidelines are those continuous elements present in the environment that allow the person with severe visual impairment to orient themselves and to maintain the direction of travel; they can be naturally present in the environment or specially built.

Examples of natural guidelines are: the continuous wall of a building, a porch, traffic noise (acoustic information).

- *Signposting*

Signage, understood as a set of coordinated signals, has the function of guiding the visitor, communicating with a universal language, made of signs, pictograms and short words, helping him to identify accesses and exits, services and routes desired.

Among the lessons learned in the design of signage with the CASP project are the major:

– avoid redundant information that can cause confusion and anxiety (the so-called "visual pollution").
– environmental signage is the most important business card, it must make the visitor feel at ease, be decipherable by as many people as possible, and be consistent, in terms of images and meaning, with all forms of communication present.
– Improving the wayfinding experience of a visitor is equivalent to improving the environmental signals that are offered to guide and convey space information.
– Analyse visitor flows and identify routes and points where information or repetition of an indication must be guaranteed, in order to offer alternative routes.
– From a graphic point of view, there are many factors which determine the effectiveness and legibility of signs. Among the most important are:

  – messages and signals must be short, legible and comprehensible;
  – the choice of colours, type and size of typefaces (fonts) and contrasts must be carefully considered;
  – signs must be visible even at distances of more than 10 m and also in motion.

– The location of signs also plays an important role. It is therefore necessary to:

  – ensure that the signals are not hidden by other temporary elements;
  – ensure that the same signals do not constitute an obstacle to the visibility of other elements or to the mobility of anyone;
  – check their readability from afar and up close;
  – verify the type of lighting present in each part of the goods;
  – avoid the use of inappropriate supports, such as reflective surfaces (glass, glossy metals, mirrors, etc.), thus favouring opaque finishes.

– From the point of view of content, signs must be clear and understandable:

  – the information should be grouped and sorted alphabetically by plan;
  – avoid placing too many messages on a single signal. Small groups of messages are more readable than a long list;
  – numbers and pictograms are more easily recognisable than words;
  – language must be clear and concise, even if brevity should not compromise understanding;
  – readability increases if the first letter is uppercase;
  – the titles and initials are more readable if the point between the initials is omitted;
  – punctuation should be used only where it is indispensable;
  – avoid abbreviations.

- Finally, the graphic efficiency of a signage system depends on the contrast between the text of the text and the background, as well as on the skillful use of colours. The colour of signage is therefore a very important and strategic factor. As it also affects the environment, the choice of color must be evaluated lighting conditions and the dominant shades of the environment, which must produce an effective contrast.
- A final consideration should be made of the maintenance of the signage system: this aspect should be considered from the design stage, possibly using solutions that provide flexibility and interchangeability, for easier assembly, maintenance and cleaning.

- *Maps*

A map is a simplified symbolic representation of the space that highlights relationships between the components of the same. The choice of the type of representation and drawing of a map depends, therefore, not only on the information that it must contain and express, but also on the recipients to whom it is addressed.

Inside the places of cultural interest, a map must be placed near the entrance, but also in other strategic points (for example near the vertical link elements, intersections, changes of direction, etc.), it is necessary to ensure the presence of a clear and accessible fixed map for the greatest number of people, including the elderly or those who have a poor habit with reading plants and floor plans.

In the case of an integrated system of maps within a site of cultural interest, attention should be paid to the homogeneity of the symbols and the consistency of the information. In the perspective of Universal Design, however, it is desirable to design and create tactile-visual maps, i.e. maps "for all", which contain additional precautions for the reading of space also by the blind: thicknesses and lines in relief, written in braille and "in black" embossed, textures recognizable to the touch.

4   Conclusive Remarks

Although many significant objectives have been achieved in recent years, much still needs to be done, considering that, to date, a large part of the sites of cultural interest continue to present objective critical issues in terms of use. Intervening on the communicativeness of space and its perceptive connotations to facilitate the wayfinding in museums has positive implications for both visitors and managers [24, 25].

Adopting a universal accessibility project, complete and respectful of the requests for protection, it can be appreciated the richness of the cultural heritage, attributing to museum structures the characters of reliability that determine its economic success (increase in the number of visitors) as well as educational-informative.

The subsequent developments of this project envisage putting them into system with all the other modalities that together will guarantee an increasing quality of the project and of the realization, within an ex post evaluation to determine the overall results.

Designing accessibility means considering not only aesthetic and formal aspects, but putting the human being and his peculiarities and needs at the center of attention: His being a man or woman that evolves from child to elderly and that in the course of life can undergo temporary or permanent changes and present characteristics different from that "normality" arbitrarily defined by conventions that often prove inadequate.

Having acquired the concept of welcoming people in a safe, easy and pleasant environment, it is equally central to strive to give information that will allow and develop the ability to choose, the possibility of autonomy and self-determination of all during the visit. By creating an accessible communication, visitors can be helped to identify an appropriate use for their physical condition, their cultural preparation and their sensitivity. Accessibility, especially to cultural goods, is, in addition to opportunities for social growth, also an economic investment, as it favors tourism and in particular cultural tourism.

References

1. Lawrence, T.B., Dover, G., Gallagher, B.: Managing social innovation. In: Dodgson, M., Gann, D.M., Phillips, N. (eds.) The Oxford handbook of innovation management, pp. 316–334. Oxford University Press, Oxford (2014)
2. Cajaiba-Santana, G.: Social innovation: moving the field forward. A conceptual framework. In: Technological Forecasting and Social Change, vol. 82, pp. 42–51 (2014)
3. Harding, R.: Social enterprise: the new economic engine? Bus. Strategic Rev. **15**(4), 40–43 (2004)
4. Brown, K., Mairesse, F.: The definition of the museum through its social role. Curator Museum J. **61**(3), 525–539 (2018)
5. Morse, N., Munro, E.: Museums' community engagement schemes, austerity and practices of care in two local museum services. Soc. Cult. Geogr. **19**(3), 357–378 (2018)
6. Cetorelli, G., Guido, M.R.: Il patrimonio culturale per tutti. Fruibilità, riconoscibilità, accessibilità. MIBACT Quaderni della valorizzazione, vol. 4. Capponi, Ascoli Piceno (2017)
7. Solima, L.: Museums, accessibility and audience development. In: Cerquetti, M. (ed.) Bridging Theories, Strategies and Practices in Valuing Cultural Heritage, pp. 225–240. EUM, Macerata (2017)
8. Vescovo, F.: Obiettivo: progettare un ambiente urbano accessibile per una "utenza ampliata. In: Paesaggio urbano, vol. 1 (2002)
9. Rappolt-Schlichtmann, G., Daley, S.G.: Providing access to engagement in learning: the potential of Universal Design for Learning in museum design. Curator Museum J. **56**(3), 307–321 (2013)
10. Walters, D.: Approaches in museums towards disability in the United Kingdom and the United States. Museum Manage. Curatorship **24**(1), 29–46 (2009)
11. Anaby, D., et al.: The effect of the environment on participation of children and youth with disabilities: a scoping review. Disability Rehabil. **35**(19), 1589–1598 (2013)
12. Levent, N., Pascual-Leone, A. (eds.): The Multisensory Museum: Cross-Disciplinary Perspectives on Touch, Sound, Smell, Memory, and Space. Rowman & Littlefield, Lenham (2014)
13. Suchy, S.: Museum management: emotional value and community engagement, in New roles and missions of museums. In: INTERCOM 2006 – International Committee for Museum Management 2006 Symposium, Edited by Chinese Association of Museums, International Council of Museums, pp. 1–10 (2006)
14. Da Milano, C., Sciacchitano, E.: Linee guida per la comunicazione nei musei: segnaletica interna, didascalie e pannelli. MIBACT Quaderni della valorizzazione, vol. 1. Capponi, Ascoli Piceno (2015)

15. De Luca, M.: Comunicazione ed educazione museale. Franco Angeli, Milano (2007)
16. Malacrino, C.: Dal passato al futuro. Il Museo Archeologico Nazionale di Reggio Calabria. Enciclopedia Italiana **10**, 28–34 (2022). ISSN 2611-8459
17. Cerquetti, M.: Dall'economia della cultura al management per il patrimonio culturale: presupposti di lavoro e ricerca. Il capitale culturale. Stud. Value Cult. Heritage **1**, 23–46 (2010)
18. Tramontana, C., Calabrò, F., Cassalia, G., Rizzuto, M.C.: Economic sustainability in the management of archaeological sites: the case of Bova Marina (Reggio Calabria, Italy). In: Calabrò, F., Della Spina, L., Bevilacqua, C. (eds.) New Metropolitan Perspectives. ISHT 2018. Smart Innovation, Systems and Technologies, vol. 101. Springer, Cham (2019). https://doi.org/10.1007/978-3-319-92102-0_31
19. Curto, R.A., Barreca, A., Coscia, C., Fregonara E., Ferrando, D.G., Rolando, D.: The active role of students, teachers, and stakeholders in managing economic and cultural value, urban and built heritage. Interdisc. J. Probl. Based Learn. **15**, 1–24 (2021). ISSN 1541-5015, https://doi.org/10.14434/ijpbl.v15i1.29626
20. Lopez, L., Pérez, Y.: Turismo, patrimonio e cultura: Verso un' educazione territoriale. Un caso di studio in Galizia (Spagna). LaborEst (21), 18–24 (2021). http://pkp.unirc.it/ojs/index.php/LaborEst/article/view/629
21. ICOM: La funzione educativa del museo e del patrimonio culturale: una risorsa per promuovere conoscenze, abilità e comportamenti generatori di fruizione consapevole e cittadinanza attiva, ICOM Italia – Commissione Educazione e Mediazione, 22 March 2021. http://www.infologsrl.it/wp-content/uploads/2015/02/LA-FUNZIONE-EDUCATIVA-DEL-MUSEO.pdf
22. Dodd, J., Sandell, R., Coles, A.: Building Bridges: Guidance for Museums and Galleries on Developing New Audiences. Museums & Galleries Commission, London (1998)
23. Onciul, B.: Community engagement, curatorial practice, and museum ethos in Alberta. In: Golding, V., Modest, W. (eds.) Museums and Communities: Curators, Collections and Collaboration, pp. 79–97. Bloomsbury, London (2013)
24. Taylor, J.K.: The Art Museum Redefined: Power, Opportunity, and Community Engagement. Springer Nature, London (2020)
25. Watson, S. (ed.): Museums and their Communities. Routledge, New York (2007)

Economic Feasibility of an Integrated Program for the Enhancement of the Byzantine Heritage in the Aspromonte National Park. The Case of Staiti

Giovanna Spatari, Immacolata Lorè, Angela Viglianisi, and Francesco Calabrò(✉)

PAU Department, Mediterranea University of Reggio Calabria, Reggio Calabria, Italy
francesco.calabro@unirc.it

Abstract. The Inner Areas of Calabria are scattered with traces of the historical-architectural heritage of the Byzantine period. Remains of fortifications and religious buildings still testify today to the wealth and exceptional value of an absolutely peculiar settlement system. The enhancement of these testimonies encounters many difficulties from a management point of view: their diffusion throughout the territory and the limited number of visitors prevent the adoption of management models that ensure their full usability, and thus increase their attractiveness. The article illustrates the economic feasibility check of a management model developed specifically for a case study, that is, the historical and cultural heritage present in Staiti, a small town in the province of Reggio Calabria, in southern Italy.

Keywords: Management models · Economic feasibility · Cultural tourism · Inland areas · Byzantines

1 Introduction

The Inland of regions whose development is lagging behind are often endowed with huge natural and cultural resources, which are currently insufficiently highlighted [1].

For these areas, the *National Strategy for Inner Areas* (*Strategia Nazionale per le Aree Interne*, SNAI) attributes a potential development role to these resources: one of the decisive issues so that the cultural and natural heritage of the Inner Areas can be adequately exploited is the economic sustainability of their management, due to the very low levels of demand [2, 3].

The territory of Calabria is strewn with traces of the historical-architectural heritage, in particular of the Byzantine period. Remains of fortifications and religious buildings still bear witness to the richness and exceptional value of an absolutely peculiar settlement system.

The work is the result of the shared commitment of the authors. However, Sects. 2, 7 and 8 can be attributed to Giovanna Spatari; to Immacolata Lorè Sects. 3 and 4; to Angela Viglianisi Sects. 5, 5.1, 5.2; to Francesco Calabrò Sects. 1 and 6. Conclusions were written jointly by the authors

F. Calabrò et al. (Eds.): NMP 2022, LNNS 482, pp. 313–323, 2022.
https://doi.org/10.1007/978-3-031-06825-6_30

The enhancement of these testimonies, however, encounters many difficulties from the management point of view: their diffusion on the territory and the relatively limited number of visitors prevent the adoption of management models that guarantee their full usability, and thereby increase their attractiveness [4, 5].

The article, after having identified the valorizing resources present on the territory and the main problems that prevent their full exploitation, illustrates the economic feasibility test of a management model developed specifically for the case study, that is the historical and cultural heritage present in Staiti, a small town in the province of Reggio Calabria, in southern Italy [6, 7].

2 The Existing Cultural and Natural Heritage

The enhancing resources present in Staiti can be substantially traced back to both typical patrimonial categories: the tangible and intangible Cultural Heritage and the Natural Heritage [8–10].

The cultural Heritage
As is well known, Calabria in the Byzantine era lived one of its most important periods, from the point of view of the geopolitical role in the Mediterranean context, due to its geographical position, and, above all, as regards the structuring of its settlement system, which profoundly marked its subsequent evolutions.

In this context, like many other places in Calabria, the territory of Staiti has hosted, probably since the sixth century AD, different forms of spirituality, even strict, such as forms of hermitism, asceticism and cenobitism: these forms of spirituality were practiced mainly by monks from the East, following the different iconoclastic struggles at first and the spread of the Muslim religion later [11, 12].

Of this past remains an extraordinary trace, consisting of the Abbey of Tridetti, which rises a few kilometers from Staiti, in a green valley surrounded by a protective blanket of mountains. It is certainly the highest and most significant monument of the visit to the village of Staiti. A Byzantine church dating back to the eleventh century located near the Badìa locality, discovered in 1912 by the archaeologist Paolo Orsi.

Paolo Orsi wrote about the discovery: "*The church is of the early Norman times but built on a Greek scheme, because it was destined for a Greek cult, as it was practiced by the Basilians.*" It is possible to see, in fact, among the most ancient architectural elements of the Church, such as capitals and other elements from classical or Hellenistic Greek architecture.

Legend has it that on the spot there was a temple built by the "Lokresi Zephiri" who in the V–VI century B.C. was raised to thank the god of the sea, Neptune, for having saved them from a storm. Between the seventh and eighth centuries on the same site the Basilian monks founded a Greek church in honor of Our Lady of the Trident, clear allusion to the divinity of the sea, then handed down in Tridetti through the dialect. The Church is mentioned for the first time in a document of 1060. Count Ruggero d'Altavilla ordered the allocation of part of the revenues to the Chapter of Bova, on which it depended.

In the church of Santa Maria de' Tridetti there is a mature fusion of western and eastern motifs, which makes one of the first examples of the three-aisled basilica typology with a transept not protruding tripartite and triabsidate, that in the twelfth century will find wide diffusion in the Greek-Calabrian and Sicilian architecture. Clearly Islamic in character is the wide use of pointed arches that, present along the naves, are repeated, in larger dimensions, on the facade of access, between central nave and presbytery and in the arch of connection between bema and central apse. Particular is the bell tower sail, in addition to the Greek architectural elements placed in the most important points of the sacred building, almost to symbolize the passage from paganism to Christianity (Fig. 1).

The monument is today a precious jewel eroded by time. The ancient church, probably, was part of a convent now disappeared. Part of the central nave, as well as part of the side walls and dome, have been lost forever. The body of the presbytery still stands, with a small hint of dome, and the facade of access to the church. It is certainly one of the best preserved examples of Byzantine architecture in Calabria.

Fig. 1. Santa Maria de' Tridetti

The close link between Staiti and the Byzantine world is also witnessed by the Museum of Saints Italo-Greeks, recently established, with its collection of Byzantine icons, the true pearl of the village. It wants to be the completion of a path of recovery of the identity traditions of the community.

Fig. 2. Museum of Saints Italo-Greeks

In accordance with the Museum of Saints Italo-Greeks there is the Path of the Byzantine Churches, an open-air museum, with bas-reliefs scattered along the streets of the village, that remind in an ideal itinerary of the most important places and events of the Byzantine history of Calabria (Fig. 2).

Evidence of the oldest roots of Staiti can be found in some typical productions, linked to the agro-pastoral activity that has characterized the economy for a long time.

This is the reason that encouraged the Municipal Administration to establish the Museum of Agro-pastoral Civilization: recently renovated but still closed to the public, presents in addition to the exhibition rooms also a kitchen where it will be possible to demonstrate the production process of typical products of the gastronomic culture of Staiti, starting from its excellence, ricotta. This dairy is often served in a very special form, warm accompanied by the serum remaining from the processing; obviously it is also used as cheese on the table or even in desserts. Among the typical Easter ones, in fact, we find the "jaluni": jewish lantern-shaped cakes filled with ricotta cheese.

The Natural Resources

The territory of Staiti is inside the perimeter of the National Park of Aspromonte and from it start routes for the use of the important natural resource [13, 14].

Near the town you can find the locality known as "Rocche du Quartu", a geosite recently equipped with panoramic binoculars, while about 3 km from the village is located Locality "Falcò", where a wooded path begins on dirt road that leads to the nearby top of Monte Cerasìa (1,023 m) or towards Campi di Bova.

3   The Critical Issues to be Addressed

From the general point of view, Staiti is the classic Municipality of the Inner Areas, affected by a depopulation continuously growing with birth rate equal to zero, which is likely to become one of the numerous "ghost" countries where the small resident population is forced to deal almost daily with the progressive closure of essential services.

There are no more schools, of any order, no medical laboratories, no ATM machines, but a Poste Italiane local open only three days a week, a little family grocery shop, a bar and a pharmacy [15].

Specifically, to promote cultural and natural resources which, as we have seen, the territory is widely equipped, it is possible to highlight some critical issues:

– one of the two existing museums is closed while the other one isn't open continuously;
– any form of receptivity is absent;
– existing resources are not known to a wide public, even if they are interested in cultural and naturalistic aspects;
– accessibility and mobility within the village are difficult, especially for people with mobility difficulties.

The following SWOT analysis focuses in detail on strengths and weaknesses, opportunities and threats [16].

Strengths	Weaknesses
– membership of the Aspromonte National Park; – Favourable climate and natural habitats; – Gastronomy and folk traditions; – Historical, architectural, religious and cultural heritage; – Bus service available	– Lack of receptivity; – Lack of essential services (health, schools, ATM); – Lack of public transport and therefore travel difficulties; – Difficulty in managing existing museums; – Carelessness of existing material cultural heritage; – Depopulation and zero births
Opportunities	Threats
– Dissemination of cultural heritage; – Digitization of existing cultural heritage; – The possibility of using disused real estate to offer accommodation; – Potential growth in tourism expenditure; – Shuttle service; – Local development through an integrated tourism supply system	– Risk of loss of tangible and intangible cultural heritage through neglect and migration; – Regional competitiveness with better tourism and marketing opportunities; – Risk of closure of the few essential structures present (post, food)

4 The Integrated Program for the Revitalization of the Village of Staiti

From the SWOT analysis it is possible to deduce the general objective the specific objectives of an Integrated Program for the revitalization of the Village of Staiti.

The general objective is: "to combat depopulation by promoting the socio-economic development of the territory through the enhancement of cultural and natural resources" [17–19].

The specific objectives are:

– Equipping the village with a reception system;
– To enable the existing cultural and natural heritage to be fully accessible;
– Encouraging the emergence of new business initiatives and the strengthening of existing ones;
– Promoting knowledge of existing cultural and natural heritage;
– Improving accessibility to and mobility inside the village.

The actions that make up the Integrated Program are:

– The construction of a reception system through the reuse of unused buildings;
– The identification of the managing entity capable of ensuring the appropriate availability of cultural and natural heritage;
– The establishment of a sustainable local mobility system;
– The implementation of communication and marketing measures.

5 The Economic Feasibility Study of the Measures

This article discusses the economic feasibility of two specific actions that make up the Integrated Program [20]:

- the construction of the reception system through the reuse of unused building stock;
- Identification of the managing entity capable of ensuring that the cultural and natural heritage is adequately accessible.

Considering the demographic and socio-economic characteristics of Staiti and the past experiences in the management of the existing museum structures, it has been hypothesized to reunify the management of both the accommodation system and the Museums in the hands of a single operator [21].

However, it is necessary to understand whether the hypothesis is really economically sustainable and what type of managing entity is most suitable for the present case.

5.1 Estimation of Tourist Flows

The diffused hotel provided by the Integrated Program consists of 8 suites and 48 double rooms. For the estimation of the expected revenues for the diffused hotel, it is necessary to estimate the potential tourist flows in advance.

It must be considered that it is the birth of a new accommodation facility in a municipality where, at the moment, none exist so it is difficult to predict which demand will be intercepted.

A reference, in theory could be the current number of visitors of cultural resources (Abbey of Santa Maria de Tridetti and Museum of Saints Italo-Greeks, in particular), but unfortunately at the moment the data is not available. In any case, a certain increase in visitors should be considered as a result of the communication and marketing campaign provided for in the Integrated Program [22].

Taking into account these considerations, to estimate the presences was used a parameter of an average figure, that is the average occupancy rate of hotel rooms in Calabria which, according to the study conducted by Assohotel in 2018, is equal to 18.5%.

5.2 Management Models and Types of Managers

Currently, the cultural resources of Staiti are difficult to use due to the scarcity of resources available for their management; on the other hand, if they are not made usable, any hypothesis of development is impracticable.

For these reasons, it is necessary to identify a manager capable of providing this service.

The Cultural Heritage and Landscape Code provides that the managing entity can be a public or private entity.

At the moment the Municipal Administration, as well as any other public entities (Metropolitan City, Superintendence, etc.), don't have human and financial resources able to allow the provision of the service.

It is therefore necessary to verify whether the conditions are met, first of all economic sustainability, to entrust the management to private entities and which model of partnership is truly sustainable.

6 The SostEc Model in Order to Verify the Economic Sustainability of Management Models

It is possible to use the SOSTEC model, which has been developed precisely to provide the necessary background information for identifying the most appropriate form of public-private partnership in this case.

Basing on the nature of the activities and of the managing entity, it is possible to hypothesize the following three models (Table 1):

Table 1. Managing models by type of subject

Managing model	Nature of entity and activities
ModeL P – Profit	Entity profit, activities profit
Model NP - Not-for-profit	Entity not-for-profit, activities not-for-profit
Model M – Mixed	Entity not-for-profit, activities profit

The SostEc model uses a typical tool in these cases, the analysis of cash flows, able to verify with relative ease whether the expected revenue flow is equal to or higher than the estimated costs or not, in relation to the different cost structures of the different types of managers.

7 The Management Models Hypothesized in the Case Study

Due to the difficulties encountered so far in the management of cultural services, it was decided to combine the management of existing museums with that of the diffused hotel, in order to optimize the use of some human resources, reducing their cost.

Usually, in the presence of limited visitor flows, such as the one in question, the most probable manager is a private non-profit entity, by virtue of the lower management costs for human resources that it has to face compared to a profit-type entity.

As will be seen, in the case in question, however, the services provided still require figures, especially those relating to accommodation, which cannot be considered voluntary as they perform tasks that are perfectly attributable to some specific categories of workers.

For these reasons, two different management models have been developed:

- the first, called Scenario 1, is attributable to the typical Model M, a non-profit entity that carries out profit activities;
- the second, called Scenario 2, is a particular type of Model P, a profit entity that carries out profit activities: in this case, the economic sustainability of the model has been verified in the hypothesis that management is entrusted to a company already existing in the area and that, in virtue of the limited flows of expected attendance, can optimize the use of human resources that it already has employees, with a certain reduction in costs for this item (Tables 2 and 3).

Table 2. Human resources Scenario 1.

Qualification	Unit	Full time	Part time
Manager of the multi-building hotel	1	x	
Receptionist	1	x	
Marketing and website management	1	x	
Tourist guide	1		x
Cleaner	1		x
Waiter	1		x

Table 3. Human resources Scenario 2.

Qualification	Unit	Full time	Part time
Tourist guide	1		x
Cleaner	1		x
Waiter	1		x

8   Feasibility of the Intervention

The following are the Project Economic Accounts associated with the two scenarios for the verification of sustainability of the intervention in the management phase (Tables 4 and 5):

Table 4. Cash flow analysis Scenario 1.

Project Income Statement (Senario 1)	
A) Value of production:	
Revenues from sales and services envisaged by the project	€ 91.896,40
Total A) Value of production	**€ 91.896,40**
B) Production costs:	
For raw materials, ancillary materials, consumables and goods	€ 10.000,00
For services (utilities; repairs; cleaning; other ordinary maintenance services)	€ 30.000,00
For personnel: a) wages and salaries; b) social security contributions; c) severance pay; d) pensions and similar; e) other costs;	€ 87.360,00
Depreciation of furniture for eating disorders center (10%)	€ 8.036,70
Provisions for risks; (2%)	€ 5.000,00
12) provisions for extraordinary maintenance	€ 10.000,00
Total B	**€ 150.936,70**
Difference between value and costs of production (A - B) -> Operating Income (RO) or (MON)	**€ −58.500,30**

Table 5. Cash flow analysis Scenario 2.

Project Income Statement (Senario 1)	
A) Value of production:	
Revenues from sales and services envisaged by the project	€ 91.896,40
Total A) Value of production	**€ 91.896,40**
B) Production costs:	
For raw materials, ancillary materials, consumables and goods	€ 10.000,00
For services (utilities; repairs; cleaning; other ordinary maintenance services)	€ 30.000,00
For personnel: a) wages and salaries; b) social security contributions; c) severance pay; d) pensions and similar; e) other costs;	€ 28.600,00
Depreciation of furniture for eating disorders center (10%)	€ 8.036,70
Provisions for risks; (2%)	€ 5.000,00
12) provisions for extraordinary maintenance	€ 10.000,00
Total B	**€ 91.636,70**
Difference between value and costs of production (A - B) -> Operating Income (RO) or (MON)	**€ 259,70**

9 Conclusions

As can be seen from the Economic Accounts above, the first scenario, in which the management would be entrusted to a non-profit entity, would generate a significant loss, while the second scenario, with the reliance on a profit entity already active in the territory, records a small asset.

Through the use of the Cash Flow Analysis, it was thus possible to verify the economic feasibility of one of the actions provided for by the Integrated Program [23].

The research continues in the direction of strengthening the reliability of forecasts, currently entrusted to information gaps and not always up to date.

Another aspect that will be deepened by the research will concern the verification of the applicability of the model in different contexts, for the purpose of its transfer in operational areas.

References

1. ICOMOS: International cultural tourism charter. managing tourism at places of heritage significance. In: Adopted by ICOMOS at the 12th General Assembly in Mexico (1999)
2. Cotella, G., Vitale Brovarone, E., Voghera, A.: I Contratti di Fiume e la Strategia Nazionale per le Aree Interne: un banco di prova per l'approccio place-based in Italia. LaborEst **22**, 21–27 (2021). https://doi.org/10.19254/LaborEst.22.03
3. Saija, L., Pappalardo, G.: From enabling people to enabling institutions. a national policy suggestion for inner areas coming from an action-research experience. In: Bevilacqua, C., Calabrò, F., Della Spina, L. (eds.) NMP 2020. SIST, vol. 178, pp. 125–134. Springer, Cham (2021). https://doi.org/10.1007/978-3-030-48279-4_12
4. Calabrò, F., Cassalia, G.: Territorial cohesion: evaluating the urban-rural linkage through the lens of public investments. In: Bisello, A., Vettorato, D., Laconte, P., Costa, S. (eds.) SSPCR 2017. GET, pp. 573–587. Springer, Cham (2018). https://doi.org/10.1007/978-3-319-75774-2_39
5. Lanucara, S., Praticò, S., Modica, G.: Harmonization and interoperable sharing of multi-temporal geospatial data of rural landscapes. In: Calabrò, F., Della Spina, L., Bevilacqua, C. (eds.) ISHT 2018. SIST, vol. 100, pp. 51–59. Springer, Cham (2019). https://doi.org/10.1007/978-3-319-92099-3_7
6. Tramontana, C., Calabrò, F., Cassalia, G., Rizzuto, M.C.: Economic sustainability in the management of archaeological sites: the case of Bova Marina (Reggio Calabria, Italy). In: Calabrò, F., Della Spina, L., Bevilacqua, C. (eds.) ISHT 2018. SIST, vol. 101, pp. 288–297. Springer, Cham (2019). https://doi.org/10.1007/978-3-319-92102-0_31
7. Calabrò, F., Iannone, L., Pellicanò, R.: The historical and environmental heritage for the attractiveness of cities. The case of the Umbertine Forts of Pentimele in Reggio Calabria, Italy. In: Bevilacqua, C., Calabrò, F., Della Spina, L. (eds.) NMP 2020. SIST, vol. 178, pp. 1990–2004. Springer, Cham (2021). https://doi.org/10.1007/978-3-030-48279-4_188
8. UNESCO: Convention concerning the protection of the World Cultural and Natural Heritage. In: Adopted by the General Conference at its 17th Session, Paris (1972)
9. UNESCO: Convention for the safeguarding of the intangible cultural heritage. In: Adopted by the General Conference at its 32nd Session, Paris (2003)
10. Council of Europe: Framework Convention on the Value of Cultural Heritage for Society, Council of Europe Treaty Series, vol. 199, Faro, Portugal (2003)

11. Campolo, D., Calabrò, F., Cassalia, G.: A cultural route on the trail of Greek monasticism in Calabria. In: Calabrò, F., Della Spina, L., Bevilacqua, C. (eds.) ISHT 2018. SIST, vol. 101, pp. 475–483. Springer, Cham (2019). https://doi.org/10.1007/978-3-319-92102-0_50

12. Campolo, D.: Greenways e velorail: una cultural route lungo gli insediamenti rupestri dei monaci greci in Calabria. LaborEst **16**(2018), 5–9 (2018). https://doi.org/10.19254/LaborEst.16.01

13. Spampinato, G., Malerba, A., Calabrò, F., Bernardo, C., Musarella, C.: Cork oak forest spatial valuation toward post carbon city by CO_2 sequestration. In: Bevilacqua, C., Calabrò, F., Della Spina, L. (eds.) NMP 2020. SIST, vol. 178, pp. 1321–1331. Springer, Cham (2021). https://doi.org/10.1007/978-3-030-48279-4_123

14. Barrile, V., Malerba, A., Fotia, A., Calabrò, F., Bernardo, C., Musarella, C.: Quarries renaturation by planting cork oaks and survey with UAV. In: Bevilacqua, C., Calabrò, F., Della Spina, L. (eds.) NMP 2020. SIST, vol. 178, pp. 1310–1320. Springer, Cham (2021). https://doi.org/10.1007/978-3-030-48279-4_122

15. Calabrò, F., Cassalia, G., Lorè, I.: The economic feasibility for valorization of cultural heritage. The restoration project of the reformed fathers' convent in Francavilla Angitola: the Zibìb Territorial Wine Cellar. In: Bevilacqua, C., Calabrò, F., Della Spina, L. (eds.) NMP 2020. SIST, vol. 178, pp. 1105–1115. Springer, Cham (2021). https://doi.org/10.1007/978-3-030-48279-4_103

16. Massimo, D.E.; Del Giudice, V.; De Paola, P.; Forte, F.; Musolino, M.; Malerba, A.: Geographically weighted regression for the post carbon city and real estate market analysis: a case study. In: Calabrò, F., Della Spina, L., Bevilacqua, C. (eds.) Smart Innovation, Systems and Technologies; New Metropolitan Perspectives. ISHT 2018, vol. 100, pp. 142–149. Springer, Cham (2018). https://doi.org/10.1007/978-3-319-92102-3_17

17. ICOMOS: The Paris Declaration on heritage as a driver of development. In: Adopted by the 17th General Assembly ICOMOS (2011)

18. European Parliament: Resolution (2014/2149(INI)), Towards an integrated approach to cultural heritage for Europe, Strasbourg, France (2015)

19. Calabrò, F.: Promoting peace through identity. Evaluation and participation in an enhancement experience of calabria's endogenous resources|Promuovere la pace attraverso le identità. Valutazione e partecipazione in un'esperienza di valorizzazione delle risorse endogene della Calabria. ArcHistoR, vol. 12(6), pp. 84–93 (2019). https://doi.org/10.14633/AHR146

20. Calabrò, F.: Integrated programming for the enhancement of minor historical centres. The SOSTEC model for the verification of the economic feasibility for the enhancement of unused public buildings|La programmazione integrata per la valorizzazione dei centri storici minori. Il Modello SOSTEC per la verifica della fattibilità economica per la valorizzazione degli immobili pubblici inutilizzati. ArcHistoR, vol. 13(7), pp. 1509–1523 (2020). https://doi.org/10.14633/AHR280

21. Mallamace, S., Calabrò, F., Meduri, T., Tramontana, C.: Unused real estate and enhancement of historic centers: legislative instruments and procedural ideas. In: Calabrò, F., Della Spina, L., Bevilacqua, C. (eds.) ISHT 2018. SIST, vol. 101, pp. 464–474. Springer, Cham (2019). https://doi.org/10.1007/978-3-319-92102-0_49

22. Ministero dei Beni e delle Attività Culturali e del Turismo: Piano Strategico di sviluppo del turismo 2017–2022. Italia Paese per Viaggiatori (2016)

23. Nesticò, A., Maselli, G.: Declining discount rate estimate in the long-term economic evaluation of environmental projects. J. Environ. Account. Manage. **8**(1), 93–110 (2020). https://doi.org/10.5890/JEAM.2020.03.007

Mobility as a Service (MaaS): Framework Definition of a Survey for Passengers' Behaviour

Giuseppe Musolino, Corrado Rindone$^{(\boxtimes)}$, and Antonino Vitetta

Dipartimento di Ingegneria dell'Informazione, Delle Infrastrutture e dell'Energia Sostenibile, Università Mediterranea di Reggio Calabria, Reggio Calabria, Italy
{giuseppe.musolino,corrado.rindone,vitetta}@unirc.it

Abstract. This paper proposes a survey framework that may be adopted in a Mobility as a Service (MaaS) context. MaaS is an integrated mobility system that considers the mobility needs of users as a central element of the transport service. Therefore, the mere definition of centralized and advanced technological systems is not enough to ensure the success of a MaaS. The paper analyses the main behavioural variables of transport users to be investigated and the methods to be adopted for the design of a survey to support the ex-ante analysis of a MaaS.

Keywords: MaaS · Transport system models · Demand analysis · User behaviour · Survey

1 Introduction

Transport and mobility are evolving towards the concept of Mobility as a Service (MaaS) [1, 2]. "MaaS is a user-centric, intelligent mobility management and distribution system, in which an integrator - the MaaS Operator - brings together offerings of multiple mobility service providers and provides end-users access to them through a digital interface, allowing them to seamlessly plan and pay for mobility" [3]. User's mobility needs are central in transport systems' planning and design. In this context, Transport System Models (TSMs) play a relevant role.

The analysis and evaluation of interventions on transport systems require, among others, knowledge about users' choices in the current and project scenarios.

The knowledge of users' choices may be carried out through existing information sources and statistical analysis. The direct estimation of users' choices starts from interviews of a sample of users, that may be supported by aggregate data (e.g. passengers' flows measured on some elements of the transport network). In addition, procedures for specification-calibration-validation of models can be used.

As far as concerns the current scenario, the surveys concern the detection of trip choices of the users, interviewed in a real or virtual context, and of the trip alternative chosen from a set of available alternatives.

As far as concerns the project scenarios, surveys may be carried out by observing users' choices in a real context, if the scenario is currently implemented, or in a virtual context, if the scenario is designed but not (yet) implemented. In the case of a real

F. Calabrò et al. (Eds.): NMP 2022, LNNS 482, pp. 324–333, 2022.
https://doi.org/10.1007/978-3-031-06825-6_31

scenario, users' choices are revealed (in literature they are defined as "revealed preferences"); in the case of the project scenario (real or virtual), users' choices are stated ("stated preferences") [4].

The detailed description of the methodologies for the survey and the estimation of travel demand is reported in [4–11].

In the surveys present in literature, MaaS scenarios are generally proposed to evaluate users' preferences with a view to customer satisfaction. These types of surveys are necessary but they are not sufficient. As matter of fact, MaaS decision-makers, planners and operators, need investigations concerning the revealed preferences of users to allow statistical analyses on users' behaviour and the specification-calibration-validation of demand models.

The main objectives of this paper are the following: i) characteristics of MaaS regarding the regulatory elements in Italy, the methodological elements for the analysis and evaluation of the transport system, the actors involved (Sect. 2); ii) consolidated survey methods concerning transport users' behaviour (Sect. 3); iii) main disaggregated and aggregated variables of the transport system to be detected for supporting the analysis of MaaS and for possible future specification-calibration-validation of demand models (Sect. 4).

According to the above objectives, the structure of the paper is the following. Section 2 describes the main characteristics of a MaaS. Section 3 presents some methods for data acquisition. Section 4 proposes a set of the data required in a MaaS survey. Some conclusions are reported in Sect. 5.

2 MaaS System

A Mobility as a Service (MaaS) system includes several actors, with different levels of involvement (Fig. 1): Local Public Authority, which is in charge of the political-administrative decisions, supporting the plan of local public transport services; Planner, who designs the transport services and supports the local public authority and the operators; MaaS Operator; MaaS transport and ICT companies (Operators); MaaS Users; other users of the transport system and citizens (not users of the transport system), who generally influence the choices of MaaS users.

2.1 Regulation

The supply of Local Public Transport (LPT) services in urban areas is regulated in Italy by specific regulations (see [12]). The local public authority, according to the mobility needs of citizens, quantifies and plans the supply of LPT services, that are classified into minimum and additional services. The supply of services is in charge of an operator, through the definition of a service contract. In recent years, due to the use of ICTs, other forms of transport services have been implemented (e.g., sharing mobility). ICT, transport and energy constitute the pillars of the Smart City according to EU [13].

The regulatory framework of LPT is different among the EU member states. For example, in Italy, the Regions and Local Authorities play an important role as they are responsible for programming services. Programming should take place on the basis of

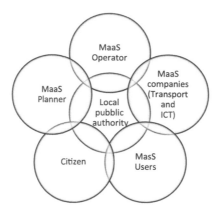

Fig. 1. Actors of the MaaS system.

users' mobility needs, who should drive the decisions concerning the organization, management, assignment and control. However, the lack of a precise regulatory framework, of consolidated financial resources are the main criticalities of the LPT in many Italian Regions. The limitations were more evident during the pandemic emergency of COVID-19. Due to the restrictions, tools for planning and programming services were necessary in order to adapt the supply of LPT services not only to the mobility needs but also to the constraints associated with social distancing.

The evolution of ICT, transport vehicles, Intelligent Transport Systems (ITS) and electric and shared mobility up to MaaS constitute opportunities to improve the supply of public transport services. However, methodological and modelling frameworks of the transport system are necessary to boost these opportunities.

2.2 Model

Decision Support Systems (DSSs) are commonly integrated inside the Intelligent Transportation Systems (ITSs), where historical and real-time information deriving from ICT interact with Transport System Models (TSMs). These tools allow, on the one hand, to monitor the conditions of the transport system (private and collective modes) and, on the other one, to predict the potential effects on mobility determined by measures concerning transport infrastructures and services [14, 15].

Modelling tools support management and control of transport services and, together with ICT tools, can be used to define and provide information for travelers in order to influence their behaviour [16]. Real-time information and data fed TSMs in order to verify and design dynamic optimal travel options of users involving a combination of different transport services [17, 18].

The necessity to analyze attitudinal and personality factors that users consider relevant for MaaS adoption is demonstrated in [19]. Then, it is important to identify the key dimensions of MaaS to highlight its possible implications [20, 21], its contributions to sustainability goals [22] and the services and prices integration [23].

TSMs and ICTs enable the possibility to design and implement MaaS solutions. As matter of fact, it is possible to analyse and model the mobility needs of people (and distribution of goods) and, on this basis, to design integrated transport services. The solutions offered should reduce the discontinuities between different transport modes (e.g., bus, bike,…) and within the same transport (e.g., coordinated time-tables). To this end, it is possible to use TSM to design transport infrastructures and services tailored to the users' mobility needs.

In a MaaS context, the design of private and collective transport services cannot therefore be separated from a correct analysis of the existing and potential travel demand both in the current state and in future scenarios.

Sample surveys are crucial to determine the characteristics of passenger mobility. The investigation concerning MaaS follows the structure of the traditional surveys carried out in the transport systems, but with some peculiarities. MaaS is not present in many cities, as some pilot experiences and some experiments showed. The MaaS analysts should propose to users a scenario that does not currently exist and that should be generated as design of existing services. It cannot be presented to the user as a completely new alternative, but as a combination of existing services proposed inside a bundle by means of ICT tools. The survey must be designed considering a new scenario, not known by the user, and generated by designing existing services.

3 Method and Survey Tools

The exact knowledge of the travel demand, in terms of trips of each user and related characteristics, requires the observation of (a sample) users in different time periods. The exact knowledge is not obtainable, nor it is necessary due to high costs and to the temporal changes of trip characteristics. Therefore, the exact knowledge would be inconsistent with the level of detail necessary for the analysis. For the above reasons, travel demand is estimated employing: disaggregated, or sample, surveys, that collect information on the travel choices of a sample of users (Sect. 3.1); aggregate surveys, that collect information on the socio-economic characteristics of an area, or aim to estimate traffic flows on selected links/nodes of the network, or on selected modes/services (Sect. 3.2). The surveys are supported by monitoring tools (Sect. 3.3).

3.1 Disaggregated Surveys

Disaggregated, or sample, surveys can be different: at home, at trip destination, along the trip, etc. The surveys detect the trip's characteristics of a sample of users. In the past, the survey was carried out by operators travelling on-board of transit vehicles (e.g. links), at terminals (e.g. nodes), or at home (e.g. at residence place).

The emerging ICTs allow today to monitor individual users while they undertake their trip in real scenarios (with the support of hardware and software systems), with semi-automatic procedures. Some characteristics of the trip that can be detected automatically (e.g. geo-location, transit service,…); other characteristics require a direct interaction with the user, as they reveal the individual perception of the service-mode used by each user, or they are associated to the individual user (e.g. trip purpose, desired departure or arrival times, socio-economic conditions, etc…).

Sampling. The survey requires to be univocally defined through the sampling unit (e.g. individual users); the extraction sampling method; the size of the sample. The sampling unit is, generally, the single user of the transport system, who must be monitored during the trip. The extraction sampling method can be the simple random sampling, the stratified random sampling, the cluster sampling. In relation to the size of the sample, each user can be observed during several days, thus carrying out a sample survey with reference to the choices made in different days (periods). The estimation depends on the sampling method, on the variable to estimate, and on statistical significance.

Attributes. Users undertake trip choices that can be modelled introducing the concept of "levels, or dimensions, of choice"; for example, whether to undertake a trip, for a given purpose and in a given time period, which destination, mode-service, path [6–9]. Each dimension of choice is defined through the perceived alternatives and the corresponding values of variables called attributes. As far as concerns the dimensions of choice commonly used for user choices, the most relevant attributes are listed below. The main attributes connected to the choice to undertake a trip for a given purpose and in a given time reference, are linked to the need or possibility to undertake the trip (e.g. the availability of cars), the trip frequency (e.g. accessibility from the origin to destinations). The main attributes connected to choice of the destination, given the purpose of the trip, depend on the attractiveness of the destination (e.g. number of shops in a zone), the transport disutility (cost) between the origin and the destination (e.g. travel time). The main attributes associated to the modal choice depend on the purpose of the trip, the level of service of the modal alternatives available (e.g. the travel time, segmented in different components), the socio-economic conditions of the user (e.g. working condition). The main attributes associated to services and routes depend on the purpose of the trip, the disutility (cost) of the available alternatives (e.g. travel time, monetary cost), the characteristics of common alternatives (e.g. common portions of paths, ….).

The above attributes are characteristics of the user, the trip, the trip alternatives (chosen and perceived), the study area. Therefore, for each user belonging to the sample in the observation periods and for each trip belonging to a trip chain, the attributes to be detected are divided into the following categories: user (level 1); trip chain (level 2); choice set of perceived alternatives (level 3.1), and chosen alternative (level 3.2).

A trip chain is characterized by a sequence of trips between an origin and a destination, with activities carried out between the trips, or with the modification of the purpose of trip [4].

3.2 Aggregate Surveys

Aggregate surveys aim to obtaining information of the study area, or of a portion of it. The information may be related to the socio-economic conditions and to flows of users and vehicles.

Socio-economic Conditions. They are indirectly representative of the mobility. For example, the number of resident population, the number of employees, the number of students in a zone influences the trip generation; while the number of workplaces, the number of school places, the number of shops in a zone influences the attraction of trips. This information can be collected during periodical census and they are present in databases (e.g. open data) provided by statistical institutes.

Flows. They may be directly measured on some elements of the network, or obtained from previous studies (e.g. passengers' flows in selected sections of transit lines; modal origin-destination matrices). Flows are essential for the study of the mobility of the sample of users, in order to capture general patterns of the mobility phenomena.

3.3   Monitoring Systems

A review of monitoring systems of vehicular traffic flow, where vehicles may be private (e.g. cars) or collective (e.g. buses), is reported below [24].

The monitoring systems of people (and goods) mobility may be subdivided into two broad categories (Fig. 2): vehicles' monitoring in a section (e.g. cars, buses): manual or automatic; peoples' (goods) monitoring along the trip.

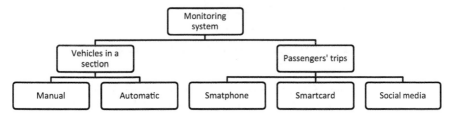

Fig. 2. Classification of monitoring systems.

Vehicle Monitoring Systems in a Section: Manual and Automatic. Manual systems are based on direct detection of flow performed by a human operator. Manual traffic measurement methods are generally used for short-term surveys. They are based on different methods: fixed observer; mobile observer. Automatic systems are based on detection through mechanical and technological systems. The main automatic systems are: pneumatic tubes; piezoelectric cables; inductive loops; magnetic-dynamic sensors; microwave sensors; infrared sensors; acoustic sensors; automatic image processing.

Peoples' (Goods) Monitoring Along the Trip (ICT). Big-data obtained, for example, from smartphones, smartcards, social media, make it possible today detailed observation of people's mobility for long periods. Despite the availability of such sources, the transport demand models commonly used are still almost exclusively built on conventional data, from travel diaries and population census. Existing research lines aim to combine the use of different data sources to increase the representative and predictive capabilities of demand models. These challenges require multidisciplinary collaboration between transport model developers and data scientists. The main systems adopted for measuring the passengers' trip are: smartphone; smartcard; GPS; points of interest.

4 Survey Design in a MaaS Context

The design and implementation of a MaaS implies the execution of: disaggregated surveys [25, 26] to estimate the mobility needs of individual users, or groups of them; aggregated surveys, to estimate the performance of a transport system (i.e. measuring users' flow through an element of the network).

The attributes to be observed in order to know the mobility and the users' choices of are described in this section, according to the tree diagram of Fig. 3.

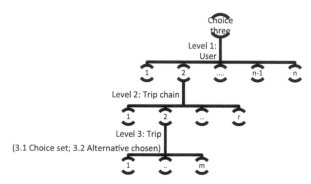

Fig. 3. Scheme of levels of data related to survey.

In Table 1, the characteristics of the user (Level 1) to detect are reported in the left column; while the characteristics of the trip chain (Level 2) and of each trip of chain within each day, which are independent from the alternative chosen by the user, are reported in the right column.

Table 1. User and trip chain characteristics.

For each user (Level 1)	For each single trip in the trip chain in the day (Level 2)
• Age • Occupation • Number of family members • Number of cars, bikes, …, owned by the family • Type of season ticket • …	• Progressive of the trip • Place of origin • Place of destination • Predominant reason or scope • Departure time constrained (and eventually time) • Arrival time constrained (and eventually time) • …

In Table 2, the characteristics of each perceived trip alternative and not chosen (Level 3.1), given the trip-chain and the single trip of the chain within each day, are reported in the left column; while he characteristics of the chosen trip alternative (Level 3.2), given the trip chain and the single trip of the chain within each day, are reported in the right column.

Table 2. Chosen alternative and choice set of the perceived alternatives in the trip chain.

For each single trip in the trip chain in the day, for the perceived alternatives in each trip (Level 3.1)	*For the chosen alternative in each trip (Level 3.2)*
• Progressive of the trip	• Progressive of the trip
• Number of alternatives perceived in each trip (For each perceived alternative)	
• Number of modes / services used (For each mode / service, read the following data)	• Number of modes / services used (For each mode / service, read the following data)
• Progressive of the mode / service used	• Progressive of the mode / service used
• Way / service	• Way / service
• Service included in the MaaS subscription	• Service included in the MaaS subscription
• If the service included, Type of fare	• If the service included, Type of fare
• Criteria considered in the choice of the alternative	• Criteria considered in the choice of the alternative
• Waiting time	• Waiting time
• Number of transfers in the same service	• Number of transfers in the same service
• Board time	• Board time
• …	• …

Aggregate surveys allow to collect (see Table 3): socio-economic data, for each traffic zone and for homogeneous groups of users (i.e. number of residents); traffic flow data (i.e. traffic flow, density, speed), for each element of the transport network (node, link, …). The aggregated data can be used for an update of the transport demand in terms of values or parameters of the models for estimating user choice behaviour.

Table 3. Aggregate data.

For homogenous group of users	For each element of supply element (link or node)
• Number of residents in the neighbourhood of origin	• Vehicular flow on a road section
• Number of active neighbourhoods of origin	• Users flow on a transit line
• Number of employees in the neighbourhood of origin	• Users flows at a bus stop
• Number of residents in the destination district	• Number of trips of a vehicle belonging to a sharing service
• Number of Active target district	• Number of vehicles parked in an area
• Number of employees in the destination district	• ……
• …	

5   Conclusions

The paper presents a survey framework for the detection of passengers' behaviour in a MaaS context. The one presented is a preliminary definition before the execution of the survey operations. The framework must be further specified to identify variables connected to the study area analysed, such as the definition of existing or project scenarios [27].

The data obtained are necessary for the analysis and evaluation steps. The analysis and the evaluation must be carried out with the support of TSMs, to quantify the variation of choices of users' groups in relation to the sample structure (i.e. the potential changes of travel choices of users related to private, collective and shared transport modes).

The proposed framework requires to be validated in an experimentation test site in the next future, also considering the trial and error process of specification-calibration-validation of demand models. Once the framework has been defined and validated, it is possible to analyze the variations of users' choices in the project scenarios in relation to a MaaS.

References

1. CIVITAS: Mobility-as-a-Service: a new transport model (2016). http://civitas.eu/content/civitas-insight-18-mobility-service-new-transport-model
2. Kamargianni, M., Yfantis, L., Muscat, J., Azevedo, C., Ben-Akiva, M.: Incorporating the mobility as a Service concept into transport modelling and simulation frameworks. In: MaaSLab Working Paper Series Paper No. 18-05 (2018)
3. ERTICO: Mobility as a Service (MaaS) and Sustainable Urban Mobility Planning. ITS Europe (editor). https://www.eltis.org/sites/default/files/mobility_as_a_service_maas_and_sustainable_urban_mobility_planning.pdf. Accessed 8 Dec 2021
4. Cascetta, E.: Transportation Systems Engineering: Theory and Methods. Springer, New York (2009)
5. Ben-Akiva, M., Lerman, S.: Discrete Choice Analysis: Theory and Application to Travel Demand. MIT Press, Cambridge (1985)
6. Birgillito, G., Rindone, C., Vitetta, A.: Passenger mobility in a discontinuous space: modelling access/egress to maritime barrier in a case study. J. Adv. Transp. **2018**, 1–13 (2018)
7. Croce, A., Musolino, G., Rindone, C., Vitetta, A.: Route and Path Choices of Freight Vehicles: A Case Study with Floating Car Data. Sustainability (2020)
8. Musolino, G., Rindone, C., Vitetta, A.: A modelling framework to simulate paths and routes choices of freight vehicles in sub-urban areas. In: 2021 7th International Conference on Models and Technologies for Intelligent Transportation Systems, MT-ITS 2021 (2021)
9. Croce, A., Musolino, G., Rindone, C., Vitetta, A.: Estimation of travel demand models with limited information: floating car data for parameters' calibration. Sustainability **13**, 8838 (2021)
10. Ortuzar, J.W.: Modelling Transport, 3rd edn. Wiley, Chichester (2001)
11. Di Gangi, M., Vitetta, A.: Quantum utility and random utility model for path choice. J. Choice Model. **40**, 100290 (2021)
12. Decreto Legislativo 19 novembre 1997, n. 422. "Conferimento alle regioni ed agli enti locali di funzioni e compiti in materia di trasporto pubblico locale, a norma dell'articolo 4, comma 4, della legge 15 marzo 1997, n. 59" pubblicato nella Gazzetta Ufficiale n. 287 del 10 dicembre (1997)
13. Russo, F., Rindone, C., Panuccio, P.: The process of smart city definition at an EU level. WIT Trans. Ecol. Environ. **191**, 979–989 (2014)
14. Anda, C., Fourie, P., Erath, A.: Transport Modelling in the Age of Big Data'. Work Report. Future Cities Laboratory, Singapore ETH Center (2016)

15. Lee, R.J., Sener, I.N., Mullins, J.A.: Emerging Data Collection Techniques for Travel Demand Modeling: A Literature Review Final Report. Texas A&M Transportation Institute (2014)
16. Nuzzolo, A., Comi, A.: Advanced public transport systems and ITS: new tools for operations control and traveller advising. In: IEEE Proceedings of the 17th International IEEE Conference on Intelligent Transportation Systems, pp. 2549–2555. IEEE (2014). https://doi.org/10.1109/ITSC.2014.6958098
17. Comi, A., Nuzzolo, A., Crisalli, U., Rosati, L.: A new generation of individual real-time transit information systems. In: Nuzzolo, A., Lam, W.H.K. (eds.) Modelling Intelligent Multi-Modal Transit Systems, pp. 80–107. CRC Press, Taylor & Francis Group, Boca Raton (FL, USA) (2017)
18. Nuzzolo, A., Comi, A.: Dynamic optimal travel strategies in intelligent stochastic transit networks. Information 12(7), 281 (2021). https://doi.org/10.3390/info12070281
19. Lopez-Carreiro, I., Monzon, A., Lois, D., Lopez-Lambas, M.E.: Are travellers willing to adopt MaaS? Exploring attitudinal and personality factors in the case of Madrid, Spain. Travel Behav. Soc. 25, 246–261 (2021)
20. Kim, S., Choo, S., Choi, S., Lee, H.: What factors affect commuters' utility of choosing mobility as a service? An empirical evidence from Seoul. Sustainability (Switzerland) 13(16), 9324 (2021)
21. Alyavina, E., Nikitas, A., Njoya, E.T.: Mobility as a service (MaaS): a thematic map of challenges and opportunities. Res. Transp. Bus. Manage. 100783 (2022, in press)
22. Hensher, D.A., Mulley, C., Nelson, J.D.: Mobility as a service (MaaS) – going somewhere or nowhere? Transp. Policy 111, 153–156 (2021)
23. Hensher, D.A., Mulley, C.: Mobility as a service (MaaS): charting a future context. Transp. Res. Part A Policy Pract. 131, 5–19 (2020)
24. Italian Government. Ministry of Transport and Infrastructures: Sistemi di monitoraggio del traffico linee guida per la progettazione (2000)
25. Ho, C.Q., Hensher, D.A., Mulley, C., Wong, Y.Z.: Potential uptake and willingness-to-pay for Mobility as a Service (MaaS): a stated choice study. Transp. Res. Part A Policy Pract. 117, 302–318 (2018)
26. Alonso-Gonzáleza, M.J., Hoogendoorn-Lanser, S., van Oort, N., Cats, O., Hoogendoorn, S.: Drivers and barriers in adopting Mobility as a Service (MaaS) – a latent class cluster analysis of attitudes. Transp. Res. Part A Policy Pract. 132, 302–318 (2020)
27. Musolino, G., Rindone, C., Vitetta, A.: Models for supporting Mobility as a Service (MaaS) design. Smart Cities 2022(5), 206–222 (2022)

Evaluation of the Structural Health Conditions of Smart Roads Using Different Feature-Based Methods

Rosario Fedele[1]([✉]) [iD], Filippo Giammaria Praticò[1] [iD], Giuseppe Cogliandro[1] [iD], and Filippo Laganà[2] [iD]

[1] Department of Information Engineering, Infrastructures and Sustainable Energy (DIIES), Mediterranea University of Reggio Calabria, Via Graziella - Feo di Vito, snc, 89122 Reggio Calabria, Italy
rosario.fedele@unirc.it

[2] Department of Civil Engineering, Energy, Environment and Materials (DICEAM), Mediterranea University of Reggio Calabria, Via Graziella - Feo di Vito, snc, 89122 Reggio Calabria, Italy

Abstract. Flexible road pavements are damaged by traffic and environmental conditions. At the same time, traditional structural health assessment and monitoring methods are often destructive and expensive. Consequently, the main objective of this study is to further develop innovative non-destructive structural health monitoring method for road pavements, which aims at identifying the presence of damages into road sections based on vibro-acoustic signature analysis. For this reason, an experimental investigation was carried out using two non-destructive monitoring systems, i.e. one traditional (based on Falling Weight Deflectometer, FWD) and one innovative (based on the aforementioned innovative method). In particular, the traditional one was used to estimate the elastic moduli of damaged and undamaged road sections, and to produce impulsive loads that, in turn, produced vibration and noise detected by the sensors of the innovative system. Signals were recorded and processed. Then, two feature-based data analysis methods (i.e., one already used in the past, herein called method 1, and one used for this application for the first time, herein called method 2) were used on the signals recorded by the innovative system mentioned above. Results show that method 2 (i.e., the chromatic method) allows obtaining better results than method 1. Indeed, thanks to both 2D and 3D graphical plots, it allows improving the ability of the innovative system to discriminate damaged road pavements from undamaged ones. This allows obtaining a solution that can be easily scaled and used in current and future smart roads.

Keywords: Smart road · Non-destructive pavement structural health monitoring · Feature-based data analysis comparison

1 Introduction

Traffic loads and environmental conditions induce stresses, strains and vibrations into pavement layers. These effects lead to the generation and propagation of several types

© The Author(s), under exclusive license to Springer Nature Switzerland AG 2022
F. Calabrò et al. (Eds.): NMP 2022, LNNS 482, pp. 334–345, 2022.
https://doi.org/10.1007/978-3-031-06825-6_32

of cracks (both surface and internal) until the failure of the road pavement occurs [1], and maintenance interventions are needed [2, 3].

In order to manage these issues, the smart city paradigm requires Intelligent Transportation Systems (ITSs), which are the combination of Internet of Things (IoT) and Information and Communication Technologies (ICTs) [4].

A smart road [5] should be able to carry out (i) active perception (i.e., collect data environmental- and traffic-related data), (ii) self-adaptation (i.e., automatic data analysis and discrimination), (iii) automatic dynamic interaction between road-to-everything, and (iv) self-powering (i.e., use self-produced energy to power itself). These objectives can be obtained using [5] advanced structural materials, smart sensor networks, information centres, and energy systems able to provide continuous energy supply. A smart road can be thought of as a "five-zero" system [5], i.e., a system where there are zero casualties, zero delays, zero maintenance, zero emissions, and zero failure. Recently [6], the concept of IoT was extended by introducing the concept of "Internet of Everything (IoE)", where information sharing between devices (IoT) was upscaled by considering Internet, devices and casual data related to road user behaviour. Sun et al. (2018) [5] reported an example of pilot smart road developed in China, which includes road that (1) are able to carry out self-healing, and energy production by mean of photovoltaic pavements (which have the ultimate goal of charging Electric Vehicles, EVs, while they are driving); (2) are equipped with optical fibres, with power cables, with sensors (temperature, traffic flow, and axle load monitoring), and self-snow melting and subgrade humidity self-regulating systems.

Both destructive and non-destructive structural health monitoring methods are currently used, and a multitude of innovative solutions are available in the literature (cf. Cafiso et al. (2020) [7]).

In the study reported in this paper, in order to further improve the efficiency of the feature-based and vibro-acoustic-based method described in [8–11], a literature review has been carried out, and the "Chromatic Clustering technique" was identified. This latter is based on three features (cf. Sect. 3), which are (i) extracted from the spectrum of the signals, (ii) are correlated each other (see equation in the Sect. 3) allowing obtain two- and three-dimensional (i.e., 2D and 3D) plots that can be very useful for clustering purposes in the method presented by Praticò et al. [6–9]. This application of the Chromatic Clustering technique to the field of transportation is innovative. Based on the above, the main objective of the study presented in this paper is to compare the efficiency of two different feature-based techniques used to analyse the vibro-acoustic signatures of four road sections that are characterized by two structural health statuses (herein called un-cracked, UC, and cracked, C) assessed through a traditional non-destructive monitoring system (i.e., a Falling Weight Deflectometer, FWD). In order to achieve the objectives mentioned above, the following tasks were carried out: Task 1: Experimental investigation description; Task 2: Methods presentation; Task 3: Method application: results and discussions; Task 4: Methods comparison: results and discussions.

Finally, it is important to underline that this research addresses some of the objectives of the on-going Italian project USR342-PRIN 2017–2022, the on-going project "SNEAK" (LIFE20 ENV/IT/000181 - optimized Surfaces against NoisE And vibrations produced by tramway tracK and road traffic), and was partially funded by the PAC Calabria 2014–2020.

2   Experimental Investigation (Task 1)

An experimental investigation on four sections (herein called Section 1–4) of an urban road in Catania, was carried out [11]. Table 1 reports the value of the elastic modulus (E_1; MPa) of the four sections for each load level (target load in N, herein called L1–L4). Based on results (cf. Table 1), Ground Penetrating Radar investigations, and visual inspections, the four section were classified as un-cracked (i.e., section 1 and 4), and cracked (i.e., section 2 and 3). At the same time, the microphone-based system mentioned above was used to gather the response of the road pavement to the loads produced by the FWD, herein called vibro-acoustic signature of the road section. Figure 1 shows: 1) one signal related to one of the un-cracked section (i.e., section 1; cf. Fig. 1.a), the related spectrum (FFT of the signal in Fig. 1.a; cf. Fig. 1.c), and the related power spectral density (PSD spectrum; cf. Fig. 1.e); 2) one signal related to one of the cracked section (i.e., section 2; cf. Fig. 2.b), the related spectrum (cf. Fig. 1.d), and the related PSD spectrum (cf. Fig. 1.f).

Table 1. Elastic modulus (E_1; MPa) from FWD measurements (cf. [9])

Sections	L1	L2	L3	L4
Section 1	1783	1815	2028	2335
Section 2	795	751	868	1216
Section 3	N/A	1246	1246	1257
Section 4	1301	1251	1365	1491

Legend. L1 = 40 kN; L2 = 56 kN; L3 = 84 kN; L4 = 120 kN; N/A = Not available.

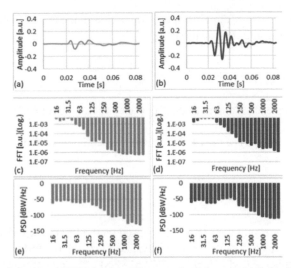

Fig. 1. Example of signals and PSD: a) signal from section 1 (un-cracked); b) signal from section 2 (cracked); c) PSD from section 1 (un-cracked); d) PSD from section 2 (cracked).

3 Methods Presentation (Task 2)

As stated above, two data analysis methods are here applied, i.e., the method proposed by Praticò et al. [8–10], herein called "Method 1", and the "Chromatic Clustering technique" [12–14], herein called "Method 2". In more details, Method 1 is based on several features extracted from time, frequency, and time-frequency domains of analysis. In this study, the following features were selected according with [11, 15]:

1. Feature 1 (herein called F1; a.k.a., the spectral centroid of the PSD spectrum in the frequency range $16 \div 2500$ Hz. This parameter aims at representing the "centre of mass" of a spectrum according to [16]. It is measured in Hz). F1 is expressed as follows:

$$f_c = \frac{\sum_{n=0}^{N-1} P_n \cdot f_n}{\sum_{n=0}^{N-1} P_n} \tag{1}$$

2. Feature 2 (herein called F2; it is the slope of the linear regression model applied on the PSD spectrum (W/Hz) in the range $16 \div 2500$ Hz. It is measured in dBW/Hz^2);
3. Feature 3 (herein called F3; it is the maximum of the PSD spectrum in $16 \div 2500$ Hz. It is expressed in dBW/Hz).

   Method 2 is based on the following features [12–14]:

4. the Energy of the FFT spectrum (E), which is measured in Joule [12–14], and is calculated as follows:

$$E = \frac{1}{2\pi} \sum_{n=0}^{N-1} f_n \cdot |FFT(sign)|^2, \tag{2}$$

where *FFT(sign)* stands for the Fourier Transform of discrete time signal, N is the number of samples (i.e., the signal length) of the discrete time signal.

- the average band (ω_c) of the FFT spectrum, which is measured in Hz [12–14], and can be calculated using the equation:

$$\omega_c = \frac{\sum_{n=0}^{N-1} f_n \cdot |FFT(\text{sign})|^2}{2\pi \cdot E}, \tag{3}$$

where f_n represents each element of the frequency vector calculated using the FFT.

- the RMS bandwidth (B) of the FFT spectrum, which is measured in Hz [12–14], and is defined as:

$$B = \sqrt{\frac{1}{E} \sum_{n=0}^{N-1} (f_n - \omega_c)^2 |Y_n|^2}. \tag{4}$$

Because of the fact that the Method 1 is based on the PSD, it was decided to use the "Chromatic Clustering technique" in two ways, i.e. (1) using the PSD instead of the FFT in the Eqs. 2–4 (this method is herein called Method 2*), and (2) using the FFT in the Eqs. 2–4 (this method is herein called Method 2). Consequently, the following features were derived applying the Method 2*:

4. Feature 4 (herein called F4; is expressed by Eq. 2 using the PSD);
5. Feature 5 (herein called F5; is expressed by Eq. 3 using the PSD);
6. Feature 6 (herein called F6; is expressed by Eq. 4 using the PSD);

while, other three features were derived applying the Method 2:

7. Feature 7 (herein called F7; is expressed by Eq. 2 using the FFT);
8. Feature 8 (herein called F8; is expressed by Eq. 3 using the FFT);
9. Feature 9 (herein called F9; is expressed by Eq. 4 using the FFT).

4   Method Application: Results (Task 3)

This section contains the results of the application of the methods 1, 2* and 2 (see Tables 2, 3 and 4). In particular, these tables report the average values of the nine features (F1–F9) for each of the four loads condition (L1–L4) generated by the FWD, and the best regression models (i.e., those with highest R-square values) derived from the relationships elastic modulus (E_1)-feature (F1–F9) from L1 to L4.

Table 2. Method 1: average features and best equations based on the best R^2.

Sections	Features	L1	L2	L3	L4
Section 1	F1	660	675	685	700
Section 2	F1	648	632	634	638
Section 3	F1	621	645	657	666
Section 4	F1	654	668	684	694
Section 1	F2	−3.2	−3.3	−3.3	−3.4
Section 2	F2	−2.6	−2.2	−2.1	−2.1
Section 3	F2	−2.3	−2.8	−2.9	−2.9
Section 4	F2	−3.0	−3.1	−3.2	−3.2
Section 1	F3	56	56	56	57
Section 2	F3	208	219	213	204
Section 3	F3	238	238	238	258
Section 4	F3	47	78	47	47
Relation	Eq. #	L1	L2	L3	L4
E_1 vs. F1	(1)–(4)	$E_1 = 8{\cdot}F1 - 3695$	$E_1 = 6{\cdot}e^{0.008F1}$	$E_1 = 13{\cdot}e^{0.007F1}$	$E_1 = 6{\cdot}10^{-9}{\cdot}F1^{4.05}$
	R^2	0.11	0.48	0.44	0.33
E_1 vs. F2	(5)–(8)	$E_1 = -559{\cdot}F2 - 210$	$E_1 = 373{\cdot}e^{-0.425F2}$	$E_1 = 545{\cdot}e^{-0.330F2}$	$E_1 = 886{\cdot}e^{-0.237F2}$
	R^2	0.30	0.58	0.44	0.25
E_1 vs. F3	(9)–(12)	$E_1 = -416.5{\cdot}\ln(F3) + 3284$	$E_1 = -386{\cdot}\ln(F3) + 3079$	$E_1 = 1880{\cdot}e^{-0.002F3}$	$E_1 = 2304{\cdot}e^{-0.002F3}$
	R^2	0.45	0.47	0.43	0.40

Legend. E1: modulus. F1 = Spectral centroid of the PSD of the signal in the range 16–2500 Hz [Hz]; F2 = Slope of the linear regression model of the signal PSD in the range 16–2500 Hz [dBW/Hz^2]; F3 = Maximum value of the PSD of the signal in the range 16–2500 Hz [dBW/Hz]; Method 1: Features 1–3; Other symbols (see Tabs above)

Table 3. Method 2*: average features and best equations based on the best R^2.

Sections	Features	L1	L2	L3	L4
Section 1	F4	25308	24041	21798	20041
Section 2	F4	19927	16994	15438	13963
Section 3	F4	22877	23248	21757	20232
Section 4	F4	21896	22381	21147	19209
Section 1	F5	910	937	957	991
Section 2	F5	888	852	858	873
Section 3	F5	823	879	908	927
Section 4	F5	858	908	956	964
Section 1	F6	2023	2013	2019	2039
Section 2	F6	2064	2061	2061	2079
Section 3	F6	2028	2050	2065	2078
Section 4	F6	2011	2021	2039	2032
Relation	Eq. #	L1	L2	L3	L4
E_1 vs. F4	(13)–(15)	$E_1 = 2 \cdot 10^{-5} \cdot F4^{1.78}$	$E_1 = 6 \cdot 10^{-4} \cdot e^{1.46F4}$	$E_1 = 4 \cdot 10^{-2} \cdot F4^{1.1}$	$E_1 = 9 \cdot F4^{0.55}$
	R^2	0.45	0.63	0.41	0.13
E_1 vs. F5	(16)–(19)	$E_1 = 4 \cdot F5 - 2494$	$E_1 = 31 \cdot e^{0.041F5}$	$E_1 = 58 \cdot e^{0.003F5}$	$E_1 = 49 \cdot e^{0.004F5}$
	R^2	0.12	0.44	0.41	0.40
E_1 vs. F6	(20)–(23)	$E_1 = 53903 \cdot e^{-0.002F6}$	$E_1 = -8942 \cdot \ln(F6) + 69426$	$E_1 = -7436 \cdot \ln(F6) + 58159$	$E_1 = 2 \cdot 10^{10} \cdot F6^{-2.14}$
	R^2	0.07	0.21	0.15	0.06

Legend. E1: modulus. F4 = Energy of the PSD of the signal in the range 16–2500 Hz [Joule]; F5 = average band of the signal PSD in the range 16–2500 Hz [Hz]; F6 = RMS bandwidth of the PSD of the signal in the range 16–2500 Hz [Hz]; Method 2*: Features 4–6; Other symbols (see Tabs above).

Table 4. Method 2: average features and best equations based on the best R^2.

Sections	Features	L1	L2	L3	L4
Section 1	F7	$8.8 \cdot 10^{-6}$	$2.1 \cdot 10^{-5}$	$7.0 \cdot 10^{-3}$	$1.4 \cdot 10^{-4}$
Section 2	F7	$1.6 \cdot 10^{-4}$	$3.2 \cdot 10^{-4}$	$1.8 \cdot 10^{-2}$	$5.5 \cdot 10^{-4}$
Section 3	F7	$1.8 \cdot 10^{-5}$	$3.3 \cdot 10^{-5}$	$5.7 \cdot 10^{-5}$	$1.1 \cdot 10^{-4}$
Section 4	F7	$2.1 \cdot 10^{-5}$	$4.6 \cdot 10^{-5}$	$1.0 \cdot 10^{-4}$	$2.0 \cdot 10^{-4}$
Section 1	F8	27	26	30	22
Section 2	F8	37	38	45	40
Section 3	F8	35	36	36	38
Section 4	F8	24	23	21	21
Section 1	F9	23	23	73	23
Section 2	F9	23	24	115	30
Section 3	F9	33	33	35	38
Section 4	F9	22	20	18	19
Relation	Eq. #	L1	L2	L3	L4
E_1 vs. F7	(24)–(27)	$E_1 = 102 \cdot F7^{-0.24}$	$E_1 = 82 \cdot F7^{-0.28}$	$E_1 = 1713 \cdot e^{-24.25F7}$	$E_1 = 2201 \cdot e^{-750.4F7}$
	R^2	0.66	0.82	0.26	0.28
E_1 vs. F8	(28)–(31)	$E_1 = 4881 \cdot e^{-0.043F8}$	$E_1 = 4224 \cdot e^{-0.04F8}$	$E_1 = 3663 \cdot e^{-0.028F8}$	$E_1 = 3671 \cdot e^{-0.03F8}$
	R^2	0.37	0.48	0.49	0.43
E_1 vs. F9	(32)–(35)	$E_1 = 1891 \cdot \ln(F9) - 4521$	$E_1 = -477 \cdot \ln(F9) + 2830$	$E_1 = 1799 \cdot e^{-0.003F9}$	$E_1 = 3146 \cdot e^{-0.02F9}$
	R^2	0.09	0.02	0.15	0.17

Legend. E1: modulus. F7 = Energy of the FFT of the signal in the range 16–2500 Hz [Joule]; F8 = average band of the FFT of the signal in the range 16–2500 Hz [Hz]; F9 = RMS bandwidth of the FFT of the signal in the range 16–2500 Hz [Hz]; Method 2*: Features 7–9; Other symbols (see Tabs above).

Table 5 reports the comparison between the three methods based on the best R^2, where symbols based on the average value of the best R^2 were used to define the efficiency of each method. Based on Table 5, it is possible to state that: (1) F3 (Method 1), F4 (Method 2*), F7 (Method 2), and F8 (Method 2) which express the maximum of the PDS-spectra, the energy of the FFT- and PSD-spectra, and the average band of the FFT of the signal in a given frequency range, respectively, provided the best correlations. (2) Methods 1 and 2 provided results better than the ones of Method 2*, confirming that the "Chromatic Clustering technique" needs the features extracted from the FFT spectra (instead of the ones derived using PSD). (3) Poor correlations were obtained using the features F6 and F9.

Table 5. Comparisons among the three methods based on the best R^2.

Load	Method 1			Method 2*			Method 2		
	F1	F2	F3	F4	F5	F6	F7	F8	F9
L1	0.11	0.30	0.45	0.45	0.12	0.07	0.66	0.37	0.09
L2	0.48	0.58	0.47	0.63	0.44	0.21	0.82	0.48	0.02
L3	0.44	0.44	0.43	0.41	0.41	0.15	0.26	0.49	0.15
L4	0.33	0.25	0.40	0.13	0.40	0.06	0.28	0.43	0.17
R^2_{AV}	0.34	0.39	0.44	0.41	0.34	0.12	0.51	0.44	0.11
	😐	😐	☺	☺	😐	☹	☺	☺	☹

Symbols. ☺ = R2 >0.4; 😐 = 0.2≤ R2 ≤0.4; ☹ = R2 < 0.2. Other symbols (see Tables above).

Figure 2 refers to feature-to-modulus relationships as a tool to isolate un-cracked, UC, from cracked, C, sections). In more detail, it is possible to see that F3 (Method 1) and F8 (Method 2) allow obtaining good discrimination between the two structural conditions under analysis, confirming the results in Table 5. Finally, low R^2 values can be attributed to a limited number of measurements and sections under analysis.

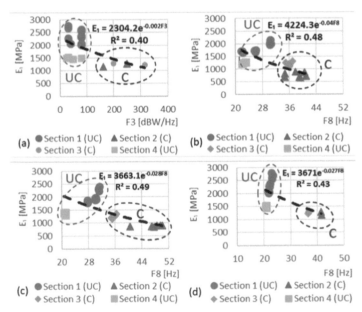

Fig. 2. Best results (clustering) obtained using: a) F3 (Method 1, load condition, LC = 4); b) F8 (Method 2, LC = 2); c) F8 (Method 2, LC = 3); d) F8 (Method 2, LC = 4).

Figure 3 reports the best results obtained using the "Chromatic Clustering technique" using 2D and 3D plots. Importantly, it was not possible to obtain clear clusters using the

Method 2*, while the Method 2 allows obtaining the results in Fig. 3. In summarizing, Method 2 provided better results than Method 1 and Method 2*. Consequently, it can be used to improve the efficiency of the vibro-acoustic-based structural health monitoring method [6–9] and its practical applications to road engineering [17].

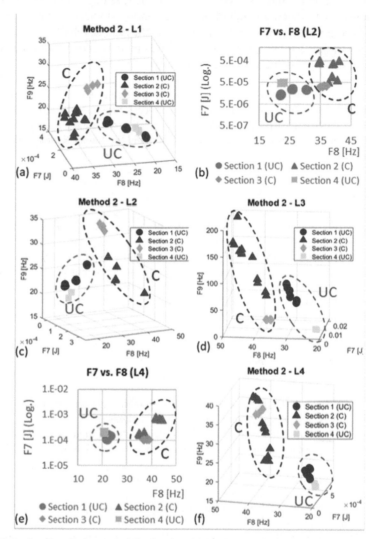

Fig. 3. Plot related to the best results related to the Method 2: a) Load condition 1 (L1), 3D plot; b) L2, 2D plot; c) L2, 3D plot; d) L3, 3D plot; e) L4, 2D plot; e) L4, 3D plot.

5   Conclusions

The assessment and monitoring of the structural health conditions of road infrastructures is a crucial task in road engineering. This study aimed at comparing the efficiency of different feature-based techniques used to analyse the vibro-acoustic signatures of road sections. Based on results, the Method 2 (i.e., Chromatic Clustering technique) provide better results in terms of discrimination between UC and C conditions. Consequently, it is possible to state that the Method 2 can be used to improve the efficiency of the Method 1. Future study will be carried out for: i) strengthening the database of signals recorded with a new version of the aforementioned innovative non-destructive method. ii) Discriminating the signals and therefore the type of vehicles that pass on the road infrastructure. iii) Implementing mathematical models capable of predicting the occurrence of structural damage to road pavements.

Funding. The authors would like to thank all who sustained them with this research, which address some of the objectives of the on-going Italian project USR342-PRIN 2017–2022, and the on-going project "SNEAK" (LIFE20 ENV/IT/000181 - optimized Surfaces against NoisE And vibrations produced by tramway tracK and road traffic), and the Italian Region Calabria (PAC Calabria 2014–2020).

References

1. Zhang, J., Yang, S., Li, S., Ding, H., Lu, Y., Si, C.: Study on crack propagation path of asphalt pavement under vehicle-road coupled vibration. Appl. Math. Mod. **101**, 481–502 (2022)
2. Bosurgi, G., Bruneo, D., De Vita, F., Pellegrino, O., Sollazzo, G.: A web platform for the management of road survey and maintenance information: a preliminary step towards smart road management systems. Struct. Control. Health Monitoring **29**(3), e2905 (2022)
3. Bosurgi, G., Pellegrino, O., Sollazzo, G.: Pavement condition information modelling in an I-BIM environment. Int. J. Pavement Eng. 1–16 (2021). https://www.tandfonline.com/doi/full/10.1080/10298436.2021.1978442?journalCode=gpav20
4. Pop, M.-D., Proştean, O.: A comparison between smart city approaches in road traffic management. Procedia Soc. Behav. Sci. **238**, 29–36 (2018)
5. Sun, L., Zhao, H., Tu, H., Tian, Y.: The smart road: practice and concept. Engineering **4**, 436–437 (2018)
6. Trubia, S., Severino, A., Curto, S., Arena, F., Pau, G.: Smart roads: an overview of what future mobility will look like. Infrastructures (Switz.) **5**, 1–12 (2020)
7. Di Graziano, A., Marchetta, V., Cafiso, S.D.: Structural health monitoring of asphalt pavements using smart sensor networks: a comprehensive review. J. Traffic Transp. Eng. (Engl. Ed.) **7**, 639–651 (2020)
8. Fedele, R., Praticò, F.G.: Monitoring infrastructure asset through its acoustic signature. In: INTER-NOISE 2019, Madrid, Spain (2019)
9. Fedele, R.: Smart road infrastructures through vibro-acoustic signature analyses. In: New Metropolitan Perspectives 2020, Reggio Calabria, Italy (2020)
10. Praticò, F.G., Fedele, R., Naumov, V., Sauer, T.: Detection and monitoring of bottom-up cracks in road pavement using a machine-learning approach. Algorithms (Switz.) **13**, 81 (2020)
11. Cafiso, S., Di Graziano, A., Fedele, R., Marchetta, V., Praticò, F.: Sensor-based pavement diagnostic using acoustic signature for moduli estimation. Int. J. Pavement Res. Technol. **13**(6), 573–580 (2020). https://doi.org/10.1007/s42947-020-6007-4
12. Dos Santos Junior, M.M., de Castro, B.A., Ardila-Rey, J.A., de Souza Campos, F., de Medeiros, M.I.M., Ulson, J.A.C.: A new acoustic-based approach for assessing induced adulteration in bovine milk. Sensors (Switz.) **21**(6), 2101 (2021)

13. Wang, X., Li, X., Rong, M., Xie, D., Ding, D., Wang, Z.: UHF signal processing and pattern recognition of partial discharge in gas-insulated switchgear using chromatic methodology. Sensors (Switz.) **17**, 177 (2017)
14. Ardila-Rey, J.A., et al.: A comparison of inductive sensors in the characterization of partial discharges and electrical noise using the chromatic technique. Sensors (Switz.) **18**, 1021 (2018)
15. Di Graziano, A., Cafiso, S.D., Severino, A., Praticò, F.G., Fedele, R., Pellicano, G.: Using non-destructive test to validate and calibrate smart sensors for urban pavement monitoring. In: BCRRA 2021 (2021)
16. Schubert, E., Wolfe, J.: Timbral brightness and spectral centroid. Acta Acust. Acust. **92**, 820–825 (2006)
17. Praticò, F.G., Ammendola, R., Moro, A.: Factors affecting the environmental impact of pavement wear. Transp. Res. Part D Transp. Environ. **15**, 127–133 (2010)

Framework of Sustainable Strategies for Monitoring Maintenance and Rehabilitation of Secondary Road Network to Guarantee a Safe and Efficient Accessibility

Marinella Giunta[✉] and Giovanni Leonardi[✉]

Department of Civil, Energy, Environmental and Material Engineering (DICEAM), University Mediterranea of Reggio Calabria, via Graziella Feo di Vito, 89100 Reggio Calabria, Italy
{marinella.giunta,giovanni.leonardi}@unirc.it

Abstract. The secondary road network represents the backbone of accessibility and mobility and constitutes the most important connection system which sustains the economies of the country and its territorial sectors, ensuring the access to markets, farm inputs, jobs, education and health services. Despite the importance of this road system for economic and social activities, the alignment and structural conditions of the roads are not such as to guarantee the level of functionality, safety and resilience required for them. A paramount component of road networks is the pavement, which provides a smooth travelling surface enabling to the vehicles of circulating with comfort and safety under various climatic conditions throughout the pavement's life cycle. However, once built, pavements suffer a deterioration over time under both traffic loads and environmental conditions. The need of maintenance has to deals with the scarcity of the financial resources available to road agencies. Considering the extension of the road network it is very important to apply low-cost techniques to detect the condition of the pavements and properly allocate interventions and resources. Different methods for the assessment of pavement condition have been developed and applied in the last years. In the paper an overview of the traditional and innovative approaches is presented, with a focus on high performance and fast ones.

Keywords: Pavement distresses · Pavement monitoring · Pavement management system

1 Introduction

The concept of accessibility is nowadays strictly linked with the concepts of economic and social development. The development of the road infrastructures, which represent the main assets to guarantee the accessibility, is generally considered vital to achieve a sustainable society and economic growth, as demonstrated by many studies in literature [1, 2].

The accessibility understood as requirement which allows to people to reach desired destination, services, activities, is the main goal of transportation system. Many factors affect the accessibility, such as:

© The Author(s), under exclusive license to Springer Nature Switzerland AG 2022
F. Calabrò et al. (Eds.): NMP 2022, LNNS 482, pp. 346–355, 2022.
https://doi.org/10.1007/978-3-031-06825-6_33

o *physical proximity*, the distance between origin and destination;
o *transport system connectivity*, density of transportation infrastructures, and quality of connections between different modes;
o *mobility*, the ability and level of ease of moving and therefore refers to the quality of travel modes.

Mobility and accessibility represent two basic concepts of a transportation system.

Accessibility can be evaluated from various perspectives: for particular users, modes, activities, locations, times and scales [3]. Regarding the scale it can be measured at neighbourhood, regional or interregional scales.

By referring to the Italian situation and considering the terrestrial road transport mode, Fig. 1 shows the development of the road network types [4].

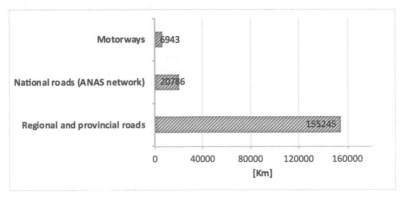

Fig. 1. Configuration of Italian road network (Source: Conto Nazionale delle Infrastrutture e dei Trasporti years 2016–2017)

It is possible to observe that there is a not balance among the different road network types since the regional and provincial road network (85%) prevails over roads of national interest (11%) and motorways (4%). The municipal roads are not counted in Fig. 2.

Responsibility for funding and management of Italian road network is divided between national (ANAS), regional and local road agencies.

The secondary road network represents the backbone of accessibility and mobility and constitutes the most important connection system which sustains the economies of the country and its territorial sectors, ensuring the access to markets, farm inputs, jobs, education and health services. This road network also constitutes an important asset of lifeline in case of extreme events such as earthquake, landslide, floods.

The roads belonging to this network are generally single carriageway and two lanes, one in each direction. The planimetric and altimetric alignment are strongly conditioned by the territorial and orographic context. In hilly and mountainous contexts, high longitudinal slopes are often associated with curves of low radius. Roads at provincial level are also characterized by low volume of traffic and speed [5, 6].

Despite the importance of this road system for economic and social activities, the alignment and structural conditions of the roads are not such as to guarantee the level of functionality, safety and resilience required of them.

The secondary road network has many elements of fragility that derive from the original materials and construction methods, and from the construction features of the structural components, such as bridges, viaducts, walls, widely present in the roads due to the morphological and hydrogeological complexity of the Italian crossed territories. However, a determining element of the current conditions is undoubtedly the lack, over the time, of ordinary and extraordinary maintenance.

As matter of the fact, the main problem for secondary roads in marginal and inner areas is the shortage of funding for maintenance and resources for rehabilitation: the road agencies and the local authorities manage funds largely insufficient to maintain the roads at good conditions and to ensure a sufficient level of services. Another problem is the technical limitation in the assessment of the current condition, the identification of the required maintenance activities and the design and execution of the interventions.

Figure 2 clearly shows as the expenditures for maintenance are negatively correlated to the extension of the Italian road networks. Expenditures for regional and provincial roads are largely less than the ones for national roads and motorways. The expenditures needed to efficiently maintain the regional and provincial roads is four time higher (average value 13,000 €/km) than the one effectively applied [4].

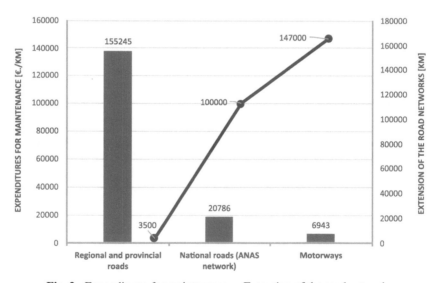

Fig. 2. Expenditures for maintenance *vs* Extension of the road networks

Literature data show that the absence of ordinary maintenance produces a reduction in service life of the infrastructures in the order of 50% and a consequent increase in maintenance costs between 30% and 60%, with peaks up to 100%.

Regarding the road safety issues, is not simple to evaluate the risk of accident due to the missing maintenance activities, however the reduction of friction in pavement surface leads to a higher probability of an accident, just as the presence of cracked floors or holes induce unexpected accelerations on the vehicle.

A paramount component of the road networks is the pavement, in asphalt concrete, which provides a smooth travelling surface enabling to the vehicles of circulating with comfort and safety under various climatic conditions throughout the life cycle [7]. However, once built, pavements suffer a deterioration over time under both ever-growing traffic volumes and environmental actions. The regular maintenance allows keeping them functional and also increases their service life. Therefore, maintenance or rehabilitation interventions throughout life span are indispensable to ensure adequate structural and functional conditions.

Maintenance and rehabilitation of pavements in secondary roads is one of the main tasks to be accomplished to restore the functionality. An estimation of the expenditures needed for the rehabilitation in Italians secondary road assign to the pavement the 49% of total expenditure followed by the road structures, major and minor, (33%), the safety barriers 8% and the road signals (5%).

In the light of the scarcity of the resources and considering the extension of the road network it is very important to apply low-cost techniques to detect pavements condition and allocate properly interventions and resources, according to a rational Pavement Management System.

2 Methodology

2.1 Pavement Condition Assessment

Two main categories of distresses can be observed in the pavement: functional and structural failures. The functional failure is often expressed by the degree of surface roughness while the structural failure is often characterized by the existence of fatigues (or what is called alligator cracks), shear developing or consolidation which is presented in one or more layer of the pavement structure. In most cases, one class of failure would potentially lead to the development of the other failure type. To evaluate the pavement condition at a certain time of its service life, different types of indicators can be assessed and used.

Among these the Pavement Condition Index (PCI), the International Rough-ness Index (IRI), and the Present Serviceability Index (PSI) are widely employed to develop pavement maintenance strategies [8].

PCI developed by the US Army and standardized by the ASTM [9]. It is a numeric index between 0 and 100 which encompasses the number of the distresses on the pavement such as: potholes, fatigue cracking, rutting, block cracking, edge cracking, longitudinal and transverse cracking, patching, shoving, bleeding, polished aggregate and ravelling and their extent. A new road has a PCI of 100, and when it deteriorates over time and reach the end of its service life PCI approaches 0.

The IRI is calculated by dividing the sum of the suspension motion of a driving vehicle by the length of the pavement section. Therefore, it is usually measured in m/km or in/mi.

PSI, introduced by the American Association of State Highway and Transportation Officials (AASHTO), is one of the most widely used pavement performance indicators after pavement condition index (PCI) and international roughness index (IRI) [10]. This performance indicator ranges between 0 and 5, 0 representing a failed pavement and 5 an excellent one. Since the PSI entails slope variance (SV), as well as Depth of rut (RD) Cracking and patching (C + P), it is correlated with performance indicators related to roughness such as IRI [11].

Correlations between the PCI and IRI have been found in many experimental studies [10, 12].

The assessment of pavement conditions which includes evaluation of friction, surface roughness, pavement structure, and existing distresses is a core task of pavement design and rehabilitation in any Pavement Management System (PMS). Most of the cost-effective Maintenance and Rehabilitation (M&R) strategies which were developed using the PMS have resulted in accurate pavement evaluation [8]. The evaluation of existing pavement conditions in terms of PCI is considered one of the main components of PMS, whose objective is to identify both degree of deterioration in the pavement sections and proper maintenance strategy, taking into account the economic constraints.

Due to the limited pavement rehabilitation allocated funds, there is usually immediate needs to prioritize any allocated funds. This prioritization is accomplished by developing systematic procedures for scheduling M&R activities to maximize the anticipated benefits of the road users and reduce the associated M&R cost. Thus, PMS would allow local agencies and engineers allocating the required budgets and funds, personnel and resources in an effective manner [13].

One of main problems to be faced by the road agencies is the detection and classification of pavement distresses and therefore assess the pavement conditions in the road network managed. Usually provincial or regional road agencies manage hundreds or thousands of kilometres of roads, so the monitoring of the pavement conditions using high-performance methodologies, techniques and devices is widely recognized as an efficient, cost-effective and sustainable approach. The introduction of new high efficiency equipment for failures detection and classification represents the new frontiers in road pavement analysis and management.

Different methods have been developed and applied in the last years. In the following an overview of the innovative approaches is presented and discusses with a special focus on the use of GIS, artificial intelligence, and image-based technologies.

2.2 Maintenance Planning Using Geographic Information System (GIS)

The Geographic Information System (GIS) is a computerized system that allows the acquisition, recording, analysis, display, return, sharing and presentation of information deriving from geographic data (geo-referred). Given its features, GIS is a useful tool in road maintenance planning since it allows to visualize the road network on a map and manage the collected related data.

Planning and scheduling of maintenance activities is an important part of any maintenance program for a large road network either at network and project level. At network level, which represents the first step of the maintenance plan, depending on pavement condition and traffic volume, it is possible to make a prioritization of the roads needing of maintenance. In the second step, at project level, for each road the more suitable repair and maintenance techniques can be identified based on the principal pavement defects, traffic volume, and climatic conditions and taking into account the environmental sustainability issues.

Figure 3 shows the data which must be collected and elaborated to perform the analyses at network and level project.

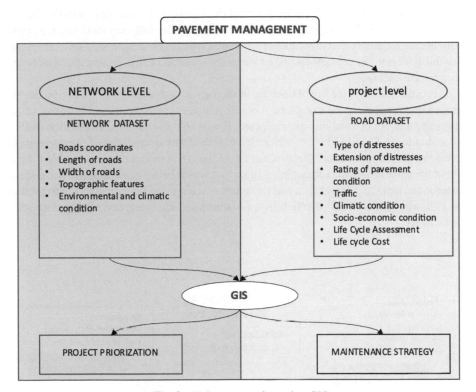

Fig. 3. Maintenance plan using GIS

According to some studies [14, 15] the pavement condition of a road (56.4%) is the most critical factor for prioritizing roads for maintenance followed by socio-economic facilities located along the road (26.2%), average daily traffic volume (11.7%), and type of connectivity provided by the road (5.7%).

The outcome of the maintenance plan using the GIS is the priority ranking along with the most suitable repair technique for each road in the network [16].

2.3 Predicting Pavement Condition Index (PCI) Using Artificial Neural Networks (ANN)

Evaluation of existing distresses and calculation of a global indicator like PCI is an essential step for pavement management and maintenance and rehabilitation programs. However, calculating the PCIs using conventional method relies on collecting of a significant sample of field data (such as distress types and severity) by visual inspection method. Starting from the collected data, the estimation of PCI values entails then a lengthy evaluation process that requires a wide technical experience. To make this process less time consuming, a relationship between distress type and severity and PCI derived from straightforward and adaptive model, can be very useful and simple to apply.

Several studies [13, 17–21] demonstrated the ability and suitability of the Artificial Neural Networks (ANN) in predicting of PCI values of the different road sections, thus reducing the required effort to estimate PCI values. Moreover, the use of ANN enables the possibility of introducing new localized variables, such as i.e. the presence of manholes in pavement sections.

According to [13] (see Fig. 4), the methodology for estimation of PCI can be developed in three phases. In the first phase, data on distress type, severity, extension, section width and number of manholes are collected. Based on these data, the calculation of PCI according to ASTM 64330-07 is performed together with an analysis of the correlation between PCI and input data. The application of the ANN model requires the identification of the suitable architecture which will be that which better generalize the correlation between the input and output data and therefore will show the better capability to predict the PCI with a high level of reliability, for training, validation and testing data sets, respectively.

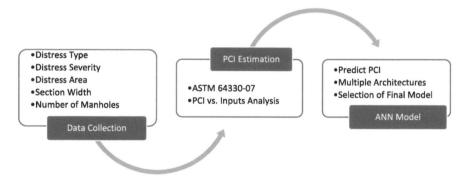

Fig. 4. Methodology for the estimation of PCI using ANN [13]

The ANN model can be integrated within an image processing platform that would automatically detect and identify pavement defects and predict the PCI value. This integration is expected to reduce the human efforts and errors when it comes to quantify the pavement conditions or defects. This can be a powerful tool for the road agencies for prioritize pavement rehabilitation programs and budget allocation.

2.4 Automated Detection of Pavement Distresses

With the aim of reducing the subjectivity of pavement distresses evaluation as well as times and costs of pavement conditions detection, in recent years many automated technologies have been developed and applied to road pavement management. These technologies are based on the use of different kind of tools and equipment.

The *image-based technologies* involve the acquisition and digitization of the image and, through an algorithms of images processing, leading to the distresses identification. Stereoscopic surveying is considered, which also includes the Structure from Motion (SfM) techniques. Photogrammetry and Structure from Motion can be applied using a single image, couples (stereo) or more images (bundle-block adjustment). Several applications of photogrammetric techniques have been realized using the stereovision approach [22].

Recent studies utilized data from *laser systems and deep learning techniques* to automate the process [23]. Laser systems have been proven to provide a high accuracy but represent a costly resource. As a result of this, image-based technologies have been widely researched to offer an alternative to laser-based systems.

An innovative approach explores the possibility to obtain information on pavement condition utilizing *drone imaginary* to replicate the roadway as a 3D model [24, 25].

Other methods for pavement distress detection rely with *radar system, acoustic techniques* [26], *friction testers, pressure-based tools.*

3 Conclusions

One of the main concerns in the service life of the secondary road network is the pavement management either for technical issues, related to its conditions assessment, and for financial issues, due to scarcity of resources for maintenance and repairs.

The overall maintenance plan of a pavement can be divided into two sub-sequent steps: prioritization of pavement maintenance within the road network and selection of the right maintenance technique for each road.

Crucial activity is the assessment of pavement condition by means the detection and classification of the existing defects. The monitoring activities through manual/visual inspections are time consuming, costly and can also cause safety concerns. The introduction of new high efficiency methodologies, tools and equipment for defects detection and classification is opening new perspective in road pavement analysis and management offering to road agencies the possibility to easily define the priorities of intervention and the more affordable and effective repair techniques and therefore leading to a more rational allocation of the available financial resources.

Technical performance is probably the first aspect to be considered in the choice of the pavement repair solution. It is fundamental in the rehabilitation to guarantee adequate surface and structural characteristics. The surface characteristics, mainly roughness and evenness, are directly related to the safety and comfort of the drivers, while the structural ones, depending on the materials and thickness applied to the layers, guarantee durability, less maintenance requirements and expected lifetime. In the framework of scarcity of resources for secondary road management the cost performance of the solution is a guiding criterion in the decision-making process. Costs must be regarded in lifetime perspective and having into account the agency and the users benefits. In this sense, repair techniques that meet the established functional and structural requirements must be compared using lifecycle cost analysis to find the most economical intervention technique.

As final note, it is worthy to highlight that the optimal maintenance and rehabilitation strategy is that which at same time ensure the meeting of several criteria such as functionality, serviceability, durability, comfort, environmental impacts, and safety as well as the lower life cycle cost.

References

1. Ng, C.P., Law, T.H., Wong, S.V., Kulanthaya, S.: Relative improvements in road mobility as compared to improvements in road accessibility and economic growth: a cross-country analysis. Transp. Policy **60**, 24–33 (2017)
2. Canning, D., Pedroni, P.: The Effect of Infrastructure on Long Run Economic Growth. Harvard University, Williams College (2004)
3. Levinson, D.M., Wu, H.: Towards a general theory of access. J. Transp. Land Use **13**(1), 129–158 (2020). https://doi.org/10.5198/jtlu.2020.1660
4. ACI: Il recupero dell'arretrato manutentorio della rete viaria secondaria una priorità per il paese: Fondazione Caracciolo (2018)
5. Praticò, F.G., Giunta, M.: Speed distribution in low volume roads: from inferences to rehabilitation design criteria. Transp. Res. Rec. J. Transp. Res. Board **2203**, 79–84 (2011)
6. Praticò, F.G., Giunta, M.: Quantifying the effect of present, past and oncoming alignment on the operating speeds of a two-lane rural road. Baltic J. of Road Bridge Eng. **7**, 181–190 (2012). https://doi.org/10.3846/bjrbe.2012.25
7. Giunta, M., Pisano, A.: One dimensional visco-elastoplastic constitutive model for asphalt concrete. Multidisc. Model. Mater. Struct. **2**(2), 247–264 (2006). ISSN: 1573-6105, https://doi.org/10.1163/157361106776240761
8. Huang, Y.H.: Pavement Analysis and Design. Prentice-Hall Inc. A Paramount Communication Company, Englewood (1993)
9. Way NC, Beach P, Materials P: ASTM D 6433-07: Standard practice for roads and parking lots pavement condition index surveys, West Conshohocken, PA (2015). https://doi.org/10.1520/D7944-15.2
10. Piryonesi, S.M., El-Diraby, T.E.: Data analytics in asset management: cost-effective prediction of the pavement condition index. J. Infrastruct. Syst. **26**(1), 04019036 (2020). https://doi.org/10.1061/(ASCE)IS.1943-555X.0000512
11. Hall, K., Muñoz, C.: Estimation of present serviceability index from international roughness index. Transp. Res. Rec. **1655**(1), 93–99 (1999). https://doi.org/10.3141/1655-13
12. Cafiso, S., Di Graziano, A., Goulias, D.G., D'Agostino, C.: Distress and profile data analysis for condition assessment in pavement management systems. Int. J. Pavement Res. Technol. **12**(5), 527–536 (2019). https://doi.org/10.1007/s42947-019-0063-7
13. Issa, A., Samaneh, H., Ghanim, M.: Predicting pavement condition index using artificial neural networks approach. Ain Shams Eng. J. **13**, 101490 (2022)
14. Nautiyal, A., Sharma, S.: Condition based maintenance planning of low volume rural roads using GIS. J. Clean. Prod. **312**, 127649 (2021)
15. Joni, H.H., Alwan, I.A., Naji, G.: Utilizing artificial intelligence to collect pavement surface condition data. Eng. Technol. J. **38**(1A), 74–82 (2020)
16. Mansour Fakhri, M., Dezfoulian, R.S., Golroo, A., Makkiabad, B.: Developing an approach for measuring the intensity of cracking based on geospatial analysis using GIS and automated data collection system. Int. J. Pavement Eng. **22**(5), 582–596 (2019)
17. Eldin, N.N., Senouci, A.B.: A pavement condition-rating model using backpropagation neural networks. Comput. Aided Civil Infrastruct. Eng. **10**(6), 433–441 (1995)
18. Yang, J., Lu, J.J., Gunaratne, M.: Application of neural network models for forecasting of pavement crack index and pavement condition rating (2003)

19. Shahnazari, H., et al.: Application of soft computing for prediction of pavement condition index. J. Transp. Eng. **138**(12), 1495–1506 (2012)
20. Amin, M.S.R., Amador-Jiménez, L.E.: Pavement management with dynamic traffic and artificial neural network: a case study of Montreal. Can. J. Civil Eng. **43**(3), 241–251 (2016)
21. Jalal, M., Floris, I., Quadrifoglio, L.: Computer-aided prediction of pavement condition index (PCI) using ANN. In: Proceedings of the International Conference on Computers and Industrial Engineering (2017)
22. Ahmed, M., Haas, C.T., Haas, R.: Toward low-cost 3D automatic pavement distress surveying: the closerange photogrammetry approach. Can. J. Civ. Eng. **38**, 1301–1313 (2012). https://doi.org/10.1139/L11-088
23. Cafiso, S., D'Agostino, C., Delfino, E., Montella, A.: From manual to automatic pavement distress detection and classification. In: 5th International Conference on Models and Technologies for Intelligent Transportation Systems - Proceedings, pp. 433–438 (2017). https://doi.org/10.1109/MTITS.2017.8005711
24. Inzerillo, L., Di Mino, G., Roberts, R.: Using UAV based 3D modelling to provide smart monitoring of road pavement conditions. Information **11**(12), 568 (2020). https://doi.org/10.3390/info11120568
25. Roberts, R., Inzerillo, L., Di Mino, G.: Developing a framework for using structure-from-motion techniques for road distress applications. Eur. Transp.\Trasporti Europei **5**(77) (2020). ISSN 1825-3997, https://doi.org/10.48295/ET.2020.77.5
26. Fedele, R., Praticò, F.G., Pellicano, G.: The prediction of road cracks through acoustic signature: extended finite element modeling and experiments. J. Test. Eval. **49**(4), 20190209 (2019). https://doi.org/10.1520/JTE20190209

Accessibility and Internal Areas - Rural Towns of Calabria and the Local Road Network

Francis M. M. Cirianni⊕, Marinella Giunta⊕, Giovanni Leonardi⁽✉⁾⊕, and Rocco Palamara⊕

Department of Civil, Energy, Environmental and Material Engineering (DICEAM), University Mediterranea of Reggio Calabria, via Graziella Feo di Vito, 89100 Reggio Calabria, Italy
{francis.cirianni,marinella.giunta,giovanni.leonardi, rocco.palamara}@unirc.it

Abstract. The morphological conformation of Calabria presents itself in a singular way compared to other Italian regions and is the reason for the plurality of landscapes ranging from the mountain and hilly type to that of flat areas divided into alluvial, valleys and typically marine terraces. The territory from a geological and hydrogeological point of view is particularly unstable and subject to erosion, landslides, floods; these events, together with seismic phenomena, make some areas particularly dangerous.

To ensure accessibility to the internal areas of Calabria and to allow the inhabitants of small towns or villages adequate services and safety conditions even in emergency conditions, it is essential to have an overall redevelopment of the secondary road network and an enhancement of the "dirt roads".

The dirt roads can represent an important resource for improving the accessibility of areas that are difficult to reach and also represent an important resource for tourist activities.

Keywords: Secondary roads · Accessibility · Unpaved roads

1 Introduction

1.1 Regional Road Infrastructure System

The regional road infrastructure system is part of this territory, whose overall development, as shown in the latest edition of the National Transport Account, consists of 288 km of motorways, 1689 km of state roads and approximately 7594 km of roads regional and provincial.

The entire road network has a total development of about 16,000 km, equal to just over 5.00% of the national network. The backbone of the Calabrian Road system is represented by the Salerno - Reggio Calabria motorway which, running along the region in the longitudinal direction, directly connects some "bearing" centers and supports the transverse connections with the two banks.

The infrastructure is conditioned both by a "historical" choice of the route, and by the geomorphological characteristics of the Calabrian territory. These two factors determine

F. Calabrò et al. (Eds.): NMP 2022, LNNS 482, pp. 356–362, 2022.
https://doi.org/10.1007/978-3-031-06825-6_34

the conditions of direct exclusion from the great viability of the small villages, located mainly in the inland areas of Calabria (Fig. 1). To partially obviate this problem, a series of transversal connections which, reaching the shores of the Tyrrhenian and Ionian rivers, partially satisfy the demand for connections, intervene with comb couplings on the motorway ridge; in fact, the secondary road network, even if sufficiently extensive, remains of a poor quality level, creating conditions of objective discomfort, diseconomy of mobility and inadequate functional levels especially in the summer period when the tourist flows highlight the numerous criticalities and maintenance limits of a road system no longer able to adequately meet the mobility needs of a region that has tourism as one of its main economic resources. The infrastructural system is unequivocally the hinge of every development process and the road equipment with its efficiency is closely related to the structural conditions that determine the dynamics of local economies [1].

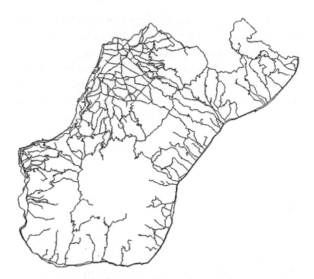

Fig. 1. Calabrian road network

The extra-urban road network therefore represents an important connection system for the economy of small Calabrian towns. These are typically roads with a single carriageway, with geometry strongly conditioned by the territorial and orographic context and by a surface hydrography characterized by ditches, engravings, waterways with typically torrential regimes that determine a high density of bridges, manholes, overpasses, bridges, guard ditches and other poorly maintained works of art.

This reduced or almost zero maintenance together with an intrinsic fragility of the infrastructures that derives from the original construction characteristics, in terms of materials and construction methods, has determined localized degradation, deterioration, subsidence and damage to the road body (Fig. 2), circumstances that have led to an evident reduction of safety and in some cases the closure of sections with consequent isolation of entire areas already suffering economically.

These difficulties in the connections, in recent decades, have favoured depopulation, with an unstoppable "demographic loss" of internal areas, and migration from the

Fig. 2. Road body collapse

villages of the hills and mountain areas in favour of the plain and costal settlements. The described phenomenon is more marked than data shows, as often the old town in the internal/mountain location and the modern settlement close to the sea and/or the main infrastructures and transport system fall within the same municipal area, and the depopulation being within the same administrative boundaries is not registered by the national census. The morphological complexity of the territory, which affects the costs of building and maintenance of the infrastructures, and the depopulation which affects the per capita public funding which can be allocated in the budgets, leads as a direct consequence that accessibility is not guaranteed adequately to large areas of the regional territory [2].

2   Local Road Network

The improvement of the secondary and local infrastructural network in terms of efficiency, safety and resilience cannot, however, disregard the protection and enhancement of the landscape, the historical and architectural heritage of the territorial context through the use of sustainable technologies and materials.

Within this objective, a strategic and often neglected role can be played by the so-called "dirt roads", which, if properly built, adapted, and maintained, can improve the overall accessibility of entire areas, especially mountainous areas, as occurs in other areas of Italy and in other European and non-European countries (Fig. 3).

In addition to the tourist value, the advantages that the enhancement of the dirt roads could determine also in terms of greater safety for the population must not be ignored; and as a matter of fact, even this class of roads can have an important "lifeline" function for areas that are otherwise poorly, if not at all, connected by ordinary roads [1, 2].

The term "lifelines" means network systems that relate and/or connect different spatial areas in order to guarantee a multitude of essential and indispensable services for the survival of the population, in particular, in the event of natural disasters (e.g. road, gas, electricity, water, IT communications networks).

Among these networks, road infrastructures represent, on the one hand, elements which are exposed to the various natural ordinary and extraordinary phenomena (as landslides, floods, earthquakes) and, on the other, infrastructures of strategic importance

Fig. 3. Dirt road in Aspromonte

both on a daily basis, and under emergency circumstances for the accessibility management, to guarantee physical access and exodus from the affected areas in the aftermath of a natural disaster [3].

In a vulnerable territory such as the Calabrian one, to guarantee a greater accessibility to areas difficult to reach through the ordinary road network, such as internal areas, mountain, and hill towns, represents a strategic goal both in terms of prevention and emergency management.

3   The Dirt Roads

Despite some recent proposals for regulatory interventions for the protection and enhancement of "white roads", better known as "dirt roads", these are often neglected by the owner administrations and, most often than not, end up in a very bad state due to lack of attention. Nevertheless, theology and materials for their stabilization have been available for a while, such as eco-compatible polymeric and/or organic anti-dust liquid binders, products that can be used with most types of natural terrain and traffic.

Alongside the stabilization interventions, an improvement in the bearing capacity and, therefore in the resistance to permanent deformations of the superstructure [4, 5], can be obtained through reinforcement operations with geosynthetics and/or geocells.

The insertion of reinforcement interlayers can increase the life span of the road or reduce the thickness of the layers (base and foundation) of the superstructure, in both cases with a consequent financial saving. The reinforcement of unpaved roads, in fact, can reduce, thanks to the "interlock" process, the differentiated yields and the formation of ruts (permanent deformations) on the road surface. These technical solutions, if used on a large scale, could improve the accessibility of areas currently isolated or difficult to reach and also represent an important resource for some types of tourist activities (e.g. hiking, trekking, cycling) (Fig. 4).

Fig. 4. Dirt road stabilization

3.1 Unpaved Road Improvement by Geosynthetics

The technique of soil-improvement using geosynthetics is extensively used in the construction of unpaved roads. When the subgrade is weak, due to its poor consistency and high compressibility, generally, a geosynthetic reinforcement is placed over the subgrade followed by a compacted granular fill layer. This technique is particularly effective because the performances of reinforced unpaved roads are enhanced by reducing permanent rut deformation for a given number of axle loads. Rutting is a permanent deformation that accumulates as the traffic loads increases. Large rutting may cause discomfort to drivers, damage to vehicles, and instability of the vehicles; therefore, excessive rutting should be avoided. For these reasons in many cases, it is used the geosynthetic like reinforcement [6, 7].

Unpaved roads reinforced with geosynthetic usually are realized by a base (made of granular material) resting on subgrade soil (typically a cohesive soil) and a geosynthetic (usually a geogrid) with reinforcing function included between the base and the subgrade. The goals of geosynthetic reinforcements are an increase of the road service life; a decrease of the construction cost by decreasing the base layer thickness (if the cost of the geosynthetic reinforcement is less than the cost of the saved base material); a decrease of the time required for the construction and of the periodic maintenance interventions. Geosynthetics used in unpaved road are essentially geotextiles and geogrids.

In the first case, in real unpaved roads geotextiles were used more than geogrids because separation in many cases was the main function required. In these cases, there isn't generally a design method.

When no design is involved, the aggregate layer thickness and the geotextile are selected based on existing projects using the same geotextile.

In unpaved roads which are the first phase of the construction of paved roads, geotextiles are used if the main concern is separation, whereas geogrids are used more than geotextiles when reinforcement is required, when it is considered important to keep deformations of the aggregate layer as small as possible. Geocells are also used, but to a limited extent. In this paper, the attention is focused on the use of geogrids as reinforcement.

The interaction due to the geogrid interlocking with aggregate minimizes lateral movement of aggregate particles and increases the modulus of the base course, which leads to a wider vertical stress distribution over the subgrade and consequently a reduction of vertical sub-grade deformations. The degree of interlocking depends on the relationship between geogrid aperture size and aggregate particle size; geogrid aperture shape; shape and stiffness of ribs instead the effectiveness of interlocking depends on the in-plane stiffness (more than strength) of the geogrid and the stability of the geogrid ribs and junctions [5, 8–10].

The geosynthetic contribution to the performance of an unpaved road is achieved through several mechanisms that take place in the road structure, which consist:

- Separation between base and subgrade;
- Lateral confinement of the base material;
- Improvement of wheel load distribution;
- Tensioned membrane effect.

The mainly geosynthetic contribution to the performance of an unpaved road is achieved through lateral confinement effect and tension membrane effect. They require different depth values of rutting in order to be mobilized.

At small permanent deformation magnitudes, the lateral restraint mechanism is developed by the ability of the base aggregate to interlock with the geogrid. As increasing of permanent deformations (which are often acceptable in unpaved roads), the tension membrane mechanism [11–13] develops. Geosynthetic reinforcement becomes increasingly effective as the dis-placements become large, so when the substantial surface rutting is acceptable, and if the geosynthetic has a sufficiently high tensile modulus, tensile stresses will be mobilized in the reinforcement, and a vertical component of this tensile mem-brane resistance will help to support the applied wheel loads [14, 15].

4 Conclusions

In a morphologically complex territory such as the Calabrian one, improving the accessibility of the internal areas cannot ignore the needs of environmental protection and enhancement of the landscape, of the historical and architectural heritage.

For this reason, it is necessary to plan maintenance and redevelopment operations which make use of innovative and eco-sustainable technologies and materials.

These technical solutions, used on a large scale, together with an adequate enhancement of dirt roads could improve the accessibility of areas currently isolated or difficult to reach and also represent an important resource for some types of tourist activities (e.g. hiking, trekking, cycling).

References

1. Cirianni, F., Leonardi, G.: A methodology for assessing the seismic vulnerability of highway systems. In: Santini and Moraci (eds.) Seismic Engineering International Conference - AIP. American Institute of Physics Conference Proceedings, 08–11 July 2008, vol. 1020 - Part one, pp. 864–894 (2008)

2. Cirianni, F., Leonardi, G., Fonte, F., Scopelliti, F.: Analysis of lifelines transportation vulnerability. Procedia – Soc. Behav. Sci. **53**, 29–38 (2012). SIIV 5th International Congress, 29–31 October 2012, Roma, Italy. Elsevier
3. Leonardi, G., Palamara, R., Cirianni, F.: Landslide susceptibility mapping using a fuzzy approach -. Procedia Eng. **2016**(161), 380–387 (2016)
4. Buonsanti, M., Cirianni, F., Leonardi, G., Scopelliti, F.: Dynamic behaviour of granular mixture solids. In: 10th International Conference FDM11, Dubrovnik (Croatia), 19–21 September 2011, vol. 488–489, pp. 541–544. Key Engineering Materials, Trans Tech Publications, Switzerland (2012)
5. Leonardi, G., Bosco, D.L., Palamara, R., Suraci, F.: Finite element analysis of geogrid-stabilized unpaved roads. Sustainability (Switzerland) **12**(5), 1929 (2020)
6. Barenberg, E.J., Dowland, J.H.Jr., Hales, J.H.: Evaluation of soil aggregate systems with Mirafi fabric. Civil Engineering Studies, Department of Civil Engineering, University of Illinois, Report No. UILU-ENG-75-2020, 52 pp. (1975)
7. Giroud, J.P., Noiray, L.: Geotextile-reinforced unpaved road design. J. Geotech. Geoenviron. Eng. **107**(9), 1233–1254 (1981)
8. Cazzuffi, D., Moraci, N., Calvarano, L.S., Cardile, G., Gioffrè, D., Recalcati, P.: The influence of vertical effective stress and of geogrid length on interface behaviour under pullout conditions. Geosynthetics **32**(2), 40–50 (2014)
9. Moraci, N., Cardile, G., Gioffrè, D., Mandaglio, M.C., Calvarano, L.S., Carbone, L.: Soil geosynthetic interaction: design parameters from experimental and theoretical analysis. Transp. Infrastruct. Geotechnol. **1**(2), 165–227 (2014). https://doi.org/10.1007/s40515-014-0007-2
10. Moraci, N., Cazzuffi, D., Calvarano, L.S., Cardile, G., Gioffrè, D., Recalcati, P.: The influence of soil type on interface behavior under pullout conditions. Geosynthetics **32**(3), 42–50 (2014)
11. Fannin, R.J., Sigurdsson, O.: Field observations on stabilization of unpaved roads with geosynthetics. J. Geotech. Eng. **122**(7), 544–553 (1996). Art. no. 10806
12. Calvarano, L.S., Gioffrè, D., Cardile, G., Moraci, N.: A stress transfer model to predict the pullout resistance of extruded geogrids embedded in compacted granular soils. In: Proceeding of 10th ICG 2014, Berlin, Germany, 21–24 September 2014, Code 110984 (2014)
13. Calvarano, L.S., Leonardi, G., Palamara, R.: Finite element modelling of unpaved road reinforced with geosynthetics. Procedia Eng. **189**, 99–104 (2017)
14. Leng, J., Gabr, M.A.: Numerical analysis of stress-deformation response in reinforced unpaved road sections. Geosynth. Int. **12**(2), 111–119 (2005)
15. Kawalec, J., Grygierek, M., Koda, E., Osiński, P.: Lessons learned on geosynthetics applications in road structures in Silesia mining region in Poland. Appl. Sci. (Switzerland) **9**(6), 1122 (2019)

Strategies and Measures for a Sustainable Accessibility and Effective Transport Services in Inner and Marginal Areas: The Italian Experience

Francis M. M. Cirianni[1] ⓘ, Giovanni Leonardi[1](✉) ⓘ, and Angelo S. Luongo[2] ⓘ

[1] Department of Civil, Energy, Environmental and Material Engineering (DICEAM), University Mediterranea of Reggio Calabria, via Graziella Feo di Vito, 89100 Reggio Calabria, Italy
{francis.cirianni,giovanni.leonardi}@unirc.it
[2] Economics and Mathematical Methods, University of Bari Aldo Moro, Largo Abbazia Santa Scolastica, 70124 Bari, Italy

Abstract. Inner and marginal areas, territories usually identified in the context of rural and mountainous regions with low accessibility levels, are characterized by low population density and depopulation. The concept of "inner" or "marginal" areas embraces a set of different settlement conditions. Italy's definition of Inner Areas focuses on rural areas characterized by the distance from the main service centres (education, health and mobility). It is clear that one of the key elements is accessibility, availability of adequate transport systems. The functional layout of the road network, heavily influenced by the orographic and geological characteristics of the territory, substantially affects the affordability and quality of the transport supply. To ensure quality of services in such areas, transport service supply has to be improved on the base of non-conventional, on demand and flexible mobility systems, to improve widespread accessibility and urban-rural connectivity. An experimental extensive plan of action interesting the inner areas of Italy is being implemented through the "National Strategy for Inner Areas", where the accessibility plays an important role and in that sense the transport network is redesigned in an integrated way with the territorial development scenario and in the perspective to guarantee access to essential services.

Keywords: Inner Areas · Sustainable mobility · Mobility on demand

1 Introduction

Inner and marginal areas, territories usually identified in the context of rural and mountainous regions but also with low accessibility level, are characterized by low-density and phenomenon of depopulation. The concept of "inner" or "marginal" areas cannot be narrowly defined. In this sense, many definitions are possible which are related with the concept of "rural" which include a generic dispersion definition of counties or regions with small villages, e.g. in the U.S.A. basic rural areas are described as "dispersed counties or regions with few or no major population centers of 5,000 or more" [1].

The Scottish Executive Urban Rural Classification is based on these two criteria of settlement size (as defined by the General Register Office for Scotland) and accessibility [2], otherwise in Italy Inner areas are defined as territories far from centers offering essential services and thus characterized by depopulation and degrade [3]. In other words, Italy's Inner Areas are rural areas characterized by their distance from the main service centers (education, health, and mobility).

These areas in socio-economic terms are meanly weak territories, for many marginal aspects, where distance and peripherality, in respect to service centers, sometimes represents an anti-performing prejudice [4].

One of the key elements, to implement sustainable development and cohesion policies, is accessibility through the availability of an adequate transport system for an adequate access to basic service.

Regular and traditional transport services in the above territories, characterized by a weak and dispersed demand, are usually characterized by a very low offer, no longer able to satisfy the mobility needs expressed by the local population and visitors.

It is therefore clear that to ensure the livability and quality of life in such territories, an integrated transport service offer should be based on a non-conventional mobility system, to improve accessibility and urban-rural connectivity. In these areas services should be designed on the specific needs of a wide selection of users, different in terms of age, trip purposes and social and economic conditions and dispersed in scattered settlements of small size.

In this regard, as shown in literature and in research, flexible transport systems (FTS) can offer an alternative to conventional public transport (fixed route -fixed schedule) bus services, filling the gaps providing essential services [5–8].

Mobility as a Service (MaaS) is an emerging concept of new integrated transport services, and although it is often developed and studied from an urban point of view, in inner and marginal areas, with sparse population, long distances, and low-capacity utilization rates it could improve efficiency by integrating different types of transportation [9, 12]. As carried out in the study of pilot cases the main opportunities on Maas implementation are related to collaboration and combining of different services, creating travel chains, and a good developed and extensive ICT infrastructure and digitalization is considered necessary [9–11, 13, 14].

The above concept and solution have been the basis of strengthening and rethinking services in the inner/marginal areas within the integrated local development projects developed in the context of the National Strategy of Inner Areas (SNAI by its Italian acronym) [3]. In a perspective of financial sustainability, the services have been redesigned to improve mobility from and within the areas, cutting the time taken to access the services available in the hubs. The rethinking process was based on the rationalization of the main local public transport services and on in the forecast of the implementation of flexible and unconventional transport services within an integrated network [12, 13, 15].

The above-mentioned National Strategy of Inner Areas, which includes 72 pilot areas situated throughout the Italian territory characterized by a great variability and differences between each other (on all levels: geographical, economic, social, cultural, eco-systemic), is currently being implemented.

In the following Sect. 2 an overview of the methods and criteria adopted to identify rural/inner/marginal areas is presented. Section 3 describes transport supply solutions suitable to ensure a financial and social sustainable mobility system for low and dispersed transport demand. More specifically the opportunities and benefits given by the implementation of not traditional and Flexible integrated transport services (FITS) are outlined. The planning and programming approach to mobility and transport solutions developed under the 'National Strategy for Inner Areas in Italy is presented in Sect. 4. Finally, conclusions present policy implications, limitations, and recommendations.

2 An Overview of the Methods and Criteria for the Identification of Inner and Marginal Areas

A preliminary step to the issue of accessibility and effective transport services in rural, inner, and marginal areas concerns the methods used to identify the areas as such.

Almost all the quantitative procedures for setting such areas also include a measure of accessibility, therefore it is useful to introduce meaningful definitions from a literature review.

Accessibility indicators can be defined to reflect both within-region/areas transport infrastructures and infrastructures outside the region which affect the region/areas [18].

The concept of accessibility can refer to the need of a person living in an area to carry out activities (shopping, work, education, recreation et sim.), which can be defined as "active accessibility", or the need of an activity located in a certain area to be reached by possible users (customers, employees, suppliers, et sim.), seen as "passive accessibility" [19].

The above definition can be summarized as "accessibility indicators which describe the location of an area regarding opportunities, activities or assets existing in other areas and the area itself, where "area" may be a region, a city or a corridor" [20].

In Relation to the structure of the transportation system and the transport services offered in the area, it is the physical accessibility that can be simply defined as the degree of difficulty that people or communities have in accessing locations to satisfy their basic social and economic needs, such as food, water, heating, education, health care, trading, and transport [21].

Transport accessibility is defined as the potential to participate in activities (or, equivalently, interacting with people) which are distributed over space. Intuitively, the more opportunities available to a person to participate in each kind of activity, the more attractive these opportunities are for engagement, and the easier it is to travel to these activity locations, the higher the accessibility [22].

Accessibility can be differentiated from connectivity on the one hand, and mobility on the other. Connectivity deals with the extent that one point (node) in a network is connected to other points (nodes) in the network. Mobility involves the actual movement of people (and goods) from point to point within the transport system. It thus represents the actualization of accessibility (i.e. the potential to travel and interact) in terms of actual trips and interactions. Connectivity deals with the extent that one point (node) in a network is connected to other points (nodes) in the network [23].

In a simple way a peripheral region is defined as a region with low accessibility. However, in addition to accessibility, many other criteria are used to delineate centers and peripheries in regional research [18].

Rural and Inner Areas accessibility means more than just reliable roads and cost-effective transportation networks [24]. Accessibility is crucial to guarantee considerable improvements in the provision of essential services (primary and secondary school and vocational training, healthcare, and medical services).

Different thresholds for the peripherality criterion have been proposed over the years with a continuous process of change and refinement of the methods used.

Most of the recent and current classification methods are essentially based on settlement size, population distributions and the quantitative criteria of peripherality/accessibility/remoteness.

The rural typology developed by the Organization for Economic Co-operation and Development (OECD) is exclusively based on population density and is applied at two hierarchical levels: the local community level and the regional level. At the first level (LAU2 level), towns with population densities lower than 150 inhabitants per km^2 are classified as rural otherwise, they are classified as urban. At the second level (NUTS3 level or NUTS2 level), a region with more than 50% of population living in rural communes is classified "predominantly rural"; if this share is between 50 and 15 it is classified "intermediate"; if lower than 15% it is "predominantly urban" [25].

To identify a region also as remote it was performed an accessibility analysis. Therefore, in this type of analysis the driving time needed for a certain percentage of the population of a region to reach a populated centre is quantified. The extended typology classifies then rural and intermediate regions as remote when 50% of the regional population needs at least 60 (45) minutes of driving time to reach a populated centre with at least 50 000 inhabitants [26].

In Italy the first step in the development of the National Strategy of Inner Areas has been the elaboration of criteria to identify territories as "Inner areas".

The methodology was developed from two main concepts (Public Investment Evaluation Unit- UVAL, 2014) [3]:

- the Italian territory is characterized by a dense and varied network of urban centers which offer a wide range of essential services like healthcare, education, and transport. These centers (defined as 'Service provision Centers') represent a "point of convergence" for people living in remote areas:
- the distance from these urban networks affects people's quality of life, and their sense of social inclusion.

A "service provision Centre" have been identified as those municipalities which offer:

- an exhaustive range of secondary schools

- at least a 1st level DEA hospital[1]
- at least a 'Silver - type' railway station (RFI)[2]

Areas have been mapped without the use of demographic criteria, according to the distance (travel-time) from these 'Service Centers' as:

- 'Belt' areas – up to 20 min far from the centers
- 'Intermediate' areas – from 20 to 40 min (Inner Areas)
- 'Remote' areas – from 40 to 75 min (Inner Areas)
- 'Ultra – remote' areas – over 75 min far (Inner Areas)

The maps of Italy's inner areas are shown in Fig. 1.

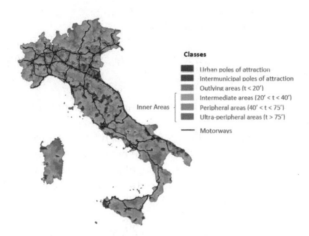

Fig. 1. Italian inner areas [3]

The leading hypothesis of the methodology employed identifies in the first place the nature of Inner Areas in terms of 'distance' from essential services. In this sense,

---

[1] Grade 1 emergency care hospitals (DEA) include a set of operational units that, in addition to Casualty departments, guarantee observation facilities, short stays, resuscitation and diagnostic-therapeutic general medical intervention, general surgery, orthopaedics and traumatology, cardiology intensive care. They are also able to provide chemical, clinical and microbiological laboratory services, medical imaging and carry out transfusions.

[2] The Italian Rail Network (RFI) classifies stations as: PLATINUM (13 large track systems): stations serving an average of more than 6,000 passengers/day and with a large average number of trains/day with a good number of high-performance trains; GOLD (103 medium/large track systems): these include medium/large systems serving a reasonably high number of passengers, providing good local and high-performance transport; SILVER (medium/small systems) include all the other medium/small systems with an average degree of uptake for metropolitan/regional services and shorter long-distance journeys than GOLD stations; BRONZE (small systems with lower uptake). These include smaller systems serving fewer passengers, providing regional services.

Inner Areas are not necessarily synonymous with "weak areas". The individual degree of remoteness rather identifies a characteristic of the areas and refers exclusively to the factors under consideration (education, health, and rail services). Only by examining the characteristics and dynamics of the demographic and socio-economic structure of the identified areas a complete reading of the various territorial development paths can be had.

The Scottish Government classifies urban and rural areas on the basis of settlement size and driving time from the centres of the Settlements with a population of 10,000 or more (i.e. Large and Other Urban Areas). Settlements were grouped into the following categories [2]:

1. Large Urban Areas - populations of 125,000 or more
2. Other Urban Areas - populations of 10,000 to 124,999
3. Small Towns - populations of 3,000 to 9,999
4. Rural Areas - populations less than 3,000

To distinguish between accessible and remote areas the following definitions of remoteness were defined:

1. Accessible – areas within a 30-min drive time of a Settlement with a population of 10,000 or more.
2. Remote – areas that are more than a 30-min drive time (6-fold classification), or areas that have a drive time more than 30 min but less than or equal to 60 min (8-fold classification) from a Settlement with a population of 10,000 or more.
3. Very Remote – areas that are more than a 60-min drive time from a Settlement with a population of 10,000 or more (8-fold classification only).

Accessibility categories were assigned to the Settlement boundary layer based upon the location of the Settlement population weighted centroids.

Recently a Espon report [27] explained the concept of the "Inner Peripherality" intended, then the simple conventional "peripherality", as a more complex, multidimensional phenomenon which compounds the effects of various socio-economic processes that cause disconnection from external territories and networks. The distinctive feature of Inner Peripheries is their degree of "disconnection" and not their geographical position in relation to the "core areas" of Europe. Inner peripheries have in common the fact that their general performance, levels of development, access to services, or the quality of life of the population, are relatively worse when compared with their neighboring territories [27]. The above definition could be applied on relatively minor spatial scales, as the regional and local scales.

3   Transport Supply Solutions for a Widespread Accessibility

Transport services in sparsely populated territories, characterized by a weak demand, face issues resulting from the low and dispersed nature of the population, which makes providing conventional (fixed-route and schedule) public transport even more complex, inefficient, and not financially viable.

The profitability of public transport in inner/marginal areas already is suffering low returns, which will continue to worsen, increasing the burden on the government to sustain the transport system. In other words, people living in rural areas have fewer options for mobility and the transport gap is increasing. In this vicious cycle, rural people have trouble maintaining their quality of life [28].

However, conventional public transport systems have a rigid framework with routes where specific locations are predefined to have a regular service which serves as much population as possible. Also, the functional layout of the road network, heavily influenced by the orographic and geological characteristics of the territory, substantially reduced the affordability and quality of the transport supply.

Not all rural/inner areas have the same territorial characteristics, structure of demographic groups and needed for moving, thus there is not a single solution to be developed and implemented but a mix of them based on flexibility and user oriented.

An integrated flexible transport systems (FTS) can be defined as a system that provides a desirable level of flexibility for passengers when choosing routes, time of travel, modes of transport, service providers, and payment systems while integrating different available services to improve the efficiency and performance of a transport service [29].

Flexible transport services (FTS) consist of a range of mobility services offering greater flexibility than regular public transport services. Whereas urban flexible transport includes shared taxis, car-pooling, and carsharing [30].

These FTS are characterized by flexible routing and scheduling of small to medium-sized vehicles operating in shared-ride mode between pick-up and drop-off locations according to passengers' needs [30].

This is achieved through demand responsive transport (DRT) for the public or more commonly through dedicated services for a specific trip (i.e., access to education, social care, and health services).

The supply adaptation to the demand needs is achieved through demand responsive transport (DRT) for the public or more commonly through dedicated services (i.e., transport for specific groups of the population, e.g., the elderly, vulnerable people, or youth) [5].

Recently to adapt the offer to the users' needs a new way of travel in rural/inner territories has been developed, named "Mobility as a service" (MaaS), which means "Put[ting]…users, both travelers and goods, at the core of transport services, offering them tailored mobility solutions based on their individual needs. This means that, for the first time, easy access to the most appropriate transport mode or service will be included in a bundle of flexible travel service options for end users" [12].

The number of good practices identified show that there is a great diversity of solutions and ways of providing public transport, adjusted to the specifics of the territories and the needs of the residents. Flexible transport services with minibuses, door-to-door DRT with "virtual" stops, shared taxis, and carpooling are only a part of the many solutions which can improve mobility for people in rural areas [31]. Figure 2 offers a clustering of a set of services which can be implemented in a rural/inner area as part of a flexible and shared mobility system.

In any case even if the above services are usually characterized by a lower operational cost compared to traditional public transport services, it is necessary to pay a flat rate

Fig. 2. The array of rural shared mobility services identified in the SMARTA Project

subsidy or block grant regardless of the low number of the passengers they transport (or are paid the subsidy if they fulfill their quota of trips within a given period) [5]. Therefore, if in one hand it is necessary to design a cost-efficient flexible integrated transport service, on the other it is still necessary for these services to be supported by public finances, given that those operated in a market situation are prone to failure.

As indicated by the European Network for Rural Development, around 200 case studies of rural mobility services, with information such as how they were implemented, what proved effective, in what context, and site contacts, are described in European projects as SMARTA, Euromontana, LAST-MILE, MAMBA, HiReach and RuMobi.[3]

The SMARTA Consortium identified a set of Good Practices in rural mobility, not only related to conventional Public Transport but also extended to innovative transport forms based on ride sharing schemes for the residents of a rural areas, vulnerable social groups and for visitors and tourists (a group that can result in highly variable demand).[4]

4   The Italian Experience

An experimental extensive plan of action interesting the inner areas the whole of the Italian territory is in the process of being implemented through the "National Strategy for Inner Areas". A detailed description of the whole process is described in a 2014 publication of the Public Investment Evaluation Unit [3] and the status of implementation of National Strategy could be inferred from the annual 2018, 2019 and 2020 reports drawn

---

[3] https://enrd.ec.europa.eu/sites/default/files/enrd_publications/smart-villages_brief_rural-mob ility.pdf.

[4] https://ruralsharedmobility.eu/good-practices/.

up the Department for Cohesion Policy. Below are in summery described the essential structure of the National Strategy with a focus of the mobility system.

The above strategy, applied to every region and macro-area of Italy, aims to contribute to the country's economic and social recovery, creating jobs, fostering social inclusion, and cutting the costs of depopulation and degrade in the 'Inner Areas', territories substantially far from centres offering essential services [3].

These areas, classified by applying the criteria set out in Sect. 2, currently cover approximately 60% of the Italian territory, the 52% Municipalities and hosting nearly 13.540 million people.

The 72 pilot areas proposed by the Regions in agreement with the Centre, that become the focus of Project Framework Agreements, as part of the National Strategy for Inner Areas. The above-mentioned pilot areas consisting of 1.077 Municipalities and hosting 2.072.718 people and each area is composed of 29.400 residents and 15 Municipalities on average.

The Strategy aims to create new income opportunities, enhance territories maintenance, and ensure that "Inner Areas" inhabitants have access to essential services (local public transport, education, social and healthcare services). The medium-term objectives are:

- supporting Inner Areas inhabitants' access to basic services through innovative delivery mechanisms. Education, Health and Transport are conceived as pre-conditions for any further development of such areas.
- fostering Local Development. Development projects aim to increase local job opportunities and profit from available local capital.
- increase local employment and work opportunities.
- recover un-valorized natural resources and territorial capital Strengthen local development factors.

In the long-term the Strategy aims to:

- reverse the demographic decline that is embodying these areas.
- improve the quality of life and wellbeing of people living in relatively isolated and sparsely populated areas of the country.
- reduce the social cost due to the progressive depopulation trend characterizing these areas (hydro-geological risk, cultural heritage losses and landscape degradation).

Considering the above objectives in each Project area a plan of action developed in 3 steeps has been set up:

1. In the first phase a first document entitled "Draft strategy" was carried out, which contains a guiding idea for essential services (Schools, Health and Mobility) and Local development. In this stage Area Coordinator, Institutions, associations, competence centres, the stakeholders and people living in the territory were involved, representing collective needs and interests and the kind of knowledge needed to carry out actions, having the incentives to propose the actions, as well as collecting the benefits.

2. Subsequently the "Preliminary Strategy Document" was drawn, in which the guiding idea is translated into expected results and actions. In this stage the Area Coordinator, the Region and the Committee on Inner Areas were involved.
3. At last, the harmonization of the ideas and the engineering phase of the projects is realized in the "Strategy document of the Area", which is submitted to the Committee on Inner Areas, to the Ministries and Regions to be evaluated, negotiated and redefined.

The set of actions planned for each project area as part of the Strategy for Inner Areas take the form of Area Projects, corresponding to the selected areas with targeted spheres of action. The tool for implementing the Area Projects and taking on specific steps (drafting the project, its functions, and deadlines) for the Regions, Towns and Provinces to follow is the Project Framework Agreement (PFA) – undersigned by the Regions, Local Administrations, the Central Coordination Administration, and other dedicated Administrations.[5] To date all the 72 Project Framework Agreements (PFA) have been signed and the project scenarios are being implemented. The status of implementation is available on the Opencoesione website.[6]

The National Inner Areas Strategy is assuming that mobility services are a precondition for development, these differ from education and health services because they provide a service function.

To avoid that remoteness turns into marginality, there is a need to increase Inner Area accessibility for those essential services which qualify the concept of citizenship, education, and health. This can be achieved through two different, not necessarily mutually exclusive, sorts of action: a) bolstering and rethinking services in these areas; b) improving mobility from and within the areas, cutting the time taken to access the services available in the hubs.

A document of the Area Strategies for mobility identifies the actions according to specific factors which characterize the mobility demand in the Inner Areas. Firstly, it relates to the movement of both goods and people. Secondly, it highlights the distinction between internal mobility in each individual area and between adjacent areas and mobility to and from Inner Areas (i.e. originating or finishing in centripetal hubs).

These preliminary considerations lead to the identification of four basic criteria to help shape a mobility strategy for the Inner Areas, also in the light of subsequent meetings with stakeholders:

- environmental sustainability of the actions.
- well-coordinated territorial programming and transport planning.
- transport demand analysis.
- where possible and economically viable, the consolidation of pilot projects already implemented and in force that have produced good results.

---

[5] Detailed information about the whole process is available by the website of the Agency for the Territorial Cohesion at the following web address https://www.agenziacoesione.gov.it/str ategia-nazionale-aree-interne/.

[6] https://opencoesione.gov.it/it/strategie/AI/.

As shown in the 2018 Report [33] most areas have directed their attentions on action choices related to the rationalization of the existing supply of transport public services and the use of the recovered financial resources for the activation of non-conventional ad flexible specific services in reference to the specific's mobility needs of each area.

The key issue of approach used by most areas lies in the effort to place into a network the transport system resources of the territory, sometimes underused. In this regard the fleets (e.g. use of school buses in the afternoon for extracurricular activities), public transit operators and staff have been involved in a multiservice perspective, as private cars for particular types of carpooling or voluntary service (e.g. community voluntary drivers).

Flexible transport services with minibuses, door-to-door DRT with "virtual" stops, shared taxis, and carpooling are a part of the defined solutions set. Particular attention is given to the travel needs groups who are additionally hindered by their physical and mental disabilities, or social "handicap" (e.g. younger and elderly people). For this user set dedicated flexible transport services must be planned and the services should be provided by community transport organizations, health sector funded organizations, or local authority departments involved in social care. Furthermore, the acquisition of the minibuses is financed, including special equipment and the implementation of ITS [32, 35, 36].

A common critical factor for the Project area has been identified by the bad road infrastructures conditions of the inner/rural network, which limit the performance of transport services, restraining access and therefore constraining development.

For most of the 72 pilot areas funding was allocated to the carry out Mobility and feasibility studies to rethink the transport system in a financial sustainable way and design the services in an integrated way and according to the specific needs for both local user and visiting clients.

Based on such criteria and considerations in accordance with the specific Guidelines,[7] innovative and original pilot-projects have been set led.

Local development interventions in the selected project areas are financed by all the available EU (ERDF, ESF, EAFRD and EMFF), national and local funds, and of the total amount of €509,37 million allocated for the essential services for 71 areas, 50 percent is for the transport system (infrastructure and services) [34].

In the draft versions of the Partnership Agreement for the 2021–2027 period for Italy, the National Strategy for Inner Areas is part of the policy objective 5 - Europe closer to citizens will assume the role of structural cohesion politics and the number of inner areas affected by the Strategy will increase.

5 Conclusions

The concept of "inner" or "marginal" areas cannot be narrowly defined but in a simple way a rural/inner area is defined as an area with low accessibility. However, in addition to

---

[7] The Guidelines for the actions on the mobility system in the Inner areas are available at the following web address: http://territori.formez.it/content/aree-interne-e-mobilita-linee-guida-interventi-aree-progetto.

accessibility, many other criteria are used to delineate centres and peripheries in regional research.

To identify a region also as remote in many cases an accessibility analysis is performed. Therefore, in this type of analysis the driving time by car needed for a certain percentage of the population of a region to reach a populated centre is quantified. This approach has a point of weakness related to the postulate that in inner/rural areas everyone has a car. Many people cannot drive, by reason of age, condition, or financial issues. When the household car(s) is(are) in use, other household members do not have access to it. Low-income households and individuals may not have a car. In such situation the measures of remoteness adopted is misleading and doesn't give the full extent of the real marginality of the areas. It should be taken in account the availability of public transport services and the travel time by bus to reach urban centers which offer a wide range of essential services like healthcare, education, and transport. Additionally, it is necessary to consider that in Italy in most rural/inner areas, although there is a well-established network of local roads the infrastructure is characterized by low functionality due low maintenance, landslides, and geological instabilities. It follows, therefore, that in the accessibility analysis the real traveling speed must be used.

Accessibility is crucial to guarantee considerable improvements in the provision of essential services (primary and secondary school and vocational training, healthcare, and medical services.

Literature and the Italian National Inner Areas Strategy assume that mobility services are a precondition for development, and this differ from education and health services, as they provide a service function.

Rural mobility presents many problems. Different issues regarding transport provision were identified: limited access and connectivity, longer distances, lack of public transport and/or alternatives to private cars, and lack of financing for mobility schemes. The specific rural land-use patterns (urban sprawl, scattered regions, lack of territorial planning for rural areas) are an important factor that leads to a high degree of car dependency in rural populations and, together with the poor infrastructure and lack of public transport supply, lead to limited accessibility and connectivity.

Re-thinking the mobility system needs for a new vision and a new setting for rural mobility policies to improve rural mobility and to bring long-term benefits. In the inner and rural areas as evidenced in many UE projects and in the National Strategy for Inner Areas implemented in Italy, securing equal mobility rights is a much more difficult task in rural areas. In less/sparsely populated areas and with weak demand there is not a single solution to be developed and implemented but a mix of them based on flexibility and user-oriented and financial sustainability.

Intermediate results show that the Italian National Strategy has produced a significant change in Inner Areas by outlining the importance of place-based solutions to place-specific issues, involving in the process local stakeholders, communities, and policymakers. The mobility system to achieve the long-term objectives is redesigned in an integrated way with the territorial development scenario and in the perspective to guarantee access to essential services.

One of the challenges is to maintain the implemented services in operation after the experimental phase ends, and the operating costs are not covered by dedicated funding.

Implication if the National Strategy for Inner Areas were to become a structural policy would be that the necessary financial resources to the transport system in addition to the National fund for Transport should be guaranteed. Cohesion and integration policies cannot be implemented without a planned and continuous financial support.

References

1. US Department of Transportation, Federal Highway Administration: Planning for Transportation in Rural Areas (2001). 04705r02 090701-10.47
2. Scottish Executive: Urban Rural Classification 2016 (2018). ISBN: 9781788516204. https://www.gov.scot/publications/scottish-government-urban-rural-classification-2016/pages/1/
3. Public Investment Evaluation Unit (UVAL): A Strategy for Inner Areas in Italy: Definition, Objectives, Tools and Governance (2014). www.agenziacoesione.gov.it/lacoesione/le-pol itiche-di-coesione-in-italia-2014-2020/strategie-delle-politiche-di-coesione/strategia-nazion ale-per-le-aree-interne/
4. Brovarone Vitale, E., Cotella, G.: Improving rural accessibility: a multilayer approach. Sustainability 2020(12), 2876 (2020). https://doi.org/10.3390/su12072876
5. Mounce, R., Wright, S., Emele, C.D., Zeng, C., Nelson, J.D.: A tool to aid redesign of flexible transport services to increase efficiency in rural transport service provision. J. Intell. Transp. Syst. 22, 175–185 (2018)
6. Hunkin, S., Krell, K.: Policy Brief on Demand Responsive Transport. Interreg Europe, Lille (2018)
7. Wright, S., Emele, C.D., Fukumoto, M., Velaga, N.R., Nelson, J.D.: The design, management, and operation of flexible transport systems: comparison of experience between UK, Japan and India. Res. Transp. Econ. 48, 330–338 (2014)
8. Jain, S., Ronald, N., Thompson, R., Winter, S.: Predicting susceptibility to use demand responsive transport using demographic and trip characteristics of the population. Travel Behav. Soc. 6, 44–56 (2017)
9. Eckhardta, J., Nykänena, L., Aapaojaa, A., Niemi, P.: MaaS in rural areas - case Finland. Res. Transp. Bus. Manage. 27, 75–83 (2018). https://doi.org/10.1016/j.rtbm.2018.09.005
10. VTT Technical Research Centre of Finland: Mobility as a Service (MaaS) in rural context On NORDIC, Road and Transport Research (2018). https://nordicroads.com/mobility-ser vice-maas-rural-context/
11. Barreto, L., Amaral, A., Baltazar, S.: Mobility as a Service (MaaS) in rural regions: an overview. In: 2018 International Conference on Intelligent Systems (IS), pp. 856–860 (2018). https://doi.org/10.1109/IS.2018.8710455
12. Alyavina, E., Nikitas, A., Njoya, E.T.: Mobility as a service (MaaS): a thematic map of challenges and opportunities. Res. Transp. Bus. Manage., 100783 (2022)
13. Hensher, D.A., Mulley, C., Nelson, J.D.: Mobility as a service (MaaS) – going somewhere or nowhere? Transp. Policy 111, 153–156 (2021)
14. Lopez-Carreiro, I., Monzon, A., Lois, D., Lopez-Lambas, M.E.: Are travellers willing to adopt MaaS? Exploring attitudinal and personality factors in the case of Madrid, Spain. Travel Behav. Soc. 25, 246–261 (2021)
15. Cirianni, F., Leonardi, G., Iannò, D.: Operating and integration of services in local public transport. In: Bevilacqua, C., Calabrò, F., Della Spina, L. (eds.) NMP 2020. SIST, vol. 178, pp. 1523–1531. Springer, Cham (2021). https://doi.org/10.1007/978-3-030-48279-4_142
16. Ministero per la Coesione Territoriale e il Mezzogiorno: Relazione annuale sulla Strategia Nazionale per le Aree Interne (2018). https://www.agenziacoesione.gov.it/wp-content/uploads/2020/07/Relazione_CIPE_2018.pdf

17. Ministero per la Coesione Territoriale e il Mezzogiorno: Relazione annuale sulla Strategia Nazionale per le Aree Interne (2019). https://www.agenziacoesione.gov.it/wp-content/uploads/2021/02/Relazione-annuale-al-CIPE-Anno-2019.pdf

18. Schürmann, C., Talaat, A.: Towards a European Peripherality Index. Final report, Berichte aus dem Institut für Raumplanung 53, IRPUD, Dortmund, Germany (2000). https://ec.eur opa.eu/regional_policy/sources/docgener/studies/pdf/periph.pdf

19. Cirianni, F.M.M., Leonardi, G.: Analysis of transport modes in the urban environment: an application for a sustainable mobility system. WIT Trans. Ecol. Environ. **93**, 637–645 (2006)

20. Wegener, M., Eskelinnen, H., Fürst, F., Schürmann, C., Spiekermann, K.: Criteria for the Spatial Differentiation of the EU Territory: Geographical Position, Forschungen 102.2, Bundesamt für Bauwesen und Raumordnung, Bonn (2002)

21. Donnges, C.: Improving access in rural areas guidelines for integrated rural accessibility planning. ASIST-AP Rural Infrastructure Publication No. 1, Bangkok (2003). ISBN 92-2-113649-3

22. Páez, A., Scott, D.M., Morency, C.: Measuring accessibility: positive and normative implementations of various accessibility indicators. J. Transp. Geogr. **25**, 141–153 (2012)

23. Miller, E.: Measuring accessibility: methods and issues. International Transport Forum Discussion Papers, No. 2020/25., OECD Publishing, Paris (2020)

24. Amed, S., Ehlund, E.: Rural accessibility, rural development, and natural disasters in Bangladesh. J. Dev. Soc. **35**(3), 391–411 (2019)

25. OECD: Creating Rural Indicators for Shaping Territorial Policy. OECD, Paris (1994)

26. Brezzi, M., Dijkstra, L., Ruiz, V.: OECD extended regional typology: the economic performance of remote rural regions. OECD Regional Development Working Papers, 2011/06. OECD Publishing (2011). https://doi.org/10.1787/5kg6z83tw7f4-en

27. ESPON: PROFECY- Processes, Features and Cycles of Inner Peripheries in Europe. Applied Research. Handbook ESPON EGTC, Luxembourg City (2017)

28. Word Economic Forum: Transforming Rural Mobility with MaaS. White Paper (2021)

29. Palmer, K., Dessouky, M.M., Abdelmaguid, T.: Impacts of management practices and advanced technologies on demand responsive transit systems. Transp. Res. Part A-Policy Pract. **38**, 495–509 (2004)

30. Nelson, J.D., Wright, S.: Flexible transport management. In: Bliemer, M.C.J., Mulley, C., Moutou, C.J. (eds.) Handbook on Transport and Urban Planning in the Developed World, pp. 452–470. Edward Elgar Publishing Ltd. (2016)

31. SMARTA consortium: Rural Mobility Matters Sustainable Shared Mobility Interconnected with Public Transport in European Rural Areas (Developing the Concept of "SMArt Rural Transport Areas") Insights from SMARTA. Smarta broschure II 08-03 (2021). https://rurals haredmobility.eu/wp-content/uploads/2021/01/Smarta-broschure-II-08-03.pdf

32. Nuzzolo, A., Comi, A.: Dynamic optimal travel strategies in intelligent stochastic transit networks. Information **12**(7), 281 (2021). https://doi.org/10.3390/info12070281

33. European Network for Rural Development (ENRD): Strategy for Inner Areas Italy. Working Document (2018). https://enrd.ec.europa.eu/sites/enrd/files/tg_smart-villages_case-study_it.pdf

34. Presidenza del Consiglio, Dipartimento per le politiche di coesione: Relazione annuale sulla Strategia Nazionale per le Aree Interne (2020). https://www.agenziacoesione.gov.it/wp-con tent/uploads/2021/11/Relazione-CIPESS-2020_finale.pdf

35. Comi, A., Nuzzolo, A., Crisalli, U., Rosati, L.: A new generation of individual real-time transit information systems. In: Nuzzolo, A., Lam, W.H.K. (eds.) Modelling Intelligent Multi-Modal Transit Systems, pp. 80–107. CRC Press, Taylor & Francis Group, Boca Raton (2017)

36. Nuzzolo, A., Comi, A.: A subjective optimal strategy for transit simulation models. J. Adv. Transp. **2018**, 8797328 (2018). https://doi.org/10.1155/2018/8797328

Optimization of Local Road Network Quality

Francis M. M. Cirianni[ID] and Giovanni Leonardi[(✉)] [ID]

Department of Department of Civil, Energy, Environmental and Material Engineering (DICEAM), University Mediterranea of Reggio Calabria, via Graziella Feo di Vito, 89100 Reggio Calabria, Italy
{francis.cirianni,giovanni.leonardi}@unirc.it

Abstract. Optimal management of a road transport network is requested to ensure adequate safety and comfort standards to users, while containing the generalized transport cost.

It is necessary that the administrations which manage roads adopt an adequate tool to monitor the network supply (DSS), giving an efficient technical support for decisions at different levels of competence.

This assumes relevance for the choice of maintenance intervention alternatives especially in areas of environmental complexity.

In the present study a methodology of analysis is presented, relating the construction of an information set to the definition of a matrix $R = r_{ij}$, $(i = 1..n; j = 1..m)$, being n the number of links of the network and m the indicators used (performance, econometric, accidents, environmental, etc.).

Finally, using a specific statistic method, we arrive to an assessment of the relative significance of each of the m vectors (of dimension n) representative of the synthesis.

Keywords: Road network · Management · Road safety

1 Introduction

1.1 The Analysis of the Functional State of a Road Network

The management optimization of a local road network is one of the key-elements to achieve the definition of the generalized transport cost and to minimize the maintenance costs and the vehicular traffic pollution [1–4].

If an infrastructure has been programmed at a decision-making level of economic policy and consequently designed and built, it is because it falls within the actions which fulfil the needs of territorial mobility of people and goods. In particular, local roads must guarantee preset performance qualities, such as given levels of safety, comfort, and service in accordance with the identified characteristics of the flow and the environmental contest [5].

From this follows the necessity that any road element is maintained, for the whole life-time, in full efficiency and functionality, so to obtain the greatest global benefits for the community.

F. Calabrò et al. (Eds.): NMP 2022, LNNS 482, pp. 377–383, 2022.
https://doi.org/10.1007/978-3-031-06825-6_36

Nowadays, thanks to the state of research in the road transport field and to available modern technologies; it is possible to perform the scheduled analysis of the demand in maintenance of the road network (urban and extra-urban).

Once defined the strategic variables, the survey data is essential to draft an effective and integrated planning of the maintenance activities. The consequent actions depend on the technical-organizational ability of the Road Administration and on the economic resources available.

For an optimal management of the road network, the analysis and monitoring of strategic variables which characterize the road network should start during the in the design phase, through construction and opening to use, to create a complete a dataset [6].

The functional contribution given by this database extends not only to the design stage of the network (projects or plans of intervention on the territory) but it assumes particular importance also for the optimization of the scheduled processes of maintenance.

It is necessary to point out that planning, realization, service, and maintenance, although all fall indifferent time phases from each other in the lifespan of an infrastructure, must be considered as successive stages in a single integrated plan, consisting in the full and satisfactory use of a public good by the community.

Within the network information system, analyzing a given data set, it is possible to identify the areas of risk, which are identified as the black spots on the road network, to draft a ranking of priorities in the requalification of the analyzed network.

1.2   The Construction of a Road Network Information System for the Integrated Maintenance

The allocation of the budget for maintenance is crucial for the management of a road network, and therefore the use of IT implementing a dynamic model allows to have the knowledge in real time of the state of use of the network, planned maintenance programs, extraordinary works and interferences with other networks (e.g. services and communications) and give an order of priority to arcs and nodes.

The proposed model is the core of a system decision management supporting technical decisions in the choice of the alternatives also in areas of environmental vulnerability [7].

As a basic requirement, the road information system requires the availability of a Geographical Information System associated with it, including the survey of the functional and operational characteristics of the different arcs and nodes of the network and the environment critical levels in particular areas [8].

The operating steps for an effective management of information [9], aimed at the process of global optimization (under the technical, economic, and environmental profile) of the network configuration, presuppose the development of the following activities:

- development of the aggregation techniques of the elements of the cartography (arcs and nodes of the road network) through the creation of a logical superstructure based on the original structure for the construction of the graphs of the same road network.
- conversion of the logical structure created in the previous point (*route system*) in the mathematical structure (road network graph) used by calculation models.

- extraction of subsets of attributes of nodes and arcs relating to the basic maps associated with elements of the analysed graph.
- realization of the environments interface "user-system" for the updating of the attributes of the network elements on the base of the territorial data.
- interpretation and evaluation of the indicators of the system (obtained by modeling the transport system) as support to the decisions.

So, the logical reference scheme is shown in the following Fig. 1:

- the system of information, constituted by a base of descriptive data (including the environmental characterizations of the area of interest) and by a base geographical-territorial data and its management system.
- the system of the models, constituted by a base of the models and its management system.
- the module of processing problems, described as "inside interface".
- the module of user interface, known in literature as "module for the generation of the dialogue".

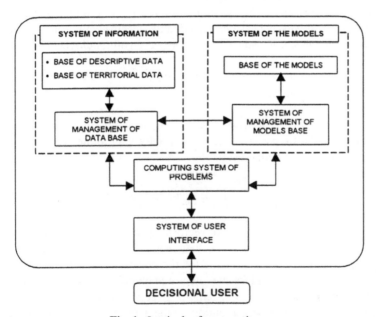

Fig. 1. Logical reference scheme

With the aim of an efficient use of the resources of the different components, particular attention must be put in the process of integration between systems, in logical-functional and operational terms [10].

2   The Mathematical Characterization of the Network Quality

From the mathematical point of view, the formulated methodological approach for the construction of the information system of the road network can be represented in the s-dimensional reference space R^s of the s vectors associated with the different classes of strategic variable observed in the analysis of network.

If, in fact, for the generic h-th arc of the network, we indicate with:

$\vec{q}_h$ the vector of the route operating characteristics and quality (geometry of the axis, level of service, safety and comfort, etc.);
$\vec{t}_h$ the representative vector of the state of efficiency in exercise of the crossing infrastructures, for the single considered typologies (bridges, galleries, etc.);
$\vec{q}_h$ the characterizing vector of the functionality and the adequacy of the minor art works to service of the road structure (roadside ditches, retaining walls, etc.);
$\vec{g}_h$ the relative vector of the typical indicators of the exercise management (generalized cost of the transport of persons and goods, indicators of pavement quality, monitoring networks, etc.);
$\vec{n}_h$ the characterizing vector of the global quality of the nodes (functionality and efficiency of the intersections, of the roundabouts, etc.);
$\vec{e}_h$ the representative vector of the existing relationships between the generic arc of the network and the considered ecosystem, included the specific configurations characterizing of the urban, architectural and landscaped aspects, anthropologized sites of elevated sensibility, etc.;
$\vec{s}_h$ the characteristic vector of the interactions between the examined h-th arc and the context of multi-modal transport system.

For all the n arcs of the network, considering all the n component of the column vectors above brought, for $h = 1, \ldots, n$, it can totally refer to the vector $\vec{v}$, union set of the s classes of previously defined variable.

In symbols we have:

$$\vec{v} = \{\vec{v}_h\}_1^s = \vec{Q} \cup \vec{T} \cup \vec{O} \cup \vec{G} \cup \vec{N} \cup \vec{E} \cup \vec{S} \tag{1}$$

Such methodological formulation, of vectorial type in the space R_S, allows a suitable and easy use of the important database that it is necessary to build for the network quality characterization. This is strategic for the optimization of the maintenance and requalification interventions and for the integrated planning of new interventions.

In such circumstance, for every specific project it will be necessary to analyze the effects generated from the new intervention on the network, defining a new vector $\vec{K}$.

The vector $\vec{K}$ keeps account of the vehicles flow variations on the arcs, of the changes in the generalized cost of transport, of the changes in the configuration of the multi-modal system, of the changes in the localized infrastructures (as an example, intermodal centers, ports, airports) and of the positive and negative variations of the impacts generated by the flow on the network, etc.

It is possible to define a different vector $\vec{v}$ m with component vectors (with $m > s$) which also consider the classes of the variable above delineated.[1]

In any case, for the considered vector $\vec{v}$, the problem can be mathematically represented by a matrix $\mathbf{R}$ constituted from column vectors of order n (equal to the number of network arcs) [11, 12].

Let us consider, then, the general case in which the vector $\vec{v}$ has components $\vec{v}_j$ with $j = 1, ..., m$. Each of these vector $\vec{v}_j \in \vec{v}$ will be formed by element r_{ij}, with $i = 1, ..., n$, relative to every observed variable j-th, with $j = 1, ..., m$.

Generally, the considered variables are expressed in different units' measure (being, as above illustrated, reported to technical, economic, and environmental aspects), so it is necessary to make the acquired date comparable to proceed to a global analysis and to individualize the eventual existing correlation between the different used indicators, eliminating those that result statistically *redundant*.

To this aim, once calculated the average $\bar{r}_j$ and the standard deviation σ_j it is possible to associate with every original element r_{ij} of v_j a new a-dimensional element r_{ij}^* given by the following formula:

$$r_{ij}^* = \frac{r_{ij} - \bar{r}_j}{\sigma_j \sqrt{n}} \tag{2}$$

In conclusion, at the matrix $\mathbf{R}$ is possible replace the matrix $\mathbf{R}^*$ of standardized variables, always of $n \times m$ order:

$$R^* \equiv \left\| r_{ij}^* \right\| \tag{3}$$

con $i = 1,..., n; j = 1, ..., m$

The following step is the transposed matrix $^T\mathbf{R}^*$ of $\mathbf{R}^*$, and therefore calculating the matrix $\mathbf{C}$ given by the product:

$$C = {}^T R^* \cdot R^* \tag{4}$$

Such symmetrical matrix (of the *deviancies* and *codeviancies*) has elements $c_{ij} = c_{ji}$ ($i, j = 1,..., m$; for $n > m$) such that:

- for $i = j$ they will be equal to 1 (degree of correlation of every variable with itself);
- for $i \neq j$ they will assume values contained in the interval $[-1, 1]$, characterizing the respective degree of correlation between the different considered couples of variables.

Starting, from such formulation of the problem, we will search a particular vector $\vec{s}$ that, through its correspondent $\vec{s}_h$, can characterize the content of information associate with it, in progressively decreasing way to the growing of h.

To this aim, we consider the roots of the equation:

$$\det(C - \lambda I_m) \tag{5}$$

---

[1] In the specific case of the construction of new q arcs, the dimension of $\vec{v}$ will suffer variations, becoming completely equal to $r = n + q$.

in which I_m is the unit matrix, that determines the *eigenvalue* λ_h ($h = 1, .., m$), with $\lambda_h > 0$, and, after, the correspondent *eigenvector* $\vec{g}_h$, obtained by the expression:

$$(C - \lambda_h I_m) \cdot \vec{g}_h = O_{m1} \tag{6}$$

where O_{m1}, is the column matrix of order n, in which every element is zero.

Ordering, now, the eigenvector $\vec{g}_h$ in such way that the respective modules will be: $g_{h-1} > g_h > g_{h+1}$, it is possible to construct the researched characteristic vector $\vec{S} \equiv \{\vec{s}_h\}_1^m \equiv \left\| g_{gh}^2 \right\|$ in which the generic element represents a content of decreasing information at h increasing.

In such way, analyzing the values assumed by the components s_{gh} it is possible to estimate the capacity of every original variable, used for the study of the quality of the network offer, and, then, to explicit the required information to have a descriptive economy of the phenomenon and the observed processes within the structure of the road information system, without, however, an appreciable loss of information.

3   Conclusions

The progressive increase of the vehicle flows on the network and the elevate levels of impact on the environment requires the adoption of suitable monitoring policies of the road network quality, to optimize the relation *"transport system-environment"*.

Within this frame a dynamic monitoring tool is proposed, which can classify the conditions of the single arcs of the road network, and prioritize ranking on parameters of quality and safety, providing that such standards are adequate, with the aim to contain, within acceptable limits, possible impacts on the ecosystem.

In the present paper, an interpretative model for the study of the quality of the local network based on vectorial algebra is proposed.

Representing the problem in the $\Re^m$ space of the observed variables (aggregated indicators of technical, economic, and environmental order) calibrated by the means of an appropriate mathematical algorithm, the more important strategic variables (with high content of information) are defined for the characteristics of the network. Analyzing the values assumed it is possible to estimate the capacity and therefore the quality of the network offer, giving information of the road information system with a full field of information, allowing the ranking of arcs and the prioritization of maintenance works.

References

1. Cirianni, F., Leonardi, G.: Environmental modeling for traffic noise in urban area. Am. J. Environ Sci. **8**(4), 345–351 (2012)
2. Friedrich, M.: Functional structuring of road networks. Transp. Res. Procedia **25**, 568–581 (2017)
3. Ma, J., Cheng, L., Li, D.: Road maintenance optimization model based on dynamic programming in urban traffic network. J. Adv. Transp. **2018**, 4539324 (2018)
4. Giunta, M., Bosco, D.L., Leonardi, G., Scopelliti, F.: Estimation of gas and dust emissions in construction sites of a motorway project. Sustainability (Switzerland) **11**(24), 7218 (2019)

5. Núñez-Alonso, D., Pérez-Arribas, L.V., Manzoor, S., Cáceres, J.O.: Statistical tools for air pollution assessment: multivariate and spatial analysis studies in the Madrid region. J. Anal. Methods Chem. **2019**, 9753927 (2019)
6. Keeney, R.L., Raiffa, H.: Decisions with Multiple Objectives: Preferences and Value Trade-Offs. Cambridge University Press, Cambridge (1993)
7. Castro, J.T., Vistan, E.F.L.: A geographic information system for rural accessibility: database development and the application of multi-criteria evaluation for road network planning in rural areas. In: Bougdah, H., Versaci, A., Sotoca, A., Trapani, F., Migliore, M., Clark, N. (eds.) Urban and Transit Planning. ASTI, pp. 277–288. Springer, Cham (2020). https://doi.org/10.1007/978-3-030-17308-1_26
8. Barrile, V., Fotia, A., Leonardi, G., Pucinotti, R.: Geomatics and soft computing techniques for infrastructural monitoring. Sustainability (Switzerland) **12**(4), 1606 (2020)
9. Jesus, M., Akyildiz, S., Bish, D.R., Krueger, D.A.: Network-level optimization of pavement maintenance renewal strategies. Adv. Eng. Inform. **25**(4), 699–712 (2011)
10. Chun-quan, Y.: Dynamic information system of real-time traffic data on Beijing road network. Commun. Transp. Syst. Eng. Inf. **3** (2002)
11. Leonardi, G.: A fuzzy model for a railway-planning problem. Appl. Math. Sci. **10**(27), 1333–1342 (2016)
12. Ji, A., Xue, X., Wang, Y., Luo, X., Zhang, M.: An integrated multi-objectives optimization approach on modelling pavement maintenance strategies for pavement sustainability. J. Civ. Eng. Manag. **26**(8), 717–732 (2020)

Knowledge and Innovation Ecosystem for Urban Regeneration and Resilience1 - Inner and Marginalized Areas Local Development to Re-Balance Territorial Inequalities

Adapting Outdoor Space for Post COVID-19

Chro Hama Radha[1]([✉]) and Sivan Hisham Taher Al-Jarah[2]

[1] Department of Architecture, Faculty of Engineering, Koya University, Koya KOY45,
Kurdistan Region, Iraq
chro.ali@koyauniversity.org
[2] Technical College of Engineering, Sulaimani Polytechnic University, Sulaimani, Iraq

Abstract. Great disasters such as earthquakes, epidemics, environmental calamities have been known to trigger transformations of the built infrastructure, environment and give rise to unpredicted or unexpected conditions of social, cultural, and economic development. The fragility of settlement systems, when faced with unexpected threats, indicates the need for re-planning and changing our perspective on the city. This paper examines the effect of COVID-19 and its related social and infrastructure changes on cities and metropolitan regions. It looks at the literature on how pandemics have affected the built environment in the past and postulates how the current pandemic might affect the structure and morphology of cities examining evidence from the imposed lockdowns. An in-depth look at the transformation of the city of Melbourne (Australia) is explored and initiatives in other cities are highlighted. Moreover, 15-min cities can cut carbon emissions and bring communities closer together.

Keywords: Pandemic · 20-min city · Complete neighbourhood

1 Introduction

Historically, epidemics and pandemics are known to have significant impacts on settlement planning. The use of urban space, the material, and construction solutions are affected by great disasters. Fracture lines that were previously unclear are usually revealed in the face of unexpected threats. Settlement systems are normally shaken and the role of architects and researchers is addressing the emergency through planning for the future. The key to defining a design approach is observing past reactions of people in the face of sudden threats. Over the years there have been many contagion diseases caused by bacteria and viruses and COVID-19 is not the first virus to affect our cities. Looking at past events and cycles, it can be deduced that COVID-19 will not be the last virus to affect our cities. During the 14th century, one-third of the population in Europe and the Middle East was killed by the Black Plagues. During the 19th-century Cholera outbreaks decimated large cities like Paris, London, Hamburg, Moscow, New York, Chicago, among other cities. The Great Flu killed over 50 million people worldwide. The current pandemic has also left profound scars in many cities. Disasters such as the 1980 earthquake in the city of Palomonte (Italy) also took many lives. Past pandemics and disasters influenced substantial political, cultural and urban design changes but non manage to change the role and functioning of large cities in society [1]. Innovation, economic growth, and creativity require the coming together of talent, face-to-face interaction, and diversity which can only be provided by cities. Despite the levels of devastation COVID-19 has caused in certain cities, it is unlikely that it will derail the economic role of cities and the process of urbanization [2]. However, even though cities are not likely to lose their role, they will be transformed.

The transformation will range from how people work, how they interact, office space design, residential homes design, physical distancing in workplaces, outdoor space design in homes and neighbourhoods. COVID-19 has resulted in social scarring which might cause people to avoid crowded places. This is likely to influence travel patterns, residence choice, and the viability of certain businesses [3]. Due to the pandemic, classrooms and workplaces have transitioned to remote and social life is being carried out mostly on digital media. Despite the success of these measures, there are signs that distanced interaction cannot fully substitute face-to-face socialization. This indicates that public infrastructure requires design changes to facilitate social distancing. Architects, planners, and designers will need to seriously consider permanent interventions which will respond to future pandemic threats. This will result in changes to urban built form. Different configurations for outdoor and indoor spaces will be required to facilitate social distancing.

2   Effect of Pandemics on Settlement Systems in the Past

Disease and pandemics have shaped the way cities have been built. Cholera caused the death of millions between 1835 and 1893. It spread through trade routes used by eastern traders. Cholera resulted in the re-design and building of sewage systems. The main aim was to have zero emissions of harmful substances and the biodegradation of sewages [4]. For instance, Cholera in London was eliminated by the redesigning of the sewage system. A central park in New York City was put in place in the wake of Cholera outbreaks in the 1850s. Authorities began to believe that open space in cities improved human and environmental health. In the 19th century, tuberculosis invaded Europe [5]. Much like COVID-19, tuberculosis is an airborne disease but is caused by mycobacteria. Tuberculosis resulted in the construction of buildings equipped with verandas, balconies, and terraces which aimed at exposing users/residents to air and sunlight. Houses were constructed with extra open spaces to be used for temporary functions. In the 19th century, buildings in Paris (France) were designed with large high-up windows to allow plenty of sunlight. They were supposed to stop the spread of tuberculosis [5].

The 1793 yellow fever outbreak killed 10% of the population of Philadelphia (USA). Over the next 50 years, cities added alleyways to allow for garbage removal.

Airborne diseases are hard to combat because physical distancing and business closures are difficult to maintain over a prolonged period of time. Cities are trying some measures such as closing some streets to cars, that allows people to walk outside whilst maintaining some distance from others. A benefit for the residents' physical and mental health [6].

Studying historical events, three strategies for the intervention on the urban landscape can be identified:

- Design and construction of new infrastructures
- Reorganization of urban accessibility
- Creation of open spaces for collective use

3 Potential Effect of the Covid-19 Pandemic on Settlement Systems

Actions and strategies to fight the Covid-19 virus can be defined by understanding what was done in the past under the same pandemic circumstances. Recent scientific literature looks at the preventive potential offered by design strategies which on top of having the capability to contain the Covid-19 pandemic, they can also be linked to the United Nations Sustainable Development Goals. Suggestions are pointing towards transition to sustainability [7]. Significant proposals are those balancing between reducing the risk of Covid-19 transmission as well as increasing other performance levels such as reducing energy needs, improving well-being, and air quality.

Historically it is known that density and connection enable civilization to strive, innovate and achieve economies of scale. However, our desire to live close to one another has resulted in creating an environment conducive for infectious diseases like Covid-19. The behaviour and lifestyle changes triggered by the Covid-19 pandemic created ideal conditions to look more into the feasibility of 15-min cities. Cities will not have to be divided into district zones for living, working, entertainment but all those needs and services which daily life will be within a 15-min radius. The effect of the Covid-19 pandemics is to magnify and hasten transitions that are already apparent [8]. There is a strong possibility that the Covid-19 pandemic might prompt a shift from functional and economic cities to healthier ones. Several researchers around the world seem to be in support of the hypothesis. physical distancing and containment measures introduced during the pandemic brought about the realization of the importance of public space. Residents in the urban area are more aware of the importance of gathering places as being essential to their psychological and physical well-being.

Numerous studies indicate that loneliness is linked to health problems such as heart disease, depression, and reduced life expectancy. Individualism and new technologies have already contributed to isolation, with the lockdown in place, it deprived people of their daily interactions. Many people were able to maintain close contacts and exchanges with neighbours through the use of balconies, alleyways, and front yards while respecting physical distancing. Public spaces such as parks helped in maintaining socialization and interaction during lockdown periods. People also used walking to exercise and escape from confinement and get exposure to sunlight and fresh air. The pandemic has shown the need for wider sidewalks and it has demonstrated the importance of parks. Cities around the world have developed an awareness of the excessive space dedicated to the automobile and are trying to take initiatives to put urban spaces at the service of people.

4 Proposed Changes Post-Covid

Cities should be designed for humans. The idea currently being used is suburban sprawl. It is the reorganization or the creation of the landscape around the requirement of auto-mobile use. Portland Oregon made decisions in the 1970s that began to distinguish it from almost every other American city. Whilst most cities were pushing for suburban sprawl, they instituted an urban growth boundary. While most cities were reaming out their roads, removing trees in order to flow more traffic, Portland invested in bicycling and walking. These changes changed the way the people of the city lived. Portland now drives 20% less than the rest of the country. The money saved from driving is being spent in other areas. The main advantage is that they spend more in their homes. The best economic strategy for cities in the post-Covid era is not the old way of trying to attract corporations but rather to become a place where people want to be. Indeed, the rapid spread of the coronavirus is creating an urban design crisis. Inactivity causes health problems, that's why the outdoor activity is so beneficial, there is no such thing as a useful walk.

The pandemic will likely accelerate already emerging trends pushing towards more humane, healthier cities which may outlive it [9]. The push is aimed at developing people focused policy, updating planning policy to allow change of use in employment zones, listening to communities and addressing their concerns, removing private vehicle storage if it contradicts any of the above. Covid 19 has brought parks and the quality of facilities to the fore. Local authorities must invest in parks and playgrounds. Improve and enhance parks and the routes used to access them. Post-Covid this may mean evaluating existing statutory instrument agreements to reallocate funding. Changes to the infrastructure should be centred on the opinions of residents. Engage with parents doing the school run, who normally spend time in the neighbourhoods. Some suggestions are to make children's lives central to changes. If it works for them, it works for everyone. Create usable amenity space outside schools, shops and around parks [10].

The slower rate of contagion in open or outdoor spaces makes it safer to shop in a public market or commercial street than in an air-conditioned area or sealed shopping mall. This could affect urban design to reduce risks of transmission. Parks and greenery proved to be fundamental and proximity to a park within a 5-min walk was the strongest predictor of satisfaction during the lockdown. A survey conducted showed that neighbourhoods composing houses score markedly better than those with apartment blocks, particularly if high rise. Residents felt that the apartment neighbourhoods served them less well during the pandemic induced lockdown. People were looking for walking and cycling space. Post-Covid, this would require the availability of less-trafficked streets. A walk score rating was developed; it measures how walkable an area is.

The most popular post-Covid proposal which takes all the discussed considerations into account is the concept of complete neighbourhoods popularly known as the 15- or 20-min city, as shown in Fig. 1. A complete neighbourhood must satisfy five of the following seven factors:

– Streets with sidewalks on at least one side
– 0.25 mile to a trail or greenway
– 0.5 mile to general service, 0.25 mile to frequent service, or 0.125 mile to a regular service
– 0.5 mile to a neighbourhood park and 3 miles to the community centre
– 0.5 mile to a store
– 0.5 mile to business or service cluster
– 1 mile to a public elementary school

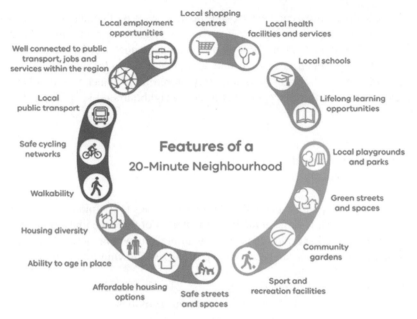

Fig. 1. Features of a 20-min neighbourhood [10].

Three of the factors are related to transit and four of them are related to provision of services. One of the requirements for a complete neighbourhood or a 15-min city is to have sidewalks [11]. This will require cities to adapt outdoor space to create wider sidewalks. Cars could be banned from some roads and the roads reserved for pedestrian use. Some cities have already started implementing such measures. In Rotterdam, cars are not allowed on certain roads after 4 pm. In California, Oakland there are streets where cars no longer have priority when it is mixed-use with pedestrians. Large urban parking lots are being transformed into farmer's markets in low-income neighbourhoods of Portland, Oregon. In the last decade, Montreal has been transforming street parking into temporary café patios. There is an observation indicating that the pandemic reinforced the trend of urban agriculture. This can act as agricultural therapy and improve on eating more healthily. Cities are rethinking their urban mobility systems. Paris has plans to make more space for bicycles by removing 72% of on-street parking space [12]. New ideas in Paris for a 15-min city aim to make all public services available within a 15-min walk to help people working from home. Indeed, Paris is removing 140,000 on-street car parking bays and stripping back roads to create more space for pedestrians and cyclists.

London has introduced its "Mini-Hollands" project where three boroughs have been given 30million pounds to create space for cyclists. Barcelona has been experimenting with "superblocks" where all interior roads within a 400 by 400-m block are closed off to all traffic except residents. 90% of Portland's residents will be able to walk or cycle to meet non-work needs by 2030. The future of cities is likely to be car less. Barcelona and Portland are also working towards it, ensuring 90% of your daily needs are within walking distance [13]. The walk score will be used in planning the cities. This looks at a neighbourhood and looks at the time it takes to reach essential services. In Copenhagen

90% of people do not use a car, yet 64% of transport space is allocated to cars, with only 26% for sidewalks and 7% for bikes. Vancouver city plans to connect walk and pathways citywide. By 2030 the city plans for 80% of trips to be made on foot, bike, or transit. A 15-min city has affordable, accessible, and adaptable housing for households. It means you can work close to home or work remotely more often. Manchester city has policy strategies that support the concept of a complete neighbourhood [11]. The policies are as follows:

- Thriving educated youth
- Economic prosperity and affordability
- Healthy connected city

One of the action plans which fall under these policies is complete neighbourhoods. Manchester city is also looking at the implementation of the 'Bee lines' project. This is a vision for the city to have a fully joined-up walking and cycling network. The goals of a complete neighbourhood can be summarized by the following four points:

- Ecology for green and sustainable city
- Proximity – to live with reduced distances to our activities
- Solidarity – to create links within people
- Participation

There are associated challenges with implementing the complete neighbourhood concept which are:

- Lack of buy-in for the schemes
- There is the risk of 'reverting to old ways after the pandemic
- Ensuring equality and accessibility

5   Proposal for the Development of Melbourne into a 20-min City

Melbourne is the capital city of Victoria and second largest city in Australia. Melbourne is a thriving cos metropolitan city of 5 million people and has been ranked as the world's most liveable city for 3 consecutive years. There is a proposal for a hypothetical Melbourne which is a 20-min city where you can easily move around in a fast flexible manner without the need for a car. The idea is presented at two scales:

- A neighbourhood – a walking or a cycling city
- A greater or metropolitan area – a commuting city

The importance of the central business district as a point of convergence is an outdated model for a resilient city. Melbourne's existing train network will not be able to cater to the future population. The new model advocates the introduction of 10 new urban centres at strategic locations on the existing train line network as shown in Fig. 2. These urban centres are identified in areas that are 30 min from the central business district. These are shown connected by a green line (railway line).

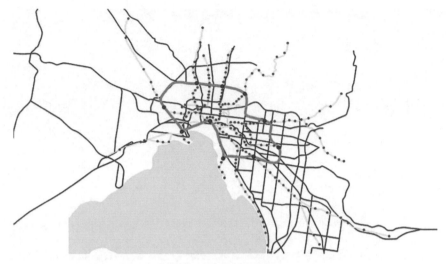

Fig. 2. New urban centres for Melbourne City [13].

These urban hubs are envisioned as a medium to high density redevelopments which will become centres for housing, education business and community. This will create opportunities for employment and will activate suburbia through the rejuvenation of suburban space for social interaction. The urban centres will be connected by rail establishing a suburban loop that capitalizes on existing infrastructure resulting in improved connectivity. At a local scale, a central point is identified in an existing suburb. When walking or cycling the 20-min city exists within a 2-km radius [14].

The 20-min radius nodes will be linked to urban centres by high-speed trains. The current lifestyle of the residents in the neighbourhood will be maintained. The sites dedicated to industrial space will continue to operate as per normal. The existing infrastructure of roads are now transformed to extend public open space within the street. A retail or commercial centre services the neighbourhood. Car parks are now identified as potential public open spaces. These spaces are now reclaimed for the community and become designated areas for interaction, urban farms, recreational parks, and an extension of the community infrastructure, as shown in Fig. 3. This modern living allows for greater flexibility, supermarkets now have the ability to grow fresh produce in once occupied car parks, streets are now free after the removal of cars. This provides greener space and expansion of the sidewalk giving more opportunity for community space and gathering. As a result, a local community is evolved into a 20-min walking and cycling city [15].

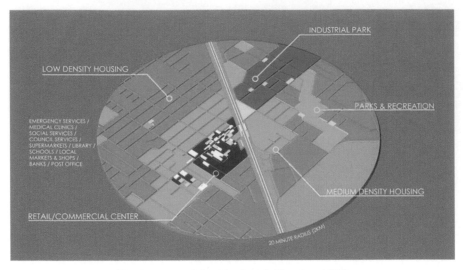

Fig. 3. Proposal for 20-min neighbourhood [13].

6   Conclusion

Complete neighbourhoods or 15-min cities can cut carbon emissions and bring communities closer together. The idea is a world away from many modern cities where all of our space is either planned around or surrendered to cars. In Copenhagen 90% of people do not use a car, yet 64% of transport space is allocated to cars, with only 26% for sidewalks and 7% for bikes. The lockdowns of 2020 and 2021 are changing how we see our cities and the idea of re-planning them is catching on. 15-min cities have the benefits of making the world a greener place and giving humanity a better chance at fighting the COVID-19 pandemic and other future airborne pandemics which might affect our cities.

7   Future Work

More studies need to be done in this field in crowded cities and different urban planning features and characteristics. Moreover, the focus could be on noise and pedestrians' thermal comfort, natural ventilation, and pollution in crowded cities.

8   Scope and Limitations

The paper looks at the literature on how pandemics have affected the built environment in the past- and postulates how the current pandemic might affect the structure and morphology of cities examining evidence from the imposed lockdowns. An in-depth look at the transformation of the city of Melbourne (Australia) is explored and initiatives in other cities are highlighted.

Acknowledgments. This research was supported by the Sustainable Buildings Research Program -SBRP- from the Research Centre at Koya University.

References

1. Sergio, C., Stephan, L., Emil, V.: Pandemics depress the economy, public health interventions do not: evidence from the 1918 flu (2020)
2. Michael, S., Anthony, J.: Buzz: face-to-face contact and the urban economy. J. Econ. Geogr. **4**, 351–370 (2004)
3. Maria, R., Serena, V., Katia, F.I., Maria, G.: Adaptive reuse process of the Historic Urban Landscape post-Covid-19. Int. J. Archit. Technol. Sustain. **5**, 87–105 (2020)
4. Bonj, S., Robert, H.: Nineteenth-century medical landscapes: John H. Rauch, Frederick Law Olmsted, and the search for salubrity. Bull. Hist. Med. **74**, 708–734 (2000)
5. Martini, M., Gazzaniga, V., Behzadifar, M., Bragazzi, N., Barberis, I.: The history of tuberculosis: the social role of sanatoria for the treatment of tuberculosis in Italy between the end of the 19th century and the middle of the 20th. J. Prev. Med. Hyg. **59**, E323–E327 (2018)
6. Lina, M., John, R.: The pandemic city; urban issues in the time of COVID-19. Sustainability **13**, 3295 (2021)
7. Sheldon, C.: Psychosocial vulnerabilities to upper respiratory infectious illness. Sage J. **16**, 161–174 (2020)
8. Violeta, B., Andrés, F., Tatsuya, A., Kelly, S., Rachel, R., Richard, A.: Urban green space use during a time of stress: a case study during the COVID-19 pandemic in Brisbane, Australia. People Nat. **3**, 597–609 (2021)
9. Anne, M.: The pandemic could force cities to build more outdoor and public spaces (2021). https://www.businessinsider.com/cities-focusing-on-building-outdoor-spaces-after-pandemic
10. Tom, R.: Ten-minute neighborhoods, 30-minute cities, Covid-19, and the future of public transit, Livable City. Org (2020). https://www.livablecity.org/ten-minute-neighborhoods-30-minute-cities-covid-19-and-the-future-of-public-transit
11. Anne, H.: Paris mayor unveils '15-minute city' plan in re-election campaign. The guardian for 200 years, International edition (2020)
12. Michael, G.: Outdoor Space Vital for Post Covid Planning (2020). https://www.built-enviro nment-networking.com/news/outdoor-space-planning-covid/
13. Richard, F., Andrés, R., Michael, S.: Cities in a post-COVID world. Urban Stud. J., 1–23 (2021)
14. Daniel, M.: The 15 minute City, Transformative Urban Mobility Initiative (TUMI) (2020)
15. Royal Town Planning Institute (RTPI): Implementing 20 Minute Neighbourhoods in Planning Policy and Practice (2021)

Reshaping Public Spaces Under Impacts of Covid-19

Sivan Hisham Taher Al-Jarah[1]([⊠]), Chro Ali Hama Radha[2],
and Rebaz Jalil Abdullah Abdullah[1]

[1] Sulaimani Polytechnic University, Sulaimani, Iraq
sivan81_univ@yahoo.com
[2] Koya University, Koya, Iraq

Abstract. (COVID-19) crisis has forced the entire world to face one of its most difficult challenges. In contemporary history, as it caused the injury of millions and the death of hundreds of people. Covid-19 cannot be described as a health crisis only, as it is a large-scale humanitarian crisis that leads to misery and the suffering of all mankind and push its social and economic well-being to the brink of collapse. Therefore, various issues have constantly been strived to shed light by the scientific community. Perhaps, as the number of infected people and death because of Corona virus have been raised, reshaping cities components and urban structure have been crucial need. This study aims to review impacts of epidemic on reshaping public spaces in both past and present time. Furthermore, propose new recommendations and strategies to generate healthy public spaces. The study result shows that each of governance and management, environmental quality, transportation and urban design and socioeconomic effects are the main impacts of COVID-19 that obligated planners and designers to rethink about the city components. Finally, COVID_19 crisis necessitates policy makers to take transformative actions to create public spaces in cities that are more sustainable and resilient.

Keywords: COVID-19 · Public space · Urban structure · City reshaping

1 Introduction

Throughout history, cities have been reshaped under impacts of pandemic [1]. The corona pandemic associated with many health crises, therefore, it has imposed a massive wave of revisions and rethinking of the modern human lifestyle and well-being [1, 2]. Additionally, COVID-19 caused economic and social consequences. The lockdown in cities and urban centers have shaped a new city image that have created disturbing experience for city dwellers who first saw their cities in a different and unprecedented figure [2, 3]. However, this is not the first time in the history that pandemic reshape cities, limited studies regarding cities and pandemic have existed before emergence of the COVID-19 pandemic [2].

The global health crisis has forced urban planners to rethink and rearrange their deeply held beliefs about what a post COVID-19 cities look like [1, 2, 4]. Investigating

© The Author(s), under exclusive license to Springer Nature Switzerland AG 2022
F. Calabrò et al. (Eds.): NMP 2022, LNNS 482, pp. 396–405, 2022.
https://doi.org/10.1007/978-3-031-06825-6_38

around impacts of pandemic on cities and how cities might be reshaped in this regard. The planning purpose is more resilient or containment to epidemics, including alternatives to urban innovation to address issues of environment, density and beauty [5]. Therefore, it is very important to reconsider the size, layout and spatial distribution of cities components including public spaces with its sidewalks, parks and open spaces as well as public facilities such as offices and community centers [6].

In this regard, it is very important to emphasis on designs and strategies that offered a healthy environment in cities [2, 7]. Understand what actions are required to minimize the impact of pandemic on urban area and its public spaces in particular [8].

This study is focused on the impact of the COVID-19 on the public space historically. The association between the elements of public spaces such as (streets, sidewalks, parks, squares, subways and other share spaces) are predominantly affected by the pandemic, therefore, many challenges have been faced by these areas that were derived from the health perspective. The study method depends on the investigations of viewpoint of planners and designers regarding urban area and health through literary review of previous studies regarding epidemic impacts on reshaping public spaces in both past and present time. Then, the paper is interpreting the primary literature on the subject in order to achieve the way that the function of public spaces can be managed and organized during the crisis. At the end the study comes up with new strategies and recommendations of the public spaces in pandemic.

2 History of Pandemics and Cities Reshaping

Throughout history pandemics caused transformation and reshaping in cities [1, 2]. Therefore, COVID-19 is not the only pandemic that have hit the world and left health crisis, economic consciousness and urban influences [9]. COVID-19 caused the worst scenario and produced a depressing image of the city due to the empty environment and closure of public spaces [1, 10]. The sudden transformation and related issues that associated to pandemics have activated the argument on vulnerability of cities. Perhaps, pandemics arising interest in the topic.

Certainly, since the early stages of the COVID-19, studies have paid special attention on cities and awareness regarding the impact of pandemic [8, 11]. Researches that always seek to unravel the original trends and pattern of the COVID-19 pandemic in cities and public spaces by highlighting on the various features of this pandemic are shedding light of the research published in recent times [6, 12, 13]. Without a hesitation, the pandemic has reawakened apprehensions about urban spaces and their possible susceptibilities to pandemics and infectious diseases [14].

Throughout history, studies regarding the impact of pandemics on urban spaces show that cities its experiences of epidemics connected to each other [15]. One of the important facts that arisen in the current crisis is that planning is not only about the materiality of the city and its physical components, but about the participation of its people in the production of the city and its features and priorities [16, 17]. The future will certainly require architects, urban designers and planners to descend from their ivory towers, where they have been holed up for decades, and to deal with the challenges of how to build society based on social interaction [18]. Even though, isolation is one of the phenomena that occurs due to the COVID_19. How can communities be involved in shaping space, space and the city? How can communities' contribution be activated to meet real needs?

The most important part for those interested in the built environment, is how to build an environment that adapted with the changes associated with post-coronavirus era. It is necessary to fully realize the importance of the social responsibility by architects, urban designers and planners [2, 15, 18]. Their design philosophy needs to depend on real community involvement so that its members contribute completely in the process of designing and planning its spaces, buildings and residences [19, 20].

It seems that it is necessary to move into an era of small and cohesive community planning rather than competition to transform the world's cities into bloated global cities in which their plan and design not suit the pandemic issues [8, 21, 22].

We return to China's experience again and are surprised by the positive consequences of the country's forced stalemate in the face of the epidemic. Carbon emissions and pollution from Chinese industry have declined since the virus first hit the country [10, 23, 24]. Only two months without production, the sky and atmosphere were cleared and people were allowed to breathe healthy again.

3   Rethinking the Form and the Function of Public Spaces

3.1   Post-pandemic Public Spaces

Cities design and planning in the past and present have been addressed by its density, public spaces, building design, public transport, street layout, green areas and parks [1, 2]. Introducing a health criterion to the public urban spaces need to be reactivated due to the pandemic consequences (see Fig. 1).

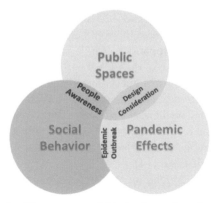

Fig. 1. Interrelationship of public spaces, social behavior and pandemic effects. [Source: By the authors]

Citizen behavior and social awareness have a significant importance in dealing with the COVID-19 era [14, 25]. Perhaps, when it comes to urban sustainability, public spaces play a critical role at all scales. They serve the city by providing social services, environmental and health scopes [26]. Also, public spaces provide excellent economic benefit to the city [27]. Public spaces and green area emphasize the importance of wellbeing and health, they used to improve health and inequality struggle [9, 13, 15].

Having access to well-maintained public urban spaces in pandemic time deliver a positive health environment and support a physiological condition and physical activities [19, 17, 22]. Even though, reducing the high population gathering is an attempt to prevent the spread of COVID-19 but public spaces can be widened and other isolated design measured can be doubted [25, 28]. Consequently, re-planning and protection of public space have become a critical issue to reconsider the size, design and spatial distribution of public space, including sidewalks, parks, and open [29].

3.2 Impacts of COVID-19 on Urban Life

A. Social Dimensions

Public spaces have a significant impact on culture identity, adding distinctiveness, and enhance sense of places among the community [13, 15]. Therefore, well- maintained and well design of open spaces foster social bonds that have been mistreated in many cities. Public spaces provide sense of belonging, recreational opportunities, support social cohesion and comprise identity features in cities. Regarding social features in the public spaces, social distancing is the most important measure that entered the social worldwide dictionary [12]. The relationship and distance of human bodies to both others decide the range of social behaviors. Four distances between people have been defined as the following Table 1.

Table 1. The hidden dimension of distance between people [8].

No.	Type	Distance	Relationship
1	Intimate	Less than 1.5 feet	Common with dear ones
2	Individual	1.5 to 4 feet	Joint with respectable friends and family
3	Social	4 to 12 feet	Common with acquaintances
4	Public	12 to 25 feet and extra	Familiar with wholly

Socially rooted, the cumulative physical distance between bodies determines these four measures, and each space form a different physical perception between people. Hall distributed interactive areas into four types (See Fig. 2).

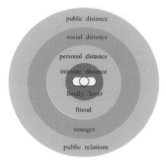

Fig. 2. Personal space & social distance [18].

Thus, the social distancing means conserving the space or distance between people to prevent the spread of the disease, or at least to slow the spread of epidemiologic diseases like COVID-19 and reduce the danger of infection [12]. In order to decrease the risk of spreading the viruses, people need to be distant 6 feet away from each other as minimum. Despite using the social distance awareness broadly but it may be carriage the incorrect message and meaning to social isolation [12].

Therefore, Physical distancing assist staying socially associated but keeping yourself away from other people physically. Moreover, it is important to maintain emotional and social communication with other, perhaps, people need to support each other but keep physical distance as far as conceivable [25]. In addition, World Human Organization (WHO) also preferred to use the phrase "physical distancing" instead of "Social Distancing" [18].

B. **Environmental Dimension**

The coronavirus pandemic has significantly affected the environment and climate in several aspects [11]. Sharp reductions in the travel and mobility of people and social and commercial activities have reduced the level of air pollution in many areas. The closures and other measures have resulted in a 25% reduction in carbon emissions, which one Earth systems scientist estimated may have saved at least 77,000 organisms over a period of two months. Despite this, the outbreak has obstructed environmental diplomacy efforts. Causing the 2020 United Nations Climate Change Conference to be postponed, and the economic fallout from it is expected to slow investment in green energy technologies.

Climate change has intensified in the world, and its negative effects have emerged clearly in various ecosystems, through climate fluctuations, frequent droughts, flash floods, and others [11]. In addition to developing practical plans and programs that help countries crystallize their development and educational projects. In accordance with the principle of precaution and comprehensive awareness of the general public [29].

It is indisputable that people staying in their homes is in the interest of the planet, which may lead to say that despite the absence of any benefits from the current epidemic, the resulting ban on unnecessary movements and the imposition of complete closure procedures in some countries of the world, made us able to monitor positive changes that occur on our planet due to the absence of its inhabitants in general for the first time ever, as is the case now [21, 27].

C. **Governance and Management**

The Corona virus crisis has posed a difficult test for many countries, which are already facing governance and institutional challenges before the emergence of the pandemic [14]. The health care systems and economies of most countries have reached the limits of their capabilities [4]. While fragile and conflict-affected states have become highly exposed, even the stable, higher-income states in the region have been significantly affected [19].

Many countries also faced difficulty in assessing and addressing the socio-economic impact of the emerging coronavirus crisis, which raised doubts about their ability to absorb shocks, whether environmental, global or societal [18, 26]. This institutional thinness, which can be attributed to inappropriate investment in basic social services, has resulted in weak capacities to provide essential jobs, a tendency to collapse when shocks and crises arrive, a slow recovery and a deterioration in state-society relations [21].

Government structures are another obstacle because they are highly centralized and poorly digitized. Service delivery can be improved by devolution of decision-making power and adoption of digital technologies, including in the health care system where digital medical records play a critical role in epidemiological surveillance [29].

D. **Economic Dimension**

The end of the year 2019 will mark the beginning of the emergence of the Corona pandemic in the state of China to gradually spread to various parts of the world [14]. So, the pandemic has taken on a global dimension similar to the current environmental problems [1, 14]. The majority countries of the world, after the spread of the Corona pandemic, have imposed a set of restrictions on economic activities, limited the movement of land, sea transport and airspace. In addition to closing unnecessary industrial establishments and adopting quarantine for the population, all with the aim of limiting the spread of the epidemic and surrounding it, or at least alleviating the severity of infections and fatalities [1, 6]. The economic activity of the public spaces became threatened as the government posed restrictions on the flow of people in public urban in order to control the pandemic and reduce its consequences [11, 21].

The Corona pandemic has caused a strong shock to the corners of the global economy, as it is expected to have an important impact on the economic sector [11]. It changings the shape of the next world because of its comprehensive wave of change after it led to millions of injuries and hundreds of thousands of deaths [21]. The impact of the pandemic on the aspects of social, medical and environmental life in individuals and societies, but it extends beyond it to other aspects such as its impact on the oil markets, stock markets, and international commodities, in addition to keeping the economic variables [2, 6].

3.3 Functional Public Spaces as a Proposal for the Post-pandemic Era

A. **Public spaces**

A practical step that should be taken immediately is to plan, protect and expand public space [6, 25]. It is very important to reconsider the size, layout and spatial distribution of public spaces, including sidewalks, parks, open spaces, and public facilities such as libraries and community centers [13]. This is especially important in crowded neighborhoods where home spaces are limited and family sizes are usually large. Although, investment in housing and private property rises with national economic development. Public space can be better allocated by the competent and empowered city authorities [13].

Urban green spaces provide great opportunities for positive change and sustainable development for our cities [7]. Public green spaces open for walking, cycling, play and other outdoor activities can improve safe mobility and provide essential services for women, the elderly and children, as well as low-income population groups, thus promoting health equity [5, 25]. Incorporating public health priorities into the development of public spaces provides a win-win approach in urban areas. A health-conscious approach when planning public green spaces can offer the potential for the greatest number of co-benefits [15, 27].

B. **Park and green area**

Parks, green spaces, and waterways are important public spaces in most cities. They provide solutions to the health and safety effects of rapid, unsustainable urbanization. Equally important are the social and economic benefits of urban green spaces, and should be seen in the context of global issues such as climate change, and other priorities set forth in the Sustainable Development Goals, including sustainable cities, public health and nature conservation [3, 20].

The scientific literature describes the various ways in which the natural environment can positively affect human health and safety, as natural areas provide opportunities for physical activity, social communication and reduced psychological stress. A growing number of epidemiological studies have demonstrated that maintaining urban green spaces has various positive health effects, including improved mental health and reduced depression; improving pregnancy outcomes; Reducing the incidence and mortality of cardiovascular disease, obesity, and diabetes [13, 19]. In addition, parks and green spaces provide people with opportunities to increase their walking, cycling and engaging in leisure time physical activity [5]. Therefore, investments in city parks, green spaces and waterways represent an effective and economical way to both improve health and mitigate the effects of climate change [5, 26].

C. **Street design & Public transport**

The new phase will impose building and enhancing confidence in the safety of travelers, and introducing new transportation methods during the phases of ending the closure and after the end of the pandemic alike [18]. Therefore, companies will have to invest in efforts to understand traveler travel paths, as well as accelerate, cancel or re-evaluate delayed projects and programs based on new global trends [13, 27].

Service providers and operators face a range of new challenges to overcome. Here, it is necessary to develop a package of solutions aimed at addressing these challenges, such as meeting the conditions for social distancing and dealing with the overall increase in travel time [2, 3]. The closures due to the Corona virus or what is known as "Covid-19" that have significantly reduced the use of roads and public transport systems, city authorities around the world, from Liverpool in the United Kingdom to Lima, the capital of the Republic of Peru, are benefiting from closing streets to cars, they opened other streets for bicycles, and widened sidewalks to help residents maintain the 6 feet distance recommended by the World Health Organization [2, 4]. It is unclear whether these urban interventions will continue once the pandemic is over. Milan plans to build 22 miles of new bike lanes and permanently expand sidewalks after the closure is lifted.

4   Conclusion

The connection of citizens to their public spaces, streets, and public facilities have changed because of restrictions on movement during the COVID-19 pandemic. Clearly public spaces must be redesign to be a part of the reaction to the virus whether to limit the spread of the virus, or to deliver ways for people to relax or carry out their livelihood.

Therefore, adequate spaces and physical distances required in the public spaces between users. The expansion of streets can ensure physical distancing is possible on pavements. Cities are temporarily or even permanently re-allocating road space from cars to provide more space for bicycles and people to move around safely, easing movement and respecting physical distancing rules. Some cities are expanding pavements to facilitate safe walking, skating and jogging. Such measures result in reduced CO2 emissions and better air quality improving people's health and well-being.

Public spaces need to be multi-functional and adaptable.

Perhaps, temporary food markets can be set up in spaces such as parking areas to decongest existing markets. Small neighborhood spaces can be transformed into pop-up community health centers areas for food distribution or food gardens. The shared use of streets and spaces can allow for organized street vending on select days or times of day or for leisure activities such as showing films or plays or holding exercise classes.

Public amenities and services can afford important services for marginalized communities during a pandemic such as providing clean water points, toilets, and appropriate cleaning products for the urban poor or those without housing. Open spaces also can provide reasonable isolated sitting area. They can also be used for Safe Street vending providing an essential living for poor families.

The design, materials used and organization and maintenance of public space is key in struggling the spread of the COVID-19 virus. Public space should be designed to tolerate for physical distancing and public space managers need to ensure that they can be cleaned frequently and thoroughly including high-touch surfaces like doors, handles, and furniture. Finally, the COVID-19 crisis has highlighted several gaps in public space including accessibility, flexibility, design, management and maintenance, connectivity and equitable distribution across a city. Going forward we need to enhance public awareness and generate a shared policy program bringing together public health, urban planning, architecture, community development and green building.

References

1. U. N. H. S. P. (UN-Habitat): Cities and Pandemics: Towards a More Just, Green and Healthy Future, United Nations Human Settlements Programme (UN-Habitat) (2021)
2. Sharifi, A., Khavarian-Garmsir, A.R.: The COVID-19 pandemic: impacts on cities and major lessons for urban. Sci. Total Environ. **749**(142391), 14 (2020)
3. Di Renzo, L., et al.: Eating habits and lifestyle changes during COVID-19 lockdown: an Italian survey. J. Transl. Med. **8**(18), 1–15 (2020)
4. Maturana, B., Salama, A.M., McInneny, A.: Architecture, urbanism and health in a post-pandemic virtual world. J. Archit. Res. **15**, 9 (2021)
5. Carriedo, A., Cecchini, J.A., Fernandez-Rio, J., Méndez-Giménez, A.: COVID-19, psychological well-being and physical activity levels in older adults during the nationwide lockdown in Spain. Am. J. Geriatr. Psychiatry **28**(11), 9 (2020)
6. Szczepańska, A., Pietrzyka, K.: The COVID-19 epidemic in Poland and its influence on the quality of life of university students (young adults) in the context of restricted access to public spaces. J. Public Health (2021). https://doi.org/10.1007/s10389-020-01456-z
7. Couclelis, H.: There will be no post-COVID city. Environ. Plann. B **47**(7), 1121–1123 (2020)
8. Hu, R.: COVID-19, smart work, and collaborative space: a crisis-opportunity perspective. J. Urban Manag. **9**(3), 276–280 (2020)
9. Eltarabily, S., Elgheznawy, D.: Post-pandemic cities - the impact of COVID-19 on cities and urban design. Archit. Res. **10**(3), 75–84 (2020)
10. LePan, N.: Visualizing the History of Pandemics. Healthcare (2020)
11. Bănică, A., Kourtit, K., Nijkamp, P.: Natural disasters as a development opportunity: a spatial economic resilience interpretation. Rev. Reg. Res. **40**, 223–249 (2020)
12. Abusaada, H., Elshater, A.: COVID-19 and "the trinity of boredom" in public spaces: urban form, social distancing and digital transformation, 8 (2021)
13. Raina MacIntyre, C., Jay Hasanain, S.: Community universal face mask use during the COVID 19 pandemic—from households to travellers and public spaces. J. Travel Med. **27**(3), 3 (2020)
14. n. authorities: Coronavirus disease 2019 (COVID-19). World Health Organization, April 2020
15. Scopelliti, M., Pacilli, M.G., Aquino, A.: TV news and COVID-19: media influence on healthy behavior in public spaces. Environ. Res. Public Health **18**(4), 15 (2021)
16. Nia, H.A.: A comprehensive review on the effects of COVID-19 pandemic on public urban spaces. Archit. Urban Plann. **17**(1), 79–87 (2021)
17. Hassankhani, M., Alidadi, M., Sharifi, A., Azhdari, A.: Smart city and crisis management: lessons for the COVID-19 pandemic. Int. J. Environ. Res. Public Health **18**(7736), 18 (2021)
18. Dreger, C., Gros, D.: Social distancing requirements and the determinants of the COVID-19 recession and recovery in Europe. Intereconomics **55**, 365–371 (2020)
19. Lunn, P.D., Belton, C.A., Lavin, C., McGowan, F.P., Timmons, S., Robertson, D.A.: Using behavioural science to help fight the coronavirus: a rapid narrative review. J. Behav. Public Adm. **3**(1), 35 (2020)
20. Kordshakeri, P., Fazeli, E.: How the COVID-19 pandemic highlights the lack of accessible public spaces in Tehran. Cities Health, 13 (2020)
21. Ang, C.: Visualized: the World's Deadliest Pandemics by Population Impact. Datastream (2021)

22. Nieuwenhuijsen, M.J.: Urban and transport planning pathways to carbon neutral, liveable and healthy cities; A review of the current evidence. Environ. Int. **140**, 105661 (2020)
23. Zevi, A.T.: The Century Global Cities-How Urbanization Changing The World and Shaping our Future. ISPI, Milano (2019)
24. Okubo, T., Chen, C., Dasgupta, S., Li, K., Liu, X., Mai, F.: COVID ECONOMICS, vol. 32, p. 246. The Centre for Economic Policy Research (CEPR) (2020)
25. Szczepańska, A., Pietrzyka, K.: The COVID-19 epidemic in Poland and its influence on the quality of life of university students (young adults) in the context of restricted access to public spaces. J. Public Health: Theory Pract., 11 (2020). https://doi.org/10.1007/s10389-020-014 56-z
26. Ferrante, G., et al.: Did social isolation during the SARS-CoV-2 epidemic have an impact on the lifestyles of citizens? Epidemiol. Prev. **44**(5–6), 353–362 (2020)
27. Cilliers, E.J., Sankaran, S., Armstrong, G., Mathur, S., Nugapitiya, M.: From urban-scape to human-scape: COVID-19 trends that will shape future city centres. Land **10**(1038), 12 (2021)
28. Sepúlveda-Loyola, W., et al.: Impact of social isolation due to COVID-19 on health in older people: mental and physical effects and recommendations. J. Nutr. Health Aging **24**(9), 10 (2020). https://doi.org/10.1007/s12603-020-1500-7
29. Buoite Stella, A., et al.: Smart technology for physical activity and health assessment during COVID-19 lockdown. J. Sports Med. Phys. Fitness **61**(3), 452–460 (2021)

Where Do the Children Play? Taxonomy of Children's Play Areas and Role of the City of Culture in Santiago de Compostela

María de los Ángeles Piñeiro-Antelo(✉) ⓘ, Lucrezia Lopezⓘ, and Miguel Pazos-Otónⓘ

Santiago de Compostela University, Santiago de Compostela, Galicia, Spain
manxeles.pineiro@usc.es

Abstract. In recent years, more and more attention has been paid to the relationship between urban spaces and children. This is the subject of the research herein carried out in the Spanish city of Santiago de Compostela, while analysing the functional transformations of the architectural complex of the City of Culture (CoC). The objectives are to devise a taxonomy of urban spaces for children in Santiago de Compostela, analyse the role of the CoC in urban spaces for children in Santiago and study whether this transformation is linked only to the end of the works of the CoC, or if CoC planners are willing to adapt this great cultural infrastructure towards the enjoyment of children and families. After a brief theoretical review, the taxonomy of the leisure offer for children in the city is analysed. From a methodological point of view, there is a brief review of the planning documents and interviews are carried out. The results point to the emergence of a new urban reality that encompasses the taxonomy of children's leisure spaces, where the CoC takes on a greater role both in the offer of outdoor spaces, and in the programming of indoor activities.

Keywords: Santiago de Compostela (Spain) · Taxonomy of urban spaces for children · City of Culture · Child-Friendly Cities (CFC)

1 Introduction

Children's relationship with their urban environment has been a research topic in children's Geography [1]; Cotterell, back in 1993, pointed to the need to pay continuous attention to the use of spaces by children, since it is linked to their development and experience of the environment [2]. Indeed, since its ratification in 1990, the *United Nations Convention on the Rights of the Child* (UNCR) has tried to guide urban planners managing urban policies by referring to the *Child-Friendly Cities* (CFC) agenda. A few years later, James et al. (1998) stated that: "the child is conceived of as a person, a status, a course of action, a set of needs, rights or differences—in sum, as a social actor" [3, p. 207]. Even today, and despite efforts to implement the UNCR, the rights of children to a quality urban life are far from being met [4]. In 2005, the Italian pedagogist Francesco Tonucci published the book *La città dei bambini. Un nuovo modo di pensare la città*,

© The Author(s), under exclusive license to Springer Nature Switzerland AG 2022
F. Calabrò et al. (Eds.): NMP 2022, LNNS 482, pp. 406–417, 2022.
https://doi.org/10.1007/978-3-031-06825-6_39

which claimed the need to affirm the role and importance that children had in the urban planning and decision-making process. As a result, he pointed out the quiet invisibility of children when dealing with urban issues, as well as other researchers, who claim the right of children to urban public spaces and their participation in formal planning processes [4–8].

This research focuses on the Spanish city of Santiago de Compostela, and more specifically on the city's play areas and the functional transformations of the architectural complex of the City of Culture (CoC), increasingly adapted as a space for children. The objectives are to devise a taxonomy of urban spaces for children in Santiago de Compostela, analyse the role of the CoC in urban spaces for children in Santiago and study whether this transformation is linked only to the end of the works of the CoC, or if CoC planners are willing to adapt this great cultural infrastructure towards the enjoyment of children and families. After a brief theoretical review that reiterates the relevance of this problem, the taxonomy of children's leisure offer in the city is presented. From a methodological point of view, a brief review of the planning documents is made, and interviews are carried out with experts in the field of play in early childhood education. The results point to the emergence of a new urban reality that encompasses the taxonomy of leisure spaces for children's offer (open air spaces, natural environments, cultural activities) where the CoC takes on an ever-greater role both in the offer of outdoor spaces, and in the programming of indoor activities.

2   Literature Review: Towards a Taxonomy of Children's Urban Spaces

Over the last few decades, the *urban entrepreneurial turn* that took place in different cities has supposed a change in terms of urban functions and priorities [8, 9]. Cities are no longer just socialising and meeting spaces, rather, they have turned into specialised and separated realities that compete among each other to reach higher international positioning [10]. In this never-ending run toward the top, urban planners forget about the weaker groups (namely children and older people) [8, 11]. In addition to this new urban approach, other issues typical of the contemporary city are emerging, among them, the privatisation of public space, new technological relationships, growing urban insecurity, parents' concern about the city as a dangerous setting [12], or the dehumanisation of the space characterised mainly by declining transit in public spaces [13]. All these problems make the urban space increasingly hostile to children and considerably reduce their freedom of movement [6]. According to Tonucci [14], children have not only seen their mobility and autonomy of movement in the city reduced, but also their free time, which is fundamental to their growth and personal development. In this sense, Krysiak [7] points out that urban design must guarantee environments that favour the active mobility of children in public spaces, boosting their health and their physical and emotional well-being.

This concept of freedom of movement is an important challenge in the construction of a *CFC*, even in the case of playgrounds, specially designed for them, which, certainly, are not spaces for children, but spaces where children are accompanied by adults who decide where and how to move [8, 15]. Therefore, playgrounds are spaces lived according to adult mediation [16], since they are designed and organised at different levels, so that adults can supervise the overall game of their children. And it is through these spaces that children build their urban imagery, their experience of the city [12], and, above all, their territoriality [17].

Currently, children are not only consumers of these public spaces planned and created *ad hoc* for them, but they are also conceiving shopping centres with indoor play areas as leisure and socialisation environments [12, 18]. Lewis [19] spoke of shopping centres as multifunctional indoor cities. Indeed, shopping centres have supplanted the city ecosystem, since inside (without cars, with streets and squares) a safe context is created for children, for whom specific and assisted spaces are designed [8, 18]. In the category of indoor spaces, toy libraries (*ludotecas*), whose origin is in the United States in 1934, are included. In 1965, these libraries first arrived in Europe, prompted by the *UN Declaration of the Rights of the Child*, and in the 1980s to Spain. Since then, the initial toy library project has been modified. Today, there are different models that respond to socio-economic, and business demands that do not necessarily prioritise the needs of children [20]. Nowadays, toy libraries are spaces that serve to solve problems, such as work and family life balance, the lack of space and playmates in homes, or the problem of children playing in unsafe streets [20]. And some authors point out that they cannot replace sidewalks, streets, squares, and gardens as play areas [3, 6, 8, 14].

Among open spaces, the city's rural surroundings should be highlighted. If the city has become dangerous and unsuitable for children's activities, natural environments continue to be the ideal and desired spaces where children can move freely and play [1, 8]. The most important child development and learning experiences are those that occur away from home, school, and adults [14]. Outdoor spaces are relevant for child development and so, activities in contact with nature are encouraged [21], thus natural environments are a critical aspect when planning CFCs [5, 22]. In general, natural spaces are among children's favourites, as they guarantee participation and safety, but especially in recent months "biosecurity", that is, the set of Covid-19 sanitary and hygienic conditions required to live a normal life.

And, finally, in this city of children, we must not forget the cultural offer for children, museums, exhibitions and cultural programs. Museums are essential places for the development of children's creativity, and for this reason, in recent years in the Western world, this model has been promoted to offer a new informal educational environment [23]. From their origins, museums have undergone functional and conceptual transformations, and today they are integrating spaces that are increasingly open to society, with a clear educational and pedagogical function [24]. Also, the museum has become a centre for education and cultural and heritage dissemination, transmitting a series of universal values to visitors. Among these visitors, the need to promote children as a cultural public is consolidating thanks to a new approach that reconsiders this group as a social actor, and as human beings capable of modifying the environment and/or interacting with it [25].

The result is a taxonomy of childhood places. First of all, they can either respond to adults' planning [8, 16] or children's place preferences [12]. Secondly, these spaces have a diverse nature, that is, they can be indoor and outdoor spaces. Third, these spaces differ according to a functional criterion [26], and not all of them are necessarily designed for children's use but are adapted according to emerging needs.

3 Methodology

To introduce such a new discourse about children's urban spaces in Santiago de Compostela, we opted to combine different methodological patterns because we believe that all of them would reinforce the trends that are being recorded.

Firstly, as far as the first aim is concerned, that is to advance a taxonomy of the children's urban spaces, based on the theoretical premises, we present and analyse the urban space of Santiago de Compostela according to the above-mentioned categories of children's space (indoor, outdoor, playgrounds, etc.). In this sense, it is worth highlighting the importance of the methodology participant observation, as field trips have been made and the city's playgrounds have been reconnoitred, paying special attention to the CoC, both indoors and outdoors. The output of this first territorial analysis is a thematic map showing their localization in the city. More detailed information on the characteristics of these urban spaces and on the CoC are offered in their proper sections, in fact, in this last case we introduce a map representing the different indoor and outdoor spaces that are undergoing a re-semantisation process.

Secondly, to investigate the growing role that the CoC is assuming as children's urban space, we review a corpus of primary sources (namely strategic and planning documents, reports and activity audits, projects and investments, related to the CoC), produced mostly at a regional and local-urban scale. These sources are analysed according to a qualitative interpretive criterion, that explores and points out action strategies referred to as cultural politics and interventions that recognise "children as social actors", thus repositioning them in the centre of their program.

Thirdly, these documents together with interviews are the main sources that we consider and use to understand the willingness of the planners of the CoC to direct this great cultural infrastructure towards the enjoyment of children and families. In relation to the interviews (Table 1), we contacted cultural programming planners, academics, and professionals in early childhood education, with years of experience in educational work with children, mainly developed in Santiago de Compostela. They also have a

Table 1 Interviews conducted and qualified opinions within the framework of this research.

Person	Organisation	Date and method
Ana Isabel Vázquez	Manager of the City of Culture Foundation	2.12.2021, Personal
Nerea Couselo	Xandobela Cultural Education	13.12.2021, Personal
Adriana Pazos	OLALAB Early Childhood Education	14.12.2021, Videocall
Pablo Meira	University of Santiago. Lecturer on Environmental Education	14.12.2021, Phone
Ana Munín	Santiago Tourism Municipal Company	14.12.2021, Personal

comprehensive knowledge of the main characteristics of play areas and the organisational component of early childhood education and the programming of cultural activities.

4   The Case Study: Santiago de Compostela and City of Culture

Santiago de Compostela was declared a World Heritage City by UNESCO in 1985. It is the only one in Galicia and one of the 15 World Heritage Cities in Spain, which are grouped together to defend common interests and address similar challenges for their citizens, including bringing these cities closer to children. From this perspective, and within the framework of the activities commemorating the 40th anniversary of the World Heritage Convention (2012), the documentary *Las Ciudades Patrimonio en los ojos de los niños* (*Heritage Cities in the eyes of children*) was produced, where children presented and shared their impressions of life in their cities: how they walk through them, how they see the streets like labyrinths, or how the stones on the ground make it difficult for them to skate. The cities also created teaching units for school children and their teachers and families. The Didactic guide of Santiago de Compostela featured the "secrets and treasures" of the city to involve children in its conservation [27, p. 3].

Specific problems in relation to mobility, and specifically traffic, have also been jointly addressed in the *White Paper on Mobility in Spain's World Heritage Cities* [28]. In Santiago, as recommended in the document, the creation of safe school paths has been promoted, and the school path pilot project (*Colecamiños*) has been launched to promote a change in the mobility guidelines of school children. The location, typology and offer of children's leisure spaces in these cities, which are shared by both visitors and residents, are strongly conditioned by mobility and traffic.

Cities must provide environments for children in public spaces, capable of promoting their health and physical and emotional well-being, focusing on games and "active mobility" [7]. And this is a challenge that some cities in Galicia have undertaken, such as Pontevedra [29], and that World Heritage Cities in Spain, including Santiago de Compostela, must follow. Santiago de Compostela, the capital of Galicia, is a city of 81,695 inhabitants within a municipality of 97,848 in 2020, with a medieval historic centre, which grew around its Cathedral. With an oceanic climate, it is among the rainiest cities in Spain, with an average annual rainfall of around 1,886 mm which conditions the use and enjoyment of open-air public spaces. Additionally, its orography conditions the use by certain population segments (elderly, disabled, children, etc.) of certain non-motorised transport modes, especially journeys on foot or by bicycle [28, p. 100]. The pedestrianisation process of the historic centre, and the construction of public parking areas in its surroundings, have made it possible to gain space for pedestrians and restrict vehicle access, mainly within the old walled enclosure of the historic centre. However, outside this perimeter, the dominance of private vehicles for urban and metropolitan mobility has negative consequences in various fields, including habitability and land tenure [28].

The CoC is located on Mount Gaiás, 3 km away from the historic centre of the city of Santiago de Compostela. The project of the (CoC) began in 1999, as a proposal of the regional government with the aim of spreading the Galicia brand internationally. Since its inception, this project has been surrounded by multiple controversies, mainly related to the opacity of its motivations, size, budget and content, which increased throughout the construction phase. In January 2011, the first two buildings of CoC were opened to the public, and in 2020 the urbanisation and humanisation works on the outdoor area, which had begun in 2010, were concluded, made up of a surface of 14 ha. At present, the Complex is comprised of 4 buildings: namely the Centre for Creative Entrepreneurship (CCE), Library and Archive, Museum and Centre for Cultural Innovation (CCI), and Fontán Building, a multifunctional building conceived to contribute to the integration of the 3 Galician public universities, hosting the headquarters of several research centres.

The functions currently carried out by the CoC could be summarised in 3: (i) tourist, due to the attraction of the architectural project by the architect Peter Eisenman; (ii) cultural, since it was conceived with the objective of being an international cultural centre of excellence as well as a socio-cultural centre for the city of Santiago; (iii) technopolis, since a large part of the complex's buildings and facilities have been turned into a technology-based place. In this work, we recognize and introduce a (iv) function, the recently developed City of Children, which gives the CoC a special role in the recreational spaces of Santiago, and which we think will be key in the coming years. The planners have chosen to create a playful, open space, of broad dimensions, suitable for families with children. The area is equipped with children's services, playgrounds, spaces for family walks, or adventure sports. The emergence of this City of Children, a space redesigned to make it family-friendly, is linked to a key aspect, which is the importance of children in the programming of cultural activities.

5   Results

5.1   Taxonomy of Urban Spaces for Children

In this work, we advance a proposal for the classification of children's play spaces in Santiago de Compostela, which we have carried out based on theoretical hypotheses, interviews and participant observation (Table 2), and distinguish between outdoor and indoor spaces. As previously mentioned, the adverse weather of Santiago de Compostela for much of the year makes the latter especially relevant. Likewise, we have differentiated between regulated spaces (initially planned for children's play) and unregulated spaces (multipurpose spaces where children coexist with other users).

Within the category of regulated open-air spaces, we find the playgrounds, which make up a small part of the play areas (Fig. 1). First of all, there are many urban playgrounds, located in the squares of city areas with high residential and traffic density. These are small spaces, with external barriers that guarantee the aforementioned function of parental surveillance [8, 15, 16]. Adriana Pazos highlights their crucial function as spaces for social contact, meeting points for families, although they are far from being the ideal spaces for play, due to the increasing dehumanisation of contemporary urban space (pollution due to traffic, noise and limited movement) [6, 12, 13]. Examples of this are the playgrounds of Galicia Square, Roxa Square and Square of Vigo. Sometimes these playgrounds may have minimal natural features that provide alternative play elements, such as Alameda Sur and San Roque.

A second typology, which increases the quality of these regulated play spaces, are the so-called "peri-urban parks", which benefit from the rich natural environment in which they are located [1, 21]. This is the case of the Granxa do Xesto, a park located on the slope of Pedroso Hill, with native vegetation, excellent views of the city, and play facilities (catwalks, zip lines, etc.) that are a novelty in relation to the equipment of conventional play facilities that we find in other playgrounds.

Table 2 Proposal for the classification of children's spaces in Santiago de Compostela

	Open-air spaces	Under cover spaces
Regulated spaces	- Urban playground: Galicia Square, Vigo Square, Roxa Square, Alameda South Park, San Roque, ... - Peri-urban playgrounds: Granxa do Xesto - City of Culture	- Toy libraries - Sociocultural Centres - Event planning companies - Playgrounds and children's spaces in shopping centres - Buildings of the Gaiás Complex
Not regulated spaces	- Historical centre (pedestrian) and rest of the city - Outdoor urban parks: Bonaval, Belvís, Galeras, etc. - Green periurban spaces: Selva Negra, Pedroso Hill	- Private homes - Associations premises

Regarding unregulated outdoor spaces, the interviewed specialists believe that there is a notable loss of the street as a place for children to play. This situation can be seen especially in the historic city, due to the tourism boom. In a pandemic scenario, and thanks to new city ordinances, squares, streets and public spaces have increasingly been occupied by terraces owned by hospitality businesses, resulting in the appropriation of pedestrianised areas in most cases. One of the implications of this has been the reduction of play spaces for children in the street. Nerea Couselo points out the loss of play spaces, especially in the historic city, such as San Martiño Pinario Square, Mazarelos Square, or the Rúa de San Pedro, the entrance of the French Way into the city.

Among the peri-urban parks, we highlight the Selva Negra, the Sarela walking path or Pedroso Hill, on the outskirts of the city. Pablo Meira, Nerea Couselo and Adriana Pazos mention the central role that these nature-dominated spaces can play in the city, due to the importance of children playing in natural environments integrated in nature.

As for play spaces under cover, they are especially necessary in a city like Santiago de Compostela. Among the regulated ones, we first highlight the toy libraries and similar spaces. Also, Adriana Pazos refers to the great role played by the Sociocultural Centres (CSC): "The City Council of Santiago has a network of sociocultural centres that integrate the family and children's fabric who are making use of these centres". CSCs are conceived as safe, regulated spaces, where it is not necessary for parents to be present.

Other spaces that have multiplied their visibility in recent years are *playgrounds* in children's event planning businesses. These are very basic spaces, in which, in addition to playing games, entertainment and catering services are of great importance. The rise of this type of businesses is directly related to the loss of the role of family homes and

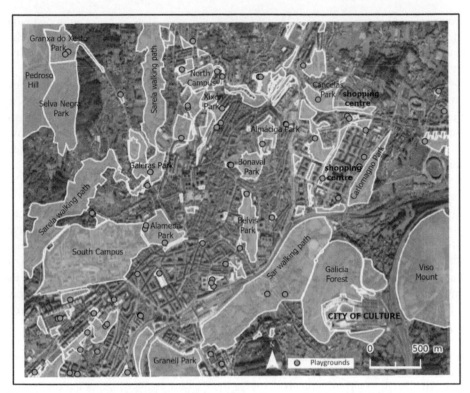

Fig. 1. Playgrounds in Santiago de Compostela. (Source: SDI Municipality of Santiago de Compostela. (Authors' own work).

public spaces as places for celebrating birthdays and children's parties in general. Hand in hand with this physical "externalisation" of play, in a rainy city like Santiago, we must also highlight the play spaces inside shopping centres [12, 18, 19], that constitute a free option for many families and open to various possibilities.

Finally, the rise of parent and child associations with a clear neighbourhood character is a new successful opportunity according to the interviewees. Communities of children who interact outside the school context and extracurricular activities emerge. According to Adriana Pazos: "the phenomenon of neighbourhood associations is working very well. They are people and families from the neighbourhoods who fulfil a need for unity, for community, which for a while has been slightly disaggregated". In Santiago, San Pedro neighbourhood, Sete Portas-Altair, Itaca and others of similar characteristics are especially active. Leisure activities take place in the association's premises, sometimes in neighbourhood association premises and, when the weather permits, open-air in the above-mentioned public spaces.

5.2 The Role of the City of Culture in Urban Spaces for Children

The facilities of the complex of the CoC constitute a category in themselves, due to their importance in the offer in Santiago and its metropolitan area. In recent years, the provision of play spaces outside Mount Gaiás has increased, with parks equipped with attractions designed for children. All this is located in an environment dominated by nature, in which special care has been taken to create a space of environmental quality, with Galicia's Forest.

The network of trails and the growing environmental importance of the complex explain the success of these facilities as play spaces for children and families. As Nerea Couselo notes, children enjoy extending the playgrounds towards open and natural spaces. They enjoy the forest, the lakes, the closeness to nature. But they also use the architectural constructions as a play space, as they climb the gently sloping warped ramps of the Einsenman buildings. In recent times, the children's offer has been completed with facilities aimed at young people: a skate park and a climbing wall in the Garden Theatre, or a BMX bicycle track.

It is also worth highlighting the great revolution that the offer of children's programming inside the buildings of the CoC has meant. The evolution of the more inclusive offer that contributes to creating new informal educational environments [23, 24] from other museum contexts can also be seen in this complex, which brings together a series of features that favour interactive education with the environment [25]. As Adriana Pazos says: "magic shows or theatre for children that did not exist in the city before are now being offered". According to the early childhood educator, the will to promote children as public explains not only the local, but also regional impact of this cultural offer.

Therefore, the CoC has acquired a leading role within the leisure spaces for children in the city. In this work we have asked ourselves to what extent this evolution with respect to the initial objectives of the CoC has been planned. To this end, the Strategic Plan 2012–2018, the II Strategic Plan 2021–2027, and the programming reports for the years 2016, 2017, 2018, 2019 and 2020 have been analyzed. The first strategic objective of the Plan 2012–2018 aims to offer citizens both activities and quality cultural programming. Regarding activities, objective 1.3 is aimed at making the CoC a space for the development of leisure activities, in contact with nature and sport [30, p. 62], which undoubtedly connects with a family audience. In relation to programming, the CoC's Cultural Action Plan encourages permeability to all types of audiences, although programming related to cultural action includes "exhibitions, concerts and various activities aimed at target audiences such as children or women. families" [30, p. 71]. Much of the programming is clearly didactic and educational in nature, with a playful approach.

The *II Strategic Plan of the City of Culture* values the diversity of activities as an "opportunity". Specifically, "the commitment to outdoor projects allows the development of activities that attract new audiences and maximise the impact of the project on society" [31, p. 12]. From our point of view, it is a clear commitment to activities for children, in relation to the use of open-air spaces for children. According to the manager, Ana Isabel Vázquez: "we will continue to segment the programming according to the audience. One of the successes of the CoC has been to have a wide range of activities and to have programming for specialists, families and the general public. We will continue along that line and have programming for a family audience".

The fact that the first full-page photograph presented in the document (of 47 pages) shows a children's show in the open-air auditorium in the centre of the CoC Complex is a declaration of intent in that respect (Fig. 2).

Fig. 2. Photograph of the Executive Summary of the II Strategic Plan of the CoC [31, p. 5].

6   Conclusions

The proliferation of regulated children's play areas, both outdoor and indoor, in the contemporary city is due to the scarcity of open spaces, where freedom of movement to play and the autonomy of the child in a safe urban environment is made possible, as stated in the Convention on the Rights of the Child.

In this context, the recent evolution of the CoC takes place, becoming and affirming itself as a *Child-Friendly meta-city*, that is, a city within another city, which brings together an offer and a safe environment that complements the offer of outdoor activities of open urban playgrounds and peri-urban parks while programming a wide, varied, and quality cultural offer. The interviews carried out confirm the commitment of the CoC to its recreational function (as a space, open and closed), with children's services, playgrounds, spaces for family walks, etc. This is turning the CoC into a more attractive space for young people, as it has been redesigned to become more family-friendly. All this is linked to a key aspect, which is the assessment of programmed cultural activities. In fact, the cultural policies in recent years and the cultural programming analysed reveal the intention to offer activities and attractive programming that generate value for society, which can be used especially by the citizens of Santiago. This architectural megaproject, with its open spaces (parks, forest, lake, tracks, etc.) and its facilities for cultural events, complements a scarce urban offer, and is giving signs of a *childrification*, that is, a social process marked by age as an identification criterion. The space of this meta-city is lived and presented [17], since the different outdoor or indoor experiences that children enjoy respond to the production schemes of a new social space [32].

The documents consulted do not allow to conclude that there is a desire in the planners of the CoC to exclusively orient this great cultural infrastructure to the enjoyment of children and families. But the objectives included in the strategic plans try to bring the complex closer to citizens, seeking to attract new audiences, and consolidate cultural programming and educational and training activities for young people and families. This objective has been achieved especially with these groups, which become the greatest beneficiaries of the cultural and leisure offer in the CoC.

References

1. Béneker, T., Sanders, R., Tani, S., Taylor, L.: Picturing the city: young people's representations of urban environments. Child. Geogr. **8**(2), 123–140 (2010)
2. Cotterell, J.L.: Do macro-level changes in the leisure environment alter leisure constraints on adolescent girls? J. Environ. Psychol. **13**(2), 125–136 (1993)
3. James, A., Jenks, C., Prout, A.: Theorizing Childhood. Polity Press, Cambridge (1998)
4. Bishop, K., Corkery, L.: Designing Cities with Children and Young People: Beyond Playgrounds and Skate Parks. Routledge, Abingdon (2017)
5. Adams, S., Savahl, S., Florence, M., Jackson, K.: Considering the natural environment in the creation of child-friendly cities: implications for children's subjective well-being. Child Indic. Res. **12**(2), 545–567 (2018). https://doi.org/10.1007/s12187-018-9531-x
6. Christensen, P.: Place, space and knowledge: children in the village and the city. In: Christensen, P., O'Brien, M., (eds.) Children in the City. Home, Neighbourhood and Community, pp. 13–28. Routledge Falmer, London (2003)
7. Krysiac, N: Where do the Children Play? Designing Child-Friendly Compact Cities, Australian Institute of Architects (2017). https://www.researchgate.net/publication/326697435_Where_do_the_Children_Play_Designing_Child-Friendly_Compact_Cities
8. Tonucci, F.: La città dei bambini. Un nuovo modo di pensare la città. Bari, Laterza (2005)
9. Harvey, D.: From managerialism to entrepreneurialism: the transformation in urban governance in late capitalism. Geografiska Annaler Ser. B Hum. Geogr. **71**(1), 3–17 (1989)
10. Evans, G.: Hard-branding the Cultural City. From Prado to Prada. Int. J. Urban Reg. Res. **27**(2), 417–440 (2003)
11. Matthews, M.H.: Living on the edge: children as 'outsiders.' Tijdschr. Econ. Soc. Geogr. **86**(5), 456–466 (1995)
12. Spencer, C., Woolley, H.: Children and the city: a summary of recent environmental psychology research. Child Care Health Dev. **26**(3), 181–198 (2000)
13. Rissotto, A., Giuliani, M.V.: Learning neighbourhood environments: the loss of experience in a modern world. In: Spencer, C., Blades, M. (eds.) Children and Their Environments. Learning, Using and Designing Spaces, pp. 75–90. Cambridge University Press, Cambridge (2006)
14. Tonucci, F: La ciudad de las niñas y los niños. In: Navarro Martínez, V., Raedó Álvarez, J., Rosales Noves, X.M. (coords.) Ludantia. I Bienal Internacional de Educación en Arquitectura para a Infancia e a Mocidade. Colexio Oficial de Arquitectos de Galicia, Pontevedra, pp. 27–32 (2018)
15. Karsten, L.: It all used to be better? Different generations on continuity and change in urban children's daily use of space. Child. Geogr. **3**(3), 275–290 (2005)
16. Sebba, R.: The landscape of childhood: the reflection of childhood's environment in adult memories and children's attitudes'. Environ. Dev. **23**(4), 395–422 (1991)
17. Raffestin, C.: Space, territory, and territoriality. Environ. Plann. D Soc. Space **30**(1), 121–141 (2012)
18. Uzzell, D.L.: The myth of the indoor city. J. Environ. Psychol. **15**(4), 299–310 (1995)

19. Lewis, G.H.: Community through exclusion and illusion: the creation of social worlds in an American shopping mall. J. Pop. Cult. **27**, 121–136 (1990)
20. Reyes Ruiz de Peralta, N.: Las ludotecas: orígenes, modelos educativos y nuevos espacios de socialización infantil. Ph.D. thesis. Granada, Editorial de la Universidad de Granada (2012). http://hdl.handle.net/10481/21771
21. Kahn, P.H., Kellert, S.R.: Children and Nature. Psychological, Sociocultural, and Evolutionary Investigations. MIT Press, Cambridge (2002)
22. Bridgman, R.: Criteria for best practices in building child-friendly cities: involving young people in urban planning and design. Can. J. Urban Plann. **13**(2), 337–346 (2004)
23. Gong, X., Zhang, X., Tsang, M.C.: Creativity development in preschoolers: the effects of children's museum visits and other education environment factors. Stud. Educ. Eval. **67**, 100932 (2020)
24. Izquierdo Peraile, I., López Ruiz, C., Prados Torreira, L.: Infancia, museología y arqueología. Reflexiones en torno a los museos arqueológicos y el público infantil. Archivo de Prehistoria Levantina **30**, 401–418 (2014)
25. Observatorio Vasco de la Cultura: La infancia como publico cultural. Financiación y gasto público en Cultura - CAE - 2018 - Conócenos I Observatorio Vasco de la cultura. Euskadi.eus (2016)
26. Heft, H.: Affordances of children's environments: a functional approach to environmental description. Child. Environ. Q. **5**, 29–37 (1988)
27. Ciudades Patrimonio de la Humanidad: Guía Didáctica de las Ciudades Patrimonio de la Humanidad. Santiago de Compostela (2016)
28. Ciudades Patrimonio de la Humanidad: Libro Blanco sobre la Movilidad en las Ciudades Patrimonio de la Humanidad de España: la problemática del tráfico en los cascos históricos y sus posibles soluciones (2016)
29. Navarro Martínez, V., Raedó Álvarez, J., Rosales Noves, X.M. (coords.): Ludantia. I Bienal Internacional de Educación en Arquitectura para a Infancia e a Mocidade. Colexio Oficial de Arquitectos de Galicia, Pontevedra (2018)
30. Cidade da Cultura de Galicia: Plan Estratéxico 2012–2018 (2012). https://www.cidadedacult ura.gal/es/system/files/downloads/2012/09/plangaias.pdf
31. Cidade da Cultura de Galicia: II Plan Estratéxico 2021–2027. Resumo Estratéxico (2021), https://www.cidadedacultura.gal/sites/default/files/ii_plan_estratexico_cidaded aculturadegalicia21-27_resumo_executivo.pdf
32. Lefebvre, H.: The Production of Space. Blackwell, Oxford (1974)

A Triadic Framework for Sustaining Metaphorical Conceptualization to Automate Urban Design Creativity

Mohammed Mustafa Ahmed Ezzat(✉)

AI4Creativity, Bursa, Turkey
mohammed.ezzat@ai4creativity.com

Abstract. *"Metaphorical generating/understanding is the monolithic tool of analysis that unilaterally spans all the known variants of knowledge, e,g, concrete sciences, and art, which are otherwise considered as distinctively disparate fields of intentionality."*

"The reality of urban constructs may be sufficiently shaped by two opposite perspectives named rational and emotional. Such reality may be reinterpreted/transformed by a creative third perspective, named visual, into unvisited novel realities that are potentially art-driven environments."

The article consolidates these two statements into a single framework for generating/understanding architectural metaphors, and whence automating architectural design creativity. The article substantiates this goal over two parallel trajectories. The first is a theoretical investigation that debates that mathematics is a mere narration of geometric spatiality. Meaning that geometric physicality is the source of metaphors for mathematics, and even time, which in turn is the metaphorical source for the entirety of concrete sciences, including computation. The goal of this investigation is to accentuate the importance of metaphorization as the general form of analysis that unilaterally encompasses continuous, discrete, and art phenomena. Additionally, it is a good preparation for the proposed framework that infuses materiality and conceptuality into a single realm of conceptualization. Such a philosophical deliberation is then coupled with the second trajectory of empirical investigation that demonstrates the computability of architectural creativity. That demonstration is an extension of a comprehensive proposition of urbanism that states that three perspectives named rational, emotional, and visual may sufficiently disclose all the variants of any given concept. The empirical investigation concludes by a space of conceptualization that asserts the fundamentality of certain concepts and discloses the design process as an endless/iterative composition of concepts similar to the way natural language manifests.

Metaphorization-ability is presented by prevalent philosophers and scientists as the main contributor to the human mind's high-level cognitive-ability. Nonetheless, no serious framework has been introduced for sustaining the metaphorical production of new concepts out of already learned ones, and the article tends to close such a gap.

Keywords: Design creativity · A unifying model for automating design creativity · Metaphorization-able machine · Urban identity

© The Author(s), under exclusive license to Springer Nature Switzerland AG 2022
F. Calabrò et al. (Eds.): NMP 2022, LNNS 482, pp. 418–431, 2022.
https://doi.org/10.1007/978-3-031-06825-6_40

1 Introduction

The article extends a three perspectival model of urbanism [1] and constructs a metaphorical framework that grounds a human-computer augmented creative design process. To have a better grasp of the contribution of this article, the chronological development of the project needs to be firstly explicated. The project in hand emerged with the single motivation of scientifically analyzing *urban identity*. Regularly, any project of urban identity's analysis starts by collecting data, using traditional observational and surveyal techniques, that is coupled with interpretative analysis. Such interpretations are variably and subjectively shaped by local conditions, and the researcher needs to cognitively cope with these variations. Consequently, data itself is not as significant as studying its various interpretations, which is the main challenge of the field of urban identity. In summary, the Interpretative biasness and the non-transferability of the findings concerning a specific locality into another were the challenges of the project's intention. Given this progression of thought, it was found that reducing the subjective versatile interpretations into something general, common, and universal is the only way for laying an objective (scientific) tool of analysis for the field that is otherwise inherently deemed psychological/philosophical and outside the reach of the scientific domain, and It turned out later that this adopted methodology of reductive generalization matches the way Husserl founded the science of phenomenology, and phenomenology is the founding movement of modern philosophy. In fewer words, objectiving the subjective is not only a prevalent inquiry of philosophy since its inception but rather the central polarity of modern philosophy for infusing the concrete and philosophical fields under a single umbrella.

For doing so, the entirety of urban theories, movements, and constructs was thoroughly analyzed and a reductive three perspectival model was rigorously debated for its sufficiency in perceiving both the material and conceptual manifestations of urban constructs [1, 2]. The three perspectives are named *rational, emotional,* and *visual*. These three perspectives, as were originally introduced, may be summarized as:

- The *rational* perspective may be exemplified by modern rationalism and the international style. It is a reductive perspective that believes that whole is fully comprehensible by its parts. The part/whole equivalency suggests purposefulness, functionality, and utalitriality, but most importantly, ethicality and beauty are related by the slogan of "truthfulness is the source of beauty".
- The *emotional* perspective may be exemplified by the sprawl and spontaneous settlements. These environments manifest as a continual interaction by their constituting agents. Such procedurality suggests values like liveness, harmony, human centeredness, and volatility.
- The *visual* perspective may be exemplified by Romanian general purpose buildings and medieval cities. A visual designer is a philosopher who philosophizes ideologies, along with related inspiring references, that shape every corner of a materialized product. Such ideology-drivenness suggests the visual designs' timeless (time-independent). The paper's proposed framework is the aiding tool that substantiates the visual designer's creativity.

The three perspectives are found to be the least unique views that are fit to structure the variants of any concept, and consequently, structure a knowledge base of those perceived variants. Nonetheless, elaborating on this three-perspectival proposal led to the findings of this article, and that is the reality of urban constructs is either rational or emotional and the visual designer's creativity is the mere reinterpretation of these realities. Additionally, the contrasting contention between the three perspectives structures the fundamental concepts found in the Spoken language, as contrastively perceived by the three perspectives in parallel, into the proposed knowledge base of conceptions. Furthermore, the whole design process may be depicted as a mere ongoing composition of these fundamental concepts. Such a lingual-like behavior is a factuality that unilaterally merges the human mind's faculties and the materialistic computability under a single umbrella, or in other words, objectively materialize art as a computational production [3–5].

Therefore, **concepts** are proposed as the basis for such a unifying paradigm. On one hand, concepts strongly manifest as the foundation of the mind's **cognitive** and other functional attributes, and from another hand, physicality may be **conceptualized** by describing it by its features. Meaning that the features of physicality are the way to bring it to the realm of conceptuality, and that is actually how physicality is internalized in any conscious cognitive agent. But if concepts may equally span the physical and the conceptual realms, then what may differentiate these two realms from each other. It turns out that physicality is always reductively composable by its constituting elements. Meaning that the whole is fully comprehensible by its compositing parts. Nevertheless, the conceptual composition is not as fortunate as that of the reductive physicality. Meaning that a conceptual composition may yield another entity that is fairly distinctive from that of its compositing concepts. Therefore, the only way to understand conceptual compositions is to sketch *analogies* of such a conceptual composition. To follow such a path of metaphorization, if classical algebra and arithmetic evolve over equality, and abstract modern mathematics around isomorphism, then the human mind models and understands the world by the tools of analogical metaphorization. In fact, this incomprehensibility of structured conceptions is the central notion that motivated the field of modern philosophy, starting from structuralism and phenomenology up to deconstructionism, existentialism, and beyond.

The article is accordingly structured by starting, in Sect. 2, with a brief theoretical investigation that sketches the idea that mathematics is a mere narration of the geometric physicality, and in turn, mathematics is the source of metaphorization for the rest of concrete sciences. Nevertheless, such spatial metaphorization is a fragmented, limited version of metaphorization that deals with a certain facet of spatiality at a time. That limited version may be encompassed by the higher level of metaphorization evident in the human mind where all the structured facts of any concept are at its disposal all the time to infinitely compose/interpret any conception as unlimited sources of metaphorization. In fact, the three rational, emotional, and visual perspectives were devised under the assumption that they are the least three unique, independent perspectives that may contrastively structure the faceted variations of any concepts, and as a consequence, would clearly speciate a rigor conceptualization of each of these three realms themselves. The sketchy theoretical section is included to familiarize the reader with the idea that metaphorization is the unilateral tool of analysis that encompasses physicality and conceptuality (science and art) all alike.

The bulk of the article is in Sect. 3, where an empirical investigation of the proposed model is conducted. In Sect. 3.1 qualities that contrastively differentiate the three perspectives are sketched. The core of this section is laid in Sect. 3.2 where a single facet of *structurality,* which is believed to be central, is deliberated to crisply differentiate between the rational and emotional realms. In Sect. 3.3, exemplar visual modifiers are discussed that may reintroduce such rational/emotional reality into a creative being [6–8]. The section concludes by establishing the design process as a mere lingual-like composition of conceptuality, and such *lingual-compositionality* is stylistically influenced by *urban identity*. In fact, an exemplar achievement of substantiating such a lingual-like conceptual-based design process is *bringing creativity into the preserved spatial places* to fulfill the *ambition of contemporary users* while *preserving* even the most historical places with the aid of computation.

For the sake of clarity, and without delving into the deep territory of discriminating the three terminologies of *art, creativity*, and *beauty*, the article adopts distinct interpretations of each term that are still alignable with many philosophical elaborations [6, 7, 9], and they are used as such throughout the paper. These interpretations are as the following:

1. *Creativity* is the production of meaningful novelty, and such meaningfulness must be attested against a certain domain/s of knowledge.
2. *Art* is a production that instructs contemplative interaction with its experiencing human agent. Meaning that it is established as a deliberative interaction. For such contemplation to manifest, *novelty* is an assumed prerequisite.
3. *Beauty* is an evaluation system possessed by any conscious being, let that be a society or an individual, that scales the artful/personal experience on an ugly-beautiful scale.

2 The Theoretical Investigation

The goal of this section is to bring into attention the idea that geometric physicality is the sole metaphorical reference for the entirety of concrete sciences, and whence, metaphorization, which is undeniably a major force of the mental attributions, is qualified to be the monolithic framework that spans computability and creativity all alike. One solid finding that may theoretically back this argument is that the seemingly continuous phenomena (quantitative) of physicality are nothing but a stateful, discrete information-driven (qualitative) phenomenon. Such discretization is evident from the field of quantum mechanics that is devised by humanity to most accurately model the physical reality is based on stateful nuclear quanta. Additionally, cognitive systems perceive physicality, which is represented as a flat image over the retina, as informational-based 3D scenes. Meaning that what we see, hear and sense, is mentally processed, information-based, qualitative reflections that virtualize reality as a continuous phenomenon, and such behavior is at the core interest of the field of Artificial Intelligence (AI).

It is not only how the mind discretely recognizes physicality, but also how it metaphorically processes it for substantiating concrete sciences that support the section's argument. For example, the known numeric systems are mere metaphors of the geometric spatiality. Using the Cartesian geometry to sustain this argument, the physical space may be accurately described by using three dimensions that are generally labeled (X,Y,Z). This coordinate system is axiomatized for modeling linear spaces (X), planar spaces (X,Y), or the observed volumetric space (X,Y,Z). The known natural numbers, e.g., integers, decimals, rationals, etc., including time as contemporarily recognized, are metaphors of the single-dimensional space (X), while the complex numbers, which are foundational for the concrete sciences' mathematical modeling of physicality, are metaphors of the planar (X,Y) spaces. Moreover, Quaternion numbers, which are widely used in graphic applied mathematics and the gaming industry, are metaphors of the volumetric 3D (X,Y,Z) space. It is important to note that these numerical systems, along with their operators, ground classical (concrete) and modern (abstract) mathematics. The same geometric attitude is prevalent in modern physics. Namely, General Relativity, which is one of the two pillars of modern physics, adopts differential geometry, which is a relaxed version of the Euclidean space, for the uniliteral representation of space and time. The second pillar of modern physics of quantum mechanics adopts Hilbert space, which unifies geometry, topology, and algebra, for the most accurate non-relativistic modeling of quantum physics. Accending from concreteness to the highest levels of abstraction, concrete sciences revolve around equations (*Equality*) for representing knowledge; abstract modern mathematics emerges over *homomorphism* between different mathematical structures to represent/infer knowledge, while the mind's *analogical metaphorization* is the most general and efficient form of conceptualization that supersedes/encompasses the equivalent/homomorphic representations.

Furthermore, the behavior on the quantum scale is believed to maintain the informational consistency between the lowest observed quantum level with that of the witnessed reality, and such behavior may suggest that the quantum scale may hold similar attributions with that of the mind's metaphorize-ability.

Such arguments theoretically support the notion that geometric physicality is the sole metaphorical reference for the concrete sciences, and as such, a metaphorical framework may monolithically underpin both physicality and conceptuality all alike. Nonetheless, the metaphorical generation and inference of new novel concepts is the main challenge of any framework that may sustain metaphorical generation/understanding. Given this context of deliberation, it is proclaimed, starting from the coming section, that depending solely on the two opposite *rational-emotional realms* and *the visual compositing modifiers* may sufficiently establish such a framework.

3 The Empirical Investigation

A three-perspectival model was rigorously debated for its sufficiency in structuring a comprehensive conceptualization of urban constructs [1, 2]. Such an assumption led this research to a better understanding of the three perspectives themselves. Namely, it is found that urban constructs realities merely reside in *rationality* and *emotionality*, while the visual perspective is the one that rephrases these realities into novel creative ones that are potentially artful. These findings are not only in close synchronization with what art is about, such as depicted by Danto as a transfiguration of the commonplace [7, 8], but also is a pillar of a prominent theory of creativity and its computational applicability [9, 10]. These consequential interpretations of the three perspectives lead to the constructability of the conceptual spaces of the fundamental concepts found in spoken language (see Fig. 1). The section starts by briefly surveying several qualities of the rational/emotional perspectives. The bulk of the section is at Sects. 3.2 and 3.3 for depicting the way rational/emotional physicality may be conceptualized that may then be reintroduced by the creative visual designer as a work of art. The section concludes by discussing the constructability of the proposed conceptualization and the way the design process may be established as lingual-like compositionality that could be stylized by the local/global spatial urban identity.

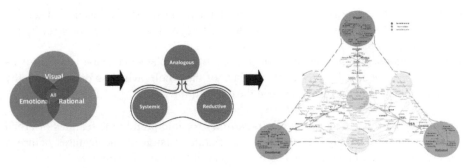

Fig. 1. The competence between the three perspectives for shaping a conceptual space of any concept.

3.1 Facts About the Three Perspectives

Figure 2 is a hierarchical structuring of the single concept of "Functionality". The multi-faceted reality of this concept is sketched at this figure, and it may even more clearly be demonstrated using the graphical representation of Fig. 1. These variants are thought to be a prerequisite transitory to Sects. 3.2 and 3.3.

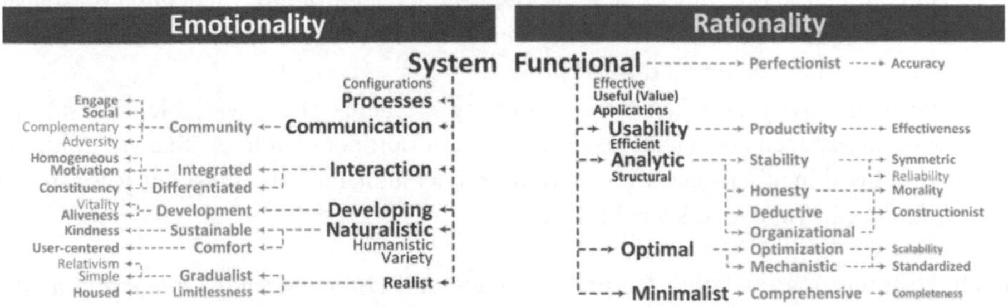

Emotionality	Rationality

System Functional ----------→ Perfectionist ----→ Accuracy

Fig. 2. Basic qualities that differentiate the two emotional and rational perspectives.

3.2 A Big Picture About the Features of the Rational and Emotional Realities

Once more, one of the contributions of the three perspectives is structuring the variants of any concepts found in natural language into a knowledge base of conceptualizations. Nonetheless, such a knowledge base of conceptualizations may be too massive, and consequentially too complex, to deliberate in a concise context like the scale of this paper. Consequently, the focus would be on a single facet of *structurality,* or in other words, how the constituting elements contribute to the manifestation of the whole. Structurality is perceived by the two opposite rational/emotional perceptions as the following:

- *The rational perspective*: the constituting elements have a single function of serving the whole. Any single element ceases to exist beyond serving the goals set by the monolithic whole.
- *The emotional perspective*: each of the constituting elements shape a world of its own. Consequently, locality exists in place of the whole, which is only sensible by the harmonious interactions between the constituting agents.

Employing this single aspect of *structurality* for interpreting the features of urban constructs may yield Fig. 3, which is a miniature version of the features of urban constructs. Based on that, Fig. 3 may be summarized as the following:

- The functionality of urban constructs may be featurized as:

  - The rational monolithic whole has the role of protecting the internal environment against the external environment's aspects of sound, mobility, solar radiation, heat, sight, and wind.
  - The interactivity between the constituting elements of the emotional environments merely delimits the internal/external spaces.

- The urban constructs' interrelated activities may be featurized by their distribution, relatedness, and flow patterns as the following:

  - The rational monolithic whole is densely, concentrically, and equally spaced arrangements. The wholistic reality is reflected by single-entry, protective indoor with a continuous, uninterrupted flow.
  - The interactivity between the constituting elements shape variable, sparse, and randomly spaced arrangements. Spaces are of outdoor characters that are infinitely housed within other spaces, e.g., nature, and as such, they are multi-entry constructs with the obstructive, dispersed flow.

- The different materializing features of shape and color may be summarized as the following:

  - The rational monolithic whole suggests monochromatic/dark-colored, simple, sharp, linear shapes.
  - The interactivity between the constituting elements suggests multi-chromatic, harmonious, light-colored, fluid/free-formed shapes.

The crucial two observations that need to be maintained by the end of this section is:

1. The two rational or emotional realities are shaped by the accumulation of the related concepts of each environment. To exemplify this notion:

   - For rationality: a wholistic single entry enclosed cube or a wholistic concentric dense clustering are two instances of pure rational environments. Additionally, the black-colored or sharp-edged materializations are reflections of the protective and reductive/productive nature of this environment as well.
   - For emotionality: delimiting, freestanding, scattered, multi-chromatic harmoniously colored, free-formed walls may represent a reality of this environment.

2. Each of the aforementioned Different products is a mere composition of certain facets distinctively belonging to each of the two environments.

Presuming the contrast-ability between the three visual, rational, and emotional perspectives made it possible to reach solid conclusions concerning the nature of each perspective from one hand, but most importantly, shaping a clear, structured, deep understanding of the two rational and emotional realities. Nonetheless, presuming that the rational and emotional perspectives reflect reality implies that they are not art-oriented on their own [7]. But rather it is the visual designer who shifts these two realities into the new unvisited realms, that are creative and potentially beautiful.

The visual designer does this job using compositional or transformational modifiers. For example, using the earlier rational metaphor of an enclosing single-entry monochromatic dark boxed urban construct:

- The visual designer may **compositionally** *replace the walls with emotional white textile-based materials, replace the single entry with multi-scattered ones (see Fig. 5 (2))*, remove walls or the roof to shift its enclosure into the emotional continuity, etc.
- The visual designer may **transformationally** curl the box, melt it down, flexibly flow it with a bath, tear it, scratch it, etc.

In both of these scenarios, the novel entity that mixes rationality with emotionality emerges and enriches the creative design space. *In the coming section,* several exemplar visual modifiers are represented in preparation of perceiving the visual designer production as a mere lingual-like composition that may be tailored according to urban or spatial identities, and whence, creativity is even brought to conservative environments.

3.3 The Visual Modifiers

The visual designer is the one who introduces creative urban constructs by reinterpreting/transforming the known rational and emotional realities into novel constructs. for doing so, a set of modifiers is needed. Figure 4 represents examples of these modifiers. The notable things about these modifiers are:

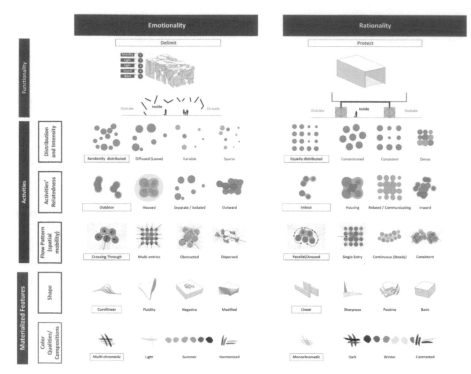

Fig. 3. Examplar opposite features of the two emotioal and rational perspectives.

1. The modifiers always belong to either the rational or the emotional realms and they coexist as opposites of each other.
2. There are four main clusters, underneath which, these modifiers may be classified. These four classes may be described as the following:

   a. As previously stated, the notion of structurality manifested over the rational integrated vs. emotional disintegrated variants, and the first two classes of modifiers belong to these two variants (Fig. 4 (1&2)).
   b. Transformative shape modifiers (see Fig. 4 (3)) change/transfigure the two emotional/rational realities.
   c. The other modifiers are derivatives from the different forms of art, e.g. painting, sculpting, poeticality, and musicality (see Fig. 4 (4)).

The consecutive application of reinterpretation-composition-transformation applied by the visual designer constructs a lingual-like design product. Reinterpretation is precedented by a metaphorical understanding of the design solution and its elements then other consecutive transformative and compositional modifiers are applied. This process may be infinitely and recursively applied. The knowledgebase proposed by the triadic model and needed for computation to metaphorically understand the fundamental concepts and their compositions is presented in Sect. 3.5.

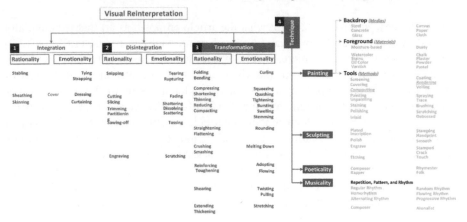

Fig. 4. Examplar modifiers used by the visual perspective for reinterpretting the emotional and rational realities.

The last thing to emphasize is the way such reinterpretation-composition-transformation process may be applied for shaping design compositions. For example, in Fig. 5, there are two exemplar creative design spaces that may be described as:

1. In Fig. 5 (1.0), the scattered natural forms that may resemble a tree are reflections of an emotional reality (1.0), but in (1.1), some of these scattered elements are condensed and floored with a black color, which are both rational facets. In (1.2.1), the forms are condensed then linearly transformed (1.2.3) then finally contained in a simple rectangular shape (1.2.4), which all are of rational modifications.
2. In Fig. 5 (2.0), a rational, protective, single-entry, monolithic, simple box is presented. Such reality is transformed by replacing the single-entry with emotional scattered multiple entries, in (2.1), that is furtherly treated with emotional curvilinearity, in (2.1.1 & 2.1.2).

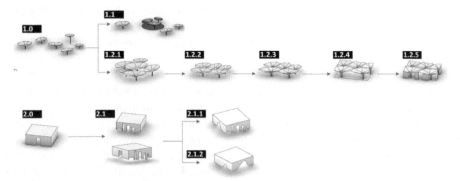

Fig. 5. An examplar design space that reinterprets emotional (1.0) or rational (2.0) variations into creative ones that rather hold joint rational/emotional traits, and as such, they are unvisited realities that are potentially artful.

Both of these alternatives exhibit what the article proposes as art that is potentially beautiful. Meaning that the original emotional or rational constructs, (1.0) and (2.0) in Fig. 5 respectively, are a mere replica of certain facets of reality, and they were considered as artful products until the visual modifiers are introduced in the rest of variants of Fig. 5. It is important to note that if different facets of the same qualities would have been selected or reinterpreted, fairly different variants of the design space would have been explored, which support the article's assumption of the sufficiency of the compositionality of the rational/emotional various facets and the visual modifiers to exhaustively experiment with creative design variants. Nonetheless, this opens the question of which of these variants, or any other if it matters, may be considered as more/less beautiful ones. The coming section tends to briefly answer this question in preparation for a more satisfactory answer by the conclusion of the article.

3.4   The Computational Utilization of the Conceptual Space for Augmenting a Creative Architectural Design Process

Constructing a knowledge base of conceptions is the indispensable starting point for computation to get involved in a creative design process. Meaning that there has to be a common conceptual space between the machine and the human agents to share during a creative design process. In fact, when an architect sits down with a client or community representative, this sharable knowledge space of conceptualization is taken for granted, or else, they wouldn't be able to communicate. Consequently, an augmented creative design process is shaped by continual interaction between the human agent and the machine. The human agent expresses his/her interests in ceratin favored composed variants of the design solutions and the computational agent extends these variants by reinterpreting/transforming the constituents elements using the visual modifiers (see Fig. 6 for an exemplar design space unfolded by these continual interactions).

3.5   The Conceptual Space

A parallel research effort has been conducted for constructing the proposed space of conceptualization using state-of-the-art Artificial Intelligence (AI) techniques. Figure 7 summarizes this effort. In brevity, a single conceptual space of the concept of "functionality" (Fig. 7 (1)) is used as a training dataset for a neural network model (Fig. 7 (2)). The trained neural model then constructs the proposed conceptualization that discloses a connected space of concepts that wholistically discloses their variants. Such a conceptual space is partly demonstrated through this paper for its essentiality in sustaining the metaphorical reinterpretation, composition, and transformation needed for any creative design process, and is the foundation of the computational augmentation to such a design process. Even so, the challenge that is still standing is the way the rational/emotional qualities along with the visual modifiers are composable in a lingual-like manner that sustains intentional creativity, which may be discussed in the coming Conclusion section.

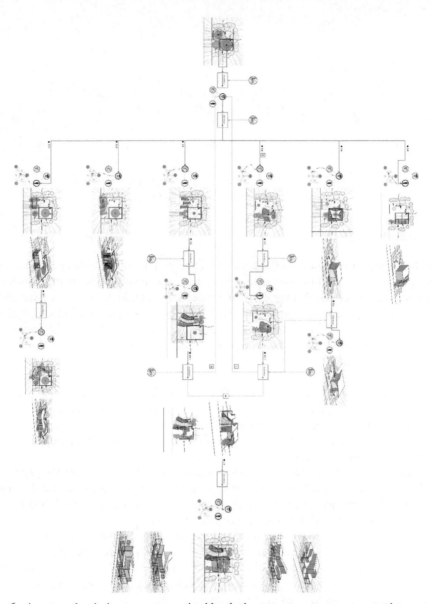

Fig. 6. An exemplar design process practiced by the human-computer agents over the proposed conceptualization.

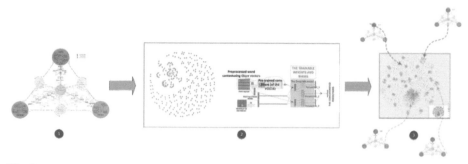

Fig. 7. The methodology adopted by the article for automating the structuring of the conceptual space of any concept found in natural language using AI techniques.

4   Results

Structuring the various materialistic/conceptual interpretations into a coherent whole (the conceptual space) is the intension of the research. Consequently, attesting the sufficiency of the three perspectives, and the detailed implied concepts, in disclosing the variations is the proper approach for concluding the efficiency of the proposed model, and the role of this section is to do so. Although, there are enormous evidence, with hardly any counterexample found yet, for asserting the article's assumption, the following three evidences are selected:

1. The three versoins of the theory of "Meaning" are fully explicable by the three perspectives. Namely, the Carnap's Truth-conditionality (the rational perspective), the Wittgenstein's Contextual Use (the emotional perspective), and the Quine's Narrated Synonymity (the visual perspective) are the three versions for sufficiently disclosing the meaning of any entity.
2. Elaborating on the three perpectives led to the framework proposed by the paper, and that is the reality of urban constructs is either rational or emotional and the visual designer's creativity is the mere reinterpretation of thcse realities. Such findings are alignable with the prominent theories of art and mind.
3. If the proposed framework is being examined over any exemplar architectural design, as the reader is encouraged in doing so, the sufficiency of the three perspectives in composing/reproducing any such design would be attested. Meaning that compositing the fundamental concepts do prove the sufficiency of the three perspectives in reproducing any envisioned urban constructs.

Nevertheless, the goal of the proposed knowledgebase of conceptions, the conceptual space, is to creatively compose concept into lingual-like designs, and that sets the trajectory for future developments that are detailed in the following section.

5   Conclusion

The proposed project started by proposing three perspectives named *rational*, emotional, and *visual* as a monolithic unifying model of urban constructs. Crystalizing the various subjective interpretations into a mere triadic ontology is only the first step for achieving a higher intended detailed conceptualization of the concepts found in natural language. The ability of the proposed model for exhausting the materialistic/conceptual variations is attested both on the high abstract level (*the three perspectives*) as much as on the detailed *concepts'* level. The design process, which is envisioned as a lingual-like composition of the detailed fundamental concepts, may not be mere accidental experimentation with various compositions of the fundamental conceptions but rather be of artful intentional involvement that may be exhaustively aided by computation, and that, until the moment, still not materialistically examined. this sets the trajectory for the future development of the project, which is as the following:

- Compositing and disentangling (understanding) the Lingual-like metaphorization by using the proposed conceptual space.
- Measuring the capability of such compositionality to sophistically and reliably establish ***communicative channels*** between computational-human agents.

References

1. Ezzat, M.: A comprehensive proposition of urbanism. In: Calabrò, F., Spina, L.D., Bevilacqua, C. (eds.) ISHT 2018. SIST, vol. 100, pp. 433–443. Springer, Cham (2018). https://doi.org/10.1007/978-3-319-92099-3_49
2. Ezzat, M.: A comprehensive conceptualization of urban constructs as a basis for design creativity: an ontological conception of urbanism for human-computer aided spatial experiential simulation and design creativity. In: Bevilacqua, C., Calabrò, F., Spina, L.D. (eds.) New Metropolitan Perspectives: Knowledge Dynamics and Innovation-driven Policies Towards Urban and Regional Transition Volume 2, pp. 580–591. Springer International Publishing, Cham (2021). https://doi.org/10.1007/978-3-030-48279-4_55
3. Evans, V.: "What's in a concept? Analog versus parametric concepts in LCCM theory. In: The Conceptual Mind: New Directions in the Study of Concepts. The MIT Press (2015)
4. Forbus, K.D., Gentner, D., Markman, A.B., Ferguson, R.W.: Analogy just looks like high level perception: why a domain-general approach to analogical mapping is right. Expt. Thfor. Artif. Intell. **10**(2), 231–257 (1988)
5. Helman, D.H. (ed.): Analogical Reasoning. Springer, Dordrecht (1988). https://doi.org/10.1007/978-94-015-7811-0
6. Carroll, N.: Theories of Art Today. The University of Wisconsin Press (2000)
7. Carrier, D.: Introduction: danto and his critics: after the end of art and art history. Hist. Theory **37**(4), 1–16 (1998)
8. Danto, A.C.: The Transfiguration of the Commonplace a Philosophy of Art, Cambridge. Harvard University Press, Massachusetts (1981)
9. Boden, M.A.: What is creativity? In: Dimensions of Creativity, pp. 75–118. MIT Press, Cambridge (1994)
10. Boden, M.A.: Computer models of creativity. AI Mag. (2014)

Sustainability Frameworks and the Recovery and Resilience Plan. Challenges from the Italian Context

Lavinia Pastore[1,2(✉)], Luigi Corvo[1,3], and Luca Tricarico[4]

[1] Open Impact, Piazza Manfredo Fanti, 22 Rome, Italy
{pastore,corvo}@economia.uniroma2.it
[2] Department of Management and Law, University of Roma Tor Vergata, Via Columbia 2, Rome, Italy
[3] Department of Business and Law, University of Milano-Bicocca, Via Bicocca degli Arcimboldi 8, 20126 Milan, Italy
luigi.corvo@unimib.it
[4] Department of Business and Management, LUISS University, viale Romania 32, Rome, Italy
ltricarico@luiss.it

Abstract. The paper analyzes the PNRR in relationship with the sustainable development frameworks and policy agendas. The aim of the research is to identify possible methodologies for assessing the social and environmental impact of this coming season of public investments that will affect Italy through the PNRR.

Keywords: PNRR · Sustainable development · Social impact · Environmental impact · Public policy

1 Introduction

The spread of COVID-19 turned the world upside down, generating health, economic, social, and governance system repercussions within each country (Aristovnik et al. 2021; Bacq and Lumpkin 2020). The coronavirus epidemic was so unprecedented that, with few exceptions, governments around the world had no experience or benchmarks to rely on (Kuhlmann et al. 2021). Given the multifaceted nature and unprecedented scale of the COVID-19 crisis, comparisons with past crises, including the 2008–2009 financial crisis, have significant limitations (OECD 2020a). Overnight, governments were forced to make fundamental decisions with the aim of protecting the health of citizens and limiting adverse consequences as far as possible (Spina et al. 2020; Cepiku et al. 2021; OECD 2020a). Measures were thus taken to support the economy, and among the productive activities that were interrupted in order to contain the virus, guarantees were recognised regarding bank loans and subsidies of various kinds were provided (OECD 2020a). The pandemic has shaken the world, setting in motion waves of change with a wide range of possible trajectories (OECD 2020b; Leal Filho et al. 2020).

In this regard, the European Union has intervened with the Next Generation EU (NGEU), a programme of unprecedented scope and ambition, which envisages a series of investments and reforms aimed at accelerating the ecological and digital transition, promoting better training for workers, and ensuring greater gender, territorial, and generational equity. The programme was approved on 21 July 2020 by the European Council with a total value of €750 billion, which will be used to finance national recovery and resilience plans to be implemented over the period 2021–2026.

At the same time, the PNRR is a plan for the recovery, resilience, and reform of the country, drawn up to deal with the pandemic crisis and its consequences, but also to take advantage of the opportunities dictated by globalisation and digitalisation, as well as by the ecological transition, in order to design a 'new Italy', with reforms to resolve inefficiencies and problems of a structural nature.

The COVID-19 global pandemic has provoked an intense debate around the (new?) role of the state and the re-centrality of public policies to balance the effects of a development model that exposes the population to risks of social and environmental crises.

While pandemic management has shown the importance of a public sector capable of responding in a timely and effective manner to largely unpredictable scenarios, the post-COVID world requires a profound rethinking of public policy.

The National Recovery and Resilience Plan (PNRR) itself represents the first signal of this change: the mere fact of having conceived a new season of public investments is a change of historic significance (both from a quantitative point of view and from the point of view of the quality of the objectives to be achieved). This first signal, however, should be followed by a rethinking of the logic underlying the construction of public policies, structurally introducing the concepts of sustainability and impact not only as moments of verification of the results achieved, but as circular processes that inspire the intentionality of policies, their management and their evaluation.

Ambitious indeed, the PNRR lacks explicit reference to all those actions and strategies implemented in the context of sustainable development within the Italian context and internationally: hence the purpose of this paper.

After analysing the 2030 Agenda signed by the UN member countries, containing the 17 Sustainable Development Goals (SDGs) set by 2030, the National Strategy for Sustainable Development, through which Italy has implemented and declined these goals at the national level, and the BES, developed in order to measure the equitable and sustainable well-being of a society by integrating its economic, environmental, and social aspects, the present work aims precisely at identifying a connection and inter-relationship between the PNRR and the three strategies oriented towards sustainable development, so as to propose a systemic framework for interpreting the PNRR, suggesting a possible measurement of the social and environmental impact of all the investments of a public nature that are envisaged.

In addition, at a later stage, it was also decided to investigate the calls issued to date, in order to verify an effective correspondence between what was planned at the level of strategies to be implemented and what was actually done.

Is it possible to reconfigure the PNRR to recognise the conditions for intentional sustainability right from the planning stage? Does the integration between the main pre-COVID sustainable development strategies and the PNRR make it possible to construct a framework in which it is possible to combine public investment and the measurement of economic, social and environmental impacts?

These are the questions that this paper seeks to answer.

The paper is structured as follows: the following Sect. 2 is a background section, with the aim of presenting the documents that will be taken into consideration during the analysis, i.e. the PNRR, and then the three sustainable development strategies, and then the 2030 Agenda at the international level and then the National Sustainable Development Strategy and the BES at the national level.

Next, Sect. 3 is dedicated to the explanation of the methods used in the analysis, which, starting from the use of different sources, aims to arrive at the understanding of a specific phenomenon. Section 4 then presents the results, which will then be discussed but which, however, are not as hoped for, highlighting a way of doing things that is still too tied to the logic of inputs and spending constraints, which does not pay attention to the impacts generated. The final Sect. 5 will then be devoted to conclusions and future lines of research.

2   Background

The purpose of this section is to offer an analysis of the Recovery and Resilience Plan (PNRR) elaborated by the Italian Government and presented in Brussels on 30 April 2021. More specifically, first we will focus on the structure of the Italian PNRR, highlighting its most salient points, then we will consider the efforts implemented by the country in terms of sustainable development, going on to examine the 2030 Agenda, the National Strategy for Sustainable Development, and then the Equitable and Sustainable Wellbeing policy. At the end of the section, from a theoretical point of view, an attempt will be made to identify potential relationships between what is envisaged within the NRP and the sustainability strategies, essentially indicating what the contribution of the Plan will be in terms of achieving the sustainability objectives already set, and in part those pursued, by the country.

2.1   The PNR

Italy is the first country to benefit, in terms of absolute value, from the two main instruments provided by the NGEU, namely the Recovery and Resilience Facility (RRF) and the Recovery Assistance Package for Cohesion and European Territories (REACT-EU). As far as the RRF is concerned, it alone guarantees an amount of 191.5 billion Euros to be used in the period 2021–2026, of which 68.9 billion are represented by non-repayable grants (NRP 2021), while the REACT-EU is conceived from a short-term perspective precisely with the aim of assisting the various countries in the first phase of the recovery of their economies.

As its name suggests, the NRP aims to be:

– A Recovery Plan, because it was drawn up to deal with the economic and social impact of the pandemic crisis and has as its objective not only to return Italy to pre-pandemic levels, but to build a 'new Italy';
– A Resilience Plan, as it presents itself as an opportunity to focus attention on crucial issues and to acknowledge and govern the transformations, dictated by globalisation and new technological frontiers, without suffering from them;
– A Reform Plan, because the planned investments are accompanied by a strategy of reforms that will modify the country, also from a structural point of view.

In line with what is shared at the European level, the Plan is developed around three strategic axes: 1) digitalisation and innovation, 2) ecological transition, and 3) social inclusion; it is divided into six specific Missions and includes four important contextual reforms, concerning public administration, justice, simplification of legislation, and promotion of competition.

The six Missions are organised as follows:

1) Digitalisation, innovation, competitiveness, and culture: the underlying objective is the digital modernisation of the country's communication infrastructure, public administration, and production system;
2) Green Revolution and Ecological Transition: This mission covers various interventions aimed at achieving a more sustainable agriculture and the implementation of the circular economy model, focusing on the exploitation of renewable energy sources and the development of sustainable mobility;
3) Infrastructures for sustainable mobility: the objective is to strengthen and extend the national high-speed rail network and upgrade the rail network at the regional level, paying particular attention to the areas of southern Italy;
4) Education and research: the focus is on young people, with the aim of boosting potential growth, productivity, social inclusion, and adaptability to the technological and environmental challenges of the future;
5) Inclusion and cohesion: the mission is divided into three components and aims at a structural overhaul of active labour policies, a strengthening of employment centres, and their integration with social services and with the network of private operators;
6) Health: this mission is divided into two components, aimed at guaranteeing the strengthening of the territorial network on the one hand and the modernisation and digitalisation of the technological equipment of the National Health Service on the other, leveraging on the Electronic Health File and the development of telemedicine.

The Plan intervenes within sectors that are crucial for the national system, aiming to promote a digital and ecological transition. With regard to digital adoption and the degree of technological innovation, Italy still lags despite recent marginal improvements, ranking 24th among the 27 EU Member States (Desi 2021)[1]. On the other hand, as far as the ecological transition is concerned, it is clear how exposed Italy is to significant

---

[1] https://digitalstrategy.ec.europa.eu/en/policies/desi-italy.

climate change and how necessary it is to accelerate the path towards climate neutrality. If it is true that today the emissions of climate-altering gases per person are lower than the EU average, it is also true that there are many delays and vulnerabilities, especially with regard to transport, and it is in this sense that the Plan is presented as a real opportunity to overcome all those barriers that have proved critical in the past, introducing advanced and integrated systems for monitoring and analysis processes and increasing investments to make the infrastructure exposed to climatic and hydrological risks more robust. Therefore, it is also a way to recover from the country's historical delays.

As regards the resources required by the Plan, a total of EUR 191.5 billion is said to be made available to the RRF, of which EUR 68.9 billion would be grants and the rest, EUR 122.6 billion, issued in loans.

According to a preliminary assessment of the potential impact generated by the Plan, it is estimated that the growth rate of the Italian economy will increase by 0.5 percentage points as a result of the higher expenditure and approximately 0.3 percentage points due to the full implementation of the planned reforms (MEF calculations 2021).

The Plan is seen as a component of a much broader development strategy which the Government will implement over the next few years and which will leverage an integrated set of funding sources and policy instruments. It is a strategy articulated in priorities and objectives, both general and specific, based on a programming process structured at various levels, which will use not only the resources made available by the Next Generation EU, but also the financing of the European Cohesion Policy for the period 2021–2027, the ordinary resources of the State budget, and additional resources dedicated to financing specific interventions complementary to the PNRR.

In line with the recommendations of the European Commission, additional checks are envisaged with respect to the ordinary administrative control established at the national level regarding the regular use of the financial resources allocated. The control activities are the responsibility of the PNRR Central Service, but then the individual administrations also have the task of verifying the actual achievement of targets and milestones and the regularity of procedures and expenditure based on risk assessment and ensuring that they are proportionate to the risks identified.

The direct and indirect impact (only of a macroeconomic nature), disaggregated by product, by productive activity and by institutional sector is quantified through the MACGEM-IT model, which is based on a Social Accounting matrix; therefore the purpose, during quantification, is to take into account possible rigidities and imperfections due to the behaviour of some operators and markets (the Public Administration and the labour market are examples). According to the Plan itself (2021), GDP is expected to grow more or less evenly over the course of the Plan's implementation, with a potential increase of 3.6% points in 2026.

2.2 Sustainable Development: Agenda 2030, National Strategy for Sustainable Development, Fair and Sustainable Welfare

The concept of sustainable development was introduced in 1987 in the Brundtland Report, released by the World Commission on Environment and Development (WCED), and is defined as 'development that meets the needs of the present generation without compromising the ability of future generations to meet their own needs' (Bruntland Report 1987).

In September 2015, the governments of the 193 UN member countries signed the 2030 Agenda for Sustainable Development, an action plan for people, planet, and prosperity. Within the Agenda appear the 17 Sustainable Development Goals (SDGs), articulated in 169 targets and more than 240 indicators, with the aim of guiding and influencing the behaviour of the countries involved over the next 15 years. Specifically, they follow up on the achievements of the Millennium Development Goals and focus on important issues such as combating poverty, eradicating hunger and combating climate change.

The peculiarity of these goals lies in the fact that they are universal, interconnected, and indivisible, therefore, taking into account the specificities of each individual territorial reality, they are applicable everywhere and globally, but also nationally and locally; but above all, their peculiarity lies in the fact that they suggest integration between the social, economic, and environmental dimensions. Referring to these three dimensions, it is possible to identify three macro categories among the 17 proposed Goals: those related to the biosphere, those related to social aspects, and those related to the economy (UN 2019; 2020).

Italy, in turn, has committed itself to defining the strategic objectives provided within the 2030 Agenda, outlining what the National Strategy for Sustainable Development (SNSvS) includes. The process of defining the Strategy has involved various stakeholders, including the main public research bodies (CNR, ISPRA, ENEA, ISTAT), civil society, and several NGOs with regard to the integration of key aspects concerning the analysis of the national context and the vision towards which to strive. It is divided into five main areas: People, Planet, Prosperity, Peace, and Partnership, each of which declined in national strategic objectives, elaborated ad hoc with regard to the Italian reality and complementary to the targets provided by the 2030 Agenda.

In the area of People, the aim is to promote a social dimension that guarantees a dignified life for the entire population, so that all human beings can develop their potential in a healthy[2] environment.

The second area, that dedicated to the Planet, aims first of all to safeguard natural resources, both land and sea, and then to guarantee the offer of an adequate flow of environmental services, dedicated to present and future[3] generations.

The Prosperity area, broken down into four different strategic choices, is aimed at the world of business and research[4].

The Peace dimension, then, aims at promoting a non-violent and inclusive society, favouring the implementation of integration and reception policies able to guarantee the rights of migrants, asylum seekers, and unaccompanied[5] minors.

Lastly, with regard to the area dedicated to Partnerships, the reference is to the promotion of cooperation for development, a crucial aspect for the application of the 2030 Agenda and that therefore must also be defined within the SNSvS. This is a particularly

---

[2] This area of the National Strategy relates to these Goals set by the 2030 Agenda: 1, 2, 3, 4, 5, 6, 7, 8, 10, 11, 13, and 16.

[3] Regarding this area, the 2030 Agenda Goals involved are 2, 6, 9, 11, 12, 13, 14, and 15.

[4] Related Sustainable Objectives are 2, 4, 5, 6, 7, 8, 9, 10, 11, 12, 13, 14, and 15.

[5] The Objectives pursued through this area are certainly 2, 4, 5, 8, 10, 15, and 16.

relevant area, so much so that it can be associated with all the Sustainable Development Goals presented in the 2030 Agenda.

In addition to the five areas just analysed, there are also the sustainability vectors, which can be conceived as transversal themes with regard to the integration of the concept of sustainability in the various policies or plans or projects, which have always been identified on the basis of the Sustainable Development Goals set by the 2030 Agenda at international level.

Still with a view to sustainable development, the BES project, developed in 2010 by Istat, in collaboration with representatives of the social partners and civil society, also deserves attention for its aim of measuring the Equitable and Sustainable Wellbeing of a society, going beyond the economic, social and environmental aspects. The purely economic indicators have been integrated with measures that take into account aspects such as people's quality of life and the effects on the environment. The BES indicators, together with other SDGs indicators, are always taken into consideration by ISTAT in order to monitor the state of implementation of the SNSvS, it is a functional monitoring for a possible update of the strategy itself, which however must be connected with the other policy documents in force, such as the National Reform Programme and the Economic and Financial Document.

2.3 The Impact of the NRP on Sustainable Development

Analysing the PNRR and the various efforts made with a view to sustainable development, there inevitably emerges overlaps, or perhaps relations, between the two guidelines currently in place within the Italian context, so it is worth trying to reason out how and to what extent one might benefit, or in any case contribute, to the success of the other. More specifically, the aim is to reason about the potential impact that the PNRR, which is due to end in 2026, will have in achieving sustainable development. However, these are relationships that are not made explicit anywhere, because, despite the fact that both guidelines are clearly developed with the same priorities in mind, in practice there is a lack of real coordination, so much so that within the Plan there is no reference to the SDGs, but only a general reference to the 2030 Agenda, when it would be appropriate for the quantitative targets to correspond to the same indicators used to monitor the progress of the Sustainable Development Goals, or that there be an alignment with what is foreseen at the international level (AsviS 2021).

According to a study carried out by the Italian Alliance for Sustainable Development (AsviS 2021), the pandemic has strongly and negatively impacted the path of sustainable development undertaken in Italy, underlining how high the risk is of not succeeding, in the remaining eight years, to achieve the 17 Sustainable Goals in the 2030 Agenda, since in the year 2019 to 2020, improvements were recorded only for three Goals (Energy System, Fight against Climate Change, Justice and Sound Institutions) and the situation remained stable with regard to three other Goals (Sustainable Food and Agriculture, Water, Innovation), but deteriorations were recorded on as many as nine Goals (Poverty, Health, Education, Gender Equality, Economic and Employment Condition, Inequalities, Condition of Cities, Earth Ecosystem, International Cooperation), while for two other Goals, namely 12 and 14, a lack of information does not allow an overall assessment (AsviS 2021).

What AsviS itself proposes, the possibility of updating the National Strategy in coherence with what is foreseen within the PNRR, is quite interesting as a measure to ensure that the theme of the younger generations, crucial within the Plan, is actually a driving force with regard to the design of all policies. Thus, the PNRR today represents not only a practically unique opportunity for the relaunch of Italy, but also a possible driving force with regard to the implementation of the 2030 Agenda.

It is clear, in fact, that the Plan itself was drawn up in favour of environmental sustainability and the reduction of inequalities (it is no coincidence that 76% of the funds are earmarked for projects aimed at combating the climate crisis and 56% are dedicated to the Mezzogiorno). In addition, the investments envisaged by the Plan are expected to reduce the Gini index of inequality by 38% as regards access to railway networks, and all the interventions aimed at transferring traffic from road to rail will contribute to improving the services offered to citizens and the competitiveness of businesses, while also making it possible to reduce climate-changing gas emissions and pollution, all interventions that drive towards sustainable development (data proposed by the Ministry of Sustainable Infrastructure and Mobility 2021).

It would be appropriate for the projects planned and financed by the PNRR to be monitored not only through the classic economic-financial analysis, but also by taking into account the environmental impact generated, and to do so it is necessary to intervene on four specific aspects (La Torre 2021):

- The construction of a new model of analysis regarding the positioning of PA with respect to sustainability indicators;
- The elaboration of a new impact assessment model, able to integrate economic variables with those of environmental and social sustainability;
- Updating the Regis system in line with the PNRR, with a view to 'integrated accounting';
- The definition of a new model of public-private partnership which, as far as the co-financing of investments is concerned, favours the rationalisation of public expenditure, enhancing the instrument of financial leverage.

Also in this perspective, a study carried out by FEEM and the Autonomous Region of Sardinia is interesting, the result of which can be seen as a proposal for a method to assess the contribution of the 2021–2027 Cohesion Policy to the 2030 Agenda (Cavalli et al. 2020). Similarly, as this methodology could be used to assess the impact of the PNRR on the 2030 Agenda, it would be possible to measure the impact of the PNRR investments on the SDGs based on the correspondences found.

For example, according to what is highlighted by Cavalli et al. (2020), Mission 1 of the Plan, aimed at revitalising the country through digitisation and innovation interventions and reducing structural competitiveness and productivity gaps, would seem to contribute not only to Goal 9 Industry, Innovation and Infrastructure, but also to Goal 8 Decent Work and Economic Growth, Goal 7 Clean and Affordable Energy, and, finally, with regard to investments for improving energy efficiency, also to Goal 13 Acting for the Climate. This kind of reasoning can be applied to all the missions of the NRP.

In general terms, it can be said that the NRP contributes most to Goals 9, 13, and 7, i.e., Industry, Innovation and Infrastructure; Climate Action; and Clean and Affordable

Energy, which are those most affected by the investments planned in the Plan's various Missions.

In any case, the regulation of the Recovery and Resilience Facility requires monitoring of the progress recorded on the Sustainable Development Goals and the various targets under the Plan, which will have to be done through accurate, effective, and efficient, as well as timely, data collection, thus ensuring transparency in the implementation of the various activities and the achievement of results.

3 Methodology

It is extremely difficult to investigate a social phenomenon without being part of it (Hammersley and Atkinson 1983, 1995). Therefore, this study is based on a specific qualitative approach based on direct observation of phenomena (Gobo and Maciniak 2011). In recent decades, this approach has been increasingly used in the social sciences, overcoming the apparent dominant use within traditional applied research fields such as social anthropology and sociology (Jones and Watt 2010).

The research presented here was carried out in an action-research context with qualitative methods over a two-year period (2020–2021). In particular, it was conducted within the support activities for the implementation of the National Strategy for Sustainable Development, as provided for in the agreement (ex-art. 15 Law of 7 August 1990, no. 241 and subsequent amendments) signed on 19 December 2019 between the Ministry of the Environment and Protection of Land and Sea (current Ministry for Ecological Transition, MITE) and the University of Roma Tre - Department of Architecture in collaboration with La Sapienza University - Department of Social and Economic Sciences and University of Rome 'Tor Vergata' - Department of Management and Law. This article describes the results of the research conducted by the working group of the University of Rome 'Tor Vergata' that has been involved to carry out an action-research path on the revision of the National Strategy for Sustainable Development with respect to the dimensions of social and environmental impact. The path, in addition to MITE, has seen the involvement of several key stakeholders such as: the Forum for Sustainable Development[6], the Board for sustainable development of the Regions, the Board for sustainable development of metropolitan cities, and the OECD.

The results of the research were subsequently submitted to and processed with CNEL using the Open Impact[7] database.

From a methodological point of view, the study is based on a specific qualitative approach that considers multiple sources of information, triangulation, which is characterised by the use of different data sources to facilitate the global understanding of a

---

[6] https://www.mite.gov.it/pagina/il-contributo-della-societa-civile-il-forum.

[7] Open Impact is a research spin-off of the University of Rome Tor Vergata. Open Impact is a platform that supports the entire lifecycle of impact, enabling decision-makers to make more informed choices, strengthening the economic sustainability of social enterprises, and facilitating the encounter between finance and social impact. Over the years, Open Impact has collected, codified, and synthesised internationally validated social impact reports (social impact assessments) available in open source, thus creating the first database for impact benchmarking. https://www.openimpact.it.

phenomenon (Patton 1999; Carter et al. 2014). The authors actively participated over the two years of the survey in internal meetings of the MITE working group and meetings with different stakeholders, and took part in public meetings on the topic of sustainable development, presenting the intermediate results of the research.

The stages of the research were as follows:

1. Documentary analysis of pre-COVID-19 sustainability strategies;
2. Focus groups and semi-structured interviews with key stakeholders;
3. Documentary analysis of the NRP;
4. Construction of an intersection database between sustainability documents and the NRP;
5. Processing of PowerBI (business intelligence software);
6. Thematic focus groups with representatives of the Forum for Sustainable Development to compare the database. Focus group with CNEL and intersection with the outcomes of the Open Impact database.
7. Monitoring of PNRR calls - Italia Domani website

As far as phase 1 is concerned, the work of analysis was carried out considering three strategic documents that have the function of orienting the public administration (and not only that body) towards a progressive recognition of dimensions and measures that intercept the social and environmental effects, as well as the directly economic ones of investments (public and private). In particular, the following were considered:

– The UN 2030 Agenda (169 targets divided into 17 Sustainable Development Goals) as a key international strategy;
– The National Sustainable Development Strategy - NSDS;
– The Fair and Sustainable Well-being (BES) developed and monitored annually by ISTAT.

Phase 2, which took place in 2020, consisted of discussions of the three strategy documents with the following key stakeholders:

– Internal MITE working group for the revision of the SNSvS, a key player in the research project as an activator;
– Forum for Sustainable Development, a collection of civil society organisations working together with MITE to revise the Strategy. The forum is organised, like the SNSvS, in the five thematic areas (Peace, Planet, People, Prosperity, and Partnership). Due to COVID-19, it was not possible to attend the plenary meetings of the individual thematic areas. The comparison was carried out through the observed participation of the meetings between MITE and coordinators and vice-coordinators of the Forum (about 15 people divided into 2/3 for each thematic area) and through the organisation of five focus groups in collaboration with the University La Sapienza with each thematic area.
– Regions, the Board for sustainable development of metropolitan cities, in addition to participating as observers in the meetings between MITE and metropolitan cities, the authors carried out 14 semi-structured interviews with the managers responsible

for the sustainable development strategy for metropolitan cities in collaboration with Roma Tre researchers.
– Table for the sustainable development of the Regions, participation in the plenary meetings between Regions and MITE.

Phase 3 consisted of studying the PNRR in both its first version - on which an initial test of linkage with the sustainable development strategies presented at the Forum in April 2021 was made - and the analysis of the final PNRR as of 30 April 2020. The PNRR was analysed by applying an impact logic that aims to identify the Theory of Change, the impact chain (Clark et al. 2004), and the conditions of intentionality, measurability, and additionality of the impact (Tiresia 2019). This phase took place from February to May 2021.

Phase 4 consisted in developing the work of linkage and interrelation between the NRP and the above-mentioned strategies. The driver of analysis is the NRP considering the four levels, to each of which were linked the targets provided by the 2030 Agenda, the domains of the BES, and the choices and strategic objectives of the SNSvS. The linkage work was carried out by three researchers independently. This phase took place between March and June 2021.

Phase 5 was the development of PowerBi from the Excel database of document analysis. PowerBi is a business intelligence platform that aggregates, analyses, visualises, and shares data. It is used to simultaneously connect and analyse different datasets by visualising grafts in the same tab. Building PowerBI was key to being able to open up the discussion on the interrelationships of strategic documents to other stakeholders in Phase 6. Phase 5 took place from April to September 2021.

Phase 6 involved, through focus groups, all the stakeholders nominated in Phase 2, to whom the link between PNRR and the strategic documents for sustainable development was presented. In addition, the PowerBi was presented at Forum PA and the Festival of Sustainable Development. This phase took place from July to December 2021.

During autumn 2021, CNEL was engaged through three focus groups in conducting an in-depth analysis of the cross-cutting aspects of the PNRR: the generation gap, gender gap, and territorial gap. The database underlying the comparison PowerBI used with MITE[8], which establishes the connections between PNRR, SNSvS, SDGs, and BES, was further related through these three aspects (youth, women, and territorial gap) with the Open Impact database and the related impact chains (an ad hoc PowerBI processing was created for the focus groups with CNEL) (Fig. 1).

Phase 7 consists of monitoring, the Italia Domani platform, updated until 13 January 2022, where all PNRR calls are published. A database has been created where the calls have been organised by mission, component, and investment and are analysed by type of recipients, territory, opening and closing date, overall budget and project budget, type of PPP envisaged, and elements of social or environmental impact envisaged.

---

[8] On this MITE public page you can view the first processing of PowerBI and focus on the context of the research. https://www.mite.gov.it/pagina/la-strategia-nazionale-lo-sviluppo-sostenibile-strumenti-di-collaborazione-istituzionale.

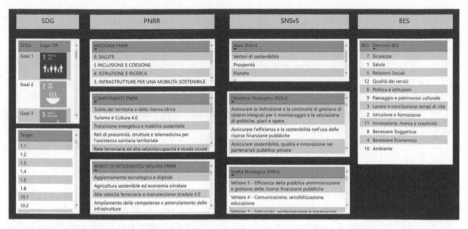

Fig. 1. Screenshot of the CNEL PowerBI (interactive and queryable visualisation) which relates SDGs, PNRR, SnSvS, and Bes.

4 Results

The first result of this analysis refers to the analysis of the intentionality of impact, and therefore of the sustainability of the NRP. The Plan does not show an explication of intentional impacts except for that portion referable to the effects of investments on GDP growth. Although the concept of sustainability is expressed and referred to several times in the official document and in the various documents linked to it, a gap of intentionality with respect to the three dimensions of sustainability is evident:

- Explicit intentionality with respect to economic-financial impact with definition of measurable objectives, targets, and timeframes;
- Reduced intentionality with respect to environmental impact, with some objectives setting expected targets and others very vague ones in terms of impact expectations;
- Very little or no intentionality regarding the social impact of the plan, with no objective expressed in terms of expected impact and no specific targets.

The observation, therefore, is that the Plan lacks a framework of intentionality of impact, exclusive of the economic-financial dimension. Therefore, returning to the first research question, 'Is it possible to reconfigure the PNRR to recognise the conditions of intentional sustainability right from the programming phase?' The answer turns out to be negative. Within the PNRR, in fact, there were no intentional, measurable, and additional impact dimensions that can be traced to a clear impact framework nor explicit links with the 2030 Agenda and consequently with the National Strategy for Sustainable Development or with BES indicators. The only impact reference provided in the NRP is of a macroeconomic nature, which aims at an increase in GDP downstream of investments.

The lack of intentionality of impact inhibits the analysis of two other key elements of the sustainability cycle: measurability and additionality. These aspects, however, are not only relevant from a theoretical point of view but also emerge as indispensable in the accountability relations between Italy and the European Commission. EC Directive

141/2021, in fact, describes Italy's social framework and highlights the transversal nature of certain particularly critical dimensions of the country, such as:

- The discomfort of the younger generations, both in terms of participation in education and in terms of job placement, which is summarised by the percentage of NEETs out of the total youth population;
- The gender differential in terms of social, work, and entrepreneurial development opportunities;
- The territorial differential between northern and southern regions and between urban and inland areas.

The reconstruction of a framework of intentionality, therefore, is essential to extend the analysis and fill the gap that emerged with the first result of this research.

Through the work of connection and interrelation between the NRP and sustainable development strategies, it is possible to reconstruct an impact framework and measurement logic that can also investigate the environmental and social impacts of the investment season. In fact, the measurability of social and environmental impacts can be identified in the taxonomies and development agendas already available (Agenda 2030, SNSvS, and BES). The answer to the second research question is therefore affirmative[9]. From the standpoint of systemic governance, these Strategies could be used as a driver to intercept (and primarily direct and stimulate) the investments envisaged in the NRP and to measure them over time. The impact framework of the PNRR, reconstructed and validated in a multi-stakeholder perspective (actors in the MITE project and CNEL), consists of five macro-areas, of which, in addition to the three referable to Directive 141/2021 and described above, the impact of the ecological transition and that of the digital transition are also considered.

In summary, then, the intentionality analysis brought out five key dimensions:

- Generational impact
- Impact on gender gap
- Impact on the territorial divide
- Digital transition
- Green transition

The first three correspond to the transversal drivers of the NRP, also understood as conditionalities, while the last two are the two vertical drivers that affect all six missions of the NRP in a diversified way. In a systemic governance, the measurement of impacts could be exercised by the SNSvS (in turn already linked to SDGs and Bes).

In fact, the lack of an ad hoc taxonomy for the NRP could turn from a critical element into a strong point that could generate opportunities. The proliferation of top-down taxonomies, in fact, does not produce improvements in the possibility of measuring sustainability performance, and the prospect of creating relational links between each element

---

[9] Does the integration between the main pre-COVID sustainable development strategies and the NRP make it possible to build a framework in which public investment and measurement of economic, social, and environmental impacts can be combined?

of the Plan (missions-programmes-interventions) with one or more objectives and targets referable to other taxonomies (SDGs, BES, SNSvS) would favour the multidimensional logic of the objectives to be achieved. The use of digital environments capable of dynamically connecting all the elements of the NRP to the existing taxonomies, therefore, can represent a perspective which, by favouring horizontal measurability, provides elements for the construction of an eco-systemic governance of the Plan.

This, in our opinion, is the crucial link and the most relevant result of this research: the possibility of building a chain of meaning between intentionality-measurability-eco-systemic governance.

In more detail, here is how this might be done. As shown in Fig. 2, the Plan can be fully read and interpreted in light of existing taxonomies, which has already been achieved as an output of this research. This initial linkage therefore allows the gaps in the Plan to be filled in terms of measurability of social and environmental impacts and to enable its execution in a framework consistent with the country's and Europe's sustainability goals.

However, this is not enough, for a number of reasons:

1) The nature of the Plan requires the verification of another key element, in addition to intentionality and measurability. That is, additionality in terms of impact, considering the extraordinary nature of the investments considered. This aspect can be defined as intertemporal complexity;

2) The guidelines of the European Commission and the vision and strategy of the Next Generation EU emphasise eco-systemic governance through the innovation of public-private relations. Alongside the existing taxonomies, therefore, it is necessary to take into account, in an integrated logic, also the emerging taxonomies, such as ESG criteria and the recent European Taxonomy. In this way, therefore, it will be possible to consider not only public investments but also mixed public-private investments and totally private investments. The framework of synthesis between these taxonomies enables a systemic reading of sustainability performance and goes in the direction of multi-stakeholder governance, from which the most innovative investment solutions such as impact investing and payment by result models based on outcomes take their cue. This aspect can be defined as relational complexity;

3) The third challenge reflects the Plan's multi-level governance perspective. Approximately 60% of the resources, in fact, will be allocated to Local Authorities, which, through investments and projects, will be able to contribute to the achievement of the objectives linked to the five key dimensions described above. This step, therefore, highlights the 'place-based' aspect of the development that the PNRR intends to generate and raises the level of complexity of the information processes. The framework of intentionality and measurability, as well as the information needed to determine the additionality of sustainability performance requires, therefore, synergetic work between central and local administrations, with a further need for coordination of data flows to be integrated in subsequent levels of aggregation and geo-location. This aspect can be defined as federal complexity.

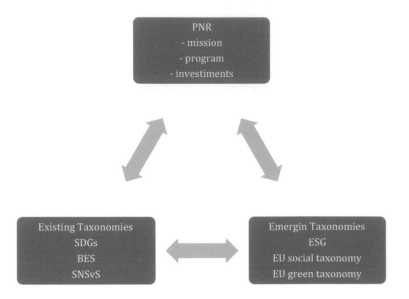

Fig. 2. PNR Integrated Reading - Existing and Emerging Taxonomies.

5   Conclusions, Limitations, and Research Perspective

First of all, this paper aims to create a systemic framework for interpreting the NRP through other sustainability strategies at the national and international level and integrating existing taxonomies with emerging ones. This opens up the possibility of measuring the social and environmental impact of planned investments, while up to now only the assessment of economic and financial impact in terms of effects on GDP growth has been made explicit.

From the point of view of implications for policy makers, this research facilitates decision-making processes that seek to take into account the social and environmental dimensions. The introduction of the latter two dimensions into the public governance discourse holds interesting implications from a research point of view.

The analysis of the three levels of complexity (intertemporal, relational, and federal) reveals on the one hand the limits of our current research, and on the other hand opens up prospects for extending the analysis towards different directions:

1) Incorporate baseline levels in the analysis in order to determine impacts in terms of additionality. To date, one of the most significant shortcomings of the overall Plan framework is represented not only by the expected targets with respect to the social and environmental dimensions, but also by the lack of baseline data describing the state of the art for each impact objective. In order to fill this gap, MITE is working on the revision of the SNSvS baselines in partnership with the OECD and the concerned local governments. This would allow two key needs to be addressed: establishing the as is for each relevant area, starting from those indicated by the European Commission with the Directive 141/2021, and determining the expected

targets according to an explicit or implicit logic. The explicit logic can be adopted if, during the implementation phase, the expected impacts are indicated in a program management logic. The implicit logic is adopted as a substitute in the absence of explicitness and foresees the adoption of international benchmarks as targets of last resort. For example, assuming for each relevant area the weighted average of the level of performance achieved by our European partners, it will be possible to consider the additional impact in terms of reduction of the gap between Italy and its partners.

2) Analyse the relational intensity of the Plan by monitoring the execution phase. A first interesting fact, found in the first phase of application of the PNRR, shows how the resources invested so far have been largely allocated through public tenders and very little space has been given so far to the construction of public-private partnerships. It will be necessary to verify, therefore, whether public-private relations can be scaled up at the territorial level after an initial allocation that expresses the transfer from central to local administrations. The risk that appears is that of a consistent increase in administrative complexity and in the proliferation of tender procedures, with the possibility of difficulties, especially on the part of municipal administrations, in superimposing on ordinary administrative management an administrative management that is completely above standard and linked to the execution of the Plan. The public-private partnerships and the tools of shared administration (co-programming, co-design, special partnership, partnership for innovation, and others) could represent an opportunity to build more efficient and performance-oriented governance solutions.

3) Monitoring the execution phase from the point of view of the administrative implementation of the Plan's impact dimensions. The first evidence, deriving from the analysis of the first calls for proposals published on the official Italia Domani portal, shows a trend that is not very encouraging. The prevailing trend, in fact, consists in the distortion of the impact objectives in constraints to the destination of inputs. By way of example, the objective of reducing the territorial gap between the territories of the centre-north and the south is expressed as a constraint to employ at least 40% of the resources in the southern regions. Similarly, the objective of favouring the empowerment of the younger generations is translated into a constraint on the implementation of activities aimed at young people for between 40% and 60% of the activities financed. The same is true for the other impact dimensions, which become, in fact, a constraint in the input-output chain and not a key element of the expected impacts, that is, of the effects produced by the use of inputs and the implementation of outputs.

The next steps of the research, therefore, will consider the three levels of complexity and will require quasi-experimental action-research methods, assuming a plural perspective that tends to closely observe the implementation process from the perspective of the different stakeholders at stake.

The contribution that this research is intended to make has a twofold purpose. From the theoretical point of view, the research is intended to be part of the critique of traditional public governance, showing how the intersection between studies of social innovation, of territorial design and of digital transformation can open up new spaces for interdisciplinary research in which we go beyond the dichotomy between quantitative vs. qualitative approaches and where the spaces of intersection between different theoretical strands create a continuum, action-research-action.

From the point of view of the implications for practitioners, the research is intended to offer a stimulus and constructive criticism to reduce the risk of the PNRR suffering the effects of a legal-administrative culture in which formal elements tend to prevail over the real generation of value and social transformation. If we want to use a representative image, we could say that the PNRR today oscillates between two possible destinations: the original one which refers to the vision of the Next Generation, or rather the possibility of opening up new development trajectories in which economic interests are included in a broader set of values and investments have a primarily transformative purpose in accompanying the players in the transition; and the conservative one which refers to the concept of recovery understood as the restoration of ex ante conditions, effectively closing the doors to a profound rethinking of the logic of government, business, and relations between social players.

References

Alleanza Italiana per lo Sviluppo Sostenibile. Il Piano nazionale di Ripresa e Resilienza #NextGenerationItalia e lo sviluppo sostenibile (2021)

Archibald, M.M.: Investigator triangulation: a collaborative strategy with potential for mixed methods research. J. Mixed Methods Res. **10**(3), 228–250 (2016)

Aristovnik, A., et al.: The use of ICT by local general administrative authorities during COVID-19 for a sustainable future: comparing five european countries. Sustainability **13**(21), 11765 (2021). https://doi.org/10.3390/su132111765

Bacq, S., Lumpkin, G.T.: Social entrepreneurship and COVID-19. J. Manage. Stud. **58**, 285–288 (2020). https://doi.org/10.1111/joms.12641

Becker, H., Geer, B.: Participant observation and interviewing: a comparison. Hum. Organ. **16**(3), 28–32 (1957)

Cavalli, L., et al.: The contribution of the European Cohesion Policy to the 2030 Agenda: an application to the autonomous Region of Sardinia. Nota di Lavoro **11**, 2020 (2020)

Cepiku, D., Marchese, B., Mastrodascio, M.: The Italian response to the economic and health crises: a budgetary comparison. J. Publ. Budg. Account. Financ. Manag. **33**, 47–55 (2021). https://doi.org/10.1108/JPBAFM-07-2020-0134

Denzin, N.K.: The Research Act: A Theoretical Introduction to Sociological Methods. McGraw-Hill, New York (1978)

Polit, D.F., Beck, C.T.: Nursing Research: Generating and Assessing Evidence for Nursing Practice, 9th edn. Philadelphia (2012)

Farmer, T., Robinson, K., Elliott, S.J., Eyles, J.: Developing and implementing a triangulation protocol for qualitative health research. Qual. Health Res. **16**(3), 377–394 (2006)

Gans, H.J.: Participant observation in the era of "ethnography." J. Contemp. Ethnogr. **28**(5), 540–548 (1999). https://doi.org/10.1177/089124199129023532

Carter, N., Bryant-Lukosius, D., Dicenso, A., Blythe, J., Neville, A.J.: The use of triangulation in qualitative research. Oncol. Nurs. Forum **41**(5), 545–547 (2014). https://doi.org/10.1188/14.ONF.545-547

Giovannini, E.: Nel Pnrr infrastrutture e mobilità sostenibili. Intervento del 4 Novembre 2021 (2021)

Governo Italiano. Presidenza del Consiglio dei Ministri. Piano Nazionale di Ripresa e resilienza (2021)

Hammersley, M., Atkinson, P.: Ethnography: Principles in Practice. Routledge, London (1995)

Hammersley, M., Atkinson, P.: Ethnograitative research. In: Principles in Practice, 2nd edn, pp. 516–529. phy, Thousand Oaks, New York (1983)

Gobo, G., Marciniak, L.T.: Ethnography. Qualit. Res. **3**(1), 15–36 (2011)

Kuhlmann, S., Bouckaert, G., Galli, D., Reiter, R., Hecke, S.V.: Opportunity management of the COVID-19 pandemic: testing the crisis from a global perspective. Int. Rev. Adm. Sci. (2021)

La Torre, M.: Pnrr e finanza sostenibile: un nuovo copione per la Pa. Il Sole 24 ore (2021)

Leal Filho, W., Brandli, L.L., Lange Salvia, A., Rayman-Bacchus, L., Platje, J.: COVID-19 and the UN sustainable development goals: Threat to solidarity or an opportunity? Sustainability **12**, 5343 (2020)

Moncavini, G., Napolitano, G.: I Piani Nazionali di ripresa e Resilienza in prospettiva comparata. Istituto di Ricerche sulla Pubblica Amministrazione (2021)

OECD. The Territorial Impact of COVID-19: Managing the Crisis across Levels of Government (2020a). https://www.oecd.org/coronavirus/policy-responses/the-territorial-impact-of-COVID-19-managing-the-crisis-acrosslevels-of-government-d3e314e1/

OECD. Strategic Foresight for the COVID-19 Crisis and Beyond: Using Futures Thinking to Design Better Public Policies. OECD Publishing (2020b). http://www.oecd.org/corona virus/policy-responses/strategic-foresight-for-the-covid-19-crisis-and-beyond-using-futures-thinking-to-design-better-public-policies-c3448fa5/

ONU. The Sustainable Development Goals Report 2019 (2019)

ONU. The Sustainable Development Goals Report 2020 (2020)

Scott Jones, J.E., Watt, S.E.: Ethnography in Social Science Practice. Routledge/Taylor & Francis Group (2010)

Spina, S., Marrazzo, F., Migliari, M., Stucchi, R., Sforza, A., Fumagalli, R.: The response of milans emergency medical system to the COVID-19 outbreak in Italy. Lancet **395**, 10227 (2020). https://doi.org/10.1016/s0140-6736(20)30493-1

Tibben, W.J.: Theory building for ICT4D: systemizing case study research using theory triangulation. Inf. Technol. Dev. **21**(4), 628–652 (2015)

Tiresia Impact Outlook (2019). http://assifero.org/wp-content/uploads/2019/11/Tiresia-Impact-Outlook-2019.pdf

Investigation on Limits and Opportunities of Rural Social Innovation in the Belluno Dolomites

Maurizio Busacca[(⊠)] 📧

Malcanton Marcorà, Ca' Foscari University, Dorsoduro 3484/D, 30123 Venice, Italy
maurizio.busacca@unive.it

Abstract. Since 2015, an increasing number of scholars associate rural development to social innovation. This paper presents the results of a participatory action-research carried out in the province of Belluno, involving various key local actors in the design of endogenous development strategies based on social and economic innovation.

1 Introduction

This paper aims at contributing to the debate on the role of social innovation as a strategy for local sustainable development and at elaborating interpretative hypothesis on the mechanisms that contribute to social innovation in rural areas. In order to do this, the article deepens the results of the pilot project "+ Resilient - Mediterranean Open REsouRcEs for Social Innovation of SociaLly ResponsIve ENTerprises", carried out in the province of Belluno and managed by the Chamber of Commerce of Treviso and Belluno. The pilot scheme took place between January and March 2021. It consisted of a participatory action-research and involved significant local actors in designing and identifying community hubs for social, professional, recreational and cultural purposes. The main goal was to define effective forms of support for social innovation in the tourism and agrifood industries. Considering that the local economy currently relies on industries, seasonal tourism and agricultural sectors were chosen as areas of potential diversification (Chamber of Commerce Treviso-Belluno Dolomites 2018). The main questions guiding the research were: how can social innovation constitute an effective strategy for local development? Which cooperation model could favour it?

We acquired data through 2 focus groups and 15 interviews. The focus groups explored social innovation in both tourism and ICT sectors and involved 18 participants. The two focus groups aimed at involving key local players to produce knowledge on the connections between social innovation processes, tourism and the mountain territory of the Belluno area. We analysed and interpreted the results of the focus groups and interviews also taking into consideration the evidence emerged during about 300 h of direct observation conducted during events, seminars and thematic courses.

We conducted 15 in-depth interviews using the Delphi Method (Landeta 2006). This approach helped us bring the answers together and co-design a Collaboration Model to implement local development strategies and social innovation, based on the local resources identified during the focus groups. We used a thematic approach to analyse and interpret the content: first, we transcribed and manually analysed the focus groups and the interviews; then, we read all the material, identifying and organizing all the emerging issues.

2   Rural Social Innovation: Theoretical Framework

Recently, the term 'Social Innovation' gained importance in scientific and political discourses. Oosterlynck et al. (2020: 5) define social innovation as «initiatives, actions and policies that aim to satisfy basic social needs (content dimension) through the transformation of social relations (process dimension), which crucially implies an increase of the capabilities and access to resources of the target group (empowerment dimensions linking the content and process dimension)».

Its acritical supporters present social innovation as an effective solution to complex and multidimensional problems and they use it as a tool to foster local development replacing the action of public authorities with citizens mobilisation (Borelli and Busacca 2020). Instead, adopting a critical approach we can enhance the transformative potential of social innovation.

In this sense, many authors see social innovation as an important engine for successful rural development (Bock 2016; Neumeier 2017), but there is very limited knowledge on how social innovation is characterized in rural areas (Noack and Federwisch 2019). The vast literature now available on Social Innovation, in fact, focuses above all on urban contexts, rich in both human and economic resources, where density (of people, knowledge, interactions) facilitates the circulation of knowledge and material resources in urban areas. On the other hand, there is still little concern on how these mechanisms might interest rural areas, where there has been a gradual shift from a purely agricultural focus towards a more diversified economy.

Now that policy and scientific discourses initiated a profound discussion on how to avoid an idealized image of rural areas and not impose top-down developmental solutions in order to offer place-shaping and networked rural development measures (Shucksmith 2018). Compared to urban contexts, in fact, the forms of mobilisation and participation required by social innovation are more complex here and hindered by the scarcity of human capital and a lower level of resources. The challenge of social innovation can be faced more effectively in contexts where there is a strong social infrastructure, therefore supporting communities involved in partnerships (Osborne et al. 2004). This is not the case in rural and mountainous areas suffering from depopulation.

In terms of resources, the difference between mountain and urban territories is crucial in explaining constraints and opportunities for social innovation. Mountain areas are defined by limited human and economic resources, reduced ability to attract investments and scarcity of long-term collaboration networks. Even so, they stand out for short-term collaboration networks and their extensive social and natural capital. Although some phenomena of neo-rurality have been changing the trend of young individuals abandoning rural territories (Orria and Luise 2017), the social and natural capital is not attractive enough to them. In fact, young individuals find greater job, training and socialization opportunities in the cities.

In order to change this trend, to facilitate and support social relations, radical interventions are required. Thus, Social Innovation appears to be an effective framework to face the challenges outlined above (Noack and Federwisch 2019; Vercher et al. 2021). We have excellent reasons to be critical and not believe in a romantic vision of social innovation (Fougère et al. 2017; Busacca 2020). However, it certainly offers an opportunity to focus on the social aspects of rural development. Social Innovation can help conceive development as a result of social interaction, commitment to collective well-being and the ability to take collective action, and not as the result of predetermined, externally imposed, top-down patterns of action. This vision of social innovation is very similar to the idea of neo-endogenous rural development and relational place-making, except for the explicit reference to innovation as a solution to the problems that hinder the functioning of rural society (Bock 2016). What unites the two lines of research and policies is the importance of governance processes in explaining socio-economic results. These are produced by the interactions between government agencies and local networks based on potential roles, trade-offs and on desirable outcomes (Vasstrøm and Normann 2019).

3   Belluno Dolomites and Their Main Challenges: Research Setting

The Belluno area is famous in the world for having determined the development of a successful industrial model, the eyewear district, an alternative to the Fordist model, based on small and medium-sized widespread enterprises (Piore and Sabel 1984).

However, by the end of the 1990s, the area was marked by the centrality of a few large local firms highly internationalised (Camuffo 2003). The initial development model was based on large endowments of social capital and dense social networks, favouring the creation of networks typical of the district system. Recently, the model assumed a different form with the vertical integration of a few globalized companies, as in the case of Luxottica, Safilo and Marcolin. However, the historical endowment of social capital has left traces in forms of mobilisations based on Local Action Groups (Da Re et al. 2017), as well as in the presence of historical local corporations such as the 'Regola' and mountain or agricultural cooperatives. This situation led to a sort of local bipolarity: on the one hand, there are few leading companies that monopolize most of the workforce; on the other hand, there is an unanswered demand for an extensive local development involving local professionals and companies. The Belluno context emerges here as an area populated by a limited number of potential social innovators where strong networks connect residents each other and where their interactions are limited and often competitive, as the result of a rapid and pervasive process of decay of local social capital over the last few decades. The Belluno territory is thus configured as an area with several structural holes (Burt 2004) with the presence of small and dense social networks. The territory appears highly fragmented and this is the main obstacle to participation in the decision-making processes invoked by local actors.

The area expresses an intense demand for local innovation, primarily to involve local actors in redesigning the trajectories of local development. The collaborative perspective outlined by this challenge presumes a vision of economic action as embedded in localized social networks (Granovetter 2017). As a result, the importance of social structure in the genesis of local socio-economic institutions is greatly emphasized. At the origin of an institution, some relationships are 'crystallized', while embeddedness accounts for the registration of economic actions in the networks of social relations. At this level, information is transmitted more easily between individuals connected by weak ties. This happens because they are defined by a lower degree of *homophilia* and therefore by the availability of more and diverse information. This implies that social and economic actions and their outcomes are influenced by the social relations between actors with other individuals and by the overall structure of the social network in which these relations are in turn inserted (Burt 1992). As a result, the economic behavior is influenced both from specific social relationships and from the aggregate effect of all relationships as a whole. This perspective is therefore characterized as a critique of both hyper-socialized and hypo-socialized views. The sociological representation of the actors as individuals determined by their social environment or, on the contrary, the idea that individuals act in an atomized way, pursuing their personal advantage according to the logic of effectiveness and efficiency, represent two apparently different perspectives. What the two have in common is the same concept of human action as a product of atomized actors: in the hypo-socialized version, to pursue one's own interest, while in the hyper-socialized version it is due to internalized patterns of behavior. In both cases, the influence of social relations would be weak. For Granovetter (2017), however, actors behave neither as atoms independent of the social context nor as executors of prescribed behaviors, but act within the framework of institutions and social relations.

From this point of view, the dynamics of local innovation become the result of individual choices, influenced by utilitarian objectives and cultural constructs, under the influence of social and political institutions and the result of localized social interactions. Being aware of these aspects can help us better understand the reasons why there is not a 'one-best-way' but multiple trajectories for local development. This kind of scenario recalls a collaborative form of local governance (Stoker 1998), where the actors continue to act individually but within a cooperative environment, giving rise to multipurpose networks where individual aims are reconciled with the objectives of the entire system. Analysing the collaborative forms of local governance from the perspective of social networks, implies observing local actors and their forms of interaction, while trying to simultaneously restore structural features. The term 'collaborative governance' is used to describe a wide range of processes and activities, where the boundaries between the public and private sectors are blurred by virtue of cooperation agreements, partnerships or networking between the public and private sectors aiming at redefining the roles that each can endorse. Still, the Government is not able to impose a structure to manage agreements or partnerships, but can only activate governance structures accountable to the parties involved. These descriptions are very similar to explanations of social innovation as a collaborative process (Murray et al. 2010; Tricarico et al. 2021). Thus, collaborative governance becomes an enabling factor for social innovation.

4   Data Collection and Analysis

The results summarized in Fig. 1, show the proposals emerged during the two focus groups in order to foster local social innovation processes.

During both meetings, all participants agreed on a high fragmentation among local actors, which limits the opportunities for collaboration and the social and economic potential for innovation and they proposed 8 actions. These actions cover 4 strategic dimensions: i) knowledge and human capital, ii) trust, iii) set of actors within the local system, iiii) communication within the area and between the area and other territories. In turn, these strategic dimensions have been traced back to two general purposes: a) enhancement of the territory and b) social and employment inclusion.

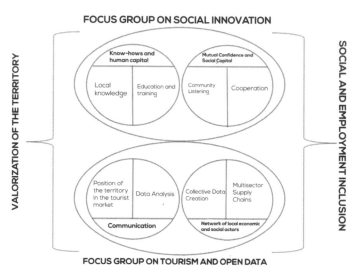

Fig. 1. Cooperation model for local social innovation.

What has emerged for the Belluno area is an endogenous strategy where human capital (knowledge, competences, etc.) and social capital (relations, trust, cooperation, etc.) are the fundamental infrastructure to create favourable conditions for innovative socio-economic action (endogenous development).

Also from the interviews, human capital and local skills have repeatedly emerged as key themes for social innovation. This underlines a general concern regarding the territory, largely attributed to the exodus of the most promising young professionals, who left the province for study reasons and are unlikely to return. Moreover, according to some stakeholders, the attractiveness of a few leading companies (e.g. Luxottica) tends to absorb most of the qualified job offers and the most promising young professionals. The lack of collaboration as well as the weakness of human capital are therefore confirmed as limiting factors for innovation. The interviewees presented numerous cases of participatory and collaborative processes promoted by local institutions in order to face these problems. However, they revealed that these attempts were a failure because the proposals were never implemented by local institutions, thus generating frustration and growing mistrust. The interviewee S.M. defined collaborative governance as «a fundamental mechanism for designing strategies and visions shared by a plurality of actors working at different institutional levels». Collaborative governance is understood here as the set of tools, knowledge and skills necessary to make the territory capable in terms of sharing the processes, information and data produced by the interaction between the actors involved. It also means exploiting the endowments of the local social capital and the opportunities offered by new technologies and main emerging trends in the tourism industry. A further deepening suggested observing and watching how the territory incubated new ideas and projects to encourage development. Incubation, unlike mere listening, provides support to ideas and a co-planning approach where responsibilities are shared among local actors. Furthermore, an incremental listening-planning process could find fertile ground in contexts where the actors highly trust each other. From these revisions, a different schematization of the work took place as a result of the focus groups.

Another part of the interview aimed at analyzing the local context where the previous considerations emerged. The focus was the incidence of endogenous and exogenous factors on the possibility of implementing the strategies and actions developed. To achieve this, all interviewees were involved in a SWOT analysis (Table 1). What emerged substantially confirmed the problems and challenges affecting the area. However, once again the results prove a clear orientation towards collaboration between actors operating at different territorial levels that therefore requires integration.

Table 1. Swot analysis for Belluno dolomites.

STRONG POINTS	WEAK POINTS
Responsiveness to a renewal of the tourism industry	The tourism sector has limited knowledge of the opportunities offered by new technologies
Strong Local Social Capital	Local actors are too self-referential
Good quality of the tourist offer	Scarcity and fragility of transports, logistics and TLC infrastructures
Increasingly aware presence of young individuals	Limited connections with other regional subsystems
Improved Territorial Marketing	Fragility of the economic fabric
Great diversification of natural and economic resources	Problematic information-gathering process
Tour operators progressively improving their skills	Disorganisation of local economies, resulting in poor hospitality
Strong local knowledge	Limited ability to listen to the territory as a whole. As a fragmented territory, each area is usually dealt with differently
	Confusing institutional structure due to the unclear role of the Province of Belluno
	Limited opportunities for lifelong learning
	Lack of innovation centers. The initiative is left to individuals
	DMO action is unbalanced in marketing actions and less active in governance processes
	Lack of strong, authoritative and shared local leadership

(continued)

Table 1. (*continued*)

OPPORTUNITIES	THREATS
The tourism sector has an acknowledged strategic potential	Systemic projects require very precise deployment times. The risk is to get them wrong and invalidate the efforts
The programs available can better qualify local actors	Depopulation is increasing in the upper regions of the Belluno area
Integrated multi-sector supply chains	Uncontrollable exogenous factors (such as
Proximity Tourism	COVID-19), could determine the definitive
Networking between small local operators	closure of some businesses
Mountain Tourism	High territorial dispersion
Deseasonalization as an opportunity to distribute and increase touristic flows	High level of dependency on the regional body
Cultural tourism - still marginally valued in the Belluno area even if rich in resources	Residues of decision-making models based on concertation and not on cooperation
Regional and National Programmes that would address many of the territory's needs	
The community is more committed in creating a vision for its local economy	
Environmental sustainability is becoming a crucial concern	
Collaboration between local actors is improving	

5 Building an Attractive Rural Territory: Participation, Integration and Connections

The results of the research were subsequently organized under three themes. In this section we use these findings in order to elaborate an interpretative hypothesis on how (and when) social innovation can work in remote, fragmented and isolated territories (i.e. inner or marginal areas).

5.1 Defining Participation

The idea prevails that local development includes the active involvement of local actors, thus idealizing network models and local coalitions that act on a territorial scale as the ideal dimension to deal with its matters. Hence, very different visions. In some cases, the scale is municipal, as in the essential services countering the depopulation of the higher valleys. In others, the functional area, as in the case of area brands that favor Destination Governance processes. In others again, it is a homogeneous area, as in the case of the Destination Management Organization. From the actors' perspective, instead, an orientation prevails towards the integration of agents operating in different sectors and at different territorial levels, without however detailing how these are integrated with each other. The role of public authorities should not only be to create suitable contexts for participation, but also, unlike many past experiences, to participate in the processes and give the actors involved the opportunity to decide and experiment with local development paths.

Although empirically fragile, this perspective presents an important aspect. Rather than producing a model similar to the pipes of an organ, where problems and solutions are compartmentalized and managed separately within the scientific method, the local development perspective considers the territory as a complex system, where several interconnected dimensions come together. These dimensions open to integrated policies as a set of actions that aim at integrating and thickening learnings that otherwise would be isolated in their respective fields of action.

5.2 The Reconstruction of Social Capital

Local social capital (Helliwell and Putnam 1995), consisting of the network of relationships of trust and civic institutions, does not necessarily imply the existence of a socio-economic infrastructure founded upon personal and organizational relationships nor it presupposes mutual acquaintance and trust as a basis for triggering local economic development and innovation (cf. competitive collective goods, Crouch et al. 2004). Personal, trusted, direct acquaintances are frequently indicated as essential elements for fostering regulatory mechanisms that imply sharing the space for action and cooperative relations between local actors. What emerges here is the absence of relationships of trust between different persons and organisations and the participants mentioned the inclination to individualism and the reform of the Province of Belluno (cfr. Graziano Delrio's Bill) as possible causes of it. The latter has eliminated an intermediary body with a fundamental role in mediating the different interests that actors and territories normally have.

Thus, the research shows that the territory is characterized by a system of relationships and trust between local actors but it is however very fragmented and presents many structural holes (Burt 2004) between small clusters of local actors in cooperative relationships with each other. This would suggest that what once was a solid social capital has recently deteriorated. Today, this capital requires targeted interventions aimed at restoring the previous conditions of trust. These solutions were outlined by the participants as in opportunities for discussion and sharing that would trigger interaction chains (Collins 2004).

5.3 Attracting People in the Area

The third key theme that emerged during the early stages of the research is the frailty of the relationship between the territory and its residents, which causes three outcomes:

1. Dispersion of skilled professionals due to the scarcity of local job opportunities, universities and post-graduate training centers. Residents move away from the territory to pursue their careers elsewhere, in so creating the conditions for non-return;
2. Except for the eyewear district, the scarcity of high-level job opportunities makes the territory unattractive. The serious lack of transport and digital infrastructures and services make it difficult to reach the territory, therefore limiting mobility and hindering digital connections;

3. The increasing depopulation of mountain municipalities at 1000 mt AMSL is strongly affecting the maintenance of essential services to citizens. As a result, these territories can only offer limited assistance to their residents.

In this perspective, innovation needs time. Along with the typical complexity that comes with innovation processes, it also takes large investments, resilience towards failures, the ability to enhance resources that are not immediately coordinated, visions and interests that may not converge, tests and models to measure the effects.

6   Conclusion

The research findings show, as recently highlighted also by Tatiana Kluvankova et al. (2020), that social innovation involves both local and external actors, but cannot develop without specific endogenous activities.

In the case of rural and marginal areas such as the Belluno mountains different needs add up. In these areas characterized by a rarefied economic structure, weak institutions and an adverse demography, innovation acquires specific traits (Mayer 2020). Here, innovators may not have immediate access to sources of information and knowledge as in urban contexts. This considered, breeding structures should be designed, built and become laboratories that welcome diversity. However, here the innovators create groups defined by strong ties, of friendship or kinship, and fewer contacts with distant and different worlds. In light of all this, innovation needs planning. The delay and rarefaction of the processes require a strong intention, more than speed.

For the innovators living in the peripheral areas, the space and spatial distribution of resources, as well as the time required to reach the locations, represent more constraints than in cities. Here, in fact, in addition to being slower, innovation is very dependent on non-codified local knowledge, which is difficult to transmit and formalize except through informal and tacit transfer systems. This requires face-to-face interactions, confidence building and the ability to understand cultural nuances.

Acknowledgements. This research was made possible thanks to the European project "+Resilient", the Chamber of Commerce of Treviso and Belluno and Ecipa scarl.

References

Bock, B.B.: Rural marginalisation and the role of social innovation; a turn towards nexogenous development and rural reconnection. Sociol. Rural. **56**(4), 552–573 (2016). https://doi.org/10.1111/soru.12119

Borelli, G., Busacca, M. (eds.): Society and the City: The Dark Sides of Social Innovation. Mimesis International, Milan (2020)

Burt, R.S.: Structural Holes. Cambridge University Press, Cambridge (1992)

Burt, R.S.: Structural holes and good ideas. Am. J. Sociol. **110**(2), 349–399 (2004). https://doi.org/10.1086/421787

Busacca, M.: The social innovation dispositive. In: Borelli, G., Busacca, M. (eds.) Society and the City: The Dark Sides of Social Innovation, pp. 33–49. Mimesis International, Milan (2020)

Camuffo, A.: Transforming industrial districts: large firms and small business networks in the Italian eyewear industry. Ind. Innov. **10**(4), 377–401 (2003). https://doi.org/10.1080/136627 1032000163630

Chamber of Commerce Treviso-Belluno Dolomites. Il tessuto economico produttivo della provincia di Belluno. Dati, tendenze e criticità nel 2017 (2018)

Collins, R.: Interaction Ritual Chains. Princeton University Press, Princeton and Oxford (2014)

Crouch, C., Le Galès, P., Trigilia, C., Voelzkow, H.: Changing Governance of Local Economies: Responses of European Local Production Systems. Oxford University Press, Oxford (2004)

Da Re, R., Franceschetti, G., Pisani, E.: LEADER and social capital in veneto: the case studies of prealpi e dolomiti and bassa padovana local action groups. In: Pisani, E., Franceschetti, G., Secco, L., Christoforou, A. (eds.) Social Capital and Local Development, pp. 305–326. Springer, Cham (2017). https://doi.org/10.1007/978-3-319-54277-5_13

Granovetter, M.: Society and Economy: Framework and Principles. Harvard University Press, Cambdrige and London (2017)

Fougère, M., Segercrantz, B., Seeck, H.: A critical reading of the European Union's social innovation policy discourse:(Re) legitimizing neoliberalism. Organization **24**(6), 819–843 (2017)

Helliwell, J.F., Putnam, R.D.: Economic growth and social capital in Italy. East. Econ. J. **21**(3), 295–307 (1995)

Kluvankova, T., et al.: Social innovation for sustainability transformation and its diverging development paths in marginalised rural areas. Sociol. Rural. (2020). https://doi.org/10.1111/soru. 12337

Landeta, J.: Current validity of the Delphi method in social sciences. Technol. Forecast. Soc. Chang. **73**(5), 467–482 (2006). https://doi.org/10.1016/j.techfore.2005.09.002

Mayer, H.: Slow innovation in Europe's peripheral regions: innovation beyond acceleration. Schlüsselakteure Reg. Welche Perspekt. Bietet Entrep. Ländliche Räume **51**, 9–22 (2020)

Murray, R., Mulgan, G., Caulier-Grice, J.: The Open Book of Social Innovation. Generating Social Innovation: Setting an Agenda, Shaping Methods and Growing the Field. The Young Foundation, London (2010)

Neumeier, S.: Social innovation in rural development: identifying the key factors of success. Geogr. J. **183**(1), 34–46 (2017). https://doi.org/10.1111/geoj.12180

Noack, A., Federwisch, T.: Social innovation in rural regions: urban impulses and cross-border constellations of actors. Sociol. Rural **59**(1), 92–112 (2019). https://doi.org/10.1111/soru. 12216

Orria, B., Luise, V.: Innovation in rural development: "neo-rural" farmers branding local quality of food and territory. Italian J. Plan. Pract. **7**(1), 125–153 (2017)

Osborne, S., Williamson, A., Beattie, R.: Community Involvement in rural regeneration partnerships: exploring the rural dimension. Local Gov. Stud. **30**(2), 156–181 (2004). https://doi.org/ 10.1080/0300393042000267218

Oosterlynck, S., Novy, A., Kazepov, Y. (eds.): Local Social Innovation to Combat Poverty and Exclusion: A Critical Appraisal. Policy Press, London (2020)

Piore, M., Sabel, C.: The Second Industrial Divide. Possibilities for Prosperity. Basic Books, New York (1984)

Shucksmith, M.: Re-imagining the rural: from rural idyll to Good Countryside. J. Rural. Stud. **59**, 163–172 (2018). https://doi.org/10.1016/j.jrurstud.2016.07.019

Stoker, G.: Governance as theory: five propositions. Int. Soc. Sci. J. **50**, 17–28 (1998). https://doi. org/10.1111/issj.12189

Tricarico, L., Bitetti, R., Buonanno, F.: Can we shape social innovation-based urban policy? Reflections on the fondo per l'innovazione sociale strategy in Milan. Econ. Lavoro **55**(1), 121–138 (2021). https://doi.org/10.7384/101088

Vasstrøm, M., Normann, R.: The role of local government in rural communities: culture-based development strategies. Local Gov. Stud. **45**(6), 848–868 (2019). https://doi.org/10.1080/030 03930.2019.1590200

Vercher, N., Barlagne, C., Hewitt, R., Nijnik, M., Esparcia, J.: Whose narrative is it anyway? Narratives of social innovation in rural areas–a comparative analysis of community-led initiatives in Scotland and Spain. Sociol. Rural. **61**(1), 163–189 (2021). https://doi.org/10.1111/soru. 12321

Culture Leading to Urban Regeneration. Empirical Evidence from Some Italian Funding Programs

Francesco Campagnari, Ezio Micelli(✉), and Elena Ostanel

Department of Culture and Arts, Università Iuav di Venezia, Dorsoduro 2206, I-30123 Venice, Italy
micelli@iuav.it

Abstract. In recent years policies fostered the leading role of culture in urban regeneration in Italy, financing and supporting not only consolidated cultural and art institutions, but also providing new economic opportunities to emerging community-based groups and associations. Research on culture-led urban interventions has however mostly adopted a micro-level of analysis, missing the possibility to grasp general trends also usable in a policy dimension. The paper aims to advance the state-of-the-art of the national debate on long-lasting culture-led regeneration processes, by offering a landscape view of the phenomenon in Italy. The research is based on the analysis of two databases of project applications for national-level funding programs on culture-led urban regeneration, containing respectively 141 and 54 processes. The paper identifies the most salient trends and common characters of these initiatives: i) the spatial dimension (location and real estate assets) ii) the scale of intervention iii) the service delivered in the regeneration effort. The paper argues that culture is neither uniform, nor alone in culture-led urban regeneration initiatives: culture is present in different shapes and in different types of activities, ranging from fruition to production; furthermore, it is often hybridised with other services, linked with local welfare provisions or commercial activities. While most of the initiatives are rooted on the regeneration of a single building, most of them adopt an area-based approach, being oriented to the revitalization of an entire urban area also through spatial and intangible actions, tending to collaborate with public authorities.

Keywords: Culture-led urban regeneration · Hybridization · Local infrastructures

1 Introduction

Two trends converge into a new approach to urban regeneration. On the one hand, bottom up community-based initiatives reuse abandoned or underused real estate properties transforming them into multifunctional venues in which market and not-for-profit activities co-exist giving birth to innovative social and economic hubs [1–3]. On the other hand, culture represents an important field of investigation in relation to the development processes of Italian cities, as it is a key driver of their regeneration [4–9].

© The Author(s), under exclusive license to Springer Nature Switzerland AG 2022
F. Calabrò et al. (Eds.): NMP 2022, LNNS 482, pp. 461–470, 2022.
https://doi.org/10.1007/978-3-031-06825-6_43

The goal of this paper is to analyse the main features of these increasingly recognized urban regeneration initiatives. In particular, the paper focuses on their diffusion throughout the country, their ability to interact with the local community and on the way culture eventually combines with other economic and social activities.

In recent years specific policies fostered the leading role of culture in urban regeneration financing and supporting not only consolidated cultural and art institutions, but also providing new economic opportunities to emerging community-based groups and associations. Through the dataset related to the important funding calls of the Ministry of Culture and of some other important national funding institutions it is possible to draw a new map of the emerging scene of culture-led urban regeneration in Italy.

The paper is structured as follows. The following section introduces the topic of culture-led urban regeneration in Italy, focusing on the role of long-lasting initiatives in the current debate. The third section describes the data sources for this research and the methods adopted to analyse them. The fourth section presents and discusses the main results of the research. The final section concludes the paper and presents the main lessons emerging from the research.

2 Culture-Led Urban Regeneration Processes in Italy: Towards a Nation-Wide Perspective

Urban regeneration is today a crucial approach for the transformation of cities [10]. Urban regeneration can be seen as a set of integrated actions leading to the resolution of urban problems and seeking to bring a lasting improvement in the economic, physical, social and environmental conditions of an area subject to change [10].

In this debate on urban regeneration, culture has been used in different forms [1–3], ranging from a "culture and regeneration" model based on a circumscribed role for culture in larger strategic plans, to a model of "cultural regeneration" where culture is fully integrated into an area strategy alongside other activities in the environmental, social and economic spheres, and to "culture-led regeneration", where cultural activity is seen as the main catalyst and engine of the regeneration [11].

Culture has been used as a tool in very different urban transformation processes, ranging from state-led top-down interventions oriented on real estate valorisation [12], reuse of abandoned real properties transforming them into multifunctional venues in which market and not-for-profit activities co-exist, to bottom-up practices [13] dedicated more to social inclusion [14].

Culture-led interventions have mostly been analysed adopting a micro-level of analysis, missing the possibility to grasp general trends also usable in a policy dimension. Research has in fact mostly focused on single or collections of case studies [6–9]. Other important analyses [15] have instead adopted a nation-wide perspective, but only focussed on the actions of specific funding institutions.

Given this backdrop, the Paper is aimed to advance the state-of-the-art of the national debate on culture-led regeneration processes, by offering a landscape view of culture-led urban regeneration processes in Italy. The paper identifies the salient trends and common characters of these initiatives: i) the spatial dimension (location and real estate assets) ii) the scale of intervention iii) the service delivered in the regeneration effort.

The Paper argues that culture is neither uniform, nor alone in culture-led urban regeneration initiatives: culture is present in different shapes and in different types of activities, ranging from fruition to production; furthermore, it is often hybridised with other services, linked with local welfare provisions or commercial activities. While most of the initiatives are rooted on the regeneration of a single building, most of them adopt an area-based approach, being oriented to the revitalization of an entire urban area also through spatial and intangible actions, tending to collaborate with public authorities. This attitude is of utmost importance if we consider that public administrations and community based initiatives have in some cases together experimented processual approaches to urban transformations that need today to be evaluated both in their spatial and social outcomes.

3   Data and Methods

The research aimed to describe the characters of long-lasting culture-led urban regeneration processes in Italy. Considering the richness and broadness of funding schemes on the topic, we chose to use as sources of data the project applications for these calls. The research aggregated multiple funding calls, in order to reduce the influence of the specificities of each one of them (such as geographical or thematic biases).

The research analysed two databases: the first was composed of project applications for calls for projects by Direzione Generale Creatività Contemporanea (DGCC, General Secretariat for Contemporary Creativity) of the Italian Ministry of Culture. The second was composed of project applications for urban regeneration calls by other funding institutions at national level, such as CheFare and Culturability (Fig. 1).

Fig. 1. Graphic representation of the research methodology for each database

The first database contained 502 project applications from three calls by the DGCC: Creative Living Lab 2018, with 201 applications; Creative Living Lab 2019, with 196 applications; Prendi Parte 2018, with 105 applications. The raw database was first reduced to 478 project applications, as 24 files were damaged and unusable.

The analysis of the clean database followed two main steps. Firstly, we explored all 478 project applications selecting projects coherent with the scope of analysis, defined as processes of culture-based urban regeneration, with a deliberate intention of establishing continuative community relations. We adopted two selection criteria: first, the processes should have the intention of being continuative and lasting on the territory, therefore excluding one-shot and temporary activities. Second, they should have a direct connection with the local spatial dimension, either by refurbishing spaces or acting on entire neighbourhoods. This criterion excluded the initiatives working only on an immaterial level, with indirect effects on space. Of the 478 project applications explored, 176 fit these criteria. Considering the submission of projects to different calls by the same initiatives, the number of processes to analyse was reduced to 141.

Secondly, these 141 processes were explored through 28 descriptive statistical variables, divided in 10 macro-themes. These themes include general information, localization features, spatial intervention types, characteristics of buildings and contracts of use of buildings, orientation of project, features of the leading organisation and its network, relations with public policies, main audiences, level of consolidation.

The same procedure was applied to the second database. Differently from the first database, the second database collected the finalists of eight calls from two granting programs, for a total of 101 project applications: CheFare, with 25 applications from three editions of the call (2012/13, 2014/15, 2015), and Culturability, with 76 applications from five editions of the call (2014/15, 2016, 2017, 2018, 2020).

Of the 101 project applications explored, 59 fit the criteria. Considering the submission of projects to different calls by the same initiatives, the number of processes to analyse was reduced to 54. These 54 processes were explored through the same descriptive statistical variables as the first database.

4   Analysis and Discussion. Urban, Local and Hybrid: The Features of Culture-Led Regeneration

The analysis of the two databases offers a good perspective on the phenomenon of culture-led urban regeneration in Italy. Aggregating the descriptive statistics, we can discuss the most relevant and important themes emerging from the analysis. We will present the results through histograms, reporting the absolute values and the horizontal (or vertical) axes show the percentage values. Following the analysis, we will discuss their relevance for culture-led urban regeneration.

The analysis of the two databases offers a good perspective on culture-led urban regeneration in Italy. Descriptive statistics allow the dataset analysis, and the discussion of the results is carried out accordingly.

The first theme refers to the spatial aspects of long-lasting culture-led urban regeneration processes. Data show how culture-led regeneration is an urban phenomenon, mainly concentrated in the North of the country, exploiting public real estate assets with an intentional relationship with the surrounding neighbourhoods.

From a macro-geographical standpoint (Fig. 2), these processes tend to be localised in Northern Italy. The MiC database shows a relative majority of application being localised in Northern Italy, while the other database sees an absolute majority.

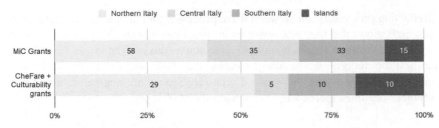

Fig. 2. Location of the projects in the Italian macro-regions

These processes tend to take place in larger urban areas (Fig. 3). In both databases over half the projects are localised in municipalities with over 100.000 inhabitants, with a considerable part of them in municipalities over 500.000 inhabitants.

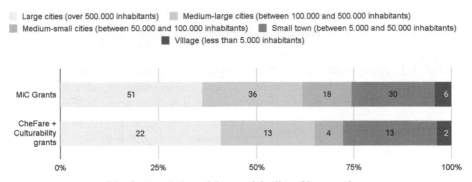

Fig. 3. Population of the municipality of intervention

The regeneration processes are distributed mostly within the urbanised areas of municipalities, with a concentration in peripheral areas for the MiC database and a more diverse distribution in the second database (Fig. 4).

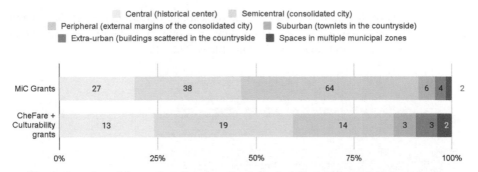

Fig. 4. Location of the projects based on the Agenzia delle entrate spatial categorization

Reflecting on the relation between these processes and their spaces of intervention, the research shows that they often base their operations in a single space for their activities. The size of these spaces is quite diverse across the two databases (Fig. 5): while in the first the majority are buildings smaller than 1000 square metres, in the second the majority is bigger than this value. Large size spaces are neglected: a minority of experiences occupies spaces whose size exceeds 5.000 sq.m.

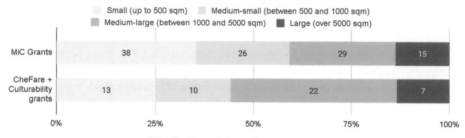

Fig. 5. Size of the real-estate assets

The real estate assets where the initiatives take place are most of the time publicly-owned. The owners are often the local municipalities or other local public administrations. The presence of private ownership is also considerable (Fig. 6).

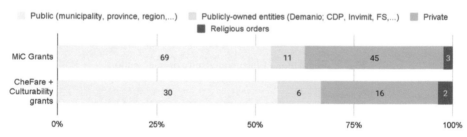

Fig. 6. Ownership of the real estate assets

While most of these initiatives are rooted on a single building, most of them adopt an area-based approach and in most cases they are oriented to the regeneration of an entire urban sector through spatial and intangible actions (Fig. 7).

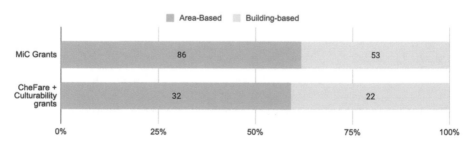

Fig. 7. The nature of the intervention: building based vs area-based

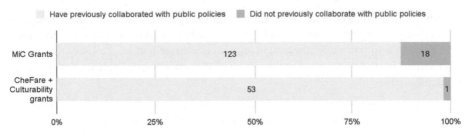

Fig. 8. Previous collaboration of interventions with public policies

The second theme is the interpretation of long-lasting culture-led urban regeneration processes as <u>proximity infrastructures</u>. As any infrastructure, these initiatives are supported by the public sector in spite of the commonplace of economic self-sufficiency. Almost all of them (and in particular the ones from the second database) have been supported in the past by public policies (Fig. 8), either for economic support, for using public buildings, or for collaborating in developing shared projects. Looking at the scale of these policies (Fig. 9), the interventions have often collaborated with policies at municipal level, with secondary quotes of national and regional policies.

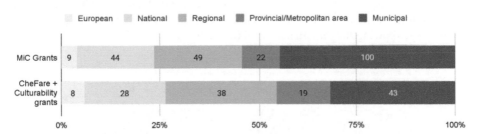

Fig. 9. Scale of public policies the initiatives have previously collaborated with

Urban regeneration acts locally. The local dimension of these initiatives is confirmed in the analysis of their main users and audiences (Fig. 10). Almost all of them are focused either on a level of proximity or municipality. This focus might also imply a similar scale of action, depending on the actual size of the municipality.

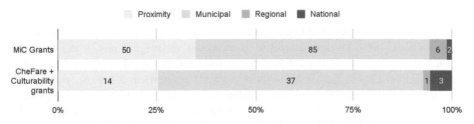

Fig. 10. Scale of the main users and audiences of the initiatives

While these processes are often led by single organisations, they are managed in connection with important networks of actors, scaling from the local to the international level (Fig. 11). In particular, these networks involve over half of the time other actors at municipal level. Other actors in the region of the leaders or in Italy are involved less frequently.

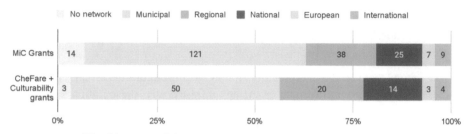

Fig. 11. Scale of the networks activated by project partnerships

The third theme refers to the hybrid nature of long-lasting culture-led urban regeneration processes. If we consider the thematic orientations of the projects (Fig. 12), we can see that most initiatives of both databases are oriented towards culture alone or a mix of culture and welfare. However, the average project of both databases present 1.6 orientations, showing the importance of mixing themes.

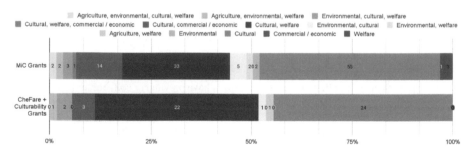

Fig. 12. Thematic orientations of the projects

Furthermore, considering the activities operationally developed in these spaces, we recognize the coexistence of multiple activities in the same projects. Figure 13 shows the number of initiatives developing each activity: we can see that while cultural activities (concerts and exhibitions) are quite diffused, many of these projects also organise cultural production activities (trainings, workshops, artist residences). Many also offer welfare services (after-school, mutual aid activities, social support) and engage in the direct sale of goods and services (bars, food and beverage services, training and consulting services). The average initiative of the first database develops 2.3 activities, while the average one from the second develops 3.1.

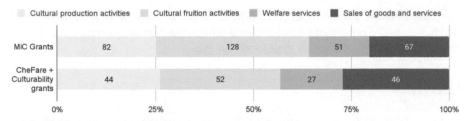

Fig. 13. Main activities developed in the process of urban regeneration

Considering the thematic orientations and the activities of these processes, we recognize that while culture is the driving force behind these experiences, they hybridise it with other services, creating social and community infrastructures and platforms [16].

These emergent themes allow a better understanding of long-lasting experiences of culture-led urban regeneration in Italy. As expected, culture is a central element of these initiatives of urban regeneration. However, it is neither uniform, nor alone: culture is present in different shapes and in different types of activities, ranging from fruition to production; furthermore, it is often hybridised with other services, linked with local welfare provisions or commercial activities. Culture is therefore at centre stage in long-lasting culture-led processes of urban regeneration, but it is surrounded by a varied cast of services and activities, which help strengthen the initiatives and their effects.

5   Conclusions

Culture-led urban regeneration is today a central paradigm of urban action, using cultural activities as leverage to kickstart urban transformations. Long-lasting experiences of culture-led regeneration have proved over the years an important role in transforming urban areas across Italy. These processes surged in the last years.

The paper explored two databases of applications to national and regional funding schemes, focusing on processes with continuative community relations and analysing them through descriptive statistical variables. This approach offered a view of long-lasting culture-led urban regeneration, sacrificing a broader understanding of other types of culture-led urban regeneration processes in Italy.

Three aspects emerge from the analysis. First, culture-led regeneration is mainly an urban phenomenon, especially diffused in northern Italy's large cities. Initiatives occur in peripheral areas where abandoned real estate assets offer potential venues at little or no cost. Second, they represent proximity infrastructures, given their solid local dimension, local audiences, and connections with local policies and actors. This character was highly visible in their direct actions of solidarity during the COVID-19 lockdowns. And third, they have a hybrid nature, mixing multiple artistic fields and services, either oriented to complement the local welfare or commercial and market-oriented ones.

These research results open new directions for further research. The integration of other databases at international, national, regional, and local level can help expand or revise these hypotheses of interpretation. Secondly, a comparative in-depth inquiry into a few paradigmatic case studies among these project applications could help understand how these characters are developed in different contexts, how they are consolidated, what benefits they bring to the initiatives and what impacts they generate through their urban regeneration processes on their local contexts.

References

1. Sacco, P., Tavano Blessi, G.: The social viability of culture-led urban transformation processes: evidence from the bicocca district, Milan. Urban Stud. **46**(5–6), 1115–1135 (2009). https://doi.org/10.1177/0042098009103857
2. Ferilli, G., Sacco, P., Tavano Blessi, G., Forbici, S.: Power to the people: when culture works as a social catalyst in urban regeneration processes (and when it does not). Eur. Plan. Stud. **25**(2), 241–258 (2017). https://doi.org/10.1080/09654313.2016.1259397
3. Sacco, P., Ferilli, G.: Tavano Blessi: Cultura e Sviluppo Locale. Verso il Distretto Culturale Evoluto. Il Mulino, Bologna (2015)
4. Lami, I.M. (ed.): Abandoned Buildings in Contemporary Cities: Smart Conditions for Actions. SIST, vol. 168. Springer, Cham (2020). https://doi.org/10.1007/978-3-030-35550-0
5. Mangialardo, A., Micelli, E.: New bottom-up approaches to enhance public real/estate property. In: Stanghellini, S., Morano, P., Bottero, M., Oppio, A. (eds.) Appraisal: From Theory to Practice. GET, pp. 53–62. Springer, Cham (2017). https://doi.org/10.1007/978-3-319-49676-4_5
6. Mangialardo, A., Micelli, E.: Grass-roots participation to enhance public real-estate properties. Just a fad? Land Use Policy **103**, 105290 (2021). https://doi.org/10.1016/j.landusepol.2021.105290
7. Moroni, S., De Franco, A., Bellè, B.M.: Unused private and public buildings: re-discussing merely empty and truly abandoned situations, with particular reference to the case of Italy and the city of Milan. J. Urban Aff. **5**, 203–207 (2020)
8. Governa, F., Saccomani, S.: From urban renewal to local development. New conceptions and governance practices in the Italian peripheries. Plan. Theory Pract. **5**(3), 327–348 (2004). https://doi.org/10.1080/1464935042000250212
9. Cerquetti, M., Cutrini, E.: The role of social ties for culture-led development in inner areas. In: The Case of the 2016–17 Central Italy Earthquake. European Planning Studies (2020)
10. Roberts, P., Sykes, H.: Urban Regeneration. A Handbook. SAGE, London (2000)
11. Evans, G., Shaw, P.: The contribution of culture to regeneration in the UK: a review of evidence. In: A Report to the DCMS. LondonMet (2004)
12. Ostanel, E.: (In)visibilizing Vulnerable Community Members: Processes of Urban Inclusion and Exclusion in Parkdale. Space and Culture, Toronto (2020)
13. Granata, E.: Placemaker. Gli Inventori dei Luoghi che Abiteremo. Einaudi, Torino (2021)
14. Ostanel, E.: Spazi Fuori dal Comune. FrancoAngeli, Milano (2017)
15. Franceschinelli, R. (ed): Spazi del Possibile. FrancoAngeli, Milano (2021)
16. Tricarico, L., Daldanise, G., Jones, Z.M.: Spazi piattaforma: quando la cultura interseca l'innovazione sociale e lo sviluppo territoriale. BDC. Bollettino Del Centro Calza Bini **20**(1), 139–165 (2020). https://doi.org/10.6092/2284-4732/7548

Strengthening Community-Based Organizations Through Social Impact Readiness: Lessons from "Periferiacapitale"

Luca Tricarico[1](✉), Luigi Corvo[2,3], Lavinia Pastore[2,3], and Arda Lelo[2,3]

[1] Department of Business and Management, Luiss University, viale Romania 32, Rome, Italy
ltricarico@luiss.it
[2] Open Impact, Piazza Manfredo Fanti, 22, Rome, Italy
{corvo,pastore,lelo}@economia.uniroma2.it
[3] University of Roma Tor Vergata, Via Columbia 2, Rome, Italy

Abstract. In 2021 the Charlmagne foundation has launched "Periferiacapitale": a social innovation program to support community-based organizations (n = 13, based in Rome municipality) with different social impact objectives in urban areas characterized by social exclusion and marginalization conditions. This set of different organizations (NPOs, networks of associations, training institutions, social cooperatives) were supported by overcoming the traditional paradigm of philanthropic intervention, focusing on social impact-based capacity building: in order to visualize, measure and communicate the social impact evaluation of their projects. With respect with this policy framework, the present contribution intends to describe the work SI^2 methodology (Social Enterprise * Social Impact) developed through research-action activities with the organizations involved in the program. The objective of the contribution is to introduce the value of *impact readiness* methodologies as a strengthening strategy with which these organizations can influence the policy making processes with which to activate new resources and activities.

Keywords: Social innovation · Community-based organizations · Social impact · Impact readiness

1 Introduction: Social Innovation, Community-Based Organization and Impact Readiness

The methodologies of social innovation are becoming diffused in various fields of policy in the light of activating mechanisms of collective intelligence and shared responsibility [1], transforming and recombining the traditional production factors through the involvement of different actors that are engaged in various roles and sectors [2]. The spread of these methodologies today seems to outline an articulated community of practices, capable of asserting a language that is becoming progressively common, pervading very different sectors, branches and fields in terms of dimension and quality, seeking actors

F. Calabrò et al. (Eds.): NMP 2022, LNNS 482, pp. 471–482, 2022.
https://doi.org/10.1007/978-3-031-06825-6_44

and approaches that until years ago seemed to be incompatible: foundations, banks, universities, nonprofit organizations, multinational, small and medium enterprises, cooperatives, social movements and individuals such as researchers of various disciplines, innovators, civil servants, freelancers.

At the same time, the pervasiveness of the concept of social innovation seems to flatten its use as a buzzword linked to vaguely defined phenomena. However, a more systematic analysis [3] allows us to divide three main sub-themes where the use of social innovation differs in territorial oriented approaches entrepreneurship, inclusion and co-production.

- Considering the *entrepreneurial* sector in particular, the social innovation phenomena is mainly connected with the distribution of individuals and organizations that in different territories and among different organizations aim to carry out projects and instances of social change within entrepreneurial processes. When present, the interconnected action of this community of organizations and professionals represent an opportunity to create social innovation ecosystems.
- The *inclusion* segment focuses mainly on the careful design and incorporation of effective governance schemes in social policies, regional development and urban policies [4, 5].
- The *co-production* subject focuses on defining parameters, markers, and decision-making processes aimed at expanding and diversifying the definition of the concept of *public value* and *general interest*. Here, in the design of services of public and general interest, arises the need to clarify the ways in which values are produced and defined [6].

In this context, the present contribution intends to observe the experience of scientific assistance carried out on "Periferiacapitale" program: a social innovation project launched by the Charlemagne foundation to support community-based organizations (n = 13, based in Rome municipality) with different social impact objectives in urban areas characterized by social exclusion and marginalization conditions. The idea behind the strategy of the program is the following: If in the second half of the last century these reactions were deliberately hidden by politics and public affairs, today we are witnessing an evident and incisive overturning of this hierarchy, which sees, on the contrary, the peripheries as active forces of change. This is due to an overall reinterpretation of the urban systems and the overcoming of a certain rhetoric use of the peripheral concept (framed as poor and marginal) instead of revealing the richness in terms of community-based developments, intensity and complexity of its social relations [7]. At the same time, there is the risk of delegating to the aforementioned citizen action issues whose management should be taken over by public policies.

According to this background, the objective of the contribution is to describe the work done in spreading *impact readiness* methodologies as a strengthening strategy to activate new resources and activities in Community-based organizations, and therefore be a crucial tool in policy making strategy to empower their role in peripheral area.

In order to discuss this issue the paper proceeds as follows: Sect. 2 describes the context of social innovation where the "Periferiacapitale" can be framed; in Sect. 3 we describe the SI^2 Theoretical Framework and implications in social innovation-based programs design; in Sect. 4 we describe the preliminary findings discussed with the data gained through the survey conducted on the 13 organizations; the last section concerns a conclusive framework about the role of *impact readiness* in social innovation programs design.

2   The Context of the "Periferiacapitale" Program

The "Periferiacapitale" program is part of the strategy of Charlemagne Foundation[1] implemented in order to sustain the activities of 13 community-based organizations in different neighborhoods of the city of Rome (Tor Sapienza, Corviale, Pietralata, Tor Marancia, Pisana, Laurentino, San Basilio, San Lorenzo, Tufello). This set of different organizations (NPOs, networks of associations, training institutions, social cooperatives) was supported by overcoming the traditional paradigm of philanthropic aid, focusing on social impact-based capacity building: in order to visualize, measure and communicate the social impact evaluation of their projects. To work on these issues, the Charlemagne Foundation has signed an agreement with Open Impact, a Roman university spin off that deals with social innovation and social impact assessment. From a methodological standpoint, the analysis conducted in the present article has been complemented with the presentation of the approaches and data collected by the authors in dealing with the "Periferiacapitale" case study, based on an action research methodology applied in the first phase of program development assistance [7–9]. Within this context, the research practice is the result of a mixture of research activities carried out in policy design processes, and interaction with stakeholders, from which a "reflection on the course of action" emerges [10].

With respect with this policy framework, the tool experimented to sustain these organization is the SI^2 methodology (Social Enterprise * Social Impact) developed through research-action activities with the organizations involved in the program. The goal of the SI^2 approach is to allow the organizations involved in the program to view, measure and communicate their social value through a new system for assessing the impact of their projects.

The ability to measure, evaluate and communicate the impact is part of the logic of a continuous improvement process which intends:

- Strengthen transparency and accountability towards current and future lenders to consolidate the relationship with them and therefore the reputational capital of the organization.
- Enhance the human capital of the organization, strengthening awareness of the value generated by the work done and the conditions of economic sustainability of the organization.
- Enable the full circularity of the social capital generated by the organization in its process of building, sharing and disseminating the common good.

---

[1] https://www.fondazionecharlemagne.org/.

3   The SI^2 Theoretical Framework and Implications in Social Innovation-Based Programs

Before presenting the results, it is necessary to deepen the theoretical and cultural background that inspired this type of approach. If the international debate on *social impact assessment* (SIA) is very developed and heterogeneous as regards the social impact of projects [8, 9] on the opposite the debate on the social value of organizations, not referable to specific projects, is still lacking today (apart from contributions mainly contextualized to community-based organizations in the central-northern european context [11]).

An interesting overview on the theme of the assessment of organizations, which goes beyond the sum of the SIA of the projects, was carried out by Zamagni et al. [12]. In the model proposed by the authors and called *Social Enterprise Impact Evaluation* (SEIE), the goal is not to quantify "the outcome of the action" (the projects) but to evaluate the model, that is the "how to do it" (the identity dimension of the organization). In this perspective, this approach intends first of all to enhance *organizational intentionality*: namely the intentional choice of building organizations as an urban common good, characterized by management regimes and accessibility of activities addressed in order to produce relational goods and mitigate the inequalities resulting from extractive urban development models [13–17]. The latter represent the enabling factor that set the conditions to generate a social impact towards people, the community and the collectivity, the way in which the organization works, the way in which it defines its presence in society and the choices for building internal and external relationships. It follows that the social value turns out to be a combination of the territorial relations that are established in the place where the organization operates and the relationship with the stakeholders and communities with which it comes into contact [18–21].

The theoretical framework just described as the cultural context in which we operate, what Giordano and Arvidsson [23] would define as "Societing", or the Mediterranean way to social innovation, has helped to develop the SI^2 model (social impact * social enterprise). A model that contains the key dimensions for "enabling the capacity" of social impact of social enterprises, as it is able to provide a picture of their capacity to adopt a governance capacity able to sustain territorial cohesion processes. The approach described, however, suffers from a criticality: the possibility of actually putting into practice a dual and non-dichotomous impact assessment between two different aspects:

- what an organization is: its identity, its way of being in society and being a part of it connected with the rest of the social fabric.
- what an organization produce in terms of activities (its projects, its programs);

4   The SI^2 Early-Stage Impact Readiness Methodology

As regards the impact readiness methodology experimented during the "Periferiacapitale" program, we can here introduce the SI^2 model (social enterprise * social impact). The SI^2 model is divided into two tools, both of which are necessary for a complete and detailed investigation on impact readiness of organization. A first tool (SI^2 Early Stage) mainly focused on perception of economic sustainability "readiness" to evaluate the social impact; a second tool (SI^2 advanced) whose purpose is the effective measurement of the social impact of the organization considering 6 dimensions.

Since the "Periferiacapitale" program is under development, in this contribution we will focus only on the first SI^2 Early-stage methodology and further discuss its preliminary findings.

SI^2 Early Stage is the tool intended for an initial assessment of an organization's readiness to evaluate. The SI^2 Early-Stage score is calculated as the arithmetic mean of the scores of the two synthetic indicators (SER and PES).

1. SER - Social Evaluability Readiness aims to outline the readiness to evaluate organizations by investigating the availability, use and management of the collected data. The first component - Social Evaluability Readiness - was conceived as an indicator of the level of evaluability of the organization with respect to the social impact and provides an estimate of the reliability of future measurements and assessments of social impact. In order to obtain the SER indicator, six items were considered which, integrated and weighted, led to the construction of the synthetic indicator. Below is the name of the 6 items with the relative weight for the determination of the index (Fig. 1).

Fig. 1. SER Items and weights

Two other components also contribute to the SER: "data collection method" (weight 15%) and "use of collected data" (weight 10%). The last two components of the index are represented by the "relationship between the use of data and organizational development" (weight: 10%) and the "ability to evaluate the impact of the organization's services/projects on public policies" (weight: 10%). By aggregating and weighing the six components of the SER, just described, it is possible to determine the readiness of the organization to assess the social impact it generates.

2. PES - Perception of Economic Sustainability aims to outline the perception of economic sustainability and the propensity to invest in the organization. This tool was instead conceived as an indicator that seeks to explore the level of perception of the organization's economic sustainability as well as the organization's propensity to invest. As was also the case for the SER, three items were taken into consideration which, after a weighting according to the principles of relevance, led to the construction of the PES indicator (Fig. 2).

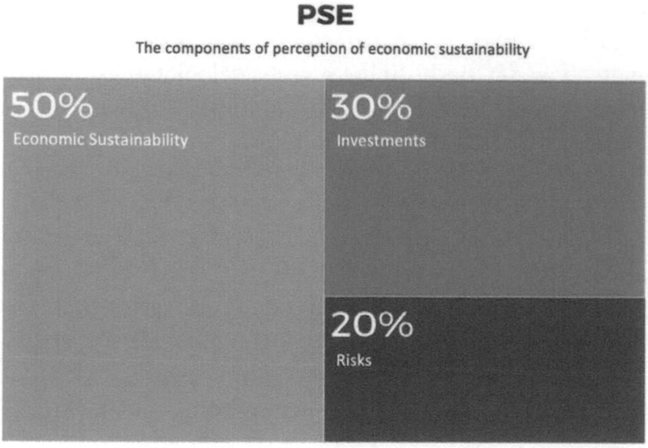

PSE

The components of perception of economic sustainability

50%
Economic Sustainability

30%
Investments

20%
Risks

Fig. 2. Pes Items and weights

Within the PES there is the component of Economic Sustainability (ES), which is not investigated in depth at this stage. In fact, the questions in the questionnaire only offer us an overview of the economic health of the organization. The ES has a very significant weight (50%), since without this it would be impossible for the organization to guarantee the performance of its activities and therefore the generation of the social impact. The perception of economic sustainability depends, for the remaining 50%, on two other factors: the share of investments and the perceived risks on economic sustainability. The investments (present and future/foreseen) in the calculation of the index have a weight equal to 30%, while the perception of the risks that the organization runs, not only in investing but also in relation to the provision of services/projects, has a weight of 20%.

More in detail, the Early Stage assessment is conducted through a 4 macro-sections survey: 1) *The organization features* and information relating to the main services and projects; 2) *The social impac*t, both as an attitude to the collection and use of data, and as the measurement process is incorporated - made its own - by the organization and to what extent this "appropriation" affects the business model; 3) *Economic sustainability*, both in terms of perception of risk today and in the future and in terms of investment propensity and how important and sustainable they are over time; 4) *The future of the organization* in the medium term, leaving room for the most diverse answers through an open question.

5   Discussing Preliminary Findings

As already mentioned in the paragraphs on the method, the SER combines 6 different components with different weights. By analyzing each of the components separately and, immediately after, by verifying the combination of these as a synthetic index, the first element we've noticed is that 5 out of 8 organizations have already built data collection processes and useful tools to start thinking in terms of impact. It is interesting to investigate the use that organizations made of data collected during the development of their community-based activities. The information obtained shows us how the predominance on aspects of communication, design and reporting were the most relevant, while to a lesser extent the data collected is intended for the use of human resource management, community management and networking activities.

Also regarding the SER tool, we've noticed that regarding the "recognition of each organization as a social enterprise" the majority of the answers were positive with the exception of only one organization. This result seems particularly interesting because it testifies an open approach to grasp the evolutions that Third Sector Entities are having, also in light of the recent innovations introduced by the Third Sector Code reform in Italy.

Another element to consider concerns the already consolidated attitude of most of them, 7 out of 13, to insert impact forecasts in the planning of activities in the medium and long term.

The topics on the sustainability of organizations and the investments necessary to ensure the production of value over time are both at the basis of the other Early-Stage indicator: the PES. There are three components that are summarized in the indicator of perception of economic sustainability:

1. analysis of the perceived conditions of economic sustainability
2. analysis of investment intentions
3. perception of the risks that the organization may have to face

A fair number of organizations (46.1%) have declared that they feel quite "confident" with respect to the truly key issue of their own economic sustainability; in fact, they declared they believe their organization to be quite sustainable, however 15% (2 organizations out of 15%) consider themselves not sustainable at all.

Regarding the topic of investments, the picture with respect to the areas does not present particularly unexpected data, in fact, as on average it happens in the third sector, we are dealing with priorities concerning training, planning, communication but also partly with an interest in managerial innovation (4 out of 13 assign high priority to this issue). Finally, as the latest figure, 58.3% of the sample perceived "the inability to manage future investments that they would need to increase their impact". In this context, however, we were not surprised by observing a very strong part the sample (almost all the organizations involved; 12 out of 13) feeling that "their organizations are running economic risks when launching new projects", an aspect to be strongly considered in the Advanced-stage path program to lever their economic solidity.

Regarding the synthesis of the two indicators on the Early-Stage path, we have implemented an average of SER and PES, in order to position the different organizations with respect to the significant dimensions of the model. The first part of the indications is a reflection of the "average" character of the SI^2 Early-Stage indicator with respect to its two macro-components, namely the SER and PES indicators. Two groups have also been created: a first group called "LOW" where all those below the average are positioned and a second called "HIGH" where all those who record values above the average are placed (Fig. 3).

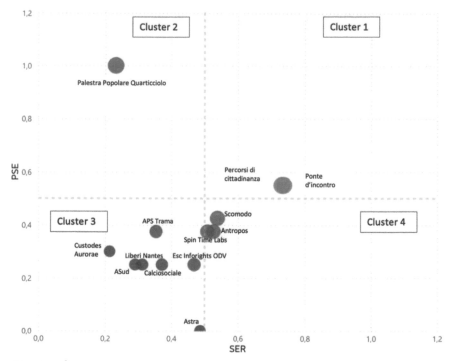

Fig. 3. SI^2 Early-Stage synthetic indicator assessment of the "Periferiacapitale" program

The main characteristics of the 4 groups are the following:

Cluster 1 (SER: high - PES:high): The first cluster is characterized by having both values, SER and PES, positive. We find within it only two organizations that report values that tend to be higher than the others involved in the process and with a strong focus on impact, from conception, planning to implementation of the activities. In accordance with a higher-than-average PES, the two have a turnover between 25,000 and 50,000 euros and both perceive themselves as a social enterprise with a strong impact orientation.
Cluster 2 (SER: low - PES: high): The second cluster includes within it those organizations with a lower SER than the average and instead a higher perception of sustainability. In this case we are dealing with a person who enjoys a second-tier turnover (25,000–50,000) and who identifies self-financing as the main source of income. As expected, the sample does not recognize itself as a social enterprise and is not particularly attentive to data collection and its use as an impact orientation. However, as demonstrated by the PES equal to 1, the perception of sustainability is high, as well as the certainty in not incurring risks for investments.
Cluster 3 (SER: low - PES:low): Cluster 3 appears to be more numerous, in fact within it we find 7 out of 13 organizations. In this quadrant are positioned the organizations

that find SER and PES values below the average. Let's see how some indications, such as turnover, appear to have little influence on SER and PES, thus denoting a long-term management and vision problem. In fact, it appears that the organizations declare a turnover that varies from the first indicated range of 0–25,000 euros up to significantly higher numbers. However, the unifying elements of the subjects have to do with a poor identification with social enterprise and almost all of them believe they are afraid of not knowing how to manage future investments. The percentage of organizations that declare that they carry out an activity linked to impact assessment is also low.

Cluster 4 (SER: high - PES:low): The fourth and last cluster finds characteristics in common with the first quadrant with respect to impact strategies and feeling like a social enterprise, although they have never entered the world of evaluation. Also, in this case, as in the third quadrant, in the face of high turnover, there is however an uncertainty in the perception of sustainability and also in the ability to control the risks deriving from new investments.

The analysis of the matrix (Fig. 3) opens both a reflection on the current positioning of the organizations involved, and an opportunity for strategic evaluation of the program. What is reported in the table highlights how the organizations are more polarized in the two quadrants 3 and 4. We can therefore think of starting a path with group 3 that allows them to develop attitudes and practices related to impact by imagining a path of strengthening skills, first with respect to its own sustainability and then to prepare the possibility of starting a concrete evaluation process by reaching the Advanced phase. The same work inherent to the VIS, can start with group 2 trying to support methods and preparatory practices to start monitoring and then evaluation processes. The goal is to be able to bring a concentration as high as possible of subjects in the first quadrant in order to increase awareness and develop the functional tools to make organizations tend towards a model of social enterprise. The ultimate goal is to ensure that their economic sustainability becomes an added value in the development of their individual activities, increasing their capacity for impact in the long term.

6   Conclusive Remarks: About the Role of Impact Readiness in Social Innovation Programs

The of dimension of economic sustainability and social impact is here analyzed in order to reveal the link found in the literature(and empirical experiences) about the connection between social value and the ability to last over time in non-profit-oriented activities.

What we've found is that the purpose of these organizations in generating social value depends on the economic sustainability: an instrumental element (therefore necessary but not sufficient) to determine the sustainability of social value over time.

This allows us to find substantial confirmation of the validity of the assumptions on in organizations based in peripheral areas, operating in territories characterized by pressing social needs and conditions of inequality that put people at risk the levels of cohesion, expressing an intrinsic demand for support from public and social infrastructures.

The state of the "economic sustainability" of the organizations involved testify two aspects:

1. Their presence on the territory and their work with the communities creates a social value that is widely recognized by multiple interlocutors[2].
2. The same does not take on strategic importance either in determining the conditions of economic sustainability or in the construction of medium-long term relationships with institutional actors.

About this aspect, the strategic answer given by the program takes on two potentially very relevant perspectives:

1. The strengthening of organizational capacity (capacity building) and the strengthening of the conditions of institutional recognition is crucial and needs to be encouraged through capacity building actions. Strengthening the organizational capacity is the element that determines the conditions of economic sustainability and that denotes the possibility that the interventions provided for by the program may change the photograph shown with Fig. 3 over time. The result that is expected to be achieved, in fact, is an average shift of organizations upwards, therefore towards more stable levels of economic sustainability.
2. The actions foreseen by the program are built with a logic that enables this type of impact: while the financial aid support activities in the short term (with the effect of reducing the pressure on the imminent difficulties) an impact readiness preparatory action, is a way to support planning, management, communication and, more generally, towards an entrepreneurial development aim to consolidate the results achieved in the medium-long term [].

Another conclusive aspect that deserves to be emphasized is the circular relationship between economic sustainability and social impact. The awareness on the necessity to gather the capacity to represent the social impact generated is considered as a way to create an effect of legitimization, where key local and city-wide stakeholders can positively influence the conditions of their economic sustainability.

For this reason the second expected strategy to be developed in the *Periferiacapitale* program concerns the improvement of the skills of strategic management of the social value generated by the organizations involved. In particular, returning to the matrix of Fig. 3, this consists in a shift of the average of the organizations towards the right part of the graph.

At the end of the first phase of the joint work between the Charlemagne Foundation and Open Impact, the following report is concluded with the next steps to follow:

*   Strengthening the territorial networks that work at the same scale where the community-based organizations work;
*   Promote community development processes in a selected number of territories to set up community foundations;
*   Promoting a joint strategy of social impact communication;
*   Stimulate the philanthropic ecosystem in order to incentivize investments in Rome through social impact methodologies supported by capacity building programs.

---

[2] As a part of the research action activities (Interviews with stakeholders) conducted during the implementation of the programs.

References

1. Mulgan, G.: Social Innovation: What it is, Why it Matters and How it Can be Accelerated. Working Paper. Skoll Centre for Social Entrepreneurship, Oxford (2007)
2. Tricarico, L., Bitetti, R., Leone, M.I.: Come disegnare politiche urbane ad impatto sociale? Un caso studio nel contesto italiano. Territorio **96**, 108–115 (2021). https://doi.org/10.3280/TR2021-096010
3. Tricarico, L., De Vidovich, L.: Imprenditorialità, Inclusione o Co-produzione? Innovazione sociale e possibili approcci territoriali. CRIOS **21**, 74–85 (2021). https://doi.org/10.3280/CRIOS2020-019006
4. Tricarico, L., De Vidovich, L., Billi, A.: Situating social innovation in territorial development: a reflection from the Italian context. In: Bevilacqua, C., Calabrò, F., Spina, L.D. (eds.) New Metropolitan Perspectives: Knowledge Dynamics and Innovation-driven Policies Towards Urban and Regional Transition Volume 2, pp. 939–952. Springer, Cham (2021). https://doi.org/10.1007/978-3-030-48279-4_88
5. Bragaglia, F.: Social innovation as a 'magic concept' for policy-makers and its implications for urban governance. Plan. Theory **20**(2), 102–120 (2021)
6. Corvo, L., Pastore, L.: Social impact assessment: measurability and data management. In: La Torre, M., Chiappini, H. (eds.) Contemporary Issues in Sustainable Finance. PSIF, pp. 247–262. Springer, Cham (2021). https://doi.org/10.1007/978-3-030-65133-6_10
7. Chevalier, J.M., Buckles, D.J.: Participatory Action Research: Theory and Methods for Engaged Inquiry. Routledge, New York (2019)
8. Greenwood, D.J., Whyte, W.F., Harkavy, I.: Participatory action research as a process and as a goal. Human Relat. **46**(2), 175–192 (1993)
9. DeLeon, P.: Participatory policy analysis: prescriptions and precautions. Asian J. Publ. Admin. **12**(1), 29–54 (1990)
10. Schön, D.A.: The Reflective Practitioner: How Professionals Think in Action. Routledge, New York (2017)
11. Cellamare, C.: Città Fai-Da-Te. Tra antagonismo e cittadinanza. Storie di autorganizzazione urbana. Donzelli, Roma (2019)
12. Grieco, C.: Assessing Social Impact of Social Enterprises: Does One Size Really Fit All? Springer, Heidelberg (2015). https://doi.org/10.1007/978-3-319-15314-8
13. Tricarico, L., Jones, Z.M., Daldanise, G.: Platform spaces: when culture and the arts intersect territorial development and social innovation, a view from the Italian context. J. Urban Affairs 1–22 (2020)
14. Kleinhans, R., Bailey, N., Lindbergh, J.: How community-based social enterprises struggle with representation and accountability. Soc. Enterp. J. **16**(1), 60–81 (2020). https://doi.org/10.1108/SEJ-12-2018-0074
15. Zamagni, S., Venturi, P., Rago, S.: Valutare l'impatto sociale. La questione della misurazione nelle imprese sociali. Impresa Sociale **6**, 77–97 (2015)
16. Tricarico, L.: Community action: value or instrument? An ethics and planning critical review. J. Archit. Urban. **41**(3), 221–233 (2017)
17. Tricarico, L.: Imprese di comunità come fattore territoriale: riflessioni a partire dal contesto italiano. Crios **11**, 35–50 (2016)
18. Tricarico, L., Vidovich, L., Billi, A.: Situating social innovation in territorial development: a reflection from the Italian context. In: Bevilacqua, C., Calabrò, F., Spina, L.D. (eds.) NMP 2020. SIST, vol. 178, pp. 939–952. Springer, Cham (2021). https://doi.org/10.1007/978-3-030-48279-4_88
19. De Vidovich, L.: Socio-spatial transformations at the urban fringes of Rome: unfolding suburbanisms in Fiano Romano. Eur. Urban Region. Stud. 09697764211031620 (2021)

20. De Vidovich, L.: The outline of a post-suburban debate in Italy. Archiv. Stud. Urban. Region **LI**(129), 127–151 (2020). https://doi.org/10.3280/ASUR2020-129006
21. Tricarico, L.: Imprese di Comunità nelle Politiche di Rigenerazione Urbana: Definire ed Inquadrare il Contesto Italiano, Euricse Working Papers 68|14 (2014)
22. Bailey, N.: The role, organisation and contribution of community enterprise to urban regeneration policy in the UK. Prog. Plan. **77**(1), 1–35 (2012)
23. Arvidsson, A., Giordano, A.: Societing Reloaded: pubblici produttivi e innovazione sociale. EGEA spa (2013)
24. Tricarico, L., Pacchi, C.: Community entrepreneurship and co-production in urban development. Territorio **87**, 69–77 (2018)
25. Calvaresi, C., Lazzarino, E.: Community hub: un nuovo corso per la rigenerazione urbana? Territorio **84**, 77–78 (2018)

Social Innovation or Societal Change? Rethinking Innovation in Bottom-Up Transformation Processes Starting from Three Cases in Rome's Suburbs

Luca Brignone[✉], Carlo Cellamare, Marco Gissara, Francesco Montillo, Serena Olcuire, and Stefano Simoncini

Dipartimento di Ingegneria Civile Edile e Ambientale, Università "Sapienza" di Roma, Rome, Italy
luca.brignone@uniroma1.it

Abstract. Among the consequences of multiple phenomena, including planetary urbanization, the long cycle of neoliberal policies, the rise of the platform economy and finally the pandemic crisis, has taken place a radical transformation of living – physical, social, and anthropological - linked to processes of increasing marginalization and peripheralization. Parallel to a decreasing ability of public authorities to regulate the economy and govern transformations, there is an evident and growing protagonism of civil society, whose very heterogeneous forms have often been unified through the 'quasi-concept' of 'social innovation'. The paper aims to test the usefulness of this concept starting from the research-action approach adopted by LabSU - Laboratory of Urban Studies 'Territori dell'abitare' (DICEA, Sapienza University of Rome), and from three cases of application of this approach in the context of the suburbs of Rome. Starting from a process perspective in the interpretation of social innovation, the three cases in question suggest a series of inductive theoretical considerations. First of all they show that the "Janus-face" of social innovation can be properly adopted and disambiguated only by understanding it as a dynamic and multidimensional process of societal change. As shown by the three cases described, characterized by a homogeneous context and very diverse processes, there are no universal formulas that have value regardless of the contingencies of contexts and processes, while it is crucial recognize the transformative capacities of the local actors, and support of forms of self-organization that they put in place. In this sense, the role of the university takes on great importance both in discriminating between different practices and in making things happen, especially promoting horizontal and vertical collaboration and setting collaborative environments and lived spaces, be they physical or digital.

Keywords: Social innovation · Urban suburbs · Local development

1 Introduction: Research-Action 'Situating' Social Innovation

The recent evolution of urban and metropolitan contexts, from a spatial, but also social and political point of view, is shaped by fundamental processes that are the starting point

© The Author(s), under exclusive license to Springer Nature Switzerland AG 2022
F. Calabrò et al. (Eds.): NMP 2022, LNNS 482, pp. 483–493, 2022.
https://doi.org/10.1007/978-3-031-06825-6_45

of the research group of the LabSU-Laboratory of Urban Studies 'Territori dell'abitare' (DICEA, Sapienza University of Rome). Firstly, according to the reflections on planetary urbanization [1], urban transformation is articulated in various forms of diffusion and dispersion of settlements, and it implies a multiplication of often indefinite agglomerations and centres of urban regionalization. It is not just about physical transformations, but also about a profound anthropological transformation of dwelling. This settlement evolution, originated with industrial capitalism [2, 3] and today globally fostered by neoliberal policies [4–6], has increased not only social, but also spatial and environmental inequalities, generating processes of marginalization and peripheralization. The result is a multiplication of suburbs, as can be seen in the emblematic case of Rome - to such an extent that it can be said that "Rome is its suburbs" [7, 8]. These are very different suburbs, not necessarily associated with a gradient of distance from the centre or with a scale of incomes. They are 'pieces of the city' which, however, often lack adequate urban conditions.

At the same time, the ways in which cities are governed have changed. The economy prevails more and more strongly over the political, and a growing distance of politics and institutions from the local contexts can be observed. Public interest itself remains increasingly uncertain and indefinite under neoliberal pressures [9].

Faced with this situation, and also because of it, forms of active citizenship and self-organization are growing in the cities. In some cases, they develop mutualistic services and volunteering practices to respond to social and local needs that do not find adequate satisfaction on the part of public policies, due to the lack of the public welfare and urban maintenance, and for this reason they often take on a subsidiary character, which can be problematic. In others, they express relationships, projects and practices capable of prefiguring, and sometimes experimenting with a paradigm shift in local development and governance [10]. Furthermore, the growing importance of the corporate digital platforms in the mediation of services and locale relationships is determining strong imbalances in local systems, further compromising the capacity of the State to regulate the economy and govern transformations [11].

In this context, combined consequences of the pandemic and global warming, highlighting and accentuating social and environmental imbalances on a planetary scale, as well as accelerating the process of digitization of the economy and society, seem to tip the balance towards a radical paradigm shift. In the face of the actual unsuccessful attempts to restore State action on both fronts, with contradictory programs of economic revitalization and decarbonization of the economy, the key role of the so-called civil society emerges even more clearly, both in a collaborative and conflictual perspective. The heterogeneous forms of this growing protagonism, in the general process of redesign and rescaling of local governance, have often been unified through the 'quasi-concept' of 'social innovation' [12]. The rise of this protagonism, and its problematic relationship with technological innovations, makes it necessary to thoroughly reconsider the usefulness of this quasi-concept.

In fact the category of social innovation is particularly controversial as an analytical tool [13], because it leads very often to abstractly taxonomic or normative approaches, and to rhetoric that sometimes tends to mask, rather than identify and counteract, the real causes of the social problems. On the contrary, this paper wants to contribute to the reflection on social innovation following research that has more systematically interpreted it from a place-based, political and process perspective [14–16].

In order to investigate more deeply the local variegations of the dominant processes of urban and local transformation briefly described above, as well as to know and in some cases support the social responses to them, LabSU research group developed its own research approach: that is interdisciplinary, 'situated' within the research-field, and choose action-research and co-research perspective. This 'situated' approach is essential for urban planning (as well as for the social disciplines that already practice it, such as applied anthropology), for several reasons. First of all, it is the only way to understand local social dynamics, practices and uses of space, but also to detect which are the main problems as well as the needs expressed by the inhabitants; moreover, looking at the future, it makes it possible to define together with the social actors which paths and tools allow to improve their living conditions. Furthermore, co-research makes it possible to enhance the point of view of the inhabitants, the knowledge they carry, which is sometimes significantly richer and more complex than the researchers' one.

Furthermore, LabSU places at the centre the processual logic of radical change, anchored to social protagonism. In this sense, it favours the development of contexts of interaction for the peer production of knowledge and co-design (in which the University plays a catalytic role), as well as collaborative networks within civil society, but with the aim of a wider involvement of different subjects, including institutional ones. The forms of collaboration between local actors and institutions always seek to transcend the occasional and opportunistic character to place themselves in a broader perspective of rethinking institutions. Upstream of this objective, there is always the need to support the formation of new political subjectivities from below, a necessary condition to guarantee a transformative effectiveness to collective action.

Based on three of LabSU's recent action-research paths, this paper intends to show the effectiveness of this approach, and for this purpose it tries to overturn the usual correlation between theoretical framework, case studies and research method. In the second, three cases of application of this approach are described, correlated to as many contexts in the Rome. Starting from these cases, in the third and fourth paragraphs, a theoretical reflection, with some conclusions, is developed to question the social innovation concept.

2 Three Action-Research Paths in Rome's Suburbs

In this paragraph we briefly report three action-research paths that LabSU have undertaken to support grassroots societal change processes. All the case studies are located in the eastern quadrant of Rome, a heterogeneous urban axis that in the past hosted important manufacturing settlements and that today but today is characterized by complete abandonment or partial gentrification. Indeed, the worst socio-economic parameters of the city are found here [17, 18], as well as the worst level of pollution and state of the built environment [19]. The description of the three cases aims at framing the type of local actors with whom LabSU works, and the different processes of urban transformation carried out.

2.1   A Bottom-Up Development Process in a Public Housing Neighbourhood

Despite its 'human-scale' dimensional and formal aspects Quarticciolo remains a neighbourhood on the fringes of Rome's intra-ring periphery, and it presents all the consequences of this location. Among them, the deliberate retreat of public initiative that leaves ample room for informal responses to daily needs, some of which are intertwined with criminal economies[1] and others by a network made of self-organising realities that carry out actions that are essential for the neighbourhood, and that we will now mention shortly.

The former Casa del Fascio hosts a housing occupation and, on the ground floor, the Red Lab social centre, closely related to the Palestra Popolare: this gym, inspired by Brazilian experiences that conceive boxing as a means of building social ties, currently occupies a neighbourhood's former boiler room that has been renovated and set up by the activists to host classes for the inhabitants (particularly, kids and adolescents).

At the beginning of 2018, the reaction of the *borgata* on the occasion of the forced eviction of a family generated the organisation of a Comitato di Quartiere (Neighbourhood Committee). The committee plays a role in monitoring the conditions of the buildings and the people who live in them and follows up on some of the most urgent requests by acting as a mediator with the managing public body (ATER). In particular, it played a fundamental role in the implementation of the recovery and renovation works of two occupied buildings (the so-called *favela*), collaborating with ATER in the census of the occupants, their socio-economic conditions and accompanying them in the process of regularisation or assignment of other accommodation.

In March 2021, after facing further difficulties generated (or exacerbated) by the pandemic period, the Comunità Educante Quarticciolo was set up: the organisation aims to deepen relations between local associations, secondary school teachers, the municipality's social services and cultural institutions, in order to combat early school leaving[2], a phenomenon that reaches significant peaks in this neighbourhood.

Like many other informal realities on which a large part of Roman welfare depends, the work of the activists of this informal network is capillary, with activities that aim to conceive and experience Quarticciolo in its collective, aggregative and mutualistic dimension. These actions make it possible to intercept different situations and have an overall image of the neighbourhood's vulnerabilities and resources, built and reshaped not on a static level, but on a dynamic process.

The network's latest project involves the recovery of a building in partial disuse to make it into a Casa di Quartiere, a sort of civic centre that shall host the many activities carried out by the network and at the same time manage to put in place a space for mediation with the managing body and local institutions.

LabSU is accompanying this project and the ongoing processes in general, facilitating relations with new partners (such as the builders' union that is supposed to support the *borgata*'s buildings renovation) and offering its expertise for the success of the project.

---

[1] Quarticciolo is considered to be one of Rome's major drug markets (especially for hashish and cocaine); [20,21]

[2] As emerges from the ISTAT data on non-completion of the secondary school cycle. Cfr. a georeferenced representation of these data at: https://www.mapparoma.info/mappe/mapparoma25-esclusione-sociale-quartieri-roma/.

2.2   Digital Social Innovation in Suburbs: East Rome Green Belt Participatory Masterplan

Centocelle belongs to the so-called historical periphery, a city full of contradictions within the V municipality, which compared to the rest of the city is the penultimate district in terms of income per capita [18], third for land consumption (with 61% of the surface consumed) [19] and last in terms of pollution.

Centocelle is now at the center of a major transformation, emblematic of the pulviscular forms with which urban extractivism manifests itself in Rome. The realization of the new subway line, the redevelopment of some squares, together with some purely immaterial features linked to its popular heritage, to the imaginary of the historical memory of the resistance, to the migrant territorialization, and to the presence of some of the most important realities of the counterculture and self-organization in the Roman scene, have generated an unprecedented attractiveness that has triggered the current transformation. This attractiveness exerts pressure on the real estate market that has given new impetus to the historical engine of urban rent, to which the city of Rome has inextricably linked its development model [22, 23]. Among the foreseeable consequences, in the absence of public policies, there has been an increase in evictions that have affected the most fragile population, exacerbating socio-spatial inequalities, and the attempt by the rent actors to saturate the few green areas that escaped edification during the years of the great expansion after World War II.

The district of Centocelle, and more generally the fifth municipality, is at the same time rich in 'values', resources and commons. Particularly relevant are the relational assets represented by the vast world of activism, self-organization and more generally of the dense fabric of associations [24]. Despite the fragmentations and political differences that cross these experiences, in November 2019 many of them came together in a variegated movement called Libera Assemblea di Centocelle (LAC). The LAC is a network born in response to some arson attacks that hit some of the commercial activities that, like the socio-cultural space of the 'Pecora Elettrica', had become a symbol of the cultural vibrancy of the neighbourhood and its countercultural spirit.

In order to promote alternative models of local development, based on environmental and social sustainability and intended to counteract criminal interests, or in any case traceable to the monoculture of rent, some activists belonging to the LAC have formed a working group called 'Environment and Territory Group' (GAT), engaged in actions to protect and enhance the huge natural and cultural heritage represented by the green areas surrounding Centocelle and its archaeological assets. The LabSU has supported the GAT in a process of critical and collaborative mapping of the district focused on the use of digital tools related to the concept of 'Digital Social Innovation' [25, 26][3] and aimed at sharing and producing bottom-up knowledge. These tools were adopted in order to create from below an open digital ecosystem, decentralized and widely shared at the urban scale, which was able to respond to the widespread need among local actors to network locally to share knowledge and planning and coordinate direct actions. The

---

[3] According to an effective definition taken from the NESTA report [25], DSI is "a type of social and collaborative innovation in which innovators, users and communities collaborate using digital technologies to co-create knowledge and solutions for a wide range of social needs and at a scale and speed that was unimaginable before the rise of the Internet".

ICT appropriated by social cooperation is supporting new social formations that seem to produce leaps of scale in the organization of practices, integrating them into networks that act in various fields in the construction of shared spatial knowledge, community visioning and innovative politics [27].

The use of these technologies, associated with the progressive involvement of groups and committees active in the protection of the various green areas of East Rome in exploration and mapping initiatives, led above all to an overall reorientation of the GAT, which first extended its interests to a wider scale, setting itself the goal of enhancing the vast "green belt" that surrounds Centocelle from below, and then operationally built a path of progressive involvement of the committees and associations of the quadrant in the mapping and co-design process.

2.3 New Public and Social Spaces for a Fragmented and Marginalized Neighbourhood

Tor Bella Monaca (TBM) district is located in the eastern outskirts of Rome, at a long distance from the city centre. The district was realized to host people suffering important economic and social problems and, with its vast open spaces defined by huge slabs and tower buildings, is clearly distinguished from the surrounding suburbs, characterized by a minute urban factory often related to spontaneous and outlaw construction processes. The area is the worst in Rome for social hardship indicators, including rates of employment, unemployment, youth concentration and school drop-out [28].

TBM is the last large public housing district built in Rome: a "piece of city" built in the early 80s of the last century on a unitary project and inhabited today by about 30,000 people. A historical example of a 'public city', still maintaining significant resources as a big public housing heritage to fight against an increasing housing emergency, as well as the opportunity of a public neighbourhood management offered by soil properties.

In a context afflicted by organized crime and social unease, these grassroots experiences offer an alternative based on the development of local projects, enhancing existing resources and promoting public participation.

Today, among the most active local actors there are: the union of the tenants of public housing, AsIA-Usb, strictly linked with the Comitato di Quartiere Nuova Tor Bella Monaca; the social center El 'Che'ntro, related to Cubolibro library and La Gabbia bicycle workshop; the association promoting rights of the disabled (Sindacato italiano per i diritti degli invalidi, SIDI), which its national headquarters; Tor Più Bella, an association recently created to involve inhabitants in taking care of public spaces in the neighborhood. LabSU, moving from an interdisciplinary workshop in 2015 [29], has been carrying out research-action activities in this district for several years [30], with the aim of intercepting and promoting ongoing grassroots practices.

Recently, LabSU has coordinated MeMo – Memorie in movimento, a project carried out by a local network including some schools and the local municipality, funded by Italian government (Mibact) and working on historical awareness of the inhabitants' struggles of 1980s, in order to address the issue of public participation and civic engagement in marginal urban areas.

Today, LabSU is coordinating CRESCO – Cantiere di Rigenerazione Educativa Scuola Cultura Occupazione, a project promoted and financed by Fondazione Paolo Bulgari to support and enhance the neighborhood's educational community.

3 Comparative Analysis and Theoretical Inductions

We have seen how these cases are diverse, in terms of contexts[4] and social agency (which entails different forms of horizontal collaboration, and therefore of organization and goals), but also in terms of different types of processes (which entail different forms and modes of vertical collaboration, and therefore a different positioning of the researchers themselves).

Despite the fact that the contexts on which the research group works obviously present some peculiarities that do not make them completely comparable experiences, we can trace some common threads. The three are neighbourhoods in which it is particularly evident how the public body (at various levels) is not only unable to meet the needs of the inhabitants, and sometimes not even to recognize them, but also how it has completely lost its imaginative and planning capacity.

The forms of social protagonism that we encounter in the districts in question are thus found to undertake mutualistic practices to overcome the daily difficulties of the neighbourhood. In addition, these subjects manage to express medium- and long-term planning, producing projects that transform their local contexts and indicate to the public actor innovative ways of 'making the city': recognizing the lesson of these bottom-up experiences is useful to rethink the way in which the public service operates, making its action more efficient and calibrated on the real needs of the inhabitants.

At the same time, however, the self-organization that underlies the local actors with which LabSU works plays a further, essential role in defining a community that understands and expresses the needs and possibilities of the territories.

In this framework, what is the role of the university institution? Working in these places requires constant self-reflexivity on the part of those who examine and support these practices, and particularly to understand how to move in the direction of the dissemination of more effective policy-making.

In the first instance, LabSU tries to collaborate and support local realities. This role allows us not only to collect their voices (although this is a fundamental aspect of field research), but also to stay inside the processes, sometimes even with a critical eye, to accompany and promote them so that they have a favourable outcome.

Often, this translates into a role of mediation (sometimes problematic) with the public body (whether it be, as in the cases we have seen, the City Hall, the Municipality, the Region, ATER), or with intermediate bodies, the realities of the private social or third sector that aspire to collaborate in the transformation of a place (in our cases, trade or tenants' unions, or private foundations). In the first case, the LabSU supports

---

[4] The three neighborhoods differ in terms of the gradient of population density (Tor Bella Monaca (88 in/h), Centocelle (176 in/h), to Quarticciolo (500 in/h); of urban fabric (from the extremely fragmented (TBM) to extremely continuous (Quarticciolo), passing through the intermediate one of Centocelle); of social mix (BM lower, Quarticciolo medium, Centocelle high), of endowment of services (TBM low, Quarticciolo medium, Centocelle high).

those who can put pressure on the institutions, the only way in which they can actually transform themselves; in this sense, it tries to bring into play its privilege as an institution, attempting a sort of redistribution of power through the support and legitimation of those who carry out the interests of the neighbourhoods (often with strongly conflicting forms). In the second case, the LabSU accompanies the bodies that intend to invest resources in peripheral urban areas, indicating the possible realities to be supported and trying to channel funds and skills to the most significant projects.

This allows us to point out a further issue: the recognition of the transformative capacities of the 'DIY city' actors [10] does not mean supporting them all indiscriminately, but understanding which ones carry out an action that meets the collective well-being. In each place where it operates, the research group makes a choice in working with some subjects rather than others, which is not easy in some contexts, such as Tor Bella Monaca, where the geographies of social bodies are particularly complex. This choice is supported by a clear political positioning, which implies to prefer active subjects who share a certain value system (defined also from the scientific knowledge acquired over the years in the field); who intend to question the systems of development and government of the dominant cities; who produce generative paths of self-representative political subjectivities.

Finally, the experience of the LabSU imposes a reflection on the social and political valence of digital technologies, or more precisely on the collaborative tools and relational environments that need to be put in place to support the development of grassroots initiatives and organisations. If it is true that it is necessary to rethink social space as a whole starting from the increasingly complex relationship between society, digital and territory, especially after the pandemic, the experience of LabSU shows that this space and relationship must be rethought starting from experiments that, framed in categories and phenomena such as Digital Social Innovation, digital commons or collaborative mapping [25, 26], aim at the construction and dissemination of digital environments, methodologies and data, whose character "open" and "horizontal" enable more advanced forms and processes of sharing and co-production of knowledge, as well as networking and collaboration between local actors, formals and informals. This is all the more necessary since the environments of the dominant digital mediation, informed by proprietary and commercial logics, not only do not favor these processes in any way, but in many cases exert deconstructive impacts on the social, cultural and productive fabrics of the city. Moreover, rethinking the new "lived spaces" [31] of society, derived from the hybridization of online interactions and territorial interactions, means building the material and immaterial conditions for the development of new social formations built on new imaginaries and values.

4   Conclusions: When Social Innovation Meets Societal Change…

Our research approach gives rise to a specific way of dealing with the problems that constitutes the declination of social innovation by the research group. First, we need to rethink the suburbs as "pieces of the city" that must acquire adequate levels of urbanity, with a diffusion of local centralities In these areas there is often a lack of services, available spaces, equipment, infrastructures, but all of this is not enough. Physical interventions must be accompanied by policies for self-organization, that is, to enhance forms of self-organization and collaborative and mutualism networks, where social protagonism develops from a public interest point of view, promotion of local contexts, local development in a social and transformative economy. Supporting forms of self-organization allows us to respond more adequately to emerging needs, to enhance social leadership (which constitutes a great investment of energy and at the same time allows the cultivation of social antibodies in the most difficult contexts, especially in those characterized by the presence of criminal organizations), to build more adequate and efficient public policies, to rethink collective action as a form of collaboration between different subjects (institutional or not, formal or informal, etc.) but who operate from a public perspective, to rethink consequently also the institutions themselves and, finally, to concretely activate the promotion of the neighbourhoods and their local development.

Moreover, the presence of engaged research allows to beware of macro-processes, particularly neoliberal trends which can use social innovation as rhetoric, e.g. pursuing competition between different contexts instead of welfare [32], or shifting public services and decision-making from democratic institutions to the private actors.

The cases in question and the framework adopted show that the "Janus-face" [33] of social innovation can be properly understood, disambiguated and usefully adopted only in the sense that denotes it as a dynamic and multidimensional process of societal change and socio-spatial transformation to improve the conditions of common living. The process perspective leads to drastically shifting the attention from the output to the input, that is, from the product, service, practice in itself taken, to the agency observed in its processual evolution, material and immaterial, and in the context to which it refers.

From the point of view of research, because of the number of variables that determine the multilevel complexity of social and urban systems, and because of the incidence of the time factor on processes, there are no possible formulas. The cases analysed and the related practices are not innovative per se. The cases analysed are innovative insofar as they succeed in producing transformative urban effects, contributing to a rethinking of institutions and to new governance processes, in which local actors on the one hand and university in its various articulations on the other, play a fundamental role. Hence, It is always necessary to observe in depth and at length the contexts in their constant change as a result of the combined pressures of internal and external forces in order to understand the innovativeness of a process. From the point of view of action, it is necessary to modulate these forces at play by firstly affecting the conditions that precede transformative actions, that is, the setting of 'lived spaces', whether physical or digital, that favor the strengthening of ties, the sharing of knowledge and horizontal and vertical collaboration. Only at the end of the action-research can we understand if there are process models or assemblages of practices capable of assuming a wider value, and therefore of constituting a useful reference for other contexts, or of being susceptible to scale up and scale out.

From our experience, the qualifying element of 'innovativeness' thus becomes the greater or lesser capacity of a practice, or of an assemblage of practices, to generate change, from a local improvement in the general conditions of life to a radical change in society, to be understood as a substantial redefinition of the relations of production and power in a given context, aimed at a more equitable and sustainable relationship between society and local resources.

References

1. Brenner, N.: Implosions/Explosions. Towards a Study of Planetary Urbanization. Jovis, Berlin (2014)
2. Lefebvre, H.: La rivoluzione urbana, Armando, Roma (1973) [original ed. La rèvolution urbaine, Editions Gallimard, Paris, 1970]
3. Choay, F.: Le règne de l'urbain et la mort de la ville. In: Choay, F. (ed.) Pour une anthropologie de l'espace. Seuil, Parigi (2006) [original ed. in: La ville, art et architecture en Europe, 1870–1993, Centre Georges Pompidou, Paris, 1994]
4. Brenner, T., Theodore, N. (eds.): Spaces of Neoliberalism. Urban Restructuring in North America and Western Europe. Blackwell, Oxford (2002)
5. Peck, J., Brenner, N., Theodore, N.: Actually existing neoliberalism. In: Cahill, D., Cooper, M., Konings, M., Primrose, D. (eds.) SAGE Handbook of Neoliberalism, pp. 3–15. Sage, Thousand Oaks (2018)
6. D'Abergo, E.: What is the use of neoliberalism and neoliberalisation? Contentious concepts between description and explanation. Partecipaz. Conflit. 9(2), 308–338 (2016)
7. Cellamare, C.: Trasformazioni dell'urbano a Roma. Abitare i territori metropolitani. In: Cellamare, C. (ed.) Fuori Raccordo. Abitare l'altra Roma, pp. 3–30. Donzelli editore, Roma (2016)
8. Cellamare, C.: Abitare le Periferie. Bordeaux edizioni, Roma (2020)
9. D'Albergo, E., Moini, G. (eds.): Il regime dell'urbe. Politica, Economia e Potere a Roma. Carocci, Roma (2015)
10. Cellamare, C.: Città fai-da-te: Tra Antagonismo e Cittadinanza. Storie di Autorganizzazione Urbana. Donzelli Editore, Roma (2019)
11. Bratton, B.H.: The Stack: On Software and Sovereignty. MIT Press, Cambridge; London (2016)
12. Jensen, J., Harrison, D.: Social innovation research in the European Union. In: Approaches, Findings and Future Directions. Publications Office of the European Union, Luxembourg (2013)
13. Moulaert, F., MacCallum, D., Mehmood, A., Hamdouch, A. (eds.): The International Handbook on Social Innovation: Collective Action. Social Learning and Transdisciplinary Research. Edward Elgar Publishing, Cheltenham-Northampton (2013)
14. Moulaert, F.: Social innovation: institutionally embedded, territorially (re)produced. In: MacCallum, D., Hillier, J., Vicari, S. (eds.) Social Innovation and Territorial Development. Ashgate, Aldershot (2009)
15. Swyngedouw, E.: Animating social change: political transformation and/or social innovation? In: Van den Broeck, P., Mehmood, A., Paidakaki, A., Parra, C. (eds.) Social Innovation as Political Transformation. Thoughts for a Better World, pp. 8–12. Edward Elgar Publishing, Cheltenham, Northampton (2019)
16. Tricarico, L., De Vidovich, L., Billi, A.: Situating social innovation in territorial development: a reflection from the Italian context. In: Bevilacqua, C., Calabrò, F., Della Spina, L. (eds.) NMP 2020. SIST, vol. 178, pp. 939–952. Springer, Cham (2021). https://doi.org/10.1007/978-3-030-48279-4_88
17. Lelo, K., Monni, S., Tomassi, F.: Le Mappe Della Disuguaglianza: Una Geografia Sociale Metropolitana. Donzelli, Roma (2019)
18. Celata, F., Lucciarini, S.: Atlante Delle Disuguaglianze a Roma. Camera di Commercio Industria Artigianato e Agricoltura, Roma (2016)
19. ISPRA. Consumo di Suolo, Dinamiche Territoriali e Servizi Ecosistemici (2019)
20. Olcuire, S.: Quarticciolo, the Perfect Dimension. Decay, Coexistence and Resistance in a Roman Ecosystem, Los Quaderno 53 (2019)

21. Olcuire, S.: Sex Zoned! Geografie del sex work e corpi resistenti al governo dello spazio pubblico, doctoral dissertation. Department of Civil, Constructional and Environmental Engineering, Sapienza University of Rome (2019)
22. Insolera, I.: Roma moderna. Da Napoleone I al XXI Secolo. Einaudi, Torino (2011)
23. Tocci, W.: Roma come se. Alla Ricerca del Futuro per la Capitale. Donzelli, Roma (2020)
24. Brignone, L., Cacciotti, C.: Self-Organization in Rome: A Map. Tracce Urbane, Giugno **3** (2018)
25. NESTA. Growing a Digital Social Innovation Ecosystem for Europe, Final Report (2015). https://media.nesta.org.uk/documents/dsireport.pdf. Accessed 10 Dec 2021
26. Anania, L., Passani, A.: A Hitchiker's guide to digital social innovation. In: 20th ITS Biennial Conference, Rio de Janeiro, Brazil. The Net and the Internet - Emerging Markets and Policies, International Telecommunications Society (ITS), Rio de Janeiro (2014)
27. Simoncini, S.: Reti sociali interorganizzative, tecnologie del sociale e autogoverno del territorio: l'avvio di una ricerca sul contesto romano. In: Gisotti, M.R., Rossi, M. (eds.), Territori e comunità. Le sfide dell'autogoverno comunitario, Atti dei Laboratori del VI Convegno della Società dei Territorialisti. Castel del Monte (BA), 15–17 November 2018, pp. 226–238. SdT, Firenze (2020)
28. https://www.comune.roma.it/web/it/roma-statistica-benessere-economico.page. Accessed 10 Dec 2021
29. AA.VV. Territorio n. 78. Franco Angeli, Milano (2016)
30. Cellamare, C., Montillo, F.: Periferia. Abitare Tor Bella Monaca. Donzelli, Roma (2020)
31. Lefebvre, H.: La Production de l'espace. Anthropos, Paris (1974)
32. Brenner, N.: 'Glocalization' as a state spatial strategy: urban entrepreneurialism and the new politics of uneven development in Western Europe. In: Peck, J., Yeung, H.W. (eds.) Remaking the Global Economy: Economic-Geographical Perspectives, pp. 197–215. SAGE Publications, London (2003)
33. Swyngedouw, E.: Civil Society, Governmentality and the Contradictions of Governance-beyond-the-State: The Janus-face of Social Innovation. In: MacCallum, D., Hillier, J., Vicari, S. (eds.) Social Innovation and Territorial Development. Ashgate, Aldershot (2009)

Pandemic, Fear and Social Innovation

Federica Scaffidi[✉]

Leibniz University of Hannover, Institute of Urban Design and Planning,
30167 Hannover, HAJ, Germany
scaffidi@staedtebau.uni-hannover.de

Abstract. Cities and regions are challenged by the impacts of the pandemic. Fear and loss of freedom are now spread, creating social conflicts. The only hope is that there will be a Post-pandemic Future. What about the period during the pandemic? This paper addresses the topic of how to face the current pandemic and create inclusive, open and innovative places. Social innovation can play a key role in mitigating the effects of this situation. It satisfies human needs and develops social benefits. Nevertheless, the "social innovation places" such as urban labs, cultural centres, social enterprises have been strongly affected by the pandemic. The paper therefore aims to discuss the importance of social innovation for cities during the pandemic and how this can reduce fears and loss of hope. Considering this purpose, the research examines specific examples of social enterprises in Europe through interviews and surveys, and how they are responding to the pandemic.

Keywords: Pandemic · Social innovation · Cities

1 Pandemic, Fear and Social Needs

Since the beginning of the Covid-19 pandemic, messages of fear and hope were spread, splitting the society in two main radical parts. There are people who live in loneliness and others with psychotic attitudes all because of their fear of being infected. On the other hand there are those who feel that their rights are being oppressed and are mourning their loss of freedom. In this name they oppose treatment and vaccine. Media contributed by spreading fear, sowing terror against neighbours, against people who come from "outside", because they could be carriers of the virus. At the same time messages of hope, mostly illusory hope, of a happy near future, without deaths, without the infected were spread as well. One of the most famous is the Italian slogan '*andrà tutto bene*'. A positive message of hope to imagine a happy near future, with no more deaths, no more infected and no more social restrictions, a Post-pandemic Future. However, this does not seem to be the case. On the threshold of 2022, according to the ECDC[1], the epidemiological curves and deaths for Covid-19 are still increasing worldwide[2]. The

---

[1] ECDC is the acronym of European Centre for Disease Prevention and Control.

[2] Although a vaccination campaign is underway in all Western countries, epidemiological curves from Covid-19 are still increasing worldwide due to the rapid genetic mutation of the virus.

F. Calabrò et al. (Eds.): NMP 2022, LNNS 482, pp. 494–500, 2022.
https://doi.org/10.1007/978-3-031-06825-6_46

longed-for post-pandemic future of unrestricted urban life, without masks, tests and social distancing, and above all without death and illness, still seems far away.

The following question arise spontaneously: will there really ever be a post-pandemic? It is necessary to ask what solutions can be adopted to break down these conflicts, to reduce the distance between these two radical positions, to create spaces for the community during the pandemic. Considering social innovation as the ability to respond to a social problem and to improve the quality of life of communities (Moulaert et al. 2005; Phills et al. 2008), it seems that it could be a good solution to solve this social problem. It would reduce loneliness and xenophobia, and reconnect these two parts of society together again, by initiating democratic processes and cooperation. The problem is that the "social innovation places" have also been affected by the pandemic. These community spaces play an important role in the urban transition, because they are the places where culture, social activities, new forms of entrepreneurships and jobs create innovation and urban dynamism. So it is becoming increasingly important to understand how to address urban resilience and to recreate the conditions of social innovation that reduce barriers, and create novelty and integration (Dorobantu and Matei 2015). During these times of uncertainty, environmental, financial and health crises, there is a need for academics to debate these topics, analyse the issues outlined above, and provide practical recommendations and innovative solutions for the cities and regions. This paper addresses the topic of social innovation to create safe and social urban spaces for all during the pandemic. Social innovation is a novel solution to a social problem and aims to enhance rights and satisfy the human needs (Phills et al. 2008; Moulaert et al. 2005). Therefore, it is important to foster processes of social innovation to reduce social isolation and xenophobia. The "social innovation places" such as community hubs, urban labs, cultural centres, social enterprises are important spaces to build the community, enhance networking skills, reduce inequality, promote new jobs, and create a more open and inclusive society (Mangialardo and Micelli 2018; Scaffidi 2019; Tricarico et al. 2020). It is important to analyse the effects of the pandemic on these centres because of their critical role in our cities. This paper looks at social innovation and the effects on the "social innovation places" in Europe, by analysing examples in Italy, Austria, Germany, and Spain. A qualitative research method has been adopted, with exploratory and dialogic surveys[3] carried out during the current pandemic where the following questions were addressed: What actions can be adopted in order to reduce fears, the sense of oppression, loss of freedom and hope? Can social innovation play a key role in addressing the pandemic? How can social innovation make cities safe and social? And how to ensure that "social innovation places" are active, safe and open?

2   Cities, Social Innovation and Pandemic Effects

Cities and regions are challenged by the impacts of the pandemic because of the exacerbation of social distances and fears. It is clear that they are experiencing a transition, and new policies, projects and research must promote innovative urban solutions to address the current pandemic (Schröder 2018). In the framework of the Green Deal, new ideas

---

[3] Specifically, participant observation and semi-structured qualitative interviews were carried out.

and innovations are proposed to ensure a long-term strategy to achieve resilience and sustainability goals. The New European Bauhaus proposes new strategies to deal with this change and brings the Green Deal to our everyday places by designing a sustainable and more inclusive future. Community spaces are more important than ever because they provide a place where dialogue can happen between people and collective needs can be discussed. Social innovation can promote this interaction and contribute to achieving these goals and ensure a long-term sustainable urban development (Carta et al. 2020). Social innovation develops more effective solutions in response to a problem and a social need, whose resulting value benefits the community (Phills et al. 2008; Clark and Wise 2018). It solves a problem of society and generates new social value through greater community involvement and the satisfaction of collective needs. According to Riccardo Maiolini (2015, 35–36): "The goal of social innovation concerns the improvement of society as a whole, through the identification of innovative practices and/or products able to improve the widespread and collective well-being of a given community". Social innovation improves the communities' quality of life and creates positive effects for the population, by analysing social actors' behaviour and encouraging them to actively participate in the development processes (Scaffidi 2021). Therefore, socially innovative actions and policies should play a key role in addressing the issues arising from the pandemic, by creating new social places, and a more inclusive and open society. In this perspective, there is a need for new urban programmes and policies with strong social and cultural impacts, where socio-cultural regeneration, community spaces and innovative organisational models are developed. Social innovation also promotes new organisa- tional forms and innovative business models where the main objective is to create social benefits and services that respond to a problem (Maiolini 2015). Social enterprises indeed are community spaces that contribute to creating cities that are more inclusive and open to diversity and dialogue. Their aim is to create a structured social impact on the site, improving the social result in relation to existing solutions. They play an important role in creating socio-cultural spaces and innovation in the cities (Micelli 2018). They pro- mote open debates, cooperation, co-working that contribute to the creation of a strong and open society. In Europe there are many examples of social enterprises that together with local governments and citizens improve the urban quality of life, and promote local development and socio-cultural regeneration. They have an important impact in the urban development and social cohesion. These could have a key role in the reduction of fears, and loss of freedom and hope. However, as the current pandemic is reshaping cities and the way to experience private and open spaces, it is also changing the use of these "places of social innovation". In the last two years cultural centres, social enterprises and community places have had significant restrictions, which have prevented activities from taking place. Periferica of Mazara del Vallo is a socio-cultural centre in Sicily. It is a community space, that organises socio-cultural events, open activities, design work- shop and visits that involve local and international people. Periferica has been closed for almost two years (Fig. 1). In an interview, Carlo Roccafiorita, founder and director of Periferica, talks about the difficulties of staying "alive" and the injustices done to the socio-cultural sector. Periferica was unable to organise its summer school again this year, despite taking additional safety measures, and introducing social distancing. It is a time of contradictions, where clubs, restaurants and discos are open, and community and cultural centres are closed (Fig. 2). Social innovation can be a response to the problems generated by the pandemic, but at the same time clear rules are needed to protect these places of openness, integration and social inclusion, which are essential for overcoming a period of fears. The findings show that other social enterprises have also had issues in dealing with the effects of the pandemic. Another case is the Real Fábrica de Cristales of La Granja near Segovia, in Spain. It is a very important cultural space for the local community and an important heritage site for the town. They organise socio-cultural activities, workshops, courses and produce the crystal of La Granja. The dialogic survey showed that this "social innovation place" also had some problems due to the current pandemic. They are open, they can continue with many activities, production, guided tours, but with fewer people and fewer services. Paloma Pastor Rey de Viñas, the director of the Real Fábrica de Cristales confirmed during an interview, that educational courses and workshops have been discontinued, creating a reduction of income within the social enterprise. It is no longer possible to have those spaces of community, of co-design, of educational workshops for children and adults. These "social innovation places", so important for the social cohesion, innovation and the expression of community needs, are badly affected by the pandemic. The exploratory and dialogic surveys also show other European examples with similar situations. The Wuk Werkstatten in Wien is a centre of culture and social activities. They hold art exhibitions, performances, educational workshops, handicraft workshops, product design courses and events. It is a place of community.

Fig. 1. Periferica is closed. Graphic by the author.

It is a social space for the city that promotes the interaction between people and creates new socio-cultural values. However, this has also been badly affected by the pandemic and many restrictions were imposed on them. A similar situation emerges in the survey in Bremen, in Germany. The Kulturzentrum Schlachthof is a socio-cultural place that has been damaged by the pandemic. It is a "social innovation place" where they organise concerts, festivals, readings, panel discussions, conferences and workshops. However, these debates and cultural events were often cancelled or postponed. This led to a drop in visits and a drop in income. It is possible to draw a parallel with another analysed case. Salinas de Añana saltworks in Spain is also a socio-cultural centre. It is a place of interaction, cultural development, social production, workshops and events. Alberto Plata Montero, the head of culture and communication of Valle Salado de Añana Foundation, stated during an interview that the Salinas de Añana saltworks are still active despite the pandemic (Fig. 3). They continue to produce salt but there has been a drop in sales and visits. The chart (Fig. 4) shows a growing trend in the number of visits, but in 2020 there was a drastic decrease in the number of visits and consequently revenue. It is important to understand how these places can continue to be active, economically stable so they can keep creating innovation and social benefits specifically, during the pandemic.

Fig. 2. Windows closed and upcoming projects. Graphic by the author.

Fig. 3. Looking at the future of Salinas de Añana. Graphic by the author.

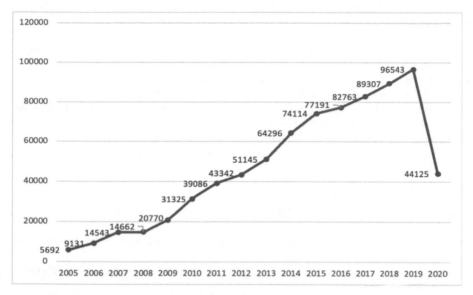

Fig. 4. Number of visitors of Salinas de Añana saltworks. Graphic by the author.

3   Conclusions

The findings show that "social innovation places" have been affected by the pandemic, with a reduction in activities, cancellation of events, restrictive measures and sometimes closures. The literature shows that social innovation promotes cohesion, integration and novelty, reduces community problems and satisfies social needs. This is necessary to promote a good quality of life in the pandemic, addressing fears, eliminating diversity, sharing needs and debating different issues and opinions, and creating new projects and actions that give hope back to communities. However, because of the health emergency and the risk of infection, these "places of social innovation" are often closed or cannot carry out the activities necessary to contribute to this development. At the same time, restaurants, clubs and stadiums are operating because they are considered to be an important socio-cultural part of our society. This leaves the questions of how to keep the "places of social innovation" active, making them safe places so they can promote resilient development. Perhaps, is it time to revaluate the true risks and benefits of different places and make the restrictions more appropriate?

Acknowledgments. The author of this article acknowledges the support provided by the research incentive program Aulet 13 of the Faculty of Architecture and Landscape of the Leibniz University Hannover. The author would also like to thank Alberto Plata Montero, Paloma Pastor Rey de Viñas, Carlo Roccafiorita, Juan Ignacio Lasagabaster, Eduardo Lomo Vadillo, Valentin Angulo, Macarena Ruiz Redondo, Pablo de Oraá, and many other actors and citizens for providing relevant information for the research.

References

Carta, M., Ronsivalle, D., Lino, B.: Inner archipelagos in sicily. From culture-based development to creativity-oriented evolution. Sustainability **12**, 1–16 (2020)

Clark, J., Wise, N. (eds.): Urban Community and Participation: Theory, Policy and Practice. Springer, Berlin (2018). https://doi.org/10.1007/978-3-319-72311-2

Dorobantu, A.D.: Matei a social economy – added value for local development and social cohesion. Proc. Econ. Financ. **26**, 490–494 (2015)

Maiolini, R.: Lo stato dell'arte della letteratura sull'innovazione sociale. In: Caroli, M. (ed.) Modelli ed Esperienze di Innovazione Sociale in Italia, Franco Angeli, Milano, pp. 23–37 (2015)

Mangialardo, A., Micelli, E.: The role of the social entrepreneur in bottom-up enhancement of Italian public real-estate properties. In: Mondini, G., Fattinnanzi, E., Oppio, A., Bottero, M., Stanghellini, S. (eds.) SIEV 2016. GET, pp. 569–577. Springer, Cham (2018). https://doi.org/10.1007/978-3-319-78271-3_45

Micelli, E.: Enabling real property. How public real estate assets can serve urban regeneration. Territorio **87**, 93–97 (2018)

Moulaert, F., Martinelli, F., Swyngedouw, E.: Towards alternative model(s) of local innovation. Urban Stud. **42**(11), 1969–1990 (2005)

Phills, J.A., Deiglmeier, K., Miller, D.T.: Rediscovering social innovation. Stanf. Soc. Innov. Rev. **6**(4), 34–43 (2008)

Scaffidi, F.: Soft power in recycling spaces: exploring spatial impacts of regeneration and youth entrepreneurship in Southern Italy. Local Econ. **34**(7), 632–656 (2019)

Scaffidi, F.: Social innovation in productive assets redevelopment: insights from the urban development scene. In: Bevilacqua, C., Calabrò, F., Della Spina, L. (eds.) NMP 2020. SIST, vol. 178, pp. 1003–1011. Springer, Cham (2021). https://doi.org/10.1007/978-3-030-48279-4_94

Schröder, J.: Open habitat. In: Schröder, J., Carta, M., Ferretti, M., Lino, B. (eds.) Dynamics of Periphery, pp. 10–29. Jovis, Berlin (2018)

Tricarico, L., Jones, Z.M., Daldanise, G.: Platform spaces: when culture and the arts intersect territorial development and social innovation, a view from the Italian context. J. Urban Affairs 1–22 (2020)

Impact Assessment for Culture-Based Regeneration Projects: A Methodological Proposal of Ex-post Co-evaluation

Maria Cerreta[1], Ludovica La Rocca[1], and Ezio Micelli[2(✉)]

[1] Department of Architecture, University of Naples Federico II, via Toledo 402, 80134 Naples, Italy
{maria.cerreta,ludovica.larocca}@unina.it

[2] Department of Culture del Progetto, Università IUAV di Venezia, Santa Croce 191 Tolentini, 30135 Venice, Italy
ezio.micelli@iuav.it

Abstract. The emergence and diffusion of social impact measurement for assessing culture-led regeneration projects underline the need for suitable approaches and tools to support public institutions and local administrations to guide their implementations. The paper introduces a methodological proposal for social impact assessment, structured and tested within the framework of the research agreement between the Directorate-General for Contemporary Creativity (DGCC) of the Italian Ministry of Culture (MiC) and the IUAV University of Venice, with the purpose to evaluate the impacts generated by the projects implemented in Italy through the public calls "Creative Living Lab" and "PrendiParte", initiatives born in 2018 to finance and support urban regeneration projects through cultural and creative activities.

Keywords: Urban regeneration · Cultural innovation · Impact assessment

1 Introduction

In recent years, Impact Assessment (IA) and Social Impact Assessment (SIA), in their various interpretations and different techniques, has also spread in Italy both thanks to scholars interested in the topic and in response to requests for new financiers (foundations and tenders promoted to support activities of social interest), as well as following the requests of the legislator, starting from Law 106/2016 and formalized in the Guidelines for Social Impact Assessment (Legislative Decree 24 July 2019). The SIA is closely related both to the reporting processes relating to the social dimensions of companies (from the reporting methods of the Corporate Social Responsibility of companies, to the social balance sheets of non-profit entities), as well as to the processes of evaluating investments, services and public policies developed to verify the efficiency and effectiveness of the process [1, 2]. For these reasons, methods and methodologies for reporting and assessing the social impact have multiplied over the years [3, 4], spreading approaches and techniques capable of measuring the value created for all stakeholders and highlighting some significant considerations: we are witnessing a fundamental cultural change that requires the need for both private and public "accountability"; the awareness is spreading that we live in a world of "scarce resources", and this leads companies to demonstrate the value created and disseminated; new regulations are emerging related to the reform of the third sector (Directive 2014/95/EU), which imposes the obligation of non-financial reporting for "large" companies; new sectors of finance are born (such as social finance) that link investments and social objectives, with the consequent need to describe, measure and evaluate the return on investments, but above all the impact determined on stakeholders; new needs for external communication and transparency of processes are emerging by third sector entities and public administrations for which the measurement of impacts assumes a significant role in assessing the quality and effectiveness of performance; new calls are structured for urban regeneration projects with innovative formats in which impact assessment is considered a necessary and complementary action to implementing services and monitoring activities. The emergence and diffusion of the issue of social impact measurement find their presuppositions in the transition phase from a Welfare State model to a Welfare Society model [5–7], two welfare systems based respectively on the principle of redistribution and on the principle of circular subsidiarity, where the second overcomes the public-private dichotomy and includes the private social sector that performs a public function, generating goods and services of general interest. The transition from the logic of production and provision of services to that of shared production with the beneficiaries of the services in co-production identifies a change of perspective and makes the evaluation process central. Therefore, if in the past it was sufficient to "report", in the sense of accounting for the use of resources, in the era of generative welfare, it is essential to evaluate and "give value", to restore a vision of value that is not only assessed from output and performance but evidence linked to the change generated by the processes activated and implemented. In this perspective, the paper introduces a methodological proposal for social impact assessment, structured and tested within the framework of the research agreement between the Directorate-General for Contemporary Creativity (DGCC) of the Italian Ministry of Culture (MiC) and the IUAV University of Venice, with the purpose to evaluate ex-post the impacts generated by the projects implemented in Italy through the public calls "Creative Living Lab" and "PrendiParte", initiatives born in 2018 to finance and support urban regeneration projects through cultural and creative activities. The research group[1] dealt with the national mapping of cultural-based urban regeneration projects and assessing their social impact, highlighting the need for an integrated ex-post co-evaluation approach.

The paper has been articulated in Sect. 2 related to Materials and Methods; in Sect. 3 that describes the methodological proposal of ex-post co-evaluation applied to the selected projects; in Sect. 4 that discusses the results and defines some conclusions oriented to the replicability of the evaluative approach.

---

[1] The research group, coordinated by prof. Ezio Micelli, is composed by dr. Elena Ostanel, dr. Francesco Campagnari and arch. Alessia Mangialardo of the IUAV, by prof. Maria Cerreta and arch. Stefano Cuntò, Ludovica La Rocca, Chiara Mazzarella, Eugenio Muccio, Stefania Regalbuto and Sabrina Sacco of the University of Naples Federico II.

2   Material and Methods

As part of the research agreement between the Directorate-General for Contemporary Creativity (DGCC) of the Italian Ministry of Culture (MiC) and the IUAV University of Venice, "National mapping of cultural-based regeneration, analysis of the calls of the Ministry of Culture and other entities and evaluation of their social impact", the research program aimed to elaborate and interpret the data relating to urban regeneration projects carried out in Italy and promoted with the support of the MiC. The research has been based on examining a sample of projects selected by some national and regional competitions relating to the reuse and regeneration of urban spaces through practices of social innovation, cultural promotion, participation and active citizenship. The study focused on the transformation paths of culture-based places to establish relationships with communities longer time and a link with urban spaces [8]. To achieve the objectives of the SIA, a methodological approach was defined in an ex-post co-evaluation perspective, to be applied through the involvement of the identified stakeholders, to obtain an orientation framework of the effects generated by the case studies selected in the related territories, as well as a comparison among the thematic clusters that cases represent. To define the most consistent procedure with the requests of the MiC, the impact assessment procedures present in the literature have been analyzed, excluding those that are sectorial, such as performance evaluation, charity evaluation and finance micro-activities. Starting from the Theory of Change (ToC) and reflections on the Impact Value Chain [9–11], an ex-post co-evaluation process has been structured (see Fig. 1).

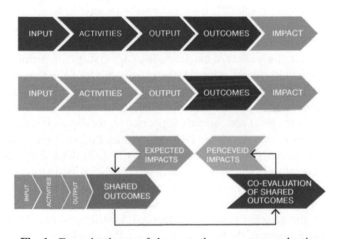

Fig. 1. From the theory of change to the ex-post co-evaluation

The ToC is a versatile approach suitable for evaluating projects that promote social change through the participation and involvement of communities. Furthermore, the ToC, through the Impact Value Chain, considers it essential to map all the conditions necessary to produce a specific impact, structuring a functional approach, especially in the ex-ante phase. The production of changes (*impacts*) is triggered by the investment of resources (*inputs*) and activities following a linear procedure. On the contrary, assessing the impacts themselves requires a backward path, which moves from long-term to intermediate (*outcomes*) and short-term (*output*) results, according to the logic of ex-post evaluation. Therefore, in the methodological proposal tested and described in this paper, the Impact Value Chain has been reinterpreted, analyzing only the impact objectives and outcomes. Outputs, activities and inputs have not been considered because they were inconsistent with the purposes of the MiC. The adaptation of the ToC, by examining the outcomes and impacts, has made it possible to assume a perspective of interpretation and evaluation of circular projects. The impacts help to trace the outcomes and, in turn, the outcomes allow to make explicit the impacts. Therefore, it has been possible to develop a versatile methodological approach suitable for evaluating projects that promote social change through the participation and involvement of communities.

3   Ex-post Co-evaluation for Culture-Based Regeneration Projects: A Methodological Proposal

The first part of the research has been aimed at examining the data extrapolated from the projects shared by the MiC, referring specifically to three editions of two calls: PrendiParte (2018) and Creative Living Lab call (2018 and 2019), with the intent to search within the dataset the processes intended to last over time in the territory and with a link with the related places (area-based, building-based, network-based).

Among the delivered projects, 141 projects were found consistent with the focus of the analysis. The main characteristics of the candidate projects emerge from the descriptive analysis of the case studies: 1. they are mainly located in Northern Italy and the peripheral areas of large urban centres; 2. they are configured as hybrid experiences, based mainly on cultural action of proximity, with services on a neighbourhood and municipality scale; 3. the leaders of the proposing partnerships are mainly associations, leading networks with different actors; 4. the candidate actions are mainly oriented to an urban area, starting from reference buildings of medium-small size, often publicly owned, with significant private shares; 5. the projects in which the actions are inserted are experiences, for the most part, already supported by local policies, therefore mainly municipal, with significant national and regional shares. For the impact assessment, six case studies were selected from the winners of the MiC calls (CLL 2018, 2019, Prendi-Parte 2018), taking into account: homogeneous geographical distribution on the national territory, selecting two cases per macro-region; criteria linked to the representativeness and diversification of the main variables that characterize the different projects.

In particular, the selected cases are: Mapping San Siro, Milan; Mercato Sonato, Bologna, Emilia Romagna; PostO, Rome, Lazio; Palombellissima/Hipnic, Ancona, Marche; #CUOREDINAPOLI, Naples, Campania; GOS Distillerie Culturali, Barletta, Puglia. Starting from these six case studies identified, an ex-post co-evaluation of the impacts generated has been carried out in the second part of the research. The main objectives of the evaluation process were: 1. return the DGCC-MiC an intuitive, easy and replicable assessment tool over time; 2. involve in the evaluation the different stakeholders who, for different reasons, have interfaced with the projects analyzed; 3. grasp the effects and, therefore, the changes triggered on the territories, using both quantitative and qualitative metrics. To achieve these objectives, it was necessary to consider a relatively short time of observation (from 2 to 3 years) between the activation of the project (2018, 2019) and that of the impact assessment (2021), in addition to the influence of the Covid-19 health emergency on projects triggered.

In the perspective of a collaborative evaluation, that allows developing a co-evaluation approach [12, 13] starting from a re-elaboration of the ToC (see Fig. 2), able to include the points of view of the main stakeholders involved in the six selected projects, three types of roles have been identified with relative three different typologies of impact concerning which everyone was invited to express:

1. Process Managers (PM), as they are the most relevant experts of the dynamics implemented by the MiC in the project evaluation phase and able to explain the expected impact of the DGCC in the evaluation phase of each project;
2. Project Managers (MP), as they are the most relevant experts of the selected winning projects and able to explain the potential impact, interpreted by the expected goals of the DGCC and potentially generated on their territory through the projects activated;
3. Users (US) are subjects directly involved in using the services and experiences offered in the area and able to express the perceived impact of the activated project.

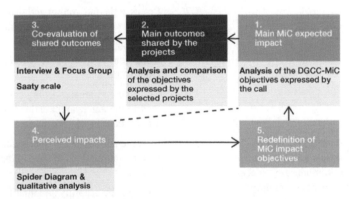

Fig. 2. The methodological approach of the ex-post co-evaluation: phases and tools

Starting from a cyclical interpretation between impacts and outcomes, the phases and related tools of the ex-post co-evaluation process have been defined: 1) selection of the main impact objectives; 2) identification of the primary outcomes; 3) co-evaluation of outcomes; 4) analysis and interpretation of the emerged data.

In the second phase, the common central outcomes of the six selected projects were identified, analyzing and comparing the objectives expressed in the tender phase, at the time of writing and applying the project to the call of interest. The purposes most shared by the projects analyzed have been grouped into thematic clusters concerning which the ex-post co-evaluation process has been developed. In this way, the outcomes were understood, in a sort of ex-post reconstruction, as sub-objectives, defined by the Project Managers and specified based on the related territories and of the main impact objectives expected by the MiC, set in the phase of writing the call. The four main impact objectives that emerged from the analysis of the purposes shared by the MiC through the "Creative Living Lab" and "PrendiParte" calls were:

1. Culture and Creativity (encourage the development and implementation of cultural and creative, innovative and experimental activities);
2. Collaborative Regeneration (encourage the development of collaborative urban regeneration processes for the reuse of disused spaces);
3. Innovative Entrepreneurship (favouring the development of new processes, business ideas and increasing the employment level);
4. Partnerships and networks (favouring the creation and strengthening of multidisciplinary territorial networks).

Table 1. The 4 main impact objectives and the 13 outcomes

Objectives	Outcomes	
1. Culture and Creativity Implement and diversify cultural, creative and social activities	1.1	Promoting the inclusiveness and accessibility of the artistic-cultural offerings
	1.2	Promote the integration and diversification of local cultural and creative actions
	1.3	Encouraging the integration of technological tools in the cultural and creative field
	1.4	Contribute to greater ecological and environmental awareness through artistic-cultural experiences
2. Collaborative Regeneration Increase the reuse and transformation of abandoned and/or underused spaces through processes of community participation and involvement	2.1	Increase and revitalize intergenerational and intercultural exchanges in regenerated spaces
	2.2	Increase the usability and quality of spaces ensuring accessibility to all
	2.3	Increase the attractiveness of places by encouraging contamination between local and temporary communities
3. Innovative Entrepreneurship Encourage the birth and/or growth of entrepreneurship in the cultural and/or social sphere	3.1	Activate enabling and professionalizing environments in the creative and cultural field
	3.2	Encourage the development of an entrepreneurial spirit in the cultural and/or social field in its territory

(continued)

Table 1. (*continued*)

Objectives	Outcomes	
4. Partnerships and Networks Encourage the establishment of multidisciplinary territorial networks and empowerment processes	4.1	Implement paths of co-design of territorial activities
	4.2	Stimulate paths of comparison and multidisciplinary learning in the artistic and cultural field
	4.3	Establish new networks, lasting over time, among local realities (e.g. associations, administration, schools, universities, etc.)
	4.4	Promote channels of collaboration between territorial networks and public administrations

The relevant outcomes extracted from the four impact objectives declared by the Project Managers in the candidate projects were connected to each of them (see Table 1). The 13 outcomes were translated into 13 questions to be submitted to the three different types of stakeholders identified for each project. In the third phase, the process of ex-post co-evaluation of the outcomes was activated through interviews and focus groups. The outcomes were explained in 13 structured questions and 2 open questions, for a total of 15 questions, submitted to the three typologies of stakeholders (PM, MP, US) for each of the six projects. The interviews were conducted online for approximately 60 min each and were facilitated by an interviewer and rapporteur. The interviewees were presented with two introductory slides on the research topics and the evaluation system, 13 slides dedicated to as many closed-ended questions and 2 slides dedicated to the respective open-ended questions, submitted to obtain different outcomes not considered and/or information on possible comparison with the Italian legislation on urban regeneration. At the beginning of each interview, the Project Managers were informed that a randomized control test could be activated on the users of their project through a Focus Group to encourage the PM to make the most objective assessment possible. The Users of a specific project were interviewed in a dedicated Focus Group, thus activating a complete control test among the six selected projects aimed at integrating the points of view of the three different stakeholders for at least one of the chosen case studies. The Users were first asked to answer the questions individually and then share their motivations as a group to reach a shared judgment. All interviewees expressed their opinion for each question using the Saaty scale [14] and commented on the two open-ended questions. During the interviews, the motivations and reflections that emerged were also recorded, collecting valuable qualitative information to complete the knowledge framework and perfect the judgment expressed with the Saaty scale. In the fourth phase, the data that emerged were analyzed and interpreted, summarizing the results in a spider diagram (named Kiviat or radar diagram too) [15, 16], drawn up for each project carried out to represent the judgments that emerged as clearly as possible and being able to compare them among projects (see Fig. 3).

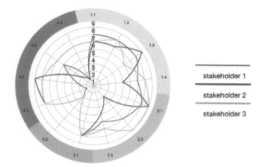

Fig. 3. Example of spider diagram elaborated for each project

The vertices of the spider diagram coincide with the 13 outcomes specified for the 4 main impact objectives. To facilitate the critical reading of the graph, convergence analysis of the points of view was set up. The average between the opinions expressed by the different stakeholders represents a "convergence index" (CI) which, based on the numerical range of belonging, implies a greater or lesser correspondence between the interviewees concerning the outcome investigated (convergent, moderately convergent, not very convergent or not convergent). In addition, text boxes have been elaborated, dedicated to interpreting the emerged qualitative data, grouped by keywords to which the interviewees showed particular convergence or divergence. The approach thus defined has been tested on the six selected case studies to be discussed and approved by the DGCC of the MiC.

4 Discussion and Conclusions

The analysis of the results of the methodological approach proposed for the social impact assessment, structured using an interpretation of the ToC, supported by an ex-post co-evaluation process, based on interviews and focus groups and summarized by spider diagrams, convergence analysis and qualitative descriptions, made it possible to understand the impacts perceived and assessed through the Saaty scale (1–9).

From the results (see Fig. 4), it is evident that those outcomes relating to new employment and innovative entrepreneurship, shared between Project Managers and Process Managers, are all last position in the ranking of satisfaction. On the contrary, community involvement is the outcome with a greater degree of shared satisfaction.

In addition, some issues that have emerged, relating to the investigation of the impacts, are particularly interesting for the MiC, such as: "supporting processes that have already started or new projects", which implies financing organizations that have already been structured over time in the reference area; "the role of technology and ecology in the cultural and artistic offer", an impact not particularly expected by the PM and MP, yet satisfied mainly in the implementation of the activities; "employment in the cultural sphere", declared not particularly satisfied from all three heard points of view. The methodological approach proposed for the assessment of social impact, defined under the requests of the MiC, appears to be a streamlined procedure, which simplifies

CULTURE AND CREATIVITY		COLLABORATIVE REGENERATION		INNOVATIVE ENTREPRENEURSHIP		PARTNERSHIPS AND NETWORKS	
Expected impacts (PM)	Value	Generated impacts (MP)	Value	Perceived impacts (US)	Value		
1.1 Inclusiveness and accessibility	8	1.3 Technological innovation	9	2.1 Community involvement	9		
1.2 Integration and diversification	8	2.1 Community involvement	9	2.2 Usability and accessibility	9		
1.3 Technological innovation	8	4.2 Multidisciplinarity	9	1.2 Integration and diversification	8		
2.1 Community involvement	8	3.1 Hard and soft skills transmission	8	1.4 Ecological awareness	8		
3.1 Hard and soft skills transmission	8	4.1 Cooperation	8	3.1 Hard and soft skills transmission	8		
1.4 Ecological awareness	7	1.1 Inclusiveness and accessibility	7	4.1 Cooperation	8		
2.2 Usability and accessibility	7	1.2 Integration and diversification	7	4.2 Multidisciplinarity	8		
2.3 Attractiveness and identity	6	1.4 Ecological awareness	5	1.1 Inclusiveness and accessibility	7		
4.3 Economic and social productivity	6	2.3 Attractiveness and identity	5	2.3 Attractiveness and identity	7		
3.2 Generativity and employment	3	2.2 Usability and accessibility	3	4.3 Economic and social productivity	7		
4.2 Multidisciplinarity	3	3.2 Generativity and employment	3	4.4 Trust in institutions	7		
4.1 Cooperation	2	4.3 Economic and social productivity	3	1.3 Technological innovation	6		
4.4 Trust in institutions	2	4.4 Trust in institutions	2	3.2 Generativity and employment	3		

Fig. 4. Rankings of the satisfaction of the outcome

the Theory of Change by focusing on the impacts generated by the projects supported by the DGCC-MiC in the medium-long term. At the same time, it is a co-evaluation tool that involves different points of view interested in projects in different ways in the assessment process [17, 18]. The proposed and tested approach is easily replicable over time since it uses a survey submitted through interviews and focus groups on obtaining comparable quantitative data, expressed through the Saaty scale, and qualitative data to grasp the reasons for the opinions expressed. Furthermore, the descriptive analyses that conclude the co-evaluation allow to identify some themes for reflection that could guide the future policy of the MiC: the predominantly local nature of urban regeneration experiences highlights how policies, target audiences and networks operate on an urban and neighborhood scale. Therefore, it becomes essential to understand how the MiC, operating at a national level, can support these experiences and how it can interact constructively with local institutions; the organizational fragility of the experiences of urban regeneration underlines how the associative form is the most widespread, but also the one that presents strong limits of scalability and consolidation over time; the hybridization of urban regeneration experiences among artistic forms, types of cultural activities (production, use), and cultural services and others highlights how culture represents the driving force behind these experiences, which interfaces with other services by creating social and community infrastructures. However, maintaining these services is difficult due to the absence of financing systems that take this integration into account. The ex-post co-evaluation process, elaborated and tested for culture-based regeneration projects impact assessment, represents the first result of a research activity oriented to answer to the MiC requests and, at the same time, improve the quality of the urban regeneration projects and their implementation and management, where the involvement of the different stakeholders is essential to facilitate the partnership activation and the mutual learning. The proposed approach considers the relevance of collaborative evaluation in the ex-post assessment to improve the ex-ante and ongoing assessments.

References

1. Vanclay, F.: International Principles for Social Impact Assessment: their evolution. Impact Assessment and Project Appraisal **21**(1), 3–4 (2003). https://doi.org/10.3152/147154603781766464

2. O'Flynn, M.: Impact assessment: understanding and assessing our contribution to change. M&E Paper **7**, 10 (2010). https://scholar.google.com/scholar?hl=it&as_sdt=0%2C5&q=O%E2%80%99Flynn+M%2C+Impact+assessment%3A+understanding+and+assessing+our+contributions+to+change&btnG=. Accessed 30 Dec 2021

3. Gray, R.: Thirty years of social accounting, reporting and auditing: what (if anything) have we learnt? Bus. Ethics Eur. Rev. **10**(1), 9–15 (2001). https://doi.org/10.1111/1467-8608.00207

4. Arena, M., Azzone, G., Bengo, I., Calderini, M.: Measuring social impact: the governance issue, May 2015. https://irisnetwork.it/wp-content/uploads/2015/06/colloquio15-arena-azzone-bengo-calderini.pdf. Accessed 30 Dec 2021

5. Skillen, A.: Welfare state versus welfare society? J. Appl. Philos. **2**(1), 3–17 (1985). https://doi.org/10.1111/J.1468-5930.1985.TB00015.X

6. Zamagni, S., Venturi, P., Rago, S.: Valutare l'impatto sociale. La questione della misurazione nelle imprese sociali. Impresa sociale **6**, 77–97 (2015). http://rivistaimpresasociale.s3.amazonaws.com/uploads/magazine_issue/attachment/7/ImpresaSociale-06-2015.pdf#page=65. Accessed 14 Dec 2021

7. Cicerchia, A., Rossi Ghiglione, A., Seia, C.: Welfare culturale, June 2020. https://scholar.google.com/scholar?hl=it&as_sdt=0%2C5&q=Cicerchia%2C+A.%3B+Rossi+Ghiglione%2C+A.%3B+Seia%2C+C.+Welfare+Culturale.&btnG. Accessed 30 Dec 2021

8. Sacco, P., Ferilli, G., Tavano, B.: Understanding culture-led local development: a critique of alternative theoretical explanations. Urban Stud. **51**(13), 2806–2821 (2014). https://doi.org/10.1177/0042098013512876

9. Lcat, D.: Theories of social change. In: International Network on Strategic Philantropy. Bertels-mann Stiftung, Washington DC (2005). https://www..scholar.google.com/scholar?hl=it&as_sdt=0%2C5&q=Theories+of+Social+Change.+In-ternational+Network+on+Strategic+Philanthropy+2005&btnG=. Accessed 30 Dec 2021

10. Vogel, I.: Review of the use of 'Theory of Change' in international development. Review Report (2012)

11. Valters, C.: Thcories of change in international development: communication, learning, or accountability. JSRP Paper **17**, 1–29 (2014). https://www.alnap.org/system/files/content/resource/files/main/jsrp17-valters.pdf

12. Cerreta, M., Elefante, A., La Rocca, L.: A creative living lab for the adaptive reuse of the morticelli church: the SSMOLL project. Sustainability **12**(24), 10561 (2020). https://doi.org/10.3390/su122410561

13. Cerreta, M., Daldanise, G., La Rocca, L.: Triggering active communities for cultural creative cities: The 'Hack the City' play ReCH mission in the salerno historic centre (Italy). Sustainability **13**, 21 (2021). https://doi.org/10.3390/su132111877

14. Saaty, R.W.: The analytic hierarchy process-what it is and how it is used. Math. Model. **9**(3–5), 161–176 (1987). https://doi.org/10.1016/0270-0255(87)90473-8

15. Centis, L., Micelli, E.: Regenerating places outside the metropolis. a reading of three global art-related processes and development trajectories. Sustainability **13**(22), 1–23 (2021). https://doi.org/10.3390/su132212359

16. Mangialardo, A., Micelli, E.: La partecipazione crea valore? Modelli di simulazione per la valorizzazione dal basso del patrimonio immobiliare pubblico. Valori e Valutazioni **19**, 41–52 (2017). https://www.siev.org/wp-content/uploads/2020/02/19_04_Mangialardo-and-Micelli.pdf. Accessed 30 Dec 2021

17. Cerreta, M., Giovene di Girasole, E.: Towards Heritage Community Assessment: Indicators Proposal for the Self-Evaluation in Faro Convention Network Process. Sustainability **12**(23), 9862 (2020)
18. Cerreta, M., Muccio, E., Poli, G., Regalbuto, S.: Verso un Modello Città-Porto Circolare: Un Sistema di Supporto alla Decisione Multidimensionale e Multiscalare Per Napoli Est. Laborest **22**, 57–63 (2021)

"As Found". The Reuse of Existing Buildings with an Identity Character as a Fundamental Element of Regeneration in the New Community Centers' Design. A Brutalist Building as a Case Study

Francesca Ripamonti[✉]

Politecnico di Milano, DAStU, via Bonardi 3, 20133 Milan, Italy
francesca.ripamonti@polimi.it

Abstract. The places of primary health and social care, at a territorial level, can and must be, through their design, elements of community building and urban regeneration. Among the essential characteristics to direct the design actions towards this goal, it was decided to investigate the reuse of existing buildings with an identity character as a preferential location for the new Community Centers.

The analysis of the case study of a building belonging to the Brutalist current, and originally inserted in a broader plan for a Civic Center, for which the re-functionalization in a Community Center is foreseen, becomes a pretext for a reasoning on the correlation between the health of the person and healthiness of the environment, on the need to think of each new project as an identity element for the citizen, as an individual and as a member of a community, and as a builder of new urban centralities, as well as on the active role that the community, with the help the designer, can have in the care of existing places and in the foreshadowing of future places.

Keywords: Urban and social regeneration · Building reuse · Sustainability

1 Premise

The proposed reflection starts from the collaboration in the research project *Coltvare-Salute.Com* [1] (funded by the Politecnico di Milano through the Polisocial Award 2020, the University's program of commitment and social responsibility [2]) as well as from the desire to deepen this study through a doctoral thesis.

The fundamental role of primary health care, made evident both in its weaknesses and in its potentialities by the pandemic crisis still underway, has become part of the civil and political debate, especially following the drafting of the NRRP (Italian National Recovery and Resilience Plan) and the funding allocated by this plan to the constitution of new health and community facilities at the territorial level, called "Community Centers" (*Case della Comunità*).

In Italy, these Centers were already established by the Ministry of Health in 2006. They were, and are still called "Healthcare Centers" (*Case della Salute*) and their objective was to have an active territorial system that would keep the health, social-health and social aspects of the person in integrated management and define, through these centers, prevention programs and health education, together with a conscious path towards autonomy and active participation of citizens [3].

It is in the pre-NRRP scenario that the research *"ColtivareSalute.Com. Cities and Healthcare Centers for Resilient Communities. The Healthcare Centers as builders of urbanity and widespread sociality in the post- COVID-19 era: new peripheral centralities in healthy and integrated cities."* started its analysis of this Healthcare system.

Having as a privileged test bench the region Emilia Romagna, albeit at the forefront of the national panorama for what concerns the Healthcare Centers, the research project started from the observation of the fragility of this system (due to the lack of a localization strategy, to the disconnection from the network of public services, to the absence of recognition of these structures by citizens…), as well as of its potential, summarized in the ability of Healthcare Centers to become nodes of urban and social regeneration.

Within a necessarily multi-disciplinary research, the reflection on the design dimension, at the architectural, urban, interior and open space scale, identifies in the Healthcare Centers an ideal symbiosis between three decisive actions: building communities; regenerating cities; implementing and communicating the network of public services to bring out the potential and articulated system in which citizens, as individuals and as a community, can be inserted.

The research aims to define guidelines for the design of these Centers, which it defined as "Community Healthcare Centers" (*Case della Salute della Comunità*), anticipating the official name of Community Centers, in an urban context, identifying national and international best practices, using selection criteria which can later be set as objectives to be achieved in the design process.

According to this criteria, a good project for a Community Center is such when: it contributes to defining urban centrality, it is the fulcrum of peripheral districts, it structures the surrounding urban public space, it is inserted in a system of green spaces, it promotes environmental sustainability, it is integrated with other public services, it guarantees the quality of the organization of the interior space and its flexibility and adaptability (especially from a pandemic point of view), and reuses existing buildings with high identity features for its community.

Starting from the observation that, just as it is not always possible to respect all these criteria (and this was evident in the selection of best practices), they are equally often closely correlated, the reflection proposed aims to focus on the last point: recovery of existing buildings with an identity character as a fundamental element of regeneration, in the project of the new Community Centers.

2 Community Centers as an Enhancing and Regenerative Element of the Existing Public Heritage

In a country like Italy, strongly characterized by a large presence of an existing – often abandoned or disused - public heritage, the use of an existing building to host any new function can be a form of economic and environmental sustainability.

As regards the location of a Community Center, this choice implies a reasoning on the settlement logic, and therefore cannot be dictated by the choice of an existing building only as it is of public property and in disuse. The need to adhere to pre-existing settlement and urban planning logics must therefore be taken into consideration, possibly implementing, through the project, any existing connections: whether they are soft mobility paths, a system of green spaces and public places, or the proximity or networking of other health, cultural, social services of interest to the community.

The choice of existing buildings and public spaces, within the heritage of unused public buildings, must keep in mind the primary objective of activating urban and social regeneration processes that respond to the now essential requirements of healthiness and sustainability.

The regeneration of a site also passes through its recognition as an urban centrality, an attractive and recognizable place for a community. For this reason, it is a great strength, that the building we chose to intervene on has an identity character, recognizable architectural and spatial characteristics, whose intrinsic vocation is compatible with the new functional program, and is already part of a collective memory.

3   The Case of the Proposed Transformation of the Former INAM Building in Paderno Dugnano, Milan, into a Community Center

However, there are cases in which the existing building, from the recovery of which the establishment of the new Community Center is expected, is as well known as it is not appreciated by the citizens.

This is the case of Palazzo Sanità, formerly Palazzo INAM (National Institute for Health Insurance, dissolved in 1974), in Paderno Dugnano - a town of nearly 50,000 inhabitants, of the Metropolitan City of Milan - for which it was made official, within the regional plan for the distribution of funds deriving from the NRRP, the conversion into a new Community Center [4].

3.1   The Building Origins: A Plan for a (Brutalist) Civic Center

The building project fits into a context that saw the typology of Civic Centers at the heart of the planning action of the new municipal administrations in the early 1960s.

At that time, architect Marco Romano was entrusted with the tasks for the design, as well as the exploration and proposal of a functional program, for the Civic Centers of three municipalities in the Milan hinterland: Bollate, Bresso and, precisely, Paderno Dugnano.

In an article for the magazine *Spazio e Società* in 1980, in which the designer reconstructs in detail the project and the administrative and political dynamics that led to its final modification, he asserts: "*What the civic centers could then consist of was not completely clear at that moment, but the bare denomination seemed beautiful and promising anyway*" [5].

Due to the urban and social structure of the town -at the time, as well as today, crumbling into multiple, more or less recent, settlements, then still even a few kilometers away from each other- the planning and design direction moved towards the definition of a single Civic Center, which would concentrate in itself all the services and functions at the municipal level which then lacked, if at all, at least centralization.

For this reason, unlike the projects of Civic Centers for the other two municipalities, which remained at the scale of the building, here the architect developed a real urban plan that provided for the gradual construction of all the aforementioned equipment, concentrated in only one place, although of urban scale, in order to *"enhance accessibility by the whole sector and allow an immediate alternative use"* [5].

Having in mind international experiences, such as the new centers of northern European *satellite cities* and *new towns* (Cumbernauld in the first place, but also Farsta, Vallinby, Stevenage) and the phenomenon of large shopping centers in the desert, such as the Taunus in Frankfurt, *"Which was believed to be soon to be built also in Italy in the wake of the prospects for neo-capitalist transformation"* [5] the project made its basic scheme, in which all the services, such as cinemas, theaters, drugstores, restaurants, etc. gravitated around a single longitudinal transport system.

The project (signed by architect Hugh Wilson) of the Civic Center of Cambernauld, Scottish New Town established in 1958 to cope with the demographic increase of the city of Glasgow, was particularly influential, with regard to the idea of keeping motorized traffic separated from pedestrian paths, maintaining the underground parking and subsequently develop the building on multiple overlapping pedestrian levels, to connect public, administrative and commercial functions in a unique, compact, urban complex.

The same layout was re-proposed by the architect Romano in the project for the Civic Center in Paderno Dugnano, developed in the winter of 1964. A dual carriageway and four lanes would have crossed it for all its 800 m in length, remaining at about 4 m below the ground level, where you would have had direct access to the underground car parks. On the ground floor, slightly raised above the ground zero level, the main pedestrian axis would have developed, along which one would have had access to the various services of the Center.

Further transversal pedestrian paths would have found space on the first floor, at intervals of 200 m, connecting the center with the external urban fabric, in particular to residential buildings, which are also part of the plan. Starting from this fixed structural scheme, the entire urban plan linked to the Civic Center would then develop over time.

The only definitive location was foreseen for the buildings of direct municipal initiative, pilot constructions of the whole system, located at the head of the Civic Center, towards the city center (Town Hall with the INAM service and the police station) and in the middle (middle School).

These buildings, or rather one of them, were finally, for reasons well described by Romano in the article for *Spazio e Società*, the only part actually built of this large and complex plan.

As memory and trace, even if on a smaller scale, of this much wider project, today only the *"First (and Last) Block of the first Lot"* [5] remains. Built in 1970, it originally included the health offices of the Municipality, the INAM offices and the Police Station.

In 1980, the designer describes this construction, in relation to the largest Civic Center, with these words: *"The political line underlying the very idea of this center is that of a Northern European social democracy of 1960s, which corresponds to the formal character of an architecture of a Calvinist cheerfulness, whose rigor is accentuated by the concern to define an iterable composite scheme for all the eight hundred meters of the future building: rigorously square meshes, constant modules, a repeatable organization of the external surfaces"* [5].

Part of a discourse that would have been broader, the building is organized around a central courtyard: the internal facades and pedestrian walkways overlook a central void that brings light to the complex, connecting visually horizontally and vertically. The user who enters on foot from the city center, via a ramp that, crossing a portal, overcomes the difference in height from the street level, thus finds himself moving in this small space of the city without crossing cars, which he can observe from above, starts of a frenetic world, parallel but divided from his. *"Everything in the project is free and viable, accessible by the people just like in the Nordic social democracies"* [5].

The visual permeability of the building is added to the physical permeability of the project, thanks to the skillful use of full-height openings, both towards the internal courtyard and towards the external fronts of the building, characterized by a continuous passage protected by a concrete balcony to which a seat is integrated. Each clinical room and office had and has this small outdoor area of relevance.

External paths designed from an urban point of view thus connect the different levels, also connected internally and via an elevator, of a building born as part of the city, created to host functions of high civic, social and health value, and therefore open and prepared to the free use of the citizen (Fig. 1).

3.2 The Building Protection Proposal

Citizens have never loved the former Palazzo Inam, now called Palazzo Sanità, seat of the territorial polyclinic of the ASST Rhodense [6] and some city associations (A.N.P.I., AVIS, Civil Protection, Red Cross, among others).

They never loved it to the extent that there is often a *"negative perception of brutalist buildings by a large part of the citizens"* [7].

The main element of disaffection, if not distrust, is given, in this, as in many other cases of brutalist architecture in exposed reinforced concrete, by the aesthetic deterioration of the materials, which have not been maintained over the years.

Even before the opening of the debate regarding the conversion of this building into the Community Center, the hypothesis of its demolition by the municipal administration had not shocked public opinion.

On several fronts, however, an interest has rekindled in this brutalist stable, well present in the lives of citizens, and a proposal for its protection has taken place: a case of active citizenship in all respects.

In the summer of 2021, it was presented by a group of citizens and professionals, led by the architect Patrizia Borghi, former manager of the Public Works sector of the Municipality, and, among others, the former mayor Gianfranco Massetti, a letter with the request for the enhancement of the property to the Superintendence of archeology, fine arts and landscape of the Metropolitan City of Milan, to protect it once and for all from the risk of demolition.

At the same time, the building was registered in the SOSBRUTALISM [8] archive, an online portal sponsored by the DAM (Deutches Architektur Museum) for the safeguarding of brutalist architectural artefacts. This operation took place thanks to the synergy between several professionals (architect Marco Romano, engineer Ottorino Pagani, arch. Giovanni Maffioletti and arch. prof. Gionata Tiengo) and the initiative and interest of a student of the latter, attending the Liceo Artistico Boccioni of Milan, who indicated the building as an object of study. In addition to fostering the objective of protecting the property, this synergy has given rise to a multidisciplinary educational project linked to its study, from a historical and social, structural and compositional point of view, and to its communication.

In the collaboration between institutions, schools, citizens and professionals and in the desire to study and communicate an artifact of identity interest for its past role (and for the sense of a historical period and of a society of which it is the spokesperson) and, consequently, for its future, an urban, social and community regeneration operation was started even before applying the first design modification.

3.3 The Reuse Project and the New Community Center

The design proposal for the new Community Center, from the recovery of Palazzo Sanità, formerly Palazzo Inam, starts from a recognition of the historical, social, urban and architectural value of this building, as well as from the awareness of how, regardless of these values, its demolition would have a negative environmental impact, due to the dispersion of carbon in the atmosphere.

The re-functionalization is facilitated by the compatibility of the new functional program with the intrinsic vocation of the existing building, formerly the seat of the ASST territorial outpatient clinic, including, among others, a vaccination center and a blood collection point, a family clinic that takes care, between other things, of prevention and health, psychological and social assistance, as well as a center for child neuropsychiatry.

With the goal of creating a building dedicated to Public Health and Wellbeing, the designer arch. Maffioletti and Eng. Pagani, with the collaboration of a group of students and former students of the Architecture School of Politecnico di Milano, have studied a careful refurbishment of the existing spaces starting from the regional guidelines on Community Centers, integrating them with a study of guidelines of other regions, in particular those of Emilia Romagna, more complete from the point of view of the dimensional indication of the functions.

The approach of the group that worked on this preliminary project was focused on the functional and performance rationalization of the building. Attention to economic and environmental sustainability has led to a minimal impact intervention, which interprets the primary vocation of the existing building, expanding its urban dimension through a reasoning on its more public spaces, from the urban access space to the building, to the rooftop garden square.

The basement, with direct driveway access from via 2 Giugno, that same low road that should have continued its course in the original project of the Civic Center, continues to see the positioning of the differentiated car access, with rationalized parking spaces for private cars and housing for existing Red Cross vehicles.

The mezzanine and first floor, with pedestrian access from Viale della Repubblica, and therefore from the city center, undergoes the minimal changes designed to accommodate the new functions or to reconfigure the existing ones according to the new legislation. The only structural intervention is the insertion of an additional access in the form of a walkway in the middle of the existing central patio, for a more rational management of the different accesses, especially as regards the compartmentalized area dedicated to the 24/7 medical guard.

The second floor, added by the project to meet all functional and dimensional expectations, always reachable both by external route and by the internal lift and staircase systems, will see the configuration of the area for prevention and health promotion programs. An open space, covered by a system of glass photovoltaic elements, will allow the use of this top floor for leisure activities, thanks to the presence of a coffee area.

The roof of this top floor will be green roof, while the eaves will host the photovoltaic system in its entirety. All around, the "garden that treats", with the vegetation that accompanies the entire perimeter of the building.

Attention to "biophilic design" connotes the entire intervention, starting from the assumption that a proximity to the natural element brings well-being to the person. The study of all the facades sees the insertion, starting with the use of the existing tanks, of small and medium-sized vegetation. Inside the courtyard, a bamboo grove will find space within a hanging circular structure, within which the new pedestrian passage is planned.

The design proposal is based on a careful inspection of the building: the picture that emerged is that of a building with a load-bearing structure in excellent condition, however characterized by aesthetic degradation as regards the concrete and by an energy-intensive behavior from the point of view of energy efficiency.

The intervention on the building, of which the designer recognizes the historical, social, urban and architectural value, proposes "retrofitting" actions, implementing a technical, spatial and functional adaptation which, without affecting the overall image of the building, leads it to comply with the highest quality standards, placing environmental and economic sustainability at the center of the design issue.

From a technical point of view, this translates into the creation of box-in-box interventions, insulated from the floor to the false ceilings. An intervention on the windows is necessary, replaced with high-performance glass and opaque parts that, however, maintain as much as possible the compositional scheme that characterized the facade of the Brutalist building. In order to ensure greater summer thermal comfort, fixed and mobile screens will be provided both for the existing facades and for those of the new floor, where they will be integrated into the gutter system that will house the photovoltaic panels, thanks to the use of which the new building has an objective of autonomy from an energy point of view.

The Community Center project does not end within the walls of the building, but extends to the surrounding public space. In particular, the reconfiguration of the road axis of via 2 Giugno responds to the same sustainability requirements as the building project. The primary objective is to reduce the flow rate of what is currently a real heat island, providing for a decrease in the road section dedicated to vehicular traffic, currently oversized, as well as a reduction to a single direction. On the side of the carriageway, new parking lots are planned on a draining pavement with recovery of waste water, a bicycle path, and a double row of trees, of different heights.

3.4 The Regenerative Value of a Community Center Project

The project for the reuse of a brutalist building, therefore arises from the impulse given by the need to give a space to a Community Center, a fundamental territorial outpost of primary care, prevention and dissemination of a culture of health that is in the first place well-being of the person, in his individual and community dimension.

Putting the theme of Health at the center, it promotes an architectural, urban, social, community regeneration that frames transformative processes in a sustainable key:

- health and salubrity as closely related elements, first of all, as demonstrated by many of the choices that define the design philosophy of the new Community Centers. A project, today, cannot be called regenerative, much less configured as successful, if it does not respond to the new environmental demands, which are no longer ignorable;
- attention to the need, for the Community Center, to become an identity element on the scale of the neighborhood as much as, in this case, of the whole city, through a restitution of meaning, through re-functionalization (or functional rationalization) and through the architectural project, which, even before its realization, has the power to prefigure scenarios of effective and qualitative use by the community;
- creation of a new urban centrality, in a context that, by its nature, is lacking that (whose only points of reference worthy of note at the urban and collective level are, to date, the new Tilane Municipal Library and public space -arch. Gae Aulenti, 2009- and the Parco Lago Nord -environmental recovery and landscape design project -arch. Maurice Munir Cerasi, 1985) which from this point of view constitutes a precedent for the city and for the imagination of citizens;
- a community and active citizenship lesson, which saw the participation of different professional figures, in order to protect an asset whose historical and social value was finally recognized, demonstrating how those communities to which, not only in the name, but intrinsically in their contents, the Healthcare Centers before and the Community today, want to connect with.

The architectural, urban, social and community regeneration is therefore finally made possible by the architectural imagination, which, certainly approaching other fundamental skills, has an essential role in foreseeing a different future for the places where the community, seeing itself put again in the center, can express itself.

4 "As Found". A Brutalist Building's Lesson

The definitions of brutalism are not unique, and are not limited to the description of reinforced concrete buildings (*beton brut*): *"In England, in 1955, Reyner Banham wrote an article entitled "The New Brutalism" in which he outlined the following characteristics: the building must be defined by the legibility of its plan, make its structure clearly visible, be immediately recognizable in its identity and should enhance the characteristics of the raw material used "as found"* [7].

The renewed interest in brutalist architecture on the part of designers, but not only (as shown by exhibitions, books, investigations and debates, pages dedicated to it on social networks...) in recent years is the result of various factors.

First of all, an affinity of socio-cultural premises, as well as design sensitivity, with the current period: the drive to build starting from what is, in the state in which it is, "*as found*", in fact, was born for *brutalism* within a climate of global recovery from the emotional and physical devastation caused by the Second World War, from which much of the post-war architecture has inserted materials of abandoned buildings in new buildings (an example is the chapel in Ronchamp by Le Corbusier (1955).

Starting from different assumptions, but witnessing "*a new kind of global devastation that is now undermining our existence - climate change - we could respond with an architecture based again on the reuse of materials*", or, in this case, of entire buildings. In the editorial entitled "*Re-evaluating brutalist architecture*", architect Jeanne Gang pleads this cause, arguing that "*a lot can be achieved by analyzing and intervening on concrete architecture; in fact, if we created new buildings by converting the existing ones entirely, we could even recover the spirit of brutalism that we so much appreciate for its distinctive trait together with its original description "as found"*" [7].

Obviously planning to modify them to meet the needs of contemporary society, "*designers can rework the existing one by correcting, synthesizing and modifying the original concrete building, while remaining faithful to the thought of its designer*" [7].

Thought that is often another factor of interest in these architectures, with "*a bit of nostalgia for their well-defined philosophical and social agenda*" [7]. Design and planning that had a well-defined ideal social model in mind, as demonstrated by the project of the Civic Center of the architect Romano, of which the current building has remained as a trace, physical and meaningful.

In a historical moment in which many of the principles on which our societies and our lifestyles have been built have been challenged, returning to interrogate existing projects that had in their constitution a great search for meaning towards the type of society that they wanted to outline, is an important lesson, especially when it comes to thinking about a very central and topical issue, such as that of the Health of the citizen and the community.

As Jeanne Gang rightly pointed out, however, "*focusing the value of these architectures solely on history restricts the discussion of the future of brutalist architecture to just a handful of buildings [...]* and we don't point out "*one of the fundamental reasons for its conservation: climate change*" [7]; and she continues "*a large amount of incorporated carbon is saved - generally between 50% and 75% - if you choose instead to recover them for a new use*" [7].

The reuse of "*as found*" buildings is an opportunity to apply the concept of circular economy to design process, to experiment with new ways of building the spaces of our cities, starting from what is there. Not only, of course, with brutalist architectures.

The choice to narrate the investigation around a product of this specific architectural movement, was an interesting pretext for a reasoning on the recovery of existing and identity buildings as one of the objectives that should be set for a design of the Community Centers. A design approach that sees very clearly the demands of the present time and therefore, necessarily, becomes a catalyst for urban, architectural, social and community regeneration.

Fig. 1. Pedestrian paths (1) courtyard (2) and internal "urban" façades (3) of existing building

References

1. ColtivareSalute.com.   http://www.polisocial.polimi.it/?s=Coltivare_Salute.Com%20/%20V incitore%20Award%202020. Accessed 20 Dec 2021
2. Polisocial. http://www.polisocial.polimi.it/it/home/. Accessed 21 Dec 2021
3. Turco, L.: Audizione sulle linee programmatiche http://leg15.camera.it/_dati/lavori/stencomm/ 12/audiz2/2006/0627/INTERO.pdf. Accessed 20 Mar 2021
4. Comune di Paderno Dugnano. https://www.comune.paderno-dugnano.mi.it/primo-si-alla-casa-di-comunita-a-paderno-dugnano/. 27 Dec Accessed 2021
5. Romano, M.: The Civic Centre of Paderno Dugnano, Milan: the first (and last) block of the first lot. Spaz. Soc. **11**, 84–91 (1980)
6. Poliambulatorio territoriale di Paderno Dugnano. https://www.asst-rhodense.it/inew/ASST/str utture/poliPaderno.html. Accessed 20 Dec 2021
7. Gang, J.: Rivalutare l'architettura brutalista. Plan **130**, 11–16 (2021)
8. Sosbrutalism. https://www.sosbrutalism.org. Accessed 16 Dec 2021

Consideration of the Potential Strategic Role of the New Community Center. Some International Case Studies: Themes and Design Practices

Stefania Varvaro[✉]

Politecnico di Milano, Piazza Leonardo da Vinci 32, 20133 Milan, Italy
stefania.varvaro@polimi.it

Abstract. A reflection is proposed on the lack of attention to the strategic potential of Community Centers in city planning. This deficit is manifested in the absence of suggestions and rules that belong to the 'art of making space', a critical factor that makes the design of these structures completely subservient to functional issues, performances, and technical logic. By accepting the PNRR proposal to focus attention on the Community, within the transition from Healthcare Centers to Community Centers, we want to put in evidence the components of an integrated design, usually not considered for these structures. In defining their physical architectural body and their relationship with the territory, they can instead become recognizable and identifiable references for the population. Through the examination of some international case studies, it becomes clear that a design aimed only at satisfying the provision of services is insufficient both to build places, between open and internal spaces, at the building, neighborhood and city scale, and to give a physical dimension to the social policies that underpin the construction of a community.

Keywords: Urban space · The 'art of space construction' · Community Center

1 The Origins of a Failed Strategy

This paper offers a reflection on some work-in-progress results of the research "*Coltivare_Salute.Com. Cities and Healthcare Centers for Resilient Communities,*" where the subtitle specifies the Healthcare Centers "*as builders of urbanity and widespread sociality in the post- COVID-19 era: new peripheral centralities in healthy and integrated cities*". A group of researchers belonging to the Politecnico di Milano proposes the drafting of guidelines for Healthcare Centers in the urban environment, adopting Piacenza as an application case.

Specifically, we assumed how, in Italy, given the regional legislation regarding this topic, there is a lack of guidance relating to a possible role for health in urban planning.

If we consider the public health system interpreted in its form of primary care, which more than any other has a direct implication with the territory, we are surprised

F. Calabrò et al. (Eds.): NMP 2022, LNNS 482, pp. 522–530, 2022.
https://doi.org/10.1007/978-3-031-06825-6_49

by the absence of indications. Since 2006, the year in which a capillary health system is promoted at a national level, through the Healthcare Centers [1], there are few regulations capable of defining the physical and architectural body of these structures and their relationship with the territory. The regional decree 291/2010 of Emilia-Romagna [2] still constitutes the most complete reference paradigm in this sense. The region which, together with a few others, represents the administrative and territorial environment where the primary care system is more structured and present in a widespread form, makes no mention of rules for the choice of locations dedicated to preventive medicine and the protection and promotion of health. What criteria are used in assigning an area or a building, whether it is better to reuse an existing building or build an entirely new one, the principles indicating where to place Healthcare Centers - none of this is clarified, neither in normative nor indicative form.

Yet primary care is associated with issues of proximity, of taking charge of the person, of active and responsible participation of the community, of relations with other health and social services. So much so that in the sixth point of Alma Ata's declaration [3] the World Health Organization writes, "It is the first level of contact of individuals, the family and community with the national health system bringing health care as close as possible to where people live and work, and constitutes the first element of a continuing health care process."

The fact that there has not been an in-depth study about the importance of the localization criterion and the strategic role that this could have in the development of city design is perhaps attributable to several factors: the too-short realization times, the constraints given by the financing calls, and more generally to a planning process subservient to practical, utilitarian, and functional logics, which, in order to respond to the service provision, has put aside the "art of building space" [4]. We propose to give value to the architectural project by considering the fundamental character of interiority that alludes to the possibility of welcoming people's lives, whether this is an open, covered, or closed space, in a multi-scale idea of an integrated project.

These reflections form a premise for the founding idea of Polisocial research which, by studying the Healthcare Centers in an urban setting, identifies them as opportunities for environmental, urban, and architectural regeneration and, in this sense, as builders of urbanity.

1.1 Note on PNRR

The pandemic has worked as an amplifier of the already existing need to implement the health system, especially in its forms of territorial primary care.

Through the transition from Healthcare Centers to Community Centers the PNRR, National Recovery and Resilience Plan, [5] made an effort to underline the collective value of these places. It is a public investment of certain significance, one of the highest recently destined for the healthcare sector, but whose too-quick timing does not allow for the tools of definition capable of combining design with multidisciplinary contributions aimed at favoring the observation of the social and physical reference context. It means giving up on a multi-scale architectural project linked to the thought that, within an urban context, the place can be important, conditioning, and that it can provide a strong base for a well-directed and future-oriented development.

The need to define a settlement strategy for the urban area of reference implies a long-term and large-scale design, with an overall vision that does not coexist with the timing that fixes the completion of the works planned and financed by European funds for 2026.

2   Analytical Methodology

Through a collection of some case studies relating to various Healthcare Centers, themes and elements of the architectural and urban project were analyzed and systematized. These studies thus provide a reference sample for a strategic definition of physical contexts, from the scale of the building to the scale of the city. We try to derive some ordering principles that relate to a general quality of the project that is not an aim in itself, but according to an intertwining of values that aim at a specific objective: that of transforming the Community Centers into urban reference points, in which citizens recognize each other and that they can also attend as active and healthy people.

Selective reading is articulated that focuses on the cardinal principle of the WHO: the need to conceive health not only as of the absence of disease but as psycho-physical and social well-being [6].

The focus is on recognizing some of the potentials of the architectural form, spaces and urban relations that transcend health care methods to focus on the characters capable of working on the habitability of a place.

The criteria that are specified below are based on the idea that providing service and doing it efficiently, whether of health, social-health, or social nature, is not enough. The urban and architectural project can add value to the construction of a new Community Center.

The goal is to transform these places into buildings for promoting health culture, rethinking their spaces and use: intended not only for health services but to become places for meeting, exchange, and interactive learning. In this vision, the Community Centers become "urban catalysts" capable of attracting different people with various interests, not necessarily linked to poor health, arranging new architectural and furnishing solutions, structuring a significant internal and external relationship, responding in a punctual way to the needs of a community.

The topics dealt with below are useful for prefiguring different aims and objectives from those pursued up to now for the Healthcare Centers. The new Community Centers can be new reference poles, nodes of urban regeneration, they must enter into a systemic relationship with the public space, the urban green system, the existing historical heritage and must express a high quality of the spatial architectural project.

2.1   As Libraries

Although in a different form, analogies can be traced to the radical change of the library building that took place starting from the 1980s, a type equally firmly linked to the cultural well-being of a community.

From exclusive and elitist places, archives, repositories and custodians of knowledge, libraries have become places of gathering and relationship, where a community finds itself, triggering a positive return on the closest context, moving people, interests, and activities.

"The creation of a library is the result of a synergy of different skills and professionalism" [7].

The same is true for the Community Centers for which a physical transformation is proposed. The aim is to make explicit and operational the whole range of activities and functions, partially disregarded, for which they were born from the beginning, where it was not a question of arranging a necessary care service, but it was about promoting a culture of health.

The figures called to give concreteness to the potential of such a complex structure, referring specifically to the fields of sociology, medicine, medical-health management and organization, its communication, can operate starting from a meticulous choice of location. They also can begin from a structure whose value of the spatial arrangement conveys a sanitary certainty together with a sense of a primary welcome of De Carli's memory [8]. It means building the space in which "the gesture of man" is welcomed and in which the architectural form guides the experience of an environment and its sequence.

On the one hand, the Community Center as a collective reference for activities and places open to the public becomes a pole of attraction; on the other hand, its location makes it the protagonist of a system in synergy with other health, social, and other public services, schools, training institutions, sports centers, associations, and the third sector. However, it should be pointed out that, while the aspects most relevant to the provision of the service (without which the Healthcare Center/Community Center would have no reason to exist) are the most tested and studied and follow the frequent changes in functional health models, the urban planning/architectural aspects have never been a basic prerogative. A critical issue that cannot be underestimated is the different speeds with which the two components, the sanitary one and the design one, move. To meet the continuous sanitary remodeling that characterizes a search for optimization of a very complex organization, the design of the spaces can work on flexibility and modularity, developing systems of adaptive relationships so that the project will not give up the interpretative component of the different contextual physical conditions. The post-war masters from Prouvé [9] to Chenut [10] and their noble experiments that tried to respond to post-war housing needs and the rapid transformations of a society and its basic nucleus, the family, could be interesting reference points, as could the dry combinable systems of Gullichsen and Pallasmaa [11] or the modulations of the Habitable Space by Munari [12], to name only a few. More consolidated spaces not subject to continuous modifications and spaces subject to change can coexist. For them it is necessary to provide extreme flexibility through suitable technologies and materials. The idea of working inside a container, whether new or existing, in different ways is implied. To follow ever-changing needs, the space can be configured through mobile and/or easily assembled devices that can be used in various ways. The mechanisms with which vertical or horizontal surfaces can be moved, folded, slipped, opened, or, light pillars or composite frames, can support, sustain define, etc., must be simple and immediate, have good solidity and require scarce maintenance (as the projects of the 50s and 60s have taught us).

3   Design Themes and Practices for a New Definition of Community Center: Examples from Some International Case Studies

Attention is focused on the components of the project because when it comes to Health Homes or Community Homes (according to the new title of the PNRR), the objectives are always primarily functional, technical and performance-related and are never accompanied by lines of thought related to the research of architectural form and the pursuit of the care of living [13].

They are places of and for the Community where a coexistence of activities and a variety of needs and people is welcomed. Danani speaks of a constitutive relationality of human beings by linking this concept to the different forms of living, suggesting once again the civil and public responsibility that a designer has in drawing the forms of hospitality.

Thinking about the Italian territorial health system and its critical issues in relation to the Health/Community Centers, we highlight those that are little or not at all dealt with. Using international case studies as examples, we underline the importance of specific issues discussed below.

3.1   Community Center as a New Centrality

Emphasis is placed on the role of the Community Center as a catalyst. Its presence can generate attractive dynamics: movements of people, various types of activities related to health-related issues - recreational-sports, cultural-educational, educational and not least commercial - and can, therefore, be defined as a recognizable new centrality. It becomes an urban piece around which to put functions and themes into a system: health, social, scholastic, cultural, green, open spaces. It becomes a pole whose location is not unimportant and should not depend solely on the logic of accessibility. These new centers must be planned considering the future development of the city, its growth and demographic composition and its possible modification over time, the population density, the presence of other public services, and the system of public and sustainable mobility (primarily).

Potentially significant locations are represented by non-central neighborhoods, outside what is considered the historic center, where the city has a harder time tracing a deep-rooted sense of belonging. Also worthy of consideration are both more densely populated as less populated urban areas, which need a meeting place, a public space that is capable of triggering a regenerative process, configuring from scratch, and/or systematizing the characters of the existing urban fabric that participate in the new definition of Community Center. It becomes a collective public space in the neighborhood, an integrated service center and a health culture center where prevention, promotion, training, information and socialization are carried out.

It becomes comparable to places that in the collective imagination refer to the market square, the oratory, the playground, a space around which the life of the local community takes shape and develops interactions between cultures, social strata and generations, fueling social cohesion.

Designed in 2016 by Steven Holl and completed in 2021, the Shanghai Cofco Cultural and Health Center aims to be a social condenser, to become a reference point for the resident community through the establishment of public spaces and the design of a linear park along a existing canal (see Fig. 1).

Fig. 1. View of the open space along the canal involved in the design of the Shanghai Cofco Cultural and Health Center (Source: https://www.stevenholl.com/)

Two buildings in dialogue structuring open public and green spaces: a health center with a space for health education, a pharmacy, consultation and examination rooms, physical therapy room, ultrasound/X-ray rooms, nursery, offices; a cultural center with an exhibition area, board games area, bar, library, gym, and areas for community and youth activities.

3.2 Community Centers and Urban Public Space

In direct connection with the idea that the Community Center is a new centrality, it is considered a living organism, which modifies what is in its close neighborhood and also in a wider context, acting on the relations between the parties. It is highlighted that the new Community Center is not a self-centered functional typology, resolved in itself and merely self-referential, which could therefore be spatialized within a building that does not interact with the context. On the contrary, the system of neighboring open spaces should become its natural completion. It means underlining that the physical margin of its volume, more or less articulated, is osmotic in its attack on the ground, in its unfolding all the elements that can set up a direct connection with the surroundings: arcades, shelters, threshold elements, access points, and in its configuration on the floors other terraces, windows and holes in general, practicable coverings.

Open space and its designed relationship with the building acquires even more value from a post-pandemic perspective. In this sense, the Maison de la Santé in Vezelay by Bernard Quirot Architecte and Associés (see Fig. 2) becomes the gateway to the town. Located on the slopes of the Vezelay hill, within the perimeter of the site classified as Unesco Heritage, it articulates, through a clever arrangement of three volumes, a public space belonging to the Healthcare Center which works with the unevenness of the land and configures a privileged point of view towards the town and its cathedral with its significant historical-architectural importance. Or consider the Healthcare Center Waldron (Henley Halebrown Architects) (see Fig. 3) in south London, designed around a new civic square with shops, a café and houses. Inside, in continuity with the surrounding open space, the volume is organized around two courtyards and a 5-story foyer with artwork by artist Martin Richman.

Fig. 2. Maison de la Santé in Vezelay (Source: photographer Luc Boegly)

Fig. 3. Healthcare Center Waldron: square and inner courtyard (Source: photographer Nick Kane)

3.3 Community Centers and Green Urban System

The relationship with the system of green spaces integrates and specifies the regenerative character of the Community House. There is a health issue, linked to all the activities that can be carried out in the open-air green spaces, such as sports, recreational, didactic and training practices. The presence of a Community Center guarantees the maintenance of a public green, equipped or not, configured as an urban park or with playgrounds. They could interpret it as an active defense that brings the population closer to the issues of environmental fragility: pollution, soil consumption, islands of heat, collection and disposal of water. The specific urban context could suggest tree species and include, as part of its green configuration, therapeutic gardens, rehabilitation paths, and social gardens.

3.4 Community Centers and Historical Existing Heritage

The use of an existing building promotes a form of sustainability by avoiding land consumption and a need to use an existing abandoned or disused public heritage that characterizes the whole of Italy in particular, but also Europe. In the choice of the building, it is important to be able to insert the single episode in an overall settlement logic in the city, foreseeing the optimal number, considering the urban structure with its characteristics and the planned urban prefigurations, in order to guide the criterion of selection without leaving only the availability of the areas and buildings and the nature of the property to contribute to the choice. A strong point is given by the possibility that the existing building has an identity character, on the one hand, that it may already be part of collective memory, and on the other hand, that it has recognizable architectural and spatial connotations, whose intrinsic vocation is compatible with the new functional program.

They can be buildings that are already iconic and recognized by the community as in the case of the church of St. Theresa in Borne, the Netherlands where the "De Poort van Borne" Health Center is set up by the Reitsmea Partners Architecten studio, or the case of the reuse of a power plant, as in the 13 kV Dordrecht Health Center designed by RoosRos Architecten.

3.5 Community Centers and Interior Space

The activities and services offered by the Community Center require particular attention to the organization and articulation of its interior spaces, essential in any project but, in this case, quite fundamental and decisive for its success. The health function, both at the hospital level and the level of Community Centers [1] has dimensional, functional, and managerial guidelines and in terms of organization of the different services, but not architectural in a broad sense. In the transition from Healthcare Centers to Community Centers, the design aims to integrate functions open to the community, articulating their hierarchy, the different levels of privacy and relationship with open space, through architectural choices that have the dimension as a physical-behavioral priority of the patient/user/citizen. This is the case of the Ballarat Community Health Primary Care Center (DesignInc) in which the structuring of the double or triple heights, the internal views, the skylights, the stairs, the furnishings constitute an integrated system, or even in the small Medical Office Renovation in Entrambasaguas (Perez - Ruiz de Apodaca) in which the shape of the space guides the progress of the paths in a game of expansion and compression which, together with wise use of materials, calibrates waiting and reception places, and organizes clinics and offices.

4   Conclusion

It is comforting to think that some international case studies show an understanding of some issues as foundations for the establishment of a Community Centers network systemic with its territory and strategic for its development. However, this type of reading is not only discussed within a disciplinary forum made up of architects and planners. Its importance is supported by a recent experience conducted by a group of researchers, sociologists, anthropologists, doctors, health and social professionals, entitled Community Express, which aims to define a concrete evaluation tool for the preparation of the new regional health and social area plan of Emilia-Romagna.

Community Express (framed within the previous Community Lab) has developed a "spatial and social observatory" through a visual method. Evaluating the potential and critical factors of territory, through the physical act of walking through it, photographing it, living it, noting the ways of life it hosts, is an analytical and, at the same time, planning formula of extraordinary interest. Keeping social policies and the places where they are expressed together is a worthy goal that invites a multidisciplinary approach.

References

1. Maciocco, G.: Cure Primarie e Servizi Territoriali. Esperienze Nazionali e internazionali, Carrocci, Roma (2019)

2. Regione Emilia-Romagna. Decreto Giunta Regionale 291, Casa della Salute: indicazioni regionali per la realizzazione e l'organizzazione funzionale (2010). https://salute.regione.emi lia-romagna.it/cure-primarie/case-della-salute/documentazione-case-della-salute/delibere. Accessed 10 Dec 2021

3. WHO. Unicef Alma Ata Declaration on Primary Health Care, p. 1 (1978). https://www.who. int/publications/almaata_declaration_en.pdf. Accessed 20 Dec 2021

4. Ottolini, G.: Forma e significato in architettura, pp. 40–42. Laterza, Bari (1996)

5. Ministero dello Sviluppo Economico PNRR Piano Nazionale di Ripresa e Resilienza, Missione 6 Salute, pp. 222–234 (2021). https://www.governo.it/sites/governo.it/files/PNRR.pdf. Accessed 14 Dec 2021

6. WHO. Constitution of World Health Organization (1948). https://apps.who.int/gb/bd/PDF/ bd47/EN/constitution-en.pdf?ua=1. Accessed 22 Nov 2021

7. Agnoli, A.: Introduzione in Muscogiuri M. Architettura Della Biblioteca. Linee Guida di Programmazione e Progettazione, p. 9. Syvestre Bonnard, Milano (2004)

8. De Carli, C.: Architettura e Spazio Primario, p. 250, Hoepli, Milano (1982)

9. Prouvé, J.: La Poetica Dell'oggetto Tecnico. Skirà, Ginevra-Milano (2007)

10. Chenut, D.: Ipotesi per un Habitat Contemporaneo. Il Saggiatore, Milano (1968)

11. Gullichsen, K., Pallasmaa, J.: https://likemyplace.wordpress.com/2013/11/16/looking-back-modular-moduli225-prototype-by-kristian-gullichsen-and-juhani-pallasmaa-helsinki-fin land/. Accessed 14 Dec 2021

12. Q Cultura Degli Interni Group Edt, Spazio Abitabile: 1968–1996/Bruno Munari. Stampa alternative, Roma (1996)

13. Danani, C.: I Luoghi e Gli Altri. La Cura Dell'abitare, Aracne (2016)

Investigating the Health-Planning Nexus in Italy: A Survey on Local and Metropolitan Plans

Luca Lazzarini[✉]

Department of Architecture and Urban Studies (DAStU), Politecnico di Milano,
20133 Milan, Italy
luca.lazzarini@polimi.it

Abstract. This contribution investigates the health-planning nexus in Italy by looking at the contents of the local and metropolitan plans of three Italian cities, Bologna, Florence, and Turin. The objective is to analyze the contribution of planning to the transition towards healthier communities, through a survey oriented to map and analyze objectives and actions related to health and wellbeing. Findings show a rhetorical understanding of how plans tackle the issues related to health. Also, scarce attention over the functioning and organization of the healthcare system and the contribution of health services to achieve healthier communities emerges in planning policies. Nevertheless, the global pandemic has initiated a change in the approach, as a growing number of plans are starting to incorporate objectives and actions related to health, with a focus on the development of primary care and community health services. The chapter comprises four sections. After a brief introduction on the context and debate, Sect. 2 presents the methodology and the criteria chosen in the survey. Section 3 highlights a discussion of the data collected concerning the five criteria previously identified. The last Sect. 4 includes some concluding remarks.

Keywords: Health · Community health · Urban planning · Metropolitan city

1 Introduction

The link between urban planning and the promotion of good health is long established and rooted in history. Modern town planning originated in the late XIX century in response to basic health problems that affected urban areas, such as poor-quality water supply and lack of ventilation, at that time characterized by the high rates of mortality and frequent epidemics [1, 2]. Despite these roots and the emergence in recent years of international movements focusing on the contribution of health in urban planning and design [3], in the last two decades the relationship between health and urban planning has remained at most unexplored by both academic and policy environments. [2] spoke about a divorce between health and town planning created by the unhealthy conditions of our habitat and the worsening of public health that many recent poorly designed urban developments have created. [4] pointed out that city planning is not currently organized to ensure that today's cities will be equitable and healthy and he claimed that the potential of healthy

F. Calabrò et al. (Eds.): NMP 2022, LNNS 482, pp. 531–540, 2022.
https://doi.org/10.1007/978-3-031-06825-6_50

city planning in eliminating the deep and persistent inequalities that plague cities has been mostly unexploited. It would not be untrue to say that until the spread of the Covid-19 pandemic the issues related to health were left off the local planning agendas. This is evident in the Italian context where a certain difficulty emerges to integrate health and wellbeing principles in local plans and planning policies [5]. Reasons lie in the scarce acknowledgment by planners and policymakers of the multiple social, economic and environmental benefits that the incorporation of the health determinants in urban planning and place-making can generate, as well as in the widespread awareness that health is just a matter for healthcare professionals [6, 7].

The Covid-19 pandemic has surely brought the issue of health back to the political and planning agenda [8, 9]. Several studies have confirmed the important role that urban planning policies and actions have, not just in addressing the short-term challenges related to the pandemic, but also in strengthening the prevention and preparedness of urban settlements to health emergencies [10]. A recent report by UN and WHO [11] highlights that planning has a great potential in improving health as it can «influence location, spatial pattern, and local design of place-based features and amenities in the built environment for the benefit of health and health equity». [12] highlight that a good neighborhood design shaped by specific planning regulations has important implications on the contribution to the physical and mental health of local communities. Moreover, the access to public amenities at short distances and time [13, 14], including not just medical, educational, and social services but also inclusive green spaces and recreation and sports facilities, is a relevant aspect responsible to improve inhabitants' health. In the same vein [15], after carrying out a systematic review on studies on neighborhood design, demonstrated that some features such as access and proximity to green spaces, walkability, access to public transport, and amenities within neighborhoods have positive impacts on health and wellbeing across all age groups.

It seems clear that the objective of enhancing public health and wellbeing requires local authorities to implement adequate policies, plans, and actions. According to [16], the integration of health issues in land-use and strategic plans is crucial for achieving health goals. Thus, the identification of tools and criteria for assessing plans concerning their impact on public health emerges as an important aspect to be investigated by researchers and policymakers.

Alongside this framework, this contribution aims at investigating the health-planning nexus in Italy, by looking at the contents of the municipal and metropolitan plans of three metropolitan cities, Bologna, Florence, and Turin. The objective is to analyze the contribution of planning to the transition towards healthier communities, through the investigation of objectives and actions related to health and wellbeing included in the plans. The chapter comprises three sections. Section 2 presents the methodology and the criteria chosen in the analysis. Section 3 highlights a discussion of the data collected concerning the five criteria previously identified. The last Sect. 4 includes some concluding remarks.

2 Methodology

The methodology is based on a screening of the municipal and metropolitan plans of three Italian metropolitan cities, Bologna, Florence, and Turin. The three cities were selected alongside the research activity carried on within the project *Coltivare_Salute.com*. The project has the objective of redefining the approach to Community Health Centers (CHCs) ("Case della Salute") in Italy: from primary care services to opportunities for urban regeneration in which "cultivating health" for treating the social vulnerability in urban areas, also during health emergencies [17]. The case studies of the research were chosen among those cities in Italy having a system of (at least two) CHCs in their municipal area. According to an official document of the Chamber of Deputies [18], there are 18 cities in Italy matching this criterion. As a second step, it was then decided to focus on a sample of three Metropolitan Cities due to the objective to investigate if and how the metropolitan planning introduced in Italy in 2014 has taken into account the healthcare system and addressed the challenges related to the organization of community health services.

The screening was combined with a discourse analysis to underly the discursive aspects of health within official planning documents and understand the particular ways in which the problem of health and wellbeing is framed and constructed in planning policies [19]. The analysis took into consideration the statutory and the strategic plans not just at metropolitan but also at the local level, in order to have a complete look at planning policies in those selected cities. At the metropolitan level, the analysis has taken into consideration two planning devices, the strategic plan, and the general territorial plan, as defined by the Italian Law n. 56/2014. While the strategic plan has a three-year duration and it aims at providing guidance for municipalities and their associations for programming the social, environmental, and economic development of the metropolitan territory, the territorial general plan is called to define the overall structure of the metropolitan territory, protecting and enhancing the natural environment, and coordinating the local planning policies in the transformation and management of the territory. Since 2014, when the new planning discipline was introduced in Italy, just in the case of Bologna, the Metropolitan City Government has adopted both a Strategic Metropolitan Plan in 2018 (a second version after a first one adopted in 2014) and a Territorial Metropolitan Plan in 2021. In the case of Florence, a Strategic Metropolitan Plan was adopted in 2018 but the process for elaborating a new Territorial Metropolitan Plan has started in 2018 but not concluded yet. In Turin, a second version of the Strategic Metropolitan Plan was adopted in February 2021 after a first version approved in 2018. As far as the Territorial Metropolitan Plan is considered, a technical proposal for a new plan was presented in 2020 but the process for its adoption has not ended yet. Even if not adopted, the research has anyway considered the technical proposals as plans *in nuce* which include all the elements needed for understanding if and how the topic of health was tackled by the planning device (see Table 1).

The screening of local and metropolitan plans was made through an analytical framework to identify those objectives and actions dealing with five criteria (see Table 2): i) the development of local community's health and wellbeing, ii) the improvement of the health system and services, iii) the forms of proximity in the configuration of public services (with specific regard to health services), iv) the spatial and functional integration between health services and other public services, and v) the relationship between health services and urban regeneration. Although referring to a specific implication of the health-planning nexus, the choice to include the fifth criteria is based upon the willingness to see if and how local and metropolitan plans explore the potentials of CHCs and, more generally, of the system of health-care services to trigger processes of urban regeneration.

Table 1. The municipal and metropolitan plans taken into consideration in the screening.

Plan	Institution	City	Year of adoption	Acronym
Strategic Metropolitan Plan (PSM)	Metropolitan City	Florence	2018	FI1
Provincial Territorial Plan (PTC)	Former Province, now Metropolitan City	Florence	2013	FI2
Structural Municipal Plan (PSC)	Municipality	Florence	2015	FI3
Strategic Metropolitan Plan (PSM)	Metropolitan City	Bologna	2018	BO1
Territorial Metropolitan Plan (PTM)	Metropolitan City	Bologna	2021	BO2
Local Plan (PUG)	Municipality	Bologna	2021	BO3
Strategic Metropolitan Plan (PSM)	Metropolitan City	Turin	2021	TO1
Territorial Metropolitan Plan (*draft proposal*) (PTGM)	Metropolitan City	Turin	2021	TO2
Local Plan (*draft proposal*) (PRG)	Municipality	Turin	2020	TO3

Table 2. Overview of the 5 criteria and their presence as objectives (o) or actions (a) in the local and metropolitan plans investigated.

Criteria	FI1	BO1	TO1	FI2	BO2	TO2	FI3	BO3	TO3
i) Development of community health and wellbeing	o	o a	o a		o	o	o	o	o
ii) Improvement of health-care system and services		o a	o a				o	a	
iii) Forms of proximity in configuration of public services		o a	o			o		a	o
iv) Integration between health services and other services	a	o a	a		a		o	a	a
v) Relationship between health services and urban regeneration					a		a		o

3 Discussion

The five criteria through which the municipal and metropolitan plans of Bologna, Florence, and Turin were investigated take into consideration both general objectives and specific actions and interventions. The distinction between objectives and actions helps to understand if the topic of health is addressed by planning devices merely in terms of identification of general issues to be tackled by plans in broad terms, or of definition of actions to be directly implemented by local administrations. At the basis of this distinction, there is also the need to see if a mismatch between rhetoric and action [20], which is typical of the Italian planning framework, is persisting and how it potentially affects the health-planning nexus.

3.1 Development of Local Community's Health and Wellbeing

Most plans show a direct reference to the issues of health and wellbeing but these aspects are subjected to a wide range of interpretations. In the case of Florence, the role of the environment in contributing to good health and wellbeing is highlighted. For instance, FI1 emphasizes the role of rural territory as a critical asset for the eco-systemic balance and the integrated development of the territory, as well as a significant component to improve the well-being and limit the problems related to air and water pollution and, more in general, promote a healthy and safe living environment for metropolitan citizens. At the municipal level, FI3 underlines the development and enhancement of the ecological network to perform multipurpose functions (plugging of microclimates, self-purification, recharge groundwater, production of oxygen, etc.) and improve the quality of life of the population. The relationship between environmental resources and health is also mentioned by TO1 which focuses in one of its strategies on making the environmental quality of the metropolitan space an enabling factor for an active, healthy, and long life, with actions ranging from urban reforestation for cooler cities and cleaner air, bottom-up air quality monitoring for citizen awareness, to high-quality food provision at the metropolitan level. The longevity of the population is a distinct objective of this plan as it explicitly sustains the active and healthy aging of inhabitants and the construction of stronger links between different generations. The social determinants of health and wellbeing are recalled by BO1 where one of the strategies refers to the need to ensure inclusion and wellbeing, especially by contrasting social, economic, and demographic fragility through processes of regeneration of the built-up environment. The plan also recognizes the social and health value of sport as a tool for realizing people's «right to health and well-being», and improving individual lifestyles. Similar is the approach taken by BO3 which operates to improve the quality of life in socially distressed urban areas and neighborhoods by implementing specific works of regeneration of public spaces and facilities.

Generally speaking, the issues of health and wellbeing are present in plans more often as general objectives than as aspects able to orient planning policies and actions, according to a prevailing approach looking at health and wellbeing as rhetorical concepts that shape the discursive part of the plan.

3.2   Improvement of the Healthcare System and Services

The reference to the improvement of health (and social)-care system and services is explicitly underlined by four plans on the ten investigated, though just the cases of the BO1 and TO1 this topic is defined both in terms of general objectives and of specific actions. In the case of Bologna, the plan underlines the need to develop clinical networks and the system of intermediate care with particular reference to CHCs and Community Hospitals, and to enhance the role of public-private partnerships in supporting consistent forms of intervention in social and health care sectors. In the case of Turin, the objective of constructing a system of decentralized health centers with advanced technology and connected to the main hospital of the metropolitan city ("Città della Salute") is mentioned in strategy n. 6.1. A list of actions and projects is also included as, for instance, the strengthening of the CHCs and their integration within a territorial network of health services, through the provision of incentives for the aggregation of general practitioners, local pharmacies, and the location of infrastructures in spaces equipped for telemedicine and easily accessible by public transport. Another relevant action is the enrollment of new community nursers and their cooperation in micro-teams with community social workers in home health services. Both BO3 and FI3 emphasize the need to shape better interdependences between hospital and territorial/community healthcare services. In the case of Bologna, these actions are framed at the metropolitan level with the urgency to rationalize the network of hospital structures and differentiate them. The objective is to qualify their role as reference points for the networks of assistance at regional, national and European levels. Moreover, the local plan mentions the contents of the plan of the local health authority, and indicates «as needed» the creation of three new CHCs in the municipal territory. A specific topic addressed in BO1 concerns the need to sustain new and existing professional roles like the so-called "basic geriatrician", the caregiver, and the community nurse in developing the services oriented to sustain elderlies and not self-sufficient individuals. As far as the community nurse is considered, the TO1 includes an action addressed to develop and extend the model of community nurses in micro-teams with community social workers to improve the quality and efficiency of home care services, also in combination with the use of new technologies like telemedicine. FI3 states the objective of setting out actions for implementing the Regional Health Plan which range from the future adaptation and renewal of existing health facilities to the development of the integration between health and social services and more effective coordination between community centers and hospitals. It is also recalled the need to check the load induced by the renewal and development of social health services on the existing and planned infrastructural system for guaranteeing adequate levels of accessibility (also for the disabled users), in terms of parking lots and public transport.

To sum up, it is important to highlight that the majority of plans emphasizes the need to strengthen the quality and performance of healthcare services, with specific attention to the primary care services such as CHCs and community hospitals. This topic was been introduced in the more recent plans (those elaborated after the pandemic) often together with the strengthening of new or existing professional roles which can provide better quality territorial and home services.

3.3 Forms of Proximity in the Configuration of Public Services (with Specific Emphasis on Health Services)

The issue of proximity to public services is mentioned by 5 up to 10 plans. The recurring objective is to increase the accessibility of public services and facilities (including health services) for achieving better territorial cohesion and equity in the whole urban or metropolitan area. In the case of Turin, starting from a peculiar territorial configuration with the metropolitan territory formed by a wide rural territory extending from the Po River valley to the alpine arch, TO2 interprets polycentrism as a «strong point of characterization of the metropolitan territory and as a lever for overcoming the asymmetry between plains and mountains [...] and creating city-mountain synergies». As a consequence, despite the identification of five levels of urban hierarchy according to which providing different public services, the plan underlines the need to support rebalancing dynamics to maintain a range of local services and ensure adequate levels of wellbeing in mountain contexts.

At the level of the TO3, the concept of polycentric city is introduced to develop and promote the livability of the different neighborhoods, counteracting the tendency towards "alienating neighborhoods", creating new public spaces for culture, sociality, leisure, and expressing equity and social cohesion in each neighborhood. In two cases (BO1 & TO1), the concept of proximity explicitly refers to the distribution of community health services such as the CHCs. This is the outcome of a harsh debate in Italy originated during the pandemic regarding the stronger capacity of territorialized and polycentric health systems and regulatory models – like those connoting some Italian regions like Veneto and Emilia Romagna – to react in a more efficient way to the diffusion of pandemics and health emergencies than the centralized and hospitalized system [21]. The first case highlights the need to develop the system of health centers for better exploiting the role of the community as the first and main subject responsible for its health and wellbeing. In the second case, the construction of a system of decentralized and technologically advanced health centers should be properly integrated into the future large metropolitan hospital complex under construction ("Città della Salute").

As a synthesis, the issue of proximity is often highlighted concerning the development of polycentrism as a condition for strengthening the territorial cohesion and guaranteeing a fair and balanced distribution of public services across the urban and metropolitan space, as also seen in other Italian and European contexts [22, 23]. Similar to the previous sections, a rhetorical understanding of proximity is prevailing with scarce critical reflections on the comprehensive reorganization of the system of public services to strengthen the conditions of accessibility and the responsiveness to social fragilities.

3.4 Spatial and Functional Integration Between Health Services and Other Public Services

The spatial and functional integration between health services and other services (social, cultural, mobility, etc.) is an aspect tackled by the majority (8) of the plans investigated. In this sense, the integration is developed at three levels. The first level relates to the professional skills working in the public services. The second concerns how the public service is managed and organized. The third level regards how the interior space of the public service is used. Most of the actions relate to the second level, while just a few have to do with the other two levels. Hence, the promotion of the integration between different existing professional skills in social and health fields is mentioned just by BO1. The importance of finding better ways for managing and organizing services opting for better integration between different typologies is recalled by several plans. For instance, in the case of FI1, this is planned through the creation of a "table of coordination" between the different stakeholders operating in the field of social policies aimed to integrate the single actions within wider strategies for achieving more effectively social cohesion. The coordination between social and sanitary services is mentioned by FI3, while in other three cases the same concept of integration is oriented to strengthen the network of services taking care of not self-sufficient individuals and elderlies (BO1 & BO2), also through the use of innovative concepts like the "caring community" (FI1). As far as the ways of use of the spaces are concerned, TO3 underlines the need to create multifunctional services, focusing on the "double use" perspective to be able to provide in the same building or facility, different services at different times.

On a wider look, the spatial and functional integration between health services and other public services is a largely unexploited topic in local and metropolitan plans. The level of integration found more frequently in the plans concerns the sharing of professional skills for tackling situations of social marginality. Thus, the sharing of spaces, users and resources between different public services is rarely mentioned.

3.5   Relationship Between Health Services and Urban Regeneration

Just three plans exhibit a focus on the potentials of the health services to promote/trigger processes of urban regeneration or, vice versa, on the opportunities of urban regeneration processes to improve the functioning and integration of health facilities in distressed urban neighborhoods. FI3 mentions that the areas subjected to transformation for establishing new public services and facilities should foster the regeneration of the surrounding context through the creation of new spatial and functional relationships. While explaining the Metropolitan Programs of Regeneration as agreements made by groups of municipalities to regenerate abandoned and underused portions of their territory, BO2 underlines that the regeneration should take place in connection with the activation or consolidation of metropolitan welfare services, especially in small municipalities and in the areas accessible by the metropolitan railway transport and close to mobility hubs. Finally, TO3 makes an explicit reference to the already mentioned new metropolitan hospital hub ("Città della Salute") as an opportunity for urban regeneration and functional reorganization of a large urban area characterized by decay and marginality. The Plan also states that the realization of the hub will provide one of the most relevant opportunities for economic development for the city in recent time, both in terms of the value generated by the functions settled down in the area and of the potentials that the areas inside the city (Molinette and Sant'Anna), emptied by the health facilities relocated in the new hub, will generate.

4 Conclusion

The screening on municipal and metropolitan plans highlights the relevance that the objectives and actions related to health and wellbeing have within municipal and metropolitan planning. As previously demonstrated, plans frame mainly in a rhetorical way the topic of health by focusing on how the transition towards more sustainable planning processes can contribute not just to fighting environmental problems and/or tackling climate change, but also to improve local community's wellbeing and health. Overall, scarce attention over the functioning and organization of the healthcare system and the contribution of health services to achieve healthier communities emerges in plans and planning policies. As also noticed by some researchers [2, 6, 7], reasons lie in the fact that urban planners often interpret the issues related to the healthcare system as out of their field of action and within other policy domains and expertise. Nevertheless, the pandemic has initiated a change in the approach. A growing number of plans are starting to incorporate objectives and actions related to the healthcare system with a focus on the development of CHCs and new professional roles that can improve the quality of primary and home-care services to communities. This direction is coherent with what stated by international organizations such as [24] and [10], which emphasize the need to strengthen the delivery of primary healthcare to make health systems more effective and responsive to vulnerabilities, as well as with the national programs of recovery and resilience implemented by governments to promote the economic and social recovery of countries in response to the pandemic crisis, such as the Italian National Recovery and Resilience Plan (NRRP) which allocated two billion-investment to supports the activation of 1.288 new CHCs [25]. The screening also underlines that strategic plans seem to be more prone to tackle the issues related to health than the territorial and land-use plans. The fact that their elaboration is often sustained by wide participatory processes which involve a range of different professional and technical skills, many of whom out of the planning field, may be a relevant aspect influencing the integration of health issues within plans and planning policies.

References

1. Mareggi, M., Lazzarini, L.: Health, an enduring theme for urban planning. In: Proceedings of the 57h ISOCARP World Planning Congress, pp. 803–812, ISOCARP, Doha (2021)
2. Barton, H., Grant, M.: A health map for the local human habitat. J. R. Soc. Promot. Public Health 126(6), 252–261 (2006)
3. Barton, H., Tsourou, C.: Healthy Urban Planning. Spon Press, London-New York (2000)
4. Corburn, J.: Healthy city planning. From neighbourhood to national health equity. Routledge, London (2013)
5. D'Onofrio, R., Trusiani, E.: Città, Salute e Benessere. Nuovi percorsi per l'urbanistica. Franco Angeli, Milano (2017)
6. RTPI Royal Town Planning Institute: Enabling Healthy Placemaking, RTPI Research paper, Royal Town Planning Insitute, London (2020)
7. Grant, M., et al.: Cities and health: an evolving global conversation. Cities Health 1(1), 1–9 (2017)
8. Martinez, L., Short, J.R.: The pandemic city: urban issues in the time of Covid-19. Sustainability 13(3295), 2–10 (2021)

9. Moccia, F.D., Sepe, M. (eds.): Benessere e salute delle città contemporanee. INU Edizioni, Rome (2021)
10. WHO World Health Organization: Strengthening Preparedness for Covid-19 in Cities and Urban Settings. https://www.who.int/publications/i/item/WHO-2019-nCoV-Urban_pre paredness-2020.1. Accessed 13 Dec 2021
11. UN-Habitat & World Health Organization (WHO): Integrating health in urban and territorial planning: a sourcebook. Geneva (2020)
12. Jackson, R.J., Dannenberg, A.L., Frumkin, H.: Health and the built environment: 10 years after. Am. J. Public Health **103**(9), 1542–1544 (2013)
13. Halpern, D.: Mental Health and the Built Environment. More than Bricks and Mortar? Routledge, London (1995)
14. Wood, L., Frank, L.D., Giles-Corti, B.: Sense of community and its relationship with walking and neighborhood design. Soc. Sci. Med. **70**(9), 1381–1390 (2010)
15. Ige-Elegbede, J., et al.: Designing healthier neighborhoods: a systematic review of the impact of the neighborhood design on health and wellbeing. Cities and Health (2020)
16. Capolongo, S., Lemaire, N., Oppio, A., Buffoli, M., Le Gall, A.: Action planning for healthy cities: the role of multi-criteria analysis, developed in Italy and France, for assessing health performances in land-use plans and urban development projects. Epidemiol. Prev. **40**(3–4), 257–264 (2016)
17. Ugolini, M.: Case della Salute: condizioni di fragilità e occasioni di rigenerazione urbana. Territorio 97 s.i. (2022)
18. Camera dei Deputati, Servizio Studi Affari Sociali: Case della salute ed Ospedali di comunità: i presidi delle cure intermedie. Mappatura sul territorio e normativa nazionale e regionale, vol. 144, 1 March 2021. http://documenti.camera.it/leg18/dossier/testi/AS0207.htm?_161919879 6640. Accessed 22 Feb 2022
19. Atkinson, R.: Narratives of policy: the construction of urban problems and urban policy in the official discourse of British government 1968–1998. Crit. Soc. Policy **20**(2), 211–232 (2000)
20. Palermo, P.C.: Urbanistica del progetto urbano: ambiguità e ipocrisie. Eco Web Town. J. Sustain. Des. **1**, 21-43 (2017)
21. Arlotti, M., Marzulli, M.: Quale rilevanza dei modelli regolativi regionali? La questione lombarda nella crisi sanitaria Covid-19. Salute e società **20**(2), 184–198 (2021)
22. Teti, E.: Italy. In: Kobel, P., Këllezi, P., Kilpatrick, B. (eds.) Competition Law Analysis of Price and Non-price Discrimination & Abusive IP Based Legal Proceedings. LCALIPUC, pp. 193–214. Springer, Cham (2021). https://doi.org/10.1007/978-3-030-55765-2_8
23. Atkinson, R., Pacchi, C.: In search of territorial cohesion: an elusive and imagined notion. Soc. Inclusion **8**(4), 265–276 (2020)
24. OECD: Realising the Potential of Primary Health Care, OECD Publishing, Paris (2020)
25. Presidenza del Consigio dei Ministri: Piano Nazionale di Ripresa e Resilienza (2021)

An Experimental Approach for the City of Health

Antonio Taccone[✉] iD

PAU Department, Mediterranean University of Reggio Calabria, Reggio Calabria, Italy
ataccone@unirc.it

Abstract. The document proposes a reflection on the role of urban and environmental planning in promoting actions aimed at improving the health and well-being of life of the inhabitants for a sustainable, safe, healthy and socially inclusive city. Today, as also envisaged by the Green New Deal, in order to support health and well-being policies, it is necessary to identify new tools and new sustainable development trajectories and medium-long term policies aimed at the specificity of local contexts that must be resilient towards the new and increasingly complex settlement geographies. Among the experiments conducted by the Lastre research laboratory, the Capacity project represented the tool capable of creating experimental research approaches to promote collective participation and sharing of project activities, placing the accent on the principle of a healthy city and eco-sustainability. In the urban and more generally territorial context. In this project, the definition of care acquires a broad meaning, not exclusively linked to the health aspect, but which broadens towards a different role of local communities that redefines urban welfare.

Keywords: Health city · Urban quality · Health equity

1 Introduction

The document proposes a reflection on the role of urban and environmental planning in promoting actions aimed at improving the health and well-being of life of the inhabitants for a sustainable, safe, healthy and socially inclusive city. Today, as also envisaged by the Green New Deal, to support health and well-being policies, new sustainable development trajectories and medium-long term policies must be identified aimed at the specificity of local contexts that must be resilient towards new and increasingly more complex settlement geographies. The well-being and health expressed by the ways of living and living in our cities are increasingly at the center of this new phase of development and in the policies of ecological transition, today supported by the second mission of the National Recovery and Resilience Plan (NRRP) drawn up by the government, which identified numerous sectors in which to intervene to support innovation and reconstruction in the post-pandemic. A very ambitious political vision project, which has developed new ways of multilevel local governance aimed at responding, through the adoption of an integrated approach aimed at promotion and local development, to the needs of territories characterized by important geographical disadvantages and competitiveness.

© The Author(s), under exclusive license to Springer Nature Switzerland AG 2022
F. Calabrò et al. (Eds.): NMP 2022, LNNS 482, pp. 541–548, 2022.
https://doi.org/10.1007/978-3-031-06825-6_51

An integrated approach that could lead to a paradigm change in the regeneration processes of our living environments also to face the challenge of climate change. In fact, research should be directed towards equilibrium between systems in order to create opportunities for regeneration that contemplate health and the quality of life in our cities, [1] also through participatory and sustainable forms of local development.

In the post-Covid era, the awareness has now been reached that it is not enough to deal with health services in the sector, but we must consider all the environmental, socio-economic and cultural factors of the city that have an influence on health. Awareness already present in the UN Sustainable Development Goals 2030 and in the various connections and links SDG 3, Good health for all, and SDG 11, Making cities and human settlements inclusive, safe, resilient and sustainable. A long work is that developed by the World Health Organization whose goal for healthy cities is to favor procedural advances that raise the quality of life and create improvements in physical and social environments in favor of communities [2]. The close connection between urban planning, medical and social sciences is the basis of the Healthy Cities project approach promoted since the 1940s and supported both by the WHO Agenda 2014–19 and by the Health 2020 policies and strategies of the European Union and then developed in Italy in the 90s through the network of "Healthy Cities" Municipalities. This approach, present in the WHO document, responds to the Sustainable Development Goals for the creation of the European network of healthy cities. The application process is outlined and the strategic objectives are specified for the promotion of policies and actions for health and sustainable development at the local level. In fact, urban areas are the engines of economic prosperity that can promote health through better access to services and by acting on cultural resources and recreational activities.

Compared to the documents of the previous phases [3], the most recent one focused on the creation of resilient communities and integrated planning tools for health and sustainable development which constitutes an innovative and strong interest also in the disciplinary debate.

In Italy, the Agenda for sustainable urban development then identified the international objectives for the healthy city: the creation of zero-carbon urban logistics systems, carbon emissions by 2030, the recovery of the delay in the provision of public transport, the promotion of intelligent mobility and the development of smart city factors for digital growth [4]. Cities need to work systematically and with a global vision of health and to intervene with integrated strategic health planning policies that involve different sectors and actors, providing shared visions starting from the values of the cities, also by putting all the written and unwritten rules in place. that concern and influence the health of citizens, who are not only simple final beneficiaries of the interventions, but must be part of the conception and promotion of the projects as subjects who know the territory [5]. In recent years, the Administrations are working precisely to be able to intercept policies capable of collecting and systematizing the results of the numerous experiences, both of existing "ordinary" and "informal" programming that are spreading in the European and National panorama.

2   New Tools for the Design of the Health City

The issues of well-being have played a central role in the world debate in recent years. A debate that has seen in particular many professional figures and institutions involved in a common reflection on the design of the physical city towards healthy activities and on the need for a renewal of the tools specifically aimed at directing the processes of redevelopment and construction of space urban taking into account the health effects. Urban health is therefore understood as the individual's physical, mental and social well-being and must be part of an integrated programming system that has public well-being as its purpose.

In Italy, the regulatory instruments of a programmatic, guiding, quantitative and functional nature, such as the plan and direction tools for territorial development policies - both at a territorial scale and locally developed - assign the control of the qualitative and formal aspects of the built project and open spaces only through regulations. Examples are the various hygiene regulations and the health risk and prevention plan that defines the approaches, activities and methodologies to be implemented aimed at intercepting health needs [6].

More and more often, the programmatic and orientative, quantitative and functional regulatory devices - the planning tools designed to direct development policies at the territorial scale or to locally govern land uses - have been accompanied by regulations dedicated to the qualitative and formal aspects of the built and open space project [7].

Some nations more attentive to the issues of urban quality, such as England, associate some texts or frameworks to the regulations, relating in particular to urban design, which are configured as an additional apparatus, not directly prescriptive but which act as a repertoire of guides and manuals drawn up both at the national scale and at that of the individual counties and districts, and which concern cities or parts thereof and where there are indicated principles and examples to be applied locally starting from a careful analysis of the context addressed to the "spatial planning for health".

The English concept has in the background more than twenty years of reflection on the themes of urban health to be obtained through structural and strategic territorial planning tools and does not intend to renounce attention to morphological, aesthetic and functional quality, rather reiterating the need to act on several levels: that of the planning of actions and that of the formulation of project requirements. *Spatial Planning for Health*, understood as an integral part of all levels and aspects of urban planning and architecture, becomes the key character of modern practice, as it does not only deal with the infrastructural project and public spaces, but integrates the interest of health to that of the urban form, from those that concern the fight against diseases to those that favor physical activity up to aesthetic ones, with a strong sensitivity to the broader themes of ecology.

In particular, the Spatial Planning for Health strategy identifies five aspects of the built and natural environment that can be designed and modeled by planners in order to promote certain health outcomes: neighborhood design; housing; healthier food; natural and sustainable environment; transport systems. In practice, it brings together in a single guidance document the city planning tools in order to fully integrate the planning process with health and wellness aspects. The guide does not take the form of mandatory regulation but provides indications and good practices to ensure compliance with health requirements in one's own territory.

The evolution of this practice has allowed experiments that are increasingly attentive to these issues. The Greater London Authority, with the Ministry of Transport, has launched the experimentation, perhaps the most advanced in Europe, of redevelopment projects that focus on the psychological and physical well-being of the population by acting on the reorganization of the London transport. This is the Healthy Street approach, a framework developed by Lucy Saunders, which puts people and their health at the center of the decision-making process and the implementation of pedestrian and cycle road network projects that discourage dependence on cars. This approach is creating good practices that can be replicated and shared with other cities around the world.

The objectives of the framework tend to encourage pedestrian or bicycle travel and the use of public transport, both by improving transport networks and by creating attractive neighborhoods and urban places. The approach is based on ten indicators, two specific on the inclusion and strengthening of walkability and cycling to encourage physical activity, and eight linked to various areas of sustainability: pollution, safety, street furniture, aesthetics and health.

Perhaps it is precisely this type of approach that should be followed in the search for an integration between urban planning and health sciences in urban centers. In a historical moment like the present one, where also in Italy strategic forms of planning are being experimented that see the transition from the classic model of the Plan into a much more complex instrument and much more dense in content, it appears fundamental identify targeted objectives for urban redevelopment that are responsive to the demands of urban, physical and social quality and responsive to the needs of well-being. An approach capable of creating innovative models and tools for the restructuring and regeneration of urban society, improving urban accessibility and safety, increasing environmental quality, management, inclusion and urban well-being [8]. These tools will be aimed at protecting and improving the state of health of the population, up to now exposed to numerous risks, linked precisely to the characteristics of the services and urban spaces, which have compromised their stability. A fundamental component must be covered by the organization of the relationships between the built environment and open spaces, by the modalities of access and circulation since the physical scheme of the city, its potential for growth and its future impacts on the structure and organization will depend on this.

3   Between Research and Urban Planning Practice

The Lastre laboratory of the Heritage, Architecture, Urban Planning Department is a university laboratory that is characterized by the method that tends to finalize the energies coming from the laboratory teaching activities, research, with the converging energies of the activities of comparison, dissemination, territorial guidance and dissemination of the results expressed in the context of the Third Mission.

Among the experiments carried out by the Lastre research laboratory of the PAU, the Capacity project represented the tool capable of creating experimental research approaches to promote the collective participation and sharing of project activities (Fig. 1), with emphasis on the principle of the healthy city and eco-sustainability in the city and more generally territorial context. In this project, the definition of care acquires a broad meaning, not exclusively linked to the health aspect [9], but which broadens towards a different role of local communities that redefines urban welfare.

The *Care Abilities and Professions for an Aggregating City* project was presented in partnership as part of the Call for Proposals of the *Urban Innovative Actions* (UIA). The area is the Pellaro district of Reggio Calabria, a semi-peripheral urban section, of low quality, which suffers from the lack of services and infrastructure. It consists of a building grouped around the original nucleus in an episodic manner that has resulted in uneven structures where the redesign and recovery work aimed at attributing centrality and recognition is also complex.

Fig. 1. Moment of confrontation with the associations and meeting with the community (Ph A. Taccone, 2018).

In fact, the lack of a healthy productive fabric, the decreasing contribution to the economy of agricultural activities and numerous other factors that urban policies alone have not been able to face have created a phenomenon that has favored the formation of these peripheral areas on the margins of the city, of low quality, without services and infrastructures. All this contributes to giving urban areas less vitality with the appearance of increasingly recurrent problems of safety, decay and abandonment.

It was found that in areas with high neighborhood services, excellent health and physical fitness conditions correspond while, in the more marginal and less served areas, cases with physical and mental pathologies are more frequent.

The project aimed at creating an integrated urban system including the fiumara, the area of the peripheral district of Pellaro with the small village of Nocille (Fig. 2) and the coastal strip starting from the forms of economy already existing in the area linked to agriculture, to the small tourist accommodation and the growing sports and leisure activities that are able to actively involve the weak and affected by the processes of urban poverty in a sustainable system at multiple levels - economic, urban, energetic, cultural, identity - structured according to a design criterion of minimum intervention/maximum profit. An important activity is that of the re-design of the wind park (Fig. 3), an important international reality for sailing sports enthusiasts.

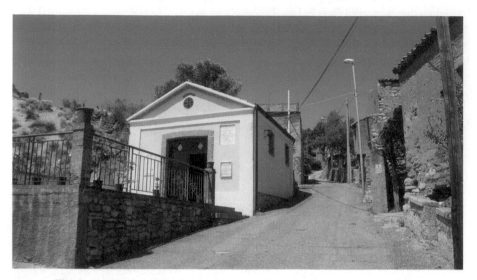

Fig. 2. Relation space in the small village of Nocille (Ph A. Taccone, 2020).

The proposal foreshadows the birth and growth of a series of social and economic initiatives which, through the enhancement and promotion of local identities, attempt to trigger an overall process of redevelopment and sustainable urban integration. In particular, the debate has widened from restricted venues to those of greater dissemination (Urban Innovative Action), which is the European field, reaffirming the need for an "urban" policy in the policies of the European Union. In fact, these policies prefigure an orientation towards an urban perspective, hoping that the reorganization of the Structural Funds and the opportunities offered by the NRRP will not limit the continuation of programs considered successful due to the reduction of the intervention objectives.

The activities, organized according to the collaboration of all partners, each with specific skills, had as their ultimate goal the involvement of the community (public administrators, natives, immigrants, economic operators, etc.) within the process of designing the space. urban and landscape, defining principles, good practices and guidelines for physical, economic and social requalification.

The approach was considered innovative because with minimal intervention in terms of transformation, material and immaterial, of the territory, a maximum economic acceleration is induced through the use of latent or underutilized natural and productive resources. With this experience it is intended to insist on a revolution in the thinking and knowledge of the community as an essential keystone for the implementation of a circular economy model, self-sufficient and lasting, essentially based on actions of inclusion not only social but also cognitive and occupational.

Fig. 3. The area of the Parco del Vento (Ph A. Taccone, 2018).

The philosophy of the approach has led to the conviction that no urban regeneration project, such as the one imagined in Pellaro, can become sustainable, lasting and replicable without a growth in thought and a substantial change in the ways of acting of the community and, more specifically, of the local production system.

Precisely as a result of a necessary rethinking of the relationship between man and the environment in which he lives, it was decided to intervene decisively in redesigning the district in order to guarantee the well-being and health of citizens. Territory as a system capable of reconnecting residential areas, public spaces and natural corridors connecting the neighborhood to the city. The open space was intended as an instrument of urban integration, as the matrix of a new structure of this peripheral area. In particular, the parks, the areas intended for outdoor sports (in particular the Parco del Vento), the green areas play an important role in the activation of internal and external eco-logical-environmental rebalancing processes. of the district, in order to give strength to the idea of ecological network which in many regions of Italy is still a concept not expressed directly or indirectly on the territory.

Furthermore, the project aimed at encouraging the coexistence of different activities and not their segregation in homogeneous and mono-functional areas, taking distances and journeys as the unit of measurement. All this to connect and configure Pellaro as part of the city, including within it a mix of uses such as to encourage different forms and levels of sociality.

The Capacity project, through the circular transition, aimed at improving human well-being, reducing emissions, protecting biodiversity and promoting equity and social justice, in line with the sustainable development goals, towards the creation of a resilient, healthy and competitive city, able to provide for all the social needs of its citizens.

A geographical context such as the outskirts of Reggio Calabria is therefore confronted with different rules, procedures, norms, customs, thus allowing to start a reflection on the meaning of the urban project free from consolidated schemes, towards the search for urban quality associated with open air, greenery, the movement for individual and collective well-being, the provision of services necessary for the protection of human health so that public policies and urban social and environmental regeneration projects can be effectively oriented, also in view of the new "opportunities" offered by post-covid recovery and resilience tools.

References

1. Fehr, R., Capolongo, S.: Promozione della salute nei contesti urbani: l'approccio urban health. Epidemiologia Prevenzione **40**(3–4), 151–52 (2016). https://doi.org/10.19191/EP16.3-4.P15 1.080
2. Edwards, P., Tsouros, A.D.: A Healthy City Is an Active City: A Physical Activity Planning Guide, World Health Organization (2008)
3. Barton, H., Grant, M.: Urban planning for healthy cities. J. Urban Health **90**(1), 129–141 (2011). https://doi.org/10.1007/s11524-011-9649-3
4. Rosa, W.: Transforming our world: The 2030 agenda for sustainable development. In: A New Era in Global Health. New York, NY Springer Publishing Company (2017). https://doi.org/10.1891/9780826190123.ap02
5. Naylor, C., Buck, D.: The Role of Cities in Improving Population Health: International Insights, King's Fund (2018)
6. Trusiani, E., D'Onofrio, R.: Città, Salute e Benessere: Nuovi Percorsi per l'urbanistica. Urbanistica 201. Milano, Italy, FrancoAngeli (2017)
7. Fallanca, C.: Gli dèi della città: progettare un nuovo umanesimo. Milano: Franco Angeli (2016)
8. Farinella, R., Dorato, E.: Paesaggi di margine e forme di vuoto. Percorsi per la costruzione della città attiva. Ri-Vista, **15**(1), 122–137 (2017)
9. Liliana, C., Ennio, R., Danilo, C., Giusi, G., Lucia, P., Rebecchi, A.: La prevenzione della sedentarietà nel Piano regionale della prevenzione lombardo 2015–2018: una strategia intersettoriale per lo sviluppo di programmi evidence-based. Epidemiologia and Prevenzione **40**(3–4), 243–48 (2016) https://doi.org/10.19191/EP16.3-4.P243.091

Toward the Development of a Planning Protocol for Public Space for Improving Health and Wellbeing of Communities

Concetta Fallanca[(⊠)] [iD] and Elvira Stagno[iD]

Mediterranean University, 89124 Reggio Calabria, RC, Italy
{cfallanca,elvira.stagno}@unirc.it

Abstract. In the field of research working on public space planning aimed at increasing the levels of health and well-being of communities, there is a significant experimental body of protocols and guidance. This study aims to identify not only the individual characteristics of the public space that influence health outcomes, already studied in the existing protocols, but also their priorities and relationships, interactions, and effects. The aim of the research is the development of a protocol that provides basic recommendations for urban planners, architects, stakeholders, and politicians interested in the design of public space aimed at pursuing the health and well-being of citizens, at the neighborhood level. We propose an approach to design aimed at increasing health through urban regeneration to considering health not only as one of the elements on which there can be positive repercussions, but as the most important element for urban regeneration for people. For pursuing this scope will be conducted a scoping review approach of protocols, evidence in literature, observations based on improving the outcomes of community health through actions on public space, and interviews to experts. The schematization and discussion of the results of the research, which is still in progress, will represent the framework for defining a planning protocol of the urban public space that affects public health at neighborhood level.

Keywords: Public space · Urban health · Protocols

1 Origin Evolution of the Debate About the Relationship Between Place and Health

The idea that "context" as a place where communities spend their lives is fundamental to individual health is not a new concept and has its roots in the holistic and Hippocratic tradition of medicine [1, 2]. In the 4th century BC Platone, in his work *La Repubblica*, described the city as a pasture, that is, a place that influences and nourishes the growth and development of the people who inhabit it, something great nourisher of society that must be organized and built to be "nourishing. and healthy". Therefore, builders and architects should also be educated and supervised to avoid the presence of weeds that poison the community. Even Hippocrates before Platone maintained that "the doctor to prescribe the right treatment must also be a small urban planner or architect to identify diagnosis and therapy"; therefore, for Platone and Hippocrates the urban planner and the architect are like doctors who build beautiful houses and beautiful cities.

The idea that the project, conceived to involve a therapeutic aspect in which good administration will have a significant impact on health of citizens, has its roots in the overall teleology that places well-being and the promotion of a qualitatively good life as the goal of any practical planning [3]. Following the Platonic assumptions, Emery supports the idea of the city project as a spatial therapy and of the political-performative dimension of architecture as a "social device". Architecture, understood as space therapy, imagines the city as a "nourishing pasture" and space as a "great nurse", according to which the living environment strongly contributes to determining our way of being [4].

Urban planning aimed at pursuing the health and well-being of communities is not just a philosophical concept but finds practical application in the many attempts to solve problems related to hygiene, within cities, over the centuries. It is interesting to note, how Leonardo da Vinci's reflections on the rehabilitation of Milan following the plague of 1484, and then the epidemiological studies in the nineteenth century that tried to shed light on the causes of the transmissible epidemics affecting the populations, found answers exhaustive in the analysis of the characteristics of the built and social environment [5]. Louis-Renè Villermé, a scientific pioneer in the field of social epidemiology, challenged traditional medical hypotheses by shifting the focus of epidemiology to understanding the social, economic, and spatial determinants of health. At the end of the 1800s, the emerging concept was that better living conditions would improve the health of city residents from a physical and psychological point of view, but it could also improve the moral condition of the population [6, 7].

2   Urban Changes and Their Influence on the Health and Well-Being of Communities

The urban environment is a highly complex and interactive socio-physical system, in which various factors associated with the built environment are directly responsible of impacts on health [8, 9]. The built environment characteristics (housing, neighborhoods, social environments, connectivity, density, land use mix, accessibility, services, and decision-making processes) and their design have an indirect impact on health and well-being, because they influence the feelings and behavior of individuals. There is a considerable amount of literature focusing on the relationship between the urban environment characteristics and health, well-being, and social cohesion, with strong evidence associating a greater disadvantages and problems of organizing and managing urban space with a worse physical health and worse mental well-being [10, 11].

Since the early 1990s there has been a notable expansion of theoretical and empirical work that investigates the role of contextual factors in the production and maintenance of health variations. Geographers and sociologists have long argued that place is relevant to health alterations because it constitutes and contains social relationships and physical resources [12–14]. The built environment is a crucial dimension of community health because as a place where people live it is the prime example of a context that affects health levels [15]. According to several studies, the design of our communities, where we spend our life, has a deep impact on our physical, mental, social, environmental, and economic well-being [16]. Some recent epigenetic studies show that there is a close correlation between bio-genetic aspects, individual lifestyles, and the environment in which we live. These are factors that must be investigated methodically and that must lead to identifying the reasons for the different health situations between cities and within them in individual neighborhoods. It appears that epidemiological research focuses not only on sanitation but on socioeconomic and spatial determinants of health [1, 17]. This represents an important advance in studies as a holistic and multisectoral approach is adopted between fields (medicine, epidemiology, planning, social sciences) that generally do not collaborate.

The World Health Organization affirms that healthy city is that in which the physical and social environment is continuously improved and in which the resources available to the community are expanded to allow its inhabitants to be able to help each other in life daily. The WHO also declares that urban growth is increasing exponentially especially in dense urban areas, estimating that in 2050 over 70% of the world population will live in metropolitan areas (7 of 10 inhabitants), bringing with it a load of pathologies that certainly represents a challenge for the world health system. Some health indicators show how the current chronic diseases could decrease if some factors of the built environment were intervened. Current health problems, compared to those strictly related to city hygiene, are more complex, as the elements that contribute to their formation and evolution are. Think of "urban diabetes", one of the most widespread chronic diseases in the industrialized world [18] which afflicts 246 million people (65%) who live in urban centers. Diabetes is only one of the Non-Communicable Diseases (NCDs) of which urbanization is a major cause [19–21].

With health as a result, asset-based approaches are closely related to the theory of "salutogenesis", which highlights the factors that create and support human health rather than those that cause disease [22]. Therefore, the change of planning for health, concerns the exclusive purpose of the intents: pursuing the exclusive purpose of increasing health through urban regeneration and not considering health as one of the elements on which there can be positive consequences. Indeed, the urban organization of functions and spaces can influence and modify the needs of urban populations, their lifestyles, and their life expectancies, since healthy and sustainable communities create the conditions that optimize physical and mental health and well-being by affecting the social determinants of health [23].

3   Ecological Perspective in Urban Planning

Health promotion involves not only educating people about healthy ways of living, but it includes efforts to change organizational behavior as well as the physical and social environment of communities. The ecological perspective is a useful framework for understanding the range of factors that influence health and well-being. Indeed, ecological frameworks can be used to integrate components of other theories and models, thus ensuring the design of a comprehensive health promotion or disease prevention program or policy approach; they are therefore based on a holistic and multidisciplinary, multilevel, and interactive approach [24].

Health implementation programs can certainly be more effective when they are designed to address the interaction and interdependence between factors within and between all levels of a health problem, highlighting people's interactions with their physical environments. and sociocultural. Historically, many fields of health have focused on individual health determinants and interventions. The new approach to health planning should therefore expand its focus to emphasize the social and physical environments that improve health. The ecological perspective emphasizes the interaction and interdependence of factors within and across all levels of a health problem. It highlights people and interactions with their physical and socio-cultural environment. Two key concepts from the ecological perspective help identify interventions for health promotion: first, behavior influences, and is influenced by, multiple levels of influence; secondly, individual behavior shapes and is shaped by the physical and social environment (mutual causality).

In accordance with ecological planning, health commitments should be directed with more conviction on prevention and health protection, rather than on exclusive disease care. Caring for individuals is essential to caring for the environment in its various social, economic, natural, landscape and spatial expressions (ACE onlus).

4   Explanation and Characteristics of the Public Space in the Urban Environment

Public spaces represent the key element of individual and social well-being. They are the places of the collective life of communities, an expression of the diversity of their common cultural and natural heritage, and the foundation of their identity (European Landscape Convention). The community recognizes itself in its public places and pursues the improvement of quality.

Public spaces are the physical network and support for the movement and parking of people; the functioning and vitality of cities depend on them. They offer valuable opportunities for recreation and exercise for all; they promote conviviality, meeting, and freedom of expression. Public space is a place of democracy, an opportunity to create and maintain over time the feeling of citizenship and awareness of the role that each of us has and can have, with daily lifestyle for the environment in which we live.

It is essential to consider urban public spaces as a continuous, articulated, and integrated system, which develops from the neighborhood relations to large environmental systems, to encourage the spread of their enjoyment to the entire community and raise urban quality. Public spaces are places accessible and usable by all and take on various spatial forms; they offer distinct benefits to the physical, mental health and well-being of individuals and communities. Issues of distribution, quality, location, access, and management of public open spaces directly affect human health and health equity.

This study investigates that typology of public space which, by function, shape, meaning, and above all in the built/non-built relationship, mainly plays the role of aggregation or social condensation. In the connecting network of these is the essence of a city [25]. We intend to investigate the possibilities in progress of those areas of public property not yet accessible or no longer accessible and or usable, considered as "potential public spaces"; a precious resource for upgrading and updating the existing public space system that contributes to urban quality. That type of small-scale public space, built, potential or disused, branched out within the neighborhood. The neighborhood, which represents the primary physical unit that makes up the urban pieces, is, for the logic of proximity, the space most experienced by all the inhabitants; it is precisely at the neighborhood level that the closest sense of community emerges. The urban system of public spaces, as a network of elective places for associated living, requires an overview that highlights the peculiarities to be maintained, enhanced, and communicated.

There are several factors that hinder the creation and maintenance of public spaces. Difficulty of many local authorities in assuming an effective role of public direction; real or perceived insecurity of public spaces, with consequent effects of poor attendance, abandonment, and decay; weakening of social cohesion and increasing frequency of vandalism [26]. The use of public spaces is an indicator of their quality, to be used throughout the entire creation-management-use cycle. Indeed, the good use of public spaces is closely linked to their mutability and adaptability, in relation to the evolution of citizens' needs. Therefore, it is advisable that local governments adopt a specific document of guidelines for the validation of the network of public spaces. And we are referring above all to local authorities, located in marginal areas, in which the skills and design structure is lacking in ideas and budget. In this sense, the construction of a planning protocol for the public space based on minimal interventions which aims to obtain a good qualitative basis of the space that responds to the needs and problems of current living becomes useful.

Considering health as an input and output would therefore mean thinking of public spaces that have those characteristics of attraction within the community and that increase the levels of health and well-being of the individuals who live there.

5   Protocols and Strategy Models for Promoting Wellbeing Through Public Space in Urban Environment

Although the general agreement on the complex interplay between the factors at the individual, family, organizational and community levels that influence community health outcomes, there is a considerable gap between research and the practical application of its results in planning for health promotion. There is a discrepancy between multiple theories and models of health promotion and the need by professionals for a more unified set of guidelines for comprehensive program planning [27]. The analysis of the literature allows us to underline how the protocols and guidelines created up until now have in common the same purpose of pursuing health through planning, but differ depending in methodology, scale of application, level of detail of the interventions, geographic region, conceptual framework.

Some protocols, those used for example in English planning, work at territorial levels and the spatial elements investigated and, the degree of detail of the proposed interventions depend on this. The *London Plan*, for example, is a national spatial development strategy on which all local development documents are based at the district level, as well as district plans promoted by the communities. These documents are arranged into policies divided by macro-topics for which targeted strategies are proposed. Among these, the *Healthy Urban Planning Checklist*, is the document that, focusing on urban planning for health and well-being, brings together the requirements and standards of planning policy that influence health and well-being, providing targeted indications to integrate the health in the planning system. Other protocols further analyze the compositional factors, therefore the socio-economic aspects within the urban environment; this is the case of the study protocol *VOORSTAND ON THE MOVE*, promoted and implemented by the municipality of Deventer (NL), in partnership with experienced researchers in community involvement for health promotion; the protocol aims to develop the skills, empowerment and learning of the population on health issues, in a disadvantaged neighborhood from a socio-economic point of view. Other protocols focus on contextual factors, therefore on the spatial factors of the built environment, such as the study *Guidelines for healthier public spaces for the elderly population: Recommendations in the Spanish context*, which provides guidelines and recommendations on public space planning to promote the well-being of the elderly population and its integration with younger age groups within the same physical space of interaction. Other protocols provide far-reaching recommendations on compositional and contextual factors together, such as *HEALTHY PLACES INCLUDED. A Guide to Inclusion & Health in Public Space: Learning Globally to Transform Locally*, promoted by the Gehl Institute in collaboration with the Robert Wood Johnson Foundation. The study focuses on the principle of inclusion to promote equity in health through public spaces. It is based on the analysis of the context and how this affects health equity, on the analysis of the processes that shape the public space by promoting civic trust, participation, and social capital, on the design and programming of a quality public space that foster health equity and support social resilience and the ability of local communities to engage in addressing change over time by promoting representation, action, and stability.

The authors and promoters of the guidelines are different as well and affect their contents. Those proposed and promoted by governments, such as the WHO, are based on a very broad approach to the topic; they address the issue of health as a global priority in the annual programming and embrace both the compositional and contextual factors of the urban environment, without providing detailed prescriptions. Furthermore, they do not refer to a specific geographic region, but consider the planning of a global space. We refer to documents such as *Health as the Pulse of the New Urban Agenda*, promoted by the WHO. In general, all protocols are based on a methodology that depends on the scale of application and the detail of the interventions, which is always based on a theoretical framework that come from the scientific literature. Some protocols provide both practical indications and method of application, some focus on one of the two aspects. The *Open Public Life Protocol*, promoted by City of San Francisco's Planning Department, Copenhagen Municipality's City Data Department, Seattle Department of Transportation, provides methodological indications and standard data collection to be used in construction of a protocol for the public space.

Considering the above, this study aims to collect data, regarding the protocols and strategy models taken so far and investigate the priorities and relationships, as well as the interactions and effects, between the elements of the public space that influence public health. Investigating the actual effectiveness, the qualitative scale, the theoretical model used and the resulting deductions, could allow the development of a synthesis protocol of the most interesting data and the most effective actions introduced so far.

6 Methodological Framework for Developing the Protocol

To date, there are several protocols, general guidelines and frameworks that prescribe practical and methodical actions on the design of public space; the aim is to provide the structure and directions needed to help neighborhood projects reach their full potential, or to provide a common framework that make up for the lack of collaboration between public health and the planning sector. Given the complexity and general order addressed in the protocols described above, this study aims to deepen the protocols, guidelines and strategic models concerning the planning of public space for health, at the neighborhood level. Providing a protocol of actions on small-scale public space could be effective mostly for small institutional and municipal entities, located on the physical and social margins, which do not have the economic resources and design structures suitable for planning the urban space that has as its input and output the pursuit of the health and well-being of the community.

The study is guided by the following research question: *How can priorities and relationships, interactions, and effects between the physical elements of the public spaces, according to existing protocols, be used to affect public health?*

In view of the large range of protocols and guidelines existing, the analysis will be conduct through the scoping review method. The scoping review allows to obtain a general framework of a potentially large and diversified body of literature, related to a vast topic, providing a descriptive overview of the material examined [28, 29]. The scoping review is guided by a search string consisting of keywords and selection criteria established through the analysis of the literature. This phase represents an important step to avoid incurring accumulations of data that are difficult to manage and dubious value. The search string uses the words "public space OR urban space", "health OR urban health", "urban planning OR city planning", "protocols OR guidelines". For the accumulation of data, was used the scientific databases such as Scopus, Google Scholar, Researchgate, Jstor.

After the collection of articles and protocols following the search string, criteria for choosing the most suitable material were established [30–32]. The articles chosen are based on the relationship between the characteristics of the urban public space and the relative influence on health and well-being; on the characteristics of the urban public space at the neighborhood level and its influence on elements such as walkability, play, rest, community behavior. The protocols chosen concern small areas of open space (pochet parks) of size under 3000 m^2, excluding urban parks. The protocols, guidelines and regulations considered, regard physical inactivity, public space, environmental aspects that had been developed since 1990.

The scoping review, which has an umbrella structure, was organized according to three themes. The first theme was defined as a practical inventory and concerns the protocols corresponding to the keywords and criteria established according to the literature; it also includes interviews with the authors of some particularly interesting protocols or those presenting problems that are not well explained and is supported by field observations. The second theme concerns the evidence in the literature that highlights the correlations between health and public space, and satisfies the criteria established thanks to the reference literature [33, 34]. The third theme is iteration, and it is based on expert opinion about the correlation between the results of the first two parts of the investigation. It represents a kind of post hoc analysis of the results. The selection of experts is based on multidisciplinary, which allows a holistic and complete approach to the topic; Italian and Dutch politicians, teachers, doctors, and planners were chosen.

This methodological part of the study was planned during a one-year internship at the Technische University of Eindhoven - TU/e - at the research unit of the Built Environment Department.

7   Early Results Achieved for a Protocol for the Developing Protocol for Public Spaces

If the purpose of planning were not for health, what could it be? (Un-habitat, 2016). There is no doubt that good planning means an increase in health, and this is a finding established over time. Once the indirect link between the physical environment in which we live and the actual and perceived state of health of individuals is established, the two elements must be correlated in terms of planning and strategies so that the increase in health becomes both an input and a result.

The first results achieved from the scoping review of this study, which discloses the results of a research still in progress, show that one of the recurring interactions in current protocols certainly concerns the approach aimed at community involvement, civic participation, and the enhancement of human and social capital. The effect that arises from the active participation of the inhabitants of a neighborhood in the design of public space increase the social contacts and, consequently, reduces the actual and self-perceived social inequalities and the sense of exclusion and loneliness, promoting the psychological well-being of individuals. Even the analysis of the context represents a periodic interaction; understanding and recognizing the community context, in which one wants to act, based on existing conditions, resources and, experiences in place, has the effect of planning interventions aimed at solving the major issues related to the lack of health and psycho-physical well-being of individuals.

For this aspect, the importance of the method and process adopted is evident. It emerges that the primary objective of the investigation method is to establish the trend of major pathologies, related to real and self-perceived health within the community. This is followed by an investigation into the perception of well-being that the living space has on the inhabitants, then, an analysis on the physical elements considered by individuals as carriers of health and well-being dynamics. The investigation tends to be conducted according to a multidisciplinary approach that involves experts from different disciplines and not only in the medical field. Other recurring interactions are those concerning comfort, safety, ease of movement, the mix of age groups and a strong presence of urban greenery, both on site and in proximity; the effect translates into an increase in the livability of public spaces, social relations, and the dynamics of attraction of space and, physical activity. Indeed, the nature of human beings is to adopt behaviors and habits that positively affect their physical and psychological health, be it real or perceived. The places that reveal health within the city can only be a strong attraction for people which will want experience them since they will satisfy his psycho and physical well-being.

Even the scientific evidence studied until now identifies the elements mentioned above, as those most positively associated with the health and well-being of individuals. Sense of safety, understood as safety and security, active community involvement, bottom-up processes, interventions on a small territorial scale, emerge as the most recurring elements associated with the planning of a public space capable of increasing the levels of effective health and self-perceived well-being of individuals.

References

1. Macintyre, S., Ellaway, E., Cummins, S.: Place effects on health: how can we conceptualise, operationalize, and measure them? Soc. Sci. Med. **55**, 125–139 (2002)
2. Meade, M.S., Earickson, R.: [BOOK REVIEW] medical geography. Soc. Sci. Med. **54**(6), 998–999 (2002)
3. Giannantoni, G.: Il primo libro della "Repubblica" di Platone. Rivista critica di storia della filosofia **12**(2), 123–145 (1957)
4. Emery, N.: Progettare, costruire, curare. Per una deontologia dell'architettura, Edizioni Casagrande s.a., Bellinzona, 1st edn. 2007 (2010)
5. Corbun, J.: City planning as preventive medicine. Prev. Med. **77**, 48–51 (2015)
6. Rosen, G.: A history of Public Health. The Johns Hopkins University Press (1993)
7. Hamling, C., Sheard, S.: Revolutions in public health: 1848, and 1998? BMJ (1998)
8. Rao, M., Prasad, S., Adshead, F., Tissera, H.: The built environment and health. Lancet **370**(9593), 1111–1113 (2007)
9. Faggioli, A., Capasso, L.: Inconsistencies between building regulations in force in Italy for indoor environment and wellness factors. Annali di igiene: medicina preventiva e di comunità **27**(1), 74–81 (2015)
10. Ludwig, J., et al.: Neighborhood effects on the long-term well-being of low-income adults. Science **337**(6101), 1535–1510 (2012)
11. Roux, A.V.D.: Neighborhoods and health: what do we know? What should we do? Am. J. Public Health **106**(3), 430 (2016)
12. Jones, K., Moon, G.: Medical geography: taking space seriously. Prog. Hum. Geogr. **17**(4), 515–524 (1993)
13. Kearns, R.A., Joseph, A.E.: Space in its place: developing the link in medical geography. Soc. Sci. Med. **37**(6), 711–717 (1993)
14. Macintyre, S., Maciver, S., Sooman, A.: Area, class and health: should we be focusing on places or people? J. Soc. Policy **22**(2), 213–234 (1993)
15. Duncan, D.T., Kawachi, I.: Neighborhood and Health, 2nd edn. Oxford University Press, New York (2018)
16. Dannenberg, A.L., Frumkin, H., Jackson, J.: Making Healthy Places: Designing and Building for Health, Well-Being, and Sustainability. Island Press, Washington, DC (2011)
17. World Health Organization, Urban population growth (2018). http://www.who.int/gho/urban_health/situation_trends/urban_population_growth_text/en/. Accessed 26 June 2019
18. Penno, G., et al.: Hemoglobin A 1c variability as an independent correlate of cardiovascular disease in patients with type 2 diabetes: a cross-sectional analysis of the Renal Insufficiency and Cardiovascular Events (RIACE) Italian multicenter study. Cardiovasc. Diabetolo. **12**(1), 1–13 (2013)
19. Barton, H., Grant, M.: A health map for the human habitat. J. R. Soc. Promotion Health **126**(6), 252–261 (2006). https://doi.org/10.1177/1466424006070466. Accessed on 22 July 2019

20. World Health Organization Measuring health gains from sustainable development. Sustainable cities, food, jobs, water, energy, disaster management. Public Health & Environment Department (PHE) Geneva, Switzerland: World Health Organization (2012). http://www.who.int/hia/green_economy/sustainable_development_summary2.pdf?ua=1. Accessed July 2019

21. Wang, H., Naghavi, M., Allen, C.: Global, regional, and national life expectancy, all-cause mortality, and cause-specific mortality for 249 causes of death, 1980–2015: a systematic analysis for the Global Burden of Disease Study 2015. Lance **388**, 1459–1544 (2015)

22. Kelly, M., Morgan, A., Ellis, S., Younger, T., Huntley, J., Swann, C.: Evidence based public health: a review of the experience of the National Institute of Health and Clinical Excellence (NICE) of developing public health guidance in England. Soc. Sci. Med. **71**(6), 1056–1062 (2010)

23. Marmot, M., Allen, J., Bell, R., Bloomer, E., Goldblatt, P.: WHO European review of social determinants of health and the health divide. Lancet **380**(9846), 1011–1029 (2012)

24. McLeroy, K.R., Bibeau, D., Steckler, A., Glanz, K.: An ecological perspective on health promotion programs. Health Educ. Q. **15**(4), 351–377 (1988)

25. INU Carta dello spazio pubblico (2013). http://www.biennalespaziopubblico.it/wpcontent/uploads/2016/12/CARTA_SPAZIO_PUBBLICO.pdf. Accessed June 2020

26. Gehl, J.: Cities for People. Island Press (2010)

27. Gehl Institute, Inclusive healthy places. A Guide to Inclusion & Health in Public Space: Learning Globally to Transform Locally (2018). https://gehlinstitute.org/wp-content/uploads/2018/07/Inclusive-Healthy-Places_Gehl-Institute.pdf. Accessed Mar 2021

28. Best, A., Stokols, D., Green, L., Scott, L., Holmes, B., Buchholz, K.: An integrative framework for community partnering to translate theory into effective health promotion strategy. Am. J. Health Promot. AJHP **18**, 168–176 (2003)

29. Pham, M.T., Rajić, A., Greig, J.D., Sargeant, J.M., Papadopoulos, A., McEwen, S.A.: A scoping review of scoping reviews: advancing the approach and enhancing the consistency. Res. Synth. Methods **5**(4), 371–385 (2014)

30. Higueras, E., Román, E., Fariña, J.: Guidelines for healthier public spaces for the elderly population: recommendations in the Spanish context. In: Martinez, J., Mikkelsen, C.A., Phillips, R. (eds.) Handbook of Quality of Life and Sustainability. International Handbooks of Quality-of-Life, pp. 35–51. Springer, Cham (2021). https://doi.org/10.1007/978-3-030-505 40-0_3

31. De Jong, M.A., Wagemakers, A., Koelen, M.A.: Study protocol: evaluation of a community health promotion program in a socioeconomically deprived city district in the Netherlands using mixed methods and guided by action research. BMC Public Health **19**(1), 1–11 (2019)

32. Gehl Institute: The Open Public Life Protocol (2017). https://gehlinstitute.org/wp-content/uploads/2017/09/PLDP_BETA-20170927-Final.pdf. Accessed Mar 2021

33. NHS London Healthy Urban Development Unit Healthy urban planning checklist (2014). https://www.healthyurbandevelopment.nhs.uk/wpcontent/uploads/2014/04/Healthy-Urban-Planning-Checklist-March-2014.pdf. Accessed Apr 2021

34. World Health Organization Health as the pulse of the new urban agenda (2016). https://www.who.int/publications/i/item/9789241511445. Accessed Apr 2020

The Eco-Neighbourhoods: Cases to Learn in the Transition Toward Urban Sustainability

Maria Fiorella Felloni[✉]

Politecnico di Milano, 20133 Milano, Italy
mariafiorella.felloni@polimi.it

Abstract. The Eco-neighbourhoods as spatially based experiences for learning the implications of urban design transition towards sustainability is the core point of the paper and considers the short retrospective of the 1990s when searching for sustainable development became a multi-targets win-win situation in the framework of the well-known United Nations reports: socially inclusive, economically viable, natural resource-conserving and dealing with climate change responsibilities.

The contribution is based on research and teaching activity where eco-neighbourhoods, chosen as cases for learning, share the common characteristics of having a recognizable unifying character due to a unitary design and having been built to represent a recognizable spatial cluster of beneficial innovations for sustainability and quality of life. To support the discourse, four characteristics – participated land use and balanced densities, mobility, open space and nature, energy and water - were adopted as evaluation criteria. Finally, some ideas for a perspective vision of eco-neighbourhoods are outlined for achieving the 2030 Agenda goals.

Keywords: Eco-neighbourhood · Urban sustainability · Urban design

1 Guidelines and Practices for a Healthy, Sustainable Urban Life

One could argue that the most challenging disciplinary goal of urban planning and design since the 20th century has been to create healthy urban spaces. This is true to the extent that the concept of health in reference to the urban environment has a dynamic rather than sectoral meaning, aimed at solving multi-scalar and multi-thematic problems. In the most recent decades, the European report on sustainable cities pointed out the following guiding concept: "Good health depends to a large extent on a healthy environment and the health of an urban population is therefore dependent on physical, social, economic, political and cultural factors related to the urban environment. Furthermore, the impact of urban processes on public health is not simply the sum of the effects of the various factors, because these factors are highly inter-related in cities" [7].

In the 1990s, guiding principles were established for European cities promoting new planning practices with an integrated approach, favouring protection of the environment, human health, greater social equity and economic viability. Sectoral environmental urban

F. Calabrò et al. (Eds.): NMP 2022, LNNS 482, pp. 559–569, 2022.
https://doi.org/10.1007/978-3-031-06825-6_53

weakness affecting health such as air and water quality, noise levels, road safety and resources consumption, are addressed as signals of an inter-related and deep-seated social and economic crisis which forced planners to rethink the modern models of organization and urban development. [6].

It is the moment in which the most internationally people-centred definition of sustainable development, coined by the Brundtland Commission, takes root: 'development that meets the needs of the present without compromising the ability of future generations to meet their own needs [17].

At the same time, it is certain that since the mid-20th century humans have had an unprecedented impact on earth's climate system and caused change on a global scale. The UNFCCC clear distinction between climate change attributable to human activities altering the atmospheric composition and climate variability attributable to natural causes established the responsibility of the economic and social models of the development through 20th century [16].

In this framework, searching for sustainable development has become a multi-targets win-win situation: socially inclusive, economically viable, natural resource-conserving and dealing with climate change responsibilities. The critical need to get beyond these challenging theoretical concepts and make sustainable urban planning and design a reality, draws the attention of various observers, professionals, academics and stakeholders to the potential of the eco-neighbourhood [1, 3, 5, 8, 9, 12].

The pioneering positions most convinced of a vision of sustainable urban development based on locality and implemented at the neighbourhood scale, focused on the concept of neighbourhood to be taken as a reference: "Neighbourhood is a residential area around which people can conveniently walk. Its scale is geared to pedestrian access and it is essentially a spatial construct, a place. It may or may not have clear edges. It is not necessarily centred on local facilities, but it does have an identity which local people recognize and value' [3].

This definition is the premise of an acceptable position that distances itself from an approach that believes this kind of neighbourhood can be created only by design and assumes the social identity and the walkability as key elements of sustainability.

However, it should be noted that in the current scenario of the 2030 Agenda for Sustainable Development, the experiences of eco-neighbourhoods deal largely with new challenging topics, as proposed in the following paragraphs, concerning participation, resources consumption, energy, nature-based solutions [2, 8, 13, 17].

2   Built to Be Eco-friendly

In the research and teaching activity that is briefly narrated in this contribution, the point is focused on sustainable neighbourhoods as spatially based experiences for learning the implications (especially techniques) of urban design transition towards sustainability. Accordingly, the eco-neighbourhoods examined have some common attributes in that they have a recognizable unifying character due to a unitary design, they are the result of the application of principles concerning urban sustainability and climate change responsibility and they have been built to represent a recognizable spatial cluster of beneficial innovations for sustainability and quality of life.

To support the research and study cases analysis, some key characteristics were adopted as evaluation criteria. The intent was to address the issue of sustainable urban settlements with climate adaptation and mitigation solutions at several levels of the urban project, namely the main physical components relating to mobility, buildings, green and open spaces, the consumption of resources – energy, water, soil - and their integration into the conceptual and strategic objective of the quality of life.

As part of the urban planning studio teaching activity for the first-year students of the bachelor's degree in architectural design, the analysis of eco-sustainable neighbourhoods has been proposed as a method of learning new urban design techniques which have implemented the guidelines and performance mentioned above.

The cases examined in the teaching activity concerned projects carried out between the end of 1980s and the present day. Here we will refer to three first generation districts built in the last fifteen years of the twentieth century and in the very first years of the twenty-first. On these it is possible to outline some thoughts also regarding achievement factors, weaknesses and the lessons learned in the perspective of the ecological transition that today globally engages urban environments.

To support the study cases analysis, some characteristics – participated land use and balanced densities, mobility, open space and nature, energy and water - were adopted as evaluation criteria. These four main characteristics are addressed in the next paragraphs. In the table below the identity card of the cases considered (Table 1).

Table 1. Identity card

Neighbourhood	Location	Size (hectares)	Goal
Bed-ZED -Beddington Zero Carbon Energy Development	Greenfield. Sutton, south London	1.7	To realize a near carbon-neutral eco-community
Egebjerggard	Greenfield. Ballerup, suburb of Copenhagen. Located in one rail corridor of the Rasmussen Five Finger Plan	37	Beyond the suburbs of the modern functional city
GWL Terrein	Brownfield, former site of the municipal water company. Amsterdam, West District. At the edge of the late 19th century city extensions	6	A totally car-free lifestyle

2.1 Participated, Mixed Uses and Balanced Densities

Functional and social mixed use is a basic mechanism for social sustainability. This principle initially expressed the desire to break with the planning practice of the function-divided city and the excluding suburban buildings of the industrialization. In the multi-functional design logic, business and social aspects coexist. People of different ages and even social backgrounds can live, socialize and work within a compact geographic area (hopefully within easy walking distance) so that social equity, security and inclusion is better realized and maintained. Compatible uses such as residential, commercial and retail businesses, form successful, healthy and attractive urban districts. In addition to this, the definition of the densities and the mix of functions must also be the result of participatory proposals by the future inhabitants of the neighbourhood.

Along with mixed uses, appropriate densities are an important factor for sustainable development. It is widely suggested that higher density development is the best design practice to use energy more efficiently, reduce the necessity of travel, increase accessibility to public transport, provide more local employment and reduce the need for private car parking [6]. Ideal sectoral gross population densities were also defined; 40 to 50 persons per hectare for example was established as a reference index for less travel planning [15]. It must be noted in any case that the density parameter in a sustainability approach, cannot be sectoral but must involve several design components. Density had to vary according to the location and accessibility of different sites, their cultural context and settings, and the different types of housing being provided. So, it is better to explore how and if densities are appropriate, case by case, with reference to location, the functions and services provided, rather than rigidly associate the concept of sustainability with high density. In the neighbourhood examined density varies significantly, even if in no case is it less than 40 persons per hectare.

In Egebjerggard, when Ballerup Municipality acquired the area in the mid-1980s, it was decided to announce a competition to get visions and ideas for a functional and social mixed urban development. One of the experimental ideas was to make an urban district catalogue collecting the suggestions of different stakeholders. It was the way to integrate future users in the experimental organizations at the earliest possible stage. The catalogue included contributions from future residents, builders, the business community and the permanent staff of schools and institutions. They were involved in planning, realization and also deciding on street names, guidelines for the design of the streets and choice of urban equipment. Functional and social mixed needs also emerged clearly in the Ten Point Programme established for the project where the goals referable to social sustainability are specifically aimed at several aspects: variation in households, mixed ownership, affordable housing, mixed land use, integration of works of art, strengthening the social structure, crime prevention and participation of future inhabitants. These goals and the Ten Point programme were later all incorporated into the Ballerup General Municipal Plan.

In BedZED, to ensure a good functional mix and give vitality to the neighbourhood, buildings were designed with working and living places together. But the 'live-work' units of the solution were difficult to market because of limited demand and because they would be treated as business space so far as local authority rating was concerned, so they were marketed as spacious homes [4]. A mix of social and market rate housing is guaranteed with 25% subsidized rent, 25% affordable home ownership and 50% for rent in open market. A specific survey explored the level of socialization with respect to the surroundings and previous homes: BedZED inhabitants knew on average 20 of their neighbours by name, while two thirds felt they knew more people living in BedZED than they did when living in their previous neighbourhoods. In the surrounding neighbourhood residents knew an average of eight neighbours by name [4].

GWL-Terrein regeneration is based on a functional and social mix that includes residences and other services for the community, typological and forms of property and rental differentiation. Shared flats, studio flats, flats for the disabled, a project of assisted living for children with multiple disabilities, commercial spaces and a neighbourhood house are also available. The water tower and the 19th-century pumping station have been restored and host restaurants and cafes (Table 2).

Table 2. Participated mixed land use and densities

Neighbourhood	Goals for mixed social and functional uses	Densities*	Participation initiative
Bed-ZED	Working and living places together Subsidized and open market housing	129	Monitoring through interviews and periodic surveys to inhabitants
Egebjerggard	Specific point programme for social mix and inclusion Children, teenagers, elderly, women, people with disabilities Neighbourhood facilities in the site	45	District catalogue with the contribution of different stakeholders integrated in the project
GWL Terrein	Social and market rate housing Commercial activities in the site Integration of historical buildings	233	Umbrella association for the site for house owners, tenants and voluntarily companies aim to guard and encourage the green character of the site

* Person per hectare

2.2 Mobility

At the local level and at the neighbourhood scale, different mobility solutions influence quality of life, economic appeal and to some extent social cohesion. At the urban and conurbation scale we know that private urban transport powered by fossil energy sources are still estimated to account for around one quarter of CO_2 emissions in Europe. Mobility solution alternatives to the traditional private and carbon-based system are a crucial focus of the pioneering experiences of sustainable neighbourhoods. Ensuring good accessibility to public transport, providing walking and cycling networks inside the neighbourhood development and with the surrounding urban environment and restraining cars parking are the principal design rules implemented in the perspective of the walkable communities and cities.

GWL Terrein development stands out among the cases studied as a successful car-free redevelopment with limited parking, car sharing provision and good transit access.

Non-motorized mobility in the development is much higher than the surrounding urban area and car use is much lower. The complex is a trial for decreasing car use and even car ownership, due to the only available parking spaces on the edge of the complex. At the same time a decrease in car use led to the car sharing mode, bike use, walking and the public transportation and so on.

In the BedZED and Egebjerggard examples, the location of car parking for the resident is outside or in the edge of residential area and the car parking place per dwelling unit is planned law (Table 3).

Table 3. Mobility

Neighbourhood	Goals and measures/rules for sustainable mobility	Car parking place per dwelling unit	Car parking for resident
Bed-ZED	Traffic free streets running between the building blocks Bicycle parking and bike storage space inside larger homes Local Car club offering a low-emissions, highly fuel-efficient hybrid car to any car club member	0.6∗	Outside the residential area
Egebjerggard	Internal paths accessible only to pedestrians and bicyclists		In external parking area of the site
GWL Terrein	Car-free eco district. Car traffic is not allowed inside the site Car sharing provision All residents signed a letter of intent stating that they were aware of the environmentally friendly and car-free character of the settlement	0.2∗∗	Outside the residential area, in a strip along the edge of the site

∗ For residents, visitors and people who work there
∗∗ Only for residents.

2.3 Nature and Open Spaces

Since the modern period, the presence of nature and open spaces for residential development have been considered as basic needs for health, social inclusion and environmental issues, along with facilities for cultural and educational purposes. Minimum mandatory quantities per inhabitant became compulsory rules in urban planning in Italy and other EU state members. In the scenario of the 1990s, green areas and open public spaces started becoming an integral part of the new urban environmental project based on nature solutions, being valued for sustainable urban drainage and climate adaptation and mitigation.

In Egebjerggard the approach to green area and open space was guided firstly by the idea of a socially and environmental integrated neighbourhood responding to different needs such as helping the elderly for socialization and providing proximity services to children and disabled people. Each housing group was equipped with clearly perceptible open spaces such as a square, a green area or a street. Street lighting style and works of art are part of the identity of the open spaces. Green spaces of different sizes, location and function were carefully designed to achieve environmental benefits. In addition to the small green areas guaranteed inside the individual housing groups, a large green belt surrounding the residential area with lakes handles ecological aspects and drainage of all rainwater on site. Landscape design is part of the project, defined by a network of paths connecting internal green spaces with the surrounding agricultural and wooded areas.

In GWL-Terrein green spaces are of similar sizes, distributed in a non-hierarchical system interspersing the built spaces, they are also used by the residents outside the neighbourhood. Green roof terraces, private gardens for ground floor apartments and kitchen gardens are the main types of solutions adopted. Residents are responsible for the condition and maintenance of public spaces and places of the settlement were unsealed and greened.

In BedZED the open space and green space were designed with technologies and solutions to contribute to the carbon free community goal (Table 4).

Table 4. Nature and open spaces

Neighbourhood	Goals and measures/rules for nature and open spaces	Nature based solutions	Functional open spaces
BedZED	Open and green spaces contribute to the driving target of the carbon free community Drainage of rainwater inside the site	Green roof Private gardens Outdoor space in the form of small gardens at ground, first or second floor level Tree-lined green ditch	Roof, roads and green area for urban drainage Open spaces and streets for neighbourliness
Egebjerggard	Topic elements of a socially and environmentally integrated neighbourhood Drainage of all rainwater inside the site Landscape design	Large green belt with lakes Small green public areas Green areas and lakes for urban drainage	Street lighting style and works of art define the identity of open spaces Traffic-free outdoor spaces for neighbourliness
GWL-Terrein	Mixed green solutions for private and public uses	Green roof Green public space Private gardens and parting hedges Kitchen gardens	Traffic-free open spaces for neighbourliness

2.4 Energy and Water

To save energy and use green energy, save water and improve the water cycle are today paradigmatic targets and rules of the economic, environmental sustainability and climate adaptation and mitigation. Already in the pioneering experiences examined, new technological solutions for water and energy occupied a significant part of residential and mixed developments. In the case of BedZed, the goal of a zero-carbon energy community was, as is well known, guiding the whole development. In monitoring activity [4] emphasis is given to how the innovative and experimental technical solutions adopted have frequently been reconsidered over time in the light of functional deficiencies or excessive maintenance costs so that these pioneer cases are useful to know what worked and what did not and which, in all likelihood, may have become the new lines of research in the technological field.

In *BedZED*, designed to be free of fossil-fuel consumption, to safeguard water by consuming less than conventional new housing development (about 87 L per person per day, - 61% of the average in Sutton in 2007), we can find the use of most of the technologies available and desirable at the end of the 20th century. Wood-burning Combined Heat and Power plant (CHP) was replaced by three conventional natural gas-fired boilers which supplied heat for the district heating scheme. In 2017 a new biomass boiler was installed burning wood pellets – a near zero-carbon fuel. Photovoltaic panels and a passive house approach implemented high levels of buildings insulation, airtightness and thermal mass and maximum use of the sun's warmth for space heating. A mini owned sewage treatment plant was realized but closed in 2005, because of its high electricity consumption and because the process needed intensive monitoring. The rainwater collection system of the original project is no longer in use because of concerns about the harvested water being contaminated from the green roofs [4].

In Egebjerggard, Skotteparken low energy housing design aims to significantly reduce energy and water consumption by incorporating a range of energy saving measures, as well as decentralized solar heating systems and a low temperature district heating scheme is used for public facilities and housing.

A passive house approach is widely used with south facing large-glazed facades to use the sun's warmth for space heating, insulation and sealed buildings. In low energy housing roofs are made of steel and solar collectors are integrated into the roof construction. Rainwater is not channelled into the sewers but runs through open gutters into small permeable lakes.

In GWL Terrein saving on water use, water recycling and drainage were strategic goals emphasized by the fact that the district has been built on the former water-company site. Rainwater is stored in an open canal in the site connected to the network of surface channels of the city. This limits the load on the sewer and surface water (Table 5).

Table 5. Energy and water

Neighbour	Energy	Water recycling and saving	Urban drainage
Bed-ZED	Combined Heat and Power plant (CHP), replaced twice with other solutions Photovoltaic panels Passive house approach	Owned sewage treatment plant (no longer in use) Rainwater collection system from green roofs (no longer in use)	Roof, roads and green area for urban drainage
Egebjerggard	Mini scale distributed Combined Heat and Power (CHP) plant Passive house approach Solar collectors integrated into the roof construction	Rainwater catching for watering gardens and other uses	Rainwater lakes
GWL Terrein	CHP Combined heat and power plant Insulation Use of passive solar energy There is no energy production from sun or wind	Special toilets to save water Use of grey (rain) water to flush the toilets Showerheads to save water and water limiters Open gutters to drain rainwater to a catchment reservoir;	About two thirds of unpaved area Rainwater canal

3 The Perspective: What Has Changed (is Changing) in the Scenario of the 2030 Agenda

The interest in eco-neighbourhoods continued to feed on projects carried out in the last two decades [5, 9, 12]. They stand out from the pioneering experiences mentioned above for the realization opportunities thanks to funding for major international events. The best known and examined cases in the planning studios are Kronsberg | Hannover EXPO 2000; Bo0 | Malmö | European housing Expo 2001; Hammarby Sjöstad | Stockholm, following the lost competition for the Olympic Games 2008.

The new generation of eco-neighbourhoods are part of large-scale transformations and regenerations: Hammarby Sjöstad | Ørestad - Copenhagen | 26.000 inhabitants; Eco-Vikki - Helsinki |17.000 inhabitants; Bo01 | Västra Hamnen |- Malmö 10.000 inhabitants. Large developments both on brownfields (Hammarby Sjöstad, Bo01) and greenfields (Ørestad, Vikki).

In some cases, sustainability concerns a radical regeneration of existing open and green spaces in large districts of the historical-consolidated city; the traditional function of these spaces is specifically adapted for further performance and functional requirements for adapting to climate changes. These are projects planned as part of municipal action plans for energy and climate.

These are examples of interest in retrospect and in perspective. In retrospect they helped to draw a critical line of evaluation of the approach we had which was sustainable neighbourhoods as spatial based experiences for learning the implications of urban design transition towards sustainability. In perspective, they help to build knowledge and understanding how local spatial design experiences had to evolve and contribute according to the new universal call of the 2030 Agenda. Here, the goals of 'Make cities and human settlements inclusive, safe, resilient and sustainable' (Goal 11) and 'Take urgent action to combat climate change and its impacts (Goal 13) are guided by a more challenging concept of sustainability based, as it is known, on the following five areas of critical importance: people, planet, prosperity, peace and partnership. The new generations of eco-neighbourhoods should perhaps become a change agent on a large scale and thus have wide influence on practice towards sustainability.

References

1. Audis, GBC Italia, Legambiente: Ecoquartieri in Italia: un patto per la rigenerazione urbana. Una proposta per il rilancio economico, sociale, ambientale e culturale delle città e dei territori (2011)
2. Baer, D., Ekambaram, A.: Integrating User Needs in Sustainable Neighbourhood Transition of the Smart City – Expanding Knowledge and Insight among Professional Stakeholders. In: Real Corp 2021 Proceedings/Tagungsband Editors (2021)
3. Barton, H.: Sustainable Communities: The Potential for Eco-neighbourhood. Earthscan, London (2000)
4. Bioregional: BedZED story (2016). https://www.bioregional.com/resources/bedzed-the-story-of-a-pioneering-eco-village Accessed 21 Dec 2021
5. Cappochin, G., Botti, M., Furlan, G., Lironi, S.: Ecoquartieri / Eco Districts - Strategie e tecniche di rigenerazione urbana in Europa / Strategies and techniques for urban regen ration in Europe. Marsilio, Verona (2014)
6. Commission of European Community CEC, Green Paper on urban environment. Brussels (1990)
7. Expert Group on the Urban Environment, European sustainable cities, European Commission Directorate General XI Environment, Nuclear Safety and Civil Protection, Brussels (1996)
8. Felloni, M.F.: Territori resilienti ai rischi climatici, Città in Controluce vol 35/36, Vicolo del Pavone, Piacenza (2020)
9. Foletta, N., Field, S.: Europe's vibrant new low car(bon) communities ITDP Institute for Transportation & Development Policy, New York (2011)
10. GWL-Terrein Home page. https://gwl-terrein.nl/bezoekers/gwl-terrain-an-urban-eco-area/
11. KCAP Homepage. https://www.kcap.eu/projects/25/gwl-terrein
12. Losasso, M., D'Ambrosio, V.: Eco-quartieri e Social Housing nelle esperienze nordeuropee, Techne 4 (2012)
13. Soma, K., Dijkshoorn-Dekker, M.W.C., Polman N.B.P.: Stakeholder contributions through transitions towards urban sustainability, *Sustainable Cities and Society* 37 -438-450 (2018)
14. Souami, T.: *Éco-quartiers, secrets de fabrication: analyse critique d'exemples européens*, Les Carnets de l'info (Eds.), Paris (2009)

15. Stead, D.: Planning for Less Travel: Identifying Land Use Characteristics Associated With More Sustainable Travel Patterns. University of London, London (1999)
16. United Nations Conference on Environment and Development UNCED, United Nation Framework Convention on Climate Change, UNFCCC, Rio de Janeiro, United Nation (1992)
17. United Nation, Transforming our world: the 2030 Agenda for Sustainable Development, United Nation (2015)
18. World Commission on Environment and Development, WCED Our Common Future, United Nation (1987)

Cultural Heritage Enhancement for Health Promotion and Environment Salubrity

Rossana Gabaglio(✉)

Politecnico di Milano, Milan, Italia
rossana.gabaglio@polimi.it

Abstract. It proposes the theme of the existing public heritage enhancement, widespread in the cities and custodian of the identifying characteristics of the places (buildings, squares, systems of open spaces), as a reading key and tool for extensive territorial diffusion facilities for health. The pandemic highlighted the need for an active and rooted network, confirmed the public and civil value of the health system, and inextricably linked the health issue to the environmental healthiness in which we live.

Community Healthcare Centers, of Anglo-Saxon origin, and Case della Salute, in Italian territory since the early 2000s, are an organizational model of social and health care services: they are capable of promoting a healthy culture, represent an opportunity to structure a coherent and necessary criterion of social, environmental, architectural and urban regeneration. Through some significant examples, let's try to understand the potentiality of cultural heritage to improve and help these structures to become, for the community, a place of health and identity.

Keywords: Cultural heritage · Enhancement · Health

1 Healthy Cities and Existing Buildings: A Relationship that Has Long Gone Beginnings

The relationship between 'hygiene issues' and the existing city is not new: it has distant origins, even if it has taken on different connotations from the contemporary ones.

Today by healthy, especially in Western contexts, we mean places where the person can take advantage of social, collective and life services that enrich and enhance the physical sphere and the psychic and social one. At the end of the nineteenth century, instead, the hygiene requirements were more related to the essential needs that were not satisfied in most situations (such as the presence of toilets in houses, the necessary distance between buildings, etc.).

1889, the year of publication *Der Stadtebau* by Camillo Sitte, marks a significant turning point in the relationship between ancient nuclei and the development of the modern city. These are the years of the Parisian model of Georges-Eugene Haussmann and the publication of the "first complete manual of urban planning" (Sica, p.33) by Reinhard Baumeister.

In this panorama, Sitte is one of the first voices that denounce the practice of destruction in favor of greater attention and respect for ancient cores. His reasons derive from considering the historic city, which he felt almost a model for modern urban planning, mainly for its 'picturesque' spatial value, and therefore closely linked to an almost romantic aesthetic vision, almost from a theatrical scene and scenographic backdrop (Giambruno 2002, pp. 17–22).

There's one element in Sitte's studies and reflections that let's still believe to be particularly significant today in addressing the theme of the relationship between healthiness and existing buildings: it's that the square is a vestibule, an essential room (Sitte, p. 28), a privileged place for socializing. Therefore the Austrian scholar anticipates the awareness of the value of the relationship between building and open space, between internal and external that collaborate in the definition of healthy spaces, in the broadest possible sense.

Even today, it is necessary to consider this principle of correlation and interconnection in regeneration processes and design of spaces for the health of those who live there.

It may also be interesting for the reflection to recall the contribution of Charles Buls, burgomaster from 1881 to 1889, who called to draw up an urban planning plan for the city of Brussels (Naretto 2016). The starting point is always that of the opposition to the evictions and demolitions considered as usual practice, claiming the awareness that historical architecture is the expression of the people who produced it, and therefore represents today, in physical form, their identity.

And here are the premises for what will be Gustavo Giovannoni's theory of thinning (Giovannoni 1931): a comprehensive interdisciplinary response, albeit still of an embryonic character - cultural, hygienic and functional - to the processes of adaptation and transformation of ancient centers about the needs of the then so-called 'modern life'.

The Belgian scholar clarifies that the question lies in reconciling respect for the ancient with the needs of modern life (Naretto 2016). For an in-depth study of contributions and an understanding of disciplinary and interdisciplinary debate developments about the epistemology of complexity as a key to interpreting the question, let's see Giambruno (2002) and Gabaglio (2008).

Today, more than a century later, we confirm that the question lies in finding a balance between conservation and respect for the existing building and its necessary transformation: between the instance of permanence and that one of mutation.

The practice of urban regeneration can become that action capable of enhancing the existing, recognizing its character and identity value, activating virtuous interdisciplinary processes (architectural but also social and environmental) of conscious re-appropriation of spaces and improvement of livability.

What role can health, or rather places of health culture, play in this process? It can become that function capably, on the one hand, of favoring the processes of enhancement and regeneration of the existing. On the other, it is precisely from the built and existing spaces to draw that necessary sense of identity that allows its practical realization of space.

2   Health as a Right and Factor of Development: Community Health Centers and Case Della Salute

Since the end of the 1970s, the concept of health has taken on an increasingly broad and interdisciplinary meaning: it is physical, mental and social well-being (World Health Organization 1978); it is a source of political and economic stability and a prerequisite for sustainable urban development (World Health Organization 2016). Its specific conditions and essential resources are peace, shelter, education, food, income, stable ecosystem, sustainable resources, social justice and equity (World Health Organization 1986). These statements define and clarify how health and society are intimately interconnected. The SARS-CoV-2 pandemic has dramatically highlighted this link, but it can also represent an opportunity to start a concrete reflection on this profound and increasingly urgent cultural change. In addition to being a fundamental individual right, health is a decisive factor for society's full and healthy development. As early as 1947, the Italian Constitution (art. 32) placed health as a fundamental right of the individual and community interest.

Therefore, it is now necessary to consider the health dimension value in the most innovative policies, in close relationship with other practices, particularly those dealing with the management and transformation of the city space (from the wide to that of detail).

The Community Health Centers, of Anglo-Saxon origin, and Case della Salute (Emilia Romagna Region 2010) or in the current definition Case della Comunità (Piano Nazionale di Ripresa e Resilienza 2021), in the Italian context, can represent interesting examples. There the health and social needs-health are resolved not only in a strictly disciplinary field but through an interdisciplinary approach and, in particular, a virtuous synergy with the disciplines of the built space. Some significant achievements of Community Health Centers and Case della Salute (presented later) can also demonstrate the added value if these health and social health structures are in existing buildings. The resulting regeneration process triggers an improvement of issues related to both health and the sense of belonging of a community[1].

2.1   Community Health Centers and Case Della Salute: Not Just a Site for Providing Services but the Place of Identity

In a nutshell, and about the food for thought proposed here, the network of Community Health Centers, present in almost all Italian regions (Brambilla et al. 2016), become the places in which disease prevention, health promotion and the dissemination of health

---

[1] The paper presents some results of research, in progress, of Polisocial Award 2020–2021 "Colti-vare_ Salute.Com. Città e Case della Salute per Comunità resilienti. Le Case della Salute quali costruttori di urbanità e socialità diffusa nell'era post COVID-19: nuove centralità periferiche in città salubri e integrate". Research group within Politecnico di Milano is composed by: M. Ugolini (scientific referent), M. Buffoli, S. Capolongo, D. Calabi, G. Costa, R. Gabaglio, E. Gheduzzi, M. Gola, E. Lettieri, M. Mareggi, C. Masella, S. Varvaro, M. Quaggiotto with F. De Luca, L. Lazzarini, A. Maturo, E. I. Mosca. Partners are: Azienda USL di Piacenza, Comune di Piacenza, Regione Emilia Romagna, Comitato Consultivo Misto (CCM), Associazione Diabetici Piacenza, Centro Sportivo Italiano (CSI) - Comitato Territoriale di Piacenza.

culture are carried out. They represent, especially in Emilia Romagna and Tuscany, one of the main tools for territorial enhancement of the health and social system (Turco 2006, pp. 4–6).

They become the physical meeting place for the population (cultural and community functions), places of potential urban regeneration also thanks to the synergy with urban public spaces with high environmental value (parks and green areas, squares, pedestrian and cycle paths, etc.).

Places, therefore, in which a community recognizes itself and feels represented; places become collective heritage in the imagination of who live there (as in the past churches, municipality, often also the schools have been ole). Not only sites of distribution of social and health services but places of identity for the community.

This essential paradigm shift allows us to reflect on the role that existing heritage can play in this necessary process of cultural change.

3 The Existing Heritage as an Opportunity in the Urban Regeneration Process

The Italian panorama, about the debate on the existing heritage and the need to protect it as an expression of civilization (AA.VV. 1967), indeed represents one of the most advanced points from a theoretical point of view. Still, it does not always find its actual realization in the reality of the facts.

The Italian territory hosts a punctual and widespread presence of existing buildings, often of small-medium size and publicly owned (school structures, small provincial hospitals, etc.), capable of representing the sense of a community, the expression of its identity. This architectonic heritage is often underused or abandoned and means, in most situations, a sustainable resource, also from the point of view of the volume available without increased land consumption.

But it is highly simplistic to consider the question only from this quantitative point of view. It is necessary to take charge of the complexity of the existing and, in particular, to assume the identity value that is affirmed in relational terms of society, space and time. This heritage is multidimensional (economic, social, cultural, physical, architectural, etc.) and multitemporal (due to the coexistence of the different stratification levels that have been deposited there).

Let's consider the regeneration process for healthy cities as a practice to encourage and trigger virtuous methods of enhancement (spatial, social, functional, etc.). It might be interesting to understand its role. Is the existing a constraint or an opportunity? (Tagliagambe 2005).

Suppose the physical and social 'empty' are also read and interpreted as conservation places of architectural values, memory and collective identity. In that case, they can become potential places of innovation and regeneration based on the principle of health.

Here the processes of underutilization and abandonment from problems can become resources: no longer as a cumbersome constraint that limits the project activities, but as an opportunity that stimulates and enriches it.

There's a relationship of mutual influence and resonance between identity and city. In reading the signs and meanings that have settled over time on the existing heritage, it is still possible to recognize, even today, the supports to which the collective identity can be anchored and to identify the substrate on which to place its transformation process. Respecting the city's identity and working in synergy with it in its regeneration does not mean considering the existing heritage as an untouchable microcosm but welcoming its inherent dynamic tension.

"Problem is the balance between the ability to innovate [...] and the ability to subordinate these changes to the preservation of a specific identity, made up of continuity of evolution and harmony between the order of the historical narrative and the order of experience" (Tagliagambe 2005, p.6).

4 Some Significant Experiences

Let's present significant examples to demonstrate how a good synergy and cooperation between the enhancement of cultural heritage and the new concept of health can positively influence the process of urban regeneration.

Here it is not possible to illustrate its complexity linked to the health and socio-health, organizational-managerial aspects, which are a fundamental prerequisite for the success of these primary care facilities. Here let's try to demonstrate the potential that the built existing makes available for the definition of identifying places for the community's health.

In the Italian context, these processes of urban regeneration often see a re-functionalization of the building (schools, hospitals, etc.) through the reorganization of services about existing spaces (for example, Casa della Salute of Bettola and Monticelli d'Ongina in Emilia Romagna Region) or more comprehensive interventions aimed at architectural recovery and completion. In Finale Emilia (Emilia Romagna Region), the old hospital, part of an interesting project to complete the entire block, becomes a vital center for the community. A new architectural volume next to the nineteenth-century building housed the hospital in Marradi (Tuscany Region), enhancing existing one architectural and functional terms and social terms.

The Casa della Salute of Sant'Antioco (Sardinia Region) and Moje (Marche Region) are interesting examples of expanding small pre-existing structures that already offered health services but have become real places for the entire community.

Let's define three general elements that represent the necessary conditions for Case della Salute to become an identity center of sociality and health. The functional mixitè (healthcare spaces, day centers, small exhibition and collective spaces, nurseries, vegetable gardens); location in identity buildings on paths where other public services and common functions are also hosted; quality architectural projects.

Broadening our gaze to an international context allows us to focus on highly significant cases both for the size of the existing architectural complex enhanced through its conservation and transformation and for the quality of the architectural project that facilitates the good result of urbanism regeneration.

The selected Community Health Centers, or similar structures, also aim to demonstrate the potential to accommodate, within the same system but in separated and recognizable areas, apparently little correlated functions: these functions contribute to making vital places and not to consider them only as spaces to go to in case of illness or for strictly health services.

Matta Sur Community Health Center in Santiago (Chile, 2018) is a meaningful example: the recovery of a school building from 1890 and a new building to complete the central open courtyard system with green spaces. The two buildings (Fig. 1) are distinct and recognizable both from the point of view of materials, language and architectural and functional composition: into the new one, there are all health and school functions (clinics, physiotherapy, study and medical rooms) while the existing one host rooms for students and families, canteen, gym, auditorium and study rooms with a library .

Fig. 1. Matta Sur Community Health Center, Santiago, Chile. Luis Vidal architect, 2018. Photos by Aryeh Kornfeld. (https://www.archdaily.com/958463/matta-sur-community-health-center-luis-vidal-plus-arquitectos).

In the past, this school building has represented, for the personal life of the inhabitants and the collective ones of the entire neighborhood, a place for the promotion of culture and has contributed to building a sense of belonging to the community: today, while hosting different functions, it can continue to be a place of promotion of culture, we would say today of health culture, and a place of identity.

There are also interesting examples of the reuse of buildings, we would almost say monumental due to their size, architectural characteristics and the identity value

Fig. 2. Health Center "De poort van Borne", Borne, Netherlands. Reitsmea a partners architects, 2017. Photos by Ronald Tilleman. (https://www.archdaily.com/891009/de-poort-van-borne-hea lthcare-center-reitsema-and-partners-architecten).

they have: Health Center "De poort van Borne" in the Netherlands (Borne, 2017, Reitsmea e partners architects) and the Repose Maternel in Gradignan (France, 2019, M. Hessamfar & J. Vérons associés).

In the first case (Fig. 2), an ex-church hosts twenty different health workers (general doctors, therapists, psychologists, masseurs, etc.): new recognizable spaces along the church's perimeter, within its majestic neo-Gothic structure with cross vaults. This intervention's extraordinarily innovative and peculiar element uses the central, unheated space, almost an internal courtyard that ends on the first floor as a multifunctional stage. It is an open but covered community space.

The recovery and expansion of the Château Lafon in Gradignan (Fig. 3), dating back to 1920, is representative for the monumental character of the entire complex and its relationship with open spaces, a large park. The architectural complex host a lot of different functions (medical center, reception center for pregnant women, early childhood support, a nursery school, offices and three lodgings for large families): while on the one hand, it does not seem to favor a daily and continuous phenomenon of appropriation of spaces, due to the specific nature of the welcome that distinguishes it, on the other hand, makes it a very significant place and also overturns the original function, of residence for wealthy classes, in a key more welcoming and caring for those who he needs.

Here are the latest examples to demonstrate how broad and articulated the category of architectural heritage can be and its identity value which, from time to time, manifests itself and is enriched concerning regeneration projects for healthy cities.

Not just school buildings, religious buildings or majestic residences: can industrial heritage also represent a place for health?

Interesting, in this regard, 13 kV Health Center in Dordrecht (Netherlands, 2020, RoosRos Architecten – Fig. 4) and Antoniny Manor Intervention in Leszno (Poland, 2015, NA NO WO Architects – Fig. 5). These experiences demonstrate the potential

Fig. 3. Repose Maternel, Gradignan. France. M. Hessamfar & J. Vérons associés, 2019 (https://www.hessamfar-verons.fr/projets/repos-maternel/).

Fig. 4. 13kV Dordecht Health Center, Dordrecht, Netherlands. RoosRos Architecten, 2020. Photo by Rene de Wit (https://www.archdaily.com/939065/13kv-dordrecht-health-center-roosros-architecten).

of the large spaces of industrial architecture in a regeneration process. Workplaces, and therefore precious and unique witnesses of the formation and development (economic, productive, etc.) of a community become not only spaces of use of health and social-health services (including pharmacy, physiotherapy, orthopedics, obstetrics center and psychiatric center) but also a place of hospitality (hotel, residences and restaurant).

Fig. 5. Antoniny Manor Intervention, Leszno, Poland. NA NO WO Architects, 2015, Photo by Marciej Lulko (https://www.archdaily.com/781567/leszczynski-antoniny-manor-intervention-na-no-wo-architekci).

5 Conclusions

Culture and places of health, existing heritage and sense of identity.

In light of the significant examples illustrated and many others that, for brevity, it has not been possible to cite, it is possible to say that it can make sense and work so that these three questions can find an effective alliance.

Undoubtedly, a cultural change capable of inverting the paradigm that has often made the concept of health coincides with the lack of a disease to reach a broader and more articulated one, which leads to well-being in all its forms (physical, social, etc.).

It is necessary to learn to consider the existing (buildings, squares, paths, etc.) not as a limit to urban transformation and regeneration but as unique and irreplaceable resources that can nourish these processes.

It is essential to contact the complexity and the profound meaning of the sense of identity, individual and collective, of a community, investigating the deep reasons that reach it and understanding that it can become an important ally for the success of interventions in the area.

And perhaps the architectural project, which deals with the nature of spaces (internal and external, and the relationship between them) and the way they are inhabited, can be a precious catalyst: a tool that allows, start from the recognition of value and identity character of the existing heritage, to define places of health where the community feels represented, welcomed and expressed itself.

References

AA.VV.: Per la salvezza dei beni culturali in Italia. Casa Editrice Colombo, Roma (1967)

AA. VV.: Piano Nazionale di Ripresa e Resilienza. Missione 6: Salute. https://www.governo.it › governo.it › files › PNRR (2021)

Brambilla A., Maciocco G.: Le Case della Salute. Innovazione e buone pratiche. Carocci Faber, Roma (2016)

Department of Health: Health Building Note 11–01 – Facilities for Primary and Community Care Service. The Stationery Office, London (2009)

Gabaglio, R.: La città tra permanenza e mutazione: l'approccio dell'epistemologia della complessità come chiave di lettura e di progetto per il costruito. Clup editore, Milano (2008)

Giambruno, M.: Verso la dimensione urbana della conservazione. Alinea, Firenze (2002)

Giovannoni, G.: Vecchie città ed edilizia nuova. Utet, Torino (1931)

Naretto, M.: Charles Buls e il restauro: antologia critica. Charles Buls et la restauration: anthologie critique. Franco Angeli, Milano (2016)

Emilia-Romagna, R.: Decreto Giunta Regionale no 291/ 2010, Casa della Salute: indicazioni regionali per la realizzazione e l'organizzazione funzionale. https://salute.regione.emilia-rom agna.it/cure-primarie/case-della-salute/documentazione-case-della-salute/delibere, (2010)

Sica P.: Storia dell'urbanistica. Il Novecento. Laterza, Roma-Bari (1985)

Sitte, C.: L'arte di costruire la città. L'urbanistica secondo i suoi fondamenti artistici. Jacabook, Milano (1980)

Tagliagambe, S.: L'albero flessibile. La cultura della progettualità. Dunod, Milano (2005)

Turco L.: Ministro della Salute Audizione alla Commissione Affari Sociali della Camera dei Deputati, 27 giugno 2006: Un New Deal della Salute. Linee del programma di Governo per la promozione ed equità della salute dei cittadini (2006). http://leg15.camera.it/_dati/lavori/ste ncomm/12/audiz2/2006/0627/INTERO.pdf

World Health Organization: Report of the International Conference on Primary Health Care, Alma-Ata, USSR, 6–12 September (1978). https://www.who.int/publications/i/item/924180 0011

World Health Organization: Constitution of the World Health Organization (1986). https://www. who.int/who_constitution_en.pdf

World Health Organization: Health as the Pulse of the New Urban Agenda (2016). https://apps. who.int/iris/bitstream/handle/10665/250367/9789241511445-eng.pdf?sequence=1)

Beyond the Official City Planning. Tirana Next Pilot for Healthier and Safer Urban Open Spaces in the Post-2020

Fabio Naselli[✉] and Klaudia Tufina

Epoka University, Tirana-Rinas Highway, Tirana, Albania
fnaselli@epoka.edu.al

Abstract. Urban open spaces are all the spaces of common use, fully accessible and enjoyable by all, at no cost and without a profit motive. They are present in our contemporary cities in various forms, including parks, boulevards, sidewalks, pathways, play areas, and also as spaces between buildings or roadsides, all of them are significant spaces for a renewed idea of commons. Due to the Coronavirus pandemic, in this post-2020, we experienced the real need to get open spaces next to our houses, often by spontaneously recovering (or re-inventing) several minor (and in many cases marginal and neglected) small and micro places. Also, the conventional uses of those spaces have been re-interpreted by giving them new values and meanings in terms of functions and features. The aim of the Tirana Next Pilot Project (TNPP) is to imagine a new-normal city where public space goes beyond the traditional concepts of a park, a square or a street, leading to a new vision in thinking about new city-forming scenarios which can promote social exchange even within the eventual need of social distancing (spatial dimension) and shortening paths (time dimension) for any daily needs.

TNPProject contains a diagnosis of the Capital City, and the greatest city of Albania, Tirana, in the past and actual conditions and through the in-force Master Plan "Tirana 030" vision. Across diverse pilot proposals, within the neighbourhoods of Tirana (Super-Blocks), the researchers want to explore the embodied experience of open spaces and the changing relations between them and the surrounding humanity and spatiality. The pilot cases are researched and analyzed by considering the physical space (hard-scape), the social life on the streets (soft-scape), the human interaction (human-scape), the virtualization process in addition to the conventional physical one (digital-scape), and the processes framework (network-scape). By investigating and analyzing the links between these elements it is expected that this research work might provide a set of practical recommendations and suggestions, pilot-cases based, for a more suitable systemic-punctual urban regeneration in the favor of Tirana neighbourhoods' quality of life, even "beyond" the in-force city planning formal process.

Keywords: Punctual urban regeneration · Public spaces · Human scale · Physical and virtual spaces · Life on the street · Cityforming process · Quality of urban life · Urban tactics · Ephemeral designs · Proxemics

© The Author(s), under exclusive license to Springer Nature Switzerland AG 2022
F. Calabrò et al. (Eds.): NMP 2022, LNNS 482, pp. 580–593, 2022.
https://doi.org/10.1007/978-3-031-06825-6_55

1   Brief About Tirana

Tirana is located at the center of Albania, within a flatland enclosed by mountains and hills, and facing a slight valley in the northwest overlooking the Adriatic Sea in distance. The city is cut by the Lana River, which crosses throughout all its east-west diameter and edged in the north side from Tirana River, also city contains two main lakes: the Artificial Lake and Farka Lake, both acting as interest points for leisure, sport, and enjoyment (Fig. 1).

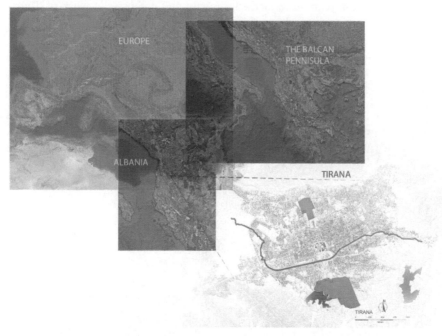

Fig. 1. Geographic location and actual map of the city of Tirana. (Source by the authors, 2021)

Tirana's growing process has been very complex and very fast, from a small city dating from 1920 until nowadays, many factors have taken part in the urbanization process of Tirana [2]. The real roots of Tirana are dated since the Byzantine age on, as a rural town, and then under the Ottoman age, it has had the first development as a commercial city. In terms of the growing process, Tirana has evolved mainly informally and organically, apart from the short Zoku I Kingdom and during the Italian period -when an official axial frame was defined to give order to the Capital- as well as significantly after the fall of the socialist regime, when democracy was raised by releasing full freedom to act, individually, for 15 years longlasting.

Despite all these deep "roots", the city is still and mostly considered a 100-year-old city, starting from 1920, when it was named as the Capital city. At the same time when the city started getting bigger and bigger from people migrating from other parts of the country. Nowadays, Tirana has the country's largest metropolitan area and the centre of the county, numbering a population of 906,166 as of 2020, making for 31.84% of the whole Albanian population.

"The sudden increase in the population of the capital city highly affected the urbanization effects on the city. Tirana grew in an organic/spontaneous way, following informal urban approaches, as a blast by the newly acquired freedom. Most dwellers are attracted to the city centre, as the most active and living part of the city" [1]. Starting from 1921, the city has gone through a huge expansion, as we can see in Fig. 2.

The city is continuing to get bigger also nowadays. As the current Mayor of Tirana, Erion Veliaj, mentions, "in a growing city, which changes with the rhythm of development, these last years have marked the hard and enthusiastic work of Tirana transformation to a European modern city".

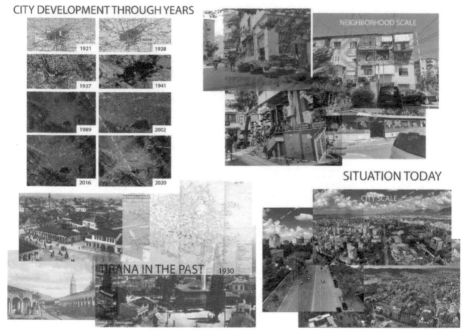

Fig. 2. Development of Tirana from 1920 until nowadays. (Source by the authors, 2021)

2 Research Issues Growing up from Reality

The main subject of this research work arose from the need to stress some of the neglected crucial topics in the ongoing urbanization and current city-forming processes. Since Tirana has gone through big extensions and transformations of the urban fabrics and building typologies through the years, the experiences of the inhabitants have faced many ups and downs with city planning. The development leaves a huge space for comments and suggestions. The below collage of newspapers (Fig. 3) gives a view of the topics and discussions for the city development and the suggestions that design professionals are taking into consideration for city organizing. This brought the inspiration to delve into details to analyze and propose inclusive urban regeneration strategies for Tirana and contribute to the current discussion in a more detail-oriented way in this regard.

Fig. 3. Motivations growing up from the reality. (Source by the authors, 2021)

As Jan Gehl reports, during the last decades our cities have been built "from above, rather from the eye-level perspective" by following an old paradigm in urban planning. This means that human needs have been forgotten (at least neglected) and the priority has been given to the buildings, the cars, the urban streets, and the highways, as they were planned and implemented to connect the city centre with the suburbs and make life 'easier'. As a result, cities have become huge, hectic megalopolis where life itself has been shoved, and in the place of human beings, we have put cars and un-human building blocks.

This state is put forward on a large scale in the city of Tirana. With the increase of population, the need for city growth has taken over urbanization, volumes of buildings have replaced and neglected the life of public spaces. The prioritization has been given to the car infrastructure to connect the centre of the city with other cities or to connect neighbourhoods. A little interest has been given to the public spaces and the needs of humans for outdoor activities. In this actual situation, we are facing dense neighbourhoods, with not enough healthy space for its inhabitants. That means that in each neighbourhood, problems of space, greenery, connections, and activities are present in everyday life.

New facts arose, led by the Coronavirus pandemic, in early 2020. When we experienced the real urge to get open spaces "proximate" to our houses, often by rethinking to several insignificant or unexisting small and micro "pockets". Also, the conventional uses of those "no-spaces" have had re-interpreted by giving them new values and meanings in terms of functions and features leading from the sudden "lockdown".

Leading us all to an evident acceleration in digital practices and processes, releasing unexpected challenges in both our social and spatial interactions and the sudden limitations in our physical spaces; changes that are marking unpredictable and deep shifts in our "old-normal". A different "new-normal" - as it has generated from the global pandemic 2020 - which set out a new mixed both physical and virtual framework of the modifications humanity undertook, being pushed into a new digital dimension; or better, as many scholars are saying, into a New Normality. Precisely into new normality in which the balance between physical and virtual interactions belong in vantage of the second one, in less than one year, by increasing, at the same time, both the quantity and the quality of exchanging digital data [3].

Furthermore, the reduction of both dimension and time necessary to evaluate those effects of the forced-on digital practices, stand in parallel to the decrease of time and the spaces dedicated to physical encounters; such as they have been reshaped for more than two years due to the recurrent waves of the global emergency conditions, and its current pandemic-related urban effects [3].

3   Healthier and Safer Urban Open Spaces: A Methodological Frame

"Cities are the places where people meet to exchange ideas, trade, or simply relax and enjoy themselves," [9] arguing that a city's public realm – the streets, squares, and parks – is the "stage" and catalyst for the abovementioned activities.

Due to the importance of social life, activities, and interactions to the general quality of life and well-being of people, it is crucial to look for enhancing the quality of our surrounding built environment in a way that can enhance the whole quality of our social experience within itself.

The overall objective is to study and investigate the quality of mixed-use public spaces that contribute to the social life of the community. Additionally, the scope of this work is to develop a design framework for the special neighbourhoods in Tirana (super-blocks) to provide future recommendations for socially sustainable urban open spaces regeneration projects at different levels such as social, cultural, and environmental. Whereas the research questions concentrate on what the main principles for a socially sustainable city on a human scale are and how to provide future design guidelines for this type of project, we aim in carrying "in": the physical space (hard-scape), the social life on the streets (soft-scape), the human interaction (human-scape), the virtualization process in addition to the conventional physical one (digital-scape), and the processes framework (grid-scape).

The focus and the main questions that we will try to answer throughout the research are illustrated in the below diagram that defines even a methodological frame (Fig. 4). It starts with the role of socio-physic-economic changes in the developing cities, how it contributes and plays out spatially in the city development. While a secondary focus lies on the key progress that can contribute to the development of Tirana in the new-normal post-pandemic stresses, in their strengths, as well as weaknesses.

The whole research process starts with the background information, then continues with the analysis section, and finally to proposals and reflections of the case. The objective is to promote Tirana as a "City for People", focusing on the human dimension by creating a safe, lively, sustainable, active, unified, socially interacted, city with standards proven by professionals and researchers of urban planning and architecture.

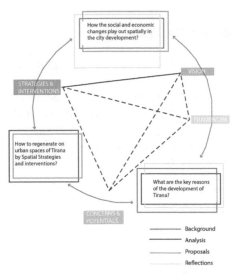

Fig. 4. Research questions methodological diagram. (Source by the authors, 2021)

The research aims at providing better solutions and experiences for the inhabitants of this city and their usual needs. How can we give these citizens a wider horizon and a variety of services and activities for a better social life? How can we revive the socio-economic condition of Tirana by spatial strategies and interventions? What are the key reasons for the development of Tirana? (Fig. 5).

Fig. 5. Key points of expected achievements through a systemic punctual urban regeneration in Tirana. (Source by the authors, 2021)

Adapting low-quality places within the superblocks and working on regenerating the micro public spaces system, promoting the circulation pathways at the neighbourhood scale, and giving priority to pedestrians and their living activities, is one of the main expected outputs of the study.

4 Beyond the Official City Plan

The process of Tirana by 2030 is defined by a set of 13 strategic projects (Fig. 6) within the local master plan TR030. The official next visions drive-by TR030 aim in solving the problems of infrastructure, mobility, public spaces, connectivity, and on natural parks and agricultural lands in the peripheral part. The project focuses mostly on a city scale, with strategic proposals for the overall city. In terms of neighbourhood scale, it recommends some minor conceptual strategies, which are very useful for future research focused on a more in-depth analysis of the city.

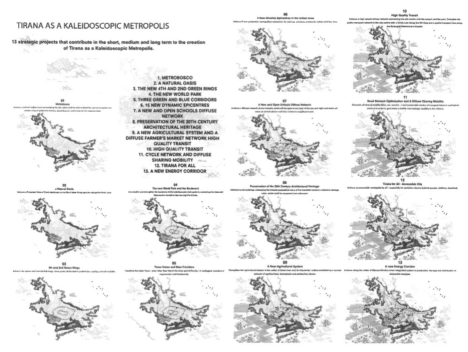

Fig. 6. The 13 strategic projects in force since 2018. (Source: Studio Boeri, 2018)

The problem is what is happening in the living city areas, behind the "perfect" strategic plans of Tirana 030. The Government plan is focused on the city scale, to solve mobility problems, infrastructure problems, take care of the main public spaces and community centres and activities in terms of city scale. What happens in the inner parts of the city, looks like a case left behind since the human dimension and needs are somehow leftover (Fig. 7).

Fig. 7. Inner-city neighbourhoods' situation. (Source by the authors, 2021)

Built structures and their functions such as buildings typologies, informal adaptations, mixed-use buildings, architectural features such as volume, shape, function, style, typology, indoor and outdoor activities, social gathering areas, greenery, and use conditions of edifices, urge to be taken into consideration.

The concern is on what happens if we zoom in to a neighbourhood scale, its infrastructure, urban fabric, built structures, architectural style, social gathering areas, and greenery. In terms of infrastructure, no plan is given in regenerating the roads, paths, pedestrian ways, car lines and bike lanes of the neighbourhoods, within the building blocks. The quality of the "inner" urban fabric is not considered at all in the government plan. Built structures and their functions, typologies, informality, ground floor uses, and their relation to the public, as well as their legacies and values, are neglected. Outdoor and indoor activities and spaces in between the built volumes within the superblocks are not analyzed and so, we do not have a regenerating plan for these fragile areas. Last, but not least, green areas, such as parks, private and public open areas are not thought at the small scale of the inner superblock (Fig. 8).

The official plan offers a wide strategy for redeveloping the city and turning it into a contemporary green metropolis. All these strategies are shareable, needed and healthy for the city, but definitely, they are not enough [10]. Many benefits are offered to the city with the Tirana 030 plan, still, we can find some weaknesses and opportunities to be further developed for the next city.

The approach of Tirana 030, approved by the Municipality in 2018, which will be processed until 2030, is a key point of this research. By the information gathered mostly from the Municipality for this development plan, studies on what Tirana aims to be in the upcoming years and conclusions of the actual/future condition of the city are clear and proper information of the city is perfectly presented. After the clear idea presented, after knowing the strengths and opportunities of the city, after knowing the city cultural heritage from the past and the future development, after knowing the potentials of the city, we aim for research and analysis on the most problematic aspects that are not covered or taken into consideration from city planning. This allowed us to study the leftover parts, which are also relevant fragments of the city, focusing mostly on the human

Fig. 8. Top view of Tirana 030 vision. (Source: Studio Boeri, 2018)

dimension and their senses, that we know for sure no one is taking after. The relationship of the new regulatory plan with the inner city, the livable city have been studied with the methodologies mentioned above. An evaluation of the outdoor spaces, territorial usage, humans' connections, interaction with the environment has been conducted with both qualitative methodologies and in-the-field urban analyses.

In conclusion, these different phases of the analysis are the basis for further development of proposals to improve the current situation, but also to enhance and promote the potentials that they comprise (Table 1).

We do lack in regenerative approaches at the smaller scale of the hidden neighbourhoods, regeneration of building typologies and green areas within the neighbourhood, but yet, this proposal gives us the room to take care of all these problems at a later stage. The project covers the most urbanized parts of the city, offering a great systemic layout and potential users' participation within the city choices. For those reasons, we focused on a more in-depth analysis of the city, to investigate its "living areas" neglected by the official vision.

Table 1. SWOT analysis of Tirana city planning

	Tirana 030	Inner-city
Strengths	Reduce land take Promoting regeneration and densification Territorial safety New urban economies Green areas proliferation Pollution reduction Biodiversity spot and connectors Bicycle lane network More efficient public transportation	Low rise-mixed rise buildings Providing citizens with all the necessary services Social cohesion Very near attractive areas of the city Mixed architectural style Cultural heritage

(continued)

Table 1. (*continued*)

	Tirana 030	Inner-city
Weaknesses	Lack of regenerating the smaller scale of the neighbourhoods Lack of regenerating building typologies Lack of green areas within neighbourhoods Lack of paths layout within neighbourhoods Lack of mobility within neighbourhoods	Lack of green infrastructure Lack of green spaces around the housing areas Wasted areas serving as parking Big scale buildings Heavy car parking Waste in potential green areas
Opportunities	Covers the most urbanized areas of the city Creates great infrastructure of the city layout Potential collaborating with the existing Infrastructure within the city Creates collaboration and social integration within neighbourhoods	Situated at the most urbanized parts of the city Near the rings Near the centre Easily accessible Close to landmarks of the city Potential infrastructure for a "walkable city"
Threats	Not connected with the livable parts Overlaps the existing situation Does not cover "inner-city" infrastructure	Limited municipal revenue sources Lack of regeneration and transformation plan Lack of organized paths

5   Study Areas and Applied Methodology

Accordingly, the areas of Selvia, Pazari I Ri, Zona e Ambasadave, Zogu I Zi, 21 Dhjetori were selected to be the pilot cases of the research investigations and proposals (Fig. 9).

The analysis phase consisted of multiple methodological methods, starting from qualitative analysis such as site observations through walker perception, photographic documentation, interviews, and discussions with citizens; to the analysis of the areas in terms of conventional urban analysis. These analyses aimed to get a clear idea of the real conditions and values, what are the physical features, their uses, accessibility, public space, and activities, what are the strengths, weaknesses, opportunities of the livable parts of the city, and on what are the real problems that its citizens face in the everyday life and what they really lack in their neighbourhoods. Case studies of relevant European practices in urban development were studied, administrative documents, historic and current data (maps, photos, statistics, projects) were collected from institutions such as INSTAT, Municipality of Tirana and other governmental bodies, which have contributed to the background set of information for this research.

Observations in terms of people's movement and use of spaces are elaborated to better understand the relationship of the people among themselves and with the environment. Public amenities, activities, services during different periods of the day are analyzed and so is analyzed the economy and their conditions also. In parallel, the current condition, structure and territorial organization, urban systems, polycentric development, solid/void relationships, spatial form and relation to the site (circulation, distances, etc.) are convicted through the researchers' investigations in the site.

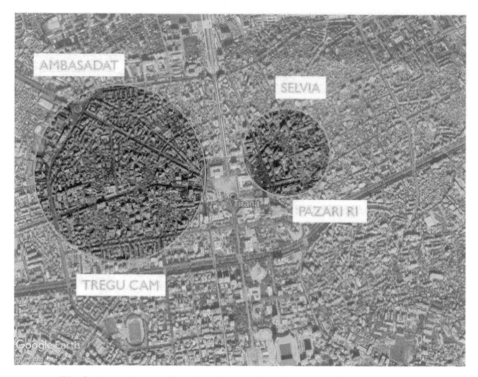

Fig. 9. Inner-city neighbourhoods' situation. (Source by the authors, 2021)

This research focuses on two of the oldest neighbourhoods of Tirana, which are near the city centre and connected with the network of main axes and rings of Tirana 030. The selection is made strategically since the zones are part of the city's most active and livable part, and are easily connected to the main landmarks. They are both considered to be very populated areas with many activities, facilities and services, and important institutional buildings. They played a significant role in the city development which contributed to the formation of the current city heritage, and way of living styles until now.

To better understand the possible impact and the real potential of these areas in shaping the city there have been site surveys and detailed analysis, through classical urban analysis at a city scale, at the neighbourhood scale (superblock), and lastly, at an eye-level perspective.

The selected areas are: "Zona e Selvia" (which covers also "Pazari I Ri") and "Zona e Ambasadave" (which covers also "Zogu I Zi" and "21 Dhjetori"). The first zone has a radius of 250 to 800 m to the city centre and the second one has a radius of 200 to 1.2 km

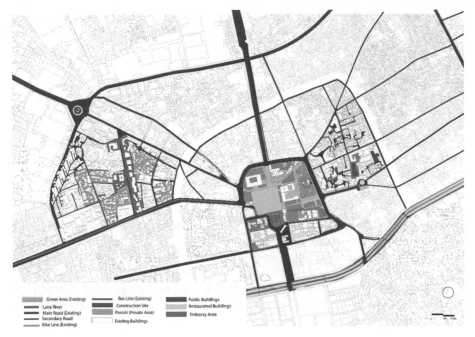

Fig. 10. Accessibility within selected case studies. (Source by the authors, 2021)

to the city centre. The two zones are also selected near each other, easily accessible from one zone to another, by walking or by vehicle. Figure 10, shows a clear understanding of the actual condition, by highlighting the main circulation network, the public and institutional spaces, and also the private spaces. As we see from the map, there is a huge public use area that covers most of the zone. This open public area is in our focus use. We see that the most active open space is at the centre of the city, near the "Skanderbeg Square" and all the green area is focused on that part. As we see from the map, it is clear that inside the selected areas, there is a lack of greenery and recreational spaces.

By understanding the current condition in this zone, we tried to create a clearer idea of what happens in the "real life between buildings". Since we do not have the "power" to transform the city into a "next city", we can focus on regenerating the people's experiences and creating a safe, healthy lifestyle for all its inhabitants. It is aspected that the regeneration concepts will give practical recommendations for the next city's development and improvement.

The intervention strategies can be used not only in the selected intervention areas but throughout the whole city of Tirana. During the research we highlighted that in-between Tirana we found:

– Low infrastructure
– Different characters within a neighbourhood
– Living areas adapted to businesses
– No design for activities and public use

– Lack of basic urban standards (uniformity, circulation, land use, public areas, green areas, play centres, outdoor activities, human-scale highly walkable)
– Self-adopted ground floors for mini business purposes.
– Replacement of low-rise private dwellings to 'superblocks'
– Bad condition buildings
– A bad condition of open space
– No public space

To answer the above-mentioned problems and improve the present condition, a few suggestions are presented throughout the research. The regeneration of the neighbourhoods and the city could start:

– Through regenerating outdoor areas within the superblocks
– Through suggesting a safe, comfortable, easily walkable, sociable neighbourhood
– Through co-creating spaces for people
– Through opening and expanding public rooms by adding the "minor" spaces
– Through refurbishment of the bad condition buildings
– Through adopting abandoned buildings to mini businesses and proximate new services
– Through enhancing the cultural/historical legacy in favour of the protection of the urban identity

This research proposal can further serve as a tool for promoting the "Humans in the city" dimension not only in the city of Tirana but also for other cities in Albania that are under the developing process.

References

1. Bulleri, A.: Back to the future. Architecture and Urban Planning for an Extraordinary Metropolis. OFL-Oil Forest League, Rionero in Vulture Italy (2018)
2. Aliaj, B., Lulo, K.: The city of Tirana, history of architecture and urban development. In: Aliaj, B., Lulo, K., Genc, M. (eds.) Tirana the Challenge of Urban Development. SLOALBA Publishers, Skofja Loka Albania (2003)
3. Bellone, C., Naselli, F., Andreassi, F.: New governance path through digital platforms and the old urban planning process in Italy. Sustainability **13**, 6911 (2021). https://doi.org/10.3390/su13126911
4. Carta, M.: Creative City. Dynamics, Innovation, Actions. Rubbettino, Palermo (2007)
5. Gehl, J.: Cities for People. Island Press, Washington (2010)
6. Jacobs, J.: The Death and Life of Great American Cities. Random House Inc., New York (1961)
7. Lydon, M., et al.: Tactical Urbanism. Short Term Action - Long Term Change. Handbook series, vol. 1–4, The street plans collaborative, New York, USA (2012)
8. Mali, F.: HAPESIRA/ZEROSPACE.. Albanian pavilion 2018 at La Biennale di Venezia. In: 16th International Exhibition of Architecture, RSH Ministria e Kultures, Tirana, Albania (2018)
9. Naselli, F., Trapani, F.: "Ephemeral Regeneration" for the Marginal Urban Spaces/Places in Enna. Planum Publisher, Rome (2019)

10. Naselli, F.: Tirana-next: a complementary development strategy (and consequent urban tactics) for the informal and historical urban fabrics within Tirana super blocks. In: Yunitsyna, A., et al. (eds.) Current Challenges in Architecture and Urbanism in Albania. The Urban Book Series, pp. 61–77. Springer, Cham (2021). https://doi.org/10.1007/978-3-030-81919-4_5

11. Pashako, F.: The legacy of informality in Albanian landscape. In: Pastore, D. (ed.) EVOKED. Architectural Diptychs, Edizioni Giuseppe Laterza srl, Bari, Italy (2016)

12. Pastore, D. (ed.): EVOKED: Architectural Diptychs, Edizioni Giuseppe Laterza srl, Bari, Italy (2016)

13. Perry, D.C., Wiewel, W. (ed.): The University as Urban Developer: Case Studies and Analysis. M.E. Sharpe, New York (2005)

14. Zeka, E., Mali, F.: Spaces to places. In: Mali, F. (ed.) HAPESIRA/ZEROSPACE. Albanian Pavilion 2018. RSH Ministria e Kultures, Tirana (2018)

Cyclical Covid Evolution and Transition Towards a Symbiosis Between Metropolitan Model and Widespread Settlement Model

Maria Angela Bedini[(✉)] [iD] and Fabio Bronzini [iD]

Polytechnic University of Marche, 60131 Ancona, AN, Italy
m.a.bedini@staff.univpm.it

Abstract. The paper starts from the possibility of a cyclical spread of the pandemic and the need for a review of urban and regional planning tools. In this perspective, the paper proposes a series of programming and planning suggestions for the protection from the risks of Covid, based on an integration between revision of times, spaces and flexible organization of the city and the regional territory. In this context, the relationship between medium / large cities and systems of small urban and rural centres of inland areas is reset. The results and conclusions highlight the close dependence between European Union objectives, consequent urban-territorial objectives, new tools and procedures for the transition from the protection of seismic risks to the global one.

Keywords: Cyclical evolution of covid and the city · Flexible transformations of urban and regional land use · New symbiosis between urban concentration areas and village diffusion areas

1 Urban Planning to Be Rethought at the Time of the Pandemic

The context of the study is a scenario of a cyclical spread of pandemic events that require a rethinking of the organization of the city and the regional territory in terms of flexibility of space use, change in the times of city use, spread of equipment on the territory, proximity services.

Some issues of the ongoing debate can be summarized as follows:

- take note of the negative impacts «on the shape of the city, on relationships and on spatial planning» [1];
- thoroughly rethink the production model of urban and rural space [2];
- consider the possibility of an escape from centralized cities to inland areas [3];
- search for more attenuated solutions of the anti-urban position [4, 5];
- ability of the city to overcome disasters with solidarity and new forms of coexistence [6];
- acknowledgment that the spread of the pandemic is encouraged by urban density, poverty and marginalization [7, 8].

F. Calabrò et al. (Eds.): NMP 2022, LNNS 482, pp. 594–602, 2022.
https://doi.org/10.1007/978-3-031-06825-6_56

In this disciplinary context, the thesis of this paper states that thousands of villages of inland areas will be called upon to carry out a propulsive function at the country level, provided that they will be able to connect with the world [9], in a difficult and unstable dynamic balance between new ways of life in the countryside and in the territories of widespread settlement and urban centers with high and medium concentration, made safe.

Assumptions:

- in inland areas with natural environments and low population density, social distancing was a specific condition of local social life [2];
- small urban settlements can become places of welcome for the population temporarily leaving the city and also for emigrants, avoiding their transformation into privileged places for the wealthy classes [10];
- a new model of urban space requires new hierarchical and functional relationships between center and periphery, between areas with a high concentration of settlements and widespread settlements, with great benefit also of interpersonal relationships of proximity [8] and for the services of a community local [11];
- the proposed settlement model for the reduction of inequalities not only requires specific European funding but also to avoid wrong policies that could increase, rather than reduce, inequalities [10];
- to tackle the social divide at urban, regional and national levels it is necessary to re-establish fair and balanced social and territorial relations between density and rarefaction, public space and space [12].

2 Objectives

The objectives of the European Union can be summarized in: ecological and digital transition, environmental regeneration, green economy, green city, reconstruction of local public health, reduction of social inequalities and imbalances between urban centers and villages scattered throughout the territory with high risk environmental, de-bureaucratization of public administrations.

The paper pursues these objectives by examining some of the most advanced solutions in the field of protection from seismic risks and coexistence with risk [13] and expands them, to also take into account the need for protection from pandemic risks.

The paper intends to search for new strategies for a different model of life in the numerous small urban and rural villages spread over large areas, to be strongly integrated with the centralized settlement model, made safe.

The objectives pursued consist therefore in some updates of the contents of the urban and territorial planning instruments capable, to some extent, of responding to the new needs of protection from global risks.

Clearly, the suggestions made to achieve these goals can only be realized by resorting to the EU Next Generation plan for large strategic global risk protection projects.

3   Methodology

Starting from the objectives set out above, some contents of the current urban planning of protection from seismic risk that have generated positive results are first reconsidered [14], to modify and extend them in order to adapt them to the new requirements imposed by global risk.

Based on a comparison between seismic risk protection solutions and pandemic risk protection solutions, a set of operational suggestions is obtained, consistent with the requests made by the European Community.

In particular, three strategic themes are first highlighted, a1, b1, c1 to address seismic risks, which are then integrated with corresponding themes a2, b2, c2, for the extension to protection from pandemic risks. Finally, three groups of qualifying actions are reported in more detail, a3, b3, c3, for the defense from global risk.

4   Results: From Seismic Risk to Global Risk

The research results can be summarized in some strategies to extend the protection from seismic risk to global risk, including pandemic risk.

First of all, the consolidated good practices for the protection from seismic risks (Fig. 1) and from natural disasters [13, 15] were evaluated, which, in a nutshell, can be exposed in three points.

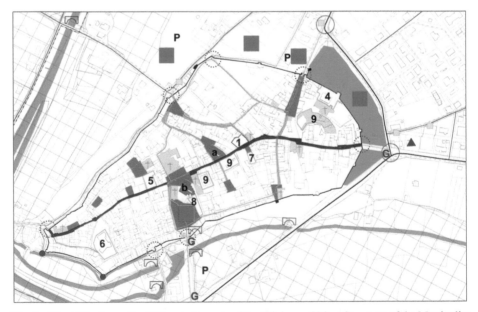

Fig. 1. The table shows the different elements of the Minimum Urban Structure of the Munipality of Bevagna (PG): structuring routes (urban crossings, links to the exterior and the main functional and strategic centres), roads to junctions and gateways to the city), escape routes, secure areas, hospitals and civil protection centres and the fire brigade. Programmatic Document of the PRG, Municipality of Bevagna.

a1. Strategies to counter territorial imbalances: enhancement of the role of the widespread settlement system and of neighborhood public services and digital transition of networked health services and business assistance.

Valorization of scattered settlements as places of reception for the population that moves away from the cities and countries affected by the earthquake [16].

b1. Strategies for urban reorganization: transformation of the rigid and deterministic uses envisaged by the Urban Plan into flexible, temporary and dynamic uses; dissemination of slow mobility within the framework of alternative traffic solutions; flexible programming of the destinations of use of paths and public spaces during the day and at night, both in case of quiet and emergency; reduction of buildable volumes in areas at risk.

Reorganization of the times, spaces and road system of the city to attenuate too high urban densities in the most fragile historical areas. Reduction of buildable volumes in zones at risk (Fig. 2).

c1. Strategies for reorganization of territorial resources: monitoring of buildings that can be used in case of emergency; reuse and securing of some inactive or underused infrastructures and services spread throughout the territory; preparation of hyper-equipped territorial centers at the service of the territorial protection system.

Realization of protected escape routes in case of emergency, places of population gathering, structures for the first reception and assistance, centers equipped at territorial level serving vast areas, easily and quickly reachable in case of danger.

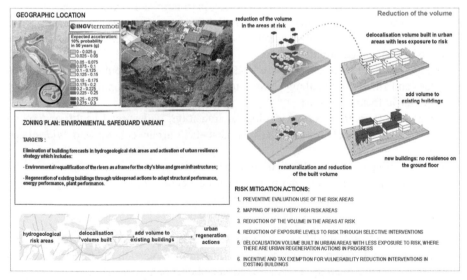

Fig. 2. Variant of environmental protection and General Land-Use Plan of Messina. First adoption in June 2018, Planner Carlo Gasparrini (graphic processing by Giovanni Marinelli).

These strategies of settlement reorganization ensure concrete answers even in case of pandemic risk, provided that the protection system is integrated, both at regional and urban level, responding to the needs imposed by the health risk. Therefore, taking into account the objectives established by the European Recovery and Resilience Plan, reported in paragraph 2, it is a question of expanding the protection strategies mentioned in the three points considered above, with attention also to pandemic risks:

a2: to dimension the population that can be accommodated in the thousands of small historic centers and villages, redesigned and requalified also to ensure isolation for quarantine and widespread health services;
b2: to make flexible, in time and space, the modalities of use of public areas, public buildings and urban roads, with possible temporary restrictions in case of emergency; to reschedule the timetables of work, public services and means of transport to reduce crowding;
c2: to identify and map, at a territorial level, widespread, stable and temporary health facilities for medical and social assistance or quarantine (including second homes that can be used).

The transition proposed here to extend the protection from seismic to pandemic risk can be expressed, in more detail, in three groups of qualifying actions.

a3:
– reduction of inequalities between centralized urban areas (encouraging forms of tele-working, social and health services at distance), peripheral areas and widespread inner areas more fragile and high environmental risk. Inland areas can thus become attractive elements in periods of social distancing, in symbiosis with the cities that will remain the engine of social and economic change;
– the scattered settlements will be able to make available, in case of emergency, thousands of houses and small historical aggregates, to accommodate the population temporarily forced to move from the cities;
– support of the green economy, ecological transition, widespread health and social assistance in a new relationship between densified urban centers and widespread settlements;
– implementation of the digital transition with the activation of broadband networks, the remote management of public and private facilities, energy control, the reorganization of the waste system, the spread of slow mobility.

b3:
– both centralized city and diffuse city will become "double cities", one operating in a normal period, and one in an emergency period. Public areas are therefore planned to be restricted, when necessary, in their function and occupancy rate (Fig. 3);

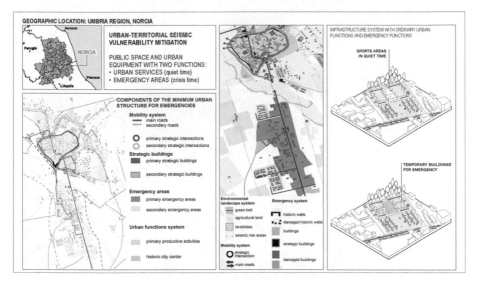

Fig. 3. Flexibility of use destinations. Norcia, Umbria Region, Master "City and Territory", Univpm, F. Malecore (graphic processing by Giovanni Marinelli).

– waiver of predetermined formal models and definition of dynamic guidelines that follow the evolution of events;
– transformation of city times (slow mobility and re-appropriation of urban spaces);
– enhancement of the "city of neighborhoods" on a human scale (Fig. 4);
– reform of the public administration (accelerating its activity, and ensuring fairness, finalization, efficiency, flexibility);
– simplification of urban planning procedures, breaking down the current bureaucracy, the approval times of the planning instrument, the approval times for authorization procedures;
– replacement of permanent zoning, with flexible zoning, to alternative destinations of use in a state of quiet or in a state of emergency (pandemic, earthquake, etc.).

c3:
– emergency destination of diffuse settlements for temporary use: second homes as places for quarantine, diffused hotels, to be used for the quarantine of individual members of family groups, villages intended for "islands" of maximum Covid-free protection, territorial hubs always accessible, equipped centres for branching out services spread over large areas;
– planning of territorial medicine, emergency health systems, local outpatient centers, home medical assistance, school services, centres for the provision of meals and beds for the needy;
– public areas for taking tampons outside pharmacies, factories, and other authorized places; spaces reserved for waiting, in conditions of distancing, facing the public transport stops, commercial and service activities.

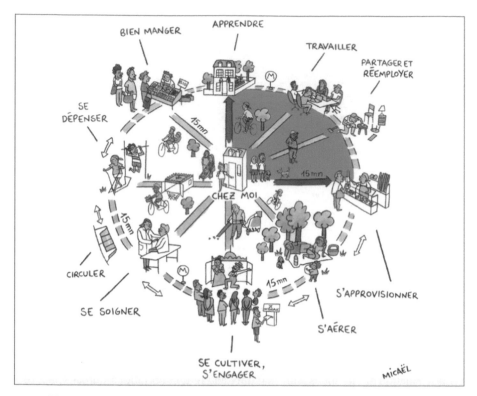

Fig. 4. Proposal of *Ville du quart d'heure* by the Mayor of Paris, Anne Hidalgo.

All the above points should be included, in graphical and normative form, in the Minimum Urban Pandemic Structure (SUMP) both at urban and regional level.

5   Conclusions

In summary, we can highlight the great difficulty to overcome to mitigate space-time inequalities and global risk without enhancing the symbiotic, unstable and dynamic relationship between scattered settlements (capable of ensuring distancing, environmental quality, qualitative neighborhood relations) and urban centers, placed in safety (maintaining their function of driving development).

Finally, it is a question of overcoming the dichotomies of the post-pandemic city by redefining the relationships between spaces and times of the city, between the contrast and coexistence with risk, between immobile city and mutant city, between centralization and diffusion, between deterministic urban planning and guidelines, between predetermined uses of city spaces and flexible, temporary and uncertain uses.

In conclusion, the results of the work lead to the awareness that the possible recurrence of pandemic cycles necessarily obliges the formalization of a pandemic plan, to be updated every few years, not only of a strictly medical-health nature but also of a territorial urban character. A pandemic urban plan at municipal, regional and national level should be aimed at organizing and managing centres equipped with reception, care and assistance, protected road systems, places of quarantine, temporary flows of inhabitants from high-density centres.

A path to pursue such reorganization of the settlement structure and of the times of work, study and leisure can develop starting from the numerous shared choices on the procedures for protection from seismic risks (or environmental disasters). The research then showed how it is possible to increase the contents of consolidated plans such as the "Minimum Urban Structures SUM" with the contents identified in the results of this work, in order to achieve urban and territorial plans defined "anti-Pandemic Minimum Urban Structures SUMP".

Finaly, a different paradigm of life and urban organization is proposed: the coexistence with global risk.

References

1. Sbetti, F.: Spazio e tempo. Urbanistica Informazioni, 3–4, 287–288. Inu Edizioni, Roma (2019)
2. Tarpino, A., Marson, A.: Dalla crisi pandemica il ritorno ai territori. Scienze del Territorio, special issue Living the territories in the time of Covid: 6–12 (2020). https://doi.org/10.13128/sdt-12369
3. Fuksas, M.: Coronavirus, Fuksas: Ridisegnare lo spazio vitale nella casa post Covid. In: Merlo, F. (ed.) La Repubblica, 18 aprile (2020)
4. Boeri, S.: Coronavirus, Boeri: "Via dalle città, nei vecchi borghi c'è il nostro futuro". In: Giovara, B. (ed.) La Repubblica, 20 aprile (2020)
5. Spada, M.: I virus passano le città restano. Urbanistica Informazioni 36, 287–288 (2020)
6. Indovina, F.: La città dopo il coronavirus. Archivio di Studi Urbani e Regionali 128, 5 (2020). https://doi.org/10.3280/ASUR2020-128001
7. Borjas, G.J.: Demographic determinants of testing incidence and Covid-19 infections in New York City Neighborhoods, National Bureau of Economic Research, Cambridg Mass Working Paper 26952 (2020). http://www.nber.org/papers/w26952. Accessed 22 Feb 2022 https://doi.org/10.3386/w26952
8. Balducci, A.: I territori fragili di fronte al Covid. Scienze del Territorio, special issue Living the territories in the time of Covid, 169–176 (2020). https://doi.org/10.13128/sdt-12325
9. Tira, M.: La pandemia come volano per il ripopolamento dei centri rurali? In: Samorì, C. (ed.) Ingenio. Informazione tecnica e progettuale (2020), https://www.ingenio-web.it/28124-la-pandemia-come-volano-per-ilripopolamento-dei-centri-rurali. Accessed 26 Feb 2022
10. Barca, F.: Ai territori serve progettualità, non sussidi e grandi opere. In: Pierro, L., Scarpinato, M. (eds.) Intervista a tutto campo al coordinatore del Forum Disuguaglianze e Diversità: sviluppo locale, aree interne, redistribuzione di opportunità e accesso alla conoscenza, ruolo degli architetti e geopolitica mediterranea, 22 Luglio (2020), https://ilgiornaledellarchitettura.com/2020/07/22/fabrizio-barca-ai-territori-serve-progettualita-non-sussidi-e-grandi-opere. Accessed 26 Feb 2022
11. Clemente, P.: Piccoli paesi nell'ondata del virus. Resistenza, democrazia, comunità. Scienze del Territorio, special issue Living the territories in the time of Covid, 44–52 (2020). https://doi.org/10.13128/sdt-12331
12. Pasqui, G.: Il territorio al centro. Urbanistica Informazioni 10–11, 287–288. Inu Edizioni, Roma (2019)

13. Bedini, M.A., Bronzini, F.: The post-earthquake experience in Italy. Difficulties and the possibility of planning the resurgence of the territories affected by earthquakes. Land Use Policy 78, 303–315 (2018). https://doi.org/10.1016/j.landusepol.2018.07.003
14. Bedini, M.A., Bronzini, F.: Old and new paradigms in pre-earthquake prevention and post-earthquake regeneration of territories in crisis. Archivio di Studi Urbani e Regionali **124**, 70–95 (2019). https://doi.org/10.3280/ASUR2019-124004
15. Bedini, M.A., Bronzini, F.: Priority in post-earthquake intervention. Territorio 96, 127–136. Milan: Franco Angeli (2021). https://doi.org/10.3280/TR2021-096012
16. Bedini, M.A., Bronzini, F, Marinelli, G.: Preservation and valorisation of small historical centres at risk. In: Gargiulo, C., Zoppi, C. (eds.) Planning, Nature and Ecosystem Services, pp. 744–756. FedOA Press, Napoli (2019). https://doi.org/10.6093/978-88-6887-054-6

The Community Health Centers: A Territorial Service in the Post-pandemic City

Marco Mareggi$^{(\boxtimes)}$ ⓘ and Michele Ugolini ⓘ

DAStU, Politecnico di Milano, Milan, Italy
marco.mareggi@polimi.it

Abstract. The global pandemic has placed health back at the center of political agendas. In particular, territorial health has become a relevant issue because it guarantees primary care close to citizens and reduces the influx of patients in large hospitals. The Community Health Center is a structure that provides these guarantees, which is being enriched with social and health-care services. The contribution presents the first results of the *Coltivare_salute.com* research, developed at the Politecnico di Milano as part of the Polisocial Award 2020. It presents the Community Health Center as a service and critically reflects upon the guidelines for its design proposed in Italy and United Kingdom. Subsequently, it highlights the issues of health in the urban plans of the city of Piacenza, the case study of the research, and develops some specific surveys. Furthermore, with respect to the case-study, the contribution discusses some criteria for the design of new Community Health Centers, conceived as opportunities for urban regeneration. In conclusion, it underlines the values and criticalities of the proposal elaborated by the research, which calls for a greater integration between health policies and territorial planning.

Keywords: Community Health Centers · Territorial health · Urban health planning and design

1 Health-Care in the Territory

Since 2020, the pervasiveness of the global pandemic has led city institutions back to discuss and deal with specific issues and services to ensure people's health.

For modern town-planning, healthiness is a constitutive issue. According to Ashworth [1] and Benevolo [2], from the second half of the XIX century in Europe the regulatory apparatus of the discipline was recognized as a derivation and evolution of the hygiene and health regulation. The end of the endemic conditions of industrial urbanization in western countries has shifted the interest of cities towards the well-being of people, to which the urban environment can make a contribution. Many were the international research and urban movements – i.e. the 'Healthy Cities' [3] and the 'Time of the City' [4] – that at the end of the XX century have served as real-world wellbeing laboratories developing and incubating innovative initiatives. With Covid-19, the health problems have returned to the fore. Many cities around the world have initiated urban

© The Author(s), under exclusive license to Springer Nature Switzerland AG 2022
F. Calabrò et al. (Eds.): NMP 2022, LNNS 482, pp. 603–611, 2022.
https://doi.org/10.1007/978-3-031-06825-6_57

transformation processes related to both the health and the well-being of people in the urban environment [5–8].

How have health and community health services recently entered the urban agenda of discussion and urban project?

This paper proposes experiences and reflections starting from a specific research project, *Coltivare_salute.com*, and in relation to a specific service, the Community Health Center (*Casa della Salute*). This service articulates its supply by moving from health to broad social and health-care services targeted to the community. These are activities that are undergoing an update, also due to the prospects opened by the National Recovery and Resilience Plan (NRRP) [9], which applies and develops the policies of the Next Generation Europe program in Italy.

Coltivare_salute.com is a research project funded by Politecnico di Milano through the Polisocial Award 2020, which supports the study and experimentation activities with a high social purpose. Interdisciplinary research (architecture, conservation, urban planning, sociology, management engineering, hygiene) aims to define guidelines for the design of new Community Health Centers in an urban context, intended as places for providing health services and at the same time social centers, capable of triggering and accompanying urban regeneration processes [10].

The essay presents the Community Health Center (CHC) as a service and critically reflects on the various indications for its design proposed in the British and Italian contexts. It then investigates the issues of health in the urban plans of the city of Piacenza, the case study of the Polisocial Award 2020 research, and develops some specific insights conducted alongside the research. Furthermore, with respect to the case study, the contribution presents some criteria for the design of new Community Health Centers, interpreted as opportunities for urban regeneration. The conclusion underlines the values and criticalities of the proposal put forward by the research.

2 Primary Health of Proximity and Planning Indications

The network of Community Health Centers represents a way to strengthen the health-care system in the area. Placed between different types of hospitals, general practitioners and social assistance services, the CHC is an intermediate service, which aims to integrate in a polycentric way the health and social assistance, the prevention and health education, making people responsible and aware of healthy and sustainable lifestyles [11]. This is a change in the approach to the primary health-care services. In fact, these structures take care of urgent and non-urgent outpatient problems, diagnostic procedures that do not require hospitalization, the management of chronic diseases, and the health prevention and promotion. They are organized according to different levels of complexity (low and medium-high) which match to the characteristics of the territories and the population density. They are open to citizens and voluntary associations [12]. In this way they attract not only the sick but a larger population. During the pandemic, it is precisely these characteristics that have highlighted the CHC as a potential effective organizational form of health services due to the variety of the offer, the proximity to people and the capillarity in the territory. This has led to strengthen their presence in the territory (in Italy through dedicated funding in the National Recovery and Resilience Plan), as well as to renew the interest in both national and local policies, not only in the health sector but also in the organization and management of the system of public services in a number of Italian cities.

The "Case della Salute" were born on the model of the British Health Centers. Subsequently, they spread to other countries, especially European ones, characterized by universal health care (Spain, France, Finland as well as the United States and Canada) [13]. In Italy they have had a differentiated implementation from region to region. This is an inherent imbalance in the national health-care system, for which each region programs health services in an independent way [14]. From the census carried out by the Italian Chamber of Deputies, in 2020 493 CHCs were active on the national territory, of which: 124 in Emilia-Romagna, 77 in Veneto, 76 in Tuscany, 71 in Piedmont, 55 in Sicily, 22 in Lazio, 21 in the Marche, 15 in Sardinia, 13 in Calabria, 8 in Umbria, 6 in Molise, 4 in Liguria, 1 in Basilicata. The other regions do not have any of them [15].

The name Health Center was introduced pioneeristically in 1920 by the Dawson report [16], a document with which the British Ministry of Health configured the new organization of health services, identifying the basic level of care in the Primary Health Centers [13]. The country thus posed itself as a historical reference for primary care. Subsequently, with the English *Health Building Notes* of 2009 [17] innovative practices were made available at international level not only for the organization of health and social services, but also with respect to the organization of the spaces in which these services are provided, with attention to the relationships between users, operators and the activities carried out. In 2013, the *Health Building Notes* [18] were enriched with a dozen best practices for the design and planning of new health buildings and for the adaptation and expansion of existing structures. The cases underline the possibilities of multi-offer integration (health, social assistance alongside libraries, gyms, spaces open to the community), multimodal accessibility (by walking and public transport; the latter less practiced in peripheral areas and at shopping center), high quality of the experience of patients and operators (intuitive orientation, enhancement of panoramic views, differences of heights and natural light), sustainable design. They thus offer complex and articulated indications with respect to the contexts, overcoming the analytical and prescriptive approach which often characterize the guidelines.

Also in Italy some regions have proposed guidelines for the design of Community Health Centers. The only ones expressed in a law are drawn up by the Emilia-Romagna Region [12]. Inspired by the British model, these aim at the homogeneity and recognizability of the structures and provide information on sizing, spatial distribution and functions, according to the tripartite division: clinical, staff and public area. For the latter, it is recommended the inclusion of comfortable recreational, meeting and refreshment spaces, for socializing, and a direct relationship with the outdoor space is suggested, encouraging its transformation from a garden belonging to the structure to an urban park. The intent is clearly to stimulate a broad involvement of the population in the use of this service, not just of those in need of care.

While the pandemic has on the one hand underlined the potential of this form of service provision, on the other it has highlighted the spatial rigidity of existing structures, starting from the reception spaces, and of their relationship with the open space. For example, the single entry required by legislation to ensure safety and control of flows is a limit [12]. Just as the lack of internal flexibility and the weak relationship with the surrounding places are obstacles to adaptability, if the need to manage the service in a differentiated way in relation to changing situations emerges. Finally, in the guidelines there are no criteria for the location of the CHCs. This highlights the lack of an idea of location of a service that activates specific urban and social relationships in the city.

3 Health in Urban Plans in Piacenza

If the legislation and the practices of the health-care system necessarily dealt with the supply of health services and their articulation in the territory, before the pandemic this was less significant in city planning, with the exception of some large-scale interventions such as the establishment of new hospital complexes. These are services increasingly planned on a supra-local scale, which drain huge public financial resources and animate local and regional debates. The case study of applied research in Piacenza is inscribed in this situation.

This city is located in Northern Italy and administratively belongs to the Emilia-Romagna Region, which has a health-care system with high quality standards and a good tradition of urban planning. The interrelation between these two fields of public action is weak. It is useful to ask whether the two areas belong to different sectoral technical rationalities and whether they have distinct decision-making arenas and programming phases.

In Piacenza, the urban plans in force at provincial and municipal level were approved respectively in 2010 [19] and in 2016 [20]. In both plans, the topic of health and health-care is treated in generic terms. Furthermore, the different types of health services and their territorial distribution are little considered and there are no specific health needs that require improvements to existing structures or new services. It is only following the reorganization of the hospital system, promoted by the regional government in 2015, that the establishment of a new provincial hospital in the capital city was proposed. Up to that date, there was no trace of this proposal either in the preliminary studies or in the planning provisions. This regional proposal was undertaken by the Municipality, in agreement with the Region, the Province and the local health authority which in 2019 started the revision of the planning regulations for the location of the new hospital structure [21].

At the same time, the competent administrations have started the definition of the new urban plans at the municipal level (general urban plan) and at the provincial one (territorial plan). Although their process started after the pandemic period, the strategic guidelines of these tools do not seem to place health issues at the center of urban planning. However an interest in the facilities where health services are provided is present in the analytical framework of the territorial plan. Specifically, the study of service facilities and their accessibility investigates the supply of health services on a territorial (hospitals, first aid, operational activities, beds, accessibility index and access times) and local scale (CHCs, general practitioners, pharmacies) [22]. The authors underline that this is an attention fostered by the administrative and scientific debate promoted by the National Strategy for inland areas. The strategy defines the marginal areas according to parameters of equipment and accessibility to transport, school and health services, which are considered fundamental for the habitability of a territory [23]. Otherwise, the studies for the new general urban plan for the city of Piacenza consider the health services in the same way as the other public services.

The *Polisocial* research, on the other hand, has focused on this specificity, analyzing the articulation of public services and equipment in Piacenza with an attention to health, socio-health and the forecasts under discussion (new hospital and other CHCs) by the various local institutions. With respect to the specific health-care supply, the research has highlighted how the city has: a hospital with pavilions, on the western edge of the historic center, built over the centuries and expanded around 1990 with a large polysurgical building; a single CHC, on the northern edge of the historic center; a widespread network of pharmacies and medical clinics (general practitioners and pediatricians); "group medicines", such as spaces shared by several practitioners or pediatricians, located where there is a concentration of demand (Fig. 1). In relation to the forecasts, however, in addition to the new hospital, the research investigates how to adjust the number of CHCs with respect to the resident population and to rebalance the territorial distribution of this health-care supply. Since the interest of the research is to put this service in relation with its potential for urban regeneration, the study investigated the provision and territorial distribution of urban and neighborhood services, open spaces and the conditions of accessibility for defining the potential settlement areas for new Community Health Centers.

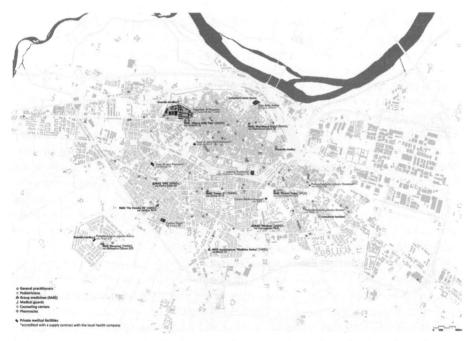

Fig. 1. The health system in the city of Piacenza, Italy. Source: Coltivare_salute.com, Politecnico di Milano, elaboration by M. Mareggi and L. Lazzarini with data provided by the local health company, 2021.

The *Polisocial* research through its studies and proposals has the objective, on the one hand, to offer a support to the ongoing municipal planning, and on the other, to provide an opportunity of reflection on the urban scale for health policies. In other words, the research aims to be an opportunity for building interrelationships between decision-making spheres that traditionally belong to different programs and that converge only *a posteriori*, when decisions produce effective transformations of the city.

4   First Guidelines for the Community Health Centers in Piacenza

Following the study of the guidelines for the design of Community Health Centers available in Italy and abroad, the research of international case studies (due to brevity not presented in this contribution), and the analysis of the health and urban planning of the city of Piacenza, and after a first discussion with local institutions (local health company and local administrations), the Polisocial research has proposed some indications and a work path to consolidate the CHCs as collective reference places for the local community.

As far as the design of the new CHCs is considered, the research identifies two peculiar characteristics: the first linked to the flexible internal organization and the second to the greater articulation of the system of external relations. On the one hand, it proposes to define new ways of organizing the interior space in order to face both emergency conditions imposed by the pandemic, and to manage a complex everyday life in which many different needs coexist. On the other hand, the possibility of triggering urban and social regeneration processes through a network of integrated health-care and social services distributed throughout the territory, which, using a careful localization strategy with respect to the context and its potential relationships, shape new identities and urban peripheral centralities, capable of substantiating an idea of healthy city and of protecting and promoting people's health as a fundamental right.

The proposal suggests to set up new CHCs in places where a vocation is recognized, that means with layered urban identities that have become manifest over time. It is a condition that overcomes performance logics and quantitative parameters and takes the form of what we can define as a localization strategy. The aim is to push the programming towards a redefinition of places, even those apparently consolidated in use, which sees the CHC as a strategic node and a place of identity recognition for the community.

This operation is not only of interest for health, but intercepts various disciplines including those of the project according to a multi-scale approach: from architecture and urban design to the dimension of internal and open spaces, to environmental design, passing through urban and territorial planning.

With a view to reducing land consumption and possibly reusing existing buildings belonging to the vast heritage of unused properties, a guiding principle invites to act within complex urban programs. This suggests placing the Community Health Centers within a framework of actions and strategies capable of constituting an articulated system of relationships that relate to: existing and new green spaces (parks, gardens, avenues), public buildings (schools, community and sports centers, gyms), and the soft and sustainable mobility networks (pedestrian, cycle, electric and public transport). Furthermore, the indications invite us to rediscover interstitial spaces of cities that express the potential to be involved in a different urban design project.

Within the idea of healthy city, the establishment of a CHC should be strengthened by a dense mix of functions, public and private, and by high-quality public spaces. The goal is not the usual practice of redeveloping open spaces, but it is a matter of imagining new urban centralities supported by a density of functions and an integral design of public space able of making these places highly attractive and recognizable for the community.

Even architecturally, the CHC should show its stable presence. It is suggested that this can happen not much through the expression of coordinated and defined images *a priori* (as if it was an urban scale logo) with the application of abstract color ranges, but rather to open relationships between architectures, capable of grasping the specificities of places, to structure complex urban relationships, both in the organization of the public open space with respect to the surrounding context, and in the definition of a flexible, adaptable and modular interior space, also with respect to its potential relationships with the buildings' open spaces.

5   Challenges to Address

Authors highlight that these suggestions for guiding the design of new Community Health Centers mark the difference of the Polisocial research project with respect to a functionalist and health-related approach, up to now implemented in the main projects in Italy.

The research project, first of all, strengthens and enhances some indications already included in the most advanced guidelines for the design of the Italian and British CHCs, for example the multifunctional articulation, the accessibility and the relations with the areas where they are located, in particular with respect to the system of open spaces.

Secondly, it critically underlines the need to strengthen the criteria to define the localization and distribution in the cities of the CHCs as a territorial socio-health service placed in an intermediate position between the general practitioners and the hospitals.

Thirdly, it reaffirms the value of the multifunctionality that the CHCs can offer and the need for these to define both spatial and social relationships with the local system of proximity services.

Fourthly, the proposal by Coltivare_salute.com pushes to invest in the design of the articulation of interior spaces that are efficient but also flexible, for example to allow a single controlled and secure access, but also multi-access in case of need.

Fifthly, they invite us to reflect upon a communication project that uses a wide range of tools, ranging from the architecture to the social interrelationships between the many institutional and non-profit entities that find or can provide services within the CHC as a collective space recognized by the community.

In conclusion, the project proposal tries to balance the important organizational needs of the health-care supply with the opportunity to give a strong urban value to a relevant service such as that of primary care, which take place in the Community Health Centers. The pandemic reminds planning authorities of the importance of a joint definition between health and territorial planning.

The authors recognize the limits of the above proposal and the difficulty of bringing together different programs and policy fields. However, the acceleration triggered by the pandemic situation makes it possible to challenge these difficulties. The investments proposed by the NRRP are in line with the indications provided. In fact, it not only economically supports the strengthening of the CHCs as proximity networks for a territorial health-care integrated with social services (with the plan to create 1,288 Centers in existing or new structures), but also renames the Community Health Centers as Community Centers, to underline their link with the local community [9].

Credits. This work is the product of a joint reflection among the authors. Sections 1, 3 and 5 are attributed to M. Mareggi. Section 2 and 4 are attributed to M. Ugolini.

References

1. Ashworth, W.: The Genesis of Modern British Town Planning. Routledge and Kegan Paul Lto, London (1954)
2. Benevolo, L.: Le origini dell'urbanistica moderna, Laterza, Rome-Bari (1963). English edition: The origin of modern town planning. MIT Press, Cambridge-London (1971)
3. Duhl, L.J., Sanchez, A.K. (eds.): Healthy Cities and the City Planning Process. A Background Document on Links Between Health and Urban Planning. WHO, Copenhagen (1999)
4. Mückenberger, U.: Local time policies in Europe? Time Soc. **20**(2), 241–273 (2011)
5. Forsyth, S.: What role do planning and design play in a pandemic? Department of Urban Planning and Design, Harvard University Graduate School of Design (2020). Accessed 7 Dec 2021
6. Leigh, G.: Re-imagining the post-pandemic city. Landsc. Archit. Australia **167**, 18–20 (2020)
7. Pisano, C.: Strategies for post-Covid cities: an insight to Paris En Commun and Milano 2020. Sustainability **12**, 5883 (2020)
8. Mareggi, M., Lazzarini, L.: Health, an enduring theme for urban planning. In: 57th ISOCARP World Planning Congress Proceedings, pp. 803–812. Doha (2021)
9. Ministero dell'Economia e delle Finanze: Missione 6: Salute. Reti di prossimità, strutture e telemedicina per l'assistenza sanitaria territoriale. In: Piano Nazionale di Ripresa e Resilienza, pp. 224–227 (2021). https://www.governo.it/sites/governo.it/files/PNRR.pdf. Accessed 4 Dec 2021
10. Ugolini, M.: Case della Salute: condizioni di fragilità e occasioni di rigenerazione sociale e urbana. Territorio **97**, 147–153 (2021)
11. Turco, L.: Un New Deal della Salute. Linee del programma di Governo per la promozione ed equità della salute dei cittadini. In: Audizione alla Commissione Affari Sociali della Camera dei Deputati, 2006/6/27 (2006). https://www.salute.gov.it/imgs/C_17_pubblicazioni_530_all egato.pdf. Accessed 8 Dec 2021
12. Regione Emilia-Romagna: Casa della Salute: indicazioni regionali per la realizzazione e l'organizzazione funzionale. Decreto di Giunta Regionale, vol. 291 (2010). https://salute.regione.emilia-romagna.it/cure-primarie/case-della-salute/documenta zione-case-della-salute/delibere. Accessed 2 Oct 2020
13. Brambilla, A., Maciocco, G.: La Casa della Salute. Innovazione e buone pratiche. Carocci, Rome (2016)

14. Della Porta, M.R., Mazzoni, E., Salerno, D., Tani, G.: Regionalismo differenziato, manifestazioni e radici. In: Da Empoli, S., Mazzoni, E. (eds.) Riportare la sanità al centro. Dall'emergenza sanitaria all'auspicata rivoluzione della governance del SSN (2020). https://www.i-com.it/wp-content/uploads/2020/09/Riportare-la-sanita-al-centro-Stu dio-I-Com.pdf. Accessed 5 Apr 2021

15. Camera dei Deputati, Servizio Studi Affari Sociali: Case della salute ed Ospedali di comunità: i presidi delle cure intermedie. Mappatura sul territorio e normativa nazionale e regionale, 144, 2021/3/1 (2021). http://documenti.camera.it/leg18/dossier/testi/AS0207.htm?_161919 8796640. Accessed 1 Dec 2021

16. Ministry of Health, Consultative Council on Medical and Allied Services: Interim Report on the Future Provision of Medical and Allied Services 1920 (Lord Dawson of Penn) (1920), Interim Report on the Future Provision of Medical and Allied Services 1920 (Lord Dawson of Penn) (sochealth.co.uk). Accessed 29 Dec 2021

17. Department of Health: Health Building Note 11-01. Facilities for primary and community care service. The Stationery Office, London (2009)

18. Department of Health; Health Building Note 11-01. Facilities for primary and community care service. The Stationery Office, Londoni (2013)

19. Provincia di Piacenza: Piano territoriale di coordinamento provinciale. Deliberazione di Consiglio provinciale, 69 (2010). https://www.provincia.pc.it/sottolivello.php?idsa=253&idbox= 40&idvocebox=165. Accessed 4 Dec 2021

20. Comune di Piacenza: Piano strutturale comunale di Piacenza. Deliberazione di Consiglio comunale, 6 (2016). https://www.comune.piacenza.it/temi/territorio/psc. Accessed 11 Dec 2021

21. Comune di Piacenza: Approvazione variante specifica al PSC vigente per la realizzazione del nuovo ospedale. Deliberazione di Consiglio comunale, 30 (2021). https://www.comune.pia cenza.it/temi/territorio/nuovo-ospedale. Accessed 11 Dec 2021

22. Provincia di Piacenza: Piano territoriale di area vasta. Quadro conoscitivo. Documento per la consultazione preliminare. Provvedimento del president, 51 (2021). https://ptavpiacenza.it/. Accessed 4 Dec 2021

23. Barca, F., Casavola, P., Lucatelli, S.: Strategia nazionale per le aree interne: definizione, obiettivi, strumenti e governance. Materiali Uval, Roma, vol. 31 (2014). https://www.agenziaco esione.gov.it/wp-content/uploads/2020/07/MUVAL_31_Aree_interne.pdf. Accessed 11 Dec 2021

Healthy Cities with Healthy Streets, Towards a New Normal of Urban Health and Well-Being

Antonio Taccone(✉) 🆔 and Antonino Sinicropi 🆔

PAU Department, Mediterranean University of Reggio Calabria, Reggio Calabria, Italy
{ataccone,antonino.sinicropi}@unirc.it

Abstract. This paper intends to illustrate the results of a line of research that explores the characteristics and aspects of the public space par excellence, the city streets, and the potential that this space has to become a third, accessible and inclusive place, generating returns in terms of 360° sociality, well-being, health, and urban well-being.

The need for public space, amplified by the restrictions of the pandemic, together with the possibility of being able to move and keep fit, is today more than ever a prerogative to guarantee the right to health enshrined in the Universal Declaration of Human Rights. Therefore, urban plans should prepare large spaces for pedestrian networks, extend them to the whole territory, and integrate them with systems and infrastructures for sustainable public/private transport (trains, trams, buses, zero-emission car sharing, cycle paths), strategically creating a multimode hub. This new way of moving would allow the transition from hypermobility, which congested urban centers mainly, to the fluidity of movements on a human scale, which can help cope with emergencies of all kinds, better health conditions, and physical well-being, and mental for people.

Keywords: Environmental quality · Healthy-city · Urban planning

1 Introduction

The Covid-19 pandemic has catalyzed essential changes in society. The main social effects are identified in the mixture of the discrete spaces of the home and work, accelerating the collapse of the border between them already eroded for years, but above all and last but not most minor, in the loss of all or most of our third places. The coronavirus pandemic has at least temporarily reconfigured city life, the relationship between work and residence and leisure, the use of public space, and the safety of transport, both public and private, and placed fundamental equity of access to resources. A smart planning scenario could not have anticipated and illustrated so vividly the far-reaching and impactful socio-spatial dimensions of the coronavirus pandemic as a whole, let alone proactively applied standard urban planning tools [1].

Measures of "social distancing", or "spatial isolation", have been the most common to contain the viral contagion. By following this approach, we must distance ourselves from others. Therefore, the most dangerous context to escape from is when intersubjective

F. Calabrò et al. (Eds.): NMP 2022, LNNS 482, pp. 612–621, 2022.
https://doi.org/10.1007/978-3-031-06825-6_58

relationships are more concentrated: the city. Again, the reason is apparent: stacking people on top of each other in buildings and offices and packing them into buses and subway cars creates an ideal breeding ground for communicable diseases.

Collective loneliness during the pandemic showed how dependent people are on each other for happiness and how interconnected they genuinely are. Healthy societies depend on the continuous interaction between people who are different in various ways. Third places are the primary places for such interactions because the shared enjoyment of their services ensures that strangers have at least one thing in common. Third places are elements that bind us to a specific place and the people who frequent it. We build a chosen community in these places, a more critical public sphere. Without them, the associations that weave a complex society will run out [2]. The coronavirus has tested our physical health and our mental, economic, and above all, social health. The urban population will have to be ready to face future challenges. Throughout history, cities have faced multiple challenges, many of which have jeopardized the safety of their inhabitants. Considering the increase in the urban population, aging, and the increase in disabling diseases resulting from unhealthy urban lifestyles, the theme of the healthy city and the healthy street is more relevant than ever and urgent to tackle. Future citizens must be ready for challenges and roll up their sleeves to face them correctly. All this will be possible only if the future citizens enjoy a good state of health and are aware of the place where they live, loving and caring for it as if it were a friend or loved one.

2 Towards Healthy-Friendly Policies of Urban Public Space

COVID-19 has given people the opportunity to reconsider the need for social interaction and to belong to the community and place. Physical distance restrictions have created opportunities for creativity in urban design. There have been radical efforts to reuse the public environment, to facilitate social interaction by precluding physical proximity around the world through actions previously deemed impractical or undesirable in otherwise automobile-centered urban centers. In North America, this has mainly included redistributing space from car lanes and converting to active modes of transportation (bicycles, scooters, pedestrians) and outdoor social activities such as outdoor restaurant seating [3–5]. In Europe, we see further expansion of already bicycle-friendly streets in cities such as Paris, Rome [4], Milan, and London.

Minor changes to the street, such as widening sidewalks and creating makeshift cycle lanes where they were previously absent, serve to illustrate what is possible when urban project priorities shift to accommodate man before the car, to broaden the public environment by abandoning the private, thus mitigating the time spent in private and individual spaces that predispose to solitude [6].

Considering that the quality of life in the city is closely linked to the level of well-being and safety perceived in the ways of moving and staying within the inhabited centers, the data are unfortunately not encouraging.

There are still many negative aspects related to the poor quality of the streets, the time lost in queuing at traffic lights, the difficulty of finding parking spaces, and last but not most minor, insufficient street safety for pedestrians, motorists, and cyclists.

The need for more excellent liveability in the city is strongly compromised by the reduced perception of architectural aesthetics linked to the excessive use of asphalt, the poor development of cycle paths, and the lack of enhancement of socialization spaces such as squares and pedestrian areas. An objective process of urban redevelopment cannot ignore the elimination of the horrible layers of asphalt, which, in addition to making our cities gray, pollute them during installation, restoration, maintenance, breakage, and during its same life. Furthermore, asphalt is a by-product of petroleum that, with its oily and hydrocarbon-rich layer, waterproofs large portions of land, contributing to the rapid desertification process of our lands.

The greater availability of public space, in addition to increasing physical activity and therefore improving health and well-being, would allow users more opportunities for socialization, enriching their knowledge and promoting differences as an added value. Furthermore, greater availability, especially in socio-economically disadvantaged neighborhoods, would allow equal access to opportunities for movement and physical activity in areas too often lacking this type of opportunity. Safe, welcoming, and attractive public spaces could increase the perception of safety and well-being in disadvantaged neighborhoods, improving their condition and potentially reducing the tendency to individualistic behaviors in the use of public space, increasing the propensity to respect the rules of coexistence. It should be added that public spaces with less vehicular traffic would allow better control of the territory by the police and reduce criminal actions in residential areas, service areas, near schools, parks, and commercial areas.

The presence of these "slow" spaces in the city should be a prerogative rather than a possibility, and the feasibility of this type of action is concrete for all types of streets and all types of neighborhoods.

We have decades behind us where cities have been dominated by cars. The car, from its advent to the entire development path, until becoming the preferred means of transport for the population, has changed the shape and structure of the city, drastically accelerating its growth, with grid planning as a mechanical repetition of elements always the same, which make it a lifeless place, suited to the needs of human "anti-men", with the streets the main enemy of man. Moreover, this primacy of the automobile has contributed to the deprivation of the elements of multi-ethnic mixing, of adaptation to human and economic diversity, generating the processes we call "sprawl", "gentrification", "ghettoization", which compromise social, cultural, and resilience of the community, growing differences between populations and conflicts within society.

The WHO in 1988 laid the foundations of the Healthy City project, which involved dozens of cities around the world. Healthy cities are places that deliver positive outcomes for people and the planet [7]. A culture of inclusion and equity is promoted. With the spread of COVID-19, Healthy Cities have been the protagonists of an effective response. The healthy cities approach assumes that cities need more green and blue spaces and a renewed healthy transport system.

The challenges faced, and that cities still have to face are manifold. The COVID-19 pandemic has shown the extent to which a rapidly spreading disease can change the life of a city, no matter how big or small, in the globalized world. Cooperation between citizens has been crucial in addressing the problems arising from the COVID-19 epidemic.

An urban population accustomed to walking, moving with ease through the city streets, and having a better health condition will respond with greater positivity to adverse events by increasing the capacity for human and urban resilience. Furthermore, the drastic reduction of passive and polluting mobility that would be obtained would lead to a significant improvement in air quality. As experienced during the first blockade of the covid-19 pandemic, the levels of fine particles in cities' air fell rapidly, and the air became breathable again. Air quality impacts the health of every person, particularly some of the most vulnerable and disadvantaged people in the community: children and people who already have health problems. Therefore, reducing air pollution benefits us all and helps reduce unjust health inequalities [8].

3   Case Studies and Best Practice

Neighborhood Slow Streets Program – Boston MA
Instead of planning and implementing changes one street at a time, the Neighborhood Slow Streets program targets an entire "area" within the boroughs of Boston, Massachusetts.

Fig. 1. Boston - North End, Harbour walkway

The evaluation considers each street within a specific area to find design problems and solutions. The entire design process occurs with a synergistic work with the entire neighborhood community. The proposals that emerged during the comparison and design phases mainly concerned traffic moderation and safety improvement elements for almost all the streets in the area (Fig. 1).

The streets have been equipped with visual and physical cues to slow drivers up to 20 mph. This new condition makes every street more inviting for people of all ages to walk, play or ride a bike. Neighborhood Slow Streets Program highlights lasting improvements in safety and quality of life by focusing on accessibility.

Neighborhood Slow Streets is a city initiative to slow down the speed of traffic and improve the safety of residential streets within a specific area, giving priority to areas with higher percentages of young, old, and people with disabilities; areas with the highest number of accidents per mile; areas that include points of interest and essential services. The goal is to support existing opportunities, plan active mobility and public transport, and implement the improvement of Boston's city.

The context of the Covid-19 pandemic made it necessary to strengthen the program, accelerating the redevelopment projects of urban areas, exceeding the number of 80 zones within the Boston neighborhoods with slow streets.

Healthy Streets Austin – Austin TX
During the pandemic, in May 2020, Walk Austin, an association of citizens who enjoy walking, running, and cycling, in partnership with the Austin city council, in Texas,

launched the Healthy Streets Austin program to prioritize streets for everyone. A program was born to help people exercise safely while respecting social distance during the pandemic. Research on best practices will bring long-term changes to city streets, helping to bring them back into authentic public spaces for all, regardless of people, age, ability, or range of motion.

Healthy streets are achieved by closing some local streets to traffic, keeping access only for residents, deliveries, and emergency vehicles. As a result, people can use these low-traffic areas to walk healthy streets, wheelchair users, run and cycle with enough space to maintain physical distance.

The streets that were the subject of the interventions became promenades frequented by all the residents, children, the elderly, up to the most avid sportsmen; protected streets to pass the mornings safely outdoors, allowing easy access to parks or points of interest. These streets have also encouraged people to exercise, walking safely within their community.

Healthy Streets – London UK

The Mayor of London's strategy, as early as 2018, sets out his plans to address unjust health differences to make London a healthier and fairer city, taking many actions to improve the environment to create a better future. One of the key points of the strategy are the Healthy Streets. The Healthy Streets approach, developed with Transport for London (TfL), creates streets and street networks that encourage car, bicycle, and public transport on foot and health problems. Using the Healthy Streets approach to prioritize human health and experience in urban planning, the Mayor wants to change London's transport mix so that the city works best for everyone.

A partnership has been established between TfL and Sustrans, implementing the Healthy Streets approach, to achieve ambitious goals of the Mayor of London's transport strategy, including achieving 80% of London journeys by foot, bicycle, or con public transport by 2041. The program addresses street risks by responding to local street safety problems and supports municipalities with local initiatives to raise public awareness of new cycle paths and discourage engine operation to a minimum.

Lucy Saunders designed the Healthy Streets implementation process for TfL. The Healthy Streets approach places people and their health at the center of decisions on public spaces' design, management, and use.

Healthy Streets officials work with some London boroughs to encourage more walking and cycling; they help community groups, schools, and businesses reduce car travel and support cycling skills training.

The approach is based on ten sound evidence-based street indicators, each describing an aspect of the human experience of being on the street. These ten must be prioritized and balanced to improve social, economic, and environmental sustainability through the way streets are designed and managed.

This approach can be applied to any street, anywhere in the world. Adopting this approach requires incremental changes in all aspects of street and transport decision-making. The streets must be welcoming places where everyone feels welcome, to spend time and interact with others, improving social interaction and health. The simplicity of using the space to reach places of interest, guaranteeing shelters and shading to favor the outdoor experience in all seasons of the year, represent a prerogative. Healthy streets are places where we can rest, with little noise and where we can feel safe, having the convenience of choosing the active mode of transport, whether on foot or by bicycle, assisted by efficient public transport.

Milan and Turin, Italy

The Covid-19 emergency has accelerated processes that were only in the embryonic stage in many Italian cities or even simple ideas of citizens' associations or local administrators.

Milan has launched an ambitious project, called "Strade Aperte", to increase the pedestrian traffic of the city streets by widening the sidewalks, where they are narrow and, where it is not possible, taking advantage of the slow and shared streets. In addition, it has increased the cycle network by a further 35 km compared to the existing 200 km, with the idea of implementing a reimbursement or rewards program for those who decide to go to work by bicycle. The goal is to decongest the city from private transport further and lighten the load on public transport. For the realizations, what was learned with the tactical urban interventions in previous years in various city streets was exploited. However, despite the administration's efforts and the projects started and completed, the demand for viable streets with active mobility in safety is still very high.

Turin was no exception. Indeed, during the pandemic, the city launched a dozen pedestrianization experiments on as many areas of the city center, experiments that will become definitive, increasing the puzzle of urban pedestrian areas and making the city highly pedestrian on a human scale and at the same time increasing the difficulty of practicability by car, with numerous bans and closed streets, making life difficult for hardened motorists, albeit with a state-of-the-art, widespread, and efficient public transport system.

4 Towards New Urban Policies for Healthy City

Closed streets and low-traffic streets have helped prevent overcrowding in public parks, footpaths, and sidewalks and have enabled people to explore their communities like never before by walking, jogging, biking, and even in a wheelchair.

The need for public space, amplified by the restrictions of the pandemic, together with the possibility of being able to move and keep fit, is today more than ever a prerogative to guarantee the right to health enshrined in the Universal Declaration of Human Rights.

Fig. 2. Boston - Rose Kennedy Greenway from "Swings by the Fountain" (early spring 2019)

Future mobility should be sustainable, accessible, fair, and safe. Even the most disadvantaged neighborhood in terms of space safety could offer opportunities for new generations to move safely on foot or by bicycle in order to access safe modes of transport and at the same time improve their physical condition and health [9].

Urban plans should prepare large spaces for pedestrian networks, extend them to the whole territory, and integrate them with systems and infrastructures for sustainable public/private transport (trains, trams, buses, zero-emission car sharing, cycle paths), strategically creating a multimode hub. This new way of moving would allow the transition from hypermobility, which mainly congests urban centers, to the fluidity of movements, on a human scale, which can help deal with emergencies of all kinds, better health conditions, and physical and mental well-being to people (Fig. 2).

Fig. 3. Boston - Rose Kennedy Greenway from "Rowes Wharf Plaza" (early spring 2019)

Ensuring priority for pedestrians or making long street routes entirely pedestrian to create an urban pedestrian network will increase perceived safety, with the necessary precautions that must be taken into consideration from the earliest stages of urban planning and design of the single block [10]. In addition, the elimination of the heavy car traffic on the streets will allow repaving with more comfortable, sustainable, and pleasant materials than the sad texture of the bituminous asphalt that covers almost all of our routes [11] (Fig. 3).

Fig. 4. Boston - Rose Kennedy Greenway (early spring 2019)

Urban pedestrian networks would allow greater accessibility and usability of public space, mainly if structured in such a way as to allow the creation of essential services, places of cultural, landscape, environmental, and above all, social interest (Fig. 4).

Cities could thus return to being not only on a human scale but above all for children, young and old, the most marginalized sections of the city's population of hypermobility. The inhabitants should be able to use the city's urban space to keep fit and healthy, but above all by the little ones to grow up in a healthy, dynamic, and stimulating environment. An urban environment that is adaptable to situations

Fig. 5. NYC - Time Square (fall winter 2019)

and usable in all conditions invites play, socialization, and creativity. The urban space, in this way, will become the third space par excellence and, perhaps, the favorite space, where young people could grow by learning from the city and learning to love it. In synergy with local administrations, the role of educational and training institutions and associations will be fundamental to foster a "very active" citizenship capable of facing future challenges. A city can only be resilient if its inhabitants are resilient and vice versa because resilience is closely linked to the balanced condition of the city.

The need for public space is now a global emergency, and the cry is even more vital in developing countries.

To ensure that the city is ready to face subsequent crises or even brief turbulences, it would be necessary to reorganize the city itself, transforming large cities into multicentric cities, thus ensuring an equitable distribution of services at a neighborhood scale and reorganizing urban street network giving greater priority and space to the pavement than the carriageway.

Like a leap into the past, the inhabitants should abandon their cars and walk again, thanks to the support of the technique and technology of our time. This return to the past could trigger a process of improvement of the city that could help reach the zero emissions goals, improve health, quality of life even in old age, rediscover belonging to an urban community, activate resilience. The possibility of moving in safe spaces, with a lower level of polluting dust, but above all the opportunity to move in welcoming and entertaining spaces, in which there is an incentive to socialize and share experiences, should be a fixed point in every action of regeneration and could lead citizens to a return to the use of proximity spaces, happily renouncing long motorized journeys to meet the needs of daily life.

The streets represent the vital space for the reconstruction of a healthy city; it is, therefore, essential to compare the size and space occupied by different street users to reveal the benefits of designing transit, cycle, and pedestrian streets. Furthermore, providing high-quality facilities for this space-efficient, convenient and sustainable mode of transport would allow the same street to accommodate more people [11]. By reducing the amount of space dedicated to the circulation and storage of private vehicles, the space available for other activities that add to the quality of the streets could be increased. By reshuffling their priorities, the post-pandemic city could become a healthier place; instead, by maintaining the position of places of consumption as the driving environment of social aggregation, thus subordinating the social bond to economic growth, we will have learned very little from the lockdowns of the Covid-19 pandemic [12]. It is necessary to rebuild the city always to make it a better means of connection; this means re-proposing public places around authentic, meaningful, and safe interactions. Streets represent the public space par excellence, but they should not be seen as a simple transit space: returning street space to pedestrians and active mobility, in general, could transform them from simple spaces to new places for the shared life of the community and places for physical activity and relaxation. To obtain safe, healthy, and social streets, it would be necessary to undertake actions which, by inserting themselves in the urban fabric and above all in the public space, could initially create disputes or tensions, precisely because they will aim at a lifestyle change. Therefore, they may find it challenging to be accepted by some parts of the population. Therefore, administrators and planners should directly involve the population, especially in areas outside the city centers, to correctly choose the streets and the actions to them. The involvement of all stakeholders in the public space will be crucial for the success of the actions (Fig. 5).

Important aspects remain open and not underestimated, such as the accessibility and inclusiveness of spaces. Although they may concern a minimum percentage of the population, the writer believes it is essential to increase research to guarantee the well-being of the urban community [13]. The future scenario of research in this sense is highly vast and transdisciplinary. However, the need to consider all the possible characteristics, deficits, and disorders of the human body and mind in the organization of public space, be it a place or a distance space, must stimulate greater interest on the part of the scientific community of the sector, of planners and urban planners, as well as of local administrations in order to be able to achieve the conditions of urban well-being for the largest share of the population. The efforts of the last decades have led to actual, but still unsatisfactory results on the motor or sensory disabilities, which often encounter obstacles in their applicability, despite the design in this sense have been regulated nationally and internationally [13]. On the approach to disabilities or disorders of the human mind in the public space and the city in general, the literature is relegated to essays or autobiographical stimuli by the scientific community of the competent medical fields [13, 14]. At the same time, the areas would require a considerable effort of urban planning and planning to achieve essential objectives, make places more livable for all and allow for better and more inclusive growth of city communities.

References

1. Banai, R.: Pandemic and the planning of resilient cities and regions. Cities **106**, 102929 (2020). https://doi.org/10.1016/j.cities.2020.102929
2. Low, S., Smart, A.: Thoughts about public space during Covid-19 pandemic. City Soc. **32**, 1–5 (2020). https://doi.org/10.1111/ciso.12260
3. Buckner, D.: Bike lanes installed on urgent basis across Canada during Covid-19 pandemic, CBC. https://www.cbc.ca/news/business/bike-lanes-covid-pandemic-canada-1.559 8164. Accessed 20 Feb 2022
4. Caballero S., Rapin P.: Covid-19 made cities more bike-friendly here's how to keep them that way. World Economic Forum. https://www.weforum.org/agenda/2020/06/covid-19-made-cit ies-more-bike-friendly-here-s-how-to-keep-them-that-way/. Accessed 20 Feb 2022
5. Schwedhelm, A., Li, W., Harms, L., Adriazola-Steil, C.: Biking provides a critical lifeline during the Coronavirus crisis. World Resource Institute. https://www.wri.org/insights/biking-provides-critical-lifeline-during-coronavirus-crisis. Accessed 20 Feb 2022
6. Heu, L.C., van Zomeren, M., Hansen, N.: Lonely alone or lonely together? A cultural-psychological examination of individualism–collectivism and loneliness in five European countries. Pers. Soc. Psychol. Bull. **45**(5), 780–793 (2019). https://doi.org/10.1177/014616 7218796793
7. World Health Organization: National Healthy Cities Network in the WHO European Region. WHO (2015). ISBN 9789289051026
8. Saunders, L.: Healthy Streets|Making Streets Healthy Places for Everyone. https://www.hea lthystreets.com. Accessed 20 Feb 2022
9. Caballero, S., Tanzilli, M.: Why the future of sustainability starts with mobility? World Economic Forum (2021). https://www.weforum.org/agenda/2021/04/future-of-transport-sus tainable-development-goals/. Accessed 20 Feb 2022
10. NACTO: National Association of City Transportation Officials. Urban Street Design Guide (2013). https://nacto.org/publication/urban-street-design-guide/
11. Kost, C., Mwaura, N., Jani, A., Van Eyken, C.: Street for walking and cycling. Un-Habitat, ITDP (2018). https://africa.itdp.org
12. Tuts, R., Knudsnen, C., Moreno, E., Williams, C., Khor, N.: Cities and Pandemic, Towards a more just, green and healthy future. Un-Habitat, Urban Impact Newsletter (2021). July
13. United Nation Centre for Human Settlements.: Improving the quality of elderly and disabled people in human settlements, vol. 1, Un-Habitat (1993). ISBN 9211312086
14. Beale-Ellis, S.: Sensing the City. Jessica Kingsley Publisher (2017). ISBN 9781784501150

Ecological Networks in the Spatial Planning of Campania Region Towards Green Infrastructures

Salvatore Losco[1](✉) ⓘ and Claudia de Biase[2] ⓘ

[1] Engineering Department, Campania University - Luigi Vanvitelli, 81031 Aversa, Caserta, Italy
salvatore.losco@unicampania.it
[2] Architecture and Industrial Design Department, Campania University - Luigi Vanvitelli, 81031 Aversa, Caserta, Italy

Abstract. *Eco-Planning* thinks nature as not only as an object of consumption and/or exclusive aesthetic enjoyment but recovers and focuses on its role as a provider of vital resources and as a mitigator of imbalances induced by uncontrolled human development. The inclusion of the Ecological Network paradigm in the spatial planning allows to design territory in an integrated way, without neglecting, on the contrary, to start from the areas of interference between anthropic and natural flows. Within this approach the EN represents the location of protection, rehabilitation, implementation of the natural space in anthropized contexts, contrasting land consumption and environmental fragmentation. The paper focuses on the analysis of the role played by EN in the planning forecasts of three spatial plans in force in Campania Region. The Regional and Provincial Ecological Networks within Campania Region Territorial Plan, Caserta and Salerno Provinces Territorial Coordination Plans will be studied and compared. The principal aim is to highlight the strategic guidelines and the prescriptions to the Communal Urban Plans to implement Municipal Ecological Network and Green Infrastructure at local scale. In a sustainable city scenario EN and the more complex GI should be considered of strategic importance for growth/development/transformation/regeneration, in the same way as grey infrastructures [1], their recognition, protection, environmental regeneration, implementation at various scales represents a real possibility both to mitigate the effects in the short term and to affect the causes in the long one.

Keywords: Eco-planning techniques · Ecological Networks · Green Infrastructures

1 Ecological Network and Green Infrastructure

The eco-centric perspective postulates that every species should have an equal opportunity to survive, but the number of endangered species has increased in recent decades. The primary cause of biodiversity (Fig. 1) decline is habitat loss, which is largely ascribed to rapid human population growth and the resulting of the natural environment anthropization.

© The Author(s), under exclusive license to Springer Nature Switzerland AG 2022
F. Calabrò et al. (Eds.): NMP 2022, LNNS 482, pp. 622–635, 2022.
https://doi.org/10.1007/978-3-031-06825-6_59

Fig. 1. Plant biodiversity: Cilento, Diano Valley and Alburni Mountains National Park. Campania Region - Salerno Province

More specifically, the protected natural heritage of Campania Region consists of 1 Unesco Geopark, 2 Unesco MAB Reserves, 2 National Parks, 5 State Reserves, 6 Protected Marine Areas, 1 Underwater Archaeological Park, 8 Regional Nature Parks, 4 Starting from the observation that it is impossible to renew environmental resources, an attitude of greater attention to the fragility of resources is becoming more and more consolidated. This attention can be implemented through the planning of an EN, which is a technique that contributes to reduce the environmental damage caused by anthropogenic pressure (decrease in biodiversity, consumption of resources, etc.) on the natural environment. The environmental continuity of a habitat is considered one of the fundamental pre-conditions to ensure biodiversity, pursuing the objective to reconstruct continuity of natural or semi-natural ecosystem units capable of playing a functional role in the regeneration of the environment. Fragmentation is a continuous process that is driven by human action and causes the isolation of natural areas, representing one of the main risks for the preservation of biodiversity. To mitigate and oppose this phenomenon it is therefore necessary to study the connectivity of natural and semi-natural areas. Connectivity can be defined as the degree to which the natural and semi-natural environment facilitates the movement of flora and fauna: it is a shared goal for ecological restoration. In biodiversity conservation strategies, it is not enough to protect specific isolated natural areas, but it is essential to network them. Urbanists are anthropocentric [2, 3] in both process and product. Man's relationship with the natural environment is aimed at managing the available resources, considered unlimited and easily renewable, and transforming nature for the economic and cultural benefit primarily of man himself.

Referring to traditional economic principles that ignore the fundamentals of environmental economics dominates urban planning and design solutions, demonstrating that the process/product is anthropocentric. Using information emerging from urban ecology and ecosystem research, urban planners and designers can develop a range of planning, design, and land management techniques that integrate eco-centric and anthropocentric values. EN and GI can contribute to this integration through a systemic and comprehensive approach involving interdisciplinary cooperation. An EN is defined [4] as a system of physically and functionally interconnected natural habitats through populations of species and ecosystems, whose biodiversity must be safeguarded, with particular attention to potentially threatened plant and animal species. A GI relates pollution, habitat, open space for recreation, and urban form [5–12]. The construction of the EN integrates the different objectives of the CUP, the improvement of the landscape (rural, peri-urban, urban), the usability and accessibility to the landscapes of rural and natural areas (routes and paths connected to EN), the enhancement of places and landscape elements of open spaces towards the implementation of GI. It can be asserted that the objectives of a MEN project can be summarized as follows: Protection, enhancement and strengthening of the existing natural system through conservation measures. Reconstruction of the EN through actions of restoration, overcoming barriers, rehabilitation, redevelopment, degraded portion of the territory or in contrast with the network projects. The EN project is an essential part of the GI and represents one of the main proposals for territorial planning and environmental protection, able to counteract the impoverishment of biodiversity by enhancing the relationships between city, natural and rural elements of the territory. The EN project aims to mitigate the effects of ecological fragmentation, through the protection of natural areas and the reconstruction of functional links between them. The EN has the capacity to perform functions related to both the conservation of biological diversity and the improvement of the human environment; the network must cover the whole territory, also including urban areas. An infrastructure is defined as a system of networked components, such as transportation infrastructure (highways, railroads, ports, airports), drinking water infrastructure (aqueducts, reservoirs), wastewater disposal and treatment infrastructure (sewers, water treatment plants), telecommunications infrastructure (data networks), and energy infrastructure (dams, power plants). In nature, there are several networks, the components of which are rivers, streams, lakes, and oceans, which together form a natural infrastructure to support ecological functions. For many plants and animals, the access to this infrastructure is necessary to survive. GI are continuous networks of corridors and spaces, designed and managed to support healthy ecosystem functions, mitigate pollution, enhance recreation, economic value, and configure new urban structure to improve the landscape and quality of life. The EN can be regarded as a subset of the GI whose purpose is to reconstruct the connections of natural and semi-natural environments of the territory. The wide area eco-connections developed in the REN and PEN projects, assume ecological and conservation interest, supporting environmental continuity, increasing biological diversity and the self-generative capacity of the same ecosystem. At the municipal level, the EN connects urban habitats and these with peripheral areas, allowing the movement of species on the territory that would otherwise be stopped by infrastructures. In the EN it is possible to identify interactions between the two main functions: ecological ie the conservation of nature, the enhancement of environmental functions and transformations for the improvement and development of habitats, and social ie the landscape and fruition aspects.

2   EN in the Campania Region Territorial Plan

Campania RTP [13], in force by the Regional Law n. 13/2008, is organized as a process-oriented and strategic framework tool to promote coordinated local actions and projects; it is aimed at implementing a wide area planning integrated with the provincial level. The RTP has no legal value as a landscape plan which schedules the protection and conservation of biodiversity and landscape in accordance with the principles of the European Landscape Convention signed in Florence in 2000. The three macro-objectives it pursues relate to territorial cohesion (to be found in the principles of sustainability), to the implementation of a polycentric scenario (aiming at the inclusion of urban systems in networks) and to co-planning (through the introduction of governance tools). The plan is divided into five Territorial Reference Frameworks (networks, settlement environments, territorial development systems, complex territorial fields, institutional cooperation), the first framework is subdivided into three sub-networks: interconnection, environmental and ecological risk. The RTP includes in the spatial articulation of the REN the elements of the Natura 2000 Network (pSCI/SAC, SPA) [14], the protected areas (national and regional natural parks, state and regional nature reserves, natural monuments, local parks of supra-municipal interest, local parks and areas designated as green by urban planning tools), categories of environmental units of intrinsic importance (forests, watercourses, lakes, wetlands and natural areas without vegetation), areas of importance for biodiversity, network nodes, corridors and ecological connections, areas of ecological rehabilitation and enhancement of degraded areas. More specifically, the protected natural heritage of Campania Region consists of 1 Unesco Geopark, 2 Unesco MAB Reserves, 2 National Parks, 5 State Reserves, 6 Protected Marine Areas, 1 Underwater Archaeological Park, 8 Regional Nature Parks, 4 Regional Nature Reserves, 8 Oases, 2 Ramsar Zones of International Interest for Bird Migration, 124 Natura 2000 Network Sites and 1 Metropolitan Park (Hills of Naples). Campania's REN configures a main connecting corridor represented by the natural park system that runs along the carbonate reliefs located on the regional longitudinal axis from north-west to south-east. This corridor represents a segment of the Apennine corridor (Project Apennine Park of Europe) [15] which extends to Calabria and the Nebrodi and Madonie mountains in Sicily. A second corridor of great strategic importance, is part of the Tyrrhenian Sea coastal corridor, climbed by migratory birds. In contrast to the first corridor, it runs along the coastal strip and is characterized by several crisis points due to high settlement pressure along the Campania coast. It is therefore a connecting corridor that needs to be regenerated. All the transversal and longitudinal corridors connecting the coastal strip with inland areas towards Apulia, Basilicata regions and the Adriatic Sea, as well as those up the Apennines towards Molise Region, must also be strengthened. The RTP relates the protection of biodiversity to the objective of guaranteeing eco-systemic services for the entire territory through a series of planned interventions and actions in areas with unique and irreproducible characteristics that must be safeguarded.

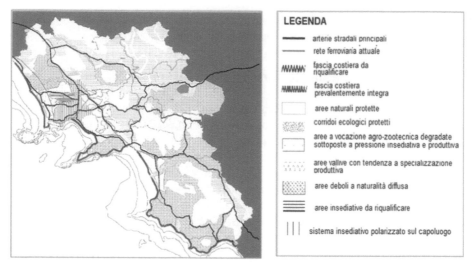

Fig. 2. Forecasts of RTP Campania Region: preferred visioning. Protected natural areas are represented in white

In the region's territory, the functionality and richness of landscape and biological diversity are heavily compromised by the fragmentation and disintegration resulting from often spontaneous and uncoordinated settlement and infrastructure development processes. The objective of safeguarding, regenerating, or creating vital connections for eco-systemic functioning and landscape continuity and usability, assumes a central role in RER planning. The RTP of Campania foresees among its strategies the realization of the RER and its configuration on the territory through the structural planning of the REP and the operational planning of the Municipal Ecological Network-MEN [16]. The REN (Fig. 2) is recognized as one of the three priority networks by the RTP and constitutes, together with interconnection and environmental risk, a strategic tool for provincial and municipal planning.

3   EN in the Caserta Province Territorial Coordination Plan

Caserta Province, to promote the development and the preservation of the naturalness of its territory, has directed its choices to enhance the naturalistic and agro-forestry resources. These areas represent almost 90% of the entire provincial territory as can be verified by consulting the regional map of land use and Corine Land Cover. The provincial territory includes four important areas of high naturalistic value that may represent the backbone of PEN planned in the reference framework of RTP networks. The first is the Domitian Coast that is characterized by formations of Mediterranean vegetation and dune pine forests, with patches of fragmented vegetation, it is particularly wet places that represent passages or even stations for species of Community interest of migratory avifauna (pink flamingo, black stork, purple heron, marsh harrier, hen harrier, etc.). The naturalistic level of these areas is very high, but it is a particularly compromised area, due to the uncontrolled expansion of illegal settlements, or even the cementing of the banks near the river mouths and the phenomenon of poaching. The second area, constituted by the fluvial-lacustrine environments that cross the entire provincial territory, is endowed with a quite high naturalistic level due to the presence of woods and vegetation patches unfortunately degraded because of anthropic interventions such as sewage, agricultural and industrial discharges. The third area concerns the mountain and sub-mountainous environments of the Campania pre-Apennines, in which the most significant types of vegetation of the southern Apennines are present with mixed forest formations, secondary grasslands belonging to the series of degradation or recovery of forest environments. The fourth and last area has a lesser presence of naturalness, it consists of semi-natural environments in which the influence of man is clear and the compromise of the natural environment is much higher than the previous areas due to a high level of habitat fragmentation. These are hilly environments or calcareous reliefs of modest elevation that occupy the areas between the coast and the inner reliefs of the Campanian Pre-Apennines. The PTCP of Caserta [17] (in force from 2012) identifies among its priority objectives the mitigation of environmental and anthropogenic risk, the minimization of soil consumption, the protection of landscape and natural values, the requalification of settlements, the mitigation of the impact of large infrastructures and the implementation of PEN. The PEN of Caserta Province is recognized in the Apennine ecological corridor and in transversal ecological corridor connecting the Provinces of Caserta, Benevento and Foggia. The mountain systems cover about 75.000 ha, equal to 31% of Province area, represent the prevalent portion (over 80%) of the natural and semi-natural habitats and form the backbone of the PEN. The province's nature areas are: 3 Regional Natural Parks (Matese, Roccamonfina and Foce Garigliano, Partenio), 1 State Reserve, 2 Regional Nature Reserves, 2 Oases, 21 Natura 2000 Network Sites (pSCI/SAC, SPA) [14] and 2 Urban Parks of regional interest (LRC n. 17/2003) Rocca d'Evandro and Riardo. The architecture of Caserta PEN is organized into central areas, corridors, buffer zones, environmental recovery zones, green belts and large territorial connectives. The province is affected by areas of maximum eco-systemic fragmentation generated by a very intense anthropization by the construction of large infrastructures and a residential building heritage, often unauthorized, by polluting discharges, water withdrawals and ecological barriers that have made, especially the coastal strip, an environmentally critical territory. To configure the PEN the plan aims at the ecological-environmental regeneration of agricultural and rural areas affected by environmental fragmentation, structural and functional deterioration of ecosystems due to pollution, reduction of biodiversity, introduction of allochthonous species, as priority areas with the function of ecological buffer and ecological corridor to increase connectivity between protected areas. The rural and open territory of the coastal ecosystem constitutes one of the most compromised portions of the provincial territory 5.000 ha comprise what remains of the dune and backdune areas, once occupied by temporary bodies of water and marshes, now reclaimed. They are hydrological discharge areas characterized by a low degree of protection of the surface water table. As a result of their high natural capacity, they are a key element of the PEN project (Fig. 3).

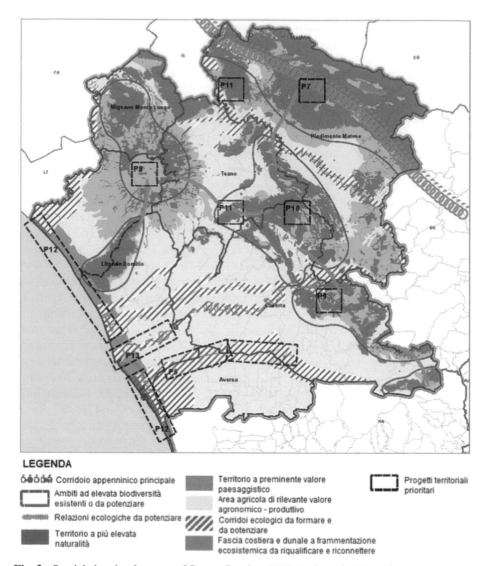

Fig. 3. Spatial planning forecasts of Caserta Province PTC: Project of PEN. Dark green represents the most natural areas, medium green the most valuable landscape areas, light green the agricultural areas of high agronomic and productive value

The rural and open territory for the ecological protection and soil protection includes the areas of pertinence of the water courses of provincial importance that extend for 32.000 ha, are characterized by high environmental sensitivity and play the important role of ecological buffer against the water courses. They constitute an important element of the PEN, as priority areas for the construction and strengthening of functional connection corridors. The re-naturalization of the Regi Lagni plays an important role in the construction of the PEN, which would transform them into an important ecological corridor between the coastal areas and the Apennines. The hilly areas characterized by a mosaic with a prevalent agricultural matrix, with the presence of discontinuous forest areas, play the key function of point areas, of ecological corridors, and sometimes of central areas of the PEN. The forest area of Roccamonfina volcano constitute an important central area of the PEN, while the agricultural areas play the key role of complementary habitats and buffer zones with respect to the more natural areas, functional connection zones between the slopes of the volcano and the adjacent plains, and multifunctional agroforestry areas in urban and peri-urban contexts. Tifatini mountains form a circle of hills around the Caserta conurbation, characterized by a highly vulnerable environmental area, which is therefore a place affected by strong anthropization. It needs to be protected to configure a continuous relationship between city and countryside. The coastal strip includes semi-natural habitats of high naturalistic, aesthetic-perceptive and recreational value, which play, within the PEN, the key role of intermediate areas in the processes of diffusion, dispersion, migration (stepping zones), but also particularly degraded areas due to the eco-systemic fragmentation; redevelopment and reconnection interventions are foreseen also through the realization of urban green belts. The interventions essentially concern: P4-Recovery of abandoned historic centers; P5-Requalification of Regi Lagni; P6-Archaeological Park of the Atellana area; P7-Mount Matese Regional Park and contiguous areas. Valorization of resources; P8-Roccamonfina Regional Park-Foce Garigliano and contiguous areas. Valorization of resources; P9-Monti Tifatini Regional Park; P10-Urban Park of Monte Maggiore; P11-Mineral Water Park; P12-Environmental requalification of the coastal dune belt; P13-Securing of the Volturno river's stem. PTCP assigns to CUPs the configuration of the MEN within the urbanized system.

4   EN in the Salerno Province Territorial Coordination Plan

The Salerno Province territory, especially in the southern and internal part, despite the increase of anthropic activities in the last decades, is constituted by a high level of naturalness and a good degree of preservation of the environment. This is due to the presence of ecosystems characterized by high biodiversity such as: prairies, shrublands, Mediterranean vegetation, wide forested mountain areas, lowlands and rivers. The substantial demonstration of the high level of naturalness is represented by the presence of rarities and particular and distinctive traits of ecological quality of flora and fauna that make the province one of the realities of strategic interest for the definition of regional and national EN. The natural and semi-natural environments of the province have undergone a deterioration that has led to the destruction and reduction of the surface extension of natural habitats, increasing the phenomenon of fragmentation. The pressure and degradation factors acting on the territory are recognizable in the: consumption of soil and natural resources, soil and groundwater pollution, seasonal forest fires, deterioration of surface and groundwater resources, increase in port infrastructure. Provincial territorial planning, therefore, has paid close attention to the essential resources of its territory to have a general picture of the problems to be addressed. The degree of naturalness depends on two variables: those that define the ecosystem community of a biotope (site) and those that define the characteristics of the context that influence the degree of naturalness of the biotope. The Naturalness Index-NI (or Ecological Value-EV) is a function of Bi = Characteristics of the biocenosis (species community of an ecosystem) and Co = Characteristics of the context and can be expressed by the following formula NI (or EV) = f (Bi, Co). The proposed naturalness indicators are identified from the agricultural land use map of Campania Region. This procedure considers the effects that the biotope context may have on the degree of naturalness. The latter can therefore change according to the context in which one finds oneself, and this makes it possible to ensure a more detailed characterization of the ecological value of the area. PTCP of Salerno [18] (in force from 2012) is focused on the improvement of the living environment of the resident populations through the conservation of environmental quality, the delimitation of areas characterized by a high level of biodiversity, the improvement of landscape quality and the formation of the PEN. In the territory of the province there are: 1 Unesco Geopark (Cilento, Vallo di Diano and Alburni), 1 MAB Biosphere Reserve (Cilento and Vallo di Diano), 1 National Park (Cilento, Vallo di Diano and Alburni), 3 Regional Natural Parks (Bacino Idrografico del Fiume Sarno, Monti Lattari and Monti Picentini), 1 State Reserve, 2 Regional Nature Reserves, 3 Protected Marine Areas, 7 Oases and 55 Natura 2000 Network Sites (SCI/SAC and SPA) [14]. The PEN, as a strategic landscape-environmental project of supra-municipal level, is based on ecological units and their interconnections, whose function is to allow the reproductive flow among the populations of living organisms that inhabit the territory, thus reducing the processes of local extinction, the impoverishment of ecosystems and the reduction of biodiversity. The starting point for the safeguard, the valorization and the conservation of the conspicuous naturalistic and landscape patrimony of the Salerno Province, is represented by the definition of the structural elements of the PEN. The framework of the PEN has been structured in: - core-areas or sources of biodiversity that include areas with a surface area greater than 50 ha, such as regional parks and nature reserves, SCI/SAC areas and SPAs [14], characterized by high levels of biodiversity, for which the plan envisages mitigation measures with a block on the construction of new infrastructures if they interfere with the PEN project - stepping zones or fragments of small environmental areas with a surface area of less than 50 ha, which constitute a valid support to the PEN because they are endowed with a high level of naturalness, for which the plan foresees the creation of reforestation and artificial wetlands; in these portions of the territory, the plan provides the creation of wildlife corridors, ie mainly linear areas that connect natural areas and allow the maintenance of the reproductive flows of living organisms; passages that constitute safeguard belts to avoid progressive building and prevent the closure of ecological corridors and the isolation of parts of the PEN - environmental regeneration areas and permeable peri-urban areas with a high degree of fragmentation in which environmental restoration processes are required with reconstruction and reconnection of the EN-areas of high naturalness that include mountainous areas and wetlands-buffer zones or zones that develop around central areas with a protective filter function, generally located in the hills and foothills that deserve protection through strategies for the conservation of ecosystems and the landscape and the establishment or extension of protected areas-second-level buffer zones or spaces located between the first-level buffer zones and the urbanized area - functional crossings of ecological corridors, river ecological corridors, infrastructural barriers or areas in which environmental upgrading and the creation of wildlife passages of vegetation planting are envisaged -

critical areas representing situations of potential conflict between the EN, the settlement system and mobility infrastructures - strategic nodes or portions of the territory that constitute fundamental elements for the continuity of the ecosystem (Fig. 4).

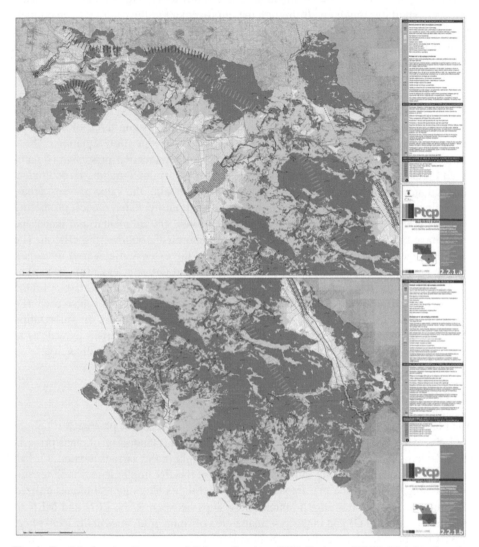

Fig. 4. Spatial planning forecasts of Salerno Province PTC: Project of PEN. High biodiversity areas are dithered in dark green, medium biodiversity and ecological connection areas in medium green, buffer zones in light green.

The plan promotes a series of actions to integrate and connect EN and urban green areas, as well as the requalification of compromised and degraded sites and the establishment of provincial parks to connect areas and implement EN in accordance with the eco-sustainable development of the provincial territory. The configuration of the network focuses on the creation of river corridors to be used as ecological connections and on the use of crossings as safeguard strips to prevent the isolation of parts of the network, with particular attention to mountain and foothill areas. Interventions concern the management of existing habitats through the improvement of the ecological functionality of lightly fragmented areas and the environmental restoration and/or requalification of moderately or highly fragmented areas using native plant species; the construction of new habitats for critical areas of eco-systemic fragmentation and defragmentation works for highly fragmented areas. These interventions must also include mitigation or compensation actions linked to the construction of new infrastructural works, such as biological bridges (overpasses) on linear transport infrastructures, faunal subways on linear transport infrastructures, fish passages (upstream ramps and thresholds), formation of permanent water flow low water riverbeds in critical water flow situations. These strategies are aimed at fostering the improvement and connection processes of natural and semi-natural ecosystems affecting the plain territory, encouraging its ecological connection with the hill and mountain territory, strengthening the value of forest areas, enhancing the function of ecological corridors carried out by watercourses, promoting actions of ecological mitigation of road infrastructures, environmental and landscape requalification of the territory through the use of naturalistic engineering criteria. The plan envisages a series of priority measures such as the extension of Sarno river park and the creation of the agricultural parks of Persano and Giffoni Sei Casali, as well as the creation of the inter-municipal agricultural parks of Alento, Fiume Tanagro, Fiume Temete and Giffoni Sei Casali. The enhancement of natural resources is also used as a tool for economic and social development and concerns investments for the valorization of natural and landscape resources to use them for sustainable tourism as well as to safeguard them from the point of view of environmental quality and biodiversity.

5 Few Concluding Remarks

The concepts of EN and GI in the vision of Eco-Planning, across plan scales, help to identify the theoretical/methodological assumptions for the elaboration of plans/projects and to extrapolate new eco-planning techniques to apply in the various territorial/urban and general/sectoral planning tools. The analysis of the EN implementation in territorial planning tools in force in Campania Region highlights planning techniques utilized recognizable in programming the interrelationships between REN, PEN and MEN to structure an efficient GI and in the systematization of some guidelines at the local scale. The recognition, forecast and implementation of EN, as an essential component of a GI, represents a mandatory choice for an urban planning that aims to integrate natural environment in the process of land transformation, to counter the impoverishment of biodiversity, to promote the relationships between urbanized natural and rural elements of the territory and to design new balances between natural and anthropized territory. Current experiences show that EN and GI must be considered fundamental components of urban and territorial infrastructures of anthropized territories, so much to require their identification and planning in territorial and urban plans, from large areas to municipalities and sub-municipalities, in a system logic between REN, PEN and MEN [19,

20], with specific functions and contents depending on the scale of intervention. The transition from REN to PEN is characterized both by a change of planning type that turns from strategies and addresses of the regional scale to structural forecasts of the provincial one thus identifying the areas affected by the main corridors of the EN. The next step, from the provincial to the municipal scale, configures a continuous and inter-connected network of environments organized with the coexistence of biodiversity and anthropization and structures a real GI that determines environmental, social and eco-nomic benefits, thus raising the level of sustainability. The planning/forecasting of nature in the plans has effects on the environment, such as the mitigation of flood risk and urban heat island phenomenon, energy saving, conservation/reconstruction of habitats for wild flora and fauna that find a suitable environment for their habitat; on the social by the diffusion of places for outdoor leisure activities with consequent improvement of health and well-being and on the economy through the increase of employment resulting from the promotion of sustainable development and smart growth. From the analysis of RTP and PTCP of Caserta and Salerno emerges that the implementation of the PEN does not constitute a macro-objective of these plans both for cultural reasons of methodological setting of the planning tools analyzed and for legal reasons since they have the legal value of general territorial plans without any landscape binding force. The analysis of the provincial plans highlighted differences between the strategies adopted to implement the PEN and a more detailed project structuring of the network within the Salerno PTCP, while Caserta focuses heavily on the recovery of agricultural areas, Salerno plans the institution of new park areas to complete the existing ones. Even though environmental issues are addressed in these plans, the analysis and design of REN and PEN represents a secondary element with respect to the other territorial networks and the GI approach is not used or delegated to the municipal scale. The strategic/structural/operational role of territorial/urban planning, at the various scales, would allow the coordination of environ-mental elements through the planning, design, implementation and management of the EN and GI, which thus branches out across the territory to increase continuity between natural and semi-natural areas, to improve their functionality, to reduce barriers and waste so that they can provide a wide range of eco-systemic services. There are many open questions, including those related to the transition towards a greener economy, which would give the right value to environmental capital, and those related to the updating and coordination of territorial, environmental and landscape regulations, which can no longer be separated, because of the most recent disciplinary acquisitions, such as the European Landscape Convention, which extends the traditional concept of landscape to the whole territory. For this purpose, an in-depth analysis of the Regional Landscape Plans with territorial validity and the Regional National and Natural Park Plans could be a contribution. If, from a scientific-technical point of view, this is a way to move the planning of the territory from an urban-centric and expansionist approach to an eco-centric and regeneration one, much remains to be done from the point of view of the economic and regulatory model to give the eco-environmental choices of planning at all scales the right market value and the prescriptive character appropriate to the superior public interest that characterizes them.

Credits.   Within this contribution, which is the result of a joint elaboration by the authors, personal contributions can be identified as specified below: 1 - Ecological Network and Green Infrastructure and 3 - EN in the Salerno Province Territorial Coordination Plan (Salvatore Losco), 2 - EN in the Campania Region Territorial Plan and 4 - EN in the Caserta Province Territorial Coordination Plan (Claudia de Biase), Abstract and Few concluding remarks (joint elaboration).

References

1. Yeang, K.: Ecomasterplanning, p. 167. Wiley, London (2009). The author describes an innovative approach to master-planning based on the ecological concept of physical planning intended as the bio-integration between the built context and natural systems and structures the natural and the man-made environment within a single system made up of four infrastructures: green, blue, grey and red. The green one is nature eco-infrastructure; The blue is water eco-infrastructure (natural drainage, water conservation systems and hydrological management in general); The grey is the engineering infrastructure, (roads, sewers, drainpipes, etc. as support systems sustainable for urban development); The red is the human infrastructure meaning the built context, including human activities and economic, legislative and social systems
2. Mell, I.C.: Green Infrastructure Planning, pp. 16–26. Lund Humphries, London (2019)
3. Austin, G.: Green Infrastructure for Landscape Planning. Integrating Human and Natural Systems, pp. 1–7. Routledge, London-New York (2014)
4. One of the most widespread definitions considers the EN as an interconnected habitat system that safeguards biodiversity, and thus focused on potentially threatened animal and plant species. Working on the EN means creating and/or strengthening a system of connections and interchanges between areas and isolated natural elements, counteracting fragmentation and its negative effects on biodiversity, see http://www.isprambiente.gov.it/it/progetti/biodiversita-1/reti-ecologiche-e-pianifica zione-territoriale/reti-ecologiche-a-scala-locale-apat-2003/cose-una-rete-ecologica
5. A definition of GI was introduced in the European Commission's White Paper on Climate Change Adaptation (2009), which states that GI is essential to mitigate fragmentation and unsustainable land use both within and outside of Natura 2000 areas and to address the need for multiple benefits to maintain and restore the ecosystem. Benedict, M.A., Machon, E.T.: Green Infrastructure: Linking Landscape and Communities. Urban Land (June), Washington DC (2006)
6. Williamson, K.S.: Growing with Green Infrastructure. Heritage Conservancy, USA (2003)
7. Town and Country Planning Association: Biodiversity by Design-Projects and Publications (2004)
8. The Environment Parthership: Advancing the delivery of green infrastructure. Targeting issues in England's Northwest, UK, Warrington (2005)
9. Natural England and Landuse Consultants: Green Infrastructure Guidance, Natural England, UK, Worcester (2009)
10. Mell, I.C.: Green infrastructure: concepts, perceptions and its use in spatial planning. Ph.D. thesis, Newcastle University (2010)
11. Landscape Institute: Green infrastructure: an integrated approach to land use, London (2013)
12. Davies, C., Macfarlane, R., Mcgloin, C., Roe, M.H.: Green Infrastructure. Planning Guide, p. 2–43. University of Northumbria, NorthEast Community Forests, University of Newcastle, Countryside Agency, English Nature, Forestry Commission, Groundwork Trusts, UK (2006)
13. http://www.regione.campania.it/regione/it/tematiche/piano-territoriale-regionale-ptr. http://www.sito.regione.campania.it/burc/pdf07/burcsp10_01_07/del1596_06all4_lineeguida.pdf. http://www.sito.regione.campania.it/burc/pdf03/burcsp08_08_03/del1543_03allegati/LG_Compatibilit%E0%20paesistica_Allegato_C.pdf. http://www.regione.campania.it/regione/it/tematiche/natura

14. Sites of Community Importance (SCI) define areas that contribute significantly to maintaining and restoring habitat types and/or maintaining the species defined in Annexes 1 and 2 of the Habitats Directive in a satisfactory state of conservation to contribute significantly to maintaining the biodiversity of its regional context. Special Protection Areas (SPAs) are located along avifauna migration routes. The regulations seek to maintain and define suitable habitats for the conservation and management of wild migratory bird populations. These areas are identified by European Union member states and, alongside the Special Areas of Conservation (SAC), constitute the Nature 2000 Network

15. Project launched in Italy in 2000 by the Inter-ministerial Committee for Economic Planning (CIPE)

16. Losco, S., de Biase, C.: Ecological network from regional to municipal scale. The case-study of San Tammaro (Ce). In: Gambardella, C. (ed.) World Heritage and Legacy Culture, Creativity, Contamination Le Vie dei Mercanti XVII International Forum, Gangemi Editore International, Roma (2019)

17. http://www.provincia.caserta.it/it/web/pianificazione-territoriale/home/

18. https://geoportale.provincia.salerno.it/page/piano-territoriale-di-coordinamento-provinciale. www.provincia.salerno.it/pagina2413_piano-territoriale-di-coordinamento-della-provincia-di-salerno.html. http://www.sito.regione.campania.it/lavoripubblici/Elaborati_PRAE_2006/salerno100.htm. www.lifecomebis.eu/privato/archivio_documenti/30.pdf. www.cilentoediano.it/

19. Guccione, M., Schilleci, F. (ed.): Le reti ecologiche nella pianificazione territoriale ordinaria. Primo censimento nazionale degli strumenti a scala locale. Rapporti 116/2010, Ispra, Roma (2010)

20. Aa, V.V.: Gestione delle aree di collegamento ecologico funzionale. Indirizzi e modalità operative per l'adeguamento degli strumenti di pianificazione del territorio in funzione della costruzione di reti ecologiche a scala locale, Manuali e linee guida 26/2003, Apat-Inu, Roma (2003)

Cultural Heritage as a Right to Well-Being and an Engine of Urban Regeneration

Chiara Corazziere[✉] [iD]

ArTe Department, Mediterranean University, Reggio Calabria, Italy
ccorazziere@unirc.it

Abstract. The post-pandemic emergency is a valuable opportunity to encourage and promote the right to culture as a driver to counteract social and environmental vulnerability and to promote well-being, new sustainable lifestyles and processes of urban and landscape regeneration. The cultural heritage not included in regimes of protection or public use can accommodate actions with an experimental and innovative character to promote the creativity of new generations and the definition of places of economic, social and cultural promotion for the disadvantaged. Conventional cultural assets, such as museums, libraries, archaeological sites and parks, archives and deposits, regardless of their administrative status, geographical location and number of users, must now, more than ever, be interpreted as principals of well-being.

This contribution analyses some recent experiences in Italy that enhance the role of culture in local development policies in pursuing not only economic but also social impact. From the study of various experiences, it can be deduced that more ready than the institutions in charge of the management of cultural heritage, seem in fact - and perhaps unconsciously - some communities of citizens who live in the places of culture and show the ability to recognize their values and potential. The thesis presented here consists, therefore, in the belief that public policies should make the places of culture look like places of well-being for the communities of inhabitants who assume a role of concrete responsibility and actively participate in the construction of new models of correct consumption.

Keywords: Right to culture · Heritage and wellness · Urban regeneration

1 A New Vision of the Right to Cultural Heritage

The spread of Covid-19 has caused the anthropocentric vision of the world to stagger and collapse, not only under the weight of the pandemic but more generally of a poor management of resources, triggering a state of alarm even with respect to the *right to culture.*

F. Calabrò et al. (Eds.): NMP 2022, LNNS 482, pp. 636–644, 2022.
https://doi.org/10.1007/978-3-031-06825-6_60

Compared to the post-pandemic state of emergency, which has only superimposed itself on the unresolved and previous state of fragility of many territories, we can seize the opportunity, now, to encourage and promote measures to counteract social and environmental vulnerability by tending towards opportunities for well-being and sustainable lifestyles, also thanks to processes of urban and landscape regeneration.

Given these as pre-conditions, the right to culture can also be expressed in the objective of an inclusive city that promotes accessibility, especially for the weakest, to quality spaces, but also of an efficient and safe city, able to prevent and treat the socio-cultural isolation and counteract the territorial fragility, thanks to cultural centres and innovative formulas for knowledge transfer and specialized training, governance of resources and by virtue of conscious and responsible communities able to propose correct ways of using the territory.

It is no coincidence that on September 23, 2020, in the midst of the pandemic, the Chamber of Deputies finally approved the ratification of the "Convention on the Value of Cultural Heritage for Society," drafted in Faro on October 27, 2005. The Italian State, which had already signed the document in February 2013, with the ratification relaunches, in fact, a vision of the right to cultural heritage that, in the post-pandemic era, can take on new contours. While not imposing specific obligations of action on signatory countries, leaving them free to decide on the most convenient means of implementing the measures provided for therein, the Convention emphasizes, in fact, the value and potential of cultural heritage «as a resource for sustainable development and quality of life» [1] and *overturns* the right to cultural life enshrined in the 1948 "Universal Declaration of Human Rights" [2], recognizing individual and collective responsibility for cultural heritage and emphasizing the importance of its preservation and its role «in the building of a peaceful and democratic society» [1]. The meaning of right *to* and not right *of* cultural heritage, indicates, in fact, the right to access cultural heritage also as a necessary resource for the formation of the identity of a citizen [3].

Less than a year later, on July 29 and 30, 2021, Italy chooses to place culture at the center of its G20 Presidency, dedicating a meeting of the ministers responsible for culture that has no precedent in history. It is certainly a strategic choice linked to the importance of the cultural sector in our country, but the global pandemic also makes us look to culture as a key to restarting, as well as a reference to the basis of post-pandemic regeneration in the broadest sense.

Even the final document of the G20 Culture Ministers "Culture Unites the World" recognizes, in fact, «the social impact of the cultural and creative sectors in supporting health and well-being, […] in amplifying behavioral change and transformation towards more sustainable production and consumption particles and in contributing to the quality of the living environment, to the benefit of the quality of life of all» [4].

2 Flexible and Permeable Collective Spaces for Social, Economic and Urban Innovation

We tend to think of the Italian cultural heritage as a collection of museums and then paintings, statues, archaeological finds, while the largest part of our heritage is composed of collective spaces that are not visited exclusively but are, first of all, containers of people, without which would be lost both the meaning of public space, of course, but also that of good used. Above all, the cultural heritage that is not part of public protection or fruition systems represents today the preferential space where to experiment urban policies at the scale of the punctual project, conceiving garrisons that contribute to the development of the territory, both at the large scale with programs that welcome the requests of local communities.

Unconventional cultural heritage and landscapes can become, especially in times of pandemic, flexible and permeable spaces for social, economic and urban innovation. Individual artifacts, entire neighborhoods, legacies of disused productive activities, can also be considered as cultural heritage if interpreted as a resource awaiting new attributions of meaning and capacity. Unconventional heritage is thus called upon to host proactive actions that promote the innovation and creativity of the new generations, that combat urban poverty and environmental vulnerability, and that encourage the definition of places of economic, social and cultural promotion, including for the weaker sections of the population.

For these places, for which fragility is not to be understood as a *status* established a priori linked exclusively to problems of physical degradation, the non-programming of processes of research of new meanings, can mean to burden the already difficult management of the territory of an additional environmental risk factor and, at the same time, to miss the opportunity to define the contours of a potential heritage capable of ensuring safe and collective spaces of daily well-being, to generate new value systems and places dedicated to new communities of work, culture, welfare.

For the *unproductive* heritage, orphan, of the original destination, in particular, fragility is not so much to be understood as a status linked exclusively to problems of physical degradation, but rather as the potential risk arising from a "non condition". This *impasse* can be overcome by initiating regeneration processes in which disused assets acquire the capacity to attract a broad community of actors according to processes whose effectiveness can be assessed in terms of urban, environmental and landscape regeneration, employment spin-offs, innovation and social inclusion, and educational and safe cities. In fact, working on disused heritage offers the advantage of being able to intervene on the container without weakening its content and can become, above all, a practice for investing in human capital and transforming problem areas of the city into opportunities for growth [5].

There are many experiences in Italy of response - formal and informal - to the official guidelines that call for a close relationship between access to culture and regeneration of living space at different scales. In most cases, these experiences have seen the consolidation, following the Covid-19 pandemic, of the principles that had substantiated their inception, exalting the role of culture in local development policies in pursuing, in addition to the economic impact, also the social one [6].

This is the emblematic case of the OGR Officine Grandi Riparazioni in Turin, protagonists of the growth of the city for about a century and at risk of demolition, following its closure in the early nineties, as established by the new Master Plan of 1995. Thus, an area of 20,000 square meters destined to become a large urban void at the center of the strategic urban quadrant called Spina 2, between the two railway poles Porta Nuova and Porta Susa, which houses the Polytechnic and its Energy Center, already characterized by a major urban reorganization, as a result of the construction of the railway link. The risk is averted thanks to a variant that allows the purchase of the area by the CRT Foundation, which starts a process of re-signification, and not just a simple recovery of spaces and structures, so that the OGR will become again the fulcrum of Turin's productivity, but this time in a cultural key and of innovation and business acceleration with an international vocation (see Fig. 1 and Fig. 2) [7].

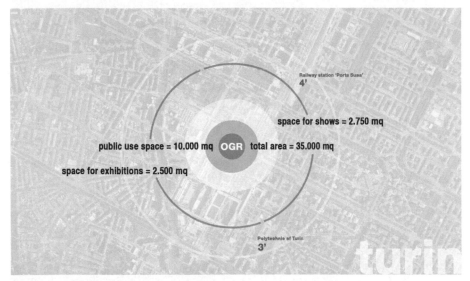

Fig. 1. Turin, OGR_Officine Grandi Riparazioni, map of the intervention area (graphic elaboration by C. Corazziere)

From the very beginning, the operation already shows the intention of not distorting the industrial essence of the complex but rather of enhancing its regenerative vocation, of ideas and no longer of machines, with social, cultural and economic spin-offs that go well beyond the physical boundaries of the area, so much so that in 2015, two years before the actual inauguration, and for a *concentrated* space that is actually a private property, it obtained the Urban Planning Award for the category "Quality of infrastructures and public spaces".

Among the many occasions of participation in the determination of common cultural references, of communities that are culturally self-defined, self-determined, even questioning consolidated past references in order to affirm the right to cultural heritage [8], an emblematic case is that of the "Miglio Sacro" in Naples which, starting from single

Fig. 2. Turin, OGR_Officine Grandi Riparazioni, the East Court, dedicated to open-air art, one of the two public squares created by the Officine re-signification intervention but independently usable by the community (ph C. Corazziere).

cultural emergencies, aims not to propose attractions for a few visitors but to regenerate an entire neighborhood, the Rione Sanità (see Fig. 3).

Fig. 3. Naples, Rione Sanità, urban heritage (ph C. Corazziere).

Characterized by deep socio-economic problems and historically labeled as an insurmountable periphery to the center of the city, in 2006 the neighborhood began to experience a process of regeneration thanks to a change of mentality towards the cultural heritage that characterizes it, interpreted as a common resource.

The cooperative that triggers the process gathers different professionals and proposes a new management - inheriting it from the Curia with which it stipulates an agreement - of sites already known but little valued, organized in a tour circuit, the Sacred Mile in fact, that crosses the neighborhood. The objective is twofold: to force the visitor to experience the widespread urban heritage - material and immaterial - that unfolds along the lines that connect the various destinations of the visit and to attract private actors in the neighborhood towards a common path of self-development.

The experiment works and in addition to obtaining significant employment effects and the gradual establishment of paths of social inclusion and cohesion in the neighborhood, it supports important investments in ordinary and extraordinary maintenance for the cultural sites that in turn motivate interventions of urban regeneration by private activities in the neighborhood, encouraged by an increasing flow of visitors. The admissions, in fact, have increased from 5,000 in 2006 to about 130,000 in 2018. Also in the same year, there are 12,100 square meters of sites recovered, 34 people permanently employed and two unused containers (former convents) transformed into accommodation facilities (see Fig. 4) [9].

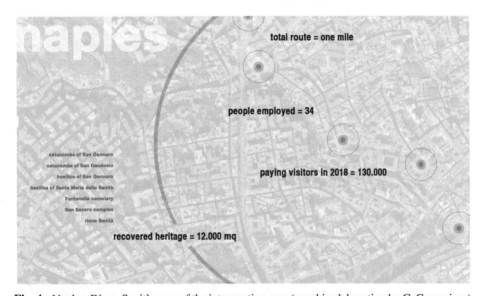

Fig. 4. Naples, Rione Sanità, map of the intervention area (graphic elaboration by C. Corazziere).

The example of the Rione Sanità, which today is considered a model of acquisition of rights to the heritage by a large community of citizens, shows that, all the more so, even the conventional cultural assets - museums, libraries, archaeological areas and parks, archives, deposits, etc., regardless of their administrative rank, geographical location and number of users - today more than ever must be interpreted as *garrisons of well-being*, understood as «public spaces of democracy where we speak to citizens and not to clients, and cultural institutions as communities of knowledge at the service of the community to which they belong» [10] and of broader strategies of qualification of the territory and of the rooting of citizenship rights.

All the traditional places of culture can become accessible spaces in which to strengthen the possibilities of restitution to the outside world of a positive and realistic image of the territory, looking at the communities as a privileged vehicle of transmission of correct and effective behavior towards common resources, including those referred to the cultural heritage.

In this sense, the National Recovery and Resilience Plan (NRRP) is also an opportunity to strengthen and support some fundamental rights, such as the basic right of access to culture, to promote measures to combat social and environmental vulnerability by moving towards sustainable lifestyles, also thanks to urban and landscape regeneration processes.

In this sense, the Regional Directorates of Museums, peripheral bodies of the Ministry of Culture, have applied for funding from NRRP, project proposals aimed at solving the difficulties related to physical and cognitive accessibility of the cultural institutions under their jurisdiction, not only in terms of infrastructural interventions of *design for all*, but also in terms of communication processes and preparation of staff responsible for reception and accompaniment [11].

In some cases, the call to enhance the identity of historic parks and gardens [9] has become an opportunity to propose the regeneration of collective spaces so that they become functional to the fruition of cultural emergencies and, at the same time, a place of life of the community of inhabitants, in degraded territories that lack of designed public spaces and where parks and gardens can become a place of well-being and sociality in the perspective also thanks to the training of local staff that can take care of them and preserve them over time.

3   Conclusion

The Covid-19 global pandemic has brought health back to the centre of the government's agendas and culture, with the National Recovery and Resilience Plan (NRRP), finally seems to be able to be the protagonist of project proposals for the regeneration of urban space and landscape that respond to the real contemporary needs of citizens, aiming at a deeper social cohesion and a renewed sense of community in the inhabitants, and that are able to attract new ones, even temporary ones.

Conceiving public policies that support the right to culture, in fact, presupposes the recognition of the value of cultural heritage for the promotion of a democratic society, and does not only mean "placing people and human values at the centre of an enlarged and interdisciplinary conception of cultural heritage" as stated in the Preamble of the Faro Convention.

This paradigm shift, therefore, requires a development policy capable of reading the territory, the economic and social dynamics that modify it, grasping its cultural specificities; a policy, therefore, that supports citizens in making their own choices regarding the right to cultural heritage. There are many effective experiences, such as those analysed, from which to draw, especially those that are informal or private.

More ready than the institutions in charge of the management of cultural heritage, it seems in fact - and perhaps unconsciously - some communities; citizens who *inhabit* the places of culture and show the ability to recognize «the scenarios of what is possible and desirable and to mobilize for its realization» [12]. It is precisely in the actions of such communities that one can grasp a sense of responsibility to contribute to a common wellbeing, to ensure not only salubrious conditions during the pandemic, but also spaces of life for those who cannot have them in their own homes or neighboring contexts.

Finally, public policies should also give structure to those *informal strategies* that generate and result in a deeper social cohesion and a renewed sense of community that interprets the quality of built space not according to exclusive aesthetic or formal canons, but according to a more correct relationship between the *design* dimension and the *use* dimension assessed in terms of *citizen well-being*.

In this direction, the NRRP could make us look at the places of culture as places of well-being for communities that, in a new vision of the right to cultural heritage, assume a role of responsibility towards the cultural resource and participate in building models of proper consumption.

Acknowledgements. This paper was developed in the context of the Italian National Research Project PRIN 2017 'Regional Policies, Institutions and Cohesion in the South of Italy' (Project code 2017-4BE543; website www.prin2017-mezzogiorno.unirc.it), financed by the Italian Ministry of Education, University and Scientific Research from 2020 to 2023.

References

1. Convention on the Value of Cultural Heritage for Society (Faro Convention, 2005). https://www.coe.int/en/web/culture-and-heritage/faro-convention. Accessed 19 Dec 2021
2. Universal Declaration of Human Rights (1948). https://www.un.org/en/about-us/universal-declaration-of-human-rights. Accessed 22 Dec 2021
3. Gualdani, A.: L'Italia ratifica la convenzione di Faro: quale incidenza nel diritto del patrimonio culturale italiano? Aedon (2020). https://doi.org/10.7390/99477. Accessed 20 Dec 2021
4. Rome Declaration of the G20 Ministers Of Culture 'Culture unites the world'. https://cultura.gov.it/g20cultura. Accessed 20 Dec 2021
5. Corazziere, C., Taccone, A.: Ri-significare il patrimonio produttivo dismesso per nuove comunità e qualità urbane. In: Mistretta, M., Mussari, B., Santini A. (eds.) ArcHistoR EXTRA 6/19_Supplemento di ArcHistoR 12/19 (2019), pp. 300–313. https://doi.org/10.14633/AHR162. Accessed 28 Nov 2021
6. OECD: Schok cultura. Covid-19 e settori culturali e creativi, 09/20. https://www.oecd.org/coronavirus/policy-responses/shock-cultura-covid-19-e-settori-culturali-e-creativi-e9ef83e6/. Accessed 28 Nov 2021
7. OGR. https://ogrtorino.it. Accessed 08 Jan 2022
8. Belotti, F.: Il diritto di partecipare al patrimonio culturale e il principio di interdipendenza dei diritti. In: Meyer-Bisch, P., Gandolfi, S., Balliu, G. (eds.) L'interdépendance des droits de l'homme au principe de toute gouvernance démocratique Commentaire de Souveraineté et coopérations. Genève, Globethics.net (2019)
9. Rione Sanità. https://www.catacombedinapoli.it. Accessed 08 Jan 2022

10. Montanari, T.: Beni culturali, 10 idee per rilanciarli. https://emergenzacultura.org/2020/04/15/. Accessed 5 Dec 2021
11. National Recovery and Resilience Plan (NRRP 2021). https://www.governo.it/sites/governo.it/files/PNRR.pdf. Accessed 17 May 2021
12. De Luca, S.: Liberare il potenziale dei territori marginalizzati. Con quali politiche? In: Bellandi, M., Mariotti, I., Nisticò, R. (eds.) Città nel Covid. Centri urbani, periferie e territori alle prese con la pandemia. Donzelli Editore (2021)

Cities and Territories Theatres of the Recovery of the Country System

Concetta Fallanca(⊠) ⓘD

Mediterranean University, 89124 Reggio Calabria, RC, Italy
cfallanca@unirc.it

Abstract. The paper proposes a reading of the NRRP with the aim of interpreting in an extensive way the interest in urban systems and territories, as irreplaceable theatres of the profound review of the country system. In the space of a quarter, local authorities will be called upon to guarantee a planning process aimed at pilot projects for the cultural, social and economic regeneration of villages at risk of abandonment, projects for Cultural and Social Regeneration, and the drafting of Integrated Urban Plans. Tight deadlines also for projects aimed at the regeneration and requalification of Italian parks and gardens of cultural interest.

The results of the rankings already approved for urban regeneration and public housing projects presented by Regions, Municipalities and Metropolitan Cities, allow for initial considerations on the risk that the objective of territorial rebalancing may not be achieved if a prudent national direction is not activated to support the territories in achieving a high level of planning.

Faced with such significant resources, the opportunity for territorial rebalancing cannot be wasted, because this would introduce a new, profound disappointment in the possibility of achieving similar opportunities for the citizens of the national territory. In order not to reduce this moment to a mere opportunity to observe phenomena, record impacts, evaluate effects, the Universities -which, with the role of the Third Mission, have guaranteed a fundamental impulse for the social, economic and cultural future of their territories of belonging- should be fully involved in the planning process, also in order to call upon and co-empower all the energies that specific territory can express.

Keywords: Urban regeneration · Integrated Urban Plan · Territorial rebalancing

1 The Transversal Opportunities of the NRRP to Redesign Urban and Territorial Landscapes

If the main mission of the National Recovery and Resilience Plan (NRRP) is to give substantial impetus to the relaunching of the competitiveness and productivity of the Country System - which presupposes "a profound intervention, acting on several key elements" [1] of the economic system - attempting to understand and promote the role that cities and territories can play in this act of redesign seems fundamental.

It becomes important to read between the lines of the NRRP and try to interpret in an extensive way any hint that may suggest an interest in urban systems and territories,

F. Calabrò et al. (Eds.): NMP 2022, LNNS 482, pp. 645–654, 2022.
https://doi.org/10.1007/978-3-031-06825-6_61

as irreplaceable theatres of these acts of re-foundation that involve a profound review of the way of relating to the geographical, economic and social context and that will determine conditions of well-being, lifestyles, within urban and territorial landscapes more appropriate and appropriate [2–4].

A first important mention that suggests long-spectrum system actions is that of the "enhancement of cultural heritage and tourism", with the correct interpretation, to avoid superficial approaches and misunderstandings of the meaning to be given to the specification "also in function of promoting the image and brand of the country". [1].

Relaunching the economic sectors of culture and tourism means giving impetus to a significant part of GDP, with interventions to enhance historical and cultural sites, aimed at improving attractiveness, safety and accessibility of places. The diffusivity of the action component can be pervasive as the interventions are intended not only to those that are defined as the "major attractions" but are also directed to the protection and enhancement of smaller sites, including "villages", "as well as the regeneration of urban peripheries, enhancing places of identity and at the same time strengthening the social fabric of the territory.

The interest and the definition of the actions of regeneration of the suburbs still seem to be defined, because at the moment they seem to be aimed at the improvement of tourist facilities and services for the redevelopment/improvement of the offer, actions inspired by a careful approach to the issues of environmental sustainability and optimization of the potential of digital and innovations offered by new technologies.

The Ministry of Culture for the *Investment 2.1 Attractiveness of the villages,* with public notice has provided two lines of action for access to resources of the National Plan Villages provided by the NRRP to win the challenge of repopulation. The first aimed at 21 villages identified by Regions and Autonomous Provinces and the second aimed at, at least, 229 villages selected through public notice addressed to municipalities.

For the 21 villages corresponding to each Region or Autonomous Province, pilot projects will be promoted with interventions considered exemplary for the cultural, social and economic regeneration of villages at risk of abandonment or abandoned. Each intervention will amount to 20 million euros and will be aimed at the economic and social revitalization of uninhabited villages or characterized by an advanced process of decline and abandonment. The projects will have to provide for the establishment of new functions, infrastructure and services in the field of culture, tourism, social or research, such as schools or academies of arts and crafts of culture, hotels spread, artist residences, research centres and university campuses, nursing homes (RSA) where to develop programs with a cultural background, residences for families with workers in smart working and digital nomads.

The first component will be implemented through public notice issued by Ministry of Culture (MiC) for the funding of proposals submitted by municipalities in single or aggregated form - up to a maximum of 3 municipalities - with total resident population up to 5,000 inhabitants. The public notice, dedicated to small historic villages, is aimed at promoting projects for the regeneration, enhancement and management of the great heritage of history, art, culture and traditions present in small Italian towns, integrating objectives of protection of cultural heritage with the needs of social and economic revitalization, revival of employment and counter depopulation.

With subsequent announcement, at the conclusion of the preliminary investigation, the funds will be assigned to companies that carry out cultural, tourist, commercial, agro-food and craft activities located in the municipalities selected for the implementation of cultural regeneration projects. For this line of action, the Regions will have to submit their proposal to the MiC by March 15, 2022, as defined in agreement with the municipality concerned.

The second line of action has a twofold soul and is implemented in two stages, it aims at the realization of local projects of cultural regeneration of, at least, 229 historic villages for proposals submitted by municipalities and, in part, is addressed to the support of micro, small and medium enterprises located or intending to settle in the villages interested in the project.

The interventions admitted to financing will have to give new life to the socio-economic fabric of these places through the redevelopment of public spaces, the regeneration of the historical-architectural heritage together with the activation of entrepreneurial and commercial initiatives that create employment in the area. In accordance with the provisions of the NRRP, 40% of the total resources will be allocated to the 8 Regions of Southern Italy and the interventions must be completed by June 2026. Finally, the investment sees an amount equal to 20 million euros designated to the intervention "Tourism of the roots" whose implementer is the Ministry of International Affairs and Cooperation.

Tight times also for the notice of the Ministry of Culture dedicated to parks and historic gardens, aimed at the regeneration and redevelopment of Italian parks and gardens of cultural interest.

The receivers of the notice are the owners, possessors or holders in any capacity - public or private - of parks and gardens of cultural interest (artistic, historical, botanical, landscape) protected under Legislative Decree n.42/2004. The interventions admitted to funding must be highly significant, capable of generating a tangible improvement in the conditions of conservation of the property, as well as a positive and high impact on the promotion of cultural, scientific, environmental, educational, economic and social development. The projects should aim at strengthening the identity of the places, improving the quality of the landscape, the quality of life and psychophysical well-being of citizens. Fundamental are the choices aimed at improving environmental conditions and the positive effects in terms of pollution reduction, regularization of the microclimate, protection of biodiversity, everything that enhances the ecosystem infrastructure. A new vision must be dedicated to the promotion of educational activities, in order to spread a renewed awareness of environmental and landscape and enhance the role of resource in terms of scientific, technical, botanical and environmental knowledge developed, tested and sedimented over the centuries, also in order to create new significant poles for cultural and tourist fruition. The financial endowment will be reserved for at least 20% to assets located in the Regions of Southern Italy since the 40% mount will be achieved with projects directly governed by the Ministry.

A further sign of interest in the "structural" and connective transformation that can make more fluid and accessible the post covid city [5] with the opportunities of the NRRP, is that provided for transversal measures in favour of senior citizens with reduced mobility and people with forms of disability. Mission 1 promotes full accessibility of Italy's cultural heritage by removing architectural and sensory barriers in museums, libraries and archives. Mission 4 envisages interventions to reduce territorial disparities in secondary schools, and Mission 5 includes an extraordinary investment in social infrastructures, with interventions on social and health services in the community and at home, in order to enhance the autonomy of persons with disabilities.

2 Urban Regeneration and the National Innovative Program for Housing Quality

The measures of the NRRP that seem most significant for an organic and integrated approach towards a post covid city transition are those intended for Integrated Urban Plans for Metropolitan Cities.

Significant funding is aimed at urban regeneration through the recovery, renovation and eco-sustainable re-functionalization of building structures and public areas to promote social inclusion, reducing marginalization and social degradation. It also co-finances projects related to smart cities, with particular reference to transport and energy consumption.

For the effects on the territory, local authorities have a key role in the success of the NRRP and are called upon to adhere to the new approach required for the realization of intermediate and long-term objectives to be achieved on individual lines of intervention, based on expenditure properly reported but especially on the performance achieved. This condition requires a higher capacity for planning and governing transformation than was previously required to access traditional structural funds. Local authorities are required to accurately measure the impact on the territory resulting from the actions planned and financed by the NRRP's implementation calls, especially for the outcomes of Mission 5: Inclusion and Cohesion, which includes the urban regeneration game and the Integrated Urban Plan.

For the projects on Integrated Urban Plans the procedure is still open, the publication of notices and calls has made known the lines that will have to be followed by the Metropolitan Cities for the presentation scheduled by March 7, 2022.

For urban regeneration and public housing projects presented by Regions, Municipalities and Metropolitan Cities, there are already approved rankings. The 159 proposals accepted have enjoyed a path that had already been initiated under the "National Innovative Program for Housing Quality" (PINQuA in Italian) and only later was hinged in the NRRP. Thus, for 159 approved urban regeneration and public housing projects in addition to the substantial funds allocated by the NRRP, residual funds from 2019 and 2020 of the PINQuA were used.

Under the fifth mission, for the strengthening of "social infrastructure", investment 2.3 provides for the implementation of the "National Innovative Program of Housing Quality". The interventions that will be carried out with this specific measure are aimed at increasing the public housing stock. In addition, projects aimed at the regeneration of urban centres and peripheries were also eligible for funding, with the priority objective of improving their accessibility, functionality and safety.

Interministerial Decree No. 395 of 16/09/2020 recalls Law 160/2019, which promotes the National Innovative Program for the quality of living in order to reduce housing discomfort with particular reference to the peripheries and to encourage the exchange of experiences between the various Regional realities. "The program is aimed at redeveloping and increasing the heritage for social housing, to regenerate the socio-economic fabric, to increase accessibility, safety and security of places and the re-functionalization of public spaces and buildings, as well as to improve social cohesion and quality of life of citizens, with a view to sustainability and densification, without consumption of new land and according to the principles and guidelines adopted by the European Union, according to the urban model of the smart city, inclusive and sustainable (Smart City)". (paragraph 437 article 1 of the recalled law of December 27, 2019, n. 160).

The Regions, Metropolitan Cities, municipalities with more than 60,000 inhabitants have submitted proposals aimed at reducing housing and settlement discomfort and increasing the quality of living, to promote regeneration processes in specific urban areas. The planning strategy must give social housing a priority role, through interventions and measures aimed at upgrading, reorganizing and increasing the residential heritage. Fundamental is the re-functionalization of areas, spaces and public and private properties also through the regeneration of the urban and socio-economic fabric and the possible temporary use of places and heritage. Actions aimed at improving the connectivity, accessibility and safety of urban places and their endowment of services and infrastructure on an urban and local scale are essential. Interventions must ensure proximity of services, aiming to reduce traffic and stress, according to the criteria of sustainable mobility, as well as increase neighbourhood ties and social inclusion.

It is considered appropriate to regenerate neighbourhoods, especially those with high housing density, increasing environmental quality and improving resilience to climate change also through the use of densification operations. Proposals must be significant and characterized by the presence of eco-sustainable solutions, elements of green infrastructure, Nature Based Solutions, depaving and ecosystemic enhancement of areas, technological innovation and typological artifacts; provide bio-architecture solutions suitable for recycling of materials, the achievement of high-performance standards, energy for seismic safety, special spaces for the management of waste collection, water recycling. Finally, it is invited to "identify and test innovative models and tools for management, social inclusion and urban welfare as well as processes of sharing and participation, including interventions of self-recovery.

Interventions and proposed measures must aim at durable solutions for the regeneration of the socio-economic fabric, improving social cohesion, cultural enrichment, quality of artifacts, places and life of citizens, with a view to innovation and sustainability, with particular attention to economic and environmental, without consumption of new land, except for any densification operations.

The maximum contribution for each proposal eligible for funding is € 15,000,000, it is ensured the funding of at least one proposal for each Region to which the proponent belongs. 34% (later increased to 40%) of the total resources must be allocated as a priority to projects located in southern Italy.

On the basis of the decree issued by the Ministry of Infrastructure and Sustainable Mobility, we know that there are 159 projects financed of which 8 are classified as high-performance pilot projects, considered to have a strategic impact on the national territory as they are oriented towards the implementation of the Green Deal and the Digital Agenda. In addition to these, there are 112 others that, despite having positively passed the evaluation of merit and being eligible, will not be funded, at least in this first phase.

With 21 funded projects, Puglia is the Italian Region with the highest number of proposals accepted. In second place we find Lombardy with 17 and in third place Lazio with 15. Among the southern Regions, Campania has 9 proposals accepted and then with the lowest number of proposals funded there are Sicily with 7 proposals, Calabria with 6, Sardinia 5 and Basilicata with only one project. Of these 159 proposals financed in the sphere of urban regeneration, 8 of them are considered "pilot projects with a high strategic impact on the national territory as they are oriented towards the implementation of the green deal and the digital agenda". In the Regions of Southern Italy three pilot projects have been approved, for Bari the reorganization of the area near the central railway station as a hinge between the historic centre and the modern urban area; infrastructural interventions for multimodal mobility; urban redevelopment and restructuring interventions with the creation of new green spaces and the increase of services; for the Metropolitan City of Messina the rehabilitation of peripheral areas through the demolition of old houses and redevelopment of the heritage intended for social housing, the recovery and regeneration of spaces and buildings, especially in areas of high population density, to improve environmental quality and resilience to climate change; for Lamezia actions to counter the phenomenon of depopulation of some neighbourhoods through the recovery of housing to be made available to families in need and improving the usability of spaces and social services, in addition to the realization of bike paths and redevelopment of the waterfront. For these particular proposals the maximum possible request amounted to 100 million euros, while for the "ordinary" ones it was 15 million.

If for the eight pilot projects a brief description is known as reported above in full, it is not known what the 151 "ordinary" projects foresee, no indication for the interventions that will be carried out in the annexes to the decree of the Ministry of Infrastructure and Sustainable Mobility, each proposal is only associated with an identification code.

The prescription of reserving 40% of the resources to the Regions of southern Italy has been respected even if it may be of some use to try to understand, beyond the differences in territorial weights in surface area, number of inhabitants and density, why some Regions such as Apulia have enjoyed the financing of 21 proposals including a "pilot", then for about 400 million euros and Basilicata only one project for an amount less than 15 million.

A first hypothesis is that some authorities, Regions, Metropolitan Cities and municipalities have been investing for some time in planning, preparing for the opportunities with seriousness, structuring efficient planning offices and calling on the expertise of professionals and the commitment of the Third Mission [6] of the Universities.

In order to understand why these results were achieved, or in any case to assume what the reasons might have been, it is worth going over the criteria for the evaluation of proposals by the High Commission appointed by the Ministry of Infrastructure and Sustainable Mobility, with the task of drawing up a list of projects to be financed.

The quality of the proposal is the dominant criterion, for the presence of innovative aspects and devices of the green economy, in particular the adherence to the Minimum Environmental Criteria (CAM in Italia), the "zero balance" of the consumption of new land through interventions of recovery and redevelopment of already urbanized areas or, if not built, however included in urban areas strongly consolidated, taking into account the significance of the interventions in terms of seismic safety and energy requalification of existing buildings, including through their demolition and reconstruction.

The extent of interventions in relation to public residential buildings, with preference for the areas of greatest housing tension, and level of integration also in terms of social mix and diversification of housing supply and related services. Ability to coordinate and involve actors in partnership for the application of measures and innovative models of management, support and social inclusion, urban welfare and activation of participatory processes.

The contiguity or proximity to historic centres or parts of the city identity. The interest in the recovery and enhancement of cultural, environmental and landscape heritage, the recovery and reuse of significant architectural evidence, connected and functional to the proposed regeneration program presented.

The involvement of private operators, including those of the Third Sector and operating in the area of intervention, also in order to activate public and private financial resources, also taking into account the possible provision of areas or properties.

Since the applications submitted were considerably higher than the resources, the priority provided by the call for proposals will be applied to those whose territory presents the highest index of social and material vulnerability (IVSM in Italian) [7].

In fact, one of the conditions of the call was that the proposals had to intervene in urban areas whose IVSM was greater than 99 or higher than the median of the territorial area. The IVSM is quite different from the poverty index and also includes a relativity linked to the conditions of the territories to which it belongs, in a logic of territorial improvement. In fact, with the same characteristics, a municipality included in a regional reality or a macro-area that is rich and socially resolved has a higher vulnerability index than a municipality that belongs to a homogeneous reality that is economically underdeveloped.

A direction aimed at offering a good level of planning to all Regions - especially those of the south, because the "southern issue" is likely to lose for the umpteenth time a good opportunity - would have been very appropriate, even given the tight schedule because the interventions admitted to funding must be tested and reported within the first half of 2026, otherwise the loss of funding itself.

3   The Integrated Urban Plans of the NRRP for the Connective Network of Well-Being of Metropolitan Cities

The decree of the Ministry of the Interior of December 6, 2021 provided Metropolitan Cities with the indications to submit Integrated Urban Regeneration Plans, which will allow large cities to start a process of transformation into smart cities, improve the quality of infrastructure services and regenerate degraded urban areas, in implementation of the project line "Integrated Plans - M5C2 - Investment 2.2" as part of the National Recovery and Resilience Plan.

The overall investment is substantial and provides for the preparation of participatory urban regeneration programs, aimed at the "improvement of degraded urban areas, regeneration, economic revitalization, with particular attention to the creation of new personal services and the upgrading of accessibility and infrastructure, allowing the transformation of vulnerable territories into smart and sustainable cities".

The Metropolitan Cities are required to identify the projects that can be financed within their urban area within one hundred and twenty days from the date of entry into force of Decree-Law no. 152/2021, therefore once again by March 15, 2022, taking into account the projects expressed by the Municipality of the Capital and the municipalities of its territory.

The projects must comply with precise conditions that allow their eligibility. They must intervene in urban areas whose IVSM is greater than 99 or higher than the average of the territorial area and have a design level not lower than the preliminary design or technical and economic feasibility study.

The main evaluation criterion concerns the degree of compliance with the conditions linked to the principle of DNSH (Do Not Significant Harm), which would seem to leave room for vagueness on the concept of "not significant" and which identifies six criteria to determine how each economic activity contributes substantially to the protection of the ecosystem, without harming any of the environmental objectives of the Green Deal [8]. A guide to assessing activities and self-assessing projects is the Taxonomy's technical annex on establishing a framework that encourages sustainable investments and refines the parameters for assessing whether different economic activities contribute to climate change mitigation by identifying sectors that are critical to effective pollution reduction. The framework defined by the Taxonomy will determine the criteria for the allocation of European resources and can be considered a useful guide to direct investments towards truly sustainable interventions.

The Operational Guide [9] is designed to assist Administrations by providing useful guidance on taxonomy requirements, the regulatory framework, and useful data and information to document compliance with DNSH requirements.

For the Administrations, this is an absolute and demanding novelty that can be supported by the provision of the technical criteria reported in the DNSH self-assessments of the NRRP, which will constitute an outlined direction in favor of the entire path of realization of investments and reforms.

For the first time, administrations are required to document and concretely guarantee that any measure does not significantly harm environmental objectives, adopting specific requirements to this effect in the main programmatic and implementation acts.

Intent that was inherent in Environmental Impact Assessment procedure, but ended up being experienced as a kind of judgment on the outcomes of a process and not as a complex process that accompanies each phase, from the preliminary design and planning.

It is now specified that the commitments made to pay attention to and respect the environmental ecosystem must be made explicit in precise ways and monitored from the first acts of programming of the NRRP measure until the final phase of the testing of the works or the certificate of regular execution of the interventions.

We are in an experimental phase and this will require as much clarity as possible to make explicit the essential elements necessary for DNSH compliance in the funding decrees and in the specific technical tender documents, in order to remove the risk of suspension of payments and the avocation of the proceedings in case of non-compliance with DNSH.

The experimentation will also concern the realization phase, which requires clear design documents, with specific indications aimed at compliance in urban planning and building regulations, in the specifications and in the specifications, so that it is possible to "trace" a detailed report on the fulfillment of the measures taken.

This is an interesting step forward in procedures and in the way of conceiving anthropic actions in the territory and in the environment, which, however, is measured against a very tight agenda foreseen by the NRRP that does not seem to offer margins for the time needed for a serious experimentation.

All calls for proposals of the NRRP that directly concern local authorities, barring possible extensions, must be launched by June 30, 2022, with the aim of starting all worksites by the end of 2023 to complete the implementation of interventions by June 30, 2026. Local authorities will be called upon to deal with a very tight schedule dictated by the European Union, and will have to prove themselves capable of measuring, in a precise and accurate manner, the impact on the territory deriving from the use of the various resources forfeited within the framework of the NRRP implementation calls. Already in the coming months they will have to provide for the quantification of the objective target relative to the square meters of the territorial basin that benefits from the intervention, to be described in the detailed report of the purposes of the intervention and the expected benefits that substantiate the project proposal.

Faced with such significant resources, never seen in recent decades and probably not to be seen again for who knows how long, the opportunity for territorial rebalancing cannot be wasted, because this would introduce a new deep disappointment on the possibility of achieving similar, if not equal, opportunities to citizens of the national territory. The results of the first distributions suggest the occasional nature of the response of the Regions and Metropolitan Cities [10] whose efficiency in managing the processes may depend on the moments of politics, on the school of thought of the decision-makers who, at that moment, determine alliances, on the relationships with the professional orders, with the Universities, with individual professionals. In some Regions, the Universities, with the role of the Third Mission, have guaranteed a fundamental direction for the socio-economic and cultural future of their territories, offering strong shoulders to the professional world and to the planning offices and obtaining the necessary listening to be able to be incisive. In order not to reduce this moment, which could be epoch-making, to a mere opportunity to observe phenomena, record impacts and evaluate effects, the University should be involved in the process, also in order to call upon all the energies that specific territory can express.

References

1. National Recovery and Resilience Plan, p. 87 (2021). https://www.governo.it/sites/governo.it/files/PNRR.pdf. Accessed 30 Dec 2021
2. Fallanca, C.: Places in the city designed for pro well-being living space To promote healthy, autonomous and active lifestyles for citizens of all ages. Upland- J. Urban Plan. Landscape Environ. Des. **5**(2), 149–172 (2021)
3. Fallanca, C.: I luoghi della città pensati per lo spazio vitale pro- benessere. In: XII Giornata internazionale di Studio Inu - 12th International Inu Study Day -Benessere e/o salute? 90 anni di studi, politiche, piani, Urbanistica Informazioni, p. 289 (2020)

4. Fallanca, C.: The city well-being. The social responsibility of urban planning. In: Bevilacqua C., Calabrò F., Della Spina, L. (eds.) New Metropolitan Perspectives. NMP 2020. Smart Innovation, Systems and Technologies. Springer, Heidelberg (2020)

5. Indovina, F.: La città dopo il coronavirus. In: Archivio di Studi Urbani e Regionali 128/2020, FrancoAngeli (2020)

6. Fallanca, C.: Didattica, ricerca e terza missione per lo sviluppo sostenibile delle città, delle comunità, del territorio. In: Supplemento di ArcHistoR 6/2019 (2019). ISSN 2384-8898

7. ISTAT Le misure della vulnerabilità: un'applicazione a diversi ambiti territoriali, https://www.istat.it/it/files//2020/12/Le-misure-della-vulnerabilita.pdf. Accessed 30 Dec 2021

8. Regolamento (UE) 2020/852 del Parlamento Europeo e del Consiglio del 18 giugno, Art 17 (2020)

9. Guida Operativa per il rispetto del principio di non arrecare danno significativo all'ambiente (DNSH). https://www.rgs.mef.gov.it/_Documenti/VERSIONE-I/CIRCOLARI/2021/32/Allegato-alla-Circolare-del-30-dicembre-2021-n-32_guida_operativa.pdf

10. Fallanca, C.: (edit by) Città Metropolitane. Linee progettuali per nuove relazioni territoriali, FrancoAngeli (2021)

Urban Regeneration and Real Estate Dynamics: A Non-linear Model of the Break-Even Analysis for the Assessment of the Investments

Francesco Tajani[1]([envelope]), Pierluigi Morano[2], Felicia Di Liddo[2], Rossana Ranieri[1], and Debora Anelli[2]

[1] "Sapienza" University of Rome, 00196 Rome, Italy
francesco.tajani@uniroma1.it
[2] Polytechnic University of Bari, 70125 Bari, Italy

Abstract. With reference to the redevelopment of brownfields areas, in the present research a non-linear model aimed at assessing the conveniences of regeneration initiatives of the parties involved has been proposed. The model is part of the assessment methods aimed at verifying the financial feasibility of an urban redevelopment intervention carried out through the public-private partnership procedure. By borrowing the operative logic of break-even analysis, the model omits the hypothesis related to the non-linearity revenues: in the specific cases in which the mentioned assumption is not valid, i.e. when the market supply significantly overcomes the market demand, the proposed model represents a useful tool to orient the renovation initiatives decisions towards effective and profitable choices.

Keywords: Break-even analysis · Brownfields · Urban regeneration interventions · Real estate dynamics · Decision-making processes

1 Introduction

Within the current urban contexts of the cities, the presence of brownfields sites constitutes a relevant issue for the definition of territorial development policies. The regeneration of degraded and abandoned areas, in fact, represents a strategy of sustainable land use and a driving force for urban revitalization in conjunction with the significant interest in environmental protection [1, 2]. According to the United States Environmental Protection Agency (EPA), the brownfield is defined as "a property, the expansion, redevelopment, or reuse of which may be complicated by the (potential) presence of a hazardous substance, pollutant, or contaminant" [3]. A research carried out by EPA in 2020, aimed at analysing environmental benefits that occur when brownfield sites are redeveloped, has found that (i) these are often "efficiently located" due to their central location and to the existing infrastructure links and (ii) their redevelopment and functional reconversion generate several benefits in economic, environmental and social terms. Haninger and Timmins have detected that cleaning-up brownfield properties leads to residential property value increases of 5–15.2% within 1.29 miles of the sites [4].

© The Author(s), under exclusive license to Springer Nature Switzerland AG 2022
F. Calabrò et al. (Eds.): NMP 2022, LNNS 482, pp. 655–663, 2022.
https://doi.org/10.1007/978-3-031-06825-6_62

The awareness of community organizations and local governments to identify, address and clean-up brownfield sites aims at a subsequent functional reconversion of the site for its safe use, in order to satisfy the needs of the community and to reduce the contaminating threats to public health and environment [5].

In general, unsafe levels of environmental contamination on a brownfield may result from former or current industrial, commercial, residential, agricultural or recreational uses and practices, causing the formation of contaminants in soil, water or air. Cleaning-up brownfield sites reduces or eliminates potential health risks to residents, workers, pets and the surrounding environment. The "size" or the typology of the rehabilitation depends on the specific contaminants found, on the extent of contamination and on the reuse modalities of the property, in order to carry out an effective brownfields land reclamation and to protect the community from potentially harmful exposures by removing or containing site contaminants.

The brownfield site redevelopment implies its transformation to satisfy the many different needs that exist within a community, by representing urban voids potentially useful for uses different from the original ones [6–9].

These polluted sites included in the urban contexts or in immediate suburbs have generally specific characteristics that allow valuable transformation and enhancement processes that are capable of producing - if they are properly managed - financial and economic benefits and new opportunities for sustainable development for the community [10].

In the context of the Next Generation EU (NGEU) [11], aimed at promoting a "sustainable, uniform, inclusive and equitable recovery" following the crisis caused by the Covid-19 pandemic, in Italy the National Recovery and Resilience Plan (NRRP) - definitively approved with the Council's Implementation Decision on 13 July 2021 – includes the issue related to the urban regeneration among the measures as a transversal and specific goal to be achieved. In particular, among the measures described in the six Missions of the plan, the measure M5C2.2 provides for 9.02 billion in loans for urban regeneration and social housing interventions [12].

Specifically, the resources are distributed with different shares in three investment lines relating to (i) investments in urban regeneration projects, aimed at reducing situations of marginalization and social degradation and improving the urban quality as well as the social and environmental context; (ii) integrated urban plans, dedicated to the suburbs of metropolitan cities which provide for participatory urban planning with the aim of transforming vulnerable territories into smart and sustainable cities, limiting the land consumption, (iii) innovative programs of the living quality that provide for the construction of new public housing structures and redevelopment of degraded areas [13].

In order to minimize marginalization and degradation situations, urban regeneration interventions intend to have significant impacts on the recovery of the most vulnerable urban tissues, i.e. peripheral and internal areas of the cities [14, 15].

In the current economic situation, the lack of funding is the most frequently cited barrier to brownfield redevelopment, i.e. a constraint compounded by uncertainty surrounding the possible cost of environmental assessments, remediation and subsequent recoverty project. The involvement of private investors represents an alternative and effective form of financing and management of the interventions [16, 17], based on the cooperation between the Public Administrations and private subjects in which the public entities are able (i) to split the resources on other interventions for the community, (ii) to restore the image of the city portion currently abandoned by introducing new functions, (iii) to carry out the redevelopment project at no cost. On the other hand, the private investor does not burden the costs relating to the purchase of the buildable area.

Furthermore, in order to guarantee the financial sustainability of the initiative, private brownfields redevelopment projects often require relevant dimensions and, therefore, economic parameters (costs and revenues), associated with significant investment risks. In these situations, reliable assessment tools, able to appropriately analyse the convenience of the subjects involved in terms of the number of units, surfaces and volumes provided for by the initiative considered, are needed.

2 Aim

The aim of the present research concerns the development of an evaluation model to support Public Administration decision processes in planning brownfields recovery initiatives to carry out through public-private partnership mechanisms. By borrowing the operative logic of break-even analysis, the proposed model intends to solve one of the main limitations of the technique in specific situations: (i) in contexts characterized by a real estate supply that already tends to absorb local demand before the realization of the initiative, (ii) where a new planning aimed at an abundant construction of new buildings and/or the possibility of re-functionalising existing disused complexes - often numerous in Italian cities – is provided. In these contexts, in which the new surfaces to be placed in the reference market are significant and abundantly satisfy the current real estate demand, the hypothesis of linearity of the increasing revenue curve - that means the invariance of the unit selling price - does not reflect the reference market: therefore, it is necessary to recall the logarithmic function of the curve of total utilities, according to which the unit selling price decreases in correspondence of an increase of the amount of gross floor area (GFA).

In this sense, the hypothesis assumed for which the unit selling price is constant whatever the quantity produced does not represent a reasonably valid assumption, as discount policies are applied in the situation in which a significant quantity of good or service are built and sold. In a context in which the supply is characterized by a greater elasticity than the reference demand, the unit selling price is characterized by a progressive decrease. In empirical terms, in fact, the unit selling prices are inversely proportional to the realized GFA, due to the reduction in the marginal utility attributed by the potential buyer in contexts of overabundant supply.

The proposed model could allow Public Administrations to orient and verify the planning decisions for the identification, under different scenarios of prices and costs, of the GFA to be built and to be sold, able to break-even the financial balance of the initiative. Furthermore, the model could allow private investors to quickly verify the financial convenience of an investment to be implemented, by neglecting the hypothesis of linearity of revenues which in some cases (real estate demand close to the available supply, construction of new buildings that involve the realization of a significant number of units, possibility of re-functionalization of existing disused complexes, etc.) constitutes a scarcely valid condition.

The paper is structured as follows. In Sect. 3 the proposed model in the research is illustrated and some considerations related to the model implementation are presented. In Sect. 4 the conclusions of the work are discussed.

3   Model

Within the financial analysis aimed at verifying the convenience of the transformation initiatives, the break-even analysis represents a technique that rapidly allow to define the GFA to be built and sold, i.e. able to determine the quantity of GFA for which the total costs and the total revenues are equal [18]. By considering in the total costs the ordinary ("normal") profit expected by the private investor, this quantity of GFA defines the condition of minimum convenience of the initiative (the "break-even" quantity): therefore, in the situations for which the total revenues are higher than the total costs, the initiative will generate an "extra-profit" (EP), that is a further compensation for the private investor or a possible maximum burden that the Public Administration could require to the private entrepreneur. As the break-even quantity of GFA indicates the minimum financial threshold for the private investor, the convenience to activate the initiative is not verified for lower quantities. In general, the implementation of the BEA in the context of the public-private partnership initiatives allows to quickly verify the financial convenience for the subjects involved through (i) the assessment of the fixed transformation costs of the initiative; (ii) the determination of the unitary variable production costs; (iii) the estimation of the unit selling prices. Starting from the mathematical relationship for which the extra-profit is equal to the algebraic difference between the total revenues and the total costs - that are constituted by the sum of the *fixed* costs (C_f), i.e. cost items defined without considering the amount of the products to be realized (e.g. acquisition of land, its environmental remediation and restoration, the urbanization and the infrastructure for mobility, the recovery of existing buildings, the establishment of spaces and equipment of collective interest), and the *variable* costs (C_v), i.e. cost items defined considering the amount of the products to be realized and sold within the initiative (e.g. energy costs, cost of raw materials directly used in the production, costs for the distribution and sale of the products, workers' salaries based on flexible contracts) – and, by assuming that the extra-profit is equal to the value zero (EP = 0), the break-even GFA represents the surface quantity to be sold, in correspondence to which the total costs (C_t - including the normal profit for the investor) are equal to the total revenues.

The developed model in the present research borrows the expression to determine the break-even GFA amount (q), that is:

$$p_u \cdot q = C_f + Cv_u \cdot q \tag{1}$$

where:
    p_u is the unit selling price of GFA [€/m²];
    q is the quantity of GFA [m²];
    C_f represents the fixed costs [€];
    Cv_u represents the unit variable cost [€/m²].

The term on the left of (1) indicates the total revenues (R_t), whereas on the right the total costs (C_t) are included. In the hypothesis of linearity of R_t and C_t functions, the formula for the determination of the break-even quantity is:

$$q = \frac{C_f}{p_u - Cv_u} \tag{2}$$

In the proposed model, the construction of the "real" trend of the marginal price curve p_u, characterized by a logarithmic and decreasing trend, is carried out. The development of this trend requires the introduction of two new variables, i.e. (i) $r =$ rate of variation, which expresses the volatility of the reference market due to changes in real estate supply/demand, (ii) $q_l =$ threshold quantity of GFA, that is the surface capable of satisfying the current demand for new real estate units, beyond which a surplus of supply is triggered that can be absorbed by the market at unit prices lower than the current market values for similar property units.

By indicating with $\overline{p_u}$ the unit market value of the new units to be built by taking into account the current supply conditions (thus *before* the introduction of new surfaces) and with $\Delta q = \frac{q - q_l}{q_l}$ the differential between the quantities of GFA generated by the investment considered and by those of saturation of the current market demand, the logarithmic function of the unit price will be:

$$p_u = \frac{\overline{p_u}}{(1 + r)^{\Delta q}} \tag{3}$$

By inserting (3) into (1) the mathematical form obtained is:

$$\frac{\overline{p_u}}{(1 + r)^{\frac{q - q_l}{q_l}}} \cdot q = C_f + Cv_u \cdot q \tag{4}$$

Having determined the starting parameters – the unit market value of the units to be built ($\overline{p_u}$), estimated with reference to current market conditions; the quantity of limit GFA (q_l), defined following a market analysis (e.g. taking into account the forecasts of current urban planning tools, the relationship between demographic trend and existing real estate stock, etc.); the fixed costs (C_f) and variable unit cost (Cv_u) of the investment; the rate of change (r), assessed by considering the geometric mean of the revaluation/devaluation rates of the half-yearly quotations published by the Real Estate Market Observatory (OMI) of the Revenue Agency for the trade area, the city and the intended use analyzed -, the amount of break-even GFA (q), through an iterative calculation, is estimated.

The model represents a highly valid tool in public-private partnership initiatives, where, in the cases of a specific request from the Public Administration - included in the amount of fixed costs -, the market conditions could not allow the private investor to satisfy in any case this burden: in this sense, the model returns a Pareto front - for different values of the variation rate r - of the optimal combinations of the surfaces to be realized (q) and the maximum request of the Public Administration (C_f).

A comparison between the "classic" graphic representation of the break-even point with the linear model – Figure (a) - and that related to the proposed non-linear model – Figure (b) - is reported in Fig. 1.

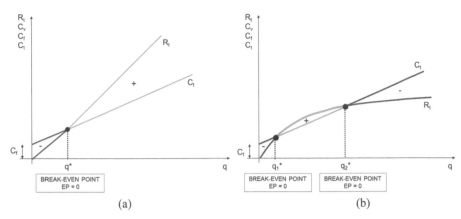

Fig. 1. Comparison between the "linear model" representation of the break-even point (Figure a) and the "non-linear model" one (Figure b)

From the analysis of the graph in Figure (a), it should be highlighted that in the linear model, starting from the quantity "0" to q^* (quantity of break-even), an absence of convenience in implementing the investment is observed, whereas from q^* an extra-profit situation for the private investor is detected. From this point, in fact, the Public Administration could require the private operator an "extra-burden" monetary sum, having already verified the convenience of the initiative for the private investor, taking into account the inclusion of a normal profit in the total costs.

The graph in Figure (b) shows that in the non-linear model the extra-profit area is limited and included between two regions, in which the investment convenience is not recorded: moreover, in the non-linear model, two break-even points, that verify the condition of minimum financial convenience for the investor (EP $= 0$), occur.

The revenue curve trend in Figure (b) highlights different conditions not only for the quantities to be produced, but also for the unit selling price. In fact, having estimated the GFA quantity to be placed on the market which saturates the current demand, a first phase in which the unit price is increasing, is identified. This condition reflects a sort of "anxiety effect" in the market demand, in which the first quantities of GFA are initially realized and the subsequent quantities planned by the project are not yet sufficiently known: a "bubble" of the unit selling price is outlined, for which a first break-even point (q_1^*) is determined with a unit selling price higher than that initially estimated. Since the normal profit is function of the total revenues/costs, observing the graph in Figure b), it is evident that the entrepreneur should increase the quantities of GFA to be realized, also generating an extra-profit that could constitute an additional monetary sum that could be required by the Public Administration. As the projected or advertised GFA quantities increase, the market bubble deflates and the unit selling price begins to decrease: at this stage for the investor it is still convenient to increase the quantity of GFA to be realized, up to q_2^*, i.e. the quantity beyond which the total revenues are lower than the total costs, and the normal profit is not guaranteed. Therefore, the private investor should aim at the q_2^* quantity, which constitutes a sort of economic optimum, capable of maximizing the normal profit, whereas the Public Administration should try to remain within the range

q_1* and q_2*, in order to both ensure the normal profit for the investor and negotiate the extra-profit.

A possible logarithmic trend of the unit selling price in the non-linear model is reported in Fig. 2.

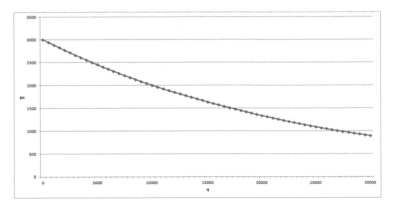

Fig. 2. Logarithmic trend of unit selling price in the non-linear model

4   Conclusions

The brownfield areas redevelopment constitutes an important opportunity to (i) recover contaminated, abandoned and/or damaged city portion [19, 20], (ii) implement larger-scale regeneration initiatives, (iii) kick-start the local economic growth [21, 22].

In the present research a non-linear model of break-even analysis for the evaluation of investments has been developed.

By borrowing the logic procedure of BEA, the model proposed does not consider the lineary of revenues (invariance of unit selling prices), i.e. the hypothesis that in contexts with a real estate demand close to the available supply or where large edification are planned, is not very consistent. The "quick" model for the identification of the minimum threshold financial convenience of the private subjects involved in the urban interventions has been applied, by starting from the "real" trend of the marginal price curve, characterized by a logarithmic and decreasing trend.

The model could be implemented in the first phases of assessment brownfields renovation investments, in order to provide a first indication of the interventions feasibility that should be completed through the implementation of more detailed evaluation techniques [23, 24].

Future insight of this research may concern the definition of a Pareto front of the optimal combinations of surfaces to be realized, able to ensure the initiative financial feasibility and the maximum request of the Public Administration. Thus, the model can be applied to an urban regeneration initiative in order to build a "Paretian optimal" range, i.e. the boundary combinations (quantity of GFA/maximum burden), able to guarantee the financial sustainability of the initiative for the private investor.

References

1. Del Giudice, V., De Paola, P., Manganelli, B., Forte, F.: The monetary valuation of environmental externalities through the analysis of real estate prices. Sustainability **9**(2), 229 (2017)
2. Del Giudice, V., Massimo, D.E., De Paola, P., Del Giudice, F.P., Musolino, M.: Green buildings for post carbon city: determining market premium using spline smoothing semiparametric method. In: Bevilacqua, C., Calabrò, F., Della Spina, L. (eds.) New Metropolitan Perspectives. NMP 2020. Smart Innovation, Systems and Technologies, vol. 178, pp. 1227–1236. Springer, Cham (2021). https://doi.org/10.1007/978-3-030-48279-4_114
3. United States Environmental Protection Agency (EPA) - Brownfields and Land Revitalization. https://www.epa.gov/brownfields/
4. Haninger, K., Ma, L., Timmins, C.: The value of brownfield remediation. J. Assoc. Environ. Resour. Econ. **4**(1), 197–241 (2017)
5. Greenberg, M., Lowrie, K., Mayer, H., Miller, K.T., Solitare, L.: Brownfield redevelopment as a smart growth option in the United States. Environmentalist **21**(2), 129–143 (2001)
6. BenDor, T.K., Metcalf, S.S., Paich, M.: The dynamics of brownfield redevelopment. Sustainability **3**(6), 914–936 (2011)
7. Bacot, H., O'Dell, C.: Establishing indicators to evaluate brownfield redevelopment. Econ. Dev. Q. **20**(2), 142–161 (2006)
8. Thomas, M.R.: A GIS-based decision support system for brownfield redevelopment. Landscape Urban Plan. **58**(1), 7–23 (2002)
9. De Sousa, C.: Brownfield redevelopment versus greenfield development: a private sector perspective on the costs and risks associated with brownfield redevelopment in the greater Toronto Area. J. Environ. Plan. Manag. **43**(6), 831–853 (2000)
10. APAT – Agenzia per la protezione dell'ambiente e per i servizi tecnici: Proposta di linee guida per il recupero ambientale e la valorizzazione economica dei brownfields. Servizio Stampa ed Editoria, Roma (2006). www.isprambiente.gov.it
11. Next Generation EU (NGEU) - Recovery Fund. www.ec.europa.eu.it
12. National Recovery and Resilience Plan (NRRP). www.mef.gov.it
13. Senato della Repubblica, Camera dei Deputati: Il Piano Nazionale di Ripresa e Resilienza (NRRP) – Schede di lettura. Dossier XVIII Legislatura (2021). www.camera.it
14. Calabrò, F., Iannone, L., Pellicanò, R.: The historical and environmental heritage for the attractiveness of cities. The case of the Umbertine Forts of Pentimele in Reggio Calabria, Italy. In: Bevilacqua, C., Calabrò, F., Della Spina, L. (eds.) New Metropolitan Perspectives. NMP 2020. Smart Innovation, Systems and Technologies, vol. 178, pp. 1990–2004. Springer, Cham (2021). https://doi.org/10.1007/978-3-030-48279-4_188
15. Mallamace, S., Calabrò, F., Meduri, T., Tramontana, C.: Unused real estate and enhancement of historic centers: legislative instruments and procedural ideas. In: Calabrò, F., Della Spina, L., Bevilacqua, C. (eds.) New Metropolitan Perspectives. ISHT 2018. Smart Innovation, Systems and Technologies, vol. 101, pp. 464–474. Springer, Cham (2019). https://doi.org/10.1007/978-3-319-92102-0_49
16. Calabrò, F., Cassalia, G., Lorè, I.: The economic feasibility for valorization of cultural heritage. The restoration project of the reformed Fathers' convent in Francavilla Angitola: the Zibìb territorial wine cellar. In: Bevilacqua, C., Calabrò, F., Della Spina, L. (eds.) New Metropolitan Perspectives. NMP 2020. Smart Innovation, Systems and Technologies, vol. 178, pp. 1105–1115. Springer, Cham (2021). https://doi.org/10.1007/978-3-030-48279-4_103
17. Tajani, F., Morano, P., Di Liddo, F., Locurcio, M.: An innovative interpretation of the DCFA evaluation criteria in the public-private partnership for the enhancement of the public property assets. In: Calabrò, F., Della Spina, L., Bevilacqua, C. (eds.) ISHT 2018. SIST, vol. 100, pp. 305–313. Springer, Cham (2019). https://doi.org/10.1007/978-3-319-92099-3_36

18. Morano, P., Tajani, F.: The break-even analysis applied to urban renewal investments: a model to evaluate the share of social housing financially sustainable for private investors. Habitat Int. **59**, 10–20 (2017)
19. Del Giudice, V., De Paola, P., Torrieri, F.: An integrated choice model for the evaluation of urban sustainable renewal scenarios. In: Advanced Materials Research, vol. 1030, pp. 2399–2406. Trans Tech Publications (2014)
20. Morano, P., Tajani, F., Guarini, M. R., Di Liddo, F.: An evaluation model for the definition of priority lists in PPP redevelopment initiatives. In: Bevilacqua, C., Calabrò, F., Della Spina, L. (eds.) New Metropolitan Perspectives: NMP 2020. Smart Innovation, Systems and Technologies, vol. 178, pp. 451–461. Springer, Cham (2020). https://doi.org/10.1007/978-3-030-48279-4_43
21. Della Spina, L.: The integrated evaluation as a driving tool for cultural-heritage enhancement strategies. In: Bisello A., Vettorato D., Laconte P., Costa S. (eds.) Smart and Sustainable Planning for Cities and Regions. SSPCR 2017. Green Energy and Technology. pp. 589–600. Springer, Cham (2018). https://doi.org/10.1007/978-3-319-75774-2_40
22. Della Spina, L.: Multidimensional assessment for "culture-led" and "community-driven" urban regeneration as driver for trigger economic vitality in urban historic centers. Sustainability **11**(24), 7237 (2019)
23. Della Spina, L.: Cultural heritage: a hybrid framework for ranking adaptive reuse strategies. Buildings **11**, 132 (2021)
24. Morano, P., Tajani, F., Guarini, M.R., Di Liddo, F.: Iniziative di riqualificazione urbana in partenariato pubblico-privato: un modello per la definizione di liste di priorità temporale. LaborEst **20**, 50–56 (2020)

An Automatic Tool for the Definition of a Sustainable Construction Investment Index

Francesco Tajani[1](✉), Lucy Hayes-Stevenson[2], Rossana Ranieri[1], Felicia Di Liddo[3], and Marco Locurcio[3]

[1] Department of Architecture and Design, Sapienza University of Rome, Via Flaminia 359, 00196 Rome, Italy
francesco.tajani@uniroma1.it

[2] Department of Architecture and Planning, University of Auckland, 26 Symonds Street, Auckland 1010, New Zealand

[3] Department of Civil, Environmental, Land, Building Engineering and Chemistry, Polytechnic University of Bari, Via Orabona 4, 70125 Bari, Italy

Abstract. With reference to a study sample related to eleven countries (Australia, Canada, Denmark, Germany, Iceland, Ireland, Italy, New Zealand, Norway, Sweden and United Kingdom), the present work intends to identify the impact of specific factors on the sustainability of the construction sector through the definition of a composite index. In particular, in order to investigate the different contributions of the factors chosen for the analysis, the data collected for the selected countries are processed by an automatic tool. The research could represent a valid reference for the Public Administrations to monitor the construction sector and to define strategies and policies able to improve the sustainability of territorial transformation interventions and built environment.

Keywords: Sustainability index · Construction sector · Automatic tool · COIN tool · Territorial interventions

1 Introduction

The Paris Agreement (2015) is the legally binding treaty focused on the climate change with the aim to reduce the temperature by 1.5° compared to the preindustrial levels [1]. During the recent years, each country involved in the agreement has produced an action plan based on a five-year cycle that has to be reviewed by the United Nation (UN) Committee [2]. The international agreement addresses the importance of reducing greenhouse gas emissions, in order to make the countries carbon neutral [3–6]. Besides, the UN have developed a specific Agenda for Sustainable Development, aimed "to achieve a better and more sustainable future for all". Among the seventeen goals settled (Sustainable Development Goals – SDGs), the goal 12 should be highlighted, that states "by 2030, substantially reduce waste generation through prevention, reduction, recycling, and reuse" [7]. However, nowadays the built environment involves a large quantity of material and energy consumption [8, 9] and the construction sector is responsible for

F. Calabrò et al. (Eds.): NMP 2022, LNNS 482, pp. 664–675, 2022.
https://doi.org/10.1007/978-3-031-06825-6_63

39% of the total global emissions [10]. In this sense, the processes aimed at the urban regeneration represent a relevant "tool" for the built environment sustainability [11–13].

In order to monitor and assess the construction sector performances, in the present research an automatic tool to calculate a composite index has been implemented. In this sense, the study aims at defining a Sustainable Construction Index (SCI) able i) to synthesize the current conditions of each country or region and ii) to address countries government's funding and policies to achieve net-zero carbon emissions. Indeed, a comparative analysis of data collected in eleven countries worldwide (Australia, Canada, Denmark, Germany, Iceland, Ireland, Italy, New Zealand, Norway, Sweden and United Kingdom) has been carried out, to understand the trends on the effects of the financial incentives, policies, and in general the built environment on the carbon dioxide emissions (CO_2e). Therefore, an automatic tool capable of assessing and ranking the index for each country selected for the case study, has been developed through the use of specific indicators. In particular, the analysis has allowed to monitor the countries' performance in CO_2e within the construction sector and the relationships between policy models and funding for existing buildings and the reduction of CO_2e. Through the trends of each performance criteria and the overall ranking of the countries, the outputs obtained have highlighted the most effective policies and funding reference models.

The manuscript is organized as follow: in Sect. 2 a brief recognition of the current literature referring the adoption of indices for the assessment of the sustainability in the construction sector is included; in Sect. 3 the description of the case study with specific regard to the tool adopted, the steps that leads to the application of the tool and the interpretation of the results obtained are carried out; finally in Sect. 4 the conclusions are presented.

2   Current State of Art Related to Composite Indices

The assessment of the sustainability of the construction sector is a complex task [14, 15]. Similarly, the measurement and the comparison of the effects of environmental funding's and policies introduced by a country are difficult [16]. A recent study has highlighted that, in the current literature, the adoption of sustainability criteria and indicators is uneven and the methodology approaches and the tools are manifold [17]. Moreover, the equivocal definition of sustainable development complicates the process of selecting the indicators to be considered for the quantitative assessment of the sustainability.

In order to support and guide the governments in the adoption of effective policies to improve the sustainability of the construction sector, during the last decades, different researches have been developed [18–23]. In particular, Ayman and Wafaa have carried out a study to establish an International Sustainability Index [24]. Bon and Hutchinson [25] have structured a framework to deepen the comprehension of the economic aspects that involve the construction sector. Furthermore, with specific reference to the highway's construction projects – that are the most energy-consuming projects [26] - a composite index has been calculated using an Analytical Hierarchy Process, that have been applied to the outputs of a survey to obtain the weights of each parameter included in the model [27]. Moreover, a composite sustainability index of a project has been defined by using a multi-criteria decision-making approach referring to the Lithuanian construction industry: sustainability criteria have been chosen and grouped on the basis of the literature analysis and a survey has been used to select and rank the most important ones [28].

In addition, given the ability of composite indices to represent complex issues in a concisely and effectively way, these have been used in multiple application fields, such as: the assessment of the resilience of territories [29], the ability of territories to react to natural disasters e.g. floods [30], or the security of the environment related to the energy sector [31].

2.1 The COIN Tool

As mentioned above, in order to assess complex and multidimensional issues such as human development, environmental performance, sustainability of the construction sector etc. – which are often difficult to define and cannot be quantitatively and directly measured – composite indices are typically used. The adoption of indices could be considered as a standard approach for policies analysis because of their ability to synthesize the performance of specific issue related to a country or a region against multiple analytic criteria. To give scientific support to the governments to develop new strategical policies, in 2019 the Joint Research Centre (JRC) of the European Commission has developed an automatic tool – named COIN Tool – aimed to (i) graphically visualize the composite indicator; (ii) analyze the relationships between the different indicators considered; (iii) check the robustness of certain assumptions [32].

Firstly, there is the need to clearly define the aspect to be measured and assessed, and then the implementation of the tool could be carried out. In particular, the main four steps of the COIN tool are [33]:

1. Selection of the indicator to be used;
2. Collection and analysis of the data;
3. Data categorization;
   a. Normalization;
   b. Weighting;
4. Execution of the tool.

2.2 The Steps Leading to the Application

With reference to the case study of the research related to eleven countries (Australia (AU), Canada (CA), Denmark (DE), Germany (GE), Iceland (IC), Ireland (IR), Italy (IT), New Zealand (NZ), Norway (NO), Sweden (SW) and United Kingdom (UK)), in order to achieve the definition of the synthetic index for the assessment of the sustainability of the construction sector, 15 national indicators have been selected (*step 1*). Thus, for each country, the most appropriate sources have been identified and data have been collected (*step 2*). The data set could have multiple levels of aggregation, with sub-indices (sn), different pillars (pn) and sub-pillars (sp$n$) to which each indicator is assigned. In Table 1 the name, the acronym, a brief description and the unit of measure of the indicators selected in the present research, and the sub-pillar to which they belong are reported.

In particular, it should be outlined that the SCI calculated in the analysis has only two levels of aggregation: in this sense, the sub-index corresponds to the index and in the tool hierarchical structure there is only one pillar and the indicators are categorized into four sub-pillars (*step 3*):

- Socio-economic aspects – referred to the factors involved both in the economic and social processes.
- Environmental aspects – all the factors related to environment with specific attention to items related to carbon emissions.
- Construction aspects – concerning the built environment and related quantitative issues.
- Funding aspects – regarding to the factors that imply the government power of each country.

Table 1. Description of the performance indicators selected

N. of indicator	Performance indicator	Description	Unit of measure	Sub-pillar
Ind. 01	Density	Density of population of each country	Pop/km^2	Socio-economic sub-pillar 01
Ind. 02	Construction Market value	Market value referred to the construction sector of each country	USD*1,000,000	
Ind. 03	Inflation	The increase in the general average level of prices of goods and services referred to year 2019 for each country	%	
Ind. 04	Real gross domestic product (GDP) per capita	The GDP per citizen for the year 2019	USD	
Ind. 05	CO_2e Growth	CO_2 emission recorded for each country between years 1990–2018	%	Environmental sub-pillar 02
Ind. 06	CO_2e Construction Sector	CO^2 emission recorded for each country referred to the construction sector	Tons	

(*continued*)

Table 1. (*continued*)

N. of indicator	Performance indicator	Description	Unit of measure	Sub-pillar
Ind. 07	CO_2e Construction waste	The percentage of CO^2 emission recorded for each country referred to the waste involved in the construction sector	Tons	
Ind. 08	Carbon Pricing	The cost applied to carbon emissions	USD per ton CO_2	
Ind. 09	Total Dwellings	The number of existing dwellings for each country	n	Construction sub-pillar 03
Ind. 10	New dwellings per year	The number of new dwellings for referred to the year 2019	n./year	
Ind. 11	Vacant dwellings	The number of vacant dwellings for each country	n	
Ind. 12	Protected buildings	The number of buildings that benefit from artistic and cultural protection measures	n	
Ind. 13	Average age of buildings	The average age of existing building at a national level	Year	
Ind. 14	Environmental funding	The presence of funding specifically referred to the protection of existing building	Point scale from 1 – very low funding to 5 – very high funding	Funding sub-pillar 04
Ind. 15	Environmental policy	The presence of specific national policies concerned the protection of existing building	Dummy indicator 1 – presence or 0 – absence of policies	

Once collected the data, the numerical values obtained have been normalized compared to their maximum value (*step 3a*). Moreover, an equal weighting for each indicator has been assigned (equal to score "1") and an arithmetic aggregation method has been implemented (*step 3b*). Furthermore, a "direction" to each indicator has been assigned in order to give some indications on the relationship between each factor and the construction sustainability (score "-1" for negative aspects and score " +1" for the positive ones). In Table 2 the characteristics of each performance indicator are summarized.

Table 2. Characteristics of each performance indicator

N. of indicator	Indicator	Sub-pillar	Weighting	Aggregation	Direction
Ind. 01	Density	sp.01	1	Arithmetic	−1
Ind. 02	Construction Market value	sp.01	1	Arithmetic	−1
Ind. 03	Inflation	sp.01	1	Arithmetic	−1
Ind. 04	Real GDP per capita	sp.01	1	Arithmetic	1
Ind. 05	CO_2e Growth	sp.02	1	Arithmetic	−1
Ind. 06	CO_2e Construction Sector	sp.02	1	Arithmetic	−1
Ind. 07	CO_2e Construction waste	sp.02	1	Arithmetic	−1
Ind. 08	Carbon Pricing	sp.03	1	Arithmetic	−1
Ind. 09	Total Dwellings	sp.03	1	Arithmetic	−1
Ind. 10	New dwellings per year	sp.03	1	Arithmetic	1
Ind. 11	Vacant dwellings	sp.03	1	Arithmetic	1
Ind. 12	Protected buildings	sp.03	1	Arithmetic	−1
Ind. 13	Average age of buildings	sp.04	1	Arithmetic	1
Ind. 14	Environmental funding for existing buildings	sp.04	1	Arithmetic	1
Ind. 15	Environmental policy for existing buildings	sp.04	1	Arithmetic	−1

2.3   Interpretation of the Results

The implementation of the COIN tool has allowed to obtain the SCI for each country selected in the present research. In Table 3 the values of the SCI, sub-pillars and indicators are reported. On the basis of the SCI calculated, the definition of a ranking has been carried out: Denmark is the country for which the best performance in terms of sustainability has been resulted. The outputs related to the socio-economic sub-pillar point out Ireland at the top and Canada at the bottom of the based ranking. Furthermore, the results regarding to the environmental sub-pillar show Italy in the first place by considering the three different categories for CO_2e and construction sector emissions. In general terms, for this sub-pillar for the European Union countries the best performance is detected compared to the other countries analyzed. The construction sub-pillar's results highlight that Denmark and Sweden present the top ranking: Denmark performs the best across all criteria with the oldest building stock and least new dwellings per year. With reference to the Sub-pillar 04, for Denmark the best performance results are assessed.

Table 3. Outputs of the implementation of the COIN tool for the case study.

Item	DE	IR	SW	IC	IT	NW	UK	GE	NZ	AU	CA
SCI	**78.33**	**70.11**	**69.01**	**67.54**	**66.55**	**60.96**	**54.21**	**41.71**	**41.30**	**36.27**	**29.07**
p.01	n/a	n/a	n/a	n/a	n/a	n/a	n/a	n/a	n/a	n/a	n/a
sp.01	78.65	66.70	67.74	67.84	71.48	57.20	60.51	49.12	35.52	32.80	21.20
sp.02	72.03	77.29	66.17	64.62	77.44	59.85	46.65	46.26	38.40	30.28	34.00
sp.03	81.89	72.50	67.66	68.18	61.81	69.07	55.66	39.19	38.93	38.29	29.10
sp.04	79.16	65.77	73.76	68.80	58.18	57.45	52.14	33.38	51.62	42.21	33.22
ind.01	79.68	69.58	67.74	65.83	68.50	59.12	57.07	43.02	38.49	32.92	25.35
ind.02	77.36	68.74	67.41	66.30	66.90	59.79	53.34	43.90	38.21	38.16	29.53
ind.03	77.21	70.72	69.31	71.10	64.88	62.49	56.20	43.78	40.95	35.41	25.34
ind.04	79.33	68.54	70.51	67.19	70.05	59.28	55.55	42.37	42.67	35.67	29.44
ind.05	78.35	73.81	65.57	69.39	66.92	61.40	50.69	37.86	45.88	35.82	28.89
ind.06	76.54	67.64	65.95	63.94	68.34	56.99	55.42	46.34	35.74	38.04	32.11
ind.07	76.54	67.64	65.95	63.94	68.34	56.99	55.42	46.34	35.74	38.04	32.11
ind.08	77.63	69.06	68.21	66.19	68.01	59.59	54.98	43.44	39.02	34.59	27.50
ind.09	77.89	69.42	68.40	66.19	65.31	63.50	54.82	41.85	39.45	36.88	29.08
ind.10	80.59	71.47	71.89	70.10	65.15	62.13	56.21	42.59	41.83	36.03	28.91
ind.11	81.06	69.58	67.72	66.84	65.15	63.50	53.96	39.28	42.63	37.75	30.27
ind.12	77.42	73.04	67.72	68.92	65.15	62.85	52.31	39.28	41.58	37.78	29.63
ind.13	81.47	69.57	76.68	72.27	62.83	62.18	57.46	40.78	45.88	37.52	29.52
ind.14	75.92	66.79	65.57	63.94	62.83	56.63	49.13	35.23	45.88	40.30	32.30
ind.15	77.67	76.15	76.68	70.95	69.85	67.74	50.30	39.32	45.88	29.19	26.45

In order to understand the further data, some individual indicators effects on the overall performance ranking of each country are reported in the graphs below (Fig. 1). The analysis has been carried out by not counting one of 15 indicators to check if some variations in the final ranking could be occurred. The changes that result on the ranking of the countries if one of the 15 indicators (Nos. 5, 6, 7, 11, 13, and 14) is not considered is summarized in Fig. 1: the blue bar of the histogram represents the original value of SCI, the orange one corresponds to the value of the SCI calculated without the specific indicator and the grey line indicates for each country the value of the "excluded" indicator.

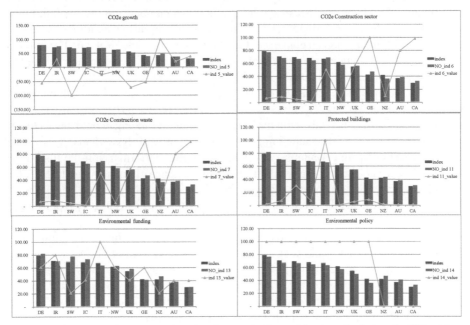

Fig. 1. Summary graphs representing the changes in the ranking by calculating the SCI without one specific indicator.

The outputs for the CO_2e growth indicator (Ind. 05) show that for New Zealand and Ireland the results affect the overall performance rating, whereas for Denmark no modifications of the ranking are found, due to the high performance levels of all other indicators.

Referring to the CO_2e Construction Sector indicator (Ind. 06), Germany has the highest emissions, so its performance rating position would increase if this indicator is excluded. For New Zealand a low value of CO_2e is measured and, consequently, the overall ranking would decrease if this indicator is excluded. The overall rankings do not almost change, except for Italy country, for which the most relevant improvement could be observed that allows its transition from the fifth position to the second one.

The Construction Waste CO_2e indicator (Ind. 07) moderately influences the ranking: the lower the CO_2e number, the higher the countries performance. The data highlight that the EU countries performance is better than non-EU ones. It should be pointed out that the UK was still in the EU at the time of the data collection (2019). Denmark has minor CO_2e from waste and, therefore, it holds the highest-ranking position. Without considering CO_2e waste, Denmark's performance would reduce while maintaining the top position. Germany has a high amount of CO_2e waste, and therefore if the waste was not included, the country performance would increase.

Aside from the protected building indicator (Ind. 11) the countries with a high number of protected building (Italy and Sweden) would decrease their performance: Denmark performs the best, whereas Canada holds the last position. The orange bar in the graphs shows the adjusted index, excluding the Protected Building indicator: Italy's overall performance would go down because it has such a high number of protected buildings, instead New Zealand's performance, with a low number of protected buildings, would go up. The overall order could change except for Germany and New Zealand. The outputs show that protected buildings are essential in a country's overall performance: the higher the number of protected buildings, the higher the performance.

Environmental funding indicator (Ind. 13) gives a point scale rating as all countries had some forms of funding. The levels drastically differ between EU and non-EU countries: the graph shows that Italy, with the highest amount of funding, would have a lower overall performance, while maintaining the position; instead, New Zealand, with the lowest amount of funding available, would have a higher performance if the funding is not considered. The graph in Fig. 1 points out the importance of financing in the SCI: for example Iceland, that is a country that uses primarily renewable energy, does not need financial incentives, and therefore the increase in position is not accurately representing the situation.

The graph on the presence of environmental policy indicator (Ind. 14) demonstrates that whether each country has a strategic specific policy, it would affect the overall performance value, while maintaining the original ranking. The existing policies in the EU countries and the United Kingdom show that programs positively influence the overall performance of the countries. Finally, the lack of policies in New Zealand, Australia and Canada strongly affects the countries performance.

3   Conclusions

The outputs obtained in the present research have outlined the relevance of developing tools able to assess and monitor the sustainability of construction sector.

In particular, the results of the implementation of the COIN tool have shown that the EU countries holds the highest ranking positions, whereas New Zealand, Canada, and Australia the lowest ones.

For socio-economic, Ireland has the best performance, followed by Denmark and Sweden. For environmental criteria, Denmark, Sweden, and Iceland have the highest performance ranking. The countries have the best performance in the carbon emissions across the construction sector, including the least amount of growth between 1990 and 2018 and the lowest emissions. Italy has the highest performance for the construction sector, followed by Sweden, Denmark, Iceland, and Ireland. The results show that the least new buildings, the oldest age of the building stock and the total number of protected buildings would increase the countries SCI performance.

Lastly, for the funding criteria, the performance ranking of each country reflects the current reference contexts: Italy has the highest performance thanks to the fiscal incentives introduction, i.e. the so-called *Superbonus* 110% aimed at supporting the energy retrofit interventions [34, 35], and New Zealand has the lowest one, with the most negligible funding and policy relating to insulation and heating questions. Again, it should be pointed out that the EU countries perform a higher standard compared to Canada, Australia and New Zealand, as the framework and policies behind the EU are more developed.

By analyzing the performance of each country in the different issues considered, it is possible to identify the Denmark, Ireland, Sweden, and Iceland policies and funding models as a valid reference to reduce emissions. In this sense, the present research could represent a useful tool for the Public Administrations to monitor the construction sector and to define strategies and policies able to improve the sustainability of territorial transformation interventions and the built environment.

References

1. United Nations. The Paris Agreement, https://unfccc.int/. Accessed 12 Nov 2021
2. Huang, L., Krigsvoll, G., Johansen, F., Liu, Y., Zhang, X.: Carbon emission of global construction sector. Renew. Sustain. Energy Rev. **81**, 1906–1916 (2018)
3. Spampinato, G., Malerba, A., Calabrò, F., Bernardo, C., Musarella, C.: Cork oak forest spatial valuation toward post carbon city by CO_2 sequestration. In: Bevilacqua, C., Calabrò, F., Della Spina, L. (eds.) New Metropolitan Perspectives. NMP 2020. Smart Innovation, Systems and Technologies, vol. 178, pp. 1321–1331. Springer, Cham (2021). https://doi.org/10.1007/978-3-030-48279-4_123
4. Barrile, V., Malerba, A., Fotia, A., Calabrò, F., Bernardo, C., Musarella, C.: Quarries renaturation by planting Cork oaks and survey with UAV. In: Bevilacqua, C., Calabrò, F., Della Spina, L. (eds.) New Metropolitan Perspectives. NMP 2020. Smart Innovation, Systems and Technologies, vol. 178, pp. 1310–1320. Springer, Cham (2021). https://doi.org/10.1007/978-3-030-48279-4_122
5. Massimo, D.E., Del Giudice, V., De Paola, P., Forte, F., Musolino, M., Malerba, A.: Geographically weighted regression for the post carbon city and real estate market analysis: a case study. In: Calabrò, F., Della Spina, L., Bevilacqua, C. (eds.) ISHT 2018. SIST, vol. 100, pp. 142–149. Springer, Cham (2019). https://doi.org/10.1007/978-3-319-92099-3_17
6. Del Giudice, V., Massimo, D.E., De Paola, P., Forte, F., Musolino, M., Malerba, A.: Post carbon city and real estate market: testing the dataset of reggio calabria market using spline smoothing semiparametric method. In: Calabrò, F., Della Spina, L., Bevilacqua, C. (eds.) New Metropolitan Perspectives. ISHT 2018. Smart Innovation, Systems and Technologies, vol 100, pp. 206–214. Springer, Cham (2019). https://doi.org/10.1007/978-3-319-92099-3_25
7. United Nations: Sustainable Development Goals. https://www.un.org/sustainabledevelopment/development-agenda. Accessed 12 Nov 2021
8. United Nation Environment Program: 2020 Global Status Report for Buildings and Construction: Towards a Zero-emission, Efficient and Resilient Building and Construction Sector. https://globalabc.org/. Accessed 12 Nov 2021
9. Del Giudice, V., Massimo, D.E., Salvo, F., De Paola, P., De Ruggiero, M., Musolino, M.: Market price premium for green buildings: a review of empirical evidence. case study. In: Bevilacqua, C., Calabrò, F., Della Spina, L. (eds.) New Metropolitan Perspectives. NMP 2020. Smart Innovation, Systems and Technologies, vol 178, pp. 1237–1247. Springer, Cham, Switzerland (2021). https://doi.org/10.1007/978-3-030-48279-4_115
10. World Green Building Council. https://www.worldgbc.org/news-media/WorldGBC-embodied-carbon-report-published. Accessed 12 Nov 2021

11. Della Spina, L.: The integrated evaluation as a driving tool for cultural-heritage enhancement strategies. In: Bisello, A., Vettorato, D., Laconte, P., Costa, S. (eds.) SSPCR 2017. GET, pp. 589–600. Springer, Cham (2018). https://doi.org/10.1007/978-3-319-75774-2_40

12. Della Spina, L.: Multidimensional assessment for "culture-led" and "community-driven" urban regeneration as driver for trigger economic vitality in urban historic centers. Sustainability **11**, 7237 (2019)

13. Della Spina, L.: Cultural heritage: a hybrid framework for ranking adaptive reuse strategies. Buildings **11**, 132 (2021)

14. Bell, S., Morse, S.: Sustainability Indicators: Measuring the Immeasurable? 2nd edn. Routledge (2012)

15. Warhurst, A.: Sustainability Indicators and Sustainability Performance Management. Mining, Minerals and Sustainable Development [MMSD] project report, 43, 129, International Institute for Environment and Development (IIED), United Kingdom (2002)

16. Botta, E., Koźluk, T.: Measuring environmental policy stringency in OECD countries: a composite index approach, OECD economics department working papers, No. 1177, OECD Publishing, Paris (2014)

17. Morano, P., Tajani, F., Guarini, M.R., Sica, F.: A systematic review of the existing literature for the evaluation of sustainable urban projects. Sustainability **13**, 4782 (2021)

18. Goel, A., Ganesh, L.S., Kaur, A.: Sustainability assessment of construction practices in India using inductive content analysis of research literature. Int. J. Constr. Manag. **21**(8), 802–817 (2021)

19. Stanitsas, M., Kirytopoulos K.: Investigating the significance of sustainability indicators for promoting sustainable construction project management. Int. J. Constr. Manag. (2021)

20. Calabrò, F., Cassalia, G., Lorè, I.: The economic feasibility for valorization of cultural heritage. The restoration project of the reformed fathers' convent in Francavilla Angitola: the Zibìb territorial wine cellar. In: Bevilacqua, C., Calabrò, F., Della Spina, L. (eds.) New Metropolitan Perspectives. NMP 2020. Smart Innovation, Systems and Technologies, vol 178, pp. 1105–1115. Springer, Cham (2021). https://doi.org/10.1007/978-3-030-48279-4_103

21. Ayman, R., Alwan, Z., McIntyre, L.: BIM for sustainable project delivery: review paper and future development areas. Arch. Sci. Rev. **63**(1), 15–33 (2020)

22. Suchith Reddy, A., Rathish Kumar, P., Anand Raj, P.: Preference based multi-criteria framework for developing a Sustainable Material Performance Index (SMPI). Int. J. Sustain. Eng. **12**(6), 390–403 (2019)

23. Kabirifar, K., Mojtahedi, M., Wang, C., Tam, V.W.Y.: Construction and demolition waste management contributing factors coupled with reduce, reuse, and recycle strategies for effective waste management: a review. J. Clean. Prod. **263**, 121265 (2020)

24. Ayman, O., Wafaa, N.: Towards establishing an international sustainability index for the construction industry: a literature review. In: First International Conference on Sustainability and the Future, Cairo, Egypt, vol. 1 (2010)

25. Bon, R., Hutchinson, K.: Sustainable construction: some economic challenges. Build. Res. Inf. **28**(5–6), 310–314 (2000)

26. Gambaotese, J.A.: Sustainable roadway construction: energy consumption and material waste generation of roadways. In: Construction Research Congress 183, ASCE, Reston, VA, pp 1–13. American Association of State Highway and Transportation Officials (AASHTO) (2008)

27. Ibrahim, A.H., Shaker, M.A.: Sustainability index for highway construction projects. Alex. Eng. J. **58**(4), 1399–1411 (2019)

28. Dobrovolskienė, N., Tamošiūnienė, R.: An index to measure sustainability of a business project in the construction industry: lithuanian case. Sustainability **8**, 14 (2016)

29. Stanickova, M., Melecký, L.: Understanding of resilience in the context of regional development using composite index approach: the case of European Union NUTS-2 regions. Reg. Stud. Reg. Sci. **5**(1), 231–254 (2018)

30. Kotzee, I., Reyers, B.: Piloting a social-ecological index for measuring flood resilience: a composite index approach. Ecol. Ind. **60**, 45–53 (2016)
31. Shah, S.A.A., Zhou, P., Walasai, G.D., Mohsin, M.: Energy security and environmental sustainability index of South Asian countries: a composite index approach. Ecol. Ind. **106**, 105507 (2019)
32. European Commission. Coin tool. https://knowledge4policy.ec.europa.eu/composite-indica tors/coin-tool_en. Accessed 21 Nov 2021
33. Becker, W., Benavente, D., Dominguez Torreiro, M., Tacao Moura, C., Fragoso Neves, A., Saisana, M. Vertesy, D.: COIN Tool User Guide, EUR 29899 EN, Publications Office of the European Union, Luxembourg (2019)
34. Italian Revenue Agency. https://www.agenziaentrate.gov.it/portale/documents/20143/233 439/Guida_Superbonus_110%25+%281%29.pdf/c11d4bd6-af26-89f9-c557-efef7c6c7452. Accessed 27 Nov 2021
35. Manganelli, B., Morano, P., Tajani, F., Salvo, F.: Affordability assessment of energy-efficient building construction in Italy. Sustainability **11**(1), 249 (2019)

Urban Regeneration Strategies According to Circular Economy: A Multi-criteria Decision Aiding Approach

Lucia Della Spina[✉] [ID]

Mediterranea University of Reggio Calabria, 89125 Reggio Calabria (RC), Italy
lucia.dellaspina@unirc.it

Abstract. As part of urban regeneration programs the implementation of policies consistent with a circular city model, social inclusion and the involvement of the local community are key factors for solving problems in urban contexts. In this context it is necessary to increase the resilience of the cities, favoring the regeneration of the existing real estate assets, in particular degraded and abandoned areas and their transformation into new junctions of interactions and actions capable of triggering local development, according to a circular economy model. Today the choices made regarding urban regeneration interventions are rarely supported by operational logics and methodologies capable of effectively rationalizing the selection processes. In the perspective of the optimal allocation of increasingly scarce resources, the proposed assessment framework is based on various methodological assessments, capable of supporting a complex, adaptive, inclusive and site-specific conflict-based decision-making process. The purpose of the study is to support decision makers in defining of alternative future scenarios capable of generating benefits in terms of cultural enhancement, social inclusion and economic development from a circular economy perspective. Through a specific set of indicators, selected considering the United Nations Sustainable Development Goals (SDGs), the scoreboard contributes to the definition of alternative scenarios, in order to lead to the choice of the shared scenario, evaluated in terms of both multi-group and multi-criteria. The methodological framework was tested for the urban regeneration of an area located in the municipality of Catanzaro in Southern Italy.

Keywords: Circular economy · Sustainable Development Goals (SDGs) · Urban regeneration · Multi-criteria decision analysis

1 Introduction

The international debate on urban regeneration consistent with sustainable development goals has become increasingly important in the implementation of circular economic models for urban policies [1–4].

The circular economy, whose mechanisms of value creation are manifold, is based on three principles: conservation and enhancement of natural capital, optimization of resources through the circulation of products, components and materials, promotion of the effectiveness of the system through disclosure and negative externalities design [3, 4].

Thanks to the circular economy processes, contrary to the current linear production-consumption-waste processes, the adoption of methods to maximize the efficient use of resources by recycling and minimizing emissions and 'waste' preserves value for as long as possible of natural and cultural resources [4–8]. This stimulates an indefinite extension of the duration of resources and their use value and promotes cooperation circuits between different actors through the improvement of the quality of life, the increase of social cohesion, the promotion of new job opportunities in the cultural sectors, and tourism, the attraction of investments, the renewal of the image of places, fostering a sense of belonging through shared and inclusive processes, etc. [9–12].

Fragments of unused, residual, abandoned, discarded landscape appear in the contemporary city. They can be recognized in brownfield sites, polluted territories, interstitial areas, waiting areas, terrain vague or drosscape [5]. These landscapes share the fact that they physically belong to the city but have been expelled from it for the end of their life cycle, for the incompatibility of their use with the urban reality or for the loss of their economic value. These "city waste" territories do not have an intrinsic negative value but a characteristic specificity, a natural consequence of the process of growth and development of the city and its metabolism [5–10].

These landscapes, hybrid urban areas on the border between urban centers and rural areas, are the natural consequence of the process of growth and development of the city and its metabolism. If properly exploited, they constitute the main potential places where it is possible to activate processes of economic, social, urban and environmental regeneration [6, 7].

In particular, in recent years there has been a proliferation of spaces dedicated to contamination and collaboration, which have represented a key tool in urban regeneration policies. Also identified as "third places" [8], they offer a neutral environment in which to develop social interactions, supporting creative and productive processes. They are fruitful places, entry points towards the circular urban economy [1–4], capable of implementing the Circular City Model (CCM) of the 2030 Agenda, social inclusion and involvement of the local community [3, 4, 9].

In the perspective of the optimal allocation of increasingly scarce resources, the purpose of this work is to define an empirical approach, capable of supporting the selection and evaluation of future scenarios in order to generate benefits in terms of cultural enhancement, social inclusion and development economic [6, 7, 12]. Through a specific set of indicators, selected considering the Sustainable Development Goals (SDGs) [4, 9–11], the preferable scenario is evaluated in terms of both multi-group and multi-criteria.

In the context outlined, a place-based and adaptive approach is proposed that uses a combination of bottom-up and top-down evaluation methods, eliciting soft and hard knowledge domains, expressed and evaluated through a series of sustainability indicators linked to the objectives SDGs [4, 9].

The operational steps of the methodological framework are described in detail in the following subsections.

2   The Case Study

The proposed adaptive multi-methodological evaluation process is applied to the target area of Catanzaro Sala (Catanzaro, Italy), selected as a case study (Fig. 1).

Fig. 1. Target area (Catanzaro Sala, Italy) 38.89603 latitude, 16.59926 longitude

The city of Catanzaro, the regional capital, is a medium-sized urban center in Southern Italy (89,065 inhabitants) which performs a significant connective and strategic function in the Calabria region. Catanzaro, strategically located at the center of the Calabrian logistic and interconnection system, between the Ionian and Tyrrhenian, is characterized as an urban area which, due to the plurality and level of functions provided, represents a functional attraction pole for the entire regional territory. It is also an important economic, commercial, university, health and cultural center, hosting important administrative functions and regional strategies. The city has undergone profound transformations in recent decades, even more marked following the economic situation that began in 2008, which has seen the progressive emptying of the administrative, managerial and commercial functions traditionally concentrated in the historic center and relocated or outside the city or in the peripheral areas. In this context, Catanzaro has been involved in the last decade by a series of urban development and regeneration programs, whose current urban development strategies see the city of Catanzaro as the "City of Hospitality, Knowledge and Innovation". The development options provide as general objectives the: (i) City specialization with a view to competitiveness; (ii) Concentration of investments on "target" areas for the requalification and regeneration of degraded urban spaces.

In particular, the area under study, in the immediate vicinity of the historic center of Catanzaro, as a whole is an area of considerable interest. There are numerous 'waiting spaces' [5] suspended landscapes, which have concluded their life cycle but which constitute a very strong regenerative potential. There is a coexistence of compromised ecosystems, of abandoned or decommissioned industrial buildings, of infrastructures (some of which are abandoned), of interstitial areas strongly hybridized with deposits, storage areas, car parks and areas for logistics.

The reuse of these underused or still unused spaces, as an alternative possibility to the state of abandonment, and the recycling of its waste areas can represent a new paradigm to escape from the crisis and reimagine the future of this urban area in a cyclical way here it is possible to trigger a regenerative strategy and a development process that takes into account its multiple domains and system of values of the stakeholders involved.

3   Methodological Framework.

In the context of the case study, the proposed methodological framework is aimed at supporting the development of development strategies according to the principles of the CCM [3] for the regeneration of urban residual space, whose planning and management strategies are particularly complex, and often conflicting, with also negative impacts.

According to the aforementioned perspective, the objective is to design a sustainable and circular development, recognizing the crucial role of an integrated multi-methodological approach capable of supporting the elaboration and selection of future scenarios, considering as a driver: (i) the observation and interpretation of the complex relationship of the target area with the surrounding urban context; (ii) the evaluation of the characteristics of identifying places and the needs of local communities.

The proposed methodological framework applies a site-specific adaptive approach, divided into three operational phases: knowledge process, development process and evaluation process (Table 1).

Table 1 Methodological framework

Step	Methodology
Knowledge	Hard analysis, soft analysis Questionnaires, interviews, storytelling SWOT analysis
Development	Strategic actions Scenario building
Evaluation	NAIADE (bottom-up approach) PROMETHEE (top-down approach) Sustainability indicators (SIs) Sustainable Development Goals (SDGs)

The following subsections provide a brief description of the different phases of the path, illustrating the objectives and methodologies used.

This paper focuses attention on the third phase of the methodological framework, relating to the application of the evaluation of alternative scenarios of urban regeneration, both in multi-group and multi-criteria terms.

3.1 Knowledge Process

The analysis of the decision-making context represents the first step of the methodological framework for the development of scenarios for regeneration of urban residual spaces in the context of metropolitan cities.

The study context is investigated through a multi-methodological approach that combines both hard and soft data and institutional analysis [13–15]. The collection of information and hard 'objective' data relevant to the decision-making context involved four areas of investigation: cultural, social, economic and urban, analyzed through a set of Sustainability Indicators (SIs) linked to the Sustainable Development Goals (SDGs) [4, 9]. At the same time, the knowledge deriving from institutional sources and the "hard dimensions" was necessarily combined with a direct investigation of the so-called "soft dimensions" [6, 7]. For soft data, we started with an Institutional Analysis [13–15], and a map of the most significant stakeholders [6, 7] useful for understanding the institutional context and the different interests that arise within complex decision-making arenas. Later, we moved on to soft data collection (online questionnaires, semi-structured interviews, storytelling) and their processing through tools such as Semantic Analysis, Social Network Analysis [16] and collaborative transformation mapping. The results of the analyzes and the definition of the decision-making context are finally structured in a SWOT matrix (Strengths, Weaknesses, Opportunities, Threats).

3.2 Development Process

In the next step of the development process, the outputs obtained from the processing of hard and soft data and the results of the institutional analysis provide a first summary of the information and a picture of the criticalities and potentials.

Through the application of the Scenario Building methodology, three possible scenarios are defined, explained in a thematic checklist (Table 2) of potential actions selected according to cultural, social, economic and urban criteria:

Scenario 1. *Cultural and creative hub of the communit*
Scenario 2. *Natural and commercial community center*
Scenario 3. *Virtual and virtuous community hub*

The construction of the scenarios starts from a vision of the future of the urban context considered, considering the potential and the development drivers of the area, oriented towards Culture-Led and Community-Led regeneration strategy [6].

The alternative regeneration scenarios are also consistent with the local development needs of the Decision Makers (DMs) and with the system of conflicting values and interests of the main stakeholders involved.

Table 2 Thematic Checklist for potential actions according circular city model

• Promote the aggregation people of all ages, social backgrounds and cultures
• Building community networks
• Encourage and accompany creative ideas and projects
• Promote multicultural integration and foster sociality and the building of meaningful relationships
• Promote culture and facilitate its accessibility to all
• Promote the sharing of knowledge and skills
• Promote and communicate to tell, be told and make reality known
• Promote new jobs for social and economic growth
• Enhance and reuse spaces and places, to restore the sense of belonging to the unused urban heritage

3.3 Evaluation Process

The complex and multidimensional nature of the decision-making problem related to development strategies for regeneration of urban residual spaces in urban contexts requires the application of multidimensional assessment operational tools [6, 7, 12] to support the evaluation of alternative scenarios.

The alternative scenarios defined are evaluated, in both multi-group and multi-criterion terms [6, 7, 12].

The structure of the decision problem required two flows of evaluation, which lead, respectively, to the elicitation of knowledge of hard and soft information, in order to respond to two main types of problems.

The first flow seeks to consider both the opportunities and threats that could arise from each of the three proposed alternatives and the needs and concerns of the people living in the area of influence of the urban study area. In this regard, the method Novel Approach to Imprecise Assessment and Decision Environments (NAIADE) [16, 17] has been selected as the most useful tool for gathering the preferences of citizens in a dynamic and fast way. The result shows the preferences of the participants with respect to the proposed alternatives.

The second flow, on the other hand, concerned the assessment of the performance of the alternative scenarios with respect to sustainability and the requirements of the CCM model and the SDGs. Furthermore, in this perspective, the preferences of the most influential stakeholders have been incorporated into the evaluation process. The multi-criteria method of Preference Ranking Organization METHod advances in operations research for Enrichment Evaluation (PROMETHEE) [18–24] was chosen, since it allows to analyze different scenarios that show the visions of the main stakeholders involved and, consequently, carry out a total ranking of alternatives based on the aggregation of all stakeholder opinions.

3.3.1 Sustainability Indicators as Tools for Choice of the Favorable Scenario

A core set of place-based indicators, relating to the Sustainable Development Goals (SDGs) [4, 9] were selected and used as operational tools to assess the performance of alternative scenarios according to the objectives of SDGs n.9: "Industry, innovation and infrastructures", SDGs n.11:" Sustainable cities and communities "and SDGs n.12:" Responsible consumption and production "[4, 9].

The selected indicators have been classified into four relevant domains for the regeneration of cities: Urban Metabolism, Landscape Quality, Society and Culture, Economic Growth and Development, further explained in nineteen SIs [4, 9] (see Table 3).

Table 3 Sustainability Indicators (SIs) list.

Domain	Indicators	Description	Goal	Unit	SDGs
Urban metabolism	Urban density	Built area on total land area	Max	%	11
	Functional mixite	N. alternative functional destinations	Max	1–5	11
	Distance from local attractors	Distance from primary public services	Min	minutes	11
	Accessibility and public transport	Pedestrian accessibility to public transport	Min	minutes	11
	Parking areas availability	Parking area extension	Max	sqm	11

(continued)

Table 3 (*continued*)

Domain	Indicators	Description	Goal	Unit	SDGs
Landscape quality	Renewable energies	Energy production from renewable sources	Max	%	12
	Environmental sustainability	Permeable surface area	Max	sqm	12
	Green areas	Presence of public green areas	Max	sqm	12
	Rehabilitation of polluted areas	Areas to be reclaimed	Max	%	12
Society and culture	Creation of Social housing	Construction of residences for social use	Max	n	11
	Gentrification	Possible phenomena of gentrification	Min	1–5	11
	Creation of attractive public functions	Attractiveness of public functions	Max	1–5	11
	Cultural and recreational activities	Recreational attraction	Max	1–5	11
	Creation of new jobs	Creation of new jobs	Max	n.	9
	Real estate market	Real estate market prices	Max	%	9
Economic growth and development	Potential of economic development	Interconnections/ synergies local economic activities	Max	1–5	9
	Investment costs	Investment costs	Min	€/sqm	9
	Management costs	Maintenance costs	Min	1–5	9
	Financial appeal for private investors	Attractiveness of private investors	Max	1–5	9

3.3.2 NAIADE: Bottom-Up Approach to Scenario Assessment

The NAIADE (Novel Approach to Imprecise Assessment and Decision Environments) multicriteria method [19] is a very flexible method used in complex decision problems in which uncertainty or blurry indeterminacy is recognized.

The NAIADE method makes it possible to evaluate social compromise solutions according to the opinion of the stakeholders, identified as relevant for the decision-making context in question. The application of the NAIADE method was very useful as, through the involvement of all stakeholders in a collaborative process, it allowed a greater awareness of the problems of the urban study context and stimulated the entire community to promote local cooperation for cultural regeneration and indicate possible development prospects.

NAIADE, applied to the case study, provides information on: (i) indicators of the distance between the interests of the various stakeholder groups, as an indication of the possibility of forming coalitions, or convergence of interests; (ii) an impact assessment for the different stakeholder groups and a ranking of alternatives or social compromise solutions for each coalition (Fig. 2).

The result of the application of NAIADE is a dendrogram of the coalition formation process, which highlights a preference towards the two Alternatives: 1. *Cultural and creative community hub of the community* and 2. *Natural and commercial community center*, which according to the different groups of stakeholder have the potential to trigger greater economic development, both in the urban area but also in the surrounding fabric.

While, among the actors G1 (Promoters) and G3 (Operators) who have a high level of credibility (0.8525), Alternative 3. *Virtual and virtuous community hub* is the least appreciated alternative from the point of view of the analysis of social conflict (Fig. 2).

EQUITY matrix	Case study Catanzaro Sala		
GROUPS ALTERNATIVES	Cultural and creative hub of the community	Natural and commercial community centre	Virtual and virtuous community hub
G1 Promoters	Perfect	Perfect	Very Good
G2 Experts	Perfect	Very Good	Good
G3 Operators	Perfect	Perfect	Very Good
G4 Users	Good	Perfect	Moderate

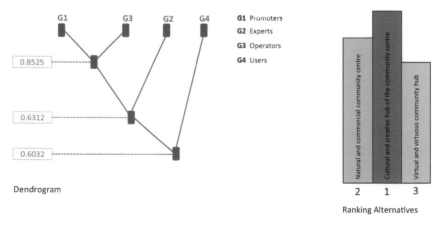

Fig. 2. Equity analysis results obtained through the NAIADE method

3.3.3 PROMETHEE: Top-Down Approach to Scenario Assessment

With the evaluation of the scenarios with the NAIADE method, emerged strongly conflicting values and interests among the stakeholders. Therefore, it was necessary to apply the multi-criteria method of outclassing of PROMETHEE (Advanced Preference Classification Organization Method in Operations Research for Enrichment Assessment) implemented through the open-source software Visual PROMETHEE [19, 24].

PROMETHEE establishes the thresholds of preference, indifference and incomparability between the alternatives with respect to each criterion [18, 25–29] and through a comparison procedure in pairs identifies the alternative that outclasses the others, based on the degree of preference of the experts from different sectors.

The assessment was then carried out by establishing the preference functions and the threshold values based on the position of each expert with respect to the four domains and related SIs (Table 3) used as a proxy to measure the performance of each alternative. To identify the weights to be attributed to criteria, in the evaluation was used the procedure of the deck of cards proposed by J. Simos (1990), revisited by Roy and Figueira, 1998 [30–32]. Influential experts from different sectors (urban planning, economic evaluation, architecture and environmental engineering) and most stakeholders agree in attributing greater importance to the aspects of Functional mixitè, with respect to the criteria relating to Gentrification and Environmental sustainability. Defined the priorities of the criteria, the final ranking of the alternative options was finally obtained by implementing the open-source software Visual PROMETHEE 1.4 [30].

The final results show that Alternatives: 1. *Cultural and creative hub of the communit* outclasses the others in all four domains, achieving positive scores in both the net outranking flows (Fig. 3) and the aggregate flow view (Fig. 4).

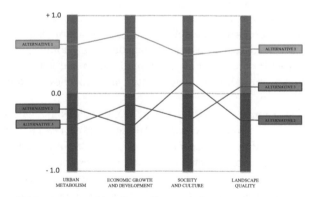

Fig. 3. PROMETHEE. Net outranking flows diagram related to the scoring of three alternatives with regard to the four scenarios.

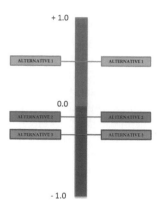

Fig. 4. PROMETHEE. Net outranking complete flow.

4   Conclusion

According to the "Report on the state of cities of the United Nations [1–3, 9–11], many cities in the world have begun to propose a renewal of the local economy by focusing on the construction of processes of cultural regeneration. Culture, a vital ingredient of the urban environment, has become a success factor for the localization of creative processes, to which the attention of local administrations is directed, motivated by the need to reposition cities in the global market and at the same time create a favorable environment for new forms of an economy based on technology, creativity, human capital and the ability to innovate [4, 8, 9].

They are complex decision-making processes, characterized by the presence of numerous variables and by a high level of uncertainty that requires integrated evaluation approaches capable of supporting the definition of shared and transparent choices.

In this context, the document proposes an integrated methodology aimed at evaluating regenerative processes for a circular city, combining hard and soft knowledge, implementing and combining multidimensional assessment approaches, Stakeholder Analysis and selection of SIs [4, 6, 7, 12].

The methodology represents a valid reference for decisions in the public and private sphere, able to effectively support the Public Administrations (PA) in defining an order of priority for actions and projects according to their multidimensional sustainability. The methodological framework defined is relevant for marginal contexts, as in the case study where the scarcity of public resources and the fragility of the economic and social systems require a correct and targeted allocation of resources. At the same time, the evaluation process, in addition to being a tool for assessing the impacts of scenarios, becomes a negotiation tool to identify values, priorities and build shared scenarios [18, 26, 33].

The places of urban waste are the privileged places from which to start to trigger regeneration processes and where it is possible to manage the circular transition, stimulating territorial productivity, economic development, and social cohesion [34–36].

In this perspective, future research directions are aimed at supporting Municipalities in regenerative programs and in circular transformation operations, building processes of dialogue and interaction between the various stakeholders, to govern the complex reconversion from urban waste spaces to places regenerative, in a perspective of recomposition and integration with the city and the surrounding territory [37–39].

References

1. Nazioni Unite: Transforming our world: The 2030 agenda for sustainable development, Resolution adopted by the General Assembly on 25 September 2015. Nazioni Unite, New York (2015). https://doi.org/10.1007/s13398-014-0173-7.2
2. Nazioni Unite: Draft outcome document of the United Nations Conference on Housing and Sustainable Urban Development (Habitat III). Nazioni Unite, New York (2016). https://doi.org/10.1257/jep.27.4.187
3. Eurocities: Full Circle, Cities and the Circular Economy. Eurocities, Brussels, Belgium (2017)
4. Ellen MacArthur Foundation: Cities in the Circular Economy: An Initial Exploration. Ellen MacArthur Foundation, Sail Loft (2017)
5. Berger, A.: Drosscape, Wasting Land in Urban AMERICA. Princeton, New York (2006)
6. Della Spina, L.: Multidimensional assessment for culture-led and community-driven urban regeneration as driver for trigger economic vitality in urban historic centers. Sustainability **11**, 7237 (2019). https://doi.org/10.3390/su11247237
7. Della Spina, L., Giorno, C., Galati Casmiro, R.: Bottom-up processes for culture-led urban regeneration scenarios. In: Misra, S., et al. (eds.) ICCSA 2019. LNCS, vol. 11622, pp. 93–107. Springer, Cham (2019). https://doi.org/10.1007/978-3-030-24305-0_8
8. Oldenburg, R.: The Great Good Place. Paragon House, New York (1989)
9. TWI2050: The World in 2050. The Digital Revolution and Sustainable Development: Opportunities and Challenges; Report Prepared by the World in 2050 Initiative; International Institute for Applied Systems Analysis (IIASA), Laxenburg, Austria (2019). www.twi2050.org. Accessed 10 Dec 2021
10. European Commission: Urban Agenda for the EU Multi-Level Governance in Action. European Union, Brussels, Belgium (2019)
11. Annoni, P. Dijkstra, L.: The EU Regional Competitiveness Index 2019. European Union, Luxembourg (2019)
12. Della Spina, L., Giorno, C., Galati Casmiro, R.: An integrated decision support system to define the best scenario for the adaptive sustainable re-use of cultural heritage in southern Italy. In: Bevilacqua, C., Calabrò, F., Della Spina, L. (eds.) NMP 2020. SIST, vol. 177, pp. 251–267. Springer, Cham (2020). https://doi.org/10.1007/978-3-030-52869-0_22
13. Checkland, P.B., Poulter, J.: Learning from Action. Wiley, Chichester (2006)
14. Checkland, P.B.: System Thinking, System Practice. Wiley, Chichester (1981)
15. Checkland, P.B.: Soft systems methodology. In: Rosenhead, J., Mingers, J. (eds.) Rational Analysis for a Problematic World Revisited. Wiley, Chichester (2001)
16. Wasserman, S., Faust, K.: Social Network Analysis: Methods and Applications. Cambridge University Press, Cambridge/New York (1994)
17. Della Spina, L.: A multi-level integrated approach to designing complex urban scenarios in support of strategic planning and urban regeneration. In: Calabrò, F., Della Spina, L., Bevilacqua, C. (eds.) ISHT 2018. SIST, vol. 100, pp. 226–237. Springer, Cham (2019). https://doi.org/10.1007/978-3-319-92099-3_27

18. Della Spina, L.: The integrated evaluation as a driving tool for cultural-heritage enhancement strategies. In: (a cura di): Bisello, A., Vettorato, D., Laconte, P., Costa, S. (eds.) Smart and Sustainable Planning for Cities and Regions. Results of SSPCR 2017. Green Energy and Technology, pp. 1–11. Springer, Heidelberg (2018). https://doi.org/10.1007/978-3-319-75774-2_40, ISSN 1865–3537
19. Munda, G.: Social multi-criteria evaluation (SMCE): methodological foundations and operational consequences. Eur. J. Oper. Res. **158**, 662–677 (2004). Sustainability 2019, 11, 7237 20 of 20 (2019)
20. Della Spina, L.: Revitalization of inner and marginal areas: a multi-criteria decision aid approach for shared development strategies. Valori e Valutazioni **25**, 37–44 (2020)
21. Munda, G.: Multicriteria Evaluation in a Fuzzy Environment: Theory and Applications in Ecological Economics. Physica-Verlag, Heidelberg (1995)
22. Ackermann, F., Eden, C.: Strategic management of stakeholders: theory and practice. Long Range Plan. **44**, 179–196 (2011)
23. Della Spina, L.: Cultural heritage: a hybrid framework for ranking adaptive reuse strategies. Buildings **11**, 132 (2021). https://doi.org/10.3390/buildings11030132
24. Mareschal, B.: Visual PROMETHEE 1.4 Manual. 2013. www.Promethee-Gaia.Net. Accessed 20 11 2021
25. Della Spina, L.: Cultural heritage: a hybrid framework for ranking adaptive reuse strategies. Buildings **11**, 132 (2021). https://doi.org/10.3390/buildings11030132
26. Carbonara, S., Stefano, D., Torre, C.M.: The economic effect of sale of Italian public property: a relevant question of real estate appraisal. In: Murgante, B., et al. (eds.) ICCSA 2014. LNCS, vol. 8581, pp. 459–470. Springer, Cham (2014). https://doi.org/10.1007/978-3-319-09150-1_33
27. Figueira, J. Greco, S. Ehrgott, M.: Multiple Criteria Decision Analysis: State of the Art Surveys, vol. 78. Springer, Berlin (2005)
28. Ishizaka, A., Nemery, P.: Multi-Criteria Decision Analysis: Methods and Software. Wiley, Hoboken (2013)
29. Ward, S., Chapman, C.: Stakeholders and uncertainty management in projects. Constr. Manag. Econ. **26**, 563–577 (2008)
30. Mareschal, B. De Smet, Y.: Visual PROMETHEE: sviluppi dei metodi di aiuto decisionale Mareschal, Bertrand, and Yves De Smet. Visual PROMETHEE: developments of the PROMETHEE & GAIA multicriteria decision aid methods. In: 2009 IEEE International Conference on Industrial Engineering and Engineering Management. IEEE (2009)
31. Roy, B., Figueira, J.: Determination des poids des criteres dans les methodes du type ELECTRE avec la technique de Simos revisee. Universite Paris – Dauphine, Document du LAMSADE 109 (1998)
32. Figueira, J., Roy, B.: Determining the weights of criteria in the ELECTRE type methods with a revised Simos' procedure. Eur. J. Oper. Res. **139**, 317–326 (2002)
33. Carbonara, S., Faustoferri, M., Stefano, D.: Real estate values and urban quality: a multiple linear regression model for defining an urban quality index. Sustainability **13**(24), 13635 (2021). https://doi.org/10.3390/su132413635
34. Carbonara, S., Stefano, D., Faustoferri, M.: Public real estate's in Italy: from decommissioning to valorization. legislative evolution and future perspectives (Il patrimonio immobiliare pubblico in italia: dalla dismissione alla valorizzazione. evoluzione della normativa e prospettive future). LaborEst **21**, 39–46 (2020). https://doi.org/10.19254/LaborEst.21.06
35. Nesticò, A., Moffa, R.: Economic analysis and operational research tools for estimating productivity levels in off-site construction (Analisi economiche e strumenti di Ricerca Operativa per la stima dei livelli di produttività nell'edilizia off-site). Valori e Valutazioni n. 20, pp. 107–128. DEI Tipografia del Genio Civile, Roma (2018). ISSN 2036–2404

36. Nesticò, A., Maselli, G.: Declining discount rate estimate in the long-term economic evaluation of environmental projects. J. Environ. Account. Manag. **8**(1), 93–110 (2020). https://doi.org/10.5890/JEAM.2020.03.007
37. Tajani, F., Liddo, F.D., Guarini, M. R., Ranieri, R., Anelli, D.: An assessment methodology for the evaluation of the impacts of the COVID-19 pandemic on the Italian Housing market demand. Buildings **11**(12), 592.C41 (2021)
38. Manganelli, B., Tajani, F.: Optimised management for the development of extraordinary public properties. J. Prop. Inv. Fin. **32**(2), 187–201 (2014)
39. Morano, P., Tajani, F., Guarini, M.R., Sica, F.: A systematic review of the existing literature for the evaluation of sustainable urban projects. Sustainability **13**(9), 4782 (2021). https://doi.org/10.3390/su13094782

Resilience of Complex Urban Systems: A Multicriteria Methodology for the Construction of an Assessment Index

Debora Anelli[1]([✉]) and Rossana Ranieri[2]

[1] Department of Civil, Environmental, Land, Building Engineering and Chemistry (DICATECh), Polytechnic University of Bari, Via Orabona 4, 70125 Bari, Italy
debora.anelli@poliba.it
[2] Department of Architecture and Design, "Sapienza" University of Rome, 00196 Rome, Italy
rossana.ranieri@uniroma1.it

Abstract. The environmental-climate changes and the Covid-19 emergency have highlighted the weakness of urban systems by raising the attention on adequate tools able to support the improvement of multi-events resilience. The social, natural and economic features that characterize the urban environment, make it a complex system that need to be comprehensively assessed for taking into account all the relevant factors that contributes on their resilience. Aim of the work is to define a multicriteria-based methodology able to create a geo-referenced Urban Resilience Index (I_{UR}) that represents the capacity of the territory to face socio-economic diseases and natural disaster. The proposed protocol consists of a *step by step* guide for creating the I_{UR} with the adoption of the Analytic Hierarchy Process technique for structuring and aggregating the system of indicators that represent the relevant economic, environmental and social contributions to the resilience of a certain territorial scale, and the geographic information system for the visualization of the different spatial distribution of the resilience. The proposed methodology can be used as a decision support tool for public-private partnership's urban intervention aimed at achieving the Sustainable Development Goals of the Agenda 2030 and the European Green Deal targets. Its flexibility makes it implementable for several sustainable urban planning decision at different scale and it can be adopted for an *ex ante* evaluation of the urban parameters from which derive the balance sheets and the pressures on the environment.

Keywords: Resilient cities · Multicriteria assessment · Real estate market · Climate change · Decision support model · Index system

1 Introduction

The 21st century poses new challenges to the city, due to the pressure generated by the increase in the number of people worldwide living in built environments and particularly within cities [1, 2]. The climate change, the pandemics disease and the economic crisis have been differently faced by the urban systems according to their structural complexity, the number of people and the level of uncertainty associated with the related exogenous events [3, 4]. City resilience describes the ability of complex socio-ecological urban contexts to adapt, change and transform in response to stresses and strains. Resilience is a term that emerged from the field of ecology in the 1970s to describe specifically the capacity of a system to maintain or recover functionality in the event of disruption or disturbance. Therefore, it is conceived to be applied to a system of different and inter-dependent elements, of which the urban context is the most exemplary representation [5, 6]. The notion of a resilient city becomes conceptually relevant when sudden shocks threaten widespread disruption, or the existent urban structure collapsed by highlighting the incapacity to deal with the changes [7, 8]. It moves away from traditional disaster risk management, which is founded on specific hazards. Resilience accepts the possibility that a wide range of disturbing events may occur and are not necessarily predictable [9, 10]. Various approaches have been taken for framing or assessing resilience. They focus either on specific urban assets or entire systems and by varying degrees, consid-ering infrastructure, natural environment, land-use management and human behavior. System-based approaches are more closely with the concept of resilience and the often long-standing notion of urban areas as "systems within system": the social layout is determined by the human behavior, which is also influenced by natural elements dis-tribution between infrastructure, services and assets in the urban environment. Based on the literature review, it is possible to identify eight critical functions that affect the resilience level: delivery of basic needs, safeguard human life, preserve and enhance assets, facilitate human relationships and identity, promote knowledge, defend the jus-tice and equity, support livelihoods, stimulate economic prosperity [11]. For measuring these factors, certain indicators appear useful: the quantification of them led the decision makers to provide a means to score, rank, and monitor progress across different strate-gies, communities, or even nations. Researchers have developed quantitative tools, prin-cipally indicators and indexes approach, to measure resilience through a wide-range of disciplinary perspectives (e.g., climate-change impacts, political economy, ecology, sus-tainability, social, environmental needs etc.), and across different spatial scales [12–14]. Fleischhauer [15] introduced several indicators that can be used to assess the resilience of spatial planning processes. Frazier and Thompson [16] conducted a study to develop a set of indicators for assessing and quantifying the spatial and temporal aspects of the resilience at the community scale in Sarasota County, Florida. They argued that indica-tors should be weighted differently in order to consider the effective contribution. Joerin et al. [17] developed a "climate disaster resilience index" for Chennai, in India, that has five dimensions: physical, social, economic, natural, and institutional. Each of these dimensions is later broken down into five parameters which in turn are composed of five variables. Although this has resulted in a reasonably comprehensive list of resilience related criteria, its focus is context-specific and there might be other criteria which are left out. Including market dynamics in the study of urban resilience can help broaden the analysis of urbanization problems. Antoniucci and Marella [18] studied the Italian housing market by finding that less dense cities have more resilient capacity. Sharifi et al. [19] highlighted the argument that advanced construction techniques and enforcement of regulations can improve urban resilience in higher density areas and thus boost eco-nomic benefits. This shows that the real estate market is a sensible marker that contains a multitude of factors that can influence urban resilience. However, there is not a consensus about what resilience means for urban areas and how introducing resilience in the urban planning process. Overall, there are still few researches that cover the multiple aspects of urban resilience and that recognize relevance to the real estate market conditions into a comprehensive assessment framework [20–22].

2 Aim

In the context outlined, this research intends to define a methodology that guides the processes of sustainable planning and urban transformation step by step in assessing the level of resilience. In particular, the proposed methodology is structured into a protocol that allows the construction of a multi-criteria index – called Urban Resilience Index (I_{UR}) – based on a system of indicators and three macro-categories of criteria, which pertain to the relevant economic, environmental and social features that affect the urban resilience level. Through the application of the Analytic Hierarchy Process (AHP) multi-criteria technique, the indicators are differently weighted and then aggregated into a spatial synthetic index, that immediately represents the level and the spatial distribution of resilience related to the urban context considered. It is possible to identify 9 phases for the construction of the I_{UR} as follows: 1) urban resilience definition; 2) territorial scale identification; 3) identification of the most relevant features that characterize the concept of resilience in the area considered and structuring of the AHP; 4) data collection; 5) normalization and correlation analysis; 6) intensity range's detection and related local weights determination; 7) indicators and criteria local weights determination; 8) aggregation of the indicators into the I_{UR} and assessment of the resilience level of the territorial scale considered; 9) sensitivity analysis and georeferencing of I_{UR} value obtained.

The use of the proposed methodology is aimed at those public and private subjects involved in sustainable urban planning decisions. The public administration can use it for identify and then monitoring the critical issues highlighted by the index. In this way, the awareness of making specific interventions necessary to improve resilience can support the achievement of the Agenda 2030 Sustainable Development Goals and the European Green Deal targets. The clear AHP's structure make the proposed protocol immediately applicable, without requiring high or specific software knowledge. It is also flexible and utilizable for different territorial scale and urban development purposes by allowing an immediate identification of the most critical resilience level for which are necessary specific urgent interventions.

The private subject can benefit from the index in the negotiation phases of the urban parameters with the public administration, as the evaluation through the I_{UR} allows to take into account the conditions of the local real estate market, also providing a measure of the risk of a potential urban transformation operation. For example, in an area with a high I_{UR}, the private subject involved in a transformation intervention can know the conditions of the local real estate market, included in the economic component of the index, requesting an appropriate premium for the risk borne by him. Furthermore, the careful calibration of the morphological and structural constraints of the intervention will be able to consider the current environmental and social conditions represented within the resulting index value.

The following structure of the work is divided into 2 sections: the first one (Sect. 3) explains and describes all the steps of the proposed methodology; the second one, instead, provides for the conclusion and future insights of the work (Sect. 4).

3   Methodology

The proposed methodology is structured into a protocol of 9 phases where, to each of them, corresponds specific operations and assessment technique to carry out for the construction of the resilience index. Figure 1 represents a summary of the entire phases that constitutes the procedure.

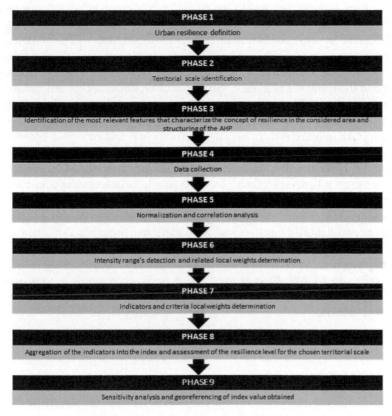

Fig. 1. Phases of the proposed methodology

3.1 Phase 1. Urban Resilience Definition

According to the existing literature on the characteristics that define the urban resilience's concept, the most relevant factors are not only related to the ecological and environmental aspects, but also the social and economic sphere play a key role. In fact, the capacity of a city, or an area pertaining to it, to resist unpredictable events of any typology defines the degree of resilience that all the elements that constitute the city are endowed. Social and economic dimensions are measured through a collection of socio-demographic data, such as household income, employment, health coverage, age or level of education. For example, dwellers who have health insurance, higher education and live in owned properties should be more resilient than those who don't. The physical or infrastructural dimension refers to the resilience of homes and services, such as electricity, water and communication systems, but also to the role of urban green infrastructures and more generally to the morphology with which the city is composed. The type of building, the age of them and the construction techniques are all relevant factors that make it possible to understand the degree of vulnerability of the assets, nevertheless, the geo-lithological fragility of the land on which they are built and the presence of green elements, jointly contribute to the resilience of a place. Given the close linkages between anthropogenic activities and the way of living the city, with the spatial distribution and the extension of the different possible uses of the land (e.g. urban, agricultural and natural), it becomes essential to take into account the social, environmental and economic components in assessing resilience. The latter component is often associated with the economic capacity or productive dynamism of a place, neglecting the key role played by the real estate market. In theory, urban resilience can increase people's confidence in consumption and investment in the real estate market. If local governments have performed well in terms of adaptation and rapid recovery after exogenous events occurred, this means that the city is strong and stable for investors and consumers and, therefore, that urban resilience is high. Furthermore, stabilizing the investment expectations of local investors attracts more foreign entrepreneurs and consumers. Finally, a dynamic real estate market means a better ability to recover the economy, and a stable economy is a cornerstone of a resilient city [23]. Therefore, it is possible to define the urban resilience as the weighted sum of three criteria:

$$\text{URBAN RESILIENCE} = \text{ECONOMIC} + \text{ENVIRONMENTAL} + \text{SOCIAL CRITERIA}$$

$$\text{URBAN RESILIENCE INDEX (IUR)} = w_1 * \text{ECONOMIC} + w_2 * \text{ENVIRONMENTAL} + w_3 * \text{SOCIAL CRITERIA}$$

where w_n are the different weights associated to each criterion that, are further composed by several indicators, chosen according to the features of the territorial context to be assessed.

3.2   Phase 2. Territorial Scale Identification

In order to take into account the relevant aspects that determine the urban resilience, the territorial scale must be chosen according to the final utilization of the index. In particular, four type of territorial scale can be individuated by varying the aims of the analysis:

1. *Neighborhood scale:* can be chosen if is intended to know the resilience level provided by the specific features of a restricted context, maybe as an *ex ante* evaluation before the regeneration and valorization of the area though a urban transformation intervention;
2. *Urban area scale:* it is wider than the neighborhood dimension and could be relevant if bigger infrastructure or services have to be realized and is necessary to understand the fragility of the urban perimeter involved for determining the sustainable pressures;
3. *City scale:* for national or regional ranking and scoring within monitoring and awareness processes could be useful to consider the entire urban systems present, in order to provide a comprehensive framework that can help the decision-makers to identify the cities less resilient. In this way the political strategy could support the improvement of the built context by promoting national urban planning and strategies;
4. *Extra-urban scale* (e.g. regional, national etc.): it appears similar to the city scale but consider the territorial administrations composed by several municipalities as unit of analysis. It can be useful for a government for monitoring the achievement of the different territorial administration in terms of resilience level.

Generally, in the sustainable urban planning decisions the neighborhood and the urban area scale are preferred by the subjects involved for the possibility to immediately identify on the urban territory the most critical areas, and therefore the ones that need more urgent interventions to improve the resilience. On the contrary, the city and extra-urban scale are useful for rapidly ranking and scoring the different territorial administration of a Country, for example, to increase political and governmental awareness of the strategic actions necessary to equally support national, regional, etc., average resilience.

3.3   Phase 3. Identification of the Most Relevant Features that Characterize the Concept of Resilience in the Considered Area and Structuring of the AHP

After having chosen the territorial scale of assessment, the identification process of the most significant economic, environmental and social features that characterize the resilience in the area considered, is the most delicate one. The existent literature on the topic, the expert and technician consultation and the scientific report on the matter could support the development of the indicators system on which is based the evaluation. The indicators must be identified due to their readability, availability, scientific sound and most of all coherence with the process and the territorial scale's peculiarities. A list of possible (not limited to) indicators related to each of the three components of the index is proposed in Table 1.

Table 1. List of possible economic, environmental and social indicators that constitute the based system for the construction of the urban resilience index.

Component	Indicator
Economic	• Trend of the selling and/or rental prices for residential and/or non-residential properties
	• Volatility of the selling and/or rental prices for residential and/or non-residential properties
	• Average sales and/or rental months • Maintenance status of the properties
	• ...
Environmental	• Land take rate
	• Average summer temperature
	• Number of minors under the age of 11
	• Blue infrastructure presence
	• Green infrastructure presence
	• Seismic risk level
	• Flood risk level
	• Urban density
	• ...
Social	• Number of old and/or young people
	• Number of young people who do not study/work
	• Number of families with more than 3 child
	• Male/Female employment rate
	• Number of metro lines and bus stops
	• ...

As can be seen in Table 1, the economic component refers also to the real estate market conditions and dynamics in order to provide a comprehensive resilience assessment. Once chosen, they must be structured within the hierarchical levels that characterize the AHP, jointly with the criteria identified, represented by the three components of the resilience concept, and the intensity ranges that can be determined according to the variations in the values of the data collected for each indicator. An example of the AHP structure with the three hierarchical levels in Fig. 2 is represented.

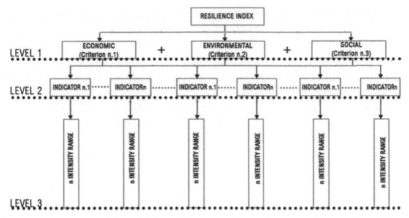

Fig. 2. General structure of the criteria, indicators and intensity ranges into the three hierarchical levels of the AHP for the index.

3.4   Phase 4. Data Collection

This phase of the proposed protocol varies according to the territorial scale considered for the construction of the index. In fact, in the case of the neighborhood or urban scale, the data collection concerning the set of the indicators will be enlarged to the entire number of neighborhoods or urban areas that are included within the municipal context considered. Instead, for the other two the data collection will regard the overall administrative subdivision within the territory to be evaluated. This passage is important to create a sample of data that constitute the system from which the index will be generated.

3.5   Phase 5. Normalization and Correlation Analysis

In order to compare and aggregate the indicators for the resilience index construction, the application of a normalization technique is helpful. The result of the resilience index can change by varying the normalization method applied, therefore it must be chosen by carefully considering if it is more or less adequate to the specific needs of the analysis (e.g. some give more relevance to the extreme values of the collected data set, others extinguish the bias deriving from anomalous values, etc.). Generally, the most used are the z-score, the min-max and the distance from a reference value [24].

The correlation analysis is performed among all the indicators that belong to the same hierarchical level, in order to ensure the lack of connection that can reduce the robustness and readability of the final results. The highly correlated indicators are removed. If for some criterion there are minor than 3 indicators, new ones have to be collected. If this condition is not respected, the AHP is not applicable.

3.6   Phase 6. Intensity range's Detection and Related Local Weights Determination

For intensity range means n buffer of variations of the indicator's values detected in the entire municipalities of administrative subdivisions, according to the territorial scale considered for the analysis. They can be different for each indicator normalized because depend on the lesser or greater sensitivity of the indicator to varying in the collected data sample. Each range has a local weight, provided by a $n \times n$ pairwise comparison matrix defined with the consultation with a panel of expert and technician of the matters addressed, ranging from 1 to 0. The judgements are provided by adopting the nine-point intensity of importance scale, modified from Schoenherr [25]. Moreover, in order to prove their robustness and coherence, the judgments of the experts must be minor of the 0.1 "Consistency Ratio" value provided by the formula below [26]:

$$\text{Consistency Ratio} = [(\lambda_{max} - n)/(n - 1)]/\text{random consistency ratio} \qquad (1)$$

where λ_{max} is the largest eigenvalue, n is the order of the matrix and the random consistency ratio is a tabulated value that depends from it.

3.7   Phase 7. Indicators and Criteria Local Weights Determination

The weights of the indicators are determined through the construction of a $m \times m$ comparison matrix, where m is the number of the final indicators that pertain to the *k-th* criterion. The same operation is conducted for the determination of the weights of the three criteria with respect to the final index, therefore a matrix of order 3 must be constructed. In both cases the panel of expert and technicians is relevant for the assessment of the weights, provided with the same nine-point intensity of importance scale, modified from Schoenherr, and used for the intensity ranges Also in this case the coherence of the judgments is verified trough the Consistency Ratio.

3.8   Phase 8. Aggregation of the Indicators into the Index and Assessment of the Resilience Level of the Considered Territorial Scale

Considering with $w_{i,j}$ the local weight of the *i-th* intensity range related to the *j-th* indicator, with $w_{j,k}$ the local weight of the *j-th* indicator that refers to the *k-th* criterion and with w_k the ones determined by the pairwise comparison of the three criteria, the aggregation of the resilience index (I_{UR}) is provided by the following formula:

$$I_{UR} = (w_{i,1} * w_{1,1} * w_1) + (w_{i,2} * w_{1,2} * w_2) + \ldots w_{i,j} * w_{j,k} * w_k \qquad (2)$$

The resilience index refers to the economic, environmental and social conditions of the territorial scale adopted for the assessment and is based on the set of indicators that refers to the features of an urban system, therefore can be considered a spatial index.

3.9   Phase 9. Sensitivity Analysis and Georeferencing of IUR Value Obtained

To verify the robustness of the resilience index obtained, the modification of the hierarchical group of local weights of the parameters considered in the AHP structure, such as for example the indicators of the economic criterion or the one related to the intensity range of the *j-th* indicator, could led to improve the results and make sure that the index take into account the real contribution of each parameter [27, 28]. For quick evidence of the value of index obtained for each of the territorial scale considered, geographic information system accessible online such as "My Maps" provided by Google, can help the final users to visualize the spatial distribution of the index into a geo-referenced map of values.

4 Conclusions

The exogenous and unexpected events recently occurred in the last decade, such as the economic crisis, the climate change and the covid-19 spread, have highlighted the fragility and the low resilience profile of the urban systems. The economic, environmental and social structures that underlay the complexity of the cities define their resilience level and, therefore, their capacity to adapt to several stresses and strains. The system-based approaches appear to be adequate and coherence with the urban resilience concept in the assessment of complex urban system's resilience, affected by several and interrelated factors that must be considered for a comprehensive evaluation.

In the context outlined, the present work is aimed at the definition of a methodology able to construct a multicriteria and spatial Urban Resilience Index (I_{UR}) that represents the vulnerability of a certain territorial scale, in terms of multi-hazards/events. The multi-criteria system on which is based the construction of the index allows to take into account the most relevant factors of the economic, environmental and social aspects of a territory that contribute to affect the resilience level. Through the AHP implementation the index takes into account all the effective contribution made by each criteria, indicators and intensity range that constitute the proposed system of evaluation from which it derives, in terms of local weights. In particular, in the economic criteria is recognized the key role of the real estate market dynamics in the determination of the resilience. In this way, the protocol that provides for the I_{UR} construction can supports the private and public subjects in: i) the sustainable urban planning decisions, ii) the monitoring process of the Sustainable Development Goals of Agenda 2030 achievement, iii) the scoring and ranking procedure, iv) the most resilient morphological composition of urban intervention, v) the determination of the balance sheets and the related risks that can occur in the territory considered, vi) raising awareness on public debate and policy-making. Future developments will regard the application of the proposed methodology to a real case study, by verifying the achievable results with the implementation of other evaluation techniques [29–31].

References

1. Dixon, J.A., Fallon, L.A.: The concept of sustainability: origins, extensions, and usefulness for policy. Soc. Nat. Resour. **2**(1), 73–84 (1989)
2. Spindler, E.A.: The history of sustainability the origins and effects of a popular concept. In: Sustainability in Tourism, pp. 9–31. Springer Gabler, Wiesbaden (2013). https://doi.org/10.1007/978-3-8349-7043-5_1
3. Wade, N.: Edward Goldsmith: blueprint for a de-industrialized society. Science **191**(4224), 270–272 (1976)
4. Purvis, B., Mao, Y., Robinson, D.: Three pillars of sustainability: in search of conceptual origins. Sustain. Sci. **14**(3), 681–695 (2018)

5. Calabrò, F., Iannone, L., Pellicanò, R.: The historical and environmental heritage for the attractiveness of cities. The case of the umbertine forts of Pentimele in Reggio Calabria, Italy. In: Bevilacqua, C., Calabrò, F., Spina, L.D. (eds.) New Metropolitan Perspectives: Knowledge Dynamics and Innovation-driven Policies Towards Urban and Regional Transition Volume 2, pp. 1990–2004. Springer, Cham (2021). https://doi.org/10.1007/978-3-030-48279-4_188

6. Spampinato, G., Malerba, A., Calabrò, F., Bernardo, C., Musarella, C.: Cork oak forest spatial valuation toward post carbon city by CO_2 sequestration. In: Bevilacqua, C., Calabrò, F., Della Spina, L. (eds.) NMP 2020. SIST, vol. 178, pp. 1321–1331. Springer, Cham (2021). https://doi.org/10.1007/978-3-030-48279-4_123

7. Massimo, D.E., Del Giudice, V., De Paola, P., Forte, F., Musolino, M., Malerba, A.: Geographically weighted regression for the post carbon city and real estate market analysis: a case study. In: Calabrò, F., Della Spina, L., Bevilacqua, C. (eds.) ISHT 2018. SIST, vol. 100, pp. 142–149. Springer, Cham (2019). https://doi.org/10.1007/978-3-319-92099-3_17

8. Del Giudice, V., Massimo, D.E., De Paola, P., Del Giudice, F.P., Musolino, M.: Green buildings for post carbon city: determining market premium using spline smoothing semiparametric method. In: Bevilacqua, C., Calabrò, F., Della Spina, L. (eds.) NMP 2020. SIST, vol. 178, pp. 1227–1236. Springer, Cham (2021). https://doi.org/10.1007/978-3-030-48279-4_114

9. Della Spina, L., Giorno, C., Galati Casmiro, R.: Bottom-up processes for culture-led urban regeneration scenarios. In: Misra, S., et al. (eds.) ICCSA 2019. LNCS, vol. 11622, pp. 93–107. Springer, Cham (2019). https://doi.org/10.1007/978-3-030-24305-0_8

10. Della Spina, L.: Cultural heritage: a hybrid framework for ranking adaptive reuse strategies. Buildings 11(3), 132 (2021)

11. City Resilience Index: City resilience framework. The Rockefeller Foundation and ARUP, 928 (2014)

12. Brooks, N., Adger, W.N., Kelly, P.M.: The determinants of vulnerability and adaptive capacity at the national level and the implications for adaptation. Glob. Environ. Chang. 15(2), 151–163 (2005)

13. Schipper, E.L.F., Langston, L.: A comparative overview of resilience measurement frameworks. Analyzing indicators and approaches (2015)

14. Parris, T.M., Kates, R.W.: Characterizing and measuring sustainable development. Ann. Rev. Environ. Resour. 28(1), 559–586 (2003)

15. Fleischhauer, M.: The role of spatial planning in strengthening urban resilience. In: Pasman, H.J., Kirillov, I.A. (eds.) Resilience of Cities to Terrorist and other Threats, pp. 273–298. Springer, Dordrecht (2008). https://doi.org/10.1007/978-1-4020-8489-8_14

16. Frazier, T.G., Thompson, C.M., Dezzani, R.J., Butsick, D.: Spatial and temporal quantification of resilience at the community scale. Appl. Geogr. 42, 95–107 (2013)

17. Joerin, J., Shaw, R., Takeuchi, Y., Krishnamurthy, R.: Action-oriented resilience assessment of communities in Chennai, India. Environ. Hazards 11(3), 226–241 (2012)

18. Antoniucci, V., Marella, G.: Small town resilience: housing market crisis and urban density in Italy. Land Use Policy 59, 580–588 (2016)

19. Sharifi, A., et al.: Conceptualizing dimensions and characteristics of urban resilience: Insights from a co-design process. Sustainability 9(6), 1032 (2017)

20. Meerow, S., Newell, J.P., Stults, M.: Defining urban resilience: a review. Landsc. Urban Plan. 147, 38–49 (2016)

21. Morano, P., Tajani, F., Anelli, D.: Urban planning decisions: an evaluation support model for natural soil surface saving policies and the enhancement of properties in disuse. Property Manage. 38(5), 699–723 (2020)

22. Della Spina, L., Calabrò, F., Rugolo, A.: Social housing: an appraisal model of the economic benefits in urban regeneration programs. Sustainability 12(2), 609 (2020)

23. Tian, C., Peng, X., Zhang, X.: COVID-19 pandemic, urban resilience and real estate prices: the experience of cities in the Yangtze River Delta in China. Land 10(9), 960 (2021)

24. Joint Research Centre-European Commission: Handbook on Constructing Composite Indicators: Methodology and User Guide. OECD Publishing (2008)
25. Schoenherr, T., Tummala, V.R., Harrison, T.P.: Assessing supply chain risks with the analytic hierarchy process: providing decision support for the offshoring decision by a US manufacturing company. J. Purch. Supply Manag. **14**(2), 100–111 (2008)
26. Saaty, T.L.: The analytic hierarchy process: decision making in complex environments. In: Avenhaus, R., Huber, R.K. (eds.) Quantitative Assessment in Arms Control, pp. 285–308. Springer, Boston, MA (1984). https://doi.org/10.1007/978-1-4613-2805-6_12
27. Orencio, P.M., Fujii, M.: A localized disaster-resilience index to assess coastal communities based on an analytic hierarchy process (AHP). Int. J. Disaster Risk Reduct. **3**, 62–75 (2013)
28. Chakhar, S., Martel, J.M.: Enhancing geographical information systems capabilities with multi-criteria evaluation functions. J. Geogr. Inf. Decis. Anal. **7**(2), 47–71 (2003)
29. Anelli, D., Sica, F.: The financial feasibility analysis of urban transformation projects: an application of a quick assessment model. In: Bevilacqua, C., Calabrò, F., Spina, L.D. (eds.) New Metropolitan Perspectives: Knowledge Dynamics and Innovation-driven Policies Towards Urban and Regional Transition Volume 2, pp. 462–474. Springer, Cham (2021). https://doi.org/10.1007/978-3-030-48279-4_44
30. Locurcio, M., Tajani, F., Morano, P., Anelli, D.: A multi-criteria decision analysis for the assessment of the real estate credit risks. In: Morano, P., Oppio, A., Rosato, P., Sdino, L., Tajani, F. (eds.) Appraisal and Valuation: Contemporary Issues and New Frontiers, pp. 327–337. Springer, Cham (2021). https://doi.org/10.1007/978-3-030-49579-4_22
31. Morano, P., Tajani, F., Anelli, D.: Urban planning variants: a model for the division of the activated "plusvalue" between public and private subjects. Valori e Valutazioni **28**, 31–48 (2021)

Evaluating the Impact of Urban Renewal on the Residential Real Estate Market: Artificial Neural Networks Versus Multiple Regression Analysis

Gabriella Maselli[(✉)] [iD]

Department of Civil Engineering, University of Salerno, Fisciano, Salerno, Italy
gmaselli@unisa.it

Abstract. Urban regeneration interventions and strategies for the restoration of eco-systemic services, in addition to generating a range of community benefits, lead to an increase in property values in the neighbourhood. This study aims to investigate how redevelopment affects residential house prices.

The aim of the paper is twofold. First the predictive potential of hedonic models is compared with that of Artificial Intelligence (AI) and Machine Learning (ML) approaches. Then, we characterise a novel Artificial Neural Network (ANN) model, which is still scarcely used to forecast how housing prices vary due to changes in the quality of the urban environment. The model includes among the variable inputs a series of environmental factors rarely considered in AI-ML approaches. The defined ANN is intended to support traditional approaches in order to provide planners and decision-makers with a more complete set of information on real estate trends.

Applications to real case studies will allow to validate the model, as well as to identify the environmental variables that most significantly influence residential property values.

Keywords: Urban renewal · Housing prices · Artificial Neural Networks · Multiple Regression Analysis

1 Introduction

Nowadays, urban regeneration strategies are a necessary response to the degradation of environmental assets caused by the uncontrolled expansion of cities. Multiple studies have shown that the fragmentation of urban ecological assets leads to dramatic consequences, such as the disruption of the hydrological system, the loss of biodiversity and the disruption of energy flows [1]. To avoid an irreversible urban deterioration, it is necessary to invest to invest more and more in the preservation and restoration of green and blue areas, so that the city itself can become a generator of resources and ecosystem services. Such investments also constitute a tangible contribution that cities can make to the United Nations' agenda on a Green Economy for the 21st century [2] and

F. Calabrò et al. (Eds.): NMP 2022, LNNS 482, pp. 702–712, 2022.
https://doi.org/10.1007/978-3-031-06825-6_66

the Sustainable Development Goals (SDGs). According to the Millennium Ecosystem Assessment (MEA) [3], urban green and blue networks can provide four different types of benefits: (i) supply services of water, food, fibre, and genetic resources; (ii) regulation services such as flooding, climate regulation, water quality and waste treatment; (iii) cultural services such as recreation, spiritual fulfilment and aesthetic enjoyment; and (iv) support services such as soil formation, pollination and nutrient cycling. The MEA results have led several authors to investigate the effects of the restoration of eco-system services in urban areas and, more generally, of urban and development renewal policies [4–7].

In this context, it is a great challenge to value the monetary benefits of ecosystem services provided by nature in urban areas [8–10]. This can be done by estimating the price increase of real estate located in the vicinity of an area affected by urban regeneration [11, 12]. Ki and Jayantha [13] show that an improved urban environment benefits nearby residents to the extent that a premium is reflected in house prices. Noor et al. [14] point out that house prices have been significantly influenced by green spaces as well as amenities and infrastructure. Nesticò et al. [15] investigate the levels of correlation between property prices and the provision of urban greenery.

In Sect. 2, traditional price prediction approaches, i.e. based on hedonic price theory, and innovative Artificial Intelligence (AI) and Machine Learning (ML) methods are compared. Section 3 defines a new Artificial Neural Network (ANN), which is still little used to assess the variation of real estate prices due to changes in the quality of the urban environment. This model, which includes environmental variables rarely considered in AI-ML approaches, is intended to be a valuable support to traditional price prediction approaches and a reference for real estate investors. Section 4 discusses conclusions and research perspectives.

2 Literature Review

2.1 Traditional Approaches to Property Price Forecasting

Several approaches for house price prediction have been tested in the literature. Among these, the Hedonic Price Method (HPM) is one of the most widely established mainly due to its relative simplicity of implementation [16, 17]. HPM is based on Lancaster's theory of consumer behaviour, according to which it is not the good itself that creates utility but its specific characteristics [18]. Under this theory, the price paid for the property can be broken down into hedonic (or implicit) prices of individual property attributes [19]. The implicit price of each property attribute is generally estimated using Multiple Regression Analysis (MRA). There are many applications that use MRA to assess environmental externalities and, more generally, the impact of urban regeneration interventions on residential property value. Just to mention a few, Zhang et al. [20] show that urban greening has a positive – and statistically significant – influence on neighbouring property values, which increase by between 5% and 20%. Jim and Chen [21] estimate the positive externalities caused by city parks that provide recreation opportunities and amenities to neighbourhoods and improve the quality of environment and life. Although HPM is a well-established and widely accepted among both practitioners and academicians, its main limitation is that it does not take into account the possible non-linearity between the dependent variable (i.e. property price) and the independent variables (i.e. property characteristics), which is a common problem in price prediction processes [17]. A second limitation is that it fails to consider possible correlation effects between variables. In addition, the sample must be large and the data uniformly distributed in space. Otherwise, the prediction theoretically could not guarantee the output accuracy. Finally, the results may be at odds with those derived from the qualitative analysis, so that it is not always possible to define a consistent regression equation [22, 23]. More sophisticated price prediction models than those analysed manage to overcome the problem of the non-linear relationship between input and output data. Among others, it is worth mentioning Evolutionary Polynomial Regression (ERP), non-linear transformable models, polynomial approach [23]. Spatial econometric approaches such as Simultaneous Autoregressive Model (SAR), Spatial Error Model (SEM), Spatial Auto-Correlation model (SAC), Geographically Weighted Regression (GWR) can account for spatial dependence and heterogeneity by including spatial "lagged" variables [24]. These models, which do not always achieve significantly better results than MRA, can be more complicated and expensive to set up.

2.2   Artificial Intelligence (AI) Approaches for Property Price Forecasting

A possible alternative to traditional price forecasting approaches is Artificial Intelligence (AI) models. AI, the broader discipline of creating intelligent machines, includes Machine Learning (ML), which refers to systems that can learn from experience. AI-ML approaches have recently found wide application in various industrial and commercial fields [24–26]. Studies claim that among the approaches to ML, the Artificial Neural Networks (ANNs) are an effective tool for property price prediction [17]. Their main advantage lies in their ability to define non-linear relationships between input and output, a condition that often occurs in the real estate sector. Several authors show that ANNs could have a great potential to predict noised and chaotic economic and financial time series [26]. ANNs are highly sophisticated modelling techniques that allow functions of a very advanced level of complexity to be designed by simulating the behaviour of the human brain. The minimum processing unit of the network is the artificial neuron, which consists of inputs, nodes, weights, and outputs. By selecting an appropriate set of connection weights and transfer functions, an ANN can learn to achieve a given task. ANNs are constituted by more layers of artificial neurons connected among them:

- the input layer, which includes the independent variables of the problem;
- the output layer, which transmits the result to the external environment;
- the hidden layer, which establishes links between the input and output layers – through the selected transfer function – and provides the generalisation of the network [27].

In summary, to build a neural network it is necessary to define:

- the architecture, establishing the number of layers, the number of neurons contained in each layer and the presence of possible feedback connections;
- the activation function, which determines the type of response that a neuron can emit and can be of different types (step, continuous linear, continuous non-linear, …);

- the learning algorithm, which has the function of training the network, going to gradually modify the synaptic weights to minimise between output and target. Two different
modes of learning can be distinguished: supervised, if it is necessary to supply to the
network possible inputs and relative outputs; not supervised, if it is necessary to supply
only the inputs [28].

With specific reference to the prediction of property values, the literature has shown
that ANNs could lead to high accuracy for different prediction settings [27]. In this
respect, experiments have been conducted with different purposes and in different geographical contexts. Morano et al. [28] characterise an ANN to predict market values of
residential properties in a neighbourhood of the city of Bari (Italy). With a method-based
housing price, Kitapci et al. [29] aim to support real estate investors by providing them
with more effective indications of property prices in Ankara (Turkey).

In the literature there are also some attempts to define advanced or hybrid AI-ML
models able to reach very high levels of performance. Abidoye et al. [30] specify a
hybrid model based on the joint use of the ANN, Support Vector Machine (SVM) models
and standard AutoRegressive Integrated Moving Average (ARIMA) to forecast out-of-
sample property values. Ho, Tang and Wong [31] compare ML algorithms – SVM, Random Forest (RF) and Gradient Boosting Machine (GBM) – for property price valuation
in Honk Kong using transaction data over an 18-year period.

The main limitation of ANNs is that the result is sensibly influenced by the input
information of the study system as well as by the training algorithms based on which
the useless data of the network are identified. A second limitation is that the elements
that constitute the network – transfer functions and number of hidden layers – must
be established a priori by the decision maker. Furthermore, the ANNs are not able to
incorporate economic laws in the learning processes [28]. The studies analysed show
that that some aspects of ANN modelling are lacking and therefore need to be further
investigated. The literature returned very different models for: (i) number of input variables; (ii) sample size; (iii) model architecture and training ratio. Regarding point (i), the
number of variables used is roughly between 6 and 40 and there is no indication on the
optimal set of input factors to be considered. There are also differing opinions on the size
of the data sample (ii): in some models, the input data are in the order of thousands; other
studies show that it is possible to achieve good results even with only a few hundred
data points [28]. Finally, the model architecture also plays a fundamental role in any
ANN: the higher the number of hidden neurons, the more complex the solution can be.
Moreover, since the architectures have been evaluated under very different conditions,
there is still disagreement on which one is able to return more accurate results.

The inclusion of environmental variables in ANN models, as well as their application
in predicting property prices following changes in the quality of the urban environment,
is another issue that deserves further investigation. In fact, in the mentioned studies
the input variables mainly concern intrinsic real estate characteristics (flat surface, floor
level, number of bathrooms, etc.) and some extrinsic or zonal characteristics (e.g. distance
from the main services). Among the few references that include environmental variables
in the ANN model, Chiarazzo et al. [32] estimate the residential properties prices in a

highly polluted urban area and identify the environmental factors that in an industrial city most influence residential location choice.

In conclusion, from the comparison between the MRA and the ANN models, it results that the latter, besides giving very performing results, can set non-linear relationships between input and output. However, implementing the MRA, it is possible to estimate the marginal prices of the single variables that influence the price function. Through ANNs, on the other hand, it is only possible to identify the most significant input variables.

3   Material and Methods

This study aims to characterise a model capable of assessing the effect generated by changes in the quality of the urban environment on the price of residential property.

Specifically, we intend to define a predictive model that: is easy to implement; returns efficient results; and allows both the identification of the most significant environmental variables and the estimation of these externalities. Thus, an innovative ANN is proposed in which environmental variables are included. In parallel, we also build an MRA model with the same input variables on which the ANN is based, to also provide indications on the marginal prices of environmental factors. In fact, it is believed that the joint use of traditional and innovative ANN models can provide the decision maker with a wider and more rigorous set of results and information on the structure of the housing market. The experimental design is described in the following 4 phases.

Stage 1: Choice of Property Variables and Data Collection. Intrinsic and extrinsic characteristics influencing the price of the property are identified. A set of factors, representing the independent variables of both ANN and regression function, is reported in Table 1. From this panel, the analyst can select the useful indicators to calibrate the models based on the specific social, economic, and environmental characteristics of the study area and of the availability of data. Once the model variables have been selected, the data collection is carried out. This step is crucial, as incorrect and inconsistent data can affect the goodness of the result. Table 1 gives the panel of property characteristics.

Stage 2: ANN Setting Up and Implementation. For this study an ANN with three layers (input layer - hidden layer - output layer) is chosen, as it is generally preferred for solving the prediction problems. So, before training the network, the decision maker must specify: the number of neurons n_i in the input layer; the number of hidden layers and the number of neurons n_h in each of these layers; the number of neurons n_o in the output layer. n_i corresponds to the number of real estate characteristics selected from the panel of Table 1. The output is given by the price of the properties, so $n_o = 1$. There is no rule to unambiguously establish n_h, so the optimal number of hidden neurons is found through an iterative process of trial and error. Each artificial neuron is constituted by n input connections (A) that receive signals from the other neurons, together with a bias adjustment (to consider the distortion of the received signal) and a set of weights for each input connection. Then, a function of transfer sums the weighed inputs and the bias to decide the value of the output deriving from that basic unit.

Table 1. Panel of the intrinsic and extrinsic features of the property.

Characteristic	Description	Data type
Intrinsic (characteristics specific to the dwelling)		
Area	Dwelling area (square meters)	Numeric
Level	Floor level (1st, 2nd, 3rd)	Numeric or Dummy
Bathrooms	Number of bathrooms	Numeric
Conservation status	Physical condition of dwelling	Dummy (bad, good, excellent)
Lift	Presence of lift	Dummy (1; 0)
Garden	Presence of garden	Dummy (1; 0)
Garage	Presence of garage	Dummy (1; 0)
Terrace	Presence of terrace	Dummy (1; 0)
Type of dwelling (DW)	Flat, Terraced, Detached	Dummy (for each type of DW)
Orientation (OR)	Sunny, Corner and Front	Dummy (for each type of OR)
Safety door	Presence of alarms	Dummy (1; 0)
Central air conditioning	Presence of central air conditioning	Dummy (1; 0)
Age	Age of the property	Dummy (assessed by age range)
Extrinsic or neighborhood (including environmental and social attributes)		
Lot density	Dwellings/Km^2 or inhabitants per area	Numeric
Services	Number of services, restaurants, retail, shopping centre, ATMs	Numeric
Schools	Presence of schools (within 500 m)	Dummy (1; 0)
Average income	Average income in the area	€
Unemployment rate	Unemployment rate in the area	%
Historical or financial centre	Distance from history. or financ. centre	m
Metro, railway, airport	Distance to main transport services	m

(*continued*)

Table 1. (*continued*)

Characteristic	Description	Data type
Bus lines	Presence of bus line within 500 m	Dummy
	Total number of lines serving the area	Numeric
Sea-coast	Distance from sea-coast	km
Pedestrian area	Distance from pedestrian area	
Green spaces	Distance from green spaces	km
Ecosystem Disservices (ED)	Distance from ED (industrial plants)	km
Degraded open spaces	Distance from degraded open spaces	km
Urban regeneration	Distance from a redeveloped area	km
Environmental quality level	Perceived environmental quality or measurement of pollutants	Discrete or $\mu g/m^3$
Traffic noise	Perceived sound level	Discrete

In the specific, the output (O_j) for the generic node j derives from the application of a transfer function φ to the sum of all the signals deriving from the single connection A_i multiplied by the value of the weight W_{ij} of the connection i entering in the node j:

$$O_j = \varphi(Sum_j) \tag{1}$$

$$Sum_j = \sum_j (W_{ji} \cdot A_i) \tag{2}$$

To solve our problem, a multilevel Perceptron network trained with the Levenberg-Marquardt back-propagation algorithm, which is widely preferred for prediction or classification problems, is set up [33]. To identify the optimal number of hidden neurons, a trial-and-error process is applied. The activation function for the hidden layer is sigmoid. The ANN is set up in MatLab using Deep Learning Toolbox.

Before training the network, the input data are normalised to [0, 1]:

$$X_i = \frac{X_i - X_{min}}{X_{max} - X_{min}} \tag{3}$$

The normalised data set is then randomly divided into three different samples: 70% is used as the training set of the network and the remaining data is equally divided between the validation set, to measure the generalisation of the network, and the testing set, to evaluate the performance of the network during and after training. It is then necessary to set the number of hidden neurons based on the performance measure of the network.

The most performing ANN is evaluated through the Root Mean Square Error (RMSE) and the regression index R^2:

$$RMSE = \sqrt{\frac{1}{n} \sum_{i=1}^{n} (P_i - \widehat{P_i})^2} \tag{4}$$

$$R^2 = 1 - \frac{\sum_{i=1}^{n} (P_i - \widehat{P_i})^2}{\sum_{i=1}^{n} (P_i - \overline{P})^2} \tag{5}$$

where: P_i is the observed price of the i-th dwelling; $\widehat{P_i}$ is the estimated price; $\overline{P}$ is the average price and n is the number of observations. The model chosen will be the one with R^2 closest to 1 and the lowest RMSE.

Finally, a sensitivity analysis is carried out to evaluate the most significant input variables. Therefore, the training phase of the ANN is repeated as many times as the property characteristics, eliminating each time one of the input variables. The significance of each eliminated input is assessed in function of the R^2 value that is achieved at the end of every training procedure. In other words, as the R^2 value diminishes, the significance of the removed variable rises [32].

Stage 3: Estimation of Marginal Prices Using the MRA. We define a regression function of the type:

$$Y = X \cdot \beta + \varepsilon \tag{6}$$

In (6): Y is a vector (n × 1) of individual house prices; X is a matrix (of size n × k) with k independent variables, represented by intrinsic and extrinsic characteristics. These variables are the same chosen by the analyst to set the ANN as said in step 1; β is the vector of k coefficients or regressors to be estimated, representing the marginal prices (MPs) of n property characteristics; ε is a vector (n × 1) of independent and identically distributed errors; n is the number of observations. Once the model has been calibrated and implemented, it is necessary both to verify the assumptions on which the multiple regression is based and to assess the acceptability of the results. Hypotheses are validated through a series of statistical tests: residual analysis; F-test; t-test. The results acceptability is verified by evaluating the indices of determination R^2, R^2_{adj}.

Stage 4: Comparison of Results. By comparing the results of phases 2 and 3, important information is obtained: (a) which model between ANN and MRA gives better results in terms of R^2 and RMSE for the neural network, and in terms R^2 and R^2_{corr} for the multiple regression; (b) the most significant input variables, with particular attention to environmental variables; (c) the estimation of MPs of each property characteristic through MRA. Figure 1 outlines the logical-operational stages to be followed.

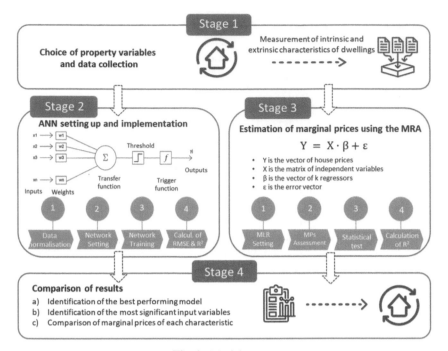

Fig. 1. Model stages.

4 Conclusion

Investment in urban regeneration, if well planned, not only give new life to run-down neighbourhoods, but also generate environmental and socio-cultural externalities on the surrounding area. These processes improve the urban quality and increase the market value of the properties in the neighbourhood affected by the redevelopment.

This study intends to propose a forecasting model able to estimate the variation of residential property prices caused by changes in environmental conditions. From literature review, it emerges that traditional regression models do not always provide effective results, as they do not consider any non-linearity between the dependent variable (the price of the property) and the independent variables (the property characteristics). In addition, these approaches fail to consider the correlation between the variables. Therefore, we analyse the predictive potential of AI-ML approaches which, in addition to defining non-linear relationships between inputs and outputs, tend to return results with high statistical precision. Thus: (a) a set of property characteristics and environmental and social indicators that influence the price function are identified; (b) an innovative ANN is defined in which environmental variables, rarely included in ML approaches, are included. This model can be a useful support to traditional MLR-based approaches which, on the other hand, make it possible to estimate the marginal prices of the individual variables of the price function. In fact, we believe that if jointly implemented, it can provide planners and urban designers with a more complete picture of real estate trends as well as a useful support to decision making.

This research represents only the starting point of the research. Applications to real case studies will allow to test the defined ANN and to return a comparison between the results deriving from the traditional approaches and those of the AI-ML models.

References

1. Assadpour, N., Melles, G.: Participation in urban renewal projects in Iran: an initial review of Mashhad, Shiraz and Tehran. Dev. Bull. Aust. Dev. Stud. Netw. **79**, 78–82 (2018)
2. UNEP: Towards a Green Economy: Pathways to Sustainable Development and Poverty Eradication - A Synthesis for Policy Makers (2001). www.unep.org/greeneconomy
3. M.E.A., Millennium Ecosystem Assessment: Ecosystems and Human Well-being: Synthesis. Island Press, Washington, DC (2005)
4. Mehdipanah, R., Marra, G., Melis, G., Gelormino, E.: Urban renewal, gentrification and health equity: a realist perspective. Eur. J. Pub. Health **28**(2), 243–248 (2018). https://doi.org/10.1093/eurpub/ckx202
5. Morano, P., Guarini, M.R., Sica, F., Anelli, D.: Ecosystem services and land take. A composite indicator for the assessment of sustainable urban projects. In: Gervasi, O., et al. (eds.) ICCSA 2021. LNCS, vol. 12954, pp. 210–225. Springer, Cham (2021). https://doi.org/10.1007/978-3-030-86979-3_16
6. Della Spina, L., Giorno, C., Galati Casmiro, R.: Bottom-up processes for culture-led urban regeneration scenarios. In: Misra, S., et al. (eds.) ICCSA 2019. LNCS, vol. 11622, pp. 93–107. Springer, Cham (2019). https://doi.org/10.1007/978-3-030-24305-0_8
7. Troisi, R., Castaldo, P.: Technical and organizational challenges in the risk management of road infrastructures. J. Risk Res., 1–16 (2022)
8. Bencardino, M., Nesticò, A.: Demographic changes and real estate values. A quantitative model for analyzing the urban-rural linkages. Sustainability **9**(4), 536 (2017)
9. Dolores, L., Macchiaroli, M., De Mare, G.: A dynamic model for the financial sustainability of the restoration sponsorship. Sustainability **12**(4), 1694 (2020)
10. Maselli, G., Nesticò, A.: L'Analisi Costi-Benefici per progetti in campo ambientale. La scelta del Saggio Sociale di Sconto. LaborEst n. 20 (2020)
11. Spampinato, G., Malerba, A., Calabrò, F., Bernardo, C., Musarella, C.: Cork oak forest spatial valuation toward post carbon city by CO_2 sequestration. In: Bevilacqua, C., Calabrò, F., Della Spina, L. (eds.) NMP 2020. SIST, vol. 178, pp. 1321–1331. Springer, Cham (2021). https://doi.org/10.1007/978-3-030-48279-4_123
12. Troisi, R., Alfano, G.: Is regional emergency management key to containing COVID-19? A comparison between the regional Italian models of Emilia-Romagna and Veneto. Int. J. Public Sect. Manag. **35**(2), 195–210 (2021)
13. Ki, C., Jayantha, W.: The effects of urban redevelopment on neighbourhood housing prices. Int. J. Urban Sci. **14**(3), 276–294 (2010)
14. Noor, N., Asmawi, M.Z., Abdullah, A.: Sustainable urban regeneration: GIS and Hedonic Pricing Method in determining the value of green space in housing area. Procedia – Soc. Behav. Sci. **170**, 669–679 (2015)
15. Nesticò, A., Endreny, T., Guarini, M.R., Sica, F., Anelli, D.: Real estate values, tree cover, and per-capita income: an evaluation of the interdependencies in Buffalo City (NY). In: Gervasi, O., et al. (eds.) ICCSA 2020. LNCS, vol. 12251, pp. 913–926. Springer, Cham (2020). https://doi.org/10.1007/978-3-030-58808-3_65
16. Ran, I., Zaman, U., Waqar, M., Zaman, A.: Using machine learning algorithms for housing price prediction: the case of Islamabad housing data. Fund. Inform. **1**, 11–23 (2021)

17. Kalliola, J., Kapočiūtė-Dzikienė, J., Damaševičius, R.: Neural network hyperparameter optimization for prediction of real estate prices in Helsinki. Peer J. Comput. Sci. **7**, e444(2021)
18. Lancaster, K.J.: A new approach to consumer theory. J. Polit. Econ. **74**(2), 132–157 (1966)
19. Capello, R.: Una valutazione di accessibilità e qualità urbana: Una stima di prezzi edonici nella città di Trento. In: Metodologie nelle Scienze Regionali. Franco Angeli, Milano, Italy (2004)
20. Zhang, B., Xie, G., Xia, B., Zhang, C.: The effects of public green spaces on residential property value in Beijing. J. Resour. Ecol. **2**(3), 243–252 (2012)
21. Jim, C.Y., Chen, W.Y.: External effects of neighbourhood parks and landscape elements on high-rise residential value. Land Use Policy **27**(2), 662–670 (2010)
22. Nesticò, A., La Marca, M.: Urban real estate values and ecosystem disservices: an estimate model based on regression analysis. Sustainability **12**(16), 6304 (2020)
23. Wang, C.: House Price Prediction Model Based on Neural Network (2021). https://doi.org/10.21203/rs.3.rs-805003/v1
24. Efthymiou, D., Antoniou, C.: Measuring the effects of transportation infrastructure on real estate prices and rents. Investigating the potential current impact of a planned metro line. EURO J. Transp. Logist. **3**, 179–204 (2013)
25. Cioffi, R., Travaglioni, M., Piscitelli, G., Petrillo, A., De Felice, F.: Artificial intelligence and machine learning applications in smart production: progress, trends, and directions. Sustainability **12**(2), 492 (2020)
26. Xu, X., Zhang, Y.: House price forecasting with neural networks. Intell. Syst. Appl. **12**, 200052 (2021). https://doi.org/10.1016/j.iswa.2021.200052
27. Karasu, S., Altan, A., Saraç, Z., Hacioglu, R.: Estimation of fast varied wind speed based on NARX neural network by using curve fitting. Int. J. Energy Appl. Technol. **4**(3), 137–146 (2017)
28. Morano, P., Tajani, F., Torre, C.M.: Artificial intelligence in property valuations: an application of artificial neural networks to housing appraisal. Adv. Environ. Sci. Energy Plann., 23–29 (2015)
29. Kitapci, O., Tosun, Ö., Tuna, M., Türk, T.: The use of Artificial Neural Networks (ANN) in forecasting housing prices in Ankara, Turkey. J. Mark. Consum. Behav. Emerg. Mark. **1**(5), 4–14 (2017)
30. Abidoye, R.B., Chan, A.P.C., Abidoye, F.A., Oshodi, O.S.: Predicting property price index using artificial intelligence techniques: evidence from Hong Kong. Int. J. Hous. Mark. Anal. **12**(6), 1072–1092 (2019)
31. Ho, W.K.O., Tang, B., Wong, S.W.: Predicting property prices with machine learning algorithms. J. Prop. Res. **38**(1), 48–70 (2020)
32. Chiarazzo, V., Caggiani, L., Marinelli, M., Ottomanelli, M.: A neural network based model for real estate price estimation considering environmental quality of property location. Transp. Res. Procedia **3**, 810–817 (2014)
33. Liang, L., Wu, D.: An application of pattern recognition on scoring Chinese corporations financial conditions based on back propagation neural network. Comput. Oper. Res. **32**(5), 1115–1129 (2005)

A Bio Ecological Prototype Green Building Toward Solution of Energy Crisis

Domenico Enrico Massimo[1](✉) ⓘ, Vincenzo Del Giudice[2], Mariangela Musolino[1],
Pierfrancesco De Paola[2], and Francesco Paolo Del Giudice[3]

[1] GeVaUL, Geomatic Valuation University Laboratory, Patrimony Architecture Urbanism
(PAU) Department, Mediterranea University of Reggio Calabria, 25 Via dell'Università,
89124 Reggio Calabria, Italy
demassimo@gmail.com

[2] Department of Industrial Engineering, University of Naples "Federico II", 80125 Naples, Italy

[3] Architecture and Project Ph.D. Program, University Rome 1 "La Sapienza", 00100 Rome, Italy

Abstract. The civil sector including construction, building management, edifice
air conditioning and heating, services, urbanism and settlement is the world's
largest consumer of fossil energy. In this structural strategy, one key factor is
the thermal high performance of GREEN buildings made possible by insulating
construction materials, better if organic or natural or forest-based, than totally
renewal and fully replenishable. In two different scenarios at least: ecological
retrofitting; brand new construction; others.

As 2020, civil sector consumes around 40% of annual world's 15,000 Mtoe
(Million tons oil equivalent) total energy, equals annual 6,000 Mtoe, an enormous
amount deriving mainly from fossil and similar materials.

Building bio ecological efficiency help in reducing significantly, up to 50%,
of the large civil consumptions of fossil energy, consequently a saving around
20% of total 15,000 Mtoe, i.e. 3,000 Mtoe worldwide, an enormous amount. This
is one strategic, perpetual and structural solution of both specific energy crisis as
well as more general climate change and consequent global warming. Research
shows that the buildings can be bio ecologically built or retrofitted at a reasonably
affordable additional initial investment cost and the cost differential pay-back is
fast, acceptable and over a short period of time. The Authors contributed equally
to the Paper.

Keywords: Appraisal · Valuation · Energy valuation · Valuation of green
building · Prototype building · Valuation of building energy · Energy
Performance Simulation Programs (EPSPs)

1 Introduction

The civil sector including construction, building management, edifice air conditioning
and heating, services, urbanism and settlement is the world's largest consumer of fossil
energy [1].

F. Calabrò et al. (Eds.): NMP 2022, LNNS 482, pp. 713–724, 2022.
https://doi.org/10.1007/978-3-031-06825-6_67

As 2020, civil sector consumes around 40% of annual world's 15,000 Mtoe (Million tons oil equivalent) total energy, equals annual 6,000 Mtoe, an enormous amount deriving mainly from fossil and similar materials.

To prevent new looming conflicts, de-escalate current wars, lower down geo-strategic slavery and dependence from fossils, trim resources prices and mitigate tragic climate change one structural permanent solution is to GET RIDE gradually of fossils and make world free of these brown, exhaustible, divisive, dirty and dark energy sources.

Specifically, for civil sector the alternative is TO DESIGN the INTEGRATION into the architecture of the several green energies, mainly: solar, pv and thermal; wind: home, onshore and remote offshore; geo thermal, domestic and remote centralized; waste treasuring; hydro power, remote; marine, remote; hydrogen, green nel electrolyzer; etc.

Generally speaking, green energies are growing and 2020 and 2021 are record years of new system development: more 270 Giga Watt of innovative green hydrogen and new generation pv solar have been deployed. In fact, thanks to innovations and new deployment, the green energy production costs (both financial as well as ecological - environmental) are progressively decreasing, compared to fossil, nuclear, bio mass and firewood [2–8].

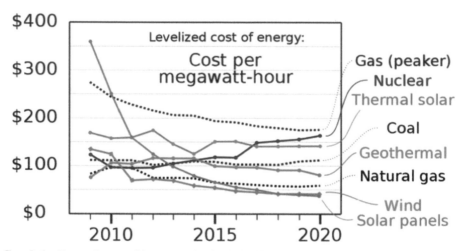

Graph 1. Costs of renewable energy declined significantly. They have a lower costs than the cheapest new fossil fuel option. Source: "Majority of New Renewables Undercut Cheapest Fossil Fuel on Cost". IEA, International Energy Agency, 2021.

Nevertheless, the best, cleanest, most convenient and cheapest alternative is the not mentioned queen of green energies: THE NOT CONSUMED one [9, 10]. I. e. the efficiency of settlements by the mean of green buildings which have outstanding climate performance compared to edifices ordinary as usual.

Building efficiency help in reducing significantly, up to 50%!, of the large civil consumptions of fossil energy, consequently a saving around 20% of total 15,000 Mtoe, i.e. 3,000 Mtoe worldwide, an enormous amount. This is one strategic, perpetual and structural solution of both specific energy crisis as well as more general climate change and consequent global warming.

In this structural strategy, one key factor is the thermal high performance of GREEN buildings made possible by insulating construction materials, better if organic or natural or forest-based, than totally renewal and fully replenishable. In two different scenarios at least: ecological retrofitting; brand new construction; others.

Thermal behaviors are thoroughly measured by Building Energy Performances Simulation\(measure) Programs BEPS(m)P made available thanks to scientific and technological innovation [11–16].

BEPS(m)Ps helps DESIGN of energy integrated INTO new architecture or retrofitting to choose the best materials and technology which lead to the highest efficiency than to the greatest energy saving (direct benefits) in green buildings and settlement leading to post carbon city and green world also thanks to significant reduction of CO_2 emission (indirect benefits) as well as minimization in the consumption of fossils. In fact, energy measures and management minimize fossil natural resource consumption and environmental deterioration while reducing residential energy use and costs.

An in progress spatial information system concerning analytic appraisal of factor costs for both ecological retrofitting as well as sustainable edifice new construction helps to optimize cost-efficiency relationship.

A long run research detected the surprising positive relationship between ecological features – energy efficiency of green buildings and their observed higher real estate market price compared to the one of no-green or brown or energy-intensive buildings, sometimes adopting highly demanding stochastic models [17–23].

A multitude of benefits occurs:

- better indoor healthiness and salubrity;
- less polluted outdoor environment due to strong abatement of CO_2 emissions;
- significant contribution to mitigation of climate change;
- smaller private consume of fossil fuel in the unit;
- lower energy bill for the unit;
- observed higher real estate market price, compared to brown buildings;

encourage to promote, incentive and foster-up self - enterprise in DESIGN toward ECOLOGICAL TRANSITION within INNOVATION ECOSYSTEM, namely Architecture-Energy integration.

Prototype building is key help in valuating architecture sustainability ex-ante before deployment of strategy at larger scale.

The preventive assessment of a new building or retrofitting strategy can and should be performed, before its implementation in real world, on an ideal small Prototype building.

The paper implements the design and analysis of Prototype in two alternative scenarios: green building versus un sustainable one.

Research shows that the buildings can be bio-ecologically built or retrofitted at a reasonably affordable additional initial investment cost and the cost differential pay-back is fast, acceptable and over a short period of time.

2   Prototype Building: A Valuation

2.1   Prototype or Reference or Sample Building

The preventive assessment of an Ecological building and Retrofitting Strategy was per-formed, before its implementation in real world case studies, on a small building (see Fig. 1); the so-called: prototype, reference or sample building.

The reference building can be easily assessed in its energy performances, under alternative scenarios, because it is small (5 × 5 × 4 m), with extremely simplified architectural characteristics, i.e., one-story or single-story cubes. It consists of a common punctiform structure in reinforced concrete (base beam, pillars, flat roof slab) and the usual buffering in common bricks.

Research includes comparative tests on the reference building without (Common Scenario) versus with (Sustainable Scenario) its envelope thermal external insulation, or coating of foundation, crawl space, walls and flat roof this latest insulated thank to natura-based bio-ecological cork panel based on local circular economy [24].

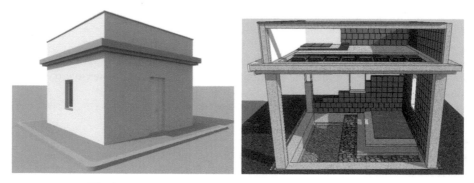

Fig. 1. Prototype/reference/sample building. Source: Authors.

2.2   Insulating Bio Materials and Bio Products

The In the Common Scenario (Business as Usual = BAS) the building is finished with popular commonly-used (external) plasters in cement-based mortar, or in industrial hydrated lime plus cement-based mortar. This plaster is made up of four to five layers including a bridge of adhesion, plaster (rustic), shaving (finishing) and putty or smooth finishing with an American metallic spatula, and with a synthetic color.

In the alternative Sustainable Scenario, the innovative external plaster is based on bio-ecological natural hydraulic lime, derived from local marlstone and mixed with expanded vermiculite (or perlite) for better insulation.

It is made up of four layers: bridge of adhesion ("aderenza"), plaster = rustic ("intonaco"), civil = shaving ("rasatura") and final colored finishing ("arenino colorato nella pasta").

There is, therefore, the addition of horizontal bio natural cork panels, derived from local cork oak forests, with a thickness of 6 cm both under the crawl space as well as above the final floor or attic. It is useful to recall the low thermal conductivity (W/m^2K) of these products:

- cork panel (0.040)
- lime-base plaster (0.066);
- and lime-base super plaster (0.029).

In the Sustainable Scenario there are also ecological windows possessing optimum thermal efficiency, involving structures based on natural wood or mineral chloride or PVC with low emission stratified double glazing.

2.3 Energy Performances and Pay Back Estimate

Building Energy Performance Simulation Programs, o BEPSPs, software provides (Table 1) the following:

- total (area × kW/m^2year) Global Primary Energy (EPgl) which demonstrates the general efficiency of the building, of the envelope and of the systems;
- total CO_2 (area × CO_2 kg/m^2 y) that the building and the systems release in the environment, as direct consequences of fossil material burning.

Estimate of energy consumption were carried out on two scenarios (Common versus Sustainable) using the EPSPs cited above, each having its own characteristics. The output is given below and is convergent to a surprising degree.

Table 1. Comparison of output concerning energy consumption and emission, from three Building Energy Performance Simulation Programs (BEPSPs).

Scenarios	#1		#2		#3	
	EPgl kW/m^2 y	CO_2 kg/m^2 y	EPgl kW/m^2 y	CO_2 kg/m^2 y	EPgl kW/m^2 y	CO_2 kg/m^2 y
01. Common (BAS)	114	24	116	11	129	15
02. Sustainable	69	15	71	8	73	9
Δ	−45	−9	−45	−3	−56	−6

The output of the three Energy Performance Simulation Programs (Table 2) was also well convergent in the percentage (%) of energy saving and sufficiently convergent in the percentage (%) of pollution mitigation of two distinct scenarios.

The valuation compares total energy consumption (kWh) and CO_2 emissions (kg) assessed by adopting the most conservative tool, Energy Plus, to test the worst scenario: 129 versus 114 and 116 kW/m^2 y.

Table 2. Percentage differential (kWh/m^2/year consumption; kg/m^2/year emissions) between Common BAS and Sustainable Bio Eco Scenarios.

	#1 Δ (%)	#2 Δ (%)	#3 Δ (%)
EPgl kWh/m^2 y	−40%	−39%	−44%
CO$_2$ kg/m^2 y	−36%	−26%	−43%

Considering just the annual saving in consumed less energy (−1400 kWh) and the statistical cost of energy for small users (€\kWh 0.42), the monetary annual saving equals: kWh 1.400 × €\kWh 0.42 = €590 (Table 3).

Table 3. Total energy consumption (kWh) and CO$_2$ emissions (kg) in 01 and 02 Scenarios per year (adopting the tool resulted most conservative tool: 129 kW/m^2 y).

Scenarios	Area m^2	EPgl kW/m^2 y	Total Annual EPgl: kWh	CO$_2$ kg/m^2 y	Total Annual CO$_2$: kg
01. Common (BAS)	25.50	129	3,313	15	382
02. Sustainable	26.21	73	1,913	9	235
Δ = differential (saving; mitigation)			−1,400		−147

Based on analytical and detailed estimates, this research forecast (Table 4) the financial costs involved in the construction of the two alternative scenarios.

Table 4. Comparison of the investment construction costs of the two prototypes. Cost differential (=Δ) of sustainability.

Prototype	Common	Sustainable	Δ = Differential	%
Tot €	37,156	40,378	+3,221	08.66
Tot €\m^2	1,456	1,540	+83	
Tot €\m^3	364	385	+20	

The difference (in both monetary amount and percentage) is very small (+3,221 € = 8.66%) compared to very large energy saving.

Given a very conservative interest rate of 4% (highly prudential until 2021) the light initial extra cost for bio ecological sustainable passivation of the building in second scenario would be paid back in a few years (Table 5). Subsequent savings, following the cost differential pay-back, represent positive added value. Passivation using biomaterials (cork panels; marlstone-based plaster) have an acceptable pay-back time of seven years.

Table 5. Pay Back (€ 3,541), in just seven years, of Differential Cost (€ 3,221).

Year	Annual saving	Anticipation Coefficient	Financial Amount	Saving Net value
n	€	$1/q^n$	€	€
1	590	0.96	567	567
2	590	0.92	545	1,112
3	590	0.89	524	1,637
4	590	0.85	504	2,141
5	590	0.82	484	2,626
6	590	0.79	466	3,092
7	590	0.76	448	3,541

3   Multiple Contextual Benefits: Healthiness, Salubrity and Other Impact Benefits

The comparison of two alternative scenarios allows for a quantitative valuation of their different energy consumption in terms of kWh, as well as CO_2 emissions, due to the above-cited low thermal conductivity of bio insulating products.

The two most immediate and visible results are the lower ecological emissions and lower energy consumption.

The positive impact of the use of bio ecological insulation in buildings is evident when compared to its non-adoption, not only from the energy and economic point of view. Also, healthiness and geo-strategic independence from oil are included in the ulterior final goals of the strategy. All the above impacts possess relevant ecologic and economic positive value. In fact, future research will outline in multidimensional terms additional "fundamental benefits" *i.e.*, healthier indoor and outdoor environments due to mitigated emissions as well as geostrategic independence from oil due to radical savings, which is the key geo-strategy of import substitution.

This research ascertained above the coherence, convergence and similar outcomes of three very different Building Energy Performance Software Programs (BEPSPs) #1, #2, #3.

All the above is will be implemented in future real-world case studies adopting one of the reliable, popular and friendly platforms, just above tested.

Furtherly, research detected that bio ecological green buildings have higher selling prices compared to other buildings, and the inferential detections are supported by robust stochastic models [25–30].

The Green Building strategy if generalized at the urban level allows to realize Post Carbon Districts [31], Historic Centers [32], Universities [33] and Cities [34] areas [35–39] which produce multiple benefits in the life cycle [39–44].

4   First Conclusions

Concerns regarding global warming, plus a Green Building exit strategy from the planet's ecological crisis and plus the reliability of Building Energy Performance Simulation Programs encourage researchers to provide help and advice to asset users, holders, contractors and all interested parties in new green building as well as in energy retrofitting.

The present research contributes TO:

- implementation approach by adopting natural, bio-ecological, historical, recyclable/renewable and local/regional materials in the framework of the Circular Economy;
- a simulation of bio-ecological protype building;
- an assessment (ecological as well as financial) of energy saving and CO_2 emission mitigation;
- an assessment of the initial investment costs involved in (ecological versus ordinary without energy enhancement) adopting a valuation based on scientific analytical techniques (with the estimate of the micro-economic production functions of indivisible works = "Lavorazioni") instead of just heuristic or empirical cost intuition;
- a forecast of the pay-back period of the additional differential initial cost of sustainable building compared with ordinary construction or Business as Usual (BAS).

The experimentation in the provided prototype or reference small building achieved the important goal of the Green Building and Post Carbon City strategies i.e. the significant enhancement of building thermal performance resulting from a few targeted keys works (= "*Lavorazioni*") and the consequent permanent structural saving of energy in those existing constructions.

Indeed, the data obtained is very encouraging because when construction or ecological retrofitting (Sustainable Scenario) is compared with ordinary mandatory usual construction or simply maintenance (Common Scenario, or Business as Usual upkeep, BAS) there are: energy efficiency, following observations can be made:

- energy saving amounts to **44%**;
- avoided CO_2 pollution is **43%**,
- the additional initial cost of sustainable works is a mere **8%** extra where this extra cost is calculated compared to the common scenario;
- the pay-back period of the additional differential cost of sustainable interventions is just two years.

All of the above show that the buildings can be bio-ecologically built or retrofitted at a reasonably affordable additional initial investment cost and the cost differential pay-back is fast, acceptable and over a short period of time.

This strategy should be generalized at urban block, districts and neighborhood [31], historic centers [32], campuses and universities [33], town and city [34] level, for revitalization projects [35, 36], and will provide benefits in the life cycle [37]. The Authors contributed equally to the Paper.

References

1. Massimo, D.E., Del Giudice, V., Malerba, A., Bernardo, C., Musolino, M., De Paola, P.: Valuation of ecological retrofitting technology in existing buildings: a real-world case study. Sustainability **13**, 7001 (2021). https://doi.org/10.3390/su13137001
2. Statistical Review of World Energy Archived 6 January 2009 at the Wayback Machine, Workbook (xlsx), London (2016)
3. Renewable Energy Employs 8.1 Million People Worldwide. United Nations Framework Group on Climate Change. 26 May 2016. Archived from the original on 18 April 2019. Retrieved 18 April 2019
4. Energy Transition Investment Hit $500 Billion in 2020 – For First Time. Bloomberg NEF. (Bloomberg New Energy Finance). 19 January 2021. Archived from the original on 19 January 2021
5. The European Power Sector in 2020/Up-to-Date Analysis on the Electricity Transition (PDF). ember-climate.org. Ember and Agora Energiewende. 25 January 2021. Archived (PDF) from the original on 25 January 2021
6. Chrobak, U. (author); Chodosh, S. (infographic) (28 January 2021). Solar power got cheap. So why aren't we using it more?. Popular Science. Archived from the original on 29 January 2021. {{cite news}}: |first1=has generic name (help) Chodosh's graphic is derived from data in "Lazard's Levelized Cost of Energy Version 14.0" (PDF). Lazard.com. Lazard. 19 October 2020. Archived (PDF) from the original on 28 January 2021
7. Majority of New Renewables Undercut Cheapest Fossil Fuel on Cost. IRENA.org. International Renewable Energy Agency. 22 June 2021. Archived from the original on 22 June 2021. Infographic (with numerical data) and archive thereof
8. Data. BP Statistical Review of World Energy, and Ember Climate (3 November 2021). Electricity consumption from fossil fuels, nuclear and renewables, 2020. OurWorldInData.org. Our World in Data consolidated data from BP and Ember. Archived from the original on 3 November 2021
9. Massimo, D.E.: Valuation of urban sustainability and building energy efficiency. A case study. Int. J. Sustain. Dev. **12**, 223–247 (2010). https://doi.org/10.1504/IJSD.2009.032779
10. Massimo, D.E.: Green building: characteristics, energy implications and environmental impacts. Case study in Reggio Calabria, Italy. In: Coleman-Sanders, M. (ed.) Green Building and Phase Change Materials: Characteristics, Energy Implications and Environmental Impacts, vol. 1, pp. 71–101. Nova Science Publishers, New York (2015)
11. Crawley, D.B., Lawrie, L., Winkelmann, F.C., Pedersen, C.O.: Energy plus: new capabilities in a whole-building energy simulation program. Build. Simul. **33**, 51–58 (2001)
12. Yilmaz, A.Z.: Evaluation of energy efficient design strategies for different climatic zones: Comparison of thermal performance of buildings in temperate-humid and hot-dry climate. Energy Build. **39**, 306–316 (2007). https://doi.org/10.1016/j.enbuild.2006.08.004
13. Crawley, D.B., Hand, J.W., Kummert, M., Griffith, B.T.: Contrasting the capabilities of building energy performance simulation programs. Build. Environ. **43**, 661–673 (2008). https://doi.org/10.1016/j.buildenv.2006.10.027
14. Rallapalli, H.S.: A comparison of EnergyPlus and eQuest whole building energy simulation results for a medium sized office building. Master's Thesis, Arizona State University, Tempe (2010)
15. Sousa, J.: Energy simulation software for buildings: review and comparison (2012). https://www.semanticscholar.org/paper/Energy-Simulation-Software-for-Buildings-%3A-Review-Sousa/b4b6593df77024a585b68d066bf2bd668838f 852. Accessed 25 May 2021

16. Malerba, A., Massimo, D.E., Musolino, M., Nicoletti, F., De Paola, P.: Post carbon city: building valuation and energy performance simulation programs. In: Calabrò, F., Della Spina, L., Bevilacqua, C. (eds.) ISHT 2018. SIST, vol. 101, pp. 513–521. Springer, Cham (2019). https://doi.org/10.1007/978-3-319-92102-0_54

17. Del Giudice, V., De Paola, P., Manganelli, B., Forte, F.: The monetary valuation of environmental externalities through the analysis of real estate prices. Sustain. Build. Environ. **9**, 229 (2017). https://doi.org/10.3390/su9020229

18. Del Giudice, V., Massimo, D.E., De Paola, P., Forte, F., Musolino, M., Malerba, A.: Post carbon city and real estate market: testing the dataset of Reggio Calabria market using spline smoothing semiparametric method. In: Calabrò, F., Della Spina, L., Bevilacqua, C. (eds.) ISHT 2018. SIST, vol. 100, pp. 206–214. Springer, Cham (2019). https://doi.org/10.1007/978-3-319-92099-3_25

19. Massimo, D.E., Del Giudice, V., De Paola, P., Forte, F., Musolino, M., Malerba, A.: Geographically weighted regression for the Post Carbon City and real estate market analysis: a case study. In: Calabrò, F., Della Spina, L., Bevilacqua, C. (eds.) Smart Innovation, Systems and Technologies; New Metropolitan Perspectives (ISHT 2018), vol. 100, pp. 142–149. Springer, Cham (2019). https://doi.org/10.1007/978-3-319-92102-3_17

20. De Paola, P., Del Giudice, V., Massimo, D.E., Forte, F., Musolino, M., Malerba, A.: Isovalore maps for the spatial analysis of real estate market: a case study for a central urban area of Reggio Calabria, Italy. In: Calabrò, F., Della Spina, L., Bevilacqua, C. (eds.) ISHT 2018. SIST, vol. 100, pp. 402–410. Springer, Cham (2019). https://doi.org/10.1007/978-3-319-92099-3_46

21. Del Giudice, V., Massimo, D.E., De Paola, P., Del Giudice, F.P., Musolino, M.: Green buildings for post carbon city: determining market premium using spline smoothing semiparametric method. In: Bevilacqua, C., Calabrò, F., Della Spina, L. (eds.) NMP 2020. SIST, vol. 178, pp. 1227–1236. Springer, Cham (2021). https://doi.org/10.1007/978-3-030-48279-4_114

22. Del Giudice, V., Massimo, D.E., Salvo, F., De Paola, P., De Ruggiero, M., Musolino, M.: Market price premium for green buildings: a review of empirical evidence. Case study. In: Bevilacqua, C., Calabrò, F., Della Spina, L. (eds.) NMP 2020. SIST, vol. 178, pp. 1237–1247. Springer, Cham (2021). https://doi.org/10.1007/978-3-030-48279-4_115

23. De Paola, P., Del Giudice, V., Massimo, D.E., Del Giudice, F.P., Musolino, M., Malerba, A.: Green building market premium: detection through spatial analysis of real estate values. A case study. In: Bevilacqua, C., Calabrò, F., Della Spina, L. (eds.) NMP 2020. SIST, vol. 178, pp. 1413–1422. Springer, Cham (2021). https://doi.org/10.1007/978-3-030-48279-4_132

24. Spampinato, G., Massimo, D.E., Musarella, C.M., De Paola, P., Malerba, A., Musolino, M.: Carbon sequestration by Cork Oak forests and raw material to built up Post Carbon City. In: Calabrò, F., Della Spina, L., Bevilacqua, C. (eds.) ISHT 2018. SIST, vol. 101, pp. 663–671. Springer, Cham (2019). https://doi.org/10.1007/978-3-319-92102-0_72

25. Manganelli, B., Morano, P., Tajani, F.: The risk assessment in Ellwood's financial analysis for the indirect estimate of urban properties. AESTIMUM **55**, 19–41 (2009)

26. Manganelli, B., Tajani, F.: Optimised management for the development of extraordinary public properties. J. Prop. Invest. Financ. **32**(2), 187–201 (2014)

27. Morano, P., Tajani, F.: Least median of squares regression and minimum volume ellipsoid estimator for outliers detection in housing appraisal. Int. J. Bus. Intel. Data Mining **9**(2), 91–111 (2014)

28. Morano, P., Locurcio, M., Tajani, F., Guarini, M.R.: Fuzzy logic and coherence control in multi-criteria evaluation of urban redevelopment projects. Int. J. Bus. Intel. Data Mining. **10**(1), 73–93 (2015)

29. Tajani, F., Morano, P., Locurcio, M., D'Addabbo, N.: Property valuations in times of crisis: artificial neural networks and evolutionary algorithms in comparison. In: Gervasi, O., et al. (eds.) ICCSA 2015. LNCS, vol. 9157, pp. 194–209. Springer, Cham (2015). https://doi.org/10.1007/978-3-319-21470-2_14

30. Tajani, F., Liddo, F.D., Guarini, M.R., Ranieri, R., Anelli, D.: An assessment methodology for the evaluation of the impacts of the COVID-19 pandemic on the Italian housing market demand. Buildings 11(12), 592 (2021)

31. Massimo, D.E., Musolino, M., Fragomeni, C., Malerba, A.: A green district to save the planet. In: Mondini, G., Fattinnanzi, E., Oppio, A., Bottero, M., Stanghellini, S. (eds.) SIEV 2016. GET, pp. 255–269. Springer, Cham (2018). https://doi.org/10.1007/978-3-319-78271-3_21

32. Bentivegna, V.: The evaluation of structural-physical projects in urban distressed areas. In: Mondini, G., Fattinnanzi, E., Oppio, A., Bottero, M., Stanghellini, S. (eds.) SIEV 2016. GET, pp. 17–36. Springer, Cham (2018). https://doi.org/10.1007/978-3-319-78271-3_2

33. Massimo, D.E., Fragomeni, C., Malerba, A., Musolino, M.: Valuation supports green university: case action at Mediterranea campus in Reggio Calabria. Proc. Soc. Behav. Sci. 223, 17–24 (2016). https://doi.org/10.1016/j.sbspro.2016.05.278

34. Musolino, M., Massimo, D.E.: Mediterranean urban landscape. Integrated strategies for sustainable retrofitting of consolidated city. In: Sabiedriba, I. (ed.) Izglitiba [Society, Integration, Education], Proceedings of the Ispalem/Ipsapa International Scientific Conference, Udine, Italy, 27–28 June 2013, vol. 3, pp. 49–60. Rezekne Higher Education Institution, Rezekne, Latvija (2013)

35. Calabrò, F., Cassalia, G., Lorè, I.: The economic feasibility for valorization of cultural heritage. the restoration project of the reformed fathers' convent in Francavilla Angitola: the Zibìb territorial wine cellar. In: Bevilacqua, C., Calabrò, F., Della Spina, L. (eds.) NMP 2020. SIST, vol. 178, pp. 1105–1115. Springer, Cham (2021). https://doi.org/10.1007/978-3-030-48279-4_103

36. Calabrò, F., Mafrici, F., Meduri, T.: The valuation of unused public buildings in support of policies for the inner areas. The application of SostEc model in a case study in Condofuri (Reggio Calabria, Italy). In: Bevilacqua, C., Calabrò, F., Della Spina, L. (eds.) NMP 2020. SIST, vol. 178, pp. 566–579. Springer, Cham (2021). https://doi.org/10.1007/978-3-030-48279-4_54

37. Mallamace, S., Calabrò, F., Meduri, T., Tramontana, C.: Unused real estate and enhancement of historic centers: legislative instruments and procedural ideas. In: Calabrò, F., Della Spina, L., Bevilacqua, C. (eds.) ISHT 2018. SIST, vol. 101, pp. 464–474. Springer, Cham (2019). https://doi.org/10.1007/978-3-319-92102-0_49

38. Barrile, V., Malerba, A., Fotia, A., Calabrò, F., Bernardo, C., Musarella, C.: Quarries renaturation by planting cork oaks and survey with UAV. In: Bevilacqua, C., Calabrò, F., Della Spina, L. (eds.) NMP 2020. SIST, vol. 178, pp. 1310–1320. Springer, Cham (2021). https://doi.org/10.1007/978-3-030-48279-4_122

39. Spampinato, G., Malerba, A., Calabrò, F., Bernardo, C., Musarella, C.: Cork oak forest spatial valuation toward post carbon city by CO2 sequestration. In: Bevilacqua, C., Calabrò, F., Della Spina, L. (eds.) NMP 2020. SIST, vol. 178, pp. 1321–1331. Springer, Cham (2021). https://doi.org/10.1007/978-3-030-48279-4_123

40. Fregonara, E., Curto, R., Grosso, M., Mellano, P., Rolando, D., Tulliani, J.-M.: Environmental technology, materials science, architectural design, and real estate market evaluation: a multidisciplinary approach for energy-efficient buildings. J. Urban Technol. 20, 57–80 (2013). https://doi.org/10.1080/10630732.2013.855512

41. Fregonara, E., Giordano, R., Rolando, D., Tulliani, J.-M.: Integrating environmental and economic sustainability in new building construction and retrofits. J. Urban Technol. 23(4), 3–28 (2016). https://doi.org/10.1080/10630732.2016.1157941

42. Fregonara, E., Ferrando, D.G.: How to model uncertain service life and durability of components in life cycle cost analysis applications? The stochastic approach to the factor method. Sustainability **10**, 3642 (2018). https://doi.org/10.3390/su13052838
43. Fregonara, E., Ferrando, D.G., Pattono, S.: Economic – environmental sustainability in building projects: introducing risk and uncertainty in LCCE and LCCA. Sustainability **10**, 1901 (2018). https://doi.org/10.3390/su10061901
44. Fregonara, E., Ferrando, D.G., Chiesa, G.: Economic valuation of buildings sustainability with uncertainty in costs and in different climate conditions. In: Bevilacqua, C., Calabrò, F., Della Spina, L. (eds.) NMP 2020. SIST, vol. 178, pp. 1217–1226. Springer, Cham (2021). https://doi.org/10.1007/978-3-030-48279-4_113

Green Building to Overcome Climate Change: The Support of Energy Simulation Programs in Gis Environment

Domenico Enrico Massimo[1](✉) ⓘ, Vincenzo Del Giudice[2], Mariangela Musolino[1], Pierfrancesco De Paola[2], and Francesco Paolo Del Giudice[3]

[1] GeVaUL, Geomatic Valuation University Laboratory, Patrimony Architecture Urbanism (PAU) Department, Mediterranea University of Reggio Calabria, 25 Via dell'Università, 89124 Reggio Calabria, Italy
demassimo@gmail.com
[2] Department of Industrial Engineering, University of Naples "Federico II", 80125 Naples, Italy
[3] Architecture and Project Ph.D. Program, University Rome 1 "La Sapienza", 00100 Rome, Italy

Abstract. The recurrent fossil energy crises and the warning signs of imminent war between countries, are jeopardizing the world's future. The increasing demand for oil, coal, uranium, lithium, biomass and firewood has caused prices to sky rocket and has triggered off dangerous conflicts all over the world. The civil sector including construction, building management, edifice air conditioning and heating, services, urbanism and settlement is the world's largest consumer of fossil energy [1]. Since 2020, civil sector has absorbed around 40% of the annual world's total energy consumption which amounts to 15,000 Mtoe (million tons oil equivalent). This comes to 6,000 Mtoe which is an enormous amount and derives mainly from fossil and similar materials. In order to prevent any new imminent conflicts, de-escalate the current wars, reduce the geo-strategic slavery to and dependence on fossils, trim resource prices and mitigate tragic climate change one structural permanent solution is to gradually GET RID of fossils and render the world free of these brown, exhaustible, divisive, dirty and dark energy sources.

As far as the civil sector is concerned, the alternative is TO DESIGN the integration of the use of the numerous green energies into architecture. The aforementioned forms of energy are mainly: solar, pv and thermal; wind: home, onshore and remote offshore; geo thermal, domestic and remote centralized; waste treasuring; hydro power, remote; marine, remote; hydrogen, green nel electrolyzer; etc. The research perform a real world experimentation of the Green Building and Ecological Retrofitting Strategy The proposed and implemented experimentation achieved the important outcome and goal of a Green Building strategy and post-carbon city framework i.e. the significant enhancement of the thermal performance of the buildings as a result of a few targeted key external works and the consequent saving of energy. The Authors contributed equally to the Paper.

Keywords: Appraisal · Valuation · Valuation of green building · Valuation of post-carbon city strategy · Valuation of building energy · Energy performance simulation programs (EPSPs) · Data base management system · DBMS

F. Calabrò et al. (Eds.): NMP 2022, LNNS 482, pp. 725–734, 2022.
https://doi.org/10.1007/978-3-031-06825-6_68

1 Introduction

The recurrent fossil energy crises and the warning signs of imminent war between countries, are jeopardizing the world's future.

The increasing demand for oil, coal, uranium, lithium, biomass and firewood has caused prices to sky rocket and has triggered off dangerous conflicts all over the world.

The civil sector including construction, building management, edifice air conditioning and heating, services, urbanism and settlement is the world's largest consumer of fossil energy [1].

Since 2020, civil sector has absorbed around 40% of the annual world's total energy consumption which amounts to 15,000 Mtoe (million tons oil equivalent). This comes to 6,000 Mtoe which is an enormous amount and derives mainly from fossil and similar materials.

In order to prevent any new imminent conflicts, de-escalate the current wars, reduce the geo-strategic slavery to and dependence on fossils, trim resource prices and mitigate tragic climate change one structural permanent solution is to gradually GET RID of fossils and render the world free of these brown, exhaustible, divisive, dirty and dark energy sources.

As far as the civil sector is concerned, the alternative is TO DESIGN the integration of the use of the numerous green energies into architecture. The aforementioned forms of energy are mainly: solar, pv and thermal; wind: home, onshore and remote offshore; geo thermal, domestic and remote centralized; waste treasuring; hydro power, remote; marine, remote; hydrogen, green nel electrolyzer; etc..

Generally speaking, the use of green energies is increasing, 2020 and 2021 have been record years for this new system of development. Over 270 Giga Watts of innovative green hydrogen and new generation pv solar have been deployed. In fact, thanks to these innovations and this new deployment, the green energy production costs (both financial as well as ecological - environmental) are progressively decreasing when compared to fossil, nuclear, bio mass and firewood [2–8].

Nevertheless, the best, cleanest, most convenient and cheapest alternative is the not mentioned queen of green energies: the NOT consumed one [9, 10]. I. e. the efficiency of green building settlements which show outstanding climate performance compared to buildings constructed using the traditional methods.

Building efficiency contributes to significantly reducing, (up to 50%!), the large civil consumption of fossil energy, consequently saving around 20% of the total 15,000 Mtoe, i. e. 3,000 Mtoe used worldwide, which is an enormous amount. This is one strategic, perpetual and structural solution to the specific energy crisis as well as the more general problem of climate change and the consequent catastrophe of global warming.

In this structural strategy, one key factor is the excellent thermal performance of GREEN buildings made possible by the use of insulating construction materials. Using insulating materials which are organic, natural or forest-based is even better than using those which are totally renewal and fully replenishable. Such materials have been used in two different scenarios at least i.e. in ecological retrofitting and in brand new construction.

Thermal behaviors have been meticulously measured by Building Energy Performances Simulation\(measure) Programs BEPS(m)P which are now available thanks to scientific and technological innovation [11–16].

BEPS(m)Ps help the DESIGN of energy integration INTO the new architecture or retrofitting of buildings. This enables the choice of the best materials and technology which lead to the highest efficiency rather than to the greatest energy savings (direct benefits) in green buildings and settlements. This leads to a post carbon city and green world and results in a significant reduction of CO_2 emissions (indirect benefits) as well as minimizing the consumption of fossils. In fact, energy measures and management minimize fossil natural resource consumption and environmental deterioration while reducing residential energy use and costs.

An in progress spatial information system concerning the analytic appraisal of factor costs for both ecological retrofitting as well as for newly constructed sustainable buildings helps to optimize the cost-efficiency relationship.

Research carried out over many years has detected a surprisingly positive relationship between ecological features – the energy efficiency of green buildings and their observed higher real estate market price compared to those values when applied to non-green, brown or energy-intensive buildings, sometimes adopting highly advanced stochastic models [17–23].

The following benefits ensue:

- a better indoor environment;
- a less polluted outdoor environment due to the strong abatement of CO_2 emissions;
- a significant contribution to the mitigation of climate change;
- a lower private consumption of fossil fuel in the unit;
- a lower energy bill for the unit;
- an observed higher real estate market price compared to brown buildings;

All these benefits encourage the promotion and development of self - enterprise in a DESIGN method aimed at ECOLOGICAL TRANSITION within an INNOVATION ECOSYSTEM, namely Architecture-Energy integration.

The Prototype building constitutes a key help in valuating sustainability in architecture ex-ante before the adoption of the strategy on a larger scale.

The preventive assessment of a new building or retrofitting strategy can and should be performed, before its implementation in the real world, on an ideal small Prototype building.

This paper implements the design and the analysis of the Prototype in two alternative scenarios i.e. green building versus unsustainable building.

Research shows that the buildings can be bio-ecologically built or retrofitted at a reasonably affordable additional initial investment cost and the cost differential pay-back is fast, acceptable and over a short period of time.

2   Case Study

The real-world Case Study is situated in the city of Reggio Calabria (the largest of Calabria, the Southernmost region of Italy), a settlement (see Fig. 1) which was rebuilt in the Liberty style after the earthquake and subsequent tsunami of 1908.

Fig. 1. "Latin Quartier". Urban Block.

The main and peculiar characteristic of the new city of Reggio Calabria, based is the fortunately small size of its Urban Blocks (about 50 × 50 m) and therefore the average footprint is around 2.500 m². This factor has created many positive effects such as the large number of streets, free street parking, sidewalks, street shop-fronts, urban tree plantation and small manageable courts.

"Latin Quarter" in the city of Reggio Calabria has been chosen as the area for the Case Study with the aim of designing a potential <sustainable neighborhood> or <green quartier> or <energy district>.

Specific experimentation is implemented on Urban Block #102, an outstanding Liberty architecture, on the Northern final stretch of main street (Corso Garibaldi) of the Liberty re-built new city of Reggio Calabria.

3   Comparative Building Energy Performance Simulation Programs (BEPSPs)

Additionally, it has been performed a comparative test of Building Energy Performance Simulation Programs (BEPSPs) through the valuation of energy consumption in kWh and CO_2 emission in kilos in the two cited different scenarios (Common *versus* Sustainable-Ecological) by means of three very different tools described below [11–16].

- Energy Plus® (Version 8.3.0) together with Design Builder (Version 4.5.0.178) is one of the best-known energy simulation software tools. It is complex software for energy diagnosis and thermal simulation in dynamic building arrangements. It has external graphical interfaces that facilitate the creation of the thermal model of the building and the inclusion of its characteristics, like Design Builder and others BIMs. Energy Plus is adopted to perform this first simple experiment on the elementary prototype edifice or reference building.
- Blumatica Energy® (Version 6.1) is user-friendly and cheap national software that allows the planner to design the thermal insulation of buildings as well as the management of their energy certification. It is interesting to compare its performance with that of more complex Energy Plus and more popular Termus.
- Termus® (Version 30.001) is one of the most popular Italian software platforms used for the assessment of energy performance of buildings. Energy certification (APE-AQE), calculation of transmittance and drafting Protocol Ithaca are some of the outputs of this software. It is the best-known standard software in Italy. It is reliable as well as friendly enough to be advised for local professionals and adopted for the present complex Case Study in this first valuation. In future research, a more complex modeling of thermal and wet transmission will be adopted to better understand and reduce the need of improvements.

4 Energy Performances Simulation Outcomes: KWh Consumption and CO₂ Emission in Alternative Common and Sustainable Scenarios

An important goal pursued in the Green Building and Post Carbon City Strategies is the strong enhancement of thermal building performance, and the consequent remarkable energy saving, with affordable bearable additional cost respect to ordinary maintenance without energy enhancement, in consequence of few targeted key works below listed as the back bone of the Ecological Retrofit approach:

- new insulating plaster (on the vertical walls) based on natural mineral Marlstone and on derived natural hydraulic lime, NHL (so called: "calce romana");
- new insulation for a flat roof, a terrace or a pitched roof, based on natural vegetal Cork derived from local cork oak forests and on an additional new slope layer based on **natural** mineral pearly-stone [24];
- new insulation for crawl space on slab intrados based on natural vegetal Cork derived from local and Mediterranean cork oak forests [24];
- efficient new windows, with optimum efficiency involving window - structures based on **natural** chloride or pvc, and low-emission stratified double glazing.

Quantitative Energy Performances Simulations (EPS) were carried out for both the Common (CS) and Sustainable Scenario (SS) interventions adopting EPSP software Ter-Mus tested and selected in previous paragraph, and in precedent researches. Each single unit of each building of Urban Block #102 have been valuated in its energy performances, deriving comparatively in both the Common (CS) and the Sustainable Scenario (SS):

annual total energy consumption in kWh and total carbon dioxide emission in CO_2 kg; annual per mq (squared meter, in International Spelling) unitary energy consumption in kWh/mq per year, and unitary emission in CO_2/mq per year.

The EPSP software has provided, among many others, the following outputs regarding:

- Envelope index (EPi, inv), i.e. the energy dispersed by the building itself;
- Global primary energy index (EPgl), which demonstrates the efficiency of both the building-plant system on heating and the domestic hot water production and distribution system;
- CO_2, i.e. the Kg of Carbon Dioxide that the Building + heating system emits into the environment.

The Energy Consumption and CO_2 emissions values of the Common (SC) and Sustainable (SS) Scenarios and differential (Δ) are shown in the Tables 1, 2, 3 below.

The clear approach of Ecological Retrofitting (based on bio ecological, natural, oil free materials and few key works) reached the goal of a strong enhancement of thermal building performance, and of a consequent remarkable energy saving in consequence of key interventions above listed.

Successful enhancement of thermal building performance is quantified and assessed thanks to a huge amount of 45 (one for each unit) detailed Building Energy Performance Simulation using Termus tool. There is remarkable energy saving around -45% of less consumed kWh ($445.266-229.729 = 215.537$), and a relevant CO_2 emission mitigation with the cut of -55% of kg dumped ($82.499-36.606 = 45.892$).

Next step is the estimate of cost differential between Common Scenario intervention and Sustainable Scenario intervention and the years needed for the pay back of this differential cost, to understand if the remarkable success in energy saving (even in a cultural relevant historic building) is also bearable in monetary and financial terms.

It is important to outline that after the additional and differential initial cost for Ecological will be paid back and off set, the perpetual and permanent energy saving in the building will be forever a positive added value [25–30], and must be taken into account at both unit and building level as well as at cumulative quartier, city and region level.

5   First Conclusions

Recent comprehensive reviews concerning global warming, plus a Green Building exit strategy from the planet's ecological crisis and plus Building Energy Performance Simulation Programs encourage researchers to provide help and advice to asset users, holders, contractors and all interested parties in building energy retrofitting who have attempted to:

- adopt natural, bio-ecological, historical, renewable/recyclable and local/regional raw materials in the framework of the Circular Economy;
- include in the energy retrofit strategy the challenging aim of conserving and restoring existing buildings;

- enhance the energy performance of such existing buildings;
- estimate the energy enhancement of such constructions as a result of few targeted external works (=Lavorazioni), examining, in particular, the initial investment costs and the longer-term multiple benefits stemming from structural energy saving as well as permanent CO_2 emission mitigation.

The present research addresses the reviews by contributing TO:

- an existing building retrofit implementation approach by adopting natural, bio-ecological, historical, recyclable/renewable and local/regional materials in the framework of the Circular Economy;
- a simulation of bio-ecological retrofitting of the prototype building such as the second Scenario;
- an assessment (ecological as well as financial) of energy saving and CO_2 emission mitigation;

The experimentation in the provided prototype or reference small building achieved the important goal of the Green Building and Post Carbon City strategies i.e. the significant enhancement of building thermal performance resulting from a few targeted keys works (="Lavorazioni") and the consequent permanent structural saving of energy in those existing constructions.

Indeed, the data obtained is very encouraging because when construction or ecological retrofitting (Sustainable Scenario) is compared with ordinary mandatory usual construction or simply maintenance (Common Scenario, or Business as Usual upkeep, BAS) the following observations can be made:

- energy saving amounts to 45%;
- avoided CO_2 pollution is 55%,

All of the above show that the buildings can be bio-ecologically built or retrofitted with a significant ecological, environmental and energy enhancement [31–44]. The Authors contributed equally to the Paper.

References

1. Massimo, D.E., Del Giudice, V., Malerba, A., Bernardo, C., Musolino, M., De Paola, P.: Valuation of ecological retrofitting technology in existing buildings: a real-world case study. Sustainability **13**, 7001 (2021). https://doi.org/10.3390/su13137001
2. Statistical Review of World Energy Archived 6 January 2009 at the Wayback Machine, Workbook (xlsx), London (2016)
3. "Renewable Energy Employs 8.1 Million People Worldwide": United Nations Framework Group on Climate Change. 26 May 2016. Archived from the original on 18 April 2019. Accessed 18 Apr 2019
4. "Energy Transition Investment Hit $500 Billion in 2020 – For First Time": Bloomberg NEF. (Bloomberg New Energy Finance). 19 January 2021. Archived from the original on 19 January 2021

5. "The European Power Sector in 2020 / Up-to-Date Analysis on the Electricity Transition" (PDF). ember-climate.org. Ember and Agora Energiewende. 25 January 2021. Archived (PDF) from the original on 25 January 2021

6. Chrobak, Ula (author); Chodosh, Sara (infographic) (28 January 2021). "Solar power got cheap. So why aren't we using it more?". Popular Science. Archived from the original on 29 January 2021. {{cite news}}: |first1=has generic name (help) Chodosh's graphic is derived from data in "Lazard's Levelized Cost of Energy Version 14.0" (PDF). Lazard.com. Lazard. 19 October 2020. Archived (PDF) from the original on 28 January 2021

7. "Majority of New Renewables Undercut Cheapest Fossil Fuel on Cost". IRENA.org. International Renewable Energy Agency. 22 June 2021. Archived from the original on 22 June 2021. Infographic (with numerical data) and archive thereof

8. Data: BP Statistical Review of World Energy, and Ember Climate (3 November 2021). "Electricity consumption from fossil fuels, nuclear and renewables, 2020". OurWorldInData.org. Our World in Data consolidated data from BP and Ember. Archived from the original on 3 November 2021

9. Massimo, D.E. Valuation of urban sustainability and building energy efficiency. a case study. Int. J. Sustain. Dev. **12**, 223–247 (2010). https://doi.org/10.1504/IJSD.2009.032779

10. Massimo, D.E.: Green building: characteristics, energy implications and environmental impacts. case study in Reggio Calabria, Italy. In: Coleman-Sanders, M. (ed.) Green Building and Phase Change Materials: Characteristics, Energy Implications and Environmental Impacts, vol. 1, pp. 71–101. Nova Science Publishers, New York (2015)

11. Crawley, D.B., Lawrie, L., Winkelmann, F.C., Pedersen, C.O.: Energy plus: new capabilities in a whole-building energy simulation program. Build. Simul. **33**, 51–58 (2001)

12. Yilmaz, A.Z.: Evaluation of energy efficient design strategies for different climatic zones: Comparison of thermal performance of buildings in temperate-humid and hot-dry climate. Energy Build. **39**, 306–316 (2007). https://doi.org/10.1016/j.enbuild.2006.08.004

13. Crawley, D.B., Hand, J.W., Kummert, M., Griffith, B.T.: Contrasting the capabilities of building energy performance simulation programs. Build. Environ. **43**, 661–673 (2008). https://doi.org/10.1016/j.buildenv.2006.10.027

14. Rallapalli, H.S.: A comparison of EnergyPlus and eQuest whole building energy simulation results for a medium sized office building. Master's Thesis, Arizona State University, Tempe, AZ, USA (2010

15. Sousa, J.: Energy simulation software for buildings: review and comparison (2012). https://www.semanticscholar.org/paper/Energy-Simulation-Software-for-Buildings-%3A-Revie-Sousa/b4b6593df77024a585b68d066bf2bd668838f852. Accessed on 25 May 2021

16. Malerba, A., Massimo, D.E., Musolino, M., Nicoletti, F., De Paola, P.: Post carbon city: building valuation and energy performance simulation programs. In: Calabrò, F., Della Spina, L., Bevilacqua, C. (eds.) ISHT 2018. SIST, vol. 101, pp. 513–521. Springer, Cham (2019). https://doi.org/10.1007/978-3-319-92102-0_54

17. Del Giudice, V., De Paola, P., Manganelli, B., Forte, F.: The monetary valuation of environmental externalities through the analysis of real estate prices. Sustain. Build. Environ. **9**, 229 (2017). https://doi.org/10.3390/su9020229

18. Del Giudice, V., Massimo, D.E., De Paola, P., Forte, F., Musolino, M., Malerba, A.: Post carbon city and real estate market: testing the dataset of Reggio Calabria market using spline smoothing Semiparametric method. In: Calabrò, F., Della Spina, L., Bevilacqua, C. (eds.) ISHT 2018. SIST, vol. 100, pp. 206–214. Springer, Cham (2019). https://doi.org/10.1007/978-3-319-92099-3_25

19. Massimo, D.E., Del Giudice, V., De Paola, P., Forte, F., Musolino, M., Malerba, A.: Geographically weighted regression for the post carbon city and real estate market analysis: a case study. In: Calabrò, F., Della Spina, L., Bevilacqua, C. (eds.) Smart Innovation, Systems and

Technologies; New Metropolitan Perspectives. ISHT 2018, vol. 100, pp. 142–149. Springer, Cham (2019). https://doi.org/10.1007/978-3-319-92102-3_17

20. De Paola, P., Del Giudice, V., Massimo, D.E., Forte, F., Musolino, M., Malerba, A.: Isovalore maps for the spatial analysis of real estate market: a case study for a central urban area of Reggio Calabria, Italy. In: Calabrò, F., Della Spina, L., Bevilacqua, C. (eds.) ISHT 2018. SIST, vol. 100, pp. 402–410. Springer, Cham (2019). https://doi.org/10.1007/978-3-319-92099-3_46

21. Del Giudice, V., Massimo, D.E., De Paola, P., Del Giudice, F.P., Musolino, M.: Green buildings for post carbon city: determining market premium using spline smoothing Semiparametric method. In: Bevilacqua, C., Calabrò, F., Della Spina, L. (eds.) NMP 2020. SIST, vol. 178, pp. 1227–1236. Springer, Cham (2021). https://doi.org/10.1007/978-3-030-48279-4_114

22. Del Giudice, V., Massimo, D.E., Salvo, F., De Paola, P., De Ruggiero, M., Musolino, M.: Market price premium for green buildings: a review of empirical evidence. case study. In: Bevilacqua, C., Calabrò, F., Della Spina, L. (eds.) NMP 2020. SIST, vol. 178, pp. 1237–1247. Springer, Cham (2021). https://doi.org/10.1007/978-3-030-48279-4_115

23. De Paola, P., Del Giudice, V., Massimo, D.E., Del Giudice, F.P., Musolino, M., Malerba, A.: Green building market premium: detection through spatial analysis of real estate values. a case study. In: Bevilacqua, C., Calabrò, F., Della Spina, L. (eds.) NMP 2020. SIST, vol. 178, pp. 1413–1422. Springer, Cham (2021). https://doi.org/10.1007/978-3-030-48279-4_132

24. Spampinato, G., Massimo, D.E., Musarella, C.M., De Paola, P., Malerba, A., Musolino, M.: Carbon sequestration by cork oak forests and raw material to built up post carbon city. In: Calabrò, F., Della Spina, L., Bevilacqua, C. (eds.) Smart Innovation, Systems and Technologies; New Metropolitan Perspectives. ISHT 2018, vol. 101, pp. 663–671. Springer, Cham, (2019). https://doi.org/10.1007/978-3-319-92102-0_72

25. Manganelli, B., Morano, P., Tajani, F.: The risk assessment in Ellwood's financial analysis for the indirect estimate of urban properties. AESTIMUM **55**, 19–41 (2009)

26. Manganelli, B., Tajani, F.: Optimised management for the development of extraordinary public properties. J. Property Invest. Finan. **32**(2), 187–201 (2014)

27. Morano, P., Tajani, F.: Least median of squares regression and minimum volume ellipsoid estimator for outliers detection in housing appraisal. Int. J. Bus. Intell. Data mining **9**(2), 91–111 (2014)

28. Morano, P., Locurcio, M., Tajani, F., Guarini, M.R.: Fuzzy logic and coherence control in multi-criteria evaluation of urban redevelopment projects. Int. J. Bus. Intell. Data Mining **10**(1), 73–93 (2015). ISSN: 1743-8187

29. Tajani, F., Morano, P., Locurcio, M., D'Addabbo, N.: Property valuations in times of crisis: artificial neural networks and evolutionary algorithms in comparison. In: Gervasi, O., et al. (eds.) ICCSA 2015. LNCS, vol. 9157, pp. 194–209. Springer, Cham (2015). https://doi.org/10.1007/978-3-319-21470-2_14

30. Tajani, F., Liddo, F.D., Guarini, M.R., Ranieri, R., Anelli, D.: An assessment methodology for the evaluation of the impacts of the COVID-19 pandemic on the Italian housing market demand. Buildings **11**(12), 592 (2021)

31. Massimo, D.E., Musolino, M., Fragomeni, C., Malerba, A.: A green district to save the planet. In: Mondini, G., Fattinnanzi, E., Oppio, A., Bottero, M., Stanghellini, S. (eds.) SIEV 2016. GET, pp. 255–269. Springer, Cham (2018). https://doi.org/10.1007/978-3-319-78271-3_21

32. Bentivegna, V.: The evaluation of structural-physical projects in urban distressed areas. In: Mondini, G., Fattinnanzi, E., Oppio, A., Bottero, M., Stanghellini, S. (eds.) SIEV 2016. GET, pp. 17–36. Springer, Cham (2018). https://doi.org/10.1007/978-3-319-78271-3_2

33. Massimo, D.E., Fragomeni, C., Malerba, A., Musolino, M.: Valuation supports green university: Case action at Mediterranea campus in Reggio Calabria. Procedia Soc. Behav. Sci. **223**, 17–24 (2016). https://doi.org/10.1016/j.sbspro.2016.05.278

34. Musolino, M., Massimo, D.E.: Mediterranean Urban Landscape. Integrated Strategies for Sustainable Retrofitting of Consolidated City. In: Sabiedriba, Integracija, Izglitiba [Society, Integration, Education], Proceedings of the Ispalem / Ipsapa International Scientific Conference, Udine, Italy, 27–28 June 2013; Rezekne Higher Education Institution: Rezekne, Latvija, vol. 3, pp. 49–60 (2013)

35. Calabrò, F., Cassalia, G., Lorè, I.: The Economic feasibility for Valorization of cultural heritage. the restoration project of the reformed fathers' convent in Francavilla Angitola: the Zibìb territorial wine cellar. In: Bevilacqua, C., Calabrò, F., Della Spina, L. (eds.) NMP 2020. SIST, vol. 178, pp. 1105–1115. Springer, Cham (2021). https://doi.org/10.1007/978-3-030-48279-4_103

36. Calabrò, F., Mafrici, F., Meduri, T.: The valuation of unused public buildings in support of policies for the inner areas. the application of SostEc model in a case study in Condofuri (Reggio Calabria, Italy). In: Bevilacqua, C., Calabrò, F., Della Spina, L. (eds.) NMP 2020. SIST, vol. 178, pp. 566–579. Springer, Cham (2021). https://doi.org/10.1007/978-3-030-48279-4_54

37. Mallamace, S., Calabrò, F., Meduri, T., Tramontana, C.: Unused real estate and enhancement of historic centers: legislative instruments and procedural ideas. In: Calabrò, F., Della Spina, L., Bevilacqua, C. (eds.) ISHT 2018. SIST, vol. 101, pp. 464–474. Springer, Cham (2019). https://doi.org/10.1007/978-3-319-92102-0_49

38. Barrile, V., Malerba, A., Fotia, A., Calabrò, F., Bernardo, C., Musarella, C.: Quarries renaturation by planting cork oaks and survey with UAV. In: Bevilacqua, C., Calabrò, F., Della Spina, L. (eds.) NMP 2020. SIST, vol. 178, pp. 1310–1320. Springer, Cham (2021). https://doi.org/10.1007/978-3-030-48279-4_122

39. Spampinato, G., Malerba, A., Calabrò, F., Bernardo, C., Musarella, C.: Cork oak forest spatial valuation toward post carbon city by CO_2 sequestration. In: Bevilacqua, C., Calabrò, F., Della Spina, L. (eds.) NMP 2020. SIST, vol. 178, pp. 1321–1331. Springer, Cham (2021). https://doi.org/10.1007/978-3-030-48279-4_123

40. Fregonara, E., Curto, R., Grosso, M., Mellano, P., Rolando, D., Tulliani, J.-M.: Environmental technology, materials science, architectural design, and real estate market evaluation: a multidisciplinary approach for energy-efficient buildings. J. Urban Technol. 20, 57–80 (2013). https://doi.org/10.1080/10630732.2013.855512

41. Fregonara, E., Giordano, R., Rolando, D., Tulliani, J.M.: Integrating environmental and economic sustainability in new building construction and retrofits. J. Urban Technol. 23, 26 (2016). ISSN: 1063-0732, https://doi.org/10.1080/10630732.2016.1157941

42. Fregonara, E., Ferrando, D.G.: How to model uncertain service life and durability of components in life cycle cost analysis applications? the stochastic approach to the factor method. Sustainability 10, 3642 (2018). https://doi.org/10.3390/su13052838

43. Fregonara, E., Ferrando, D.G., Pattono, S.: Economic – environmental sustainability in building projects: introducing risk and uncertainty in LCCE and LCCA. Sustainability 2018, 10 (1901). https://doi.org/10.3390/su10061901

44. Fregonara, E., Ferrando, D.G., Chiesa, G.: Economic valuation of buildings sustainability with uncertainty in costs and in different climate conditions. In: Bevilacqua, C., Calabrò, F., Della Spina, L. (eds.) NMP 2020. SIST, vol. 178, pp. 1217–1226. Springer, Cham (2021). https://doi.org/10.1007/978-3-030-48279-4_113

Attractiveness and Problems in a Rural Village Restoration: The Umbrian Case of Postignano

Marco Pizzi[1,2(✉)], Paola de Salvo[1,2], and Cristina Burini[1,2]

[1] University of Perugia, Umbria, Italy
marco.pizzi@studenti.unipg.it
[2] Political Science Department, Via Pascoli 20, 06123 Perugia (PG), Italy

Abstract. The hamlet of Postignano is located in Umbria, into the Municipality of Sellano, in a rural marginal peripheral area. The relaunch project for this area is the main research object, in this paper. In 1992, two Architects bought the whole hamlet, investing a great amount of money and human and professional resources to make it live again after its abandonment, which took place in the 1960's. The research is the result of a series of interviews to selected key-actors and to people who bought a flat in the hamlet. The interviews are aimed at understanding what made the hamlet attractive and how the owners' approach to the settlement management has been changed during the last decades. The project, indeed, has crossed crucial decades for the international laws about the rural development, facing new challenges like the global economic crisis in 2008 and local problems which forced the owners to reinvent deeply their approach to local development. The redevelopment process of Postignano confirms the necessity to build a touristic plan involving the local communities and being respectful through the sustainable development concepts.

Keywords: Rural renaissance · City planning · Local development

1 Introduction

Many rural and marginal areas of the European countries, today, are lived and governed as loisir and consumer places rather than spaces to be developed. The commodification of places [1], the *turistification* [2] and the outcomes of many top-down decisions about local development [3], indeed, pushed a lot of fragile areas into the socioeconomic profile of the mere show-territory [4] making them increasingly entrusting with the touristic field to gain back the lost wealth. Although such approach has become the mainstream, at least in Italy, many public initiatives have been started with the aim to promote the most efficient, participated and affective development process.

Recent Italian public policies – such as the National Strategy for Inland Areas (SNAI) – started to manage local development of such areas trying to subvert the *urbanocentric* approach which characterized the territorial interpretation from the second postwar period. The major importance given to the cities by scholars and politicians, however, is a cultural asset far to fall, overall considering the urban trends expected for

© The Author(s), under exclusive license to Springer Nature Switzerland AG 2022
F. Calabrò et al. (Eds.): NMP 2022, LNNS 482, pp. 735–745, 2022.
https://doi.org/10.1007/978-3-031-06825-6_69

the next years. According to the "World Cities Report 2016: Urbanization and Development–Emerging Futures" [5] indeed, the 70% of world's population will live inside cities and conurbations within 2050, increasing up to 9 billion units. This awareness should encourage reflections not only on the potential urban problems, threats, strengths and opportunities, but also on all the residual non-urban spaces that will inevitably change their structure facing this upcoming future scenario. Waiting for the outcomes of the SNAI, however, it would be advantageous also to focus on case studies based on different features and premises, with the aim to sharpen the analytic tools required to the upcoming scenario.

Some of the SNAI most significant premises to place restoration are the participation of local communities, the self-management of the development projects, the fostering of a local self-project capability, the improvement of the quality of life [6]. The development processes based on these assumptions are realized in target areas where a local community is engaged through years, but this, naturally, implies the presence of a community in the territory to redevelop. A characteristic of many marginal Italian areas, nevertheless, is just the lack of a cohesive community concentrated in a specific place. The SNAI approach to local redevelopment, thus, cannot be considered as the only applicable to all the marginal contexts.

The exploration of new case studies is needed to understand if a development approach not based on local participation is possible. Such case studies should be able to analyse the restoration process of a marginal area under the same conditions guaranteed by SNAI under the economic and expertise points of view, but different under the social profile. We can suppose, indeed, that a place distinguished by an important cultural, architectonical and historical heritage and a quality surrounding landscape, but fragmented under the social point of view, would have different attractive points and different needs to be restored.

The aim of this paper is to provide a description of a case study fitting with this research needs, underlining what are the distinguishing characteristics of a rural restoration processes lead not by institutions and in a deserted locality.

A fundamental intuition for this paper is that all these topics can be effectively addressed observing non-urban or rural local restoration processes lead by private citizens rather than public institutions.

Such kind of initiatives, indeed, have some specific characteristics that make them interesting study objects [7]:

- The elaboration of territorial development ideas is less complex and articulated than the Public Authorities' one, and conducted with less cognitive means;
- The entrepreneurship governance approach is more often characterized by a top-down approach rather than a participatory one;
- Following more strictly many marketing criteria, the enterprises involved in territorial restoring processes as leader need to directly follow the market gait.

The case study analysed in this paper is the Umbrian hamlet of Postignano. An hamlet totally owned by an entrepreneur since 1992, whose restoration process gives important opportunities to address the topics named before.

2   The Umbrian Case of Postignano

2.1   Case Study Description

Postignano is a medieval hamlet situated in Umbria, under the Municipality of Sellano. This village, founded in the 13th century, thrived thanks to an economy based on herding and iron files production. Due to this activity, it reached its maximum expansion in the 16th century, with a population of 400 inhabitants. The demographic thread remained stable until the 20th century, when the basilar dynamics of the desertification phenomena intervened as in many other villages of the Apennines [8]. The hamlet fell in total desertification by the end of the '60s, when a sudden evacuation has been imposed suspecting that an earthquake would have taken place soon after a slight landside. The hamlet remained desert until 1992.

In that year, a venture lead by an architect started the acquisition of the whole village, buying each estate from the heirs of the ancient owners. The idea of the architect was to give a second life to this place restoring it entirely and reselling it after doing it. This company still owns the hamlet, today, which is used as a widespread hotel for a half and is intended to the retail sail for the other half. Some of the real estate that the hamlet is composed by, indeed, have been already sold to buyers interested to own a house in a picturesque setting, enjoying the meticulous restoration work done.

The restoration effort consisted in a real reconstruction work and was not intended as a superficial operation. Nevertheless, the 1997 earthquake caused new heavy damages to Postignano just few years after its acquisition, during the first years of '90s: a building situated on the top of the village dropped triggering a chain collapse. Ten years have been taken by legal proceedings and a new projecting phase, until when the work resumed, between 2007 and 2008.

The nation in which Postignano came back to life, though, was deeply different from that of 1997. In the meantime, in fact, affective socioeconomic changes took place. The most important among them has been the 2008 global economic crisis. This event, as a matter of fact, forced the new hamlet owners to widely rethink their action plans. If, on a first time, the intent was to complete a mere architectural project to sell the entire settlement – to an institution or a corporation that could be able to revitalize the locality – after the economic crisis, when the restored hamlet had become very hard to sell, the owners had to take care about many aspects about the life of the village, that are no more connected to the simple architectural side. Owning a small town and developing a long-term revitalization process is totally different than to complete an architectural restoration to sell back the property letting other people to revitalize it. On first instance, then, the project sustainability concerned only the architectural field: buildings had to be rebuilt so that they lasted over time, resisting earthquakes being comfortable as much as possible; in the new after-2008 scenario, instead, the restoring choices taken ended with determining the new inhabitant's community entry criteria.

What makes interesting the case of Postignano, then, is the deep change happened in the management of this place, which became a social engineering experiment from a simple architectural operation.

2.2 Methodology

To understand the choices made in the reinvention of the village, key-players were interviewed: individuals who contributed substantially and directly to determining the choices that made this settlement what it is today. The 15 subjects to interview in-depth with semi-structured interviews made by videocall (due to COVID-19 pandemic) have been chosen combining the rational sampling technique with the snowball-one, asking to the first interviewee who were the key-subject to talk with. In this way, some key subjects whose names recurred during the interviews were selected. They are the architect who created the project as well as the founder of the "new Postignano" and owner of the company that bought the village; the architect co-founder of the project and, today, director of the hotel in the village; the communication manager of the village and a former consultant for the promotion of the village who served the company for a few years starting in 2010.

This methodological choice was made to give as much prominence as possible to the interviewees' point of view and to their network of relationships, according to the personal life history and environmental historic comprehension methodological frame [9, 10]. In this way, in fact, were interviewed only people who they believe to be important with respect to the project design and who they believe to be in continuity with their own point of view, namely that of project managers. During the interviews, some discrepancies emerged in the narration of the facts regarding the restoration and different points of view, but all confirmed the incidence of the other interviewees in influencing the course of work on the village. This survey process is articulated on the level of the perception of space and time of the interviewees to compare their story with the tourist, cultural, social and economic context described in the introduction. The aim is to understand what the strengths and the weaknesses of the project are and what the territory of Postignano is communicated through. These elements, indeed, could be able to make understand better some aspect of the "return to villages" phenomenon.

Another branch of interviewees (10 people), on the other hand, is the one in which all those who have bought a house in Postignano find themselves. Basically, what was asked of them concerns the reasons for their choice and the intentions on how to use the property in future years.

3 Results

The answers of the interviewees will be reworked below to highlight some issues of particular interest for the purpose of analysing the recovery process of the village of Postignano.

What Aspects of the Redevelopment Project Needed to Be Implemented to Make It Sustainable in the Long Term?
Regarding this topic, all the interviewees articulated many opinions about it. Returning exhaustively the positions of all, however few, would be impossible also due to the complexity of the subject and the amount of statements made. If we want to clarify the overall vision that the creators of the project have regarding this topic, however, we could start by creating an index of the main issues that emerged from the interviews.

On the theme of the social sustainability of the project, the interviewees expressed their thoughts on the relationship with the local community and its level of entrepreneurship; on the services offered by the village and its vitality in everyday life; on the dynamics of short-break tourism; the lack of territorial coordination by the institutions and the relationship with them; on the intention of not musealizing the village; on the protection and maintenance of the territory; on the subject of population aging. Here it will not be possible to report the complete statements of the interviewees commenting on them in detail. Instead, some of the salient contents of the interviews will be reported in the form of a list, which we will return to in the conclusions.

- **Relationships between different communities**. The social, economic and cultural space of Postignano, today, is the result of the interaction between three different communities distinguished by different time and spaces. The old Postignano community, which built the village and created its main cultural asset; the present community composed by tourists and by the new inhabitants (people who bought some of the renewed apartments); the surrounding community, which lives in the enclosing territory.

  On first instance, it can be observed that even if some efforts to involve the surrounding community – that was connected to the history of the hamlet, in some ways – in the regeneration process, today the hamlet is perceived as something quite extraneous. After centuries of craftmanship and rural lifestyle, the hamlet hosts a wealthy, high-educated and polyglottal enclave of new inhabitants. They are enjoying the great work done on the architectural side aimed at giving value to the village qualities, choosing a type of real estate that is quite countertrend respect to the local mainstream, according to which more and more locals are choosing the near cities to live.

  The cultural activities intended for enliven the hamlet, moreover, are felt by the people who live around Postignano as an additional buffer factor, sometimes.

  This perception cleavage emerged during the interviews, but it doesn't have to be considered as the only way in which the hamlet is seen. It is an interesting and recurring topic present in the interviews, that is reflected in some moments of hostility by local elites, whose institutions have not always favored the project for Postignano.

- **Sense of community and the role of the new community of Postignano**. "You cannot sell Postignano without a community" said the communication manager of the village. Therefore, many strains of the owners of the hamlet are meant to create a sense of living community inside it. Many public spaces have been refurbished with particular care. The hamlet, in fact, is now fitted out with a theater, a library, a museum, a square. The staff is always present, even in winter, when the hamlet is deserted, with the explicit intention to give the impression that the hamlet is always populated to potential residents who could decide to come during the low season. As told before, in addition, also the cultural activities are projected to create a stimulating setting.

  Listening to the new inhabitants interviews, all these initiatives seem to have been effective. They, indeed, declare to feel a sense of community coming back to the hamlet also thanks to these initiatives. In their narration it is a place in which people can really met and know each other and in whom everyone is free to chose if enjoy a more intimate routine rather than a more sociable one. The aspect of village sociability emerges in the interviews as one of the main reasons of their choice to buy an apartment in Postignano.

- **Entrepreneurship and environmental maintenance around Postignano**. The surrounding community constitutes, according with the board of the village, a potential richness and a problem at the same time. The lack of entrepreneurship of the neighboring population is a node to be solved to ensure a long life to the results of the redevelopment of this village, which needs to be insert inside an economic network to be completely lived. Furthermore, emerges in the stories of all the interviewees, who note a high rate of aging, a low cultural level, the presence of few young people and a certain reluctance to accept job offers within the village or to start businesses that could collaborate there. "We need an integrated, long-term plan, in concert with entrepreneurs, local and regional administrations, from various political and social parts which creates serious opportunities for young people". All the actors involved in the local socioeconomic development, then, should work to create am economic and social fabric to make the hamlet survive and to exploit the potential richness that it is able to bring in the territory.

The locals disaffection through the territory would clarify also why the environmental context is in bad conditions – from the point of view of the hamlet owners. Who is currently managing the village explained, in the interviews, how in many occasions they founded more interest in cultural and eco-friendly initiatives from foreign people rather than from the locals, arguing that it is a problem not only for Postignano, but also for the entire territorial context.

The Territorial Branding Dynamics in the Postignano Project

The first evidence that emerged with respect to this topic is that, thus far, "there is still no integrated communication plan for Postignano". The communications manager explains how "the exercise of sitting around a table and answering the following questions has not yet been done: what is Postignano? What do we want to communicate?". This situation appears connected to the underlying indecisions and a lack of a complete vision about the implications of the new structure of the village, but it could also have economic causes. "The investment in the website is strong, also because it costs to keep it up to date," says the co-founder. If the hypothesis on the causes of this delay in terms of communication also consider evaluations on the management strategy, however, it is because already in 2010 by the executives "there was no idea of what to sell; sell the project? A product? The rooms? The flats? The widespread hotel? There could not have been a communication since there was not a product, yet: there were only renovated walls - so the former consultant for the promotion of Postignano - they called us to promote it outside when did not yet know how to do the interiors, how to loft, how subdivide, how to do". Before the opening of the village in 2013, in fact, it seems that the consultant was hired without precise instructions on the promotion actions to be taken. "We made Postignano presentations around the world, in Milan with presentations, speeches, press conferences, different types of releases, newspapers. They asked us for help in promoting the village in events like these, creating contacts with investors, decision makers, personalities from the world of finance... Communication was a mix. We had to present Postignano as a real estate opportunity in various ways, […] but we have never taken care of an organic plan for them".

Today, however, the need for an organized communication plan is felt. The idea of the communications manager is that this should be articulated on several levels, involving subjects also outside the village: "[On the part of the surrounding municipalities, tour operators and local companies] there is a lack of updated, complete and clear sites, adequate communications, publications, signs. All this is missing because people are no longer fond of the territory, but also because there is no local public coordination, but above all regional and national". As for Postignano itself, the question to ask, says the communication manager, is "How do I communicate a place like this?". "From an aesthetic point of view, it has been recovered, but the life inside has changed. You cannot bring back the original life and the one you can bring back risks being artificial and false. It is a real risk, to rebuild places like this and make something of them for the exclusive use of the elite - the only ones who can enjoy a certain type of expense and atmosphere. From a marketing point of view, the idea of a luxurious place they had in the nineties could make sense at the time. The market, however, has changed in the meantime. Today, tourists are looking for an experience. Experiential tourism is growing as travelers seek a certain type of life. How do you make this natural and not artificial? How do you put your life back, to restart a process connected with a centuries-old history like this? […] Today, the Postignano project cannot be sold without a community […] otherwise the product becomes a sort of Disneyland and the territory dies". The founder himself acknowledges that "All in all, Postignano remains an artificial village", but - even without an organic strategic plan - what the managers of Postignano are working on are precisely the experiences shared with the inhabitants of the surrounding communities, the recognizability of the village and its insertion into the territorial fabric through the proposal of cultural events, the job offer and the inclusion programs of the younger generations. It is no coincidence that the co-founder argues that "It is essential for us to work so as not to be perceived by the population as an extraneous object".

"These are social issues, but no less than tourism and marketing communication" declares the communications manager. "These are issues of a social nature, but no less than tourism and marketing communication" declares the communications manager. It is important to make yourself recognizable on the web through the site, according to everyone, "without which no one would find us" (cit. Founder) and through the activation of social channels, but it would be equally important to intervene with signs in the area, "which we plan to install".

"We want Postignano to become a living reality. We know what we want to communicate, but we still don't know exactly how" is the summary of the communication manager.

The management of the village, therefore, recognized the need and the usefulness of branding the village and the surrounding area, but at the same time it also focused on the importance of creating a complex performative tourist proposal, which cannot be done of images only, but which requires the restoration of a stable local community and the resumption of a life lived within the village.

The Village Elements of Attraction

This section of the interviews display also includes the answers given by those who bought an apartment in Postignano and spend part of the year there. The exposition will be articulated by first presenting generic elements of attraction: ownership of the village or the service proposal aimed at new residents; then reference will be made more specifically to the community dimension.

The former consultant for the promotion of the village claims to have always insisted, in his activity, on the aspect of "the high architectural quality of Postignano". The co-founder himself argues that this same element is fundamental: "respect for housing traditions has been observed by furnishing the apartments and rooms with local furniture used or redone in the same style as in the past, by installing underfloor heating systems under the original trampling so as not to put the radiators, painting the walls as they did at the time when it was abandoned… in short, taking care of many details connected to the memory and history of the place". One of the village main attractions, perhaps is its architecture, also due to the anti-seismic safety. The architecture of the village is perhaps one of its attractions also from the side of anti-seismic safety by which it is characterized. The founder explains that "we have been able to carry out anti-seismic work impossible to repeat in other contexts. We have secured fragile and ancient buildings by doing work from their foundations upwards. This is an operation that would require the transfer of entire families in order to be implemented in an inhabited area ". The money and the efforts made have been rewarded, at least in some cases, in terms of image as one of the buyers interviewed emblematically stated: "Knowing that the area was at risk, earthquake safety was essential for me. I was convinced in October 2016, when I experienced the strongest earthquake I've ever felt and not a single brick collapsed, I didn't see a single crack. It was then that I realized it was a safe place".

"An important step was the opening of the restaurant" says the founder of the project. "The culinary offer is linked to local productions and seasons" underlines the co-founder and this seems to be of particular interest to buyers who declare themselves in love with Italian food and the village restaurant. Along with the culinary specialties, artisanal ones are also offered in small shops, displaying typical products made with local materials. These two aspects are often presented, both by the founder and by the co-founder, combined with the presence of native - or Umbrian - staff who make up 60% of the active forces of the structure.

The topic of staff introduces that of the feeling of community that is intended to be created within the village. The staff is also selected for their ability to relate to the public. "Places are made not only of stones, but also of people - says the creator of the project - if the people inside them know how to give something special, we are at a good point. Anyone who comes to Postignano feels at home, feels welcomed, but not in a hotel sense: I mean welcomed into a sort of community. I can't explain this concept, maybe it's indefinable, but I think it helps to convey the idea". Community is the word that even those who have bought a house use to describe the feeling of welcome they feel when returning to the village: "It is the right word to describe the ties we have established there - says a couple of interviewees - both with the staff and with the other people who bought a house in Postignano. In addition, there is often the opportunity to meet people who come and go thanks to the hotel". Another home buyer says: "People in Postignano find themselves in a community where there is a desire to be with others but with respect for personal spaces, if you want to be alone, but if they don't want to be alone. alone there are possibilities to meet many interesting people, who talk about life, philosophy, culture. There are exchanges of lives that are such a treasure of contact". Another relates an episode to explain what he means by community: "I'll give you an example: when I arrive in Postignano by car I always have a lot of things to bring, because we don't come often and I take a lot of things with me around the house. I always leave the car in the parking lot in front of the church… It takes two minutes for all my things to be in front of the door because everyone comes to help and it's so nice that I don't even have to ask for anything: [imitates, with an Italian accent, greetings from people who call her to welcome her and go say hello], I arrive and hop! They are all there. The warmth of this community is really unique, fantastic". The founder and co-founder are often present in the village and involve guests by talking to them, proposing activities and spending time with them. There are those who speak of "networks of friends that have started from Postignano and reach the whole world, now, also remaining outside the village". The place is very much appreciated for the "Italian hospitality, for the *energy* (a word that occurs often in the interviews, to describe the atmosphere of the hamlet), for cultural, educational, health events, for the restaurant and the bar". Someone goes so far as to talk about long-term prospects: "I have moved my residence here and I will come to spend my retirement years. It will be a perfect setting to continue to carry out my professional projects and to bring some cultural proposals to the village".

4 Conclusions

From what was told by the managers of the village redevelopment project and their collaborators during the interviews, it is possible to conclude that the approach to its management has had to evolve in the face of continuous changes: from the destruction of the 1997 earthquake, to the advent of the internet, the 2008 crisis, the return of the 2016 seismic phenomenon that affected the reputation of the whole area. The main change in the management style of the village involved the birth of a new awareness on the part of the administrators, that of having to build a profound, multifaceted, performative tourist offer, which offers an experience to the tourist who enters in Postignano, but also to who buys an apartment inside. A normal real estate sale process would probably involve focusing only on the transfer of ownership of a property from one owner to another, but, in this case, in addition to the structures, the social aspects are also part of the transactions founding architects of the project build from scratch.

All this implies a profound change for a project that was born as an "architectural challenge" (cit. Founder) to be concluded with the sale of a luxury real estate asset to an entity that should have taken on it entirely. According to the initial idea, in fact, the owner of the village should not have taken care of the design of a new community. "You cannot sell Postignano without a community" said finally the communication manager of the village, but one of the peculiarities of this case is that the village is reborn without any inhabitant and is restored according to criteria that design the future possible community upstream. What is described here is the transitional identity of Postignano, the one that must be understood to realize the actions of the village's rebirth [4]. This is a situation that will soon evolve depending on the work of those who run this place, but also on the institutions and the surrounding local community [7]. In the passivity of the district, in the lack of coordination by the institutions, in the excess of branding of the village, there is a concrete possibility that the direction conceived for Postignano is that of the territory-landscape-show [4], "In front of which the look passively disposes itself", in which the imagination of the place prevails over the life lived in it, as in a museum [4]. To avoid this, it is necessary to act precisely on the "negotiation" and mutual inclusion between the parties that Battaglini [11] indicates as a route for the sustainable development of a territory.

The leaders of the village understand with different degrees of awareness that they have to act on different fronts to create a "practiced landscape", really lived despite an open-air museum: involve the surrounding municipality in the life of the settlement, foster links between this and that of new buyers, but also foster maximum cohesion between the new inhabitants. Another front on which it is necessary to work, according to the leaders of the village, is that of the relationship with local institutions, which should contribute in terms of coordination between actors operating in the surrounding area to favour its development, as well as work in the direction to improve the response in case of large-scale emergencies.

The many aspects that make up the life of a practiced landscape, however, cannot be communicated for promotional purposes through marketing operations that still aim at transmitting flat, postcard-like images of the territory or, as they have been defined, images of a "landscape- show". From the interviews it was possible to record a certain degree of satisfaction from the new buyers, who found a positive correspondence between the promotional images that had been proposed to them and the life they had the opportunity to experience in the village. This happy correspondence, however, depends on the survival of a single company that guarantees continuity to the life of the village even in the months when no one lives there, which leads tourists to revive the life of the settlement through the hotel business, which builds the entire cultural offer of the area. It is a fragile balance, which will have to reinvent itself in order to last and be sustainable.

References

1. Celata, F.: Platform capitalism and the new logics of commodification of places, Franco Angeli, Milano (2018)
2. Gainsforth, S.: Oltre il turismo. Esiste un turismo sostenibile? Eris, Torino (2020)
3. Zenker, S., Erfgen, C.: Let them do the work: A participatory place branding approach. J. Place Manage. Dev. 7(3), 225–234 (2014)
4. Baule, G., Calabi, D.A., Scuri, S.: Narrare il territorio: dispositivi e strategie d'innovazione per gli spazi percepiti. In: 5th STS Italia Conference. A Matter of Design: Making Society through Science and Technology, Politiecnico di Milano, Milano (2014)
5. Un-Habitat: World cities report 2016: Urbanization and development–emerging futures. United Nations Human Settlements Programme (2016)
6. Lucatelli, S., Monaco, F., Tantillo, F.: La Strategia delle aree interne al sevizio di un nuovo modello di sviluppo locale per l'Italia. In: Rivista economica del Mezzogiorno, (3–4) (2019)
7. Labrianidis, L.: Fostering entrepreneurship as a means to overcome barriers to development of rural peripheral areas in Europe. Eur. Plan. Stud. 14(1), 3–8 (2006)

8. Biasillo, R.: Dalla montagna alle aree interne. La marginalizzazione territoriale nella storia d'Italia. In: Rivista di storia e storiografia onlile, vol. 47 (2018)
9. Battaglini, E.: Sviluppo territoriale. Dal disegno della ricerca alla valutazione dei risultati, Franco Angeli, Milano (2014)
10. Teti, V.: Il senso dei luoghi, Donzelli, Roma (2004)
11. Battaglini, E.: Urban heritage conservation and development. In: Encyclopedia of UN Sustainable Development Goals. Springer (2019)

Fostering the Renovation of the Existing Building Stock. Operational Models and Evaluation Tools

Fabrizio Battisti[1]([✉]) [iD] and Orazio Campo[2]

[1] Department of Architecture, University of Florence, 50121 Florence, Italy
fabrizio.battisti@unifi.it
[2] Department of Planning, Design, and Technology of Architecture,
Sapienza University of Rome, 00196 Rome, Italy
orazio.campo@uniroma1.it

Abstract. This paper analyses the 3 main forms of regeneration incentives adopted in the European Union: that is, direct funding, tax incentives and density bonus. Nowadays (2021), density bonus is the least used among the three, as only few Member States have included it in their framework. However, this tool has already been experimented in countries outside Europe, in order to foster building stock regeneration without increases in public expenditure. This incentive consists in the attribution of reward building rights following the execution of building interventions aimed to achieve objectives defined by public policies. In the European countries where it is adopted (Italy, France), density bonus is attributed for the achievement of specific energy efficiency standards. However, the use of density bonus produces significant impacts on landscape, environment and infrastructural demand; therefore, its applicability should be opportunely evaluated. Hence, this contribution aims to outline possible evaluation tools to adopt in combination with this incentive to support the renovation of the existing building stock.

Keywords: Building renovation · Urban regeneration · Density bonus

1 Introduction

In European Union, the renovation of public and private buildings is now considered a key action to promote energy efficiency in the building sector. This also contributes to the achievement of the objectives established by the European Law on climate (Regulation 2021/1119/EU) and indicated in the Green Deal, aimed to reduce emissions by 55% by 2030, and to eliminate them completely by 2050 [1–3].

Even though Next Generation EU [4] accounts for direct funding to support interventions for energy efficiency and seismic safety, additional measures can contribute significantly to foster diffuse actions [5]. Some incentives have already been established by regional and local governments since many years: these include tax incentives in proportion to the expenses for building renovation, and density bonus, that is the attribution

F. Calabrò et al. (Eds.): NMP 2022, LNNS 482, pp. 746–753, 2022.
https://doi.org/10.1007/978-3-031-06825-6_70

of incremental building rights, following energy or seismic retrofits and/or demolition and reconstruction interventions [6].

A very wide part of the European building stock can be subjected to these measures. Around the 75% of the existing 260 million building units [7] cannot be considered energy efficient, according to the requirements of EU Directive 2018/844, which states that an efficient building is considered as such when at least the 50% of thermal and electric energy demand comes from renewable sources. In addition to issues related to energy efficiency, European buildings also suffers from technical and quality lacks: according to Eurostat, 6,9% of the building stock has deep deficiencies that do not allow winter heating; 1,6% has no sanitary systems; as much as 12,7% of the building stock has significant technical and structural problems on the roof [7].

Hence, urban regeneration represents a crucial mission in EU, especially considering the objective of the European Commission to reach the regeneration trend of 5 million units per year by 2030.

The renovation of the existing building stock raises issues concerning repartition modalities for direct funding in support of renovation interventions, and regarding other possible measures such as tax incentives and density bonus.

While direct funding and tax incentives negatively affect public expenditure, density bonus does not. However, increases in building rights may produce significant impacts on urban dynamics, which require accurate evaluations.

After analyzing tools for regeneration incentives, this paper outlines a focus on the density bonus; finally, it contributes to the individuation of evaluation tools that allow assessing the applicability of density bonus itself.

2   Incentive Tools for Urban Regeneration

2.1   Direct Funding

The most common form of regeneration incentives is the direct subsidy (usually, a refund of expenses) to the building owner or user for specific interventions, generally aimed to the achievement of design standards [8].

In Europe, fundings are supplied directly by EU for specific projects that follow the policies of the Community; additionally, a form of "shared management" is used, that is a financing system where the European Commission manages funding jointly with national and regional authorities (this represents around 80% of EU budget).

Within EU, several funding streams can be used to support building refurbishment. Some of the most important ones are European Structural and Investment Funds (ESIF), the European Fund for Strategic Investments (EFSI) and the ELENA facility. Moreover, additional specific funding streams will be developed between 2021 and 2027 within InvestEU. Other related funding instruments include the LIFE programs, divided into 4 sub-programs; one of them concerns the clean energy transition. Commission policies regarding energy efficiency funding are specifically focused on buildings, as demonstrated by the Smart Finance for Smart Buildings facility, adopted within the Clean Energy for all Europeans package. Building refurbishment, in particular when performed on residential and public buildings, has been regularly discussed in the Sustainable Energy Investment Forums. These meetings are aimed to promote dialogue and share best practices in the development of projects and investment programs on sustainable energy [3, 4]. The renovation strategy of Green Deal will be supported a consistent funding, in order to allow large-scale investments.

This framework is compounded by fundings managed by national and regional authorities.

2.2 Tax Incentives

An additional incentive for urban regeneration is represented by tax incentives, which take the form of deductions to renovation costs. The practice of tax incentives has also been recognized by the World Bank as a tool to absorb private capitals for urban regeneration [8]. Unlike others (density bonus, up-zoning, transferable development rights), tax incentives involve an indirect exchange of economic resources between the public and private sector. In other words, tax incentives are compensated by an increase in Gross Domestic Product within a taxed sector, which is produced by the application of the incentive itself.

USA and United Kingdom have been using tax incentives since the 80s in order to perform urban regeneration actions, aiming also to increase occupation in the construction sector.

2.3 Density Bonus

Density bonus is an increase in building rights on a real estate, following an expense with positive effects on the achievement of public goals [8–12].

The concept of density bonus can be found ante litteram in the urban planning tools introduced since the '80s of the 20th century (Neighborhood Agreements, Urban I, Urban II, Urbact), some of which have been included in the framework of the Member States. Within such requalification programs for decayed territorial contexts, new projects could derogate from the building ratio index, generating economic increases (land value increase) that served as economic driving forces for the transformations [13].

Conversely, this approach has already been used for a long time in the United States with a specific codification. In its American applications, density bonus represents an opportunity for local administrations to foster private investments for the urban regeneration of specific areas. This tool works best in cities with a strong market demand and a limited availability of free spaces, or for project and locations where economic incentives produce a higher profitability for densification than for other options of real estate development. Density bonus has been used to promote the realization of social housing and public spaces, but also environmental safeguard, as this tool reduces land consumption.

The applicability of this tool is limited to solid markets, where project developers can afford the cost of realization of additional housing units. Moreover, it can be used with a social purpose, to foster the creation of communities with heterogeneous incomes. For example, it can allow local authorities to incentivize the construction of a public asset (for example, additional social housing at affordable prices), by compensating construction costs. In addition to experiences with density bonus in USA, there have been significant experimentations in Brazil, Hong Kong, Japan, Singapore and France, too. The following paragraph provides an outline of possible evaluation tools to support the application of density bonus.

3   Evaluation Tools for Density Bonus

Evaluating density bonus requires understanding its future impacts before its application. This operation is a necessary to assess its applicability.

First, it must be considered that density bonus has huge financial effects. Hence, it is vital to define a fair and calibrated value for it, as it requires counterbalancing.

Concerning this, it is possible to use Highest and Best Use through Cost Revenue Analysis, Discounted Cash Flow Analysis, Cost Volume Profit Analysis or Balance Sheet Model, also in a combined way.

Highest and Best Use (HBU) allows estimating the market value of a real estate as a function of its transformation scenarios (including changes of intended use). In particular, the implementation of HBU estimates the value of each hypothesized transformation of the real estate [14]. In this sense, HBU contributes to the individuation of the best possible use that the destiny bonus can produce.

Various evaluation methods can be implemented in HBU: in particular, Cost-Revenue Analysis (CRA) or Discounted Cash Flow Analysis [15].

CRA allows evaluating the results of the production process (from the owner/manager's standpoint) at time zero (the time when the estimate is performed) in financial terms, expressed through specific performance criteria, consists in the following phases:

1. Estimation of all costs and incomes of the production process, quantified in detail.
2. Cash flow creation (referred to the duration of the production process) with calculation of discount rate. After estimating costs and incomes of the production process, these must be calculated for the whole duration of the intervention. This allows estimating the financial balance for each year of the intervention; then, this can be back-discounted by using an appropriate discount rate.
3. Calculation of financial performance indicators – net annual value (NAV) and internal rate of return (IRR) – to drive decision making in relation to the convenience of the proposed intervention(s). If CRA considers cash inflows and outflows in accounting terms, it is reduced to a traditional Discounted Cash Flows Analysis (DCFA).

CVPA allows relating the main financial and dimensional parameters of a production process in the real estate sector, highlighting their interrelations, and favoring their calibration. Hence, it is possible to perform an accurate analysis of the influence of the financial composition of costs and incomes on the success of the intervention [15, 16]. The implementation of CVPA requires, in first place, the estimation of intervention costs – divided into fixed and variable – and incomes. The second step data processing in order to determine the break even point, contribution margin, extra profit margin and the operating leverage of the intervention. If the technical and financial characteristics of the intervention (costs and incomes for each scenario) are known, the break-even point allows individuating the quantity of building product that should be realized and sold in order to produce an economic balance for the transformation. The total contribution margin is the available financial amount to sustain the fixed costs of the intervention; if (fixed and variable) costs include ordinary profit, the surplus with respect to fixed costs represents an extra profit. The stability of the intervention of building production can be tested by using the operating leverage coefficient: its higher values indicate safer operations.

The Balance Sheet Model (BSM) is widely used to verify the conditions of economic and financial balance for a business [17–19]. It allows representing the general balance and the specific incomes from single business activities. At the same time, the limits of the algorithmic description of dynamic, open and non-mechanistic system – as a business is – are well-known. However, these limits are less restrictive for a business activity consisting in a single renovation intervention on a building [20, 21].

Hence, CVPA and BSM can effectively contribute to determine the consistence of density bonus.

Density bonus produces multi-dimensional effects, as the increase in building volumes has impacts on landscape, environment and infrastructural demand; therefore, the evaluation of the financial aspects must be integrated with a multi-criteria decision analysis (MCDA) or a cost-benefit analysis (CBA).

MCDA techniques allow performing evaluations with both quantitative and qualitative criteria/sub-criteria. Moreover, they allow considering different (stakeholders') standpoints. Alternatives can be ranked through MCDA techniques, by identifying the most suitable one, according to the goals set by the decision makers; additionally, it is possible to use them to perform compliance evaluations in relation to specific law parameters and/or best practice benchmarks.

According to the literature on MCDA [22–26], when using these methodologies, it is necessary to select the most appropriate one among the various tools developed over time (among the most significant ones: WSM, AHP, ELECTRE, EVAMIX, TOPSIS, MACBETH), in relation to the characteristics of the evaluation. The mentioned MCDA techniques are the most discussed in the scientific literature, but it is not excluded that also MCDA techniques of lesser diffusion can effectively contribute to solve decisive decision nodes for the success of the regeneration processes.

On the other hand, CBA allows forecasting the effects of a project/program/investment, not only from the owner/manager's standpoint, but also from the communitarian one. It is structured into two phases [27, 28]:

1. financial analysis (basically like the one outlined for CRA)
2. economic analysis, which performs the following operations on the data considered for the calculation of the financial return of an intervention: tax adjustment, conversion of market prices into shadow prices; evaluation of non-market impacts and correction for externalities (European Commission, 2014).

As in CRA, the performance of the intervention can be evaluated by adopting an appropriate discount rate and using NAV and IRR as economic indicators.

In order to assess the applicability of density bonus, it seems opportune to consider the effects deriving from the diffuse application of this incentive tool over time. Scenario analysis is particularly useful in this respect [29, 30], as it performs predictions according to tendencies and events recorded in the past, by introducing the concept of conditional hypotheses at the base of the constant changes in society.

4 Conclusions

Density bonus produces "densification" effects with significant impacts on environment, landscape and infrastructural demand, even if they are correctly managed within a building project. Regeneration is aimed to generate an overall (infrastructural, landscape and environmental) improvement in urban areas, by enhancing livability, building quality and resource efficiency. This requires reaching a polysemic and multi-disciplinary vision of interventions, beyond the notions of 'recovery', 'reuse' and even 'requalification'. Regeneration is a public and/or private action that produces an increase in economic, cultural and social values [31], assuming an inclusive meaning that encompasses multi-scalar, multi-cultural and multi-disciplinary approaches [32, 33]. Hence, the renovation of buildings and districts has a wider and comprehensive meaning, as it includes all actions that restore integral quality in contexts. This refers to the functional, distributive and technological aspects of houses and common spaces, but also to energy efficiency, representing one of the key factors for environmental quality.

When fostering regeneration through density bonus, it is then vital to consider the effects produced by it: in addition to the economic ones, those on environment, landscape and infrastructural demand must be analyzed. A possible hypothesis could be to regulate the attribution of density bonus through specific (urban planning) tools for the management of building renovation processes, possibly supported by a combined application of the evaluation tools proposed.

The conjunct use of these evaluation tools seems necessary in relation to the heterogeneity of the effects of density bonus, also considering the wideness of the concept of urban regeneration and the current (2021) relevance of building renovation policies within EU. The use of appropriate evaluation techniques can provide useful cognitive elements to public decision-makers to establish appropriate policies and initiatives of regeneration, being able therefore to resort to a proper use of incentive tools discussed in this paper [34].

References

1. European Commission: The European Green Deal. Brussels, 11.12.2019 COM (2019) 640 Final. https://ec.europa.eu/info/sites/default/files/european-green-deal-communication_en.pdf. Accessed 28 Dec 2021
2. Eurostat: Energy Data 2020 Edition. https://ec.europa.eu/eurostat/documents/3217494/110 99022/KS-HB-20-001-EN-N.pdf/bf891880-1e3e-b4ba-0061-19810ebf2c64?t=159471560 8000. Accessed 28 Dec 2021
3. European Commission: A renovation wave for Europe—greening our buildings, creating jobs, improving lives. Brussels, 14.10.2020 COM (2020) 662 Final. https://eur-lex.europa.eu/legal-content/EN/TXT/PDF/?uri=CELEX:52020DC0662&from=EN. Accessed 28 Dec 2021

4. European Commission: Recovery plan for Europe. https://ec.europa.eu/info/strategy/rec overy-plan-europe_en Accessed 28 Dec 2021

5. Bertoldi, P., Economidou, M., Palermo, V., Boza-Kiss, B., Todeschi, V.: How to finance energy renovation of residential buildings: Review of current and emerging financing instruments in the EU. WIREs Energy Environ **10**, e384 (2021)

6. Battisti, F., Campo, O.: The assessment of density bonus in building renovation interventions. the case of the city of Florence in Italy. Land **10**, 1391 (2021)

7. Eurostat. Quality of Housing. https://ec.europa.eu/eurostat/cache/digpub/housing/bloc-1c. html?lang=en. Accessed 28 Dec 2021

8. World Bank: Urban regeneration. https://urban-regeneration.worldbank.org/. Accessed 28 Dec 2021

9. Conticelli, E., Proli, S., Tondelli, S.: Integrating energy efficiency and urban densification policies: two Italian case studies. Energy Build. **155**, 308–323 (2017)

10. Beghelli, S., Guastella, G., Pareglio, S.: Governance fragmentation and urban spatial expansion: evidence from Europe and the united states. [Governance-Fragmentierung und urbane räumliche Expansion: Erkenntnisse aus Europa und den USA]. Rev. Reg. Res. **40**, 13–32 (2020)

11. Paetz, M.M.D., Pinto-Delas, K.: From red lights to green lights: town planning incentives for green building. In: Proceedings of the Talking and Walking Sustainability International Conference, Auckland, New Zealand, 20–23 February 2007

12. Amoruso, F.M., Sonn, M.-H., Chu, S., Schuetze, T.: Sustainable building legislation and incentives in korea: a case-study-based comparison of building new and renovation. Sustainability **13**, 4889 (2021)

13. Guarini, M.R., Battisti, F.: Benchmarking multi-criteria evaluation: a proposed method for the definition of benchmarks in negotiation public-private partnerships. In: Murgante, B., et al. (eds.) ICCSA 2014. LNCS, vol. 8581, pp. 208–223. Springer, Cham (2014). https://doi.org/ 10.1007/978-3-319-09150-1_16

14. Fattinnanzi, E., Acampa, G., Battisti, F., Campo, O., Forte, F.: Applying the depreciated replacement cost method when assessing the market value of public property lacking comparables and income data. Sustainability **12**, 8993 (2020)

15. Battisti, F., Guarini, M.R.: Public interest evaluation in negotiated public-private partnership. Int. J. Multicriteria Decis. Mak. **7**, 54–89 (2017)

16. Morano, P., Tajani F.: Break even analysis for the financial verification of urban regeneration projects. In: Applied Mechanics and Materials. Trans Tech Publications Ltd, pp. 1830–1835 (2013)

17. Bezemer, D.J.: The economy as a complex system: the balance sheet dimension. Adv. Complex Syst. **15**, 1250047 (2012)

18. Kulikova, L.I., Garyntsev, A.G., Gafieva, G.M.: The balance sheet as information model. Procedia Econ. Financ. **24**, 339–343 (2015)

19. Rudd, A., Siegel, L.B.: Using an economic balance sheet for financial planning. J. Wealth Manag. **162**, 15–23 (2014)

20. Chlodnicka, H., Zimon, G.: Balance sheet model for small economic entities. Econ. Soc. Dev. Book Proc. **6**, 243–250 (2019)

21. Dichev, I.D.: On the balance sheet-based model of financial reporting. Account. Horiz. **22**, 453–470 (2008)

22. Guarini, M.R., Battisti, F., Chiovitti, A.: Public initiatives of settlement transformation: a theoretical-methodological approach to selecting tools of multi-criteria decision analysis. Buildings **8**, 1 (2017)

23. Ishizaka, A., Nemery, P.: Multi-criteria Decision Analysis: Methods and Software. Wiley, Hoboken (2013)

24. Roy, B.: Méthodologie Multicritére d'Aide à la Décision. Economica, Paris (1985)
25. Guitoni, A., Martel, J.M.: Tentative guidelines to help choosing an appropriate MCDA method. Eur. J. Oper. Res. **109**, 501–521 (1998)
26. Vincke, P.: L'aide Multicritère à la Décision. Université de Bruxelles, Bruxelles (1989)
27. Pearce, D.W.: Cost-benefit analysis. Macmillan International Higher Education (2016)
28. Layard, P.R.G., et al.: Cost-Benefit Analysis. Cambridge University Press, Cambridge (1994)
29. Rescher, N.: Predicting the Future. An Introduction to the Theory of Forecasting. State University of New York Press, Albany (1998)
30. Makridakis, S., Wheelright, S.C., Hyndman, R.J.: Forecasting. Methods and Applications. Wiley, New York (1998)
31. Murtagh, B.: Urban regeneration and the social economy. In: The Routledge Companion to Urban Regeneration, pp. 219–228. Routledge, London (2013)
32. Capolongo, S., Sdino, L., Dell'Ovo, M., Moioli, R., Della Torre, S.: How to assess urban regeneration proposals by considering conflicting values. Sustainability **11**, 3877 (2019)
33. Daniels, B., Zaunbrecher, B.S., Paas, B., Ottermanns, R., Ziefle, M., Roß-Nickoll, M.: Assessment of urban green space structures and their quality from a multidimensional perspective. Sci. Total Environ. **615**, 1364–1378 (2018)
34. Battisti, F., Campo, O.: The assessment of density bonus in building renovation interventions. the case of the city of Florence in Italy. Land **10**(12) (2021)

Towards a More Sustainable Use of Land.
A Comparative Overview of the Italian Regional Legislation

Donato Casavola$^{(\boxtimes)}$ (iD) and Giancarlo Cotella (iD)

DIST | Politecnico di Torino, Viale Mattioli, 39, 10125 Turin, Italy
{donato.casavola,giancarlo.cotella}@polito.it

Abstract. The contribution explores how the different spatial planning laws that characterise the Italian regions approach sustainable land-use issues and what implications for instruments and practices they entail. At first, the national situation on soil consumption is presented, drawing on existing studies. Then, the analysis shifts its focus to how sustainable land-use issues are considered within the 21 regional/provincial legislative frameworks that define the functioning of spatial governance and planning in the country. To this end, a number of characterizing elements have been identified and used to assess the different spatial planning laws. From the analysis, it is possible to reflect on the legislative, economic-productive and morphological factors that explain the differences observed among the various Italian regions in terms of soil consumption.

Keywords: Sustainable land-use · Spatial planning · Italy

1 Introduction

Since the second half of the last century, urbanisation processes have become more and more intense, leading to the progressive recognition of the finite nature of the land. More recently, the COVID-19 crisis has further warned us about the importance that a present and future sustainable built, as well as natural, environment, could have in facing unexpected emergencies more resiliently [1, 2]. In this light, it is essential to make careful decisions on urbanization and land-use management, approaching the latter not only as a political and technocratic decision but as one that affects our society's well-being and quality of life [3, 4]. This perspective is well acknowledged at the European level, with the European Union (EU) which, through time, has introduced a growing number of policies and actions aiming at promoting a more sustainable approach to development and urbanization [5–7]. In particular, the EU is trying to halt excessive land transformation with its objective to achieve zero net land take by 2050 and, more recently, the European Green Deal has stressed the need to make Europe climate neutral by 2050 [8, 9]. As a result, in the last few years, policy and decision-makers at all territorial levels have started to dedicate increasing efforts to pursue urbanization and land-use models that are more sustainable, thus leading to the consolidation of an increasingly

F. Calabrò et al. (Eds.): NMP 2022, LNNS 482, pp. 754–763, 2022.
https://doi.org/10.1007/978-3-031-06825-6_71

heterogeneous set of interventions and practices aiming at this direction [10, 11]. This process has been reinforced by the acknowledgement, within the SDG framework, of the role that spatial governance and planning could play in the process [12], leading to action in the context of both urbanized and depopulated remote rural areas [13–15]. In particular, sustainable land-use seems to depend both on the socio-economic processes that trigger spatial development and on the effectiveness of the instruments that aim at steering and regulating these processes. In turn, both variables are directly influenced by the high heterogeneity that characterizes the European continent in terms of socio-economic development, administrative culture and spatial governance and planning [16, 17]. Acknowledging the above, this contribution develops a comparative analysis of the Italian regional spatial planning legislation, aiming at assessing its suitability to orient territorial development and land-use in a more sustainable way. After this brief introduction, the heterogeneous landscape for spatial governance and planning that characterise Italy and its regions is presented, together with some information concerning the national and regional situation on soil consumption. Then, the analysis shifts its focus to how sustainable land-use issues are considered within the laws that regulate the functioning of spatial governance and planning activities in the 19 Italian regions and the 2 autonomous provinces of Trento and Bolzano. To this end, a number of characterizing elements have been identified and used to assess the different spatial planning laws. The results of the analysis allow us to reflect on the main factors that may explain the differences observed between the various Italian regions in terms of soil consumption. Finally, a concluding section rounds off the contribution, summarizing its main findings and sketching out a number of avenues for future research on the matter.

2 The Fragmented Italian Spatial Planning Landscape and Its Impact on Land-Use

The Italian spatial governance and planning system is based on the national Law 1150/1942, and a number of amendments to the latter that have been approved through time. However, it experienced increasing regionalisation since the 1970s, when regions were created and provided with legislative competencies in the field of *urbanistica* [18]. All regions started to approve their spatial planning laws, leading to an increasing heterogeneity and divergence of regional spatial planning systems and practices [19, 20]. Nowadays we witness a situation in which some regions still operate spatial governance and planning on the basis of legal frameworks approved in the previous century, while others have profoundly renewed their action as a consequence of internal and external stimuli, among which the EU has certainly played a relevant role [21]. More specifically, most of the spatial governance and planning regional laws that have been adopted in the last 20 years, albeit with different interpretations, have sought to address topical issues such as the normalisation of innovative renewal and regeneration programmes at the local level, the introduction of communicative and participatory processes in planning, the involvement of private stakeholders in territorial transformations and service provision, the introduction of the ex-ante and ex-post evaluation of plans and programmes with various processes, from environmental assessment to integrative approaches, which aim to assess the impact of spatial transformation on the territorial system. As a consequence, the Italian regions feature significant differences in terms of spatial governance and planning instruments, procedures, objectives and functions, which also depends on the time when each law was developed and approved. The evolution of legislative planning competencies to the regional authorities have led to the consolidation of a mosaic characterised by the cohabitation of 21 different regional spatial governance and planning systems (19 regions + the two autonomous provinces of Trento and Bolzano). Despite the described regional innovation, or perhaps as a consequence of the latter, a legislative void persists at the national level concerning sustainable land-use and soil consumption. In recent years, a number of legislation proposals have been under discussion, ranging from those specifically dedicated to soil consumption to those mainly dedicated to urban regeneration, of which the most notable are (i) the law proposal n.63 'Disposizioni per il contenimento del consumo di suolo e per il riuso dei suoli edificati', (ii) the law proposal n.86 'Disposizioni per la riduzione del consumo di suolo nonché delega al Governo in materia di rigenerazione delle aree urbane degradate' and (iii) the law proposal n.164 'Disposizioni per l'arresto del consumo di suolo, di riuso del suolo edificato e per la tutela del paesaggio'. None of them has however been approved yet. In the meantime, the regions have adopted in their spatial governance and planning laws

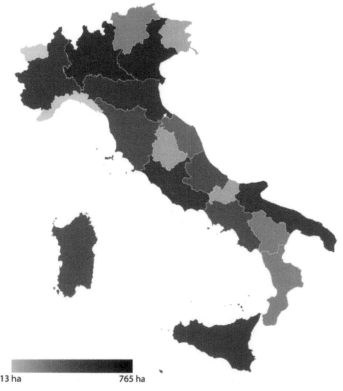

Fig. 1. Map of net soil consumption in hectares at regional level between 2019 and 2020. Source: Own elaboration on ISPRA data

several devices and amendments which, in one way or another, try to fill this gap, leading to a further increase of heterogeneity. The discrepancy between the various regional laws turns out to be a limit since the regions, feeling autonomous on the matter, very often allows derogations concerning territorial transformation interventions, in so doing leading to increasing soil consumption. As a consequence, from a quantitative point of view, soil consumption in Italy continues to grow. According to the last report published by ISPRA (Istituto Superiore per la Protezione e la Ricerca Ambientale) [22], the new artificial coverings in 2020 concern 56.7 square kilometres of land, that is, on average, about 15 hectares per day. A transformation speed is in line with that recorded in the last years and which concerns almost 2 square meters of soil which have been irreversibly lost every second. At the regional level, the highest shares of consumed soil are recorded in Lombardy (12.08%), Veneto (11.87%) and Campania (10.39%), followed by Emilia-Romagna, Apulia, Lazio, Friuli-Venezia Giulia and Liguria, with values above the national average and between 7 and 9%. Valle d'Aosta is the region with the lowest percentage (2.14%). The largest increases since 2020 have occurred in the regions of Lombardy (+765 hectares), Veneto (+682), Apulia (+493), Piedmont (+439) and Lazio (+431) (Fig. 1).

3   Methodology of Analysis

In order to compare and assess the 21 regional/provincial spatial planning legislation that characterises the country *vis-à-vis* their suitability to promote a more sustainable use of land, we adopted the following methodological steps.

At first, we explored whether the regional/provincial laws included any reference to the concepts of sustainable land-use, sustainable urbanization, urban sprawl, soil consumption containment, zero-balance soil consumption, urban regeneration, and all those concepts that make up the conceptual framework on the subject under investigation. After this initial screening, and drawing on the latter, we identified a number of characterising elements, whose presence in a regional legislative framework outlines a certain degree of attention to the subject under investigation (Table 1). A score from 1 to 5 was attributed to each characterizing element, symbolising its relative importance concerning the achievement of more sustainable use of land. To operate this 'weighting' of the identified elements, we took direct inspiration from the results of the ESPON SUPER project, which had surveyed and assessed the effectiveness of over 230 initiatives aiming at sustainable urbanisation and land-use in as many as 39 European countries [23].

More specifically: the regional regulations featuring an organic and specific regulation on the containment of soil consumption were given 5 points (the most impacting among the identified characterizing element); those laws identifying a threshold or an overall quota of eligible soil consumption scored 2 points; 3 points were awarded to legislations setting a fixed perimeter of their urbanized area; ecological compensation actions were awarded 1 point, and the same goes for the activation of programs explicitly dedicated to urban regeneration; 2 points were assigned to the presence of explicit regulations on the monitoring of land consumption as well as to those regulations that include reward measures and incentives; 3 points were awarded for the existence of specific devices that regulate and favour the temporary use of buildings and 1 point was

Table 1. Characterizing elements and related scoring

Characterizing elements	Score
CE1. Soil consumption and regeneration in legislation as a principle	1
CE2. The existence of one or more ad hoc articles on the topic of land-use	5
CE3. Regulations on limiting land consumption through regeneration programmes for the existing building stock	1
CE4. The existence of regulations that impose an overall quota of eligible land consumption	2
CE5. The existence of legislation that prescribes the perimeter of the urbanized area	3
CE6. The explicit presence of references to take actions aimed at ecological compensation	1
CE7. The existence of explicit regulations on the monitoring of land consumption and the creation of any lists of disused buildings to be redeveloped	2
CE8. The use of urban regeneration programs	1
CE9. The existence of legislation defining the temporary use of certain properties	3
CE10. The existence of regulations that include reward measures and incentives for urban regeneration and de-sealing of the land	2
CE11. The presence of a bill under discussion within which the issue of sustainable land-use and the containment of soil consumption is discussed	1

assigned to all those regions that have soil consumption and regeneration in legislation as a principle, a regulation on limiting land consumption through regeneration programmes and/or which are currently discussing a legislative bill on these themes. Once the above characterizing elements were identified and weighted, a table was produced, showing the performance of all regional and provincial spatial governance and planning legislative frameworks in relation to each of them. In so doing, it was possible to compare the 21 legislative frameworks under investigation and to assess them in relation to the higher or lower attention that they dedicate to sustainable land-use and soil consumption.

Finally, the results of the assessment were compared to the actual levels of soil consumption that characterizes the different Italian regions/provinces, to reflect on the actual relations between spatial governance and planning and sustainable urbanization and land-use.

4   Comparative Analysis of Regional Laws *vis-à-vis* the Regional Performance in Terms of Soil Consumption

From the comparison and assessment of the various regional/provincial regulations concerning spatial governance and planning, it is possible to confirm that the attention to sustainable urbanisation and land-use varies rather much from region to region (Table 2).

Table 2. Comparative table of regional regulations

	CE1	CE2	CE3	CE4	CE5	CE6	CE7	CE8	CE9	CE10	CE11	TOT
Abruzzo	1		1					1		1		**4**
Basilicata			1									**1**
Calabria	1	5	1			1		1				**9**
Campania	1		1					1		1		**4**
Emilia Romagna	1	5	1	2	3	1		1	3	2		**19**
Friuli Venezia Giulia	1		1			1		1				**4**
Lazio	1		1			1		1				**4**
Liguria	1		1			1		1		2		**6**
Lombardy	1		1	2		1		1				**6**
Marche	1		1					1				**3**
Molise										2		**2**
Piedmont	1	5	1	2	3	1	2	1			1	**17**
Apulia	1		1			1		1		2		**6**
Sardinia	1		1					1			1	**4**
Sicily	1	5	1	2		1		1				**11**
Tuscany	1		1		3	1		1				**7**
Umbria	1		1					1			1	**4**
Valle d'Aosta	1		1					1				**3**
Veneto	1	5	1	2		1		1	3	2		**16**
Bolzano	1	5	1		3	1		1				**12**
Trento	1	5	1					1				**8**

The Italian regions that have proved most virtuous and attentive to the containment of land-use are Emilia-Romagna, Piedmont, Veneto and, just one step behind, the Autonomous Province of Bolzano and Sicily. On the other hand, the regional contexts that seem to feature the more backward legislation on the matter are Basilicata, Marche, Valle d'Aosta and Molise. Overall, the regions that have legislated more carefully are located in the northern part of the Italian peninsula with numerous regions that have updated their regulations in recent years. The regions of Valle d'Aosta and Friuli-Venezia Giulia seems to represent an exception to this positive picture, as they still feature rather incomplete legislation for what concerns the promotion of a more sustainable use of land. When it comes to southern Italy, Sicily and Calabria rank as the most advanced regions, with the recent update of their respective regional regulations that have included soil containment and a more sustainable use of land among the most relevant territorial development priorities. The regions located in the central area of the country are more inertial in their legislative activity and, despite some recent updates, mainly limit themselves to include rather general principles focusing on the reduction of soil consumption, to be achieved in most cases through urban regeneration.

When comparing the results obtained from this assessment to the data concerning soil consumption at the regional/provincial level, several findings emerge that may appear paradoxical at a first glance. First of all, among the regions that currently consume more soil, there are some of those that feature a more complete and more comprehensive legislative framework. Veneto and Emilia-Romagna are respectively the second and the fourth Italian regions showing the highest share of soil consumption in 2020, but at the same time Emilia-Romagna also features the most comprehensive legislative framework to fight this phenomenon, and Veneto follows shortly after, with a number of recent updates that aim at the further integration of sustainable land-use interventions in its territorial policy. On the contrary, among the regions featuring the lowest values in terms of soil consumption, we find Valle d'Aosta, Basilicata, Sardinia and Molise, i.e., those regions that have apparently dedicated fewer efforts to the integration of sustainable land-use principles and instruments into their legislative frameworks.

All this leads us to consider that the presence of higher legislative attention to the theme of sustainable land-use does not always coincide with lower values concerning soil consumption. On the one hand, this is due to the fact that the latter very much depends on the actual pressure over land that is determined by intertwined socio-economic dynamics of land demand and supply [24], and on which spatial governance and planning activities can impact only to a certain extent [17]. On the other hand, it reflects the complex dynamics that concerns the causal relations between territorial development phenomena and the predisposition of instruments and devices aimed at their steering and regulation. For instance, the region of Veneto, in recent years has been the one that has consumed most soil, and this has triggered several legislative adjustments to limit these phenomena, in so doing leading to the consolidation of a legislative framework that is now among the most comprehensive in relation to the promotion of more sustainable use of land. Even if to a lesser extent, the same goes for the Piedmont region, which stands above the national average in terms of soil consumption in the years 2019–2020, while featuring a rather high attention to the issue in its legislative framework. Probably these regions, and in particular, the respective regulations, need time to show repercussions on the territory. A reverse argument may apply to those regions that, having until now presented a rather low degree of soil consumption did not yet deem necessary to fine-tune their legislative frameworks in order to increase the attention dedicated to sustainable land-use (this is the case of Basilicata, Sardinia, Valle d'Aosta and Molise). At the same time, when relating the various regional regulations and the data relating to the increase in land consumption between 2019 and 2020, Some additional insights can be individuated. For example, in the Calabria region, the low soil consumption that has characterised the region in recent years may be a direct consequence of the implementation of the updated regional spatial governance and planning legislation, that explicitly aims in this direction.

5 Conclusive Remarks and Future Research Perspectives

The chapter presented a qualitative analysis of the Italian regional/provincial legislation concerning spatial governance and planning, and of the extent to which the latter incorporate elements aimed at the limitation of soil consumption and the overall promotion of a more sustainable use of land. On this basis, it may be possible to advance some preliminary hypotheses on the possible relations occurring between spatial governance and planning regulations, land consumption values and other economic-productive and morphological factors. In this sense, for example, in regions such as Lombardy and Veneto, the high consumption of soil may have been influenced by both economic factors, due to the high value of private investments landing on these regions, and morphological factors, due to the presence of a predisposed soil to undergo transformations destined for production plants or settlements. The influence of these factors could also justify the results identified for the regions of Molise and Valle d'Aosta. Although these regions feature legislative frameworks that appear still inadequate in relation to the limitation of soil consumption and the promotion of a more sustainable use of land, their low values in relation to soil consumption values may depend on a lower volume of investments as well as on a morphological configuration that is less prone to major transformations. Be that as it may, a general trend may be observed that, in recent years, has led many regions to reform their spatial governance and planning legislation to incorporate elements that, in one way or another, aim at limiting soil consumption and organising land-use in a more sustainable way. However, this fertile legislative activity remains highly fragmented and uncoordinated, in the absence of any dedicated legislative framework at the national level. To fill this regulatory gap, a central attempt to draft of a single text that could encompass the various draft laws under discussion would be more than welcomed and perhaps a useful step towards the upscale and further diffusion of the innovative elements that have been introduced by the regions at various stages. Whereas it is certainly meritorious that several regions, in the absence of national regulations, have tried to legislate on the matter, the analysis also showed that land-use trends do not necessarily follow suit with this legislative impetus. The identified discrepancies highlight the fact that land-use is very much influenced by socioeconomic and morphological reasons. At the same time, they also warn us about the potential ineffectiveness or inadequacy of the existing spatial governance and planning system that characterise our country (and regions) to pursue the objectives it sets. This argument is supported by various authors in the literature, highlighting how conformative spatial planning systems, due to their rigidity, often fail to provide the public authority with effective control over spatial development dynamics [17, 25]. The various bills under discussion and the most significant elements that were included through time in the legislation of the various regions could constitute a sound basis upon which to find the development of a new national spatial governance and planning framework law, focusing on sustainable land-use, the containment of soil consumption and urban regeneration, and at the same time attempting to solve some of the traditional drawbacks that have characterised Italian spatial governance and planning since the Second World War [19]. This document should establish definitions, fundamental principles and guidelines in so doing serving as a blueprint for the consolidation and fine-tuning of the various regulations at the regional level. Despite its preliminary nature and several limits, the undertaken qualitative analysis has allowed for preliminary individuation of those components that may constitute a useful inspiration for the development of this national reference framework. However, additional research is needed to explore how, in the different regional contexts, the identified components influence the development of spatial planning tools and their practical implementation and how the latter manage to translate the objectives and principles included in the legislation into the envisaged results on the ground. This additional level of analysis will require the development of local case studies, that could also provide relevant insights concerning the actual capacity of spatial governance and planning instruments to steer and regulate land-use towards a more sustainable configuration.

References

1. Cotella, G., Vitale Brovarone, E.: Questioning urbanisation models in the face of Covid-19. TeMA-J. Land Use Mobil. Environ. 105–118 (2020). https://doi.org/10.6092/1970-9870/6913
2. Cotella, G., Vitale Brovarone, E.: Rethinking urbanisation after COVID-19. What role for the EU cohesion policy? Town Plann. Rev. **92**(3), 411–418 (2021). https://doi.org/10.3828/tpr.2020.54
3. Solly, A., Berisha, E., Cotella, G., Janin Rivolin, U.: How sustainable are land use tools? A Europe-wide typological investigation. Sustainability **12**, 1257 (2020). https://doi.org/10.3390/su12031257
4. Solly, A., Berisha, E., Cotella, G.: Towards sustainable urbanization learning from what's out there. Land **10**, 356 (2021). https://doi.org/10.3390/land10040356
5. Atkinson, R.: The emerging "urban agenda" and the European spatial development perspective: towards an EU urban policy? Eur. Plan. Stud. **9**(3), 385–406 (2001). https://doi.org/10.1080/713666487
6. Adams, N., Cotella, G., Nunes, R.: Spatial planning in Europe: the interplay between knowledge and policy in an enlarged EU. In: Adams, N., Cotella, G., Nunes, R. (eds.) Territorial Development, Cohesion and Spatial Planning: Knowledge and Policy Development in an Enlarged EU (pp. 29–53) Routledge (2012)
7. Cotella, G.: The urban dimension of EU cohesion policy. In: Medeiros, E. (ed.) Territorial Cohesion. TUBS, pp. 133–151. Springer, Cham (2019). https://doi.org/10.1007/978-3-030-03386-6_7
8. Cotella, G., Crivello S., Karatayev M.: European union energy policy evolutionary patterns. In: Lombardi P., Gruning M. (eds.) Low-carbon Energy Security from a European Perspective. Elsevier, Amsterdam, pp. 13–42 (2016). https://doi.org/10.1016/B978-0-12-802970-1.00002-4
9. Sikora, A.: European green deal – legal and financial challenges of the climate change. ERA Forum **21**(4), 681–697 (2020). https://doi.org/10.1007/s12027-020-00637-3
10. Bottero, M., Caprioli, C., Cotella, G., Santangelo, M.: Sustainable cities: a reflection of potentialities and limits based on existing eco-districts in Europe. Sustainability **11**(20), 5794 (2019). https://doi.org/10.3390/su11205794
11. Rotondo, F., Abastante, F., Cotella, G., Lami, I.M.: Questioning low-carbon transition governance: a comparative analysis of European case studies. Sustainability **12**(24), 10460 (2020). https://doi.org/10.3390/su122410460
12. Berisha, E., Caprioli, C., Cotella, G.: Unpacking SDG target 11. a: what is it about and how to measure its progress? City Environ. Interact. 100080 (2022). https://doi.org/10.1016/j.cacint.2022.100080

13. Paniagua, A.: Geographies of Differences (and Resistances) in Urbanized and Depopulated Remote Rural Areas. In: Leal Filho, W., Azul, A.M., Brandli, L., Lange Salvia, A., Wall, T., (eds.) Life on Land, pp. 1–11. Springer, Switzerland (2020). ISBN 978-3-319-71065-5. https://doi.org/10.1007/978-3-319-71065-5_136-1
14. Vitale Brovarone, E., Cotella, G.: Improving rural accessibility: a multilayer approach. Sustainability **12**, 2876 (2020). https://doi.org/10.3390/su12072876
15. Vitale Brovarone, E., Cotella, G., Staricco, L.: Rural Accessibility in European Regions. Routledge, London (2021). https://doi.org/10.4324/9781003083740
16. European Commission: Directorate General for Regional and Urban Policy. In: My Region, My Europe, Our Future: Seventh Report on Economic, Social and Territorial Cohesion; Publications Office: Luxembourg (2017)
17. Berisha, E., Cotella, G., Janin Rivolin, U., Solly, A.: Spatial governance and planning systems in the public control of spatial development: a European typology. Eur. Plan. Stud. **29**, 181–200 (2021). https://doi.org/10.1080/09654313.2020.1726295
18. Cotella, G., Berisha, E.: Inter-municipal spatial planning as a tool to prevent small-town competition: the case of the emilia-romagna region. In: Bansky J. (ed.) The Routledge Handbook of Small Towns, pp. 313–329. Routledge, London (2021). https://doi.org/10.4324/978100309 4203-27
19. Vettoretto, L.: Planning cultures in Italy – Reformism, Laissez-Faire and Contemporary Trends. In: Knieling, J., Othengrafen, F. (eds.) Planning Cultures in Europe, pp. 189–204. Ashgate, Farnham (2009)
20. Gelli, F.: Planning systems in Italy within the context of new processes of "Regionalization." Int. Plan. Stud. **6**(2), 183–197 (2001). https://doi.org/10.1080/13563470123858
21. Cotella, G., Rivolin, U.J.: Europeanization of spatial planning through discourse and practice in Italy. disP-The Plann. Rev. **47**(186), 42–53 (2011). https://doi.org/10.1080/02513625.2011. 10557143
22. Munafò, M.: Consumo di suolo, dinamiche territoriali e servizi ecosistemici. Edizione 2021. Report SNPA 22/21 (2021). ISBN 978-88-448-1059-7
23. ESPON SUPER – Sustainable Urbanisation and Land-use Practices in European Regions. Final Report. Luxembourg: ESPON EGTC (2020). Available at: https://www.espon.eu/super. Accessed 01 Mar 2021
24. Kroll, F., Müller, F., Haase, D., Fohrer, N.: Rural–urban gradient analysis of ecosystem services supply and demand dynamics. Land Use Policy **29**(3), 521–5350 (2012). https://doi. org/10.1016/j.landusepol.2011.07.008
25. Janin Rivolin, U.: Conforming and performing planning systems in Europe: an unbearable cohabitation. Plan. Pract. Res. **23**(2), 167–186 (2008). https://doi.org/10.1080/026974508023 27081

Minimum Environmental Criteria, Estimation of Costs and Regional Prices: Preliminary Considerations

Laura Calcagnini$^{(\boxtimes)}$, Fabrizio Finucci , and Mariolina Grasso

Department of Architecture, Roma Tre University, 00153 Rome, Italy
{laura.calcagnini,fabrizio.finucci,mariolina.grasso}@uniroma3.it

Abstract. Several European Directives over the last 15 years addressed the issue of environmental sustainability towards building design process and considering the public sector this led to a regulatory transposition of some requirements in the procurement phase. The European requests in this sense are implemented in Italy also with the introduction of the Minimum Environmental Criteria (MEC) which must be preliminarily set in the feasibility study and guaranteed in the definitive and executive design phases. After a brief overview and introductions about MEC the paper proposes an evaluation based on the regional price lists where the cost estimation is linked to public works. First of all, the evaluation (rapid, intuitive, and replicable) focuses on the analysis of the regional price lists verifying the presence of MEC materials or processes and highlighting the quantity with respect to the total of the price list items. A second comparative evaluation focuses on the comparison between two types of thermal insulators (considering three national geographical areas such as north, center and south) by relating two aspects: a) the price differential between the two materials (MEC versus traditional ones); b) the relative price differential relating to the processing and installation of the same materials made to vary in quantitative terms due to the same performance response to the three different climatic zone.

Keywords: MEC · Price list · Cost evaluation

1 Introduction

In Italy, the discipline of the conservation of the historical built heritage has been characterized by different approaches and continues to be a matter of great debate for cultural, social, economic and environmental reasons. Awareness of the importance of pursuing models of sustainable development also through the realization of architectural projects capable of limiting environmental impacts, of responding to new housing needs linked to contemporary lifestyles and of guaranteeing the quality of urban environments, increasingly requires the introduction of policies aimed at promoting and guiding the transformation of the existing building heritage, from ancient to more recent expansions [1, 2].

© The Author(s), under exclusive license to Springer Nature Switzerland AG 2022
F. Calabrò et al. (Eds.): NMP 2022, LNNS 482, pp. 764–773, 2022.
https://doi.org/10.1007/978-3-031-06825-6_72

The issue is particularly relevant if we consider that the country is affected by the phenomenon of urban shrinkage, i.e., by that set of dynamics concerning not only a significant demographic decline but also an overabundance of empty and obsolete buildings. This condition requires the definition of strategies for the reuse of the built environment and the regeneration of contexts at different scales.

A recent study carried out by CRESME (Centro Ricerche Economiche, Sociologiche e di Mercato nell'Edilizia) in collaboration with CNAPPC (Consiglio Nazionale Architetti, Pianificatori, Paesaggisti e Conservatori) reveals two distinctive elements in the history of Italian architectural production [3]:

Out of a total stock of 15 million buildings in Italy (as of 2016), 11.9 million are for residential use or a mix of residential and economic activities;
the most common type of size is the single-family house which accounts for 76% of residences.

CRESME's analysis provides a picture of a country hit by an urbanization of small residential buildings spread across the territory, which has contributed to increasing land consumption at a high rate. During 2019, more than 57 km^2 of land was taken away by new artificial cover corresponding to an average of about 16 ha per day. It is no coincidence that the CRESME XXVII Joint and Forecast Report on the 2020 construction market [4] clearly states that in the current environmental and socio-economic phase, the future of Italian construction will have to converge more towards quality than quantity.

The 110% superbonus (sismabonus + ecobonus) is a benefit provided for by "*Decreto Rilancio 2020*" which raises to 110% the deduction rate for expenses incurred from 1 July 2020 to 30 June 2022 for specific interventions related to the energy and structural upgrading of the building sector. The context of the building stock on which the incentives will be applied suggests a potentially more favorable scope as it is characterized by more 'historic' residential buildings on average. Despite the improvement in recent years (in the period 2016–2019 buildings with high energy performance increased from 7% to 10%) there is still much to be done: over 60% of the Italian building stock is in the least efficient energy classes (F-G). To obtain the facilitation in the energy efficiency works of the building thermal insulation materials compliant with MEC must be used.

The minimum environmental criteria and circular construction propose a design approach aimed at satisfying quality requirements for the whole building and for the transformations underway [5]. This is a revolutionary method because it involves an architecture that goes beyond the quantitative and dimensional limits of the current regulatory framework. The objectives pursued lead to replacing the dimensional indicators defined by distances, heights, parking facilities, etc. with the fulfilment of requirements in terms of quality and positive effects of the design process [6].

The study carried out (considering the context described above) aims at estimating the costs of interventions on the external envelope of the building by analyzing the incidence of MEC items.

2 State of the Art

2.1 Overview and Introductions to MEC

The impacts of the construction industry on the natural system and the co-built environment are crucial and widely debated particularly in the last two decades. The construction sector is believed to be responsible for 50% of the natural resources derived by humans and 25–40% of the total energy used [7]. The set of arguments related to these aspects fully flows into the debate of sustainability in architecture. The measurement of the reduced environmental impact of buildings has (for some years now in Italy) been the subject of technical-legislative actions referring to the process and design of public works. Since 2004, European Directives (2004/17/EC and 2004/18/EC) have aimed to include appropriate environmental considerations in the building process to define technical specifications useful for awarding contracts. Ten years later, new EU Directives (2014/23/EC, 2014/24/EC, 2014/25/EC) aim to pursue common strategic objectives at European level, requiring (once again in public procurement) the adoption of suitable instruments aligned with Green Public Procurement (GPP). The European requests in this sense are implemented in Italy with the introduction of the Minimum Environmental Criteria (MEC) with the Ministerial Decree of 24.12.2015 and subsequently with the "*Nuovo Codice dei Contratti*", Legislative Decree 50/2016. The legislation has in fact implemented the direction of national and supranational policies in pursuing the objectives of saving and reusing resources. In fact, the presence of these instruments in public works was already evident fifteen years ago with provisions governing the environmental issue in various sectors, such as the mandatory use of recycled materials to an extent equal to 30% of the annual requirement of manufactured goods and goods (products) [8] and, more recently, with the significant transposition of the European directive of 2008, with the provision aimed at reducing the negative impacts of waste production and management to achieve 70% of the level of recycling of waste in the building industry by 2020 [9].

The condition of compulsory use of recycled material for public works is finally sanctioned by the regulations deriving from the so called "connected environment" [10] and which is applied in the Minimum Environmental Criteria (MEC) Building: through them, the environmental requirements for the realisation of public works have been defined. The Building MECs contain indications addressed to all actors in the building process and with reference to the different phases of the process, from design to project co-construction. They include the definition of minimum environmental characteristics and requirements that the project and the process must respect, requirements that are higher than those provided by national and regional laws in force [10]. Since the entry into force of the Construction MECs in 2017 public works cannot be designed and built without the environmental measures adopted in them being quantified and qualified. Within the Building MECs the restrictions on the use of recovered and recycled materials in the project [11] are part of the wider framework of tools to control the environmental sustainability of the project which include systems (although not compulsory) aimed at the possibility of environmental certification of both buildings based on the definition of criteria and requirements to be met (such as LEED, ITACA protocol), and products (such as EPD environmental product labels). The actions related to the absorption of MEC in building are configured as an urban strategy of sustainable development in line with the Leipzig Charter. With reference to the articles of the law on the technical specifications of building components, the designer must indicate "the environmental information of the chosen products and provide the technical documentation that allows the criteria to be met". The content of recovered and recycled material is regulated by criteria common to all building components [11] with emphasis on the requirements of:

– disassembly of components for "at least 50% by weight of building components and prefabricated elements only" (therefore excluding plants) which must "be subject to selective demolition at the end of their life and be recyclable or reusable" with specific evidence that of this percentage, a minimum of 15% must be "made up of non-structural materials" [11];
– presence of recovered and recycled material in the materials used for "at least 15% by weight evaluated on the total of all the materials used", where 5% is the minimum amount of recovered and recycled material that must be present in the non-structural materials [12].

With regard to some materials only the MEC Building Regulations specify minimum percentages of recovered material. The prescriptions indicated on the recycled material shall be specified by the designer first and ensured by the contractor later through an environmental declaration or a product certification [13]. In summary, with these prescriptions, building MEC is a comprehensive and uncomplicated tool for project control of public works and, to date, can be applied voluntarily in private works.

3   MEC in Regional Price List

In 2019, there was still a lack of literature regarding the application of MEC in construction [14]. For the effective practical application of MEC in addition to literature references the analytical estimation of construction costs during the final and executive design phase is crucial. In fact, the implementation of these criteria is often limited in the economic calculation due to the lack of clear and explicit references. Many professionals complain about the poor information content in the different regional price lists to which (according to art. 23 of "*Codice dei Contratti*") public contracting authorities are bound in determining the costs of works, equipment and products. It may be useful to understand how the price lists of the different regions are adapting to the MEC perspective in design. To this end, a preliminary survey of the most recent official regional price lists has been set up as follows. Any price item in each official regional price list that complied with the MEC regulation was isolated. In addition to this information, those items were isolated and counted which included the use of recycled material or recyclable material at the end of its life cycle. To avoid overlaps, these latter items had to be different from the MEC materials within the item. In order to construct a synthetic index able to provide a snapshot that would make the different scenarios of the national context comparable, each total was divided by an estimate of the total items of which each price list is composed. As it is impossible to count each item or article individually the estimation was made by counting the average number of workings for a significant number of pages, multiplied by the total number of pages containing the workings (excluding the descriptive, introductory and methodological notes). The ratio between the two values gives back the percentage of items able to satisfy MEC with respect to the total number of price list items. The operation has not been easy because of the heterogeneity with which the issue is addressed in each single price list and it is affected by a possible degree of approximation in the counting of the various elements; for the purposes of this contribution, for which we want to provide a general reference

Table 1. Distribution of MEC items in official regional price lists

Region	Year	MEC items (%)	Recycled/recyclable items (%)	MEC item (N.)	Recycled/recyclable items (N.)
Lombardia	2021	8,75%	0,63%	1.620	117
Toscana[a]	2021	8,22%	0,42%	1.236	63
Campania	2021	8,03%	0,00%	1.130	0
Piemonte	2021	5,39%	0,07%	1.995	38
Emilia-Romagna	2021	3,01%	0,18%	365	23
Umbria	2019	1,28%	0,06%	215	10
Calabria	2020	0,96%	0,15%	58	9
Molise	2021	0,40%	0,13%	48	16
Liguria	2020	0,23%	0,20%	18	16
Sardegna	2019	0,18%	0,48%	23	61
Basilicata	2020	0,15%	0,19%	64	82
Friuli Venezia giulia	2021	0,07%	0,25%	6	22
Veneto	2021	0,07%	0,31%	9	41
Sicilia	2019	0,01%	0,35%	1	27
Abruzzo	2021	0,00%	0,15%	0	17
Lazio	2020	0,00%	0,12%	0	14
Marche	2021	0,00%	0,08%	0[b]	13
Trentino-Alto Adige	2021	0,00%	0,56%	0[c]	92
Valle D'Aosta	2021	0,00%	0,63%	0	34

[a] For ease of reference the survey in the Tuscany price list was carried out in the section concerning the Province of Florence. [b] According to the initial methodological note Marche MEC price list is generally referred to in each work or item but there is no explanation in the items. [c] Trentino-Alto Adige price list declares in the methodological note that it deliberately does not take into account MEC, referring the assessment of the compliance with the requirements within the single item and an adjustment or integration of the description to the project. [d] The fragmentation and separation of the continuous updates of Puglia price list did not allow a direct survey.

framework, this approximation is acceptable. In this way the index (of simple composition) represents the percentage of MEC items with respect to the total number of items in the price list, so it can be measured in any national and international price list. The result of this quantitative operation is reported in the following Table 1.

Considering the importance of the MEC price list item since it can be directly used by the designer in the analytical calculation phase the reference scenario needs to be strongly improved. A first group of 5 regions is characterized by a better presence of MEC items than the others, but this is limited to 8% (Lombardia, Campania and Toscana) and then decreases to 5.4 and 3% in Piemonte and Emilia-Romagna. These regions are still in the process of adapting these aspects, as can be seen from the initial methodological notes. Other lists simply do not take them into account. It is clear that, even with a view to increasing these items, there is no ideal value to strive for, but the adjustment process still needs to improve, especially if one considers that 70% of Italian regions (14 out of 20) present a derisory number of items (around 1%) and half of them practically none. Even independently of MEC compliance, the contribution of items containing recycled or recyclable materials is a very small part, almost non-existent, of the total number of items in the official price lists. The fact that the price lists with a better relative performance should also be improved is evident from the methodological note of the Piemonte Region: in fact, the inclusion of MEC products in the price list is subject to the acquisition of documentation showing compliance with the product requirements. The results of the survey carried out by *Regione Piemonte* with the support of a technical committee specifically set up for this purpose show that the delay is also ascribable to the market which is still not fully adapted to the new regulations also with respect to the different types of suitable products and already potentially applicable to the design but not yet certifiable. However, the price list is a tool that cascades (in economic terms) the technological changes of the construction characteristics widespread in the building market and certainly cannot anticipate them. This is evident if we consider that in 2019, out of 900 companies contacted by *Regione Piemonte* to request MEC modules in various product sectors, only 6% responded (50 companies); of these, only 12 declared a product suitability by transmitting the relative certification, i.e. 1.3%.

4   Concluding Remarks on the Costs of MEC Materials vs. Traditional Materials

In order to understand the incidence of MEC materials compared to traditional materials, a comparison was made between the first three regions, i.e. Lombardy, Tuscany and Campania (Table 2), which have more MEC items in their price lists.

This quick, intuitive and easily replicable comparison was carried out by considering two types of insulating material (of mineral origin, rock wool and of fossil origin, expanded polystyrene). This choice was motivated by the wide use of these materials due to the possibility for private subjects to have access to the incentives related to the ecobonus.

In Table 2 and Table 3 the unit prices of the minimum thicknesses of the items of the price list of the insulating materials are indicated.

Table 2. Comparison between MEC item and traditional item (rock wool)

Insulation material of mineral origin - rock wool

Price list		Material	Price
Campania 2021	No MEC	E.10.010.070.a - Thermal and acoustic insulation with rigid rock wool panels - Thickness 5 cm	7,28 €/mq
	MEC	E.10.010.070.a.MEC - Thermal and acoustic insulation with rigid rock wool panels - Thickness 5 cm	9,82 €/mq (+34,9%)
Toscana 2021	No MEC	TOS21_PR.P18.022.001 - Rock wool mats (MW) […] - 50 mm	7,28 €/mq
	MEC	TOS21_PRCAM.P18.022.021 - Rock wool mats (MW) […] - 50 mm	9.59 €/mq (+31.7%)
Lombardia 2021	No MEC	MC.10.250.0050.b - Rock wool board with thermosetting resins, […] thickness 60 mm	7,71 €/mq
	MEC	MEC - MC.10.300.0030 - Rock wool board with thermosetting resins, […] thickness 60 mm	9,47 €/mq (+22,83%)

Table 3. Comparison between MEC item and traditional item (expanded polystyrene)

Insulation material of fossil origin - expanded polystyrene

Price list		Material	Price
Campania 2021	No MEC	E.10.010.020.c - Sintered expanded polystyrene panel […] 5 cm thick	n.a.
	MEC	E.10.010.020.c.MEC - Sintered expanded polystyrene panel […] 5 cm thick	6,45€/mq
Toscana 2021	No MEC	TOS21_PR.P18.017.001 - Closed-cell sintered expanded polystyrene (EPS) panels […] 50 mm thick	3,58 €/mq
	MEC	TOS21_PRCAM.P18.017.073 - Closed-cell sintered expanded polystyrene (EPS) panels […] 50 mm thick.chiuse (EPS) […] 50 mm thick	4,38 €/mq (+22.35%)
Lombardia 2021	No MEC	MC.10.050.0040.b - Sintered expanded polystyrene panels, 50 mm thick. […]	8,79 €/mq
	MEC	MEC - MC.10.300.0020.b - Sintered expanded polystyrene panels, 50 mm thick. […]	11,51 €/mq (+30.9%)

As far as rock wool panels are concerned (the minimum thickness considered in the price lists is 5 cm) both Campania and Toscana showed an increase in the unit price of about 30% of the MEC item compared to the traditional item (no MEC). In Lombardia on the other hand the gap in terms of unit prices between the two items is less marked and it is set on 23%. As far as the expanded polystyrene panel is concerned, the Campania Region only considers the MEC item in the price list. While Toscana and

Lombardia between MEC and traditional items showed an increase in the unit price of about 20 and 30% respectively. A further analysis was carried out on the cost of processing (considering MEC items) thermal insulation using rock wool or expanded polystyrene panels with the minimum thickness necessary to guarantee the required energy performance. Since 2005, Legislative Decree 192 on the energy performance of buildings has regulated the limit values for thermal transmittance of the closures of the building body in public works, which have been progressively updated in the regulations with more restrictive values. The limit values of the thermal transmittance of the closures are different for the different climate zones: lower for the climate zones that classify the coldest municipalities, i.e. with more degree days, and higher for the hottest municipalities, i.e. with fewer degree days, as classified according to Presidential Decree 412/03 and subsequent updates. Annex E of the implementation decree of the Ministry of Economic Development "ecobonus requirements" of 6 August 2020 is also in line with this practice and regulates the maximum transmittance values allowed for access to deductions for private works. In particular, the thermal transmittance (calculated in accordance with UNI EN ISO 6946 standards) of the technical elements on which work is carried out must, following the intervention, have higher values than, with reference to opaque vertical closures, 0,30; 0,26 and 0,23 $W/m^2 K$ for the three prevailing climatic zones in the area, i.e. zones C, D and E respectively.

The prevalence of the climatic zones can also be read according to the different regions. If we want to compare the energy requalification of opaque vertical closures in Campania, we can assume climate zone C to be prevalent; in Tuscany, climate zone D; in Lombardy, climate zone E. In order to compare the cost of thermal insulation according to the legal requirements for an opaque vertical closure, a closure consisting of an empty box wall with 30 cm thick perforated bricks with a 5 cm air gap and a thermal transmittance of 1,01 $W/m^2 K$ (1) is simulated.

For climate zone C and to reach the limit value of 0,30 $W/m^2 K$, 9 cm of rock wool insulation or 8 cm of expanded polystyrene are required.

For climate zone D and to reach the limit value of 0,26 $W/m^2 K$, 11 cm of rock wool insulation and 10 cm of expanded polystyrene are required.

For climate zone E and to reach the limit value of 0,23 $W/m^2 K$, 13 cm of rock wool insulation and 12 cm of expanded polystyrene are required (Table 4 and Table 5).

With regard to the processing of rock wool panels, the data from the Lombardy Region stand out, where the cost of processing, despite the greater thickness, is comparable to the cost of processing present in the price list of the Campania Region, even though it refers to a lower thickness. On the other hand, analyzing the thermal insulation work using expanded polystyrene panels, it emerges that the cost of work in the Campania region, although considering a lower thickness than in the other regions, is clearly lower than the average. These preliminary data and comparisons make it possible to understand the current situation where most regional price lists needs to implement the items related to MEC, especially given the recent obligation to use these materials also within the private work context and especially for those who want to access the incentives of the ecobonus.

Table 4. Insulation cost using rock wool (MEC item)

Rock wool panel_ minimum thickness to ensure energy performance
Climatic zone C: Campania (thickness 9 cm)
Campania 2021 price list E.10.010.070.d.MEC - Thermal and acoustic insulation with rigid rock wool panels - 9 cm thick **22,10 €/mq**
Climatic zone D: Toscana (thickness 11 cm)
Toscana 2021 price list TOS21_01CAM.D01.001.040 – Insulation with an "overlay system" on external opaque vertical straight surfaces […] with rock wool panel insulation - thickness 100 mm **83,87 €/mq**
Climatic zone E: Lombardia (thickness 13 cm)
Lombardia 2021 price list 1C.10.250.0050.b – Thermal insulation on walls or perimeter cavities made of rigid rock wool panels […] - thickness 130 mm **19,50 €/mq**

Table 5. Insulation cost using expanded polystyrene (MEC item)

Expanded polystyrene panel_ minimum thickness to ensure energy performance
Climatic zone C: Campania (thickness 8 cm)
Campania 2021 price list E.10.010.020.e.MEC - Thermal and acoustic insulation made of 8 cm thick sintered expanded polystyrene panels **19,93 €/mq**
Climatic zone D: Toscana (thickness 10 cm)
Toscana 2021 price list TOS21_01CAM.D01.001.011 – Insulation with an "overlay system" on opaque vertical external surfaces […] with insulation in sintered expanded polystyrene panels (EPS) - thickness 100 mm **68,20 €/mq**
Climatic zone E: Lombardia (thickness 12 cm)
Lombardia 2021 price list 1C.10.300.0020.f – Thermal insulation system made of sintered expanded polystyrene sheets […] - thickness 120 mm **68,02 €/mq**

It is hoped that the production system will rapidly implement the production of MEC materials and that these will be absorbed in the regional price lists, without anomalous prices, and that, moreover, in compliance with the traditional construction systems most widespread in the regions, economies and diseconomies derive in the future from the different performance levels required by the specific localization and construction systems. What is proposed in this contribution, in addition to a first reference framework, can provide (improved and extended to other materials) a mechanism of comparison on which to set up a monitoring system.

References

1. Baratta, A.F.L., Finucci, F., Magarò, A.: Regenerating Regeneration: augmented reality and new models of minor architectural heritage reuse. VITRUVIO – Int. J. Archit. Technol. Sustain. **3**(2), 1–14 (2018). https://doi.org/10.4995/vitruvio-ijats.2018.10884
2. Acampa, G., Grasso, M.: Heritage evaluation: restoration plan through HBIM and MCDA. IOP Conf. Ser. Mater. Sci. Eng. **949**(1), 012061 (2020). https://doi.org/10.1088/1757-899X/949/1/012061
3. www.cresme.it/it/studi-e-ricerche/. Accessed 27 Dec 2021
4. Baratta, A., Calcagnini, L., Finucci, F., Magarò, A., Molina, H., Quintana Ramirez, H.S.: Strategy for better performance in spontaneous building. TECHNE J. Technol. Architect. Environ. **14**, 158–167 (2017). https://doi.org/10.13128/Tcchnc-20797
5. http://www.cresme.it/it/congiunturale-cresme.aspx. Accessed 27 Dec 2021
6. Baratta, A.F.L., Finucci, F., Magarò, A.: Generative design process: multi-criteria evaluation and multidisciplinary approach. Technè **21**, 304–314 (2021). https://doi.org/10.13128/Tec hnc-20797
7. Ministerial Decree 203/2003 set out the rules for the regions to adopt the above provisions and to allocate them to companies with predominantly public capital. The provisions, which have been partly disregarded [Legambiente, 2015], do not refer to the specific field of construction but to the whole of goods and buildings
8. EU 2008/98/CE e D.lgs. 205/2010
9. Art. 34 of Law 221/2015 (Environmental Amendment) clarifies, through the pre-disposition in contracts for public procurement, the fulfilment of obligations for contracting authorities in the field of energy and environmental sustainability, through the definition of minimum environmental criteria, also in the field of "the awarding of design services and works for the new construction, renovation and maintenance of buildings and for the management of public administration worksites"
10. As regulated by the Ministerial Decree of 24 December 2015 and the subsequent Ministerial Decree of 11 October 2017
11. Art. 2.4, D.M. 11 ottobre 2017
12. Exceptions are granted where specific performances are required (e.g. for materials for storm water protection or durability requirements or other regulatory constraints)
13. The Environmental Product Declaration must be of Type III (EPD), compliant with UNI EN 15804 and ISO 14025, such as EPDItaly© or equivalent; product certification issued by a conformity assessment body certifying the recycled content through the explanation of the mass balance, or self-declared environmental declaration, compliant with ISO 14021. The standard also accepts as certification an inspection report issued by a conformity assessment body in accordance with ISO/IEC 17020:2012 stating the recovered or recycled content of the product
14. Bassi, A., Ottone, C., Dell'Ovo, M.: I Criteri Ambientali Minimi nel progetto di architettura. Trade-off tra sostenibilità ambientale, economica e sociale. Valori e valutazioni **22**, 35–45 (2019)

An Integrated Model to Assess the Impact on World Heritage Sites. The Case Study of the Strategic Plan for the Buffer Zone of the UNESCO Site "Pompeii, Herculaneum and Oplontis"

Alessio D'Auria[1][(✉)] and Irina Di Ruocco[2]

[1] University Suor Orsola Benincasa, via Suor Orsola 10, 80132 Naples, Italy
alessio.dauria@docenti.unisob.na.it
[2] University Insubria, Via Ravasi, 221100 Varese, Italy

Abstract. The paper intends to point out the case study of the UNESCO site "Pompeii, Herculaneum and Oplontis" and its buffer zone, listed in the World Heritage List since 1997 for its exceptional cultural, artistic, and scientific value. In 2014 the Italian Ministry of Cultural Assets prepared a new version of the site management plan to ensure a more effective protection of the values recognised therein. The update also suggested to extend the buffer zone to a vast, complex, and densely populated area, one of the widest in Italy. A strategic development plan was drafted for this area, whose objectives are in apparent contradiction with those outlined in the site's management plan. We thus propose an integrated assessment model characterised by an incremental approach to plan choices, which can balance the conflicts between conservation and development issues.

Keywords: Strategic plan · World Heritage Site · Integrated assessment

1 The UNESCO Site "Pompeii, Herculaneum and Oplontis" and Its Buffer Zone Between Conservation and Development

The archaeological site "Pompeii, Herculaneum and Oplontis" is probably the most famous in the world regarding ancient Roman history and is constantly under the spotlight of the mass media due to its huge artistic and cultural value, which is so hard to maintain and manage. On the basis of such unquestionable value, it was inscribed into the World Heritage List (WHL) in 1997, addressing the criteria III, IV and V, with the following justification: "considering that the impressive remains of the towns of Pompei and Herculaneum and their associated villas, buried by the eruption of Vesuvius in AD 79, provide a complete and vivid picture of society and daily life at a specific moment in the past that is without parallel anywhere in the world."

Following the collapse of the *Schola Armaturarum* (housing an important military-style association), which occurred on 6th November 2010, UNESCO imposed to Italy a

F. Calabrò et al. (Eds.): NMP 2022, LNNS 482, pp. 774–785, 2022.
https://doi.org/10.1007/978-3-031-06825-6_73

more efficient protection policy and a complete revision of the Management Plan (MP) of the archaeological site. Therefore, in January 2014, the new MP was submitted to the UNESCO advisory board. The renewed plan contains some very significant innovations. Among all, the most fruitful of implications is probably the extension of the buffer zone from an area that is presently less than 25 hectares, to a new one that is significantly wider, of about 77 square kilometres, and that is inhabited by 380,000 people, including all the territories of the municipalities from Portici to Castellammare di Stabia (except for areas falling in the Vesuvius National Park), with the aim to rebuild an identity of the Vesuvius landscape.

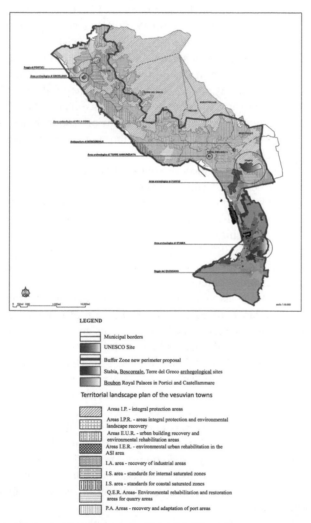

Fig. 1. Extension of the buffer zone of the "Pompei, Herculaneum and Oplontis" UNESCO site – UGP, 2013.

It should be noted that, while the concept of "world heritage", had its own specific UNESCO definition as documented since 1972, year of the International Convention on World Heritage, the same cannot be said for the concept of "buffer zone", whose identification or modification must be deemed necessary in the proposal for inclusion of an asset in the list. The buffer zone was treated in all the drafts of the *Operational Guidelines for the Implementation of the World Heritage Convention* since their first release in 1977. Initially the buffer zone was presented as an optional requirement in the application for inclusion in the WHL and its purposes were not precisely defined.

In recent years, although UNESCO and ICOMOS have always confirmed that the identification of a buffer zone is not mandatory, an increasingly in-depth analysis has been carried out on this topic and on the specification of the buffer zone functions related to guaranteeing the integrity of the OUV for World Heritage Properties [1]. According to Michael Turner: "Buffer zones cannot, by definition, exist alone. They can be part of a system, which involves areas of sustainability or areas of concern, and have been defined as a set of problems that a given project is intended to address. These areas are identified not only from inside-out but outside-in. They will include the areas of concern as defined by the various community interests. Finally, their determination is both normative and technical. Technically, the issues of sustainability are the balance between the various interests in time and place through the economics of the mutual benefits within well-defined constraints" [2].

The definition is far from the idea of a buffer zone intended as a separation or as a barrier and is closer to the idea of a management and control tool that must deal with the transition between the site and its surroundings through an adequate protection, while preserving those aspects of the asset related to the OUV, but without neglecting the interests of the local communities [3].

The buffer zone of the UNESCO site "Pompeii, Herculaneum and Oplontis" certainly meets this new policy and approach, to overcome and heal conflicts and contradictions between protection and development.

For this area, which is vast and very sensitive, the Italian Government has issued a special law that foresees a strategic plan with the aim of pursuing economic boost, environmental and landscape recovery (for waterfronts and the numerous brownfields) and an overall enhancement of the accessibility to cultural sites.

The Strategic Plan of the Buffer Zone (SPBZ) of the UNESCO site "Pompeii, Herculaneum and Oplontis" is an innovative and unique tool of its kind, born principally from the need to extend the buffer zone to the entire area of the coastal Vesuvian municipalities, for the protection and enhancement of the site and its relationship with the Vesuvian landscape. This extension was contemplated by the Site Management Plan, approved on 23rd December 2013 by the Minister for Cultural Heritage.

The "new" buffer zone is a large and complex area (77 sq. km), characterised by a high urbanisation and population density (380,000 inhabitants), densely innervated by infrastructures, and marked by brownfields and degraded areas alternating with landscape contexts of absolute relevance.

We are faced with the coexistence of two levels of planning: a management plan for the core zone, which is the archaeological site *intra moenia*, with the aim of identifying the procedures for protection, conservation and enhancement of heritage, and a strategic plan for the buffer zone, which is the wide territory *extra moenia*, identified and delimited with the purpose to ensure the integrity of the former. This strategic plan foresees the identification of shared strategic goals and their subsequent transfer into specific actions through the active involvement of public and private actors in order to enhance the development of tourism in the area.

The management plan and the strategic plan have only apparently contradictory goals. As a first goal, the management plan aims to develop strategies not only to preserve the property, but also to make it usable both by the present and future generations for purposes of cultural and economic development. In this way, the management plan may represent a new model of territorial development with the ultimate intention of combining conservation with economic development, by integrating the promotion of cultural heritage in programming and planning. In this sense, the management plan must also have a strategic dimension and not only an operative one, as also emerges clearly in the guidelines prepared by ICCROM [4].

Meanwhile, the coexistence of different planning models with divergent targets, affirms that governance and management of UNESCO sites appear today more than ever, in the scientific and cultural debate, as the necessary point of synthesis of the integration between instances of protection of cultural sites with socio-economic and tourist development needs of its territories. UNESCO site management must face these issues and therefore adapt itself to a situation where, even with fewer economic resources available, it is more than ever necessary to implement the best possible strategies for economic, social, and cultural development of the local contexts involved in the process [5].

The necessary hybridisation between the management plan and the strategic plan requires the development of an innovative governance model, capable of integrating the protection of cultural heritage with urban and territorial policies aimed at environmental and landscape restoration, and of economic assets based on tourism exploitation.

The approach proposed in the management plan of the UNESCO site "Pompeii, Herculaneum and Oplontis" is a multiscale approach and includes not only action plans for the medium term (5–10 years) but also for the long term (30 years), foreseeing the need for a shared vision among all the stakeholders involved [6]. This necessarily means that the management plan must take on an aspect that is strategic and operational at the same time, so that it can be configured as a "diagonal" tool between the exceptional and outstanding values that it aims to protect, on the one hand, and its informal nature, linked to the typical constructions of "visions" and participatory strategic plans, on the other hand [7].

At the same time, the deep awareness that the effects induced by UNESCO's attractors, as facilitators for the development of the entire Vesuvius area, cannot be separated from the improvement of the context, has led the Italian Government with the Law Decree n° 91/213 (converted, with amendments, into Law n° 112/2013 and subsequently amended and integrated by Law 106/2014) to identify the "strategic plan" as a tool for defining a homogeneous plan, aimed at social and economic revitalisation, environmental and urban redevelopment and at the overall increase in the attractiveness of the entire area including all the municipalities included in the management plan of the UNESCO site, or buffer zone. The law also identified the Greater Pompeii Unit as the advisory technical board in charge of drawing up the plan.

The identification of a large area of protection around the World Heritage Site requires an extensive protection and coordinated action and can certainly be the basis and instrument for the social and economic development of this territory [8].

2   The Strategic Plan for the Buffer Zone: Motivations and Articulation of the Plan

The territory identified and bounded by law as the buffer zone is a heterogeneous geographical area, full of contradictions, in which the following coexist: urban and residential settlements; small businesses and very high quality production chains; abandoned productive areas confining to wide portions of land to be reclaimed, and a dense network of infrastructures that develop across a landscape of high-value and across historical, cultural and environmental areas, which have been stratified over the centuries within a very complex system, often not well governed and managed.

The scenario that emerges, therefore, is a dense interweaving of resources and activities, characterised by unique assets of universal value that coexist with contemporary realities, woven over a wide historical span and often deviated by the distortion of the rules. Therefore, the development strategy, which the legislator entrusts to the strategic plan, must take into account that the significant interventions that will derive from it will aim above all at the attenuation of these always problematic dynamics, and at the proposition of a new model of territorial development, based on the enhancement of cultural heritage "in a broad sense" [9].

Certainly, the elaboration (and subsequent implementation) of a strategic plan for a wide and sensitive area, such as that in question, seems not only a good choice, but even compulsory, if the plan will be effectively integrated with political and social issues and does not limit itself to an approach merely aimed at territorial transformations.

The plan is structured into four strategic lines (three oriented towards a territorial transformation of the area and a fourth one focusing on intangible actions), which represent the goals that the plan aims to achieve:

— *strategic line 1*: improving the roads of access and interconnections to the archaeological sites, which, in turn, is divided into 4 actions: accessibility by the rail network; accessibility by sea; accessibility by tar roads; interchange and connections with the archaeological sites;
— *strategic line 2*: environmental recovery of the degraded and compromised landscapes primarily through the recovery and reuse of the brownfield sites; it is also divided, in turn, into 3 actions: recovery and reuse of the brownfields; environmental recovery of the landscape of the coastal strip; recovery of the peri-urban agricultural landscape;
— *strategic line 3*: rehabilitation and urban regeneration, which, in turn, is divided into 2 actions: urban regeneration of the functional axes and their surrounding context to guarantee the accessibility to the cultural sites; recovery, reuse and exploitation of disused volumes for tourism, commercial or craft purposes.

— *strategic line 4*: promotion of charitable donations, sponsorships, public-private partnerships, engagement activities of non-profit organisations in the promotion of cultural heritage.

The SPBZ does not consist of a discipline of land use and is aimed at the construction of a "project programme", starting from some interventions considered invariant and necessary to trigger sustainable territorial development processes.

The Greater Pompeii Unit, starting from the first months of 2015 with a preliminary and timely survey of the project actions in progress, considered the strategies and interventions proposed by the central and peripheral institutions as well as by other local stakeholders, and the actions envisaged by other existing planning tools, to carry out an integration and rationalisation rather than a further addition of proposals to better respond to the needs identified for the buffer zone.

Following this preliminary stage, at the beginning of 2016, the advisory board published the list of the most relevant projects, which were chosen among the ones submitted by the local municipalities and proposed by the regional government and other public bodies.

The list includes, for the strategic line 1: the new high speed railway station in Pompeii (that represents the "flag-project"); a new station near the Herculaneum ruins; a network of electric buses linking the most important archaeological and cultural sites in the buffer zone; the improvement of accessibility to the Vesuvius National Park. For the strategic line 2 the projects are the following: the reconversion of the railway line between Torre Annunziata (where the archaeological ruins of Oplontis are located) and Castellammare di Stabia into a light and speed tram line, with the urban and environmental regeneration of the whole waterfront; the transformation of the abandoned railway line between Torre Annunziata and Boscotrecase into a linear park with bike lanes; the recovery of the rural landscape in the surroundings of the archaeological ruins; the enhancement of the archaeological sites that have not been included yet in the WHS (namely, Torre del Greco and Castellammare di Stabia). The Unit has selected the following projects for the strategic line 3: the restoration of the axes connecting the ports and the stations to the main cultural assets in the area; the enhancement and reuse of the available real estate heritage complexes.

The Plan was approved by the Management Committee (formed by the Ministry of Culture, the Ministry of Territorial Cohesion, and the local and regional government bodies involved) on 20th March 2018.

Due to bureaucratic delays, which postponed the beginning of the application at the end of 2019 and the consequent pandemic crisis that has embraced the last two years and which, unfortunately, is yet to be resolved, the plan to date is substantially dissatisfied, despite the enormous potential for territorial development linked to the possibility of derogation from territorial urban planning.

The opportunity offered by React-EU funds and the National Recovery and Resilience Plan should allow the effective and rapid implementation of the projects envisaged in the strategic plan. An integrated assessment of the plan, therefore, becomes crucial.

3 An Integrated Assessment Model for the Strategic Plan of the Buffer Zone

The integrated assessment of the strategic plan therefore acquires importance for its ability to model and structure design choices, also to overcome the somewhat hasty approach adopted in the bureaucratic drafting of the environmental report based on the Strategic Environmental Assessment (SEA), towards a perspective of "integrated decision-making process".

The mutual integration of evaluation in the formation of plan choices is a prerequisite for ensuring coordination and coherence between the sectoral policy objectives and those of regional development and is also one of the major challenges for achieving an effective integration among environmental, social, cultural areas and economic issues [10].

From this point of view, the proposed assessment model must necessarily go beyond a merely regulatory and instrumental approach, already ensured by the SEA and by the other evaluation phases envisaged by the regulations, to verify the feasibility of individual projects and moves in the exploratory perspective, previously described [11]. This approach favours the interaction between decision makers and stakeholders, faced with visions that do not always coincide to pursue sustainable development strategies of the territory, due to the growing complexity of perspectives, interests and preferences that are often in conflict and of different forms of capital (physical, cultural, economic, social, human, etc.) and their mutual and delicate interrelationships [12, 13].

In this perspective, the integrated decision support approach is undoubtedly able to generate results that are more effective than sectoral approaches and, at the same time, offers the possibility to enter a multidimensional and intersectoral decision area [14]. Integration is a complex concept, characterised by several dimensions that need to be defined and explored. According to Lee [15], "integrated assessment" involves vertical and horizontal processes, which can be spread over different aspects of spatial development, rather than being rigid, hierarchical, unilinear [16]. The integration of evaluative approaches therefore means considering the interaction between different contextual dimensions, capable of combining existing relationships and of exploring the potential to build new relationships [17].

Therefore, the model here proposed, is articulated according to the following scheme:

1. *Relevance assessment*: relevance indicates the adequacy of the strategic guidelines of the plan, to be considered as real strategic objectives with respect to objectives of sustainable local development, expressed in the territorial government plans at the local level as well as in non-mandatory tools but with implications significant in the development of territories, such as action plans for sustainable energy. The most appropriate evaluation method is the Electre [18, 19], capable of obtaining an ordering and of assigning weights to the alternatives. Operationally, it makes a comparison between all the options with respect to all the criteria in order to understand how much each option outperforms the other, also with reference to some threshold parameters, which can be defined for each criterion.
2. *Priority assessment* refers to an essential step in the evaluation process, aimed at building a hierarchy of priorities (or better preference) of the projects that each local municipality has proposed with respect to the strategic lines of the strategic

plan. This operation shall be interpreted as the construction of a hierarchy of values characterised by different priorities [20]. The most appropriate method for carrying out the analysis of priorities is the Analytic Hierarchy Process, developed by Thomas Lorie Saaty [21–23]. The method, as is known, is based on the pairwise comparison between criteria (to assign weights) and between alternatives (to deduce an order of priority), using the calculation of the main vector of the matrices of pairwise comparisons, thus obtaining a synthetic index (expressed on a scale from zero to one), which expresses the overall preferability of each alternative with respect to the level of higher rank.

3. *Coherence assessment*, which refers both to the internal aspects of the strategic plan (internal coherence), analysing the links among the projects (for which a priority assessment has already been carried out) and the strategic objectives of the plan, and to the external projection of the plan (external coherence). In this second stage, the degree of conflict existing between the vision (i.e., the strategic objectives that generated the strategic lines) and the complex strategic directions outlined in the higher-level government (i.e. the Regional Territorial Plan or the Territorial Coordination Plan of the Metropolitan City) or in policies at the regional level (such as the Regional Operative Programme funded by the European Regional Development Fund) but also at the national and European level (e.g. the European Landscape Convention) will be assessed. From a methodological point of view, the method of expected value based on ranks, developed by Schlager in 1968, appears appropriate to evaluate the coherence of the strategic plan [24].

4. *Efficacy assessment*, intended as a verification of compliance of the impacts expected from the implementation of the plan strategies with respect to planning expectations: that is, an opinion on how much the strategic plan will be able to meet the needs of the stakeholders. This evaluation phase is necessary to answer the following questions: "to what extent are the objectives of the strategic processes justified with respect to the requests, to the issues at stake?" And above all: "are the effects and impacts (expected and unexpected) satisfactory from the point of view of direct and indirect beneficiaries?" Operationally at this stage it is appropriate to resort to the Community Impact Evaluation (CIE) [25], whose evaluation scheme helps to deduce "a common vision" for territorial development. The main aim of the CIE is therefore to evaluate the impacts that a conservation/redevelopment policy or a project will have on all groups interested in the project. Basically, this evaluation method is implemented through the chain of changes triggered by the project and the effects of these changes (the so-called impact chain), which can be economic, environmental, social, direct or induced and induce positive or negative impacts. First, the CIE identifies in parallel all groups of people likely to be affected by the impact, and then evaluates the impacts on these groups and in what proportion each group would consider the impacts beneficial or negative. The CIE highlights the contribution of individual project proposals and alternatives to social welfare, first of all identifying all the different groups that are likely to suffer or enjoy these effects and analysing the "social desirability" of the changes, highlighting the interrelations that existed before the change [26].

5. *Environmental compatibility assessment*, which must include the Environmental Impact Assessment (EIA) procedure, where required by current legislation, and referring to the programme of interventions with respect to existing constraints (environmental, geological, landscape) in the *extra moenia* area, which are governed by Territorial Landscape Plans and the Hydrogeological Structure Plan. This evaluation phase should be complemented with an appropriate methodology, such as the threshold method [27, 28]. The approach envisaged by the Ultimate Environmental Threshold represents an evolution of the threshold analysis that seems appropriate for the definition of the transformability thresholds of the territory [29, 30]. A further evolution of this methodology is the Land Suitability Assessment, which determines the suitability of a territory for a defined use [31, 32], usually among several, competing uses. The Land Suitability Assessment is a multi-criterion and context-dependent evaluation of the development capacity of the territory, based on the opinion of experts who define the most desirable factors and their optimal values and weights for this purpose [33, 34]. First, it is necessary to detect and locate in the territory those structuring values and non-negotiable invariant values with respect to which the transformations must conform.

6. *Economic feasibility assessment*, which requires quantitative methods, such as the cost–benefit analysis (CBA), to assess the actual feasibility of the proposed projects based on the economic evaluation of the return on investment. The CBA is a well-known method, as it is the technique most-widely used by all public actors in the world to evaluate the effects of a public investment, verifying whether, with the implementation of the intervention, the company obtains a benefit or a net cost. It is a tool to support the public decision-making process as, by calculating the benefits and costs associated with its implementation, it allows you to choose the best proposal among different design alternatives. Since the CBA bases its judgment of adequacy not only on accounting-financial criteria but also on cost-effectiveness and social convenience criteria, calculated from the results of the financial analysis through appropriate corrections to derive social costs and benefits, it is also the best way to express the economic surplus on cultural heritage due to the implementation of public projects [35, 36].

7. *Cultural impact assessment*, to adequately assess the potential impacts of the project on the Outstanding Universal Value (OUV) expressed by the site; to this end, the UNESCO World Heritage Committee has proposed to the States to conduct Heritage Impact Assessments (HIA – ICOMOS 2011) to large-scale projects included in the territories of the sites included in the UNESCO World Heritage list. The HIA evaluates the impacts that are directly connected with the attributes of OUV, adopting a global approach to assets, in particular relating to the protection of the values for which the sites have been recognised as a UNESCO World Heritage site.

A temporal and/or logic sequence of the different evaluation phases must be devised, as the evaluations relating to the strategic lines of the SPBZ must be carried out before deepening the project programme (refer to Fig. 2 for more details).

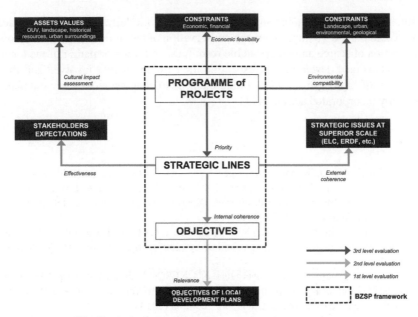

Fig. 2. Articulation of the integrated assessment process

4 Conclusions: Integrated Strategic Assessments for Complex Values. Towards a Holistic Assessment of Sustainability

The proposal of an integrated evaluation model for the strategic plan of the buffer zone of the UNESCO site "Pompeii, Herculaneum and Oplontis" made it possible to clarify the potential interactions between planning and evaluation, and to explore the field of methodologies and tools of "integrated assessment" [37, 38].

Within an integrated approach to evaluative and decision-making processes, reasoning on values means including a multi-dimensional perspective, considering the tangible and intangible aspects of hard and soft values, objective and subjective values, using values as independent use, as well as intrinsic values and their synergistic and complementary relationships.

The implementation of the proposed procedure would improve the transparency of the decision-making process as a whole: it could clarify possible conflicts of interest, allowing more advantageous decisions to be made for all those involved, developing win-win alternatives [39].

This model can integrate, in compliance with the principles of sustainability, technical choices with those of a political nature, with reference to complex value systems, inserted in conflicting and changing contexts [40]. In this sense, the proposed methodology should be interpreted as a "holistic assessment of sustainability", which is a systematic process to consider ways in which the plan can contribute to the improvement of environmental, social and economic conditions, as well as a means of identifying and mitigating any potential negative effects that the plan might otherwise have.

But, most importantly, the integrated evaluation model, articulated as indicated above, would pursue a balance between conservation of cultural heritage, understood as the protection of scarce and unrepeatable resources, and transformation of the territory, understood as local development, respecting the complex values involved and the different forms of capital, finally underlining the need to consider evaluation and planning as mutually incorporated activities [41].

References

1. Capitanio, C.: Lo studio preparatorio per il progetto Buffer Zone per il Centro Storico di Firenze. In: Bini, M., Capitanio C., Francini C. (a cura di), Buffer Zone. L'area di rispetto per il sito UNESCO Centro Storico di Firenze. Dipartimento di Architettura DIDA, Università degli studi di Firenze, pp. 25–79 (2015)
2. Sommaruga, G.: Introduction. In: Sommaruga, G. (ed.) Formal Theories of Information. LNCS, vol. 5363, pp. 1–12. Springer, Heidelberg (2009). https://doi.org/10.1007/978-3-642-00659-3_1
3. D'Auria, A., Pugliese, S.: The Governance of UNESCO Cultural Landscapes between universal values and local identity: the case of Campania. In: Volume V for IPSAPA – Interdisciplinary Scientific Conference. Udine, pp. 189–200 (2013)
4. Feilden, M.B., Jokilehto, J.: Management Guidelines for World Cultural Heritage Sites, Rockwell and Lawrence Editors for ICCROM, Roma (1998)
5. D'Auria, A.: Urban cultural tourism: creative approaches for heritage-based sustainable development. Int. J. Sustain. Dev. 12(2/3/4), 275–289 (2009)
6. Bonazzi, G., Lagi, A., Bonini, A.: Piano di gestione sito UNESCO "Aree Archeologiche di Pompei, Ercolano e Torre Annunziata". MiBACT, Roma (2013)
7. D'Auria, A.: I piani di Gestione delle World Heritage Cities: una proposta operativa per Napoli. In: Metropoli In-Transizione, Urbanistica Dossier n.75, INU Edizioni, Roma, pp. 463–464. (2004)
8. Unità Grande Pompei: Piano Strategico per lo sviluppo delle aree comprese nel Piano di gestione del sito UNESCO "Aree archeologiche di Pompei, Ercolano e Torre Annunziata", Documento di orientamento (2013)
9. Valentino, P., Musacchio, A., Perego, F.: La storia al futuro: beni culturali, specializzazione del territorio e nuova occupazione. Associazione Civita/Giunti, Firenze (1999)
10. Cerreta, M., De Toro, P.: Integrazione della VAS nei processi di pianificazione: il PTCP di Benevento. Scienze Regionali – Italian J. Region. Sci. 12(2), 15–46 (2013)
11. Fusco, Girard L., Cerreta, M., De Toro P., Forte, F.: The human sustainable city: values, approaches and evaluative tools. In: Deakin, M., Mitchell, G., Nijkamp, P., Vreeker, R. (eds.) Sustainable Urban Development. The Environmental Assessment Methods, Vol. 2. Routledge, London, pp. 65–93 (2007)
12. Kirdar, Ü.: A better and stronger system of human governance. In: Fusco, Girard L., Forte, B., Cerreta, M., De Tor, P., Forte, F. (eds.) The Human Sustainable City (2003)
13. Fusco, Girard L., Cerreta,, M., De Toro P.: Integrated assessment for sustainable choices. In: Adaptive Evaluations in Complex Contexts, Scienze Regionali, vol. 13, n.1, pp. 111–142. (2014)
14. Wiek, A., Walter, A.: A transdisciplinary approach for formalized integrated planning and decision-making. Complex Syst. Eur. J. Operat. Res. 197(1), 360–370 (2009)
15. Lee, N.: Bridging the gap between theory and practice. in: integrated assessment. Environ. Impact Assessm. Rev. 26(1), 57–78 (2006)
16. Allmendinger, P., Tewdwr-Jones, M.: Territory, Identity and Space: Planning in a Disunited Kingdom. Routledge, London (2006)

17. Cerreta, M., De Toro, P.: Integrated spatial assessment for a creative decision-making process: a combined methodological approach to strategic environmental assessment. Int. J. Sustain. Dev. **13**(1/2), 17–30 (2010)
18. Roy, B.: Classement et choix en présence de points de vue multiples (la méthode ELECTRE)", La Revue d'Informatique et de Recherche Opérationelle (RIRO) no. 8, pp. 57–75 (1968)
19. Roy, B.: Decision science or decision-aid science? Eur. J. Operat. Res. **66**, 184–203 (1993)
20. D'Auria, A.: Un modello valutativo per verificare coerenza, efficacia e fattibilità delle strategie di sviluppo nelle aree protette: una sperimentazione nel Parco del Cilento e Vallo di Diano. In: AISRe, Impresa, mercato, lealtà territoriale, Atti della XXVII Conferenza Scientifica Annuale, Pisa (2006)
21. Saaty, T.L.: Exploring the interface between hierarchies, multiple objectives and fuzzy sets. Fuzzy Sets Syst. **1**, 57–68 (1978)
22. Saaty, T.L.: Multicriteria Decision Making - the Analytic Hierarchy Process. RWS Publications, Pittsburg (1992)
23. Saaty, T.L.: Fundmentals of Decision Making and Priority Theory with the Analytic Hierarchy Process. RWS Publications, Pittsburg (1994)
24. Fusco Girard, L.: Risorse architettoniche e culturali: valutazione e strategie di conservazione. Una analisi introduttiva. Franco Angeli, Milano (1987)
25. Lichfield, N.: Community Impact Evaluation. UCL Press, London (1996)
26. Fusco Girard, L., Nijkamp, P. (eds): Le valutazioni per lo sviluppo sostenibile della città e del territorio, FrancoAngeli, Milano (1997)
27. Forte, F. (ed.): Progettazione urbanistica e territoriale attraverso la teoria e l'analisi della soglia, FrancoAngeli, Milano (1976)
28. Forte, F.: La pianificazione paesistica in Basilicata. In: Stanghellini S. (edited by) Valutazione e processo di piano, Alinea, Firenze (1997)
29. Kozlowski, J.: Threshold Approach. In: Urban, Regional and Environmental Planning: Theory and Practice, University of Queensland Press, St. Lucia, Queensland (1986)
30. Senes, G., Toccolini, A.: Sustainable land-use planning in protected rural areas in Italy. Landsc. Urban Plan. **41**, 107–117 (1998)
31. Steiner, F., McSherry, L., Cohen, J.: Land suitability analysis for the Upper Gila River Watershed". Landsc. Urban Plan. **50**, 199–214 (2000)
32. Joerin, F., Thériault, M., Musy, A.: Using GIS and outranking multicriteria analysis for land-use suitability assessment. Int. J. Geogr. Inf. Sci. **15**(2), 153–174 (2001)
33. Stoms, D., McDonald, J.M., Davis, F.W.: Fuzzy assessment of land suitability for scientific research reserves". Environ. Manage. **29**, 545–558 (2002)
34. Cerreta, M., De Toro, P.: Urbanization suitability maps: a dynamics spatial decision support system for sustainable land use. Earth Syst. Dynam. **3**(2), 157–171 (2012)
35. Ramalhinho, Ana R., Macedo Filomena, M.: Cultural heritage risk analysis models: an overview. Int. J. Conservat. Sci. **10**, 39–58 (2019)
36. Tišma, S., Mileusnić, Škrtić, M., Maleković, S., Jelinčić, DA.: Cost–benefit analysis in the evaluation of cultural heritage project funding. J. Risk Financ. Manage. **14**(10), 466 (2021)
37. Golub, A.L.: Decision Analysis: An Integrated Approach. Wiley, New York (1997)
38. Therivel, R.: Strategic Environmental Assessment in Action. Earthscan, London (2008)
39. Mondini, G.: Valutazione e complessità. In: Bottero, M., Mondini, G. (eds), Valutazione e sostenibilità. Piani, programmi, progetti. Celid, Torino, pp. 17–22 (2009)
40. Fusco Girard, L., Cerreta, M., De Toro, P.: Integrated planning and integrated evaluation. theoretical references and methodological approaches. In: Miller D., Patassini D. (eds.), Beyond Benefit Cost Analysis. Accounting for Non-Market Values in Planning Evaluation. Ashgate, Aldershot, pp. 173–203 (2005)
41. Alexander, E. R. (ed.): Evaluation in Planning. Evolution and Prospects. Ashgate, Aldershot (2006)

Preliminary Approach for the Cost-Benefit Analysis in the Building Envelope: Study and Comparison of Actions

Giovanna Acampa[1,2] , Fabrizio Finucci[3] , Mariolina Grasso[3(✉)] ,
and Antonio Magarò[3]

[1] Faculty of Architecture and Engineering, University Kore of Enna, 94100 Enna, EN, Italy
giovanna.acampa@unifi.it
[2] Department of Architecture, University of Florence, 50121 Florence, FI, Italy
[3] Department of Architecture, Roma Tre University, 00153 Rome, Italy
{fabrizio.finucci,mariolina.grasso,antonio.magaro}@uniroma3.it

Abstract. The architectural envelope plays a fundamental role in shaping the antropic space allowing man to adapt to the environment, to colonise it and therefore to evolve. The primordial needs linked to the instinct of self-protection establish, par excellence, the archaic function of the shell, as a mediator between the anthropic and the natural Within the present research work, the practice of *remodelage* on the envelopes seems to be the one that best defines the type of interventions and the strategy we intend to pursue: the typological modification by addition and volumetric expansion and performance improvement. In this study different types of interventions on the building envelopes have been analyzed and the basis for the development of cost-benefit analysis has been laid.

Keywords: Building envelopes · Cost-benefit analysis · Built heritage

1 Introduction

The issue of building and urban 'regeneration' is fundamentally based on the choice of 'demolish and rebuild' or 'recover and transform'. The prevailing model in most European countries follows a transformation logic that on the one hand gives the consolidated building fabric a cultural value and on the other hand addresses the problem of development in relation to environmental sustainability (disposal costs, scarcity of available resources, etc.). In this scenario, addition is an intervention strategy that has all the potential to radically transform large buildings and to provide effective solutions in terms of function (interior spaces), performance and the building's image. This is an important opportunity for our suburbs where most of the multi-family housing is located. Built over twenty-five years, often in speculative situations, without rules, with poor materials, today in conditions of serious technological and energy, social and architectural degradation. In Italy, according to the latest census, 75% of citizens live in the suburbs and 48% of these live in buildings constructed in the 1960s and 1970s

F. Calabrò et al. (Eds.): NMP 2022, LNNS 482, pp. 786–794, 2022.
https://doi.org/10.1007/978-3-031-06825-6_74

(27 million people). The interventions in most cases have been resolved with responses limited only to maintenance works for the renovation of the facade finishes and the adaptation of the plant. Requalifying instead means modifying the performance of the building to make them satisfactory in relation to the new needs of contemporary living of the individual and the community as a whole. It emerges then the need to provide an answer able to interpret the new user profiles (singles, young couples, elderly, non-EU) that are far from the standards of family-living units that defined the housing types of the '60s and '70s. Among the redevelopment interventions those involving large residential complexes are undoubtedly among the most complex because unlike other cases must consider the presence of inhabitants, and consequently the need to overcome all the technical and logistical difficulties with greater efforts to innovate on construction processes and safety measures of the site. Moreover, this heritage is the result of a stratification lasted decades, to the point that many of these buildings have exceeded what is the concept of useful life, and without interventions of performance improvement inevitably pour in conditions of degradation and obsolescence.

For this reason, because it is from the residential building stock that comes most of the polluting emissions into the atmosphere, and because the problems related to land consumption require that we intervene more on the existing than on the new, it is increasingly urgent the need for a strategy of urban and architectural regeneration.

The main lines of intervention, to date found, are two and in sharp contrast between them:

– the "scrapping" of entire portions of the city in order to operate a massive building replacement;
– the "mending" that provides targeted and punctual interventions, although too often hinged in operations to improve the thermal performance of the envelopes.

It is possible to imagine that this operation is carried out directly by the property, which could be an option in the case of intervention on public property carried out by the Public Administration, but similarly the social costs arising from a building replacement operation involving the relocation of a few thousand people if carried out intensively are not considered. Within this discussion, this approach represents a path not to be pursued.

Since the late 1990s, experiments in remodeling, in terms of improving the performance of envelopes, have been carried out in several European countries, to the point that this strategy is more complex and codified than it seems.

It takes the name of *remodelage*, a term coined by the French architect Roland Castro to indicate the complex of actions that lead to the substantial modification not only morphological, but also typological, of a building, and that reverberate on the urban surroundings. The *remodelage* is composed of the application, even not contemporary, of the following actions:

– add parts of the building, external areas, internal partitions, more or less advanced functional layers, etc.;
– subtract through targeted demolitions, technological units or even entire floors to implement changes of typological character;
– replace infills, window frames, cladding, etc.

Within the present research work, the practice of *remodelage* on the envelopes seems to be the one that best defines the type of interventions and the strategy we intend to pursue: the typological modification by addition and volumetric expansion and performance improvement.

With the processes of redevelopment using additions, not only is it possible to plan the intervention entirely from the outside, thus avoiding the relocation of inhabitants, but it is also possible to sensitively modify the image of the building to return in terms of architectural quality benefits to the entire community. In this study different types of interventions on the shells of the built heritage have been analyzed and the basis for the development of cost-benefit analysis has been laid, identifying the costs of interventions and the main benefits resulting from the interventions made.

2   The Architectural Envelope Between Habitability and Adaptability

The architectural envelope plays a fundamental role in shaping the antropic space allowing man to adapt to the environment, to colonise it and therefore to evolve. The primordial needs linked to the instinct of self-protection establish, par excellence, the archaic function of the shell, as a mediator between the anthropic and the natural. The functional aspects have allowed the attribution to the envelope of a series of requirements, such as the control of the internal temperature, while the formal aspects of representativeness or of creating a filtering space between private and public space have always been important [1]. Underlying the design of any envelope is the dialogue between mass and void, corresponding to the choice between separating and correlating, between full and empty, between light and shadow [2]. Therefore, the wall envelope can also become, episodically, an inhabited place. Medieval military and fortified architectures are an example of this: among all, Castel del Monte, in the Murge plateau. Although the octagonal envelope appears to consist of two walls, their mass, in relation to their close spacing, configures them as a single mighty wall in which eight passages are excavated. A similar arrangement is found in the basement of Palazzo Farnese in Caprarola (Jacopo Barozzi da Vignola): the fortress is built on a pentagonal wall in which there are rooms that pass through to form a circular courtyard. The Renaissance was an experimentation in the study and recovery of ancient architecture: in 1535 the elevation of the Theatre of Marcellus in Rome (Baldassare Peruzzi) was completed and transformed into Palazzo Savelli (later Palazzo Orsini). In truth, it had already lost its original function in the 14th century, and the arched shell typical of the Roman theatre was host to barrels and small dwellings. n the same year, but diatopic, is the façade of Palazzo dei Banchi in Bologna (Jaco-po Barozzi da Vignola). With the aim of unifying the façade and giving dignity to the eastern edge of Piazza Maggiore, the architect creates an advanced classifying façade in which the basement level is transformed into a commercial portico denouncing the volumetric extensions for the upper floors. The urban regeneration this intervention entails is comparable to that of the Libreria nella Piazzetta in Venice (Jacopo Tatti Sansovino). Set against the Doge's Palace, it unfolds over 16 bays that unify the building behind it and determine its extension. Subsequently, the concept of the inhabitable shell corresponds to the stratification of the city. Urban transformations are the fruit of this, where the roof gives way to an elevation and the overhang is closed to create a service space. It was not until the 1950s that the habitability of envelopes was systematically applied. In 1955, Le Corbusier completed Notre Dame du Haut in Ronchamp. The wall of the chapel facing south is archetypal, flared in plan and raised, with a strong thickness denoted by the powerful splays of the openings, habitable as caves in the rock. For the next ten years, many of Louis Kahn's buildings took up the concept of the habitable wall/window, translated into modernity by the fascination with Scottish castles. Moreover, he is one of the few architects who mentions the use of the invariate poché, taken from the hotels particuliers, characterised by public or private service spaces excavated in the wall structure. The baton is being picked up in the contemporary world by tectonic architects such as Souto de Moura, who have been called on to definitively consecrate the shell as a living space: one example is the Das Bernardas Convent, converted into a residence with 78 dwellings. The Portuguese Pritzker's intention is to continue in the wake of the transformations, while retaining some of the site's characteristics, such as the niches with seats carved into the wall. Just as the habitability of the envelopes is characterised by the totality of the façade and the roof, there is no biunivocal correspondence between façade and envelope. One of the most recent definitions of the envelope is that it is the set of "technical elements of closure and frontier, vertical, horizontal or inclined, of a building system, designed to perimeter, separate, protect and put in relation and interaction material and immaterial space inside of this system than outside. The continuous relationship between inside and outside reaffirms the concept of adaptability for which the envelope is that set of technical elements able to ensure the maximum level of environmental comfort with the minimum energy requirements achievable by the architecture confined in it [3]. One of the first definitions of adaptivity is by Giuseppe Ciribini, who, addressing his students, defines the architectural project as an "adaptive dynamic system" and adaptivity as "the ability of a system to naturally assimilate to different realities with consequent changes in state". From a historical point of view, some definitions anticipate the meaning of envelope adaptivity. Already in the 1960s, Banham referred to the minimum endowments of living, such as curtain and fire, highlighting how a building should be able to change its boundaries and thermal properties according to the surrounding conditions. In the 1970s, at the beginning of the computer revolution, architectural spaces intersected with digital spaces: enclosures took on a different meaning in relation to an environment that was not only real but also virtual. The concept of a responsive environment is emerging, for which interior and exterior play an active role in the dynamics of reciprocal changes, which can be defined with a complex mathematical function that can only be resolved through the use of information technology. In the 1980s, research into materials capable of responding to a complex exigential framework theorised the possibility of creating an envelope capable of modifying its performance. This is the case of the Polyvalent Wall, which is able to control energy flows between inside and outside through a succession of thin poly-performance layers. In the 1990s, the writer Stewart Brand [1995] theorises the modifiability of the envelope in relation to environmental conditions, schematising the envelope in concentric layers related to each other according to their attitude to change. Disassociating adaptivity from automatic mechanisms of movement, Frei Otto links adaptivity to the constructive lightness of involuclosures, giving the latter the possibility to modify themselves, not only morphologically but also in position and place. Currently, the most accepted definitions are that adaptive envelopes, separating interior and exterior, are able to change their function, behaviour and characteristics, over time, depending on the boundary conditions, to improve the overall performance of the building. However, it is necessary that the goal is to achieve the best comfort level through the lowest energy expenditure in comparison to the static envelope.

In accordance with this semantic framework, adaptive envelopes are those that provide an adequate response to changes occurring in both the external and internal environment, with the aim of maintaining or improving performance in terms of heat flow, air and vapour permeability, closure against atmospheric agents, protection from solar radiation and noise, fire behaviour, stability, as well as formal and aesthetic performance.

3 Main Methods of Economic Valuation of Built Heritage Envelopes

As built heritage is a social product, its analysis and evaluation must express the complexity of analysed elements.

Analyses often focus on historical-morphological changes of spatial characteristics in order to avoid excessive simplification in the presentation of space, which is typical for modern space planning [4].

The multicriteria evaluation models contain evaluations from different disciplines, and these evaluations should preferably be articulated as special modules. The evaluation of heritage requires special analyses and techniques, e.g. relating to architecture or landscape and, in these models, the aspects of analysis and evaluation can not often be differentiated from one another. The models for evaluation of individual elements of built heritage consider a single element (or a complex), while evaluations focusing on landscape comprise all significant characteristics of landscape (hydrological systems, space enclosing elements, trees, animals, etc.). Thus landscape evaluation models analyse various features of heritage: obsolescence, duration, scarcity, artistic creativity, connection with historic persons or events, recognition, registration, conservation, interpretation, loss, and passage of time.

Kalman's method [4] is based on the prepared table with indicators and sub-indicators, which are attributed appropriate values in accordance with a predefined scale. These indicators are: architecture (with sub-indicators: style, construction, age, architect, design, interior), history (with sub-indicators: person, event, context), environment (with sub-indicators: continuity, setting, landmark), usability (with sub-indicators: compatibility, adaptability, public, services, cost), integrity (with sub-indicators: site, alterations, condition). The same cards can be used for the analysis of the extant, for evaluation of the designed state (including comparison with other projects), and for monitoring. This method has been developed for the evaluation of architecture, and is hence less adequate for the evaluation of open spaces.

Campeol's pyramidal model was developed during the work on the UNESCO project Urban Development and Freshwater Resources: Small Coastal Cities.

The model is based on multicriteria analysis where data about space are grouped according to "qualities" and "damage" within the matrix model, and then the type of project is defined – conservation, valorisation, preservation, renewal, regulated use, new use [5, 5]. This model enables synthetic approach to the issue of heritage.

Lichfield defines the approach called "heritage value for money" [7], which is based on the analysis of costs and benefits. Project effects and efficiency are identified and priority activities are defined taking into account the budget allocated for the project. The method is used to analyse distribution of benefits and costs throughout the life cycle of a structure/facility, discount rate, loan costs and distribution of impacts on social sectors. The method comprises data about individual cultural assets and is especially efficient when used with other types of evaluations that estimate features of heritage [8]. The model involving the entrepreneur approach briefly describes activities that can be made with respect to heritage, taking into account the existing resources. This approach makes use of a table organised in fields such as cost (small, medium, and high cost) and type of intervention (maintenance, renewal, and change of use). For each combination of resources and intervention type, the approach gives indications about possible interventions (valorisation, maintenance, forming minimum conditions for subsequent interventions, change of use for innovative initiatives, etc.). Although the characteristic part of the evaluation is synthetic, in a part of the analysis this model checks different financial resources, social needs, results of services, market and context, management methods, financial sustainability both in the initial phase and during use. The model is based on economic analysis but is specially adjusted to built heritage evaluation [9, 9].

4 Case Studies and Classification of Action on Building Envelopes: An Approach for Assessing the Sustainability of Envelopes

In the analysis of habitable envelopes, are classified some European example cases, within a period that covers the last two decades. Residential buildings are considered, as well as those that acquire this use destination following recovery interventions. This study proposes an analysis data sheet that can become a general work tool. The types of Actions that characterize each intervention are identified. These actions have been defined to describe the general intervention strategy applied to the envelope (Table1).

Table 1. Actions on building envelopes and associated example

Action	Action descriptions	Example
Wrapping	Intervention in which the new envelope completely covers the existing building	Casa a Morchiuso
Densyfying in continuity	Intervention aimed at increasing volume and the number of inhabitants. In the same way it generally refers to an elevation of an existing building, with modification of the roof and an increase in performance of this element	Torre a Soriano nel Cimino

(*continued*)

Table 1. (*continued*)

Action	Action descriptions	Example
Densyfying in discontinuity	Intervention aimed at increasing volumes and the number of inhabitants. In the same way it generally refers to an elevation of an existing building, with modification of the roof and an increase in performance of this element	Lude House
Saturating	intervention on the envelope, it is proposed to saturate the urban fabric characterized by a voidthat interrupts its continuity	Dovecote studio;
Integrating	It expresses a collaboration between thebuilding and the new envelope, which is generally relative only to a portion of the pre-existence	Heliotrope
Grafting	The action of grafting aims to consider the existing building as a host, in which a new, generally small, body is inoculated. This new organism acts as a parasite, while the pre-existence is the host organism	Neo Leo
Redefining	This is an intervention aimed at balancing urban fronts that have not developed, generally in height, presenting inhomogeneities. Also, such interventions on the casings are often related to consolidated fabrics, for which modifi cations to the casings in height are envisaged	Surélévation in Rue Daumier
Reshaping	It involves a radical replacement of the envelope with a consequent morphological and typological modification. The other substantial difference concerns the fact that the re-modelling action requires that the new envelope works in place of the previous one, without juxtaposition, but in adherence with the existing building, like a new skin	Villa Rotterdam
Selecting	This action aims, in a surgical manner, toperform specifi c services, often localized in single portions of the envelope. Therefore, the action of selecting is relative to what needs to be changed, as to what can be maintained	La ruina habitada
Superfetation	This action, unlike grafting, requires that the neworganism is completely autonomous respect to the one on which it is juxtaposed	Arichinger House

Following this classification the next step was to identify the main benefits related to the 3 dimensions of sustainability (**environmental:** generally understood as the ability to generate not only work, but also income and increased value; **social**: aimed at ensuring conditions of well-being, linked to different parameters, such as security, health, education, democracy, justice and participation; **techno-economic:** whose objective is to maintain the quality and reproducibility of the resources) considering the actions mentioned above. Table 2 lists the main benefits related to the 3 dimensions of sustainability.

Table 2. Benefit related to sustainability dimension

SUSTAINABILITY DIMENSION BENEFIT		
ENVIRONMEN TAL	SOCIAL	TECHINCAL-ECONOMIC
Global warming potential	Improvements in personal safety	Construction cost
	Improvements in social cohesion	Maintenance cost
	Fabric densification	Savings on energy expenditure over the lifetime
Ozone Depletion potential	Increase in personal services	
	Increase in urban level services	

	Costs				
	Env.	Social		Ecomomic	
Actions	Environmental cost of building production	Increased urban load	Discomfort of the construction site	Construction cost	Maintenance cost
Wrapping					
Densyfying in continuity					
Densyfying in discontinuity					
Saturating					
Integrating					
Grating					
Redefining					
Reshaping					
Selecting					
Superfetation					

	Benefit									
	Environmental		Social					Economic		
Actions	Global Warming reduction	Ozone Depletion Potential	Improvements in personal safety	Improvements in social cohesion	Fabric densification	Increase in personal services	Increase in urban level services	Increase in the total market value of the property	Increase in the unit market value of the property	Savings on energy expenditure over the lifetime
Wrapping										
Densyfying in continuity										
Densyfying in discontinuity										
Saturating										
Integrating										
Grating										
Redefining										
Reshaping										
Selecting										
Superfetation										

Fig. 1. Cost-benefit/action relation forms

Afterwards, two forms (Fig. 1) were set up to be filled in order to indicate for each of the actions listed in Table 1 the presence (or absence) of the identified cost and benefits. If the action (e.g. Wrappring) results in a cost (environmental, social, economic) and/or in a benefit, it is suggested to enter the value 1 in the forms, otherwise 0.

5 Conclusions

This study addresses the topic related to the classification of action on building envelopes and subsequently aim to lay the foundations for a cost-benefit analysis to define from which action derives the best performance (in terms of benefits) at the lowest possible cost. Both costs and benefits are divided into to three categories (Environmental, Economic and Social) and are grouped into a form to be filled in order to carry out the analysis. The future development of this research will focus on the development of cost-benefit analysis, so far approached in a theoretical way. In particular we will look for a method to attribute in a coherent and objective way a value between 1 and 0 to the different types of costs and benefits expected in order to produce reliable results.

References

1. Serra, O., Fiorensa, R.: L'energia nel progetto d'architettura. In: Battisti, A., Tucci, F. (eds.) (a cura di) Ambiente e Cultura Dell'abitare. Dedalo Edizioni, Roma (2000)
2. Cellucci, C., Di Sivo, M., Santi, G.: Architettura Del Vano Murario. ETS Edizioni, Pisa (2018)
3. Magarò, A.: Involucri Abitabili Adattivi. Metodologia sistemica di rigenerazione urbana, Tesi di Dottorato (XXXII ciclo), Dipartimento di Architettura, Università degli Studi Roma Tre, Roma (2020)
4. Baratta, A.F.L., Finucci, F., Magarò, A.: Generative Design Process: Multi-criteria Evaluation and Multidisciplinary Approach. Technè n. 21|2021, pp. 304–314 (2021). https://doi.org/10.13128/Techne-20797
5. Acampa, G., Campisi, T., Grasso, M., Marino, G., Torrisi, V.: Exploring European strategies for the optimization of the benefits and cost-effectiveness of private electric mobility. In: Gervasi, O., et al. (eds.) ICCSA 2021. LNCS, vol. 12953, pp. 715–729. Springer, Cham (2021). https://doi.org/10.1007/978-3-030-86976-2_49
6. Nijkamp, P., Medda, F.: Integrated assessment of urban revitalization projects (chapter). In: Fusco Girard, L., Forte, B., Cerreta, M., De Toro, P., Forte, F. (eds.) The Human Sustinable City: Challenges and Perspectives from the Habitat Agenda, pp.417–428. Ashgate Publishing, Aldershot, Burlington (2003)
7. Lichfield, N.: Economics in Urban Conservation. Cambridge University Press, Cambridge (1988)
8. Baratta, A.F.L., Finucci, F., Magarò, A.: Regenerating regeneration: augmented reality and new models of minor architectural heritage reuse. VITRUVIO – Int. J. Architect. Technol. Sustain. 3(2), 1–14 (2018). https://doi.org/10.4995/vitruvio-ijats.2018.10884
9. Acampa, G., Grasso, M.: Heritage evaluation: restoration plan through HBIM and MCDA. IOP Conf. Ser. Mater. Sci. Eng. 949(1), 012061 (2020). https://doi.org/10.1088/1757-899X/949/1/012061
10. Baratta, A., Calcagnini, L., Finucci, F., Magarò, A., Molina, H., Quintana Ramirez, H.S.: Strategy for better performance in spontaneous building. TECHNE – J. Technol. Architect. Environ. 14, 158–167 (2017). https://doi.org/10.13128/Techne-20797

Implementation Tools for Projects Cost Benefit Analysis (CBA)

Giovanna Acampa[1,2](✉) and Giorgia Marino[1]

[1] Kore University of Enna, 94100 Enna, Italy
giovanna.acampa@unikore.it, giorgia.marino001@unikorestudent.it
[2] University of Florence, 50121 Florence, Italy

Abstract. The paper focus on the integration between digitalization technologies and evaluation tools. in particular how Building information Modeling (BIM) and the Geographic Information System (GIS) can be integrated to support Cost-Benefit Analysis in the first step of the project evaluation process as set forth in the National Recovery and Resilience Plan (PNRR) and the Guidelines for the project's technical and economic feasibility issued by the Italian Ministry of infrastructure and sustainable mobility (MIMS). BIM and GIS, along with BIM plug-ins to be further developed, provide useful tools to the calculation of NPV as requested in the above guidelines.

Keywords: Evaluation · BIM · PNRR

1 Introduction

1.1 The National Recovery and Resilience Plan (PNRR) and the Guidelines for the project's Technical and Economic Feasibility Issued by the Italian Ministry of Infrastructure and Sustainable Mobility (MIMS)

The pandemic, and the ensuing economic crisis, pushed the EU to formulate a coordinated structural response, in particular with the launch in July 2020 of the Next Generation EU programme (NGEU). The amount of resources provided to support growth, investment and reform is estimated at EUR 750 billion. NGEU aims to promote a powerful recovery of the European economy through green transition, digitisation, competitiveness, training and social, territorial and gender inclusion.

The main component of the NGEU programme is the Recovery and Resilience Facility (RRF), running for six years, from 2021 to 2026. Italia Domani [1], the National Recovery and Resilience Plan (PNRR) [2] presented by Italy, includes a project of reforms concerning public administration, justice, simplification of legislation and promotion of competition. Reforms in public administration includes steps to improve administrative capacity at both central and local levels; strengthening the processes of selection, training and promotion of civil servants; and encouraging the simplification

This paper was prepared with the active support of Mariolina Grasso.

and digitalisation of administrative procedures. A key point is the extensive use of digital services, with the aim of reducing bureaucratic procedures and optimising the costs and time currently burdening businesses and citizens.

PNRR has sixteen parts, grouped into six Missions [3]:

- Mission 1: Digitalisation, innovation, competitiveness, culture and tourism;
- Mission 2: Green revolution and ecological transition;
- Mission 3: Infrastructure for sustainable mobility;
- Mission 4: Education and research;
- Mission 5: Cohesion and inclusion,
- Mission 6: Health.

Focusing on Mission 3, it is divided into two main sections: the first containing *investments on the railway network*, dedicated to the completion of the main high-speed railway axes and making the entire railway network safe; the second section, *intramodality and integrated logistics*, provides for measures to support the modernisation and digitalisation of the logistics system. Overall, the planned investments are consistent with the national mobility strategy of the MIMS, articles 44 and 48 of Decree-Law no. 77 of 31 May 2021, converted into Law no. 108 of 29 July 2021. It defines an accelerated procedure for "major works" on the basis of a project technical and economic feasibility plan (PFTE).

According to the PFTE guidelines, the design and execution of works can be entrusted on the basis of the project's technical and economic feasibility plan, divided in two steps, to which the basic documentation on the relationship between the geometric-spatial layout of the infrastructure and its environmental components should be attached.

Step 1 "WHAT": setting the requirements framework related to the economic and social needs and their performance goals and indicators. It includes an analysis of:

- the general objectives to be achieved through the planned interventions, associated with specific performance indicators;
- the needs of the community, or of the specific users for who the intervention is intended, as reference basis for the works;
- the qualitative and quantitative needs of the commissioning authority and of the specific users to be satisfied through the planned works; Idea on integrated energy control system in public hospitals

On the basis of the requirements framework, the feasibility document of the project alternatives (DOCFAP) develops a comparative analysis between project alternatives and the decision support system to select the best option is considered to be the Cost Benefit Analysis (CBA).

Step 2 "HOW": after identifying the overall "preferable" design alternative, in the second phase the design guidance document (DIP) regulates the preparation of the technical and economic feasibility plan (PFTE) and contains the performance requirements to be achieved [4]. In the PFTE a comparison among different technological solutions is carried out using multi-criteria analysis (Fig. 1).

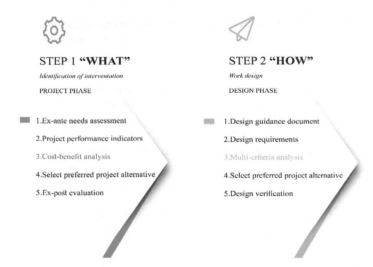

Fig. 1. Step 1 and Step 2 of PFTE.

The guidelines above outlined can be applied to any kind of project, public or private and the Cost Benefit Analysis (CBA) is time again considered as the main evaluation tool.

The paper will focus on the integration between digitalization technologies and evaluation tools. In particular how Building information Modeling (BIM) [5] and the Geographic Information System (GIS) can be integrated to support Cost-Benefit Analysis in the first step of the project evaluation process. The aim is to create the prerequisites for the optimization and organization of data useful to perform a cost-benefit analysis. At urban scale, this means sharing and organizing information collected from various sources. For this purpose, GIS models are destined to acquire, manage and view georeferenced information and at the same time it is fundamental to have the specific data referred to the specific structure and the BIM models allows to link these data to 3D geometric features of generic construction.

1.2 Economic Appraisal Tools Cost-Benefit (CBA) and Multi-criteria Analysis (MCA)

The main methodological tool to support selection between project alternatives in the guidelines is *Cost-Benefit Analysis* (CBA) [6]. Basically, on the one hand costs (investment, maintenance and operation, plus negative impacts) are assessed, and on the other, benefits (revenues, shadow prices, time savings, etc.) are all transformed into monetary values using appropriate standards. Thus getting final indicators – such as the economic *Net present value* - NPV i.e. a method to determine the current value of all future cash flows generated by a project, including the initial capital investment, or the *Internal Return Rate* - IRR, i.e. the value of the discount rate which cancels the NPV - are calculated [7, 8].

On the other hand, Multi-Criteria Analysis (MCA) could also be used along with the CBA, especially in step 2 of the evaluation process, once the best alternative is selected by means of the CBA. MCA is a methodology for assessing the different impacts of each alternative, not only economic but also social, environmental and territorial, using an appropriate measurement system. This analysis is therefore particularly suitable when several judgements which focus on a multiplicity of criteria (e.g., economic, territorial, performance, etc.) have to be considered. Evaluation consists of comparing and ordering a set of alternatives using appropriate criteria and relative weights. In practice, the results of each project alternative are measured against the set objectives and a score is attributed according to the importance that decision-makers attach to each objective. This quantitative (results) and qualitative (weights) mix is summarised in a matrix (decision matrix, the rows of which are the various alternatives and the columns of which are the various judgement criteria linked to the objectives) which, depending on the weights assigned to the various objectives, leads to a ranking of the various solutions. This method is transparent, as highlights the individual results and the final winning solutions in relation to the analysis criteria and their weights.

2   Materials and Method

2.1   The Role of BIM Within the PFTE Guidelines and Its Possible Future Developments in Public Projects with GIS Data Integration

The aim of this paper is to apply the guidelines explained above, which are generally applicable, also in the feasibility analysis of project for the construction of new buildings or the built environment transformability relying on BIM. BIM supports the management, monitoring, and determination of intervention priorities through CBA and the integration with GIS would allow Public Administrations to provide plans for the regeneration of their assets.

So, the PFTE Guidelines are also aimed at defining the essential contents of the documents, of the digital information models (if any) and of the documents and documents required by the Tender Authorities for the award on the basis of the PFTE. Therefore, the essential contents of the PFTE referring to BIM are defined with reference to what is regulated by Article 23, subsections 5 and 6 of the Italian Contracts Code (Legislative Decree no. 50/2016 [9]) and, on an optional basis, by subsection 13 of the said article ("electronic methods and tools"). In order to support the projects and speed up the procedures benefiting from PNRR funding, it is envisaged that the Tender Authorities awarding contracts may provide, in the call for tenders or the letter of invitation, for the allocation of a bonus score for the use of specific electronic methods and tools in the design (DM 312 of 2 August 2021).

In the case of the use of digital information modelling, the guidelines establish the use of the informative specification (Capitolato Informativo - CI/EIR, according to the UNI 11337 [10] and/or UNI EN ISO 19650 standards), as provided for by the D.M. 560/2017, regarding electronic methods and tools. This document integrates the IC contained in the DIP and regulates the digital processes, information modelling, technological and management choices also regarding ACDat (Data Sharing Environment) for the subsequent phases of the process, both in the design and construction fields, with a view to the final digital management of maintenance aspects. BIM modelling allows the optimisation of planning, construction and management of buildings by means of software in which all relevant data are collected, combined and linked, so as to generate a "digital twin" of the future work, on which all conditions, reactions and behaviour in geostatic, structural and energy terms can be calculated and simulated.

In this context it is important to emphasise how this type of approach can be used also in feasibility projects about reuse, transformability and design of public buildings through cost-benefit analysis [11, 12]. We will outline a methodology to support Public Administrations in using MIMS guidelines in the realm of real estate development and, consequently, urban redevelopment/regeneration projects, while integrating GIS spatial data into BIM to optimize the data for carrying out a CBA [13, 14].

2.2   Tools for CBA

The methodological process for CBA proposed in the MIMS guidelines, is focused about the calculation of NPVs. The formula is the following [15]:

$$\mathrm{NPV}_0 = -I_0 + \sum_{t=1}^{n} \frac{C_t}{(1+i)t} + \frac{R_n}{(1+i)n} \tag{1}$$

where:

I_0: Investment at the starting point ($t = 0$);

n: Duration in years;

t: Time interval;

Ct: Cash flow;

i: Discount rate in %;

Rn: Residual value.

A project is cost-effective if its NPV is positive; between two projects, the one with the higher NPV is preferable (Table 1).

Table 1. Values of $ENPV_0$.

Current value	Valuation
$NPV_0 > 0$	Profitable investment
$NPV_0 < 0$	Unprofitable investment
$NPV_0 = 0$	The investment offers no advantage over a risk-free investment bank

In the real estate realm, it is advisable to use BIM software programs. They support in calculating automatically construction costs which are a main component of the I_0. Another key element to this end, is the calculation of the land value strictly connecting to the real estate market and thus to the localization of the building. For this reason, it is appropriate to create a link between BIM and databases on land values through GIS [16]. While BIM by itself is a repository of data of a building considered as a product, it lacks a reference to the positioning of the building, which can be obtained with GIS data. GIS delivers territorial data information which are crucial to assess the land value market, but also the expected revenues coming from a building's rent which are necessary to the calculation of Ct.

Utilizing the BIM's 3D geometric documentation process is a well-established process for integrating geographic information system (GIS) thematic maps [17]. In general, the use of GIS platforms is widely accepted in a vast number of disciplines and scientific fields, but in the field of evaluation, it is still a research area that has many aspects to unfold. Within this information system, a multilateral database can be created, incorporating diverse data such as investment costs, that can be projected through the thematic maps [16, 18].

Appropriate BIM plug-ins are being developed to implement such a connection.

On the other hand, another integration useful to calculate the NPV, especially as far as the selection of the discount rate i according to transparent and objective criteria, is the connection with the quality of the construction. To this end, appropriate plug-ins can be developed according to the building's main quality features, and we have already developed a first tool to this kind, referring to public housing [19].

At this point, we could consider including in the BIM software the automatic calculation of NPV which is a main indicator in CBA.

3   Expected Results and Conclusions

The PNRR promotes, therefore, a new approach to the design, construction and management of an infrastructure, focusing on sustainability and innovation, extending this principle and attention to the various phases of the related process.

A crucial element in the methodology proposed by PNRR is CBA, which time again is considered a crucial evaluation tool for infrastructural projects.

This approach can be extended to the construction sector that can greatly benefit by implementing the methodology showed in PNRR [20]. When data are reliable and accessible in a real estate database, it is possible to implement automated valuation methods like the one proposed. The flow of information from the building model can be conveyed into a computerized database containing economic information about the location characteristics through the interoperability languages represented by the IFC. For example, real estate companies and investors use the tools in ArcGIS to research markets, identify new opportunities for growth and expansion, and manage their investments at the market and neighborhood levels. Real estate professionals can use mobile data collection tools to gather property information directly from the field and analyze and share insights across their organizations in real time.

In particular, geographic information systems (GISs) allow to work with spatial data modeling accurately, as well as with environmental, territorial, and locational factors, while building information models (BIMs) manage the complexity of a 3D building structure with its intrinsic features. Our future work will focus on testing the proposed workflow in order to demonstrate the potential of the BIM-GIS integration. In particular, the idea is to entrust the BIM model with the real estate valuation and to integrate it with the support of geographic information systems technology, useful for analyzing territorial characteristics, in order to allow the automation of the calculation of the cost of the NVP. Thanks to 3D GIS evolution, a BIM model can be integrated to a 3D GIS prototype in order to add some factors and all data about the project or investment. This will be led to a refined property valuation CBA method.

References

1. Home - Italia Domani - Portale PNRR. https://italiadomani.gov.it/it/home.html. Accessed 30 Dec 2021
2. Il Piano Nazionale di Ripresa e Resilienza (PNRR) - Ministero dell'Economia e delle Finanze. https://www.mef.gov.it/focus/Il-Piano-Nazionale-di-Ripresa-e-Resilienza-PNRR/. Accessed 30 Dec 2021
3. PNRR: Il testo delle Linee guida sulla fattibilità tecnica ed economica per l'affidamento di contratti pubblici Available online: https://www.ingenio-web.it/31568-pnrr-approvate-le-linee-guida-sulla-fattibilita-tecnica-ed-economica-per-laffidamento-di-contratti-pubblici. Accessed 30 Dec 2021
4. Lgs, D.: Linee Guida per la Valutazione degli Investimenti in Opere Pubbliche nei settori di competenza del Ministero delle Infrastrutture e dei Trasporti (2011)
5. Fattinnanzi, E., Acampa, G., Forte, F., Valutazioni, F.R.-V.E.: The overall quality assessment in an architecture project. siev.org (2018). Undefined
6. Prest, A.R., Turvey, R.: Cost-benefit analysis: a survey. Surv. Econ. Theory **1966**, 155–207 (1966). https://doi.org/10.1007/978-1-349-00210-8_5
7. Guide to Cost-Benefit Analysis of Investment Projects (2014). https://doi.org/10.2776/97516
8. Beria, P., Maltese, I., Mariotti, I.: Multicriteria versus Cost Benefit Analysis: a comparative perspective in the assessment of sustainable mobility. Eur. Transp. Res. Rev. **4**(3), 137–152 (2012). https://doi.org/10.1007/s12544-012-0074-9
9. Generali, I.: Decreto Legislativo 18 aprile 2016, no. 171, pp. 18–19 (2016)
10. UNI UNI 11337-7: 2018 Edilizia e opere di ingegneria civile - Gestione digitale dei processi informativi delle costruzioni - Parte 7: Requisiti di conoscenza, abilità e competenza delle figure coinvolte nella gestione e nella modellazione informativa (2018)
11. Drèze, J.: Economics, N.S.-H. of public. The theory of cost-benefit analysis. Elsevier (1987). Undefined
12. Layard, R., Glaister, S.: Cost-benefit analysis (1994). ISBN 9780521466745
13. PNRR: i progetti di fattibilità e la metodologia BIM. https://www.ingenio-web.it/33026-pnrr-i-progetti-di-fattibilita-e-la-metodologia-bim. Accessed 30 Dec 2021
14. Acampa, G., Marino, G., Ticali, D.: Validation of infrastructures through BIM. In: Proceedings of the AIP Conference Proceedings, vol. 2186. American Institute of Physics Inc. (2019)
15. Attualizzazione dei Flussi di Cassa Futuri - Cloud Finance. https://www.cloudfinance.it/attualizzazione-dei-flussi-di-cassa-futuri.html. Accessed 30 Dec 2021
16. Acampa, G., Battisti, F., Di Pietro, G., Parisi, C.M.: City information model for the optimization of urban maintenance cost. In: AIP Conference Proceedings, vol. 2343 (2021). https://doi.org/10.1063/5.0047779
17. Tsilimantou, E., Delegou, E.T., Nikitakos, I.A., Ioannidis, C., Moropoulou, A.: GIS and BIM as integrated digital environments for modeling and monitoring of historic buildings. Appl. Sci. **10**, 1078 (2020). https://doi.org/10.3390/APP10031078
18. D'Amico, F., Calvi, A., Schiattarella, E., Di Prete, M., Veraldi, V.: BIM and GIS data integration: a novel approach of technical/environmental decision-making process in transport infrastructure design. Transp. Res. Procedia **45**, 803–810 (2020). https://doi.org/10.1016/j.trpro.2020.02.090

19. Campo, O., Battisti, F., Acampa, G.: Integrated multi-criteria assessments in support of the verifying the feasibility of recovering archaeological sites: the case of Portus-Ostia Antica. In: Bevilacqua, C., Calabrò, F., Della Spina, L. (eds.) NMP 2020. SIST, vol. 178, pp. 1952–1961. Springer, Cham (2021). https://doi.org/10.1007/978-3-030-48279-4_184
20. Acampa, G., Diana, L., Marino, G., Marmo, R.: Assessing the transformability of public housing through BIM. Sustain. **13**, 5431 (2021). https://doi.org/10.3390/SU13105431

Photovoice and Landscape: Participatory Research-Action to Led Young People to Monitor Policies and Landscapes

Pietro Bova[✉]

Department of Architecture and Territory (dArTE), Università Mediterranea di Reggio Calabria,
Reggio Calabria, Italy
pietro.bova@unirc.it

Abstract. This paper proposes the use of the landscape as a key to the interpretation of the territory during participatory action research (PAR). The "landscape" collects and tells the effects of policies: on territory, on environment, on cultural and architectural heritage of the community. In addition, the landscape indicates past practices and may indicate future sustainable policies. In this document is tested a method of PAR called photovoice. The aim of the application is to test if the photovoice can be useful to activate civic monitoring of projects with European cohesion funding.

Photovoice is a methodology born in the 90s of the last century and brings together the visual and mnemonic perception of the community to obtain useful elements for a dialogue with local decision makers. The perception of the landscape itself is one of the best reading keys to be provided to the involved communities, meanwhile the community itself is involved doing photovoice. From the landscape elements it is possible to identify the connections between territory, perceived landscape heritage, and territorial management.

In the case study reported in this paper, the application of photovoice takes place in the inner area of Calabria called «Grecanica» or Greek. The elements identified by young people in the community are linked to local projects resulting from European cohesion policies (according to open-cohesion data and Monithon). Finally, the application verifies the usefulness of the photovoice method in reading the landscape in order to activate social innovation processes, bring the community to perform civil monitoring, write civic monitoring reports and upload them to the Monithon platform (in this case study).

Keywords: Action research · Photovoice · Landscape

1 Problem Statement – Participatory Action Research to Outline Policies in Inner Areas

The methodologies of participatory action research, started in the 70s, have been outlined to respond to a need that in the last century has been felt in the academic environment: the need to involve communities and make them co-protagonists in the search

© The Author(s), under exclusive license to Springer Nature Switzerland AG 2022
F. Calabrò et al. (Eds.): NMP 2022, LNNS 482, pp. 803–814, 2022.
https://doi.org/10.1007/978-3-031-06825-6_76

for knowledge. The involvement of the communities allows the researcher-activist to be able to compose a more complete knowledge (Borda 2006) [3]. According to Borda, the objective of the participatory actions researches (PAR) was to better outline policies for communities and territories, trying to involve and give voice to the most disadvantaged parts of communities. A "fake" participation of the community was not enough to obtain the result sought by the researchers-activists. By "fake" community participation is intended here to all those steps of the ladder of involvement [1] that see the community only as a passive subject useful to verify data in case studies [1, 13–16].[1] The aim of the researcher-activist is to achieve the empowerment of the community, with particular attention to the empowerment of those who are at a disadvantage for socio-economic reasons. How was this intended to be achieved? From the 70s different methods of participatory action-research were born and developed – structured interviews, semi-structured interviews, *photovoice*, etc. – that were as scientifically rigorous as possible in the collection of quali-quantitative data. The PAR methodologies are still evolving, perfectible and declinable according to specific case studies.

The application of PAR methodologies serves to empower a part of the community by really giving it a role in local decision-making. Most of the time the intent of a PAR is to make the community dialogue with local decision makers and stakeholders. To achieve this dialogue, the activist researcher must act as a facilitator (of dialogue), and dialogue must stimulate the writing of policies and actions on the territory.

The researcher-activist facilitator is an intermediary figure and has among its tasks to use and decline the PAR methodologies. In addition, the facilitator must provide the keys to understanding the territory and policies in it, so that the community consulted is guided in the PAR. About the involvement of the community, this paper aims to suggest the use of the landscape as a key to the interpretation of policy results. The landscape as a space lived by the community and the individual, a reading key rarely used in PAR [8] for example in landscape architecture research [5]. The landscape does not see territorial boundaries between provinces or municipalities but rather looks at the area of which the community feels part. This spatial definition, based on the perception of the community and individuals, is functional in order to outline strategies and policies for "area" without necessarily observing municipal boundaries. Moreover, this kind of PAR are useful to pursue some goal of the UN 2030 Agenda: sustainable cities and communities, reduced inequalities, etc.

2 Proposal for PAR in Inner Areas – Reading the Landscape to Read Policies During a PAR (and Suggest New Policies)

Get a qualitative assessment of policies by asking communities to describe their landscape. This type of analysis of the landscape is functional in reading the effects of policies. This statement is supported by some valuable landscape definitions including: a definitions from Rosario Assunto [2]; a definition of "space" and "history" by Fernand Braudel [4]; the definition of landscape written in the European Landscape Convenction [6] in its art. 1; further conceptions debated nowadays.

---

[1] Yin, 2012, p. 4.

Rosario Assunto [2] had described the landscape as the synthesis of natural and human history, it can be deduced that man is not only the observer necessary for the very existence of the landscape and its stimmung[2] [11] but he is also the creator of the shape of the landscape. In addition, if one takes the action of individuals, the actions of a community or also the ones of a civilization, it can be added that man will modify the landscape according to a certain ethics [13]. However, this paper will leave aside the ethical aspect that is behind the management of an environment or a territory, rather it will be held in mind that landscape is the overlap of human actions and policies. An example of landscape as an overlap of human actions and events is the Mediterranean (and European) landscape narrated by Fernand Braudel [4], or the particular history of the Italian agrarian landscape [10]. Both Sereni and Braudel can be evaluated as proof of the usefulness of the landscape as a key to understanding past policies (or human actions). The perception of the landscape can be useful in the PAR for its definition according to the ELC [6]: landscape is the set of territorial and environmental aspects as perceived by the community. Asking the community to describe its landscape is a survey of past policies to suggest new ones. To test the affinity between the perception of the landscape and the investigation of the effects of policies, among the methods that contemplate the visual perception, the *photovoice* method [14] is the methodology of PAR taken in examination by this paper.

3 The Use of the Landscape as a Key to Interpreting the Territory Using the *Photovoice* Method

The *photovoice* method was born in the nineties, conceived by Wang & Burris [14] and differs from a mere photo reportage. Unlike a photo reportage, the *photovoice* collects in photos (and videos) the perception of some socio-economic (and also territorial) problems through the eyes and thoughts of the community. In addition, the peculiarity of the photovoice is to bring to a dialogue the communities and the local decision makers, in this case the dialogue is enhanced thanks to the photographic production. The creators of the methodology described its key points and objectives in 1997 (Fig. 1):

> "Photovoice is a process by which people can identify, represent, and enhance their community through a specific photographic technique. As a practice based in the production of knowledge, photovoice has three main goals: (1) to enable people to record and reflect their community's strengths and concerns, (2) to promote critical dialogue and knowledge about important issues through large and small group discussion of photographs, and (3) to reach policymakers." [14]

---

[2] *Stimmung* is a characteristic of the place that acts as a unifying and organizing principle of the reality otherwise simply perceived (Simmel, ed. 2006).

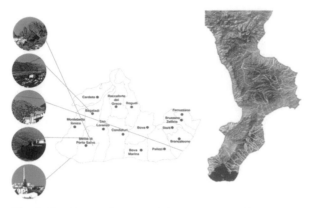

Fig. 1. Landscape elements of the Greek Area of Calabria. The Greek Area is highlighted in blue on the right of the image. The municipalities of this area are 15 in accordance with the Italian National Strategy for Inner Areas (SNAI) (2021).

The different applications of *the photovoice,* in different places in the world, have led to its natural evolution and correction. Wang himself dwells on reporting further indications for the involvement of young people in the communities [15]. In 2017 Fantini puts the method to a test [8] reporting the successes and failures of various PAR projects with *photovoice* method to investigate water management. Fantini also reports an interesting, albeit rare, use of the landscape as a key to reading the territory during some PAR. Fantini also said something about the potential given by reading the Landscape during the *photovoice* method: "The contribution of participatory visual methods to study […] perceptions of landscapes or waterscapes deserves further study." [8]. Following the suggestions of Fantini, below are the results obtained through the reading of the landscape according to the photovoice method. The case study reported is the landscape (and the territory) of the Calabrian *Grecanica* (or Greek) Area located in southern Italy. In this case the photovoice methodology has been expanded and linked to a good civic monitoring practice called *Monithon.*[3]

Fig. 2. One of the 1st "round" of photovoice with one of the groups. Ph. (n.1) by: Prof. A. M. Ermidio

4   The Structure of the Photovoice Method Used + *Monithon*

To start the experimentation I proceeded in accordance with the 9 steps outlined by Wang [15] for the involvement of young people. The 9 steps of the *photovoice* method are summarized here in 4 steps + 1 for simplification (Fig. 2):

---

[3] All the info on monithon can be found at: https://www.monithon.eu/about-english/ (consulted on 7/12/2021).

1. 3 groups of students were selected to be involved in the PAR: the optimal number of participants per group is between 7 and 10 young people according to Wang [15]. In this case there were 3 groups: a group of 7 students from the "Euclide" school for surveyors in Bova Marina (RC); two groups of students from the "Liceo linguistico Ten. With the. G. Familiari". The total amount of students is 23. Once the young people were involved, it was possible to plan the series of meetings for the PAR.

2. A series of meetings was agreed according to availability and school schedules. During the first meeting were introduced and explained: the photovoice method to the groups involved; the objective of dialoguing with decision makers; the objective to suggest policies to decision makers by reading the landscape. Moreover, unlike the procedure outlined by Wang, time was spent at this stage to give young people the knowledge of some conceptions of landscape. The conceptions of landscape – including those extrapolated from Rosario Assunto [2] Emilio Sereni [10] Micheal Jakob [9] and others – have given the basis for better reading the territory lived and photographing its. This type of reading then made it possible to identify the demands of young people connected to landscape elements: the basis for a dialogue with the decision makers.

3. Deadlines have been set for sending photographic and videographic material. After the introductory meeting participants were given time to take photos during their daily lives and to write down related comments for each photo or video. Below, there were organized meetings to discuss photographic and videographic production. This step can led to other rounds of photovoice. According to Wang, 1 to 3 rounds of *photovoice* are required, the number of rounds is to be decided according to the requests of the participants.

4. The last step is underway, there will be a dialogue with the decision makers in January 2022. In fact, this paper wants to report the results of the photovoice method obtained before the dialogue with local decision makers. The fourth step involves the organization of an event to achieve dialogue between the community and local decisionmakers. In this particular case study, there will be a debate where the students will expose their requests related to the landscape elements reported.

5. The fifth step goes beyond the photovoice method, it is peculiar to this case study and is made possible thanks to a good italian practice: *Monithon*. *Monithon* is an initiative born in 2013 with the aim of combining a "participatory" monitoring with "official" monitoring. The "official" monitoring - on *Open Coesione* data [18] - and the evaluation of public policies is accompanied[4] by a PAR. *Monithon* uses the energies and expertise of the final beneficiaries of the policies, with the aim of proposing and outlining improvements for projects financed with European funds (2021).

Below is the list of face-to-face meetings, from step 2 to the 5.

---

[4] Open platform for transparency within projects with European cohesion funds.

Table 1. Calendar of activities.

Meeting	Title of the meeting	Step (from 1 to 5)
1st – October 2021	Participatory action research: Landscape & Photovoice	Step 2
2nd – November 2021	Photovoice results. Round 1. Everyday landscape elements	Step 3
3rd – November 2021	Photovoice results. Round 2: Link between landscape elements and European, regional and local policies	Step 3
4th – November 2021	Photovoice results. Round 3. Discussion of new Material, *Monithon* and organization of meeting with decision *makers*	Steps 3 and 4
5th – January 2022	Photovoice results: discussion Public event/discussion with decision makers	Steps 4 and 5

5   The Link Between the Landscape Reported with the *Photovoice* Method and the Civic Monitoring Through *Monithon:* The Case Study of the Calabrian Greek Area

The calendar of activities (in Table 1) is useful for the replication of this photovoice methodology plus civic monitoring, especially on Italian soil with the possibility of relying on *Monithon* for civic monitoring. Please note that the reasons that have led to the choice of testing the photovoice method in the Greek area, involving and leading young people to do civic monitoring, are more than one: the objective of this activity with students is one of the sub-objectives not achieved by the interventions of the National Strategy (SNAI) [19] for this internal area[5]; the depopulation of the area reported by the data of the Italian Statistical Institute (ISTAT)[6] [17]; this activity has been recognized as a laboratory activity that connects school and university.[7]

The link between the landscape elements reported by the working groups and the civic monitoring naturally comes out of the discussions within the *photovoice* method. This is because social innovation and civic monitoring initiatives aim to bring arguments into a dialogue with decision makers [7, 12].

In the specific case of the Greek Area of Calabria, an area formed by 15 municipalities and overlooking the east coast of Calabria, the groups involved were formed with students between 16 and 18 years old. The choice of working groups derives from an old issue of depopulation of the area, with a strong youth abandonment due to several complex phenomena. In any case, this paper wants to deal with only one of the aspects that leads young people to migrate: the absence of dialogue between young people and decision makers. The young people of the Greek Area suffer from not being questioned

---

[5] The goal to create social innovation and lead the community to do civic monitoring was written inside the intervention A.1 – "Laboratorio di Sviluppo Locale e Innovazione Sociale dell'Area Grecanica" in annex 2.a of the National Strategy for Inland Areas (SNAI) – "Grecanica" (2021). Source: http://www.snaigrecanica.it/download/Allegato_2A_RelazioniTecnicheSinteti cheInterventi.pdf (on 20/12/2021).

[6] Source: https://gis.censimentopopolazione.istat.it/apps/opsdashboard/index.html#/e8e6eccf2 6f34bb6b734899354d13928/ (on 20/12/2021).

[7] In Italy this kind of activity are under the name of "Percorso per le Competenze Trasversali e l'Orientamento" (PCTO): a training path that can be integrated into the school program, with the aim of guide students into the choose of a work or of a university study path.

or considered enough in what is the implementation of local policies. Among the working groups, many of the students see themselves as incapable of influencing policies. They do not see themselves in a position of dialogue with decision makers. More than half of the participants believe that they cannot influence local policies in any way. In addition, there are seven subjects who believe it is useless to even try to dialogue with local decision makers and who have expressed this thought by giving no answer. However, those who responded to the interviews said they were confident in the projects being worked in their area. It should be noted that before the interview the projects were illustrated (Fig. 3).

Fig. 3. Trust in policies and institutions (among young people involved in this PAR) from 1 to 10.[8]

However, this trust in the projects derives from the very fact of knowing the projects. Many of the students were not aware of the 7 projects (in Table 2) covered before responding to the interview. In fact, before the start of the activities, 20 out of 23 (if we count the 8 abstentions) were not aware of the projects in progress in the area they lived (Fig. 4).

Fig. 4. Only 3 people were aware of the 7 projects covered during the face-to-face meetings.[9]

---

[8] Source: https://forms.gle/7jy5v5S1okooGPGe6 (on 20/12/2021).

[9] Source: https://forms.gle/hAHByRamgbKKSm5P9 (on 20/12/2021).

Subsequently, the working groups photographed and filmed on video the landscape elements they experienced in the Greek Area. During the discussions they were able to expose their considerations and their requests related to the territory they live. For privacy, only the initials of the students who produced the photo shoot will be reported (Fig. 5).

In photo n. 2 it is possible to see the water-front of Palizzi Marina (RC). During the discussion with the author of photo n. 2 more than one request came out: the necessary maintenance of the place together with a better management of urban waste. The request was linked to the landscape element without hiding a certain nostalgia for an idealized (but possible) past landscape.

Fig. 5. Geolocalization of photographies

In photo n. 3 it is possible to observe an effect of the current management of urban waste, but here the author (N.I.) also wants to denounce a certain level of incivility that could be reduced with a better management of the recycling chain. In photo n. 4 is instead reported by C.M. one of the remains that report the hope of past decades: it was common practice to leave the upper floor unfinished and then build it and leave it to the sons. In photo n. 5 is instead placed an element of landscape that could be better valued and that otherwise risks being "erased" by neglect: the bell tower of the church of *San Sebastiano* in the municipality of Condofuri. In photos n. 6 and n. 7 we note some last useful elements to summarize the points most treated in the dialogue that took place in the photovoice rounds: the added value that infrastructures can give to the landscape if inserted in a plastic and non-destructive way; the possibility of rethinking old and new infrastructures in relation to the local economy and the potential tourism of some almost unreachable place; a lack of interest and appreciation (from outside) of the rich heritage of legends and local knowledge - including knowledge of the Greek language that derives from past Greek colonies - related to certain places in the "Grecanica" Area (Figs. 6 and 7).

Fig. 6. From left to right: photo n. 2; photo n. 3; photo n. 4

After a first dialogue on the photographic and videographic material held for each individual student, we proceeded by better investigating the 7 projects in progress in the Greek Area (with European cohesion funds). After a brief investigation of the seven

Fig. 7. From left to right: photo n. 5; photo n. 6; photo n. 7

projects, of which none have had a civic monitoring with reports on the *Monithon* platform, the effects that these projects could have on the landscape aspects photographed were discussed. The next step was to have the working groups vote on a project that they would monitor. Finally, with me as a facilitator for civic monitoring, it was decided to start a report on *Monithon*. In addition to the classic goal of photovoice, the dialogue with decision makers, there was added a civic monitoring.

Table 2. Project ready for monitoring using *Monithon*.

Name of the project[a]	Location or municipality in the Greek Area of Calabria (Italy)	Cost in the[b] programme: POR FESR FSE CALABRIA	Spent (on data "open coesione"[c])	Monitored using *Monithon*[d]
Project to enhance the municipal solid waste management service	Motta San Giovanni (RC)	142,475.60€	0€	No
RC 018A/10 COMUNE DI MONTEBELLO JONICO[e]	Montebello Jonico (RC)	1,000,000€	584.374,54€	No
Interventions for the Protection of the "Pantano" Oasis of Saline Joniche (in Montebello Jonico)"	Montebello Jonico (RC)	50,470.55€	0€	No

(*continued*)

Table 2. (*continued*)

Name of the project[a]	Location or municipality in the Greek Area of Calabria (Italy)	Cost in the[b] programme: POR FESR FSE CALABRIA	Spent (on data "open coesione"[c])	Monitored using *Monithon*[d]
Conservation and monitoring of the habitat of the oily scrub -juniperetum turbinatae- area sic "Fiumara Amendolea" and area "Capo San Giovanni" - repopulation/ restoration of species in conservation interest of the oily scrub - juniperetum	Bova Marina (RC)	59,715.06€	33,536.86€	No
Integrated management of urban waste, assimilated and urban hygiene - Municipality of Condofuri	Condofuri (RC)	449,855.39€	0€	No
Restoration and enhancement of the castle of Palizzi	Palizzi (RC)	300,000€	101.000€	No
Enhancement of the Sea turtle recovery center in Brancaleone	Brancaleone (RC)	59,313.99€	59,313.99€	No

[a] Name of the projects are translated by the author.
[b] Data available on: https://opencoesione.gov.it/it/progetti (on 20/12/2021).
[c] Data available on: https://opencoesione.gov.it/it/progetti (on 20/12/2021).
[d] Data available on: https://projectfinder.monithon.eu/ (on 20/12/2021).
[e] Project objectives in brief: construction of new infrastructure; adaptation of the infrastructures to climate change; prevention and management of climate-related risks.

6 Considerations on the Use of the *Photovoice* Method for Social Innovation and Civic Monitoring

The considerations on the PAR project conducted up to the date of writing of this paper are for points that can be summarized: 1) asking the working groups to read their "own" landscape through the *photovoice* methodology is functional thanks to the link between inhabitants and "landscape"; 2) the *photovoice* methodology was functional in the creation of social innovation and community engagement; 3) the results obtained with this declination of the *photovoice* method can be transferred into civic monitoring reports to make communities and decision makers dialogue.

In the specific case, different needs and proposals of the young inhabitants came out of the dialogue: better management of waste and its recycling; a more careful action for the maintenance and restoration of the aspects of the landscape; the creation and adaptation of infrastructure for sustainable tourism and in response to the risks posed by climate change. From the considerations and proposals of the participants it was then possible to move on a civic monitoring action (using *Monithon* platform). The photovoice method has therefore proved to be suitable for outlining the basis for various civic monitoring actions and in particular for the monitoring of European regional cohesion policy projects.

After the application of the photovoice method presented in this paper, the students were finally intrigued by a project for the protection of the ecosystem and are now (December 2021) guided by me in the civic monitoring of this project. The project selected by the students is: "Conservation and monitoring of the habitat of the oily scrub -juniperetum turbinatae- area sic "Fiumara Amendolea" and area "Capo San Giovanni" - repopulation/restoration of species in conservation interest of the oily scrub - juniperetum."

The ongoing civic monitoring serves to monitor different aspects of the landscape that the students were interested in during the PAR: protection of places, ecosystem, tourism, culture, work. The Phoenician juniper, the kind of endangered tree focused by the project monitored, is linked to local history and culture but also guarantees the life of a rare ecosystem suitable for the nesting of turtles "*caretta caretta*".

During the last step of the PAR students will have the opportunity to dialogue with the decision makers, with me as a facilitator, and with an enhanced knowledge of local policies and projects. The PAR method also helps the facilitator, as those who lead a PAR get a deeper knowledge of community and territory. Finally, the aim of the final dialogue will be to propose improvements about the management of the projects.

A final consideration goes to the availability of decision makers in participating in the dialogues. Fantini [8] reports that photovoice projects do not always succeed in reaching a dialogue with decision makers and that there is a risk of creating dialogues for their own sake. Adding a further step to the classic photovoice method, which involves the creation of civic monitoring reports, could be the right conclusion of the method. Adding civic monitoring, giving the community the tools to write facilitated reports such as those on *Monithon*, can lead to a more lasting dialogue between communities and decision makers.

References

1. Arnstein, S.R.: A ladder of citizen participation. J. Am. Inst. Plann. **35**(4), 216–224 (1969). https://doi.org/10.1080/01944366908977225
2. Assunto, R.: Il paesaggio e l'estetica. Novecento, Palermo (1973)
3. Borda, O.F.: Participatory (action) research in social theory: origins and challenges. In: The SAGE Handbook of Action Research: Participative Inquiry and Practice. SAGE, London (2006)
4. Braudel, F., De Angeli, E.: Il mediterraneo: lo spazio, la storia, gli uomini, le tradizioni. Bompiani, Milano (1985)
5. Bruns, D., Münderlein, D.: Visual methods in landscape architecture research. Presented at the September 11 (2016)
6. Council of Europe: European Landscape Convenction (2000)
7. Derr, V., Simons, J.: A review of photovoice applications in environment, sustainability, and conservation contexts: is the method maintaining its emancipatory intents? Environ. Educ. Res. **26**(3), 359–380 (2020). https://doi.org/10.1080/13504622.2019.1693511
8. Fantini, E.: Picturing waters: a review of Photovoice and similar participatory visual research on water governance: Photovoice on water governance. Wiley Interdiscip. Rev. Water. **4**(5), e1226 (2017). https://doi.org/10.1002/wat2.1226
9. Jakob, M.: Il paesaggio. Il Mulino, Bologna (2017)
10. Sereni, E.: Storia del paesaggio agrario italiano. Laterza, Bari, Roma (1962)
11. Simmel, G.: Saggi sul paesaggio. Armando Editore, Roma (2006)
12. Den Broeck, V., et al.: Social Innovation as Political Transformation – Thoughts for a Better World. Edward Elgar Publishing, Cheltenham (2020)
13. Venturi Ferriolo, M.: Etiche del paesaggio: il progetto del mondo umano. Editori riuniti, Roma (2002)
14. Wang, C., Burris, M.A.: Photovoice: concept, methodology, and use for participatory needs assessment. Health Educ. Behav. **24**(3), 369–387 (1997)
15. Wang, C.C.: Youth participation in photovoice as a strategy for community change. J. Community Pract. **14**(1–2), 147–161 (2006). https://doi.org/10.1300/J125v14n01_09
16. Yin, R.K.: Case study methods. In: Cooper, H., et al. (eds.) APA Handbook of Research Methods in Psychology, Vol. 2: Research Designs: Quantitative, Qualitative, Neuropsychological, and Biological, pp. 141–155 American Psychological Association, Washington (2012). https://doi.org/10.1037/13620-009
17. IstatDashCens.   https://gis.censimentopopolazione.istat.it/apps/opsdashboard/index.html#/e8e6eccf26f34bb6b734899354d13928/. Accessed 28 Dec 2021
18. OpenCoesione – Home. https://opencoesione.gov.it/it/. Accessed 28 Dec 2021
19. Strategia Nazionale Aree interne - Area Grecanica. http://www.snaigrecanica.it/. Accessed 28 Dec 2021

The Mediation Role of the University Tutor to Promote Student Empowerment: A Student Voice Survey Through Documentary Writing

Viviana Vinci[(✉)] [iD]

Mediterranea University of Reggio Calabria, Reggio Calabria, Italy
viviana.vinci@unirc.it

Abstract. The most recent educational policies emphasise the need to promote the empowerment of students, who are recognised as protagonists and active partners in the learning and improvement processes of university teaching. In this direction, universities are committed to consolidate guidance, tutoring and placement services. Tutoring, in particular, is a strategic tool to accompany students in their academic career. A pilot survey carried out with student tutors at the Mediterranea University of Reggio Calabria is presented. The survey included the analysis of reflective documental devices structured to accompany the tutoring process. The results of the survey underlined the role of the Tutor as mediator and suggested several actions to improve the tutoring service. The experimentation demonstrates the power of narrative and writing devices. It also suggests enhancing the student-Tutor point of view in research on the improvement of university services and Faculty Development.

Keywords: Student empowerment · University tutor · Documentary writing

1 Theoretical Framework

The national and international policies of recent decades in the field of higher education highlight the need for young people's guidance as support for decision-making processes, for identity building and for the development of autonomy, awareness and planning throughout their lives: these are essential skills to be able to cope with the transition phases from a degree course to the professional world (which is constantly changing) and to be able to respond effectively to ongoing social changes, through the realisation of one's own life and professional project [1, 2]. It is about promoting the empowerment of the student, recognised as a protagonist and active partner in the teaching-learning process and in the quality improvement processes [3–5].

The OECD has recently pointed out some unresolved issues [6, 7]: a mismatch between the aspirations of young people, who have poor knowledge of the labour market and the reality of employment; the marginality of career guidance actions in the educational offer. In Italy, following Ministerial Decree 1047/2017 and the issuing of 'Linee Guida per i Piani di Orientamento e Tutorato' (Guidelines for Guidance and Tutoring Plans), Universities have been strongly committed to the consolidation of guidance,

F. Calabrò et al. (Eds.): NMP 2022, LNNS 482, pp. 815–825, 2022.
https://doi.org/10.1007/978-3-031-06825-6_77

tutoring and placement services, with a shift from predictive, directive and specialised models to training models enhancing the subjectivity, autonomy, responsibility and active participation of students, who gradually develop greater control and power over their own lives and choices.

Tutoring, which has been established in Italian universities since Law 341/1990, consists of global and integrated support actions aimed at the academic accompaniment of students; it covers multiple dimensions, not only academic, but also personal and professional, so it seems difficult to find a common univocal definition [8, 9]. There are many types of activity in university tutoring: of a planning nature, aimed at defining the academic and personal pathway of the student for integration into the workplace and society; of an educational nature, such as orientation actions and strategies for the development of the person as a whole; of a didactic-formative nature, as accompanying actions for the acquisition of both generic and disciplinary skills; of a psycho-educational nature, as processes for enhancing learning styles for understanding, reflection and responsibility [10]. Promoting personalised learning is now at the heart of most HE institutional missions and many universities are currently reviewing their strategic and operational tutoring infrastructure [11, 12]. On the basis of these premises, it seems of fundamental importance for educational research to assume the student-tutor's point of view on tutoring [13, 14].

2   Methodological Design

The results of a survey involving 15 qualified university tutors, launched in the academic year 2019–2020 at the Mediterranea University of Reggio Calabria, are described. The survey is part of a pilot experimentation[1] aimed at improving orientation and tutoring services through the use of narrative devices as tools for reflective accompaniment between school and university. The experimentation has foreseen different phases: 1) training activity addressed to the students on the Tutor's role at University; 2) monitoring of the tutoring service through the compilation of a register and a final report; 3) structuring of a reflective documentary device functional to assume the Tutor's point of view on tutoring (Table 1) designed to accompany the Tutors' reflection and improve the effectiveness of the tutoring pathway.

The investigation has included the analysis of documentary devices in order to verify whether reflective writing can facilitate learning processes and well-being in university students, encouraging their empowerment and preventing dropout.

---

[1] The pilot experimentation was designed as an activity carried out within the framework of a departmental delegation assumed by the Author from a.y. 2019–20 for Orientation and Pathways for Transversal Skills and Orientation. The experimentation included two investigations that move from the relevance of documentary writing for the university student's training and the improvement of the Orientation and Tutoring service: 1) an orientation pathway with a narrative background that involved 25 students of a high school institute and that included the use of different narrative devices: autobiographical writing deliveries; mind-map writing deliveries (individual/group); collaborative writing sessions of a project [15]; 2) the structuring of a documental dispositive for 15 qualified university tutors (senior students selected for the tutoring service) in the areas of Statistics-Mathematics, Economics, Law, whose results are here described.

Table 1. Documentary device for university tutors.

Biographical data	Tutor's role	Tutor's experience	Dropout: the tutor's viewpoint	Self-assessment and proposals for improvement
Personal data, department, start date of tutoring contract	Tutor functions and competences	Motivation for choice, activities carried out, interactions, difficulties, role in pandemic emergency	Prevailing student support needs, reasons for university dropout, difficulties in the transition from high school to university	Self-assessment, lessons learnt, emotions felt, expectations, training needs, proposals for improvement of the tutoring service

The research hypothesis underlying the design of the research protocol looks at narrative practices as possible paths capable of: promoting, in the student, essential skills for the success of university education, in the logic of self-direction (reflexivity, communication, capacity for analysis and choice, autonomy, self-regulation, self-efficacy); giving back, to the University, useful elements for the improvement of its services, enhancing the "student's voice", often neglected; contributing to rethink, in a formative direction, the orientation paths and the training of qualified Tutors supporting first year students, enhancing reflective and identity building practices.

Specific objectives of the survey are: a) to examine and understand the tutoring practice from the "point of view" of the Tutor's narrative; b) to introduce the Tutor to the documentary writing through a first pilot experience and to promote the narrative and explicitation competence; c) to promote in the Tutor the ability to reflect on his/her identity, role and functions; d) to validate a device of documentary writing useful to collect what said and unsaid within the training; e) to assume, from the analysis of the text corpus, some indications for the improvement of the tutoring service.

The entire text corpus was subsequently analysed by means of a Qualitative Data Analysis (QDA) procedure [16, 17] divided into the three phases of *open coding* (first conceptualisation into meaningful text units and labels), *axial coding* (identification of frequent macrocategories) and *selective coding* (hierarchization of macrocategories and emergence of core categories).

3   Data Analysis

From the analysis of the documents compiled by the qualified tutors, an interesting profile of the university tutor - considered a facilitator and mediator between students and the university context - emerged, starting from the outline of his/her competences (Table 2):

Table 2. University Tutor's role and competences: emerging core categories.

Axial coding	Selective coding	Core category
- Mastery of the chosen subject matter - Consolidate knowledge of the subject matter - Adequate knowledge of the subject matter - The ability to facilitate learning - Sufficient preparation at teaching level - Teaching and expression skills (simplifying and clarifying concepts)	Mastery of the subject matter Didactics: being able to clarify and simplify concepts	Mediation with the subject cultural matter
- An overall knowledge of the university, i.e. knowledge of the location of all the different offices and of the procedures for submitting documents or taking courses; - A good knowledge of how the university environment works - Supports and assists students on their academic path, having some extra years of university experience behind them - It should be a point of reference for those students who are lost, who perhaps do not know how to approach a subject, or the university in general - It should deal with inefficiencies, if any, in degree courses, and bureaucracy in secretariats	Ability to "orientate" in the university context Knowledge of the university environment Previous academic experience	Mediation with the university system
- It should deal with misunderstandings between teachers and students - Acts as a mediator between the student and the teacher - Communicating and supporting communication	Relationship management Communication	Mediation with Professor

(*continued*)

Table 2. (*continued*)

Axial coding	Selective coding	Core category
- Have a good understanding of the individual's difficulties and a good dose of empathy - Listen to and understand requests for help - Provide psychological and educational assistance to students during their studies - Skills to understand the personality of the student with whom they relate - Honing their empathy and their communication and listening skills - Empathy, availability and listening skills - Being able to identify a student's potential, stimulate and encourage him/her, monitoring individual progress - Helping the student to be actively engaged in the individual education process and guiding him/her to overcome obstacles and difficulties that may arise during his/her academic career - Give great importance to the individual needs of each person	Welcoming, listening, empathy, and context analysis skills Motivation and guidance Differentiation and personalisation	Mediation with students

A further noteworthy finding concerns the motivations for the choice, the description of the activities carried out and the difficulties encountered (Tables 3 and 4), made explicit thanks to the heuristic and formative power of writing [18, 19], which allows to narratively explain one's own identity and bring to light the implicit meanings embedded in the experience:

The survey also highlighted the reasons that university tutors attribute to the phenomenon of student dropout, showing a multiplicity of interrelated factors (Table 5) concerning both the difficulties in coming to terms with the academic world (criticality of the transition from high school to university; not being able to adapt, loneliness, disorientation and distance from family and friends), and more personal-psychological factors such as complex construction/management of one's own learning pathway, lack of a study method, absence of encouragement, heterogeneity in basic preparation, low

Table 3. Tutor's experience (activities): core categories.

Main activities	
Selective coding	Core category
- Help with the written test - Advice on textbooks - Support in the approach to studying the subject - Explanations of a didactic nature - Assistance in carrying out exercises - Administration of quizzes on the subject - Laboratory exercises (e.g. recognising plant slides in the laboratory, identifying plant species in the herbarium) - Supervising entry tests, with responsibility for correcting the tests - Listening to students on prepared topics - Preparing summaries of the main topics of the subjects - Providing disciplinary explanations	Disciplinary support (study, exams, activities)
- Collaboration with the teaching secretariat - Uploading and checking data on the university portal - Support in a wide range of activities	University system support
- Front-office work with students - Welcoming incoming students	Welcoming, information
- Training activities on the role of the tutor	Training

Table 4. Tutor's experience (motivations, difficulties): core categories.

Motivations		Difficulties	
Selective coding	Core category	Selective coding	Core category
Passion, self-improvement, 'challenging', personal gratification	Self-centred	Low student turnout; succeeding in motivating; creating a relationship of trust and exchange	Users
Relating to students, helping those in difficulty, ensuring that students experience the difficulties they personally experience in the absence of reference figures	Other-centred	Difficulty in interfacing at distance - online; publicising the tutoring activity	Provision
		Working alone, having few tools at own disposal; reconciling tutoring activities with lessons and study	Working methods

self-esteem, poor self-regulation skills, poor time management, poor relationship management with teachers: in both cases, the Tutor's role assumes a central function as mediator and facilitator.

Table 5. The tutor's view on university dropout: core categories.

Causes of student dropout		Needs	
Selective coding	Core category	Selective coding	Core category
Critical issues in the transition from high school to university; Failure to cope with the university environment; failing to adapt; distance from family and friends; heterogeneity in the knowledge base; past failures; difficulty in managing greater 'freedom', with little autonomy	School-university transition	Information about examination methods and type of test; support in accessing information	Informative needs
		In-depth study of the study content to be addressed; disciplinary explanations; support in the organisation of study for the preparation of the examination	Formative needs
Distance between disciplines and student's personal interests and aptitudes; fear of choosing the 'wrong' university address; error in the choice of academic path; low/no motivation	University choice	Structuring of the study plan; selection of lessons to be attended; how to relate to teachers and/or technical-administrative staff	Organisational needs
Loneliness, disorientation; lack of encouragement; comparison with better prepared students; conflictual relationship with the teacher; low self-esteem; lack of emotional support	Emotional difficulty		

(*continued*)

Table 5. (*continued*)

Causes of student dropout		Needs	
Selective coding	Core category	Selective coding	Core category
Lack of learning methods; unable to organise their own workload; unable to self-regulate; unable to manage time; lack of logistical/economic support; lack of a continuous and daily relationship with university context	Didactic-organisational difficulties		

Furthermore, from the analysis of the data some proposals for improvement and suggestions for the university tutoring service emerged, which can be summarised as follows: greater accessibility of the service on the website and more extensive advertising of the service through every possible channel; receptions by appointment and not 'free' (over-the-counter); evaluation of students at the end of the tutoring activities (feedback on their work); independent spaces for the provision of the tutoring service (i. e.: Tutor rooms); extension of the tutoring service to more disciplines than those selected in the call for proposals (e.g. at least one Tutor for each Degree Course); anticipating the start of the service (with the beginning of the academic year or with the lessons of the first semester); spending more time on tutor training and defining clear and detailed roles for the tutor.

4   Results

The results of the survey revealed a clear profile of the Tutor as a facilitator and mediator. This mediation [20] concerns the university system (i.e. the connection with information, services and the didactic and administrative components of the university system), the Professors (as facilitator of communication and support in accessing information), the cultural object (in its formative and scaffolding function in learning and metacognition processes, such as the personalisation of the cognitive pathway, the refinement of the study method, the transposition of didactic practices, etc.), and the motivation and sense of self-efficacy of the student (who needs a 'place', albeit symbolic, for listening, counselling and be enhanced).

The needs of the students are in fact plural and are of an informative nature (requests for information about the university context), of a formative nature (to strengthen or develop specific knowledge or skills for self-determination and autonomy in the learning process), and of a counselling nature (specific advice and support needs at an individual, psychological and motivational level) [21]. The Tutor's skills that emerged from the data analysis are several: mastery of the subject, flexibility, motivational capacity, organisational capacity, listening skills, adaptation, mediation (between student and teacher, between student and secretariat, between student and object of study), facilitation, differentiation, personalisation, welcoming, communication, exemplarity, context analysis (Fig. 1). They evoke the UKAT Professional Framework for Advising and Tutoring [22], that sets out the core competencies that personal tutors and academic advisors need to effectively support student success.

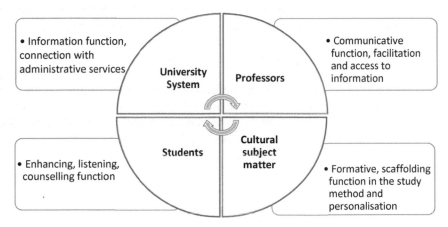

Fig. 1. Multiple mediation functions of the university tutor.

The suggested improvements have been taken up by the Mediterranea University of Reggio Calabria, which has launched a process of reorganisation of the tutoring service, starting with a rescheduling of the time in advance of its provision, the implementation of a greater number of tutoring grants (extended to more disciplines and courses of study), the organisation of the reception also in online mode and with a space specifically dedicated to communication on the University website; a student questionnaire on the provision of the service is also being implemented, in the light of the importance of feedback and peer evaluation [23] as a tool for monitoring and assessing the quality of the service. The survey described demonstrated the scaffolding power of narrative [24–28], a valuable accompanying tool, for the construction of interpretative and planning processes useful to facilitate university student learning and well-being. It is writing that enables narrative, identity and therefore orientation work, starting from the awareness of the sense of self and the construction of one's own identity. It is writing that enables the promotion of certain transversal skills that are indispensable in the delicate transition to the university and professional contexts [18]; it is writing that brings to light the implicit meanings embedded in experience [29].

Memory and narration are privileged ways of accessing knowledge: in the act of narrating oneself, the subject perceives the self as another, doubles up, being both 'here and now' and 'there and then', looking at himself from the outside and from the inside [25]. The perspective outlined here sees writing - and its heuristic and formative power [19] - as a tool for teaching school-university accompaniment [15].

5 Implications

In the transition from the school to the university context, it is necessary to think of a teaching method capable of accompanying the student, i.e. capable of following him/her, promoting empowerment and the ability to self-direct by strengthening some transversal skills central to university study and the transition from the school to the academic world. Although in an exploratory, pilot form, to be replicated with a higher number of Tutors, the survey conducted allows us to interpret the practice of tutoring by taking it from the "point of view" of the Tutor and to assume some useful indications for the improvement of the tutoring service. In this sense it fully suggests to enhance the point of view of the student-Tutor (considered as an active partner in the processes of qualification of teaching and services to promote student participation at University) also in the research on the improvement of university services and Faculty Development, i.e. that set of activities of a formal or informal nature, undertaken at individual or group level, that universities implement to encourage innovation [30, 31]. The student "voice" - often ignored by teachers and policy makers [32] - can be able to offer significant information for the improvement of university teaching and teaching innovation services.

References

1. Klemenčič, M.: From student engagement to student agency: conceptual considerations of European policies on student-centered learning in higher education. High Educ. Pol. **30**, 69–85 (2017)
2. McIntosh, E.A., Troxel, W.G., Grey, D., Van Den Wijngaard, O., Thomas, L.: Academic Advising and Tutoring for Student Success in Higher Education: International Perspectives. Frontiers Media SA, Lausanne (2021)
3. Weimer, M.: Learner-Centered Teaching. Five Key Changes to Practice. Jossey-Bass, San Francisco (2013)
4. Grion, V., Cook-Sather, A.: Student Voice. Prospettive internazionali e pratiche emergenti in Italia. Guerini, Milano (2013)
5. Kinash, S., et al.: Global graduate employability research: a report to the Business20 Human Capital Taskforce. Bond University, Gold Coast, QLD Australia (2014)
6. OECD: Youth Aspirations and the Reality of Jobs in Developing Countries. OECD Publishing, Development Centre Studies, Paris (2017)
7. Musset, P., Kurekova, L.M.: Working it out. Career Guidance and Employer Engagement. Education Working Papers, No. 175, OECD Publishing, Paris (2018)
8. Álvarez Pérez, P.R., González, M.: Los planes de tutoría en la Universidad: una guía para su implantación. Servicio de Publicaciones de La Universidad de la Laguna, San Cristóbal de la Laguna (2008)
9. Yale, A.: The personal tutor-student relationship: student expectations and experiences of personal tutoring in higher education. J. Furth. High. Educ. **43**, 533–544 (2019)
10. Da Re, L.: Favorire il successo accademico: il Tutorato Formativo fra ricerca e intervento nell'esperienza dell'Università di Padova. Formazione & Insegnamento **XVI**(3), 185–199 (2018)
11. Lochtie, D., McIntosh, E., Stork, A., Walker, B.: Effective Personal Tutoring in Higher Education. Critical Publishing, Cheshire (2018)
12. Thomas, L.: What works 2, student retention and success programme final report. Higher Education Academy, York, United Kingdom (2017)

13. Raby, A.: Student voice in personal tutoring. Front. Educ. **5**, 64–72 (2020)
14. Ghenghesh, P.: Personal tutoring from the perspectives of tutors and tutees. J. Furth. High. Educ. **42**, 570–584 (2018)
15. Vinci, V.: Dispositivi narrativi per una didattica dell'accompagnamento: una ricerca per il miglioramento dei servizi di orientamento e tutorato universitario. In: Zago, G., Polenghi, S., Agostinetto, L. (eds.) Memoria ed educazione. Identità, narrazione, diversità, pp. 213–225. Pensa Multimedia, Lecce (2020)
16. Charmaz, K.: Grounded theory in the 21st century. In: Denzin, N.K., Lincoln, Y.S. (eds.) Handbook of Qualitative Research. Sage, Thousand Oaks (2005)
17. Strauss, A., Corbin, J.: Basics of Qualitative Research: Grounded Theory Procedures and Techniques. Sage, Newbury Park (1990)
18. Perla, L.: Scritture professionali. Metodi per la formazione. Progedit, Bari (2012)
19. Perla, L.: Document-AZIONE=Professionalizz-AZIONE? Scritture per la co-formazione in servizio degli insegnanti. In: Magnoler, P., Notti, A.M., Perla, L. (eds.) La professionalità degli insegnanti. La ricerca e le pratiche, pp. 69–87. Pensa MultiMedia, Lecce (2017)
20. Damiano, E.: La mediazione didattica. Per una teoria dell'insegnamento. FrancoAngeli, Milano (2013)
21. Bertagna, G., Puricelli, E.: Dalla scuola all'Università. Orientamento in ingresso e dispositivo di ammissione. Rubbettino Editore: Sovaria Manelli (CZ) (2008)
22. UK Advising and Tutoring (UKAT): The UKAT Professional Framework for Advising and Tutoring (2019). https://www.ukat.uk/professional-development/professional-framework-for-advising-and-tutoring/. Accessed 22 Dec 2021
23. Restiglian, E., Grion, V.: Valutazione e feedback fra pari nella scuola: uno studio di caso nell'ambito del progetto GRiFoVA. Ital. J. Educ. Res., 195–221 (2019). Special Issue, Year XII. https://doi.org/10.7346/SIRD-1S2019-P195
24. Bruner, J.: The narrative construction of reality. Crit. Inq. **18**, 1–21 (1991)
25. Demetrio, D.: Raccontarsi. L'autobiografia come cura di sé. Raffaello Cortina, Milano (1996)
26. Demetrio, D.: La scrittura clinica. Consulenza autobiografica e fragilità esistenziali. Raffaello Cortina, Milano (2008)
27. Clandinin, D.J., Connelly, F.M.: Narrative Inquiry: Experience and Story in Qualitative Research. Jossey-Bass, San Francisco (2000)
28. Craig, C.: Story constellations: a narrative approach to contextualizing teachers' knowledge of school reform. Teach. Teach. Educ. **23**, 173–188 (2007)
29. Perla, L.: Didattica dell'implicito. Ciò che l'insegnante non sa. La Scuola, Brescia (2010)
30. Sorcinelli, M.D.: Fostering 21st century teaching and learning: new models for faculty professional development. In: Lotti, A., Lampugnani, P.A. (eds.) Faculty Development e valorizzazione delle competenze didattiche dei Docenti nelle Università Italiane, pp. 19–25. GUP Genova University Press, Genova (2020)
31. Perla, L., Vinci, V.: Didattica, riconoscimento professionale e innovazione in Università. FrancoAngeli, Milano (2021)
32. Fielding, M.: Beyond student voice: patterns of partnership and the demands of deep democracy. Revista de Educaciòn **359**, 45–65 (2012)

'Work'/'World of Work' and Primary Teacher Training

Laura Sara Agrati[✉] [iD]

University of Bergamo, via Pignolo n. 123, 24121 Bergamo, Italy
laurasara.agrati@unibg.it

Abstract. 'Work', as general concept, has been defined the activity that fully involves the human faculties. The work is however considered a specific form of human experience and is assumed as a founding aspect for teachers and learners.

For this reason, work has often been a specific focus into the training curricula for new generations, since primary school. After recalling some old and new perspectives about the 'work' and the 'world of work', as training interventions within the primary school curriculum, the contribution focuses attention on how these must be able to become topics of primary teachers' training. The method and the results of the first phase of analysis of a SoTL investigation, conducted within the Didactic course at the bachelor's degree in Primary Education at the University of Bergamo, are presented.

The investigation aimed to know the future teachers' representations of 'work' and 'world of work', as concepts, in general, and as themes to be integrated into the school curriculum. The analysis highlighted the ability of student teachers to choose activities suited to the characteristics of the hypothetical pupils but also the difficulty of translating these themes into curriculum learning objectives.

Keywords: 'Work' · 'World of work' · Primary teacher training

1 'Work' and 'World of Work' at Primary School - Old and New Perspectives

'Work', as general concept, is defined the activity that fully involves the human faculties (perception, movement, design and ideation etc.). It is intentionally aimed at product means for survival. Over the years, various definitions have been proposed [1], among which Arendt's well-known distinction emerges [2]. The philosopher distinguishes *work*, as a general condition of the human being, from *labor* (the production of the 'artificial' world) and *action* (the ability to move in a public political space).

Form the perspectives of different attributed meanings, work is however considered a specific form of human experience [3–5] and the work experience is assumed as an unavoidable instance for teachers' educational intervention [6]. For this reason, work has often been at the center of the training curricula for young generations since primary school: examples are the various proposals for intervention inspired by different theoretical frameworks.

F. Calabrò et al. (Eds.): NMP 2022, LNNS 482, pp. 826–839, 2022.
https://doi.org/10.1007/978-3-031-06825-6_78

H. Pestalozzi [7] was one of the first authors who considered work as a form of training for the human being, for this reason, to be encouraged from primary schools. Work - he argued - is dehumanizing if done in an oppressive sense and if practiced in a servile way, as a sacrifice; instead, if understood as the most natural human activity, any kind of work, especially the practical one, becomes a means of integral education [8, 9]. The specificity of work would be to develop the thought through physical exercise and the involvement of the psychological and emotional-affective dimensions [10]. Theoretical reflections on the so-called 'good' work [11–13] nowadays attach more and more importance to this latter aspect.

From the Kerschensteiner's pedagogical perspective [14] - inspired by Dewey, Rickert, Natorp and Pestalozzi - work is not only an intellectual and aesthetic activity but also the ability to transform own self with commitment and attention into a well-determined objective product ('objectivity'). Also for this reason, work would be characterized by the criterion of utility. In the so-called 'schools of work', inspired by Kershensteiner's perspective [15], children were not taught specific trades, but rather developed personal creative skills, combined with qualities such as self-control, sense of collaboration and precision. These abilities and skills would have contributed to the building of human society, the active development of people, in its ethical and civil sense [16].

Maria Montessori [17] also highlighted the educational role of work within the development processes of child's personality and social identity, beyond its utilitarian value linked to specific contexts. By her perspective, the 'hand' is the psychic organ par excellence, 'engine of curiosity', understood as the need to explore in order to learn to adapt to the environment [18]. She defines, therefore, work as 'a vital instinct, which must accompany all stages of the child's growth process to help him build the necessary cognitive and emotional tools, in adolescence, then as an adult, to guide himself in society' [9, p. 38].

The related 'world of work' construct is understood as the set of organizational forms of work activities (occupations, jobs, employers, employees, paychecks, promotions, etc.), subject to transformations due to social, political and anthropological contexts [19, 20]. Even the 'world of work' has often entered the school curricula, mainly in secondary school, less in primary school. With respect to the latter in particular, the experience of the 'school as a research center' [21] should be remembered. It developed in the mid-70s and was influenced by the tradition of Italian activism, which is inspired by Personalistic intellectual stance and the Dewey' thought. In the training proposal for primary schools, work is placed both among the 'educational goals' towards which the teaching activity tends, as well as among the 'detailed themes of social research', which the methods of investigation resulting from the various disciplines are inspired by - cf. Table 1. These research topics serve as 'critical examination of the organization and functioning of the society in which we live, addressing its constituent elements and structures in relation to human, individual and collective needs' [21, p. 285].

A deep reading of the school curriculum brings out an idea of 'work' that combines both the dimension of the *labor* and that of the *activity*. Furthermore, in the references to the organizational forms of the 'world of work' (e.g. unemployment and exploitation as effects of unbalanced work relationships), it is possible to recognize a sensitivity to

Table 1. 'Work' as educational goal. Adapt.: [21, pp. 68, 286–287].

Educational goal	• Working to make explicit the natural human dimension of realization (non-stressful, non-anxiety-inducing, non-alienating work) • Operational acquisition of the idea that the work is expressed in reference to specific technical structures • Rediscovery of the social value of human work (our society is founded on work). Respect for work • Rediscovery of human work as a 'service' and as a 'collaboration' • Human work fits responsibly into reality • Manual activities aimed at making tools • Subordination of production to the actual needs and real needs of the social group and the child • How human work can produce authentic well-being
Detailed topic of social research	Work in our environment • Artisan, agricultural (including fishing and breeding) and industrial work *(typologies, working conditions and safety at work, relationship between work and local economic activities, techniques and education, consequences of work - employment, well-being, pollution, etc.)* • Local economic resources *(raw materials and actual resources; economic resources and infrastructures; relationships between economic activities)* • Specific human conditions - commuters, laborers, unemployed *(causes, consequent human situation, remedies)* • Strike

the social conditions typical of the work activities that characterized the Italian nation in the 1970s (protection of workers, balance between products and well-being) [22].

The concept of 'work' and the related construct of 'world of work' find above all today a new redefinition and placement within the educational proposals also thanks to the innovations introduced in the ONU's Agenda 2030 [23, 24]. This recognizes work as an indispensable necessity to be guaranteed in an equitable manner and, for this reason, as an unavoidable theme for the education of the citizens of tomorrow [25–29] (Table 2).

The transversal analysis of the 17 objectives of the 2030 Agenda highlights an idea of 'work' in which the *activity* dimension prevails, less that of the labor. The need for decent working conditions for wider types of human beings - men, women, disabled people, immigrants - is reaffirmed, in addition to the fight against forms of forced labor. The 'world of work', on the other hand, is considered in a functional way to increase the production of goods and services, also thanks to technologies. Fayard's words can be shared therefore: 'the nature of work is evolving, with the emergence of new forms such

Table 2. 'Work' as topic into the Agenda 2030 17 SDGs. Adapt.: [23].

Goals	Description	Target and interventions
5. Gender equality	Achieve gender equality and empower all women and girls	5.4: Recognize and value unpaid care and domestic work through the provision of public services, infrastructure and social protection policies and the promotion of shared responsibility within the household and the family as nationally appropriate 5.b: Enhance the use of enabling technology, in particular information and communications technology, to promote the empowerment of women
8. Decent work and economic growth	Promote sustained, inclusive and sustainable economic growth, full and productive employment and decent work for all	8.2: Achieve higher levels of economic productivity through diversification, technological upgrading and innovation, including through a focus on high-value added and labour-intensive sectors 8.3: Promote development-oriented policies that support productive activities, decent job creation, entrepreneurship, creativity and innovation, and encourage the formalization and growth of micro-, small-and medium-sized enterprises, including through access to financial services 8.5: By 2030, achieve full and productive employment and decent work for all women and men, including for young people and persons with disabilities, and equal pay for work of equal value 8.7: Take immediate and effective measures to eradicate forced labour, end modern slavery and human trafficking and secure the prohibition and elimination of the worst forms of child labour, including recruitment and use of child soldiers, and by 2025 end child labour in all its forms 8.8: Protect labour rights and promote safe and secure working environments for all workers, including migrant workers, in particular women migrants, and those in precarious employment

(*continued*)

Table 2. (*continued*)

Goals	Description	Target and interventions
12. Responsible consumptions and production	Ensure sustainable consumption and production patterns	12.b: Develop and implement tools to monitor sustainable development impacts for sustainable tourism that creates jobs and promotes local culture and products

as open innovation and crowdsourcing, freelancing and the gig economy and artificial intelligence, and robotics' [30, p. 213].

An original perspective that responds to new emergencies is the one with which the OECD has addressed the theme of 'work' and the correlated 'world of work' in relation to the representations of primary school children [28]. 'Envisioning the Future of Education and Jobs: Trends, Data and Drawings' publishes the project promoted by the OECD based on data from the 2018 edition of 'Education at a Glance' [31, 32] and the study 'Drawing the Future' [33].

First of all, there is a need to face the challenges presented by the 'fourth industrial revolution', the digital one, in order to better equip themselves and help the younger generations to prepare for their future, through 'concerted actions by educators and entrepreneurs who foster the development of skills necessary to take advantage of the digital revolution' [28, p. 4]. It is also necessary to invest in skills related to the artificial intelligence and cognitive intelligence, as well as to the social and emotional capacities of human beings. It becomes equally necessary to communicate and make these skills understandable to the younger generations 'as soon as possible' [28, p. 4]. Therefore, current and future work trends to be communicated to the younger generations are indicated (Table 3):

Attention is focused on this latest trend, in particular, on a new task entrusted to the school. It should be noted that among the 7–11 year olds:

- 59% said they heard about their favorite job from parents/guardians or other family members; 56% heard about it via TV/cinema and social media. 'Less than 1% heard of it from someone who did the work and visited their school' [28, p. 14];
- career aspirations have little in common with expected demands from the labor market and confirm that their aspirations reflected a narrow view of the world of work [34].

The actions that the school should take are then to:

- allow to 'give all children, regardless of their social background, where they live or the work their parents do, the same opportunity to meet people - if not in person, then via the Internet - who do all types of jobs, to help them understand the wide range of opportunities open to them' [28, p. 16];

Table 3. Work trend lines in the OECD study Adapt.: [28].

Trend lines	Topics of description
Globalization – *It's a small world*	Air transport, international migrations, global mobility
Environmental security - *A change in the weather*	Climate change, use of renewal energy, awareness about environmental issue
Digitalization – *Data makes the world go round*	Social network, online services, civic participation, access to information, media and digital literacy skills
The 4th industrial revolution – *New opportunities*	Computer-based and remote working, 'gig economy', healthcare, transport and environment sectors, STEM and gender stereotyping
Lifelong learning - *Learning to work, working to learn*	Longer life expectancy, financial sustainability of pension systems, longer working lives, continuous learning
Collaboration and co-operation - *Working together to create new opportunities*	Multicultural accommodation, diverse workplaces, social skills
Get ready - *Linking the world of school with the world of work*	Career aspirations and labor markets demands, wider view of the world of work, employers and educators collaboration

- encourage the meeting between employers and educators 'to help broaden the horizons of young people, raise their aspirations and provide them with the vital work-related knowledge and skills that will help them in the transition from school to work' [28, p. 16].

The final aim becomes - according to the document -, then, to 'contribute to reducing the mismatch between the aspirations of young people and the needs of the labor market, thus guaranteeing a workforce capable of guaranteeing our economic prosperity in the future' [28, p. 16].

2 'Work' and 'World of Work' as Topics of Teachers' Training – an Investigation

If the concept of 'work' and the related construct of 'world of work' find new meanings within the debate on today's educational challenges [28, 35], we should then ask ourselves how they can be translated into a training interventions to be integrated with the current professional profile of the primary teacher - e.g. as abilities to enhance and promote the concrete experience of pupils and pupils or to raise awareness of general issues and new professional profiles to be carried out. Even more, we need to ask ourselves what place do the concept of 'work' and the related construct of the 'world of work' can realistically occupy within the training curricula of future primary school teachers.

2.1 Objectives and Investigation Questions

The place that can be granted to the concept of 'work' and to the correlated construct of 'world of work' with respect to the profile of the future primary teacher is the focus of the investigation experience carried out at the Bachelor's Degree in Primary Education of the University of Bergamo in the first semester 2020–21. The investigation had a dual purpose: a. to obtain useful evidence for monitoring the course and the training program, in general; b. to learn more about the basic professional learning processes that student teachers carry out in their academic classrooms and in other educational contexts [36, p. 23]. It asked the following research questions:

Q1 - *what representation of 'work' and of the 'world of work', in general, did the student-teachers involved develop?*

Q2 - *what representation of 'work' and 'world of work', as themes to be integrated into the school curriculum, have the student teachers involved developed?*

2.2 Context and Population

The Bachelor's Degree in Primary Education at the University of Bergamo has about 750 students from the neighboring areas of the Lombardy region and the province of Bergamo. It is at the sixth year of activation and has already graduated about 56 students with the qualification of infant and primary teacher. Its articulation is inspired by the pedagogical model of 'alternance training' [6, 26]. This provides for a strong integration between the various training activities - lectures, laboratories and internships - through frequent references to contents provided and exchange of experience activities that students develop in a university and school environment. The 'Alternance training' module of the course of Didactics III was inspired by such pedagogical model. The module is located in the third year of the degree course; it recalls the syllabi of the first and third year courses and provides internal laboratory activities to encourage the students involved to connect the internship experiences carried out in schools and theoretical knowledge provided at lecture.

The investigation involved 30 students with diversified socio-professional characteristics - mainly women, with an average age of about 22, with 1 year of teaching responsibility already matured, with at least 1.5 years of internship in childhood and primary schools.

2.3 Methodology. SoTL as Case-Study

The investigation was inspired by the 'scholarship for teaching and learning' (SoTL) method [36, 37] - i.e. scholars advancing teaching practice in higher education by making investigation results public [36] - referring to the 'micro' (individual) and 'meso' (departmental) level [39, 40]. This method allows university teachers to use the investigative practices and skills commonly carried out within their research work and apply them to their work as teachers. The purpose is to verify, for example, the effectiveness of a method, or a course, and make the results public. For this reason, the teacher involves the students who are stimulated to ask questions, collect evidence, draw conclusions [38] with reference to both the university (e.g. lecture) than real (e.g. internships at school) context [37]. Specifically, the SoTL-inspired investigation followed the case study method [38, 41, 42] carrying out the process illustrated in Table 4. In the 'after class' phase, document analysis [43] was used to evaluate the materials produced by the students and the thematic analysis [43] for the analysis of the transcripts of the group discussions. This proposal illustrates the process and the outcome of the documentary analysis of the products made by the students.

Table 4. Case study method followed. Adapt.: [41].

a. Define and design	b. Collect, prepare and analyze	c. Analyze and conclude
Review/develop theory Select case Design data collection protocol	Make video-lesson Transcribe video-lesson Qualitative data analysis 1. Document analysis 2. Transcriptions notes	Analyze and formulate the findings Write case study report Draw conclusion

3 Document Analysis of Materials

A 6-h thematic laboratory was set up inside the course. An ergative phase of illustration/deepening of the materials was carried out - some excerpts from the work as experience [6] and the synthesis of the OECD survey on the representations of primary school pupils [28]. After that, the student-teachers involved (n. 30), in working groups, were asked to produce didactic material that exemplified how they would present the concept of 'work' and 'world of work' to a hypothetical primary school class. The student-teachers could take inspiration from the illustration/in-depth materials to choose the definitions to be given, the examples to be done, how to stimulate connections with known/unknown working realities.

The didactic material supporting the learning of the hypothetical pupils had to respect specific content (definition of work, reference to experience, reference to the issues of the 'world of work') and formal (self-consistency, modularity, availability in repositories, reusability, interoperability) criteria. The materials were analyzed in reference to quality criteria of the learning objects [45] - two criteria proposed by Nesbit and colleague [46]- Table 5 -, integrated with the didactic-documentary ones, taken from the Indire-BDP GOLD protocol [47] - Table 6.

The material produced was first cataloged on the basis of macroscopic characteristics - Table 7.

Then the material was analyzed according to the chosen criteria - Tables 5 and 6. Figures 1, 2 and 3 represent the frequency of each indicator, differentiated by criterion.

Table 5. Analysis criteria of the material I. Adapt.: [46].

Criteria	Indicators
Content quality	Veracity Accuracy Balanced presentation of ideas Appropriate level of detail
Alignment to	Learning goals Activities Assessments Learner characteristics

Table 6. Analysis criteria of the material II. Adapt.: [47].

Criteria - Quality:	Indicators
Didactic-documentary	Integration with the curriculum Adequacy of contents to objectives Adequate instructions Presence of examples of didactic use

3.1 Results

The analysis of the documents produced by teacher students within the thematic laboratory followed the content quality, alignment and didactic-documentary criteria - see Tables 5 and 6.

The analysis highlighted that, with respect to the content quality criterion (Fig. 1), the documents fully respected the indicator of veracity (10/10) and almost fully that of balanced presentation of ideas (9/10); the indicators of accuracy (6/10) and appropriateness of details (6/10) are also sufficiently respected.

With respect to the alignment criterion (Fig. 2), the analysis highlighted that the documents produced fully complied with both the indicator relating to the choice of activities (10/10) and the correspondence to the pupils' characteristics (10/10). On the other hand, the indicators of alignment with the learning objectives (4/10) and with the tools and methods of evaluation (3/10) are not sufficiently respected.

As regards, instead, the didactic-documentary criterion (Fig. 3), the documents responded well to the indicators of adequacy with respect to content (8/10), adequacy of instructions (8/10) and choice of didactic examples (7/10). The indicator of the proposal with respect to the curriculum integration, instead, is very low (1/10).

The macroscopic characteristics - Table 7 - and the results of the documentary analysis of the material produced by student-teachers - Figs. 1, 2 and 3 - allow to answer the two research questions.

Table 7. Macroscopic characteristics of materials.

Id. No	n. stud.	Title	Profile/sector of work	References to the 'world of work'	Target
1	2	The world of bees	Beekeeper	n.a.	2nd grade
2	7	A life as a gardener	Gardener	n.a.	3rd kindergarten
3	1	Hands in the dough	Baker	Required employment	5th grade
4	4	In the role of …	Mayor, doctor, tailor, pizza chef, veterinarian, painter, company manager, theater director	n.a.	5th grade
5	2	A day as a fireman	Fireman	n.a.	n.a.
6	2	The magical world of the lute maker	Lute maker	Limited employment	5th grade
7	2	n.a.	Fisherman	n.a.	n.a.
8	7	n.a.	Farmer	n.a.	4th grade
9	1	From corn kernels…	Farmer Miller	Required employment	n.a.
10	1	The world of work… in my country	Carpenter	Limited employment	4th grade

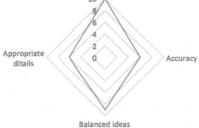

Fig. 1. Analysis of the material - 'Content quality'.

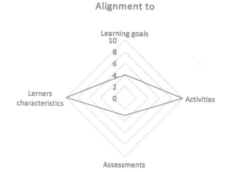

Fig. 2. Analysis of the material - 'Alignment'.

Fig. 3. Analysis of the material - 'Didactic-documentary quality'.

Q1 - *what representation of 'work' and of the 'world of work', in general, did the student-teachers involved develop?*

On the basis of the profiles and work sectors, chosen by the student-teachers as examples to be illustrated in the material produced (Table 7), representations emerge regarding work activities operating mainly in the agricultural (beekeeper, gardener, farmer, fisherman, miller), in the artisan (baker, lute maker, carpenter, also tailor, pizza maker) and in the services (fireman, also mayor, doctor, veterinarian, company manager, theater director) sectors, as well as references to the artistic sector (painter). These are mainly described through the actions carried out and the tools that are used. This method returns a 'true' and fairly accurate representation (Fig. 1).

In the material produced there are few direct references to the 'world of work': these concern, on the one hand, the job profiles defined as 'in great demand' (baker and farmer), on the other, less requested profiles, 'to be rediscovered' (carpenter) or, even, to be protected (lute maker) (Table 7).

Q2 - *what representation of 'work' and 'world of work', as themes to be integrated into the school curriculum, have the student teachers involved developed?*

From the documents produced it was also possible to know how student-teachers manage to insert 'work' and the 'labor market' within a hypothetical educational intervention program. From the analysis (Figs. 2 and 3) a double trend emerges. Student-teachers, on the one hand, are able to adapt the theme of 'work' above all to the characteristics of the learners and the activities to be carried out. They are also capable of choosing communicative content and writing instructions appropriate to the topic, as well as taking explanatory didactic references. On the other hand, however, they find it more difficult to explain the connections of the 'work' theme with the learning objectives and assessment tools and to place it within the school curriculum.

4   Discussion

The first phase of SoTL-type investigation through documentary analysis is highlighting highly realistic representations of future primary school teachers regarding working sectors such as agriculture and craftsmanship. These representations would seem more linked to a predominantly local reference context - cf. curriculum from the 'school as a research center' [21] - Table 1 -, not very 'global', in contrast with the trend lines indicated by the OECD [28].

The investigation also revealed that such realistic representations of work activities are, on the one hand, easily 'didactized' [48] in educational activities and paths adapted to the characteristics of the hypothetical students, on the other hand, not so easily 'curriculated' [49], with reference to the learning objectives of the disciplines.

These early results and, specifically, a double trend referred to properly didactic activities - if confirmed by the second phase of the analysis - would translate into concrete questions to be posed to the references of the training programs of future teachers - both at 'micro' (the responsible for the didactic course) and 'meso' (the degree course coordinator) level [39, 40].

In the perspective of didactic course responsible, it would be appropriate to train future teachers in seeking the connections between the 'work' and the 'world of work' issues within the school curricula. Some disciplines would be epistemologically easier in carrying out such exercises - e.g. History, Geography, Civics, Science [50]. However, it would be desirable to ask future teachers to practice planning as interdisciplinary as possible. The degree course coordinator, also, could try to define an 'under-track' path that allows student teachers to tackle the 'work' and the 'world of work' issues through a greater connection between courses of different years, between lectures, laboratories and internship.

As regards, however, the orientation to be given to the themes of 'work' and 'world of work' - eg. follow the tradition of the local socio-economic context or interpret the trend guidelines as indicated by the OECD -, it should fully involve the university policy.

The results of SoTL investigations often lead to new pedagogical experiments, which deserve further investigation, are mainly used by teaching and learning scholars to examine the quality of teaching practices and ask themselves 'how could we do better'.

References

1. Cukier, A.: Qu'est-ce que le travail? Vrin, Paris (2018)

2. Arendt, H.: The Human Condition, 2nd edn. The University of Chicago Press, Chicago (1958)
3. Rousseau, J.J.: Emilio o dell'educazione. La Scuola, Brescia (2016). [1762]
4. Benjamin, W.: Erfahrung und Armut, Die Welt im Wort, vol. I, p. 10 (1933)
5. Weil, S.: La condition ouvrière. Gallimar, Paris (1951)
6. Potestio, A.: Alternanza formativa. Radici storiche e attualità di un principio pedagogico. Edizioni Studium, Roma (2020)
7. Pestalozzi, J.H.: Educazione professionale elementare. In: Pestalozzi, J.H. (ed.) Popolo, lavoro, educazione, pp. 263–269. La Nuova Italia, Firenze (1974). (Original work published 1807–1808)
8. Becchi, E.: Proposta di lettura. In: Pestalozzi, J.H. (ed.) Popolo, lavoro, educazione, pp. 7–22. La Nuova Italia, Firenze (1974)
9. De Serio, B.: Il 'buon' lavoro nella storia della pedagogia. Un breve excursus storico sull'alternanza scuola-lavoro. Metis VII 1(17) (2017)
10. Scaglia, E.: 'Il cigno canta ancora?': note sulla pedagogia dell'amorevolezza di Johann Heinrich Pestalozzi. Formazione Lavoro Persona – Pestalozzi nella modernità VII(21), 59–76 (2017)
11. Costa, M.: Capacitare l'innovazione. La formatività dell'agire lavorativo. FrancoAngeli, Milano (2016)
12. Rossi, B.: Sviluppo professionale e processi di apprendimento. Nuovi scenari lavorativi. Carocci, Roma (2005)
13. Goleman, D.: Working with Emotional Intelligence. Bantam Books, New York (1998)
14. Kerschensteiner, G.: Avvertenza alla prima edizione di"Il concetto della scuola di lavoro. Marzocco, Firenze (1959)
15. Bertuletti, P.: Il lavoro nella scuola popolare. Pestalozzi e Kerschensteiner a confronto. Formazione, Lavoro, Persona XI(34), 126–143 (2021)
16. Chiosso, G.: Novecento pedagogico. Brescia, La Scuola (1997)
17. Montessori, M.: Il cittadino dimenticato. Vita dell'Infanzia. Rivista Mensile dell'Opera Nazionale Montessori I(1), 3–8 (1952)
18. Montessori, R.: La vita pratica come base spirituale della Casa dei bambini. Vita dell'Infanzia 11(12), 56–60 (2006)
19. International Labour Organization (ILO): World Employment and Social Outlook 2021: Trends 2021. ILO, Geneva (2021)
20. Gershon, L.: A World of Work: Imagined Manuals for Real Jobs. Cornell University Press, Ithaca (2015)
21. Giunti, A.: La scuola come centro di ricerca. La Scuola, Brescia (1973/2021)
22. Musso, S.: Storia del lavoro in Italia. Marsilio, Venezia (2002)
23. United Nations (UN): Resolution adopted by the General Assembly on 25 September 2015, Transforming our world: the 2030 Agenda for Sustainable Development (A/RES/70/1) (2015)
24. Grange, T.: Qualità dell'educazione e sviluppo sostenibile: un'alleanza necessaria, una missione pedagogica. Pedagogia Oggi. Rivista SIPED, anno XVI, no. 1, pp. 19–32 (2018)
25. Bertagna, G.: Lavoro e formazione dei giovani. La Scuola, Brescia (2011)
26. Bertagna, G.: Dall'esperienza alla ragione, e viceversa. L'alternanza formativa come metodologia dell'insegnamento. Ricerche di Psicologia 3, 319–360 (2016)
27. D'Aniello, F.: Le mani sul cuore. Pedagogia e biopolitica del lavoro. Fano, Aras (2015)
28. Schleicher, A., Achiron, M., Burns, T.: Envisioning the Future of Education and Jobs: Trends, Data and Drawings. OECD Publishing, Paris (2019)
29. Kohler, R.: Possiamo cambiare il mondo. L'educazione civica raccontata ai bambini. Mondadori, Milano (2018)
30. Fayard, A.L.: Notes on the meaning of work: labor, work, and action in the 21st century. J. Manag. Inq. 30(2), 207–220 (2019)

31. OECD: Trends Shaping Education 2019. OECD Publishing, Paris (2019). https://doi.org/10.1787/trends_edu-2018-en
32. OECD: Education at a Glance 2018: OECD Indicators. OECD Publishing, Paris (2018). https://doi.org/10.1787/eag-2018-en
33. Chambers, N., Rehill, J., Kashefpakdel, E.T., Percy C.: Drawing the Future. Exploring the career aspirations of primary school children from around the world. Education and Employers, London (2018). www.educationandemployers.org/wp-content/uploads/2018/01/DrawingTheFuture.pdf
34. Mann, A., Massey, D., Glover, P., et al.: Nothing in common: the career aspirations of young Britons mapped against project labour market demand, 2010–2020. Education and Employers and UK Commission for Employment and Skills, London (2013)
35. OECD: Education at a Glance 2015: OECD Indicators. OECD Publishing (2015)
36. McKinney, K.: Enhancing Learning through the Scholarship of Teaching and Learning. The Challenges and Joys of Juggling. Jossey-Bass, San Francisco (2007)
37. Felten, P.: Principles of good practice in SoTL. Teach. Learn. Inq. ISSOTL J. 1(1), 121–125 (2013)
38. Hutchings, P., Shulman, L.S.: The scholarship of teaching: new elaborations, new developments. Change: The Magazine of Higher Learning, vol. 31, no. 5, pp. 10–15 (1999)
39. Eaton, S.E.: Understanding Academic Integrity from a Teaching and Learning Perspective: Engaging with the 4M Framework (2020). http://hdl.handle.net/1880/112435
40. Roxå, T., Mårtensson, K.: How effects from teacher-training of academic teachers propagate into the meso level and beyond. In: Simon, E., Pleschová, G. (eds.) Teacher Development in Higher Education: Existing Programs, Program Impact, and Future Trends, pp. 213–233. Routledge, London (2012)
41. Mauffette-Leenders, L.A., Erskine, J.A., Leenders, M.R.: Learning with Cases, 4th edn. Richard Ivey School of Business, London (2007)
42. Simmons, N., Eady, M.J., Scharff, L., Gregory, D.: SoTL in the margins: teaching-focused role case studies. Teach. Learn. Inq. 9(1), 61–78 (2021)
43. Bowen, G.A.: Document analysis as a qualitative research method. Qual. Res. J. 9(2), 27–40 (2009)
44. Braun, V., Clarke, V.: Thematic analysis. In: Cooper, H., Camic, P.M., Long, D.L., et al. (eds.) APA Handbook of Research Methods in Psychology, Vol. 2. Research Designs: Quantitative, Qualitative, Neuropsychological, and Biological, pp. 57–71. American Psychological Association (2012)
45. Garavaglia, A.: La valutazione dei materiali didattici per l'apprendimento. In: Galliani, L. (ed.) L'agire valutativo. Manuale per docenti e formatori, pp. 167–180. La Scuola, Brescia (2015)
46. Nesbit, J.C., Belfer, K., Leacock, T.L.: LORI 1.5: Learning Object Review Instrument (2004). https://laeremiddel.dk/wp-content/uploads/2012/07/Learning_Object_Review_Instruments_LORI_1.5.pdf. Accessed 20 Dec 2021
47. Indire-BDP: Griglia di valutazione dei documenti e software. Protocollo GOLD. http://gold.indire.it/gold2/
48. Damiano, E.: La mediazione didattica. FrancoAngeli, Milano (2013)
49. Perla, L.: L'insegnamento dell'educazione civica: prodromi educativo-didattici e 'prove tecniche' di curricolo. Nuova Secondaria 10, 222–238 (2020)
50. Agrati, L., Massaro, S., Vinci, V.: Il bene comune come 'sapere da insegnare'. La ricerca-formazione. cittadinanza, costruzione identitaria e cultura del rispetto. MeTis. Mondi educativi. Temi, indagini, suggestioni 7(2), 600–637 (2017)

Developing Students' Empowerment Through Health Education for Future-Oriented Curricula and Sustainable Lifestyles

Stefania Massaro[(✉)] [iD]

University of Bari Aldo Moro, Bari, Italy
stefania.massaro@uniba.it

Abstract. Within the OECD framework, health literacy, educational processes and empowerment are actually interconnected to embed well-being and values of equity and social justice in school curriculum. Health literacy has an impact on educational outcomes such as learning and critical thinking, enabling students to become empowered individuals as well as ethically and socially responsible citizens. Within a multi/transdisciplinary study oriented to grasp the transition processes to the phenomenon of obesity and promote health education and healthy lifestyles, the present contribution presents the steps of an empowerment and game-based approach in primary school settings in Apulian region, for the development of critical and evidence-based thinking on students' health choices.

Keywords: Health education · Health literacy · Empowerment

1 The Interplay Between Citizenship Education, Health Literacy and Empowerment

The covid-19 pandemic has shown need to support community building, and resilience, placing well-being and participation at the forefront of dealing with perspective health and environmental crisis [1]. As lesson learnt from the pandemic education systems have to implement an education response that supports equity, quality and well-being developing a vision that acknowledges the crisis and restates commitment to these key educational principles with curriculum implementation strategies [2].

Education systems have to design and implement future-oriented curricula appropriate and relevant to rapidly changing social contexts ensuring the quality of students' learning but also well-being [3]. Students must be prepared to tackle future societal challenges and take responsible action towards sustainability and personal and collective well-being. Well-being involves more than access to material resources, such as income and wealth, being related to quality of life, health, civic engagement, education and adoption of a sustainable lifestyle [4].

The core foundations of the entire curriculum includes physical and mental health and well-being: the core knowledge, skills, attitudes and values of the curriculum include cognitive, emotional and social well-being and health, as well as literacy and numeracy,

F. Calabrò et al. (Eds.): NMP 2022, LNNS 482, pp. 840–848, 2022.
https://doi.org/10.1007/978-3-031-06825-6_79

being health a cross-curricular content built across boundaries of subject area emphasizing interdisciplinary knowledge. Aims are to provide opportunities to bridge equity gaps and empower students to become responsive citizens within critical thinking and practical learning experiences on health.

For the curriculum to be inclusive, it must be a stimulus for the students to be 'responsible citizens', together with 'successful learners', 'confident individuals' and 'effective contributors' [5]. Over last years citizenship education has been emerging in its task related to health education, according to WHO definition of 'health' as complete well-being (1948) and with consideration of reformulations emphasizing individual's resources to cope and self-manage [6]: citizenship education at present stresses citizen's capacity to participate on one's own health with skills of inquiry, critical thinking, problem solving and evaluation, cultivating personal and social responsibility. Schools health promotion interventions and programmes are actually built on capacity building function along with political function of more inclusive and equitable educational policies and scientific function of producing data about policies and practices [7].

The dynamic relationships between education and health are expressed in the conceptual development and empirical evidence of health literacy, considered among the thematic 'pillars' of a holistic approach to health promotion. Addressing children's and adolescents' health literacy is fundamental for sustainable development, societal growth, and it must be understood as an important educational dimension of schools in the 21st century as a promoter of resilience, fairness and inclusion.

The World Health Organization addresses health literacy as "social and cognitive skills that determine the motivation and ability of individuals to gain access, understand and use information in order to promote and maintain good health", arguing that health literacy means improving access of people to information on health and the ability to use it effectively, according to an extended definition that goes beyond the boundaries of literacy, to open up to the concept of empowerment [8].

Health literacy will enable children and adolescents to:

- access and navigate health information environments,
- understand health messages,
- think critically about health claims and make informed decisions about health,
- acquire health knowledge and use it in new situations,
- communicate about health topics and concerns,
- use health information to promote their own health, others', and environmental health,
- develop healthy behaviours and attitudes,
- engage in healthy activities and avoid unnecessary health risks,
- become aware of their own thinking and behaving,
- identify and assess bodily signals (e.g. feelings, symptoms),
- act ethically and socially responsible,
- be a self-directed and life-long learner,
- develop a sense of citizenship and be capable of pursuing equity goals,
- address social, commercial, cultural, and political determinants of health.

Health literacy therefore becomes a tool to promote health education developing skills that allow to retrieve information, evaluate its reliability, exercise greater control over the determinants of one's own health and make informed, conscious and optimal choices. Starting from a traditional definition strictly connected to the basic concept of literacy, the scope is extended to empowerment as a tool for the foundation of a modern citizenship [9]. The ability to understand health information is therefore closely linked to the ability to decide freely and consciously about it, in a connection of empowerment, health literacy and educational processes towards the goal of overcoming psychological, cultural and social barriers. This means addressing equity and social justice through curricular innovations embedding values into curriculum.

Nutbeam [10] refers to functional health literacy, interactive health literacy and critical health literacy, with curriculum content of respectively communication of information, development of personal skills and personal and community empowerment to address social and economic determinants of health, and achieve policy and/or organizational change, showing how different levels of literacy progressively allow for greater agency and personal empowerment. This goal can be attained by creating a democratic atmosphere at school for physical, social and mental well-being while the teacher should be seen as a key agent for developing health literacy approaches that are tailored to curriculum and its core educational tasks, practices and goals.

When students are empowered they are equipped with agency to shape their own lives and contribute to the lives of others. Students' agency is defined as the ability, will and beliefs to positively influence their own lives and the world around them. It entails having the capacity to set a goal, reflect and act responsibly to effect change. It is about acting rather than being acted upon, shaping rather than being shaped, and making responsible decisions and choices, rather than accepting those determined by others. Students need to exercise purpose and responsibility in their pursuit of learning and the transition to adulthood. Future-ready students need to exercise agency, in their own education and throughout life. Agency implies a sense of responsibility to participate in the world and, in so doing, to influence people, events and circumstances for the better. The principle of student agency is especially relevant to ensuring effective implementation and contributing to equity, but also to embedding values in curricula.

A curriculum granting agency incorporates learning experiences that engage students in promoting values that are of personal interest and have relevance for them in relation to their goals and aspirations. Such curricula are also self-tailored to the specific needs of the individual and support students to become increasingly self-directed over time, allowing them to gain confidence in their ability to complete learning tasks, self-evaluate and build the skills they need to monitor, review and reflect on their progress [11].

Thus implementing promotion of health means working on personal resources, within the relationship with the context, to build lifelong learning experiences generating control and well-being. Just as in life-long learning, the matter is not leading to learning a health behavior but rather to promote reflective skills to activate changes involving people, communities, the environment. The complexity of the problem of the citizen involved in the processes of construction of health implies more and more decisively the formation of an interpretative thought oriented to a daily reflexivity and committed to problematize the personal relationship with the context through the enhancement of personal experience. An ideal reference to the values of equity and justice must implies an intervention on mental, cognitive, affective and socio-cultural processes to be effective.

2 The Transdisciplinary Study on Obesity Prevention

Within these frameworks with the Citel Interdepartmental research centre in Telemedicine at University of Bari we have proposed a multi/transdisciplinary study, oriented to grasp the transition processes and complications related to the phenomenon of obesity and promote health education pathways within a virtuous network of industry, research and training institutions. Referring to health as 'complex problem' urging transdisciplinary research perspectives at present, this research-based project addresses obesity prevention through the interdisciplinary paradigm of telemedicine including medicine, education, biomedicine and computer science.

Within a holistic and sustainable approach to health literacy composed by key elements as:

- the salutogenic approach to health [12];
- the concept of health 'development' as the interaction between the individual and his environment;
- the areas of action of the Ottawa Charter for health promotion interventions
- the newest culture of citizen participation in health services as the enhancement of empowerment, with a transition from traditional methods of participation (charter of services, informed consent) towards more complex capacities including critical understanding, judgment, discussion within cognitive processes enabling an active and responsible citizenship

Aim of the project is to define experimental clinical pathways for the treatment of patients with obesity, and the prevention of obesity meant as empowerment of students and families towards healthy lifestyles [12, 13].

While the European Union with the SDGs has placed the objective to diminish of a third the premature deaths from not communicable diseases within 2030, the prevalence rate of obesity continues to rise, with up to 90% in European countries, with a higher incidence on the poorest population and vulnerable groups, and social inequalities evident especially towards women. At European level, the need to change the status quo with multidisciplinary and sustainable approaches is stressed: promote a change in the narrative on obesity traditionally focused on the person and his lifestyle and therefore on individual responsibility, in order to adopt a shared responsibility paradigm based on key elements as telemedicine, transdisciplinary care, intervention on social phenomena of weighting discrimination and promotion of health literacy or obesity health literacy [14].

In the context of citizenship education meant to promote knowledge and participation of the person in the context with aim of well-being and quality of life, research questions are: how to make the school experience more inclusive against health inequalities? How to improve lifestyles of students and prevent obesity and its complications? How can practices be transformed effectively to support well-being of young people in the context of rapid societal changes?

Educational objective is therefore engaging students in best practices on obesity and the promotion of life-long capacities of informed decision and self-care at schools as primary prevention settings.

Though not a novelty, school-based approaches are strategical when tackling childhood obesity:

- allowing complex reshaping of environmental but also social elements often triggering obesity in youngsters towards more inclusive contexts, training student to participation and initiative with empowered-based approaches
- providing children with learning experiences that bind together cognitive, affective and moral dimensions to shape more aware and responsible identities and behaviors in the domain of citizenship education
- allowing to reach families as well as vulnerable groups, disadvantaged socio-economic groups and get in contact with public entities, being schools nowadays part of bigger networks connected to their own territories in systems of educative alliances, built to develop inclusion and democratic systems
- focusing the attention on SDG 3 (Good Health and Well-being for People).

Within an obesity project centered on the topics: empowerment - AI-based enabled education and well-being as a dynamic state to develop students potential and build positive relationships, the educational specific objectives identified are:

- development of a research-based educational model to prevent childhood obesity and promote healthy lifestyles
- creation of online learning environments and experimentation of videogame-based learning to support and empower students
- involvement of students in the design and testing of innovative educational software to implement a strategic vision that sees the user at the center of the innovation process
- simulations of self-regulatory behaviors through immersive learning techniques in virtual and augmented reality learning environments
- promotion of health literacy and digital health literacy to improve well-being and quality of life
- dissemination through videogames of national and international guidelines and recommendations on nutrition and physical activity to reduce the gap between scientific knowledge and people's life choices
- structuring of knowledge-building communities and learning communities as e-learning educational solutions enhancing community-based knowledge production
- educational design of activities to combat social discrimination and prevent eating disorders
- engagement and empowerment of families
- promotion of collaborations to implement a stable integration between research, health and territory.

3   The GA4POPS - Gamification for Prevention of Obesity and Promotion of Healthy LifestyleS Project

With a focus on Apulian schools we set up a protocol in the framework of Health Promoting Schools model [15] enhancing health literacy within a holistic school approach and introducing game-based learning as composed by:

- videogame-based learning to integrate technology in the curriculum and develop trial-and-error deductive learning and team working
- game-based activities to promote critical thinking to access, understand, appraise and apply reliable health information

The research protocol was designed to engage primary schools in a collaborative research/training activity to be further developed on a European scale, as composed by:

Step 1: survey and in-depth understanding of students healthy lifestyles engagement through qualitative analysis

Step 2: develop EU guidelines and recommendations on healthy lifestyles through videogames for the promotion of students' evidence-based healthy behaviors

Step 3: implement the primary school resources set up by the international IHC - Informed Health Choices - Network [16] identified as a set of Key Concepts that students need to understand and apply to assess treatment claims and make informed health choices, with an approach consistent with the principles of a spiral curriculum.

For some years now videogame-based learning, with the use of elements of game design in non-playful contests such as health, has been associated with a fully accepted teaching methodology relying on some main objectives:

- manage an active interest in the message to be communicated
- stimulate active and measurable behavior
- give the user continuous feedback, transforming the path towards a distant goal, difficult to achieve, into a path made up of small and continuous steps achieved.

As adolescents are technology frontrunners, digital health interventions appear to be a practical modality for dietary behavior change interventions for the prevention of obesity [17] and there is evidence that supporting effective engagement in digital interventions is a critical factor in the adoption of healthy dietary behaviors in adolescents within the current digital world [18].

In order to promote a have a very 'hands-on approach', combining elements of the theory, group work and discussion, students are also introduced to concepts and models of game design, providing them with the full experience of designing a game (phases of concept, design, prototype, playtest) drawing contents from different fields to create their first game.

The IHC Key Concepts are standards for judgment, or principles for evaluating the trustworthiness of treatment claims and for making treatment choices, being a treatment something that is done for one's own health. As there are endless claims about treatments in the mass media, in the advertisements, and in everyday personal communication and some of them are true and some are false, many are unsubstantiated (we do not know whether they are true or false) through a cartoon, that is a story with words and pictures put together, students can learn about 'the questions you should ask when someone says something about a treatment' or 'the questions that you should ask when you are choosing whether to use a treatment' [19].

What is your 'health'? What is a 'treatment'? What is an 'effect' of a treatment? are the basic questions.

The Key Concepts that are highlighted in the IHC primary school resources [20] are:

1. Recognising claims about the effects of treatments that have an unreliable basis

- Treatments may be harmful
- Personal experiences or anecdotes (stories) are an unreliable basis for assessing the effects of most treatments
- Widely used treatments or treatments that have been used for a long time are not necessarily beneficial or safe
- New or more expensive treatments may not be better than available alternatives
- Opinions of experts or authorities do not alone provide a reliable basis for deciding on the benefits and harms of treatments
- Conflicting interests may result in misleading claims about the effects of treatments

2. Understanding whether comparisons of treatments are fair and reliable

- Identifying effects of treatments depends on making comparisons
- Apart from the treatments being compared, the comparison groups need to be similar at the beginning of a comparison
- If possible, people should not know which of the treatments being compared they are receiving
- Small studies in which few outcome events occur are usually not informative and the results may be misleading
- The results of single comparisons of treatments can be misleading

3. Making informed choices about treatments

- Decisions about treatments should not be based on considering only their benefits

These descriptors are closely linked to learning to learn, as an attitude of to make informed health-related decisions and distinguish reliable and unreliable health information. Students need to understand the dangers of trusting and sharing false information on health, since it may compromise medical advice or cause unjustified alarm.

4 Conclusions

Further development of the project will require the construction of tools to assess students and also parents engagement on obesity health literacy and healthy lifestyles. Within this contribution we want to highlight the role of critical thinking in mastering today evidence-based choices, in order to weigh risks and benefits and to continually question one's own beliefs or traps called bias or prejudices that we have seen spreading during the covid-19 pandemic and which we should get rid of to rely on experiments and data. It is therefore a question of taking the opportunity to link an inquiry-based methodology to problems concerning health, in order to have an advantage both on health and in the dissemination of a scientific culture, especially nowadays when the acceleration of scientific knowledge requires correct interpretations for the development of collective well-being.

Promoting citizens' evidence-based healthy behaviors is at the core curriculum of a school interested in the development of a democratic and knowledge-based society, traditionally oriented towards common good and human rights protection. This relationship has been challenged during pandemic, when society has shown mistrust towards scientific research by crediting mostly a misinformation disconnected from objective investigation while society, as composed by citizens, should build an understanding of situations referring to scientific knowledge in order to make informed decisions.

References

1. Lauriola, P., et al.: On the importance of primary and community healthcare in relation to global health and environmental threats: lessons from the covid-19 crisis. BMJ Glob. Health **6**, e004111 (2021)
2. OECD Adapting curriculum to bridge equity gaps. https://read.oecd-ilibrary.org/education/adapting-curriculum-to-bridge-equity-gaps_6b49e118-en#page4. Accessed 23 Dec 2021
3. OECD Education responses to covid-19: an implementation strategy toolkit. https://www.oecd-ilibrary.org/docserver/81209b82-en.pdf?exres=1640313237&id=id&accname=guest&checksum=E2F4518FB9B2888E814C1D576A6A66E7. Accessed 23 Dec 2021
4. European Commission, LifeComp The European Framework for Personal, Social and Learning to Learn Key Competence. https://publications.jrc.ec.europa.eu/repository/handle/JRC 120911. Accessed 28 Dec 2021
5. OECD Future of Education and Skills 2030. https://www.oecd.org/education/2030-project/. Accessed 23 Dec 2021
6. Huber, M., et al.: How should we define health? BMJ **343**, d4163 (2011). https://doi.org/10.1136/bmj.d4163
7. SHE Schools for health in Europe, Factsheet 4 School health promotion, https://www.schoolsforhealth.org/sites/default/files/editor/fact-sheets/she_factsheet_no_4_2018.pdf. Accessed 24 Dec 2021
8. WHO Health Promotion Glossary. https://www.who.int/publications/i/item/WHO-HPR-HEP-98.1. Accessed 28 Dec 2021
9. Kickbusch, I., Maag, D.: Lo sviluppo della Health Literacy nelle moderne società della salute. Franco Angeli, Milano (2007)
10. Nutbeam, D.: Health literacy as a public health goal. A challenge for contemporary health education and communication strategies into the 21st century. Health Promot. Int. **15**, 259–267 (2000)
11. OECD Curriculum (re)design. A series of thematic reports from the OECD Education 2030 project. https://www.oecd.org/education/2030-project/contact/brochure-thematic-reports-on-curriculum-redesign.pdf. Accessed 28 Dec 2021
12. Antonovsky, A.: The salutogenic perspective: Toward a new view of health and illness. Advances **4**(1), 47–55 (1987)
13. Massaro, S., Perla, L.: Studio esplorativo sulla prevenzione e cura dell'obesità attraverso la Telemedicina: orizzonti transdisciplinari del lavoro educativo. In: 10th SIRD Proceedings, Ricerca e Didattica per promuovere intelligenza comprensione e partecipazione, pp. 635–642 (2021). https://www.pensamultimedia.it/pensa/prodotto/ricerca-e-didattica-i-tomo/
14. Massaro, S., Perla, L.: A competency model for obesity prevention and healthy lifestyles education through the interdisciplinary and sustainable paradigm of telemedicine. In: 2nd International Proceedings Journal Scuola Democratica Reinventing Education, vol. 3, pp. 309–317 (2021), https://www.scuolademocratica-conference.net/proceedings-2/

15. OPEN-EU's Manifesto Changing the status quo in obesity. https://obesityopen.org/open-eu/open-eu-news-and-updates/eu-manifesto/. Accessed 28 Dec 2021
16. Schools for Health in Europe, Health literacy in schools State of the art. https://www.schoolsforhealth.org/sites/default/files/editor/fact-sheets/factsheet-2020-english.pdf. Accessed 26 Dec 2021
17. IHC Network, The Informed Health Choices (IHC) Key Concepts 2019. https://www.informedhealthchoices.org/wp-content/uploads/2019/12/IHC-Key-Concepts_Health_2019.pdf. Accessed 28 Dec 2021
18. Rose, T., et al.: A systematic review of digital interventions for improving the diet and physical activity behaviors of adolescents. J. Adolesc. Health (2017). https://doi.org/10.1016/j.jadohealth.2017.05.024
19. Gibson, A.A., Sainsbury, A.: Strategies to improve adherence to dietary weight loss interventions in research and real-world settings. Behav. Sci. **7**, 44 (2017). https://doi.org/10.3390/bs7030044
20. The Informed Health Choices Group: The Health Choices Book: Learning to Think Carefully About Treatments A Health Science Book for Primary School Children. Norwegian Institute of Public Health, Oslo (2016)
21. Oxman, A.D., Chalmers, I., Dahlgren, A., Informed Health Choices Group: Key Concepts for assessing claims about treatment effects and making well-informed treatment choices (2019). https://www.informedhealthchoices.org/wp-content/uploads/2019/12/IHC-Key-Concepts_Health_2019.pdf. Accessed 29 Dec 2021

Identification, Validation and Certification of Previous Skills to Support Vulnerable Worker in Post Pandemic Dynamics

Daniela Robasto[✉]

University of Turin, Via Verdi, 8, 10124 Torino, Italy
daniela.robasto@unito.it

Abstract. With the publication by ANPAL of the "Guidelines on the management of financial resources assigned to inter-professional joint funds" (Article 118 of Law 388/2018), it is established, for the Interprofessional Funds for Continuing Training, that "Training [...] must be designed for knowledge and skills, including for the latter suitable assessment activities aimed at issuing the student with a transparent and expendable certificate of the acquired learning". This contribution presents, in summary, the system developed by the an Interprofessional Fund in agreement with the Department of Philosophy and Educational Sciences of the University of Turin, following a Research Training e underlines the importance of using IVC systems to enhance the profile of the vulnerable worker, providing a redefinition of this worker in the post-pandemic era.

Keywords: Skills · Retraining · Certification of skill · Worker

1 Redefine the Vulnerable Worker in the Post-pandemic Time

The legislation linked to the COVID-19 pandemic situation first paid attention to the "vulnerable person" and then to the "vulnerable worker", defining this person as a carrier of current or previous pathologies that make them susceptible to particularly serious consequences in the event of contagion, also called hypersensitive. Hence the construct "vulnerable worker", when the vulnerable person is in the professional context, thus defining the vulnerable worker as "the worker in possession of the recognition of characteristics with a connotation of gravity [...] certifying a condition of risk [...]". Nonetheless, outside the pandemic context, it would be appropriate for a vulnerable worker to be circumscribed using principles other than health, which would equally lead him to be defined as susceptible to particularly serious consequences and in a condition of risk, even if in this case the health consequences are not applicable, but rather those related to difficulties to keeping a job, psychophysical well-being and competent action.

F. Calabrò et al. (Eds.): NMP 2022, LNNS 482, pp. 849–857, 2022.
https://doi.org/10.1007/978-3-031-06825-6_80

Within a post-pandemic framework, the brittle worker could be reclassified using some of the following criteria:

- Low schooling or possession of unrecognized qualifications;
- Lack of professional qualifications or certifications in the sector;
- Being an immigrant or belonging to other disadvantaged, protected or weak social categories: (disabled, ex-prisoners; ex-drug addicts; non-EU citizens, the elderly, working mothers, women victims of violence, foreign women with children, elderly people, psychiatric, etc.);
- Have a seasonal, atypical or fixed-term employment contract;
- Have a low contractual status (laborer, assistant, unskilled worker, etc.)
- Have a low salary.

In fact, there are several studies (Istat 2020) that specifically highlight some fragilities and risk situations in the professional sphere given by being a foreigner, of a female gender, within a large family, with a fixed-term contract or seasonal, without qualifications or professional specializations, with an uncertain and precarious income situation. Some co-occurrences of fragility are well highlighted in the Annual Report on foreigners in the labor market (2021), in the ISTAT surveys and in the reports of the *Statistical Information System of Compulsory Communications* (2021).

According to *XI Annual Report - Foreigners in the labor market in Italy* (2020–2021), the foreign population living in Italy as of January 1, 2021 amounts to 5,036 million. In the year of the pandemic and the consequent economic crisis, individuals in absolute poverty exceed 5.6 million (9.4%), an increase compared to 2019. If we then look at the incidence of family poverty by disaggregating it on the basis of citizenship of members, in 2020 the families of only foreigners continue to record the highest values and see their condition worsened. The family typology shows how larger families are more exposed to discomfort; families with 5 or more members of foreigners show three times higher incidence values than those with Italian members only (38.8% against 13.8%); moreover, among families with three or more children, the incidence of absolute poverty reaches 36.6% against 13.6% of Italian-only families. When minors are present, the incidence grows rapidly up to 37.0% of families with 3 or more minor children (compared to 15.7% of Italian-only families).

Furthermore, having a job is not a sufficient condition to move away from the risk of poverty. If it is true that poverty is highest among those who are looking for employment (18.1%), it is in particular among those who have a job that poverty increases: families with one person working, whether they are composed of Italians foreign only families, the incidence of absolute poverty increases, respectively from 3.3% in 2019 to 4.4% in 2020 for Italians only and from 20.0% to 26.2% for families with foreigners.

There are also some gender differences. In the EU27 and the United States, immigrant men and women recorded a relatively similar but overall reduction in the employment rate compared to natives.

In Italy, immigrant women have suffered from the crisis much more than their male counterparts, but with a twice reduction in the employment. On the other hand, among people born in Italy, the reduction in the employment rate affected men and women equally (Eleventh Annual Report - 2021, p. 3 and 4).

2 Vulnerable Worker in the Agricultural Sector

Among the sectors that involve a higher number of vulnerable workers, there is undoubtedly agriculture. According to *the Annual Report of the EBAN Observatory on Agricultural Work* (2021), in 2019, the employees employed in the Italian agricultural sector amounted to 1,095,308. The largest share (97%) is made up of blue-collar workers, flanked by a residual percentage (3%) of office workers, middle managers and executives. Furthermore, In Italy, the number of employment relationships activated annually in agriculture reflects the significant presence of fixed-term labor, even for short periods during the year (Eban 2021). Work in agriculture is also characterized by a massive presence of foreign workers. At the macro-area level (North, Center, South Italy), we find significant differences. The North has a percentage of foreign workers equal to 47%, the Center 17% and the South 36%. Another significant aspect is that in the last 10 years there has been a notable growth in non-EU foreign workers (+77). Consequently, non-EU workers are the majority (62% of total foreigners) compared to EU workers.

On the other hand, by questioning the employment relationships of employees activated in agriculture by qualification (CICO Integrated Sample of Compulsory Communications - Ministry of Labor 2019), it is clear that 49% of workers have no qualifications and 91% of the total reach at least to the middle school (cumulative percentage of no qualification, elementary school certificate and middle school certificate). In addition to this, the questioning on the employment relationships of employees activated in agriculture by professional qualification shows that over 87% of temporary workers are hired as unskilled agricultural workers. It should also be said that the previous studies mainly based on the EBAN Agricultural Labor Observatory Annual Report (2021), on ISTAT data and the Ministry of Labor, deal with the so-called "regular" economy, but the Report itself concludes with some insights relating to the "irregular" economy. By focusing on the "underground business" only, its incidence on the added value of total economic activities is equal to 12.0%. In the case of agriculture, this share rises to 17.1% (ISTAT "National Accounting", 2018).

The profile of the worker in agriculture is therefore particularly vulnerable, not only for education, qualification, level and contractual duration but also for a more likely risk of coming into contact with the "underground" economy. For these reasons, when you are in the not taken for granted of starting training for a basically vulnerable worker, it is essential not only to have full awareness of the andragogic challenge you are facing, but it is also necessary that the training process leads to: a) an increase in the motivation of the worker and the company to participate in training sessions; b) the full achievement of the learning objectives, specifically modulated on the characteristics of the worker; c) the recognition of the *learning acquired* in a perspective of transparency and usability within the national qualifications system that could make it more protected within the labor market.

3 Identification, Validation and Certification of Previous Skills to Support Brittle Worker

The *National Skills Certification System*[1], being part of the broader national process for the individual right to lifelong learning, was for years (2012–2021) in a condition of operational limbo, at least until January 5, 2021, date of adoption of the Guidelines[2]. These Guidelines represent the provision that makes the recognition and certification of skills, acquired by the individual in formal, non-formal and informal contexts operational. The target is to encourage and support a concrete increase in the participation of people in training, as well as a usability of the skills acquired even in informal and non-formal contexts within the labor market.

The identification, validation and certification of skills (IVC) services, as stated in the guidelines themselves, are of particular importance in order to: a) reduce the percentage of the population with low levels of qualification, increasingly exposed to marginalization and exclusion from the labor market, also as a result of technological innovations and digitization; b) to increase the participation of adults belonging mainly to the weakest groups of workers in training activities; c) reduce the youth unemployment rate and promote generational relay conditions; d) reduce the condition of skill mismatch between workers with low qualifications and workers with high qualifications. The National Skills Certification System is in fact based, in a nutshell, on a three-phase process (Identification, Validation, Certification) which could allow any citizen to get the own skills recognized in several contexts and, therefore, to achieve professional qualification more easily. The subject of IVC are therefore the skills acquired by the individual through various types of experience (professional, educational, personal), in different moments of life and contexts (Article 3 of Legislative Decree 13/2013). However, this is not a global balance of skills, as frequently and erroneously understood, but the reconstruction of well-defined experiences gained by the user on a specific and circumscribed work process. Precisely for this reason, the validation and possibly certification process must be supported by documentation aimed at supporting and highlighting the skills acquired by the individual the collection of such evidence, on vulnerable workers, risks to be very complex.

---

[1] The Italian legislative decree is available for consultation here (GU Serie Generale n.39 del 15–02-2013) https://www.gazzettaufficiale.it/eli/id/2013/02/15/13G00043/sg. At the European level, see Commission of the European Communities, Recommendation of the European Parliament and the Council of 9 April 2008 on the establishment of the European Credit System for Vocational Education and Training (ECVET), Brussels, 9.4.2008. https://eur-lex.europa.eu/LexUriServ/LexUriServ.do?uri=OJ:C:2009:155:0011:0018:EN:PDF.

[2] The guidelines for the interoperability of IVC systems are available for consultation here https://www.lavoro.gov.it/documenti-e-norme/normativa/Documents/2021/DI-del-05012021.pdf.

4 Research - Training with Interprofessional Fund

Those who today design training interventions in the agricultural sector and, more generally in precarious work contexts, cannot therefore refrain from wondering what the peculiarities of training aimed at the vulnerable worker should be, what elements of enhancement of previous experiences can be triggered and with what results.

In this regard, some passages of the system developed by the ForAgri Interprofessional Fund[3] in agreement with the Department of Philosophy and Educational Sciences of the University of Turin are presented below, following a Research_Training conducted in relation to the experimentation envisaged by the announcement 2/2018. Below we illustrate only some developments from life long learning, which sets itself the goal of combining "canonical" (T) training planning with possible service based on Identification - Validation- Competence Certification (IVC). In fact, in the present experimentation an attempt has been made to achieve this aim without losing sight of the central focus: the quality of the learning and teaching processes (Lichtner 1999) and their outcome even with a vulnerable target.

The research program was developed by the Department on the basis of what was agreed with the Fund and envisaged the following research objectives:

– to analyze the training proposals and continuing training activities financed on a specific notice (2/2018) by providers which were awarded the grant;
– to analyze the tools used by providers for the analysis of training needs.
– to determine the feasibility of IVC processes connected to funded training courses to highlight the critical processes that could compromise their outcomes.

From the point of view of the training objectives, on the other hand, on the basis of the criticalities identified, it was intended to start training on models for the design, evaluation, validation and certification of the *acquired learning*.

Due to the dual nature of the objectives, research and training, the Department and the Fund have agreed to adopt a Research_Training (R-T) model deemed suitable for exploratory and implementation purposes expressed by the Fund (Grange 2017). In the research approach adopted, priority objectives of change or nomothetic intentions were not defined in the first instance. Exploratory needs, of first understanding and sharing (Dewey 1938; Lewin 1949) of the problems and practices implemented, in order to contribute to the training of a reflective professional (Schön 1987, 1991; Grange 2017; Colucci 2008). The details of the R-T phases (Robasto 2021), recalibrated on the needs of the research context and on the needs expressed by the Fund, is shown in the following table (Table 1).

---

[3] Interprofessional funds in Italy are 21. The share of the paycheck with which they are fed is decidedly lower than in other European countries (0.30%). ForAgri is the main funded training fund in the agricultural sector, in Italy.

Were set at first methodological protocol: is the result of synergistic work between the Fund's representatives, researchers, methodologists who are experts in IVC processes and didactic references or coordinators of the implementing subjects participating in the experimentation. Some representatives of the Ministry of Labor and ANPAL also participated in some planning or restitution meetings envisaged by the cycle.

The methodological protocol was applied to twelve training plans funded by the lawyer 2/2018 which voluntarily expressed an interest in participating in the ForAgri experimental system; 302 workers and 79 companies operating in the Agriculture sector were involved in the "experimental" training plans.

Coming to the results of the research, in relation to the objective of detecting the criticalities connected to an IVC process within the funded training, some phases of the planning and teaching process it has been found that require a rethinking of the practices in place in providers.

The challenging situation is given by becoming able to design, deliver and evaluate training interventions that are not only suitable for the company and the worker but also aimed at achieving legible, transparent, expendable learning according to recognizable standards. Specifically, the "critical" processes detected during the investigation concern:

- The analysis of needs and the definition of learning objectives with a view to recognition within the national system (Standard)
- Taking charge of previous skills (transparency with respect to a specific profile)
- Verification of ongoing learning (with training value and greater support for the recipient of the training, especially if vulnerable)
- Summative assessment of the achievement of learning objectives designed and conducted according to specific policies and specific docimological criteria.

With the awareness that the training need should be understood as a descriptive element of the state of relations between the individual, organization and economic-social context and that its detection can help identify a potential area of training intervention (Quaglino and Carozzi 2004; Cini and Catarsi 2003; Robasto 2019), we believe that the detection of training needs within an T-IVC process must consist of at least four steps: 1) Detection of professional needs; 2) Detection of training needs; 3) Taking charge of the participants' previous skills on a specific work process or area of activity; 4) Micro-design of learning objectives with a target to recognition. In this case, it is a question of focusing, with the greatest possible rigor, the specific learning objectives on which the training actions will have to focus, moving from a more general learning goal, in line with the elements that make up the standard taken as a reference to specific learning objectives tailored to the corporate and personal needs. The qualifications of the NQF related to the Activities, in fact, make it possible to identify, only at a general level, what are the resources - in terms of knowledge, skills, competences (standard descriptors provided for by the NQF) - that allow to carry out the performance (s). However, we cannot stop at identifying a standard goal in terms of market recognition, as is frequently the case, this would compromise the design and delivery of a tailor-made training intervention built on specific professional and training needs.

Table 1. Summary of the phases of research training conducted

Phase	Description
1	Exploratory phase: interviews and contacts with providers and start of the first document collection (analysis of funded training plans and low-structured interviews with implementers)
2	Detection of criticalities related to the IVC (present in the training plans or interviews conducted) and definition of the training objectives
3	Co-design of tools and operational protocols to support the IVC process within the funded training (Methodological Protocol v.1 and design support sheets Sheet A)
4	First pilot administration of the methodological protocol (v.1) and construction of support sheets for the construction of evaluation tools (Sheet B)
5	Detection of the results of the first pilot administrations, elaboration and critical analysis of the first administrations of the protocol v.1
6	Back talk and review of the methodological protocol (v.2) and of the support sheets and Construction of additional support sheet for tracing the summative evaluation results (Sheet C)
7	Training of providers e use of the revised methodological protocol (v.2) + implementation of the support sheets (A, B, C)
8	Generalization of the Protocol to experimental plans and extended training
9	Dissemination of the research results and dissemination of the methodological protocol

5 Some Considerations for Future Applications T-IVC

The identification of the critical elements has made it possible to develop the awareness that an T-IVC service cannot afford any indeterminacy or abstract formulations of the learning objectives. The redefinition of both didactic and evaluation practices is necessary, specially so in those contexts in which the recipient of the training had some characteristics of fragility.

Probably it was found that within a framework of planning rigidity (estimate = final balance), moreover frequently present within the funded training, a significant part of the changes made in itinere would not have taken place and this, in our opinion, would have involved at least three outcomes: a lowering of the qualitative level of the training path (often designed starting from a general survey of the learning goals); a company dissatisfied with specific and tailored professional needs that remained unmet; the impossibility of proceeding with a truly transparent and expendable certification of the learning outcomes (i.e. punctually legible and traceable to reference standards) or their recognition, precisely on an already particularly disadvantaged worker profile. A reflection must therefore also be advanced on the need to reinterpret and update, in particular in the system of inter-professional funds, the concept of monitoring and effectiveness of training (Corsini and Sanzo 2009), not so much intended exclusively as a slavish achievement of pre-established training objectives and quantitative result indicators (in terms of hours provided and recipients trained), but rather as the "ability of a given action to produce a desired change" (European Commission 2013, p. 99; Visalberghi 1955).

To this, other relevant issues are added in relation to the actuators: it is a question not only of having internal figures adequately trained in the evaluation and IVC processes, but also of being in the conditions, also in terms of economic and temporal resources, to be able review the practices and timing of planning, training delivery, evaluation.

Finally, it is crucial to address the question of the relationship between the provider, the company and the recipient of the training (Cameron and Bobby 2014). In training financed with Interprofessional Funds, the company is both a beneficiary and a "client" of the training and exerts a great influence both on provider and on the recipient of the training. It is therefore necessary to establish an alliance pact that on the one hand protects the company in giving availability to train workers during working hours in order to achieve training objectives that are actually useful for the company and the objectives, on the other hand the palt should allow the training objectives is located within the individual right to lifelong learning and their achievement can be spent within a standard system.

References

AA.VV, Ministero del Lavoro. Il mercato del lavoro. Una lettura integrata (2020). https://www.lavoro.gov.it/documenti-e-norme/studi-e-statistiche/Pagine/default.aspx

AA.VV, OECD. Report Adult Learning in Italy. What role for training funds? (2019). https://www.oecd.org/italy/adult-learning-in-italy-9789264311978-en.htm

ANPAL. XIX Rapporto sulla formazione continua in Italia (2020). https://www.anpal.gov.it/-/disponibile-online-il%C2%A0xix%C2%A0rapporto-sulla-formazione-continua-in-italia

Cameron, B.: Using responsive evaluation in strategic management. Strateg. Leadersh. Rev. **4**, 22–27 (2014)

Corsini, C., Sanzo, A.: Dalla valutazione alla valutazione d'impatto. Un modello possibile: le ricerche sul Valore Aggiunto in educazione, in AA.VV., Una prima rassegna sulla valutazione di esito ed impatto formativo (2009)

Commission of the European Communities, Recommendation of the European Parliament and the Council of 9 April 2008 on the establishment of the European Credit System for Vocational Education and Training (ECVET), Brussels, 9April 2008. https://eur-lex.europa.eu/LexUriServ/LexUriServ.do?uri=OJ:C:2009:155:0011:0018:EN:PDF

Decreto Legislativo 6 gennaio 2013, n. 13 Definizione delle norme generali e dei livelli essenziali delle prestazioni per l'individuazione e validazione degli apprendimenti non formali e informali e degli standard minimi di servizio del sistema nazionale di certificazione delle competenze (GU Serie Generale n.39 del 15 February 2013). https://www.gazzettaufficiale.it/eli/id/2013/02/15/13G00043/sg

European Commission. EVALSED. The resource for the evaluation of Socio- Economic Development. Evaluation guide. Brussels: European Commission (2013). https://ec.europa.eu/regional_policy/en/information/publications/evaluations-guidance-documents/2013/evalsed-the-resource-for-the-evaluation-of-socio-economic-development-evaluation-guide

Istat, Il Mercato del Lavoro 2020, Una Lettura Integrata (2020). https://www.istat.it/it/files/2021/02/Il-Mercato-del-lavoro-2020-1.pdf

Lichtner, M.: La qualità delle azioni formative, FrancoAngeli, Milano (1999)

Linee Guida per l'interoperatività degli enti pubblici titolari del sistema nazionale di certificazione delle competenze. https://www.lavoro.gov.it/documenti-e-norme/normative/Documents/2021/DI-del-05012021.pdf

Osservatorio Stranieri INPS, Banca Dati (2019). https://www.inps.it/nuovoportaleinps/default.aspx?itemdir=54479

Palumbo, M.: Il processo di valutazione, FrancoAngeli, Milano (2001)

Rapporto Annuale ISTAT 2021. https://www.istat.it/it/archivio/258983

Rapporto Annuale dell'Osservatorio EBAN sul Lavoro Agricolo (2021). https://www.enteeban.it/studi-e-ricerche/

Robasto, D.: L'agire formativo nella formazione continua. Uno studio esplorativo sui fondi interprofessionali, Excellence and Innovation in Teaching and Learning, vol. 1 (2019)

Robasto, D.: Politiche della formazione: l'innovazione dei sistemi formativi e valutativi. Lifelong Lifewide Learning (LLL) **38**, 107–115 (2021)

Schön, A.: Educating the Reflective Practitioner: Toward a New Design for Teaching and Learning in the Professions. Jossey-Bas, San Francisco (1987)

Schön, A.: The Riflective Turn Case Studies in and on Educational Practice. Teachers College Press, New York (1991)

Sistema Informativo Statistico delle Comunicazioni Obbligatorie, Rapporto Annuale sulle Comunicazioni Obbligatorie, 2021, Undicesimo Rapporto Annuale - Gli stranieri nel mercato del lavoro in Italia (2021). https://www.lavoro.gov.it/documenti-e-norme/studi-e-statistiche/Pagine/default.aspx

Stake, R.: Responsive evaluation. U.S. Department of Health, Education (1972)

Van Eerde, W., Simon Tang, K.C., Talbot, G.: The mediating role of training utility in the relationship between training needs assessment and organizational effectiveness. Int. J. Hum. Resour. Manage. **19**(1), 63–73 (2008)

Visalberghi, A.: Misurazione e valutazione nel processo educativo. Edizioni di Comunità, Milano (1955)

Students' Role in Academic Development: Patterns of Partnership in Higher Education

Anna Serbati[2(✉)], Valentina Grion[1], Juliana Elisa Raffaghelli[1], and Beatrice Doria[1]

[1] University of Padua, Padua, Italy
[2] University of Trento, Trento, Italy
anna.serbati@unitn.it

Abstract. Academic development becomes a central strategy to help universities be suitable settings for XXI century education. However, academic development itself needs to adjust to a new post-pandemic reality and new ways of learning. In this regard, engaging students has been targeted in several areas of research on educational quality, assessment and evaluation, as well as institutional change in Higher Education.

This paper focuses more specifically on the students' role in academic development. Firstly, we aim at offering an overview of current patterns found in the literature, of good practices of the student-teacher partnership in academic development in HEIs, and how different approaches might be integrated. Secondly, we propose a model for student-teacher partnerships in academic development as a sustainable and inclusive approach towards participatory democracy in higher education.

Keywords: Academic development · Students' role · Partnership · Higher education · Transformation

1 Introduction: An Era of Transformation in Higher Education

The EU2020 benchmark in Higher education (40% of young people in the EU with university-level qualification by 2020) was set a decade before 2020 became the year of the pandemic, with its unexpected consequences for all human activity. A "Modernisation Agenda for Higher Education" provided an overarching policy framework for national and EU policies to lead institutional changes towards the EU2020 benchmark. Some of the innovations required were competency-based approaches, flexible, personalised, diversified and inclusive learning pathways, better-informed evaluation, close relationships with society and the labour market, and global visibility of the learning offer [1]. This changing landscape was and still is challenging for the academic profession.

In such a setting, academic development becomes a central strategy to help universities to be suitable providers of XXI century education. However, academic development itself needs to adjust to the new post-covid reality and new ways of learning. Firstly, there is a need to reconsider scholarship in the digital world as contextualised within the

F. Calabrò et al. (Eds.): NMP 2022, LNNS 482, pp. 858–867, 2022.
https://doi.org/10.1007/978-3-031-06825-6_81

modernisation of higher education, focusing on the actual training needs of academic staff at several stages of advancement of their careers [2]. Secondly, it becomes important to emphasise the relevance of professional learning contexts, where institutional strategies and vision, important projects and careful support can make a difference.

In this context, many authors [3] believe that a crucial step to building an institutional culture of teaching and learning that searches for excellence is to put the student at the centre. How can this important goal be achieved? The real challenge is to value students' contribution not only within the classroom but to consider students a valuable resource (like academics and staff) to co-construct our universities as an enlarged educational community.

In this regard, the studies focusing on students' evaluation of teaching [4], student generated content [5], and peer assessment [6] are some examples of the richness embedded in engaging students in university's organizational change.

This paper deals with on students role in academic development. Firstly, we aim at offering an overview of current patterns in the literature and universities' good practices of the student-teacher partnership in academic development and how different approaches can be possibly integrated. Secondly, we would like to offer our own proposal for the student-teacher partnership in academic development as a sustainable and inclusive approach towards participatory democracy in higher education.

2 The Student-Teacher Partnership as Lived Democracy

Since we launched our Manifesto for Partnership in 2012, we've seen unprecedented strides forward in higher education in developing student engagement in teaching and learning, quality enhancement and institutional governance. We have consistently argued that higher fees and marketisation will not lead to improvements in quality, but rather honest conversations and constructive engagement with students [7, p. 3].

For many years, many claims have been made to recognise that education, and in particular higher education, is one of the main principal vehicles for social and economic development [8–11], and an instrument for the promotion of a sustainable future [12]. However, Higher Education Institutions (HEIs) today appear to be stuck on a market-driven path that has lost sight of these wider social aims in education [13]. Many countries have seen a progressive implementation of policies designed to increase competition among universities for both public and private funding [14], in what has been characterised as "managerialism" in Higher Education [15].

Numerous institutional bodies that deal with higher education and various academics all agree that a real change is needed and that the role of students within universities, through a renewed idea of partnership can be a central piece of the HEIs modernisation [3, 16]. The relationship between teaching staff and students can no longer remain linked to the "customer-service relationship" model. Rather, it should be grounded on the principles of respect, reciprocity and responsibility, realising itself as a "student-faculty partnership" [3] with the co-responsibility of the students in all aspects of educational processes. This perspective proposes a partnership between students and staff, which is about investing students with the power to co-create not just knowledge or learning, but the higher education institution itself [17].

Assuming such a vision, the Higher Education Academy (now Advance HE) states

"Partnership is a process for developing engaged student learning and effective learning and teaching enhancement. At its heart, partnership is about applying well-evidenced and effective approaches to learning, teaching and assessment with a commitment to open, constructive and continuous dialogue. Partnership involves treating all partners as intelligent and capable members of the academic community" [16].

While partnership approaches remain largely still under-theorised, student-as-partner (SaP) practices are emerging in today's universities as a means to offering a more participatory agenda and transforming institutional cultures within an increasingly economically driven higher education context [18]. From within the Student Voice movement, [19] Cook-Sather and Luz (2015) see the partnership as a *threshold* concept. As [20] Meyer and Land say (2006), threshold concepts are "conceptual doors" or "portals" which, once crossed, lead to a transformational internal view of an object, of a subject's landscape or even to a different vision of the world. Crossing a threshold leads not only to new ways of knowing but also to new ways of being. In this light, [19] Cook-Sather & Luz (2015) assert that introducing student-staff partnerships means pursuing a truly democratic education. An essential focus of the Student Voice movement is precisely that of democracy through education [21–23]. Within this perspective, some authors believe that democracy should be lived in daily experience to become a "mental habitus" of each one. [23] Fielding (2012) stresses that democracy is much more than a collaborative mechanism. It is mainly a way of living and learning, at the basis of which there is a common commitment to freedom, equality, mutual respect, and solidarity. Therefore, we need educational settings in which the concepts of authority and participation need to be reviewed, and offered students the space to share leadership, a space in which young people can express what they consider to be significant in their own education [24].

As Angus explains,

"In democratic organisations—indeed in any organizations in which there is genuine leadership rather than merely managerial coercion—such organisational shaping is never just a top-down process but is an engaged process involving all organisational players. The dialectical, relational view of leadership as a process incorporates the human agency of all members of the organization. [...]. Such leadership arises not from coercion and manipulation, but from relational collaborative, participatory processes" [24, p. 372].

Building on Fielding's thoughts (2006; 2012) [23–26], we believe that in order for universities to foster more democratic learning environments, students must be empowered as active and participatory agents and work in partnership with academics and administrators. According to the author [23, p. 53], the most genuine partnership between students and staff is named *Intergenerational learning as lived democracy*, a transformative relationship in which a joint commitment to the common good is put into practice and where there are occasions and opportunities for an equal sharing of power and responsibility. This is the best pattern of relationship between students and teachers to build a democratic fellowship, to teach and learn democratic citizenship, to promote democracy as a way of living and learning together.

3 Patterns of the Student-Teacher Partnership in Academic Development

Let us now reconsider the way in which student-teacher partnerships can be introduced and thus impact academic development. Many authors [23, 27] have noted that the students are able to actively contribute to the academic community by working with academics in designing courses and curricula through the adoption of participatory and collaborative methods [27, 28]. This approach has many advantages: on the one side, students feel part of the community, and this can sustain their motivation and increase the likelihood of them engaging in deep learning, developing hard and soft skills while confronting the challenging task of contributing to curriculum design. This might have very positive impacts on students' employability, success, and adaptability. On the other side, teachers have a unique chance to create a less hierarchical learning environment [27, 29] with more active learners who can make an effective contribution in the complex tasks of planning courses and curricula, which can foster their own motivation toward professional development.

The role of students as key actors and co-creators is obviously relevant in teaching and learning and curriculum design, but it becomes even more important in planning academic development initiatives for teachers to improve their pedagogical competences.

Academic development "aims to enhance the practice, theory, creativity and/or quality of teaching and learning communities in higher or post-secondary education" [30]. The literature and international practices worldwide have developed over the years a large variety of methods between formal and informal approaches as well as individual and group models [31]. All these strategies have the common aim to support professors in their professional path for improving the quality of teaching and learning; and who are better actors than the students to help guide this process? Learners can have relevant information, can share their perspectives and difficulties, can unveil points of view not yet considered, can bring suggestions, new ideas to the discussion that go beyond traditional teacher-driven ways of interpreting teaching and learning.

As Bovill, Cook-Sather & Felten [27, p. 142] suggest, it is important to carefully analyse the academic context, and identify appropriate co-creation opportunities. In fact, there are different approaches [32] for including students' voice in faculty development initiatives, from models where learners are heard and consulted as significant actors of educational change towards more complex ones in which students become co-creators and experts, with a more active role as drivers of the change.

Among others, we present here four well developed approaches of students' participation in academic development processes, that imply different intensities of the student role and that might be applied to university contexts depending on the characteristics of each specific situation. They all have in common the opportunity for mutual exchange between learners and teachers by drawing on the resources of both and building a community where they work together towards the improvement of the whole academic experience [33–35]. Our aim is, first of all, to offer an overview of current patterns in the literature and universities' good practices of student-teacher partnerships in academic development, highlighting different ways in which young students and academics can work together to face the complex challenges of higher education. Secondly, we also aim to interpret these models within a common framework, by offering our view on how different approaches can dialogue with each other and how they can possibly be integrated. Finally, we would like to offer our own proposal for the student-teacher partnership in academic development as a sustainable and inclusive approach towards participatory democracy in higher education.

A first approach that is already well-known and disseminated is called *Hearing the Student Voice*. It aims at collecting and using feedback from students in order to develop and thus improve the quality of the courses and curricula. Students are engaged in reflective processes for continuous educational change as relevant and legitimated stakeholders that can provide meaningful information for academics to decide how to act to enhance quality. For students to feel safe in sharing their comments and truly belong to the community it is necessary to create an environment in which dialogue and mutual exchange between students and teaching staff can take place in a constructive and effective way, to be then transferred into concrete actions [37]. For students, being heard means becoming an active part of the academic community and contributing to their motivation and engagement [37]. There are a variety of methods to listen to students' voices such as questionnaires, online discussions, focus groups, meetings, blogs and reflective commentaries, etc. For this approach to work, it is important to help teachers relinquishing control over pedagogical planning [27] and find a new balance in working with students, as happens, for example, in research with master's and graduate students.

Another approach is called *Students as Learners and Teachers* (SaLT), where the students serve as consultants for academics and professionals who work within the academic context, in order to foster dialogue and collaboration between members of the university community [38]. In detail, the model aims to facilitate a process through which students and teachers collaborate to generate dialogue about teaching and learning through meetings, seminars, in-class observations and scheduled briefings [33]. Student advice and concrete suggestions for improvement are therefore considered fundamental to the pedagogical development of the whole academic context [39]. Each experience can become good practice also for other colleagues and other contexts; all experiences are discussed within curricular meetings organised by each program, becoming a shared culture of teaching and learning.

A third approach interprets *the learner as a researcher and instructor* in academic development programs. According to this perspective, the students have the ownership of designing and carrying out research projects to investigate specific issues and problems of their institution. Those projects that achieve creative and sustainable solutions are then included in professional development activities devoted to academics; in these cases, students play the role of instructors sharing their research outputs and informed suggestions to professors. Therefore, their role becomes central not only for academic development but also to foster the academic system as a whole [28]. This approach emphasises the research process as a means by which the student promotes the link between research, learning and teaching [40].

A final approach, which we consider as a sort of synthesis of the previous approaches, proposed by Healey, Flint & Harrington (2014) [41], is called *Partnership Learning Communities*. It is presented as the union and overlapping of four macro-areas of student engagement and student voice: a) learning, teaching and assessment; b) subject-based research and inquiry; c) scholarship of teaching and learning; d) curriculum design and pedagogic consultancy. The first area, "learning, teaching and assessment" concerns collaboration and active involvement of students in their own learning, and is the most common form through which participation can be promoted [42]. This implies the use of inductive methods, active learning strategies and approaches based on experiential learning, transformative learning, self-directed learning, often helped by technology-enhanced environments. Relevant methods can be used to facilitate participation, also in assessment practices, such as peer assessment, self-assessment and the use of feedback. The second area, "subject-based research and inquiry", concerns student involvement in the research process, allowing them to learn autonomously while developing collaborative skills [42]. Student involvement in research calls for learners being active not only in their learning but also in a collaborative effort to inquire and discover new knowledge, developing linkages between research and teaching. The third area is the well-known "Scholarship of Teaching and Learning (SoTL)" based concept of [43] Boyer (1990) and involves researching and theorising teaching and learning within a discipline and then communicating and disseminating the findings [44]. Felten et al. (2013) [45, p. 63] call for expanding an inclusive approach to student engagement in SoTL by "encouraging a diversity of student voices to engage in co-inquiry with faculty. Inclusive engagement has tremendous potential to enhance student and faculty learning, to deepen SoTL initiatives, and to help redress the exclusionary practices that too often occur in higher education". Finally, the fourth area is "curriculum design and pedagogic consultancy" and concerns the least developed partnership. This goes beyond involving students in course evaluations and in departmental staff-student committees to engage students as partners in designing the curriculum and giving pedagogic advice and consultancy [42].

4 RE-FL-EC-T Innovation: A Proposal for an Inclusive Sustainable Approach of the Student-Teacher Partnership in Academic Development

As the final aim of this paper, we make a proposal for the student-teacher partnership in academic development towards a sustainable and inclusive approach and participatory democracy in higher education. Such a proposal has been developed through our own teaching contexts. In fact, in our experience academics express some concerns and difficulties in implementing approaches where students become real change agents and where initiatives are student-led. Even when teachers are willing to relinquish some of their control to establish a more democratic collaboration, they see new approaches as time consuming in their already busy schedule. They also believe there is a strong need for professionals, such as academic developers, to build a bridge between students and themselves. In other words, it seems that the establishment of fully student-led initiatives require resources and long planning.

Indeed, in our experience we tried to find a balance between student agency and sustainability. We developed a model that can be implemented by every teacher in every course with small amounts of time and resources. A very helpful tool is a website/learning platform where teachers and students can share their perspectives and where students' anonymity can be assured.

The approach is called *RE-FL-EC-T INNOVATION* and is divided in four steps.

- *RE*calling practices by teachers: every academic in their own course starts with recalling what happened during the teaching and learning process, reflects on events which occurred and highlights some thoughts in writing (a sort of auto-ethnography).
- *FL*asback scaffolded by teacher narratives: these narratives are shared, so students can reflect upon the teacher's experience, integrate their opinions and build a joint commentary about the process of teaching and learning within the course. This phase requires "all" students to participate in commenting on the teacher's narrative in a shared space.
- *EC*hoing the students' perspectives: teacher and students all read this common text and prepare for a discussion.
- *T*eaching *innovation*: students and teacher jointly inquire and analyse these narratives and co-construct improvement of learning, teaching, assessment and, in general, the overall experience to generate new practices.

The *RE-FL-EC-T INNOVATION* approach is flexible and can be applied at different stages of a course, i.e. in the middle of it, or towards the end of it or after the course has ended. Depending on when it is implemented, the improvements can be applied immediately or in the following year.

The model is built on a collaborative inquiry but does not require a lot of time and effort, nor does it imply particular training for academics and students. This is because in our view it is *sustainable* and implementable in every course.

Moreover, if *RE-FL-EC-T INNOVATION* is implemented in every course of a programme, results can be shared and become a source for curriculum design and enhancement.

Moreover, the simple actions outlined above require a strong student-teacher partnership. Though the model is initially teacher-led with the teacher sharing his/her vision, this action can be seen as an offer, a gift, opening a wide space to all the students to integrate and propose their own initiatives. The very act of asking students to write (whether done anonymously or not) rather than speak aloud, allows all students (and perhaps not only the most motivated ones or the least shy) to share their opinions, so it offers an inclusive approach. In the end, the teacher learns from this exchange, with a focus on his/her practice: this is the precise moment in which professional development can move further, in a balanced action with the improvement of teaching as part of higher education INNOVATION.

As Blanchet (2018) [46] recalls, bringing students into professional development offers several advantages: to help teachers to set clear goals and tasks to collect students' voices; to create relationship between teacher and students towards the good of the academic community; to scaffold students' engagement, collaboration, autonomy and responsibility; to better tailor academic development thanks to students' feedback.

Our hope is that the empirical evidence we are working on, and which we invite others to work on, demonstrates how this model can become a source of joint (professional) learning and growth, which also sets the basis for the democratic construction of university life. There, the students can have the freedom to become critical thinkers [47] and to really contribute to educational change.

References

1. High Level Group on the Modernisation of Higher Education: Report to the European Commission on New modes of learning and teaching in Higher Education, Luxembourg (2014)
2. Steinert, Y., et al.: A systematic review of faculty development initiatives designed to enhance teaching effectiveness: a 10-year update: BEME Guide No. 40. Med. Teach. **38**(8), 769–786 (2016)
3. Cook-Sather, A., Bovill, C., Felten, P.: Engaging Students as Partners in Teaching and Learning: A Guide for Faculty. Jossey-Bass, San Francisco (2014)
4. Spooren, P., Christiaens, W.: I liked your course because I believe in (the power of) student evaluations of teaching (SET). Students' perceptions of a teaching evaluation process and their relationships with SET scores. Stud. Educ. Eval. (54), 43–49 (2017)
5. Galloway, K.W., Burns, S.: Doing it for themselves: students creating a high quality peer-learning environment. Chem. Educ. Res. Pract. **16**(1), 82–92 (2015)
6. Serbati, A., Grion, V.: IMPROVe: Six research-based principles on which to base peer assessment in educational contexts. Form@re **19**(1), 89–105 (2019)
7. National Student Union, Comprehensive Guide to Learning and Teaching. https://www.nus connect.org.uk/resources/comprehensive-guide-to-learning-and-teaching-2015
8. Altbach, P.G.: The emergence of a field: research and training in higher education. Stud. High. Educ. **39**(8), 1306–1320 (2014)
9. Benneworth, P., Cunha, J.: Universities' contributions to social innovation: reflections in theory & practice. Eur. J. Innov. Manag. **18**(4), 508–527 (2015)
10. Ghislandi, P.M.M., Margiotta, U., Raffaghelli, J.E.: Scholarship of teaching and learning: per una didattica universitaria di qualità. Scholarship of teaching and learning for a quality higher education. Formazione & Insegnamento. Eur. J. Res. Educ. Teach. **12**(1), 1–289 (2014)
11. Salas Velasco, M.: Do higher education institutions make a difference in competence development? A model of competence production at university. High. Educ. **68**(4), 503–523 (2014). https://doi.org/10.1007/s10734-014-9725-1
12. Axelsson, H., Sonesson, K., Wickenberg, P.: Why and how do universities work for sustainability in higher education (HE)? Int. J. Sustain. High. Educ. **9**(4), 469–478 (2008)
13. National Student Union, Sustainability in Education. A manifesto for Partnership. https://www.iau-hesd.net/organizations/2626-national-union-students.html
14. Capano, G., Regini, M., Turri, M.: Changing Governance in Universities. Italian Higher Education in Comparative Perspective. Palgrave, London (2016)
15. Peters, M., Tze-Chang, L., Ondercin, D.: Managerialism and the neoliberal university: prospects for new forms of 'Open Management' in higher education. In: Peters, M., Tze-Chang, L., Ondercin, D. (eds) The Pedagogy of the Open Society. Brill, London (2012)
16. Advance HE. https://www.advance-he.ac.uk/guidance/teaching-and-learning/student-engage ment-through-partnership

17. National Students Union. A manifesto for partnership. https://wisewales.org.uk/wp-content/uploads/2018/09/NUS-A-Manifesto-for-Partnership.pdf. Accessed 16 May 2022
18. Gravett, K., Taylor, C.A., Fairchild, N.: Pedagogies of mattering: re-conceptualising relational pedagogies in higher education. Teaching in Higher Education (2021)
19. Cook-Sather, A., Luz, A.: Greater engagement in and responsibility for learning: what happens when students cross the threshold of student–faculty partnership. High. Educ. Res. Dev. **34**(6), 1097–1109 (2015)
20. Meyer, J., Land, R.E.: Overcoming Barriers to Student Understanding: Threshold Concepts and Troublesome Knowledge. Routledge, Oxon (2006)
21. Fielding, M.: Transformative approaches to student voice: theoretical underpinnings, recalcitrant realities. Br. Edu. Res. J. **30**(2), 295–311 (2004)
22. Fielding, M., Moss, P.: Radical education and the common school: a democratic alternative. Rev. J. Childhood Stud. **39**, 92–96 (2014)
23. Fielding, M.: Beyond student voice: patterns of partnership and the demands of deep democracy. Revista de Educación **359**, 45–65 (2012)
24. Smyth, J.: Educational leadership that fosters 'student voice.' Int. J. Leadersh. Educ. **9**(4), 279–284 (2006)
25. Angus, L.: Educational leadership and the imperative of including student voices, student interests, and students' lives in the mainstream. Int. J. Leadersh. Educ. **9**(4), 369–379 (2006)
26. Fielding, M.: Leadership, radical student engagement and the necessity of person-centred education. Int. J. Leadersh. Educ. **9**(4), 299–313 (2006)
27. Bovill, C., Cook-Sather, A., Felten, P.: Students as co-creators of teaching approaches, course design, and curricula: implications for academic developers. Int. J. Acad. Dev. **16**(2), 133–145 (2011)
28. Seale, J.: Doing student voice work in higher education: an exploration of the value of participatory methods'. Br. Edu. Res. J. **36**(6), 995–1015 (2010)
29. Fielding, M.: Radical collegiality: affirming teaching as an inclusive professional practice. Paper presented at the British Educational Research Association Conference, Brighton (1999)
30. IJAD editorial team, 12 October 2021. https://think.taylorandfrancis.com/special_issues/our-academic-development-stories/
31. Steinert, Y.: Faculty development for teaching improvement: from individual to organizational change. In: Walsh, K. (ed.) The Oxford Textbook of Medical Education, pp. 711–721. Oxford University Press, Oxford (2013)
32. Burkill, S., Dunne, L., Filer, T., Zandstra, R.: Authentic voices: collaborating with students in refining assessment practices. In: Presentation at ATN Assessment Conference, RMIT University (2009)
33. Cook-Sather, A.: Teaching and learning together: college faculty and undergraduates co-create a professional development model. Improve Acad. **29**, 219–232 (2010)
34. Huston, T., Weaver, C.L.: Peer coaching: professional development for experienced faculty. Innov. High. Educ. **33**(1), 5–20 (2008)
35. Cox, M.D., Sorenson, D.L.: Student collaboration in faculty development. In: Kaplan, M. (eds.) To improve the Academy, vol. 18, pp. 97–106. Anker, Bolton (2000)
36. Schmolitzky, A.W., Schümmer, T.: Hearing the student's voice - patterns for handling students' feedback. In: Proceedings of EuroPLoP (2009)
37. Campbell, F., Eland, J., Rumpus, A., Shacklock, R.: Hearing the student voice. Involving students in curriculum design and delivery. Edinburgh Napier University, Edinburgh, Scotland (2009). http://www2.napier.ac.uk/studentvoices/curriculum/download/StudentVoice2009_Final.pdf. Accessed 4 Dec 2021
38. Dickerson, C., Jarvis, J., Stockwell, L.: Staff–student collaboration: student learning from working together to enhance educational practice in higher education'. Teach. High. Educ. **21**(3), 249–265 (2016)

39. Mihans, R., Long, D., Felten, P.: Power and expertise: student-faculty collaboration in course design and the scholarship of teaching and learning. Int. J. Scholarship Teach. Learn. **2**(2), 1–9 (2008)
40. Walkington, H.: Students as Researchers: Supporting Undergraduate Research in the Disciplines in Higher Education. The Higher Education Academy. Heslington, York (2015)
41. Healey, M., Flint, A., Harrington, K.: Engagement Through Partnership: Students as Partners in Learning and Teaching in Higher Education. HEA, York (2014)
42. Kuh, G.D.: The national survey of student engagement: conceptual and empirical foundations. New Directions Inst. Res. **141**, 5–20 (2009)
43. Boyer, E.: Scholarship Reconsidered: Priorities of the Professoriate, 1st edn. Josey-Bass, New York (1990)
44. Healey, M.: How to Put Scholarship into Teaching. Times Higher Educational Supplement, 4 Feb 2000. https://gdn.glos.ac.uk/confpubl/thes.htm
45. Felten, P., et al.: A call for expanding inclusive student engagement in SoTL. Teach. Learn. Inq. ISSOTL J. **1**(2), Special Issue: Writing Without Borders: 2013 International Writing Collaborative (2013)
46. Edutopia (2018). https://www.edutopia.org/article/bringing-students-professional-development
47. Freire, P.: Pedagogy of the Oppressed. Penguin, London (1990)

How Can University Promote Eco-Literacy and Education in Environmental Sustainability? A Third-Mission Best Practice at the University of Bari

Alessia Scarinci[✉] [iD] and Alberto Fornasari [iD]

University of Bari, Bari, Italy
{alessia.scarinci,alberto.fornsari}@uniba.it

Abstract. The systemic and complex nature of the environmental problem and sustainable development that sees political institutions, educational and training institutions, communities and individual citizens interconnected, brings out the need to invest in the transformative power of knowledge and knowledge, along throughout his life. Educational and training institutions, especially universities, have the task of promoting eco-literacy and education in environmental sustainability by educating to critical and responsible consumption to improve lifestyles and increase the sustainability of human activities on ecosystems. The university must coordinate, together with other stakeholders, different intervention strategies in order to address the objectives of the 2030 Agenda and the local sustainability challenges. The place-based approach, capable of linking individual and social needs to the needs of the territory, is configured as a possible approach for defining the university as a hub of sustainability on the territory.

Keywords: University · Sustainability · Place-based approach · Third Mission

1 Education Scenarios for Sustainable Development

For years, thanks to technological development, man has governed the environment and exploited the resources present in the ecosystems of which he himself is a part, thus becoming independent from nature [1] but at the same time has altered the balance nature leading to an unprecedented environmental crisis. The environmental emergency brings about a necessary change in lifestyles, production and consumption of environmental resources and the preparation of a new paradigm for development. The survival of our planet depends on the recognition of the consequences that the choices we make today can have on the future of the environment and of humanity itself [1, 2]. In the Brundtland report (1987), a document that has given international importance to the concept of sustainable development, this interdependence between our daily decisions and the effects on the entire system emerges, as also underlined by Goleman [3].

F. Calabrò et al. (Eds.): NMP 2022, LNNS 482, pp. 868–877, 2022.
https://doi.org/10.1007/978-3-031-06825-6_82

Development must, in fact, concern the needs of the present, guaranteeing "the best possible conditions for its longevity" [4], without, therefore, compromising the possibility for future generations to be able to meet their own needs. of growth and development [5]. This model is based on the concept of generational equity (inter and intra-generational), i.e. on a preventive economic system that considers ethical principles of social justice and solidarity [6]. This implies a common responsibility that is expressed in the active participation of all citizens and local communities for the protection of the environment, natural systems and human well-being; recalls the dimension of otherness, the willingness to see the other, the concern for guaranteeing future generations equal opportunities for growth and being responsible towards those who live with us on Earth [7, 8]. As underlined by Malavasi, Iavarone and Mortari [9] "we are all called to make a decision and stand up to fight against degradation, poverty, in favor of the environment, equality, in order to implement a culture of peace and of hope for the companies and all their members". In fact, as Jonathan Safran Foer argues in his novel "We can save the world before dinner" [10], when we talk about the environmental "crisis" we refer to a decision: it is a question of "deciding" which side to take, what action to take, what to give up in order to make the necessary change to save us and our common home. Environmental and social change can occur if everyone participates in the collective action of taking care of "a common good in the general interest" [5, 10]. We recall the principle, already present in the documents of the World Commission for Environment and Development, known as the Brundtland Report of 1987, and of the UN conference on environment and development held in Rio De Janeiro in 1992, of "acting locally and thinking globally "which is based on processes of involvement and participation from below for the promotion of sustainable actions that start from a local dimension of social responsibility and then arrive at interventions at a global level [4]. For the implementation of sustainable development policies, therefore, the fundamental role of the local community is recognized and the need to implement a process shared by all the players in the area from a systemic perspective (Agenda 21). This sense of common responsibility involves a rethinking of citizens who are activated and become capable of "taking care of the common good in the general interest" [5] and which leads to new relationships between citizens and the territory. To participate in global citizenship and in the path of sustainable development it is necessary that everyone is guaranteed the right to a quality education capable of facilitating the generative and transformative path of the citizen and therefore forming the basis for achieving the objectives set by the Agenda. 2030 [11].

"It must be strongly reaffirmed that education, in all contexts, constitutes the first and fundamental right, the first common, tangible and intangible good" [5] and is central to the development of the person and the community and to the formation of environmental awareness [12]. Environmental problems must be tackled starting from a review of culture and in particular of schools and universities as places for the promotion of culture and then investing in the transformative power of knowledge and knowledge throughout life [13, 14]. As underlined in goal 4 of the 2030 Agenda, it is the task of education and training "to transmit to all students the knowledge and skills necessary to promote sustainable development" (Goal 4.7), to train the new generations more attentive to causes that generate impacts rather than the techniques to reduce them, moving the problem from the resolution of the effects to the elimination or reduction of the causes.

Educational and training institutions must promote ecological literacy and education for environmental sustainability by educating to critical and responsible consumption to improve lifestyles and increase the sustainability of human activities on ecosystems; reflect on cultural, anthropological and social values to foster the development of a critical and conscious environmental awareness as an expression of an active citizenship and of one's life in the world. This requires an integrated action between education, pedagogy and politics capable of promoting attitudes and behaviors that can meet the needs of intra and intergenerational equity and allow individual and collective actions to be taken on local, national and global urgencies [15]. Important are the international and national political provisions on sustainability which identify in knowledge and training throughout the life span "the transversal element for change" [16] of attitudes, behaviors and values and a "planetary fraternity" [9, 12]. "Education must raise awareness of ecological and ethical issues among the population, ensure effective decision-making participation of the public in the positions taken and consolidate values and behaviors compatible with sustainable development" (Chapter 36 of Agenda 21).

Therefore, a new pedagogical-didactic project is necessary capable of educating through reflection, collaboration and participation in social life to the provision of the best conditions, both environmental, cultural and emotional, to favor human growth [17]. Arrange interventions that allow you to experience being in relationship with nature, taking care (taking care and worrying), the complex intertwining of nature and culture and understanding the impact of one's actions on the territory [2].

2 Universities and Policies for Sustainability

The transformative value of education, as discussed above, is the founding element on which sustainable development policies are based and which guide social and cultural change. As stated in the declaration of objective 4 of the 2030 Agenda or in the UN General Assembly Resolution 72/222 (2017), education is not just a lifelong learning opportunity for the acquisition of tools and knowledge necessary for full participation and active in social life, but it is the basis, the necessary condition for the achievement of sustainable development [11, 15]. All countries are, in fact, invited to strengthen education and improve the implementation of education for sustainable development by strengthening educational and training policies.

In order to give substance to sustainable development, it is necessary to involve at the local, national and regional level, in order to reach the global level, of interested parties such as institutions, schools, universities and communities, in a systemic and integrated perspective promoted by the 2030 Agenda. Educational and training structures (schools and universities) are also involved in the Sustainability Education Plan [18] for the implementation of initiatives aimed at promoting knowledge and practices for sustainable development in collaboration with the community room in which students and people work [15, 19].

The university, the place for the construction of the mindset and the educating community, has the task of promoting knowledge and training future teachers and educators, as well as citizens, through the activation of paths aimed at developing a feeling of intergenerational responsibility and a ecological awareness that allows us to understand how much our behavior and our choices affect the planet and how strong is the danger that comes from not considering the close interdependence that exists between our daily decisions and the effects on the whole system [2, 19]. The university, therefore, as a space for the "promotion of values and ideals" [20] as well as a place of social community in which to make change. The university context creates the conditions, thanks to its ability to forge relationships with the social, institutional and territorial fabric, for the development of pro-environmental attitudes and behaviors, for the practice of education for sustainable development throughout the life span [19].

Sustainability in its three dimensions (environmental, social and economic) is one of the university's missions from a perspective of transversality and interdisciplinarity. The implementation of this mission is expressed on three levels: organizational, pedagogical-educational and macrosocial. At the organizational level, the sustainable dimension should lead to an internal reorganization of the bodies and management that takes into account the environmental, social and economic impact of the activities [20]. At the pedagogical-educational level, reference is made to the university's own purposes, also referred to in the Sustainability Education Plan, namely:

– training, understood as education to sustainability through teaching (degree courses, masters, training courses, specific courses);

– research, promotion of research projects that broaden the knowledge of environmental problems and propose solutions;

– the dissemination of knowledge, through the dissemination of research results and the university's commitment to the "Third Mission" which relates training and development of the person and which leads to the cultural and economic growth of the social fabric in which it is inserted through sharing knowledge and strategies with local stakeholders.

Finally, at the macrosocial level, the synergistic work of the university with policy makers, administrations and communities for the promotion of sustainable actions and initiatives is highlighted [19, 20]. The mission of sustainability in the university context declined in these three scenarios emphasizes the assumption of a social responsibility of the educational and training institution and therefore as an agent of change.

3   University and Territory: The Value of the Third Mission in Light of the 2030 Agenda

Sustainability, climate change, energy and energy efficiency, green consumption and sustainable urban development are globally recognized challenges for the 21st century. But all the explicit indications risk remaining an end in themselves if they do not materialize also in formative experiences that stimulate real behaviors in this direction. Facing these challenges requires a strong commitment and specific skills on the part of businesses and private sectors, governments, agencies, NGOs and universities.

Citizenship education and education for sustainable development are in fact fundamental themes for the education of the citizens of tomorrow but working, as Santelli argues, on a vision of the world which, by protecting itself from an eco-sustainable model, also brings with it the right to peace in positive ways. of coexistence in diversity and social justice. The role of the Third Mission of the universities appears in this important intertwining between training and the central territory. By university Third Mission (TM) we mean "the set of activities with which universities enter into direct interaction with society, providing a contribution that accompanies the traditional missions of teaching and research" [21] but also the "propensity of structures to open up to the socio-economic context, exercised through the enhancement and transfer of knowledge" [21]. Unlike research and teaching, which are the institutional duty of every teacher, the Third Mission is an institutional responsibility to which each university responds in a differentiated way, according to its specific characteristics, where an important variable is constituted by being generalist universities versus Polytechnics and Specialization Schools.

The Third Mission has also been recognized as an institutional mission of universities only in recent times and with a still incomplete regulatory provision. It seems appropriate to remember that the Third Mission of the University should not be understood in its purely economic dimension [22] but rather it should be understood in its socio-cultural mission as a tool for enhancing knowledge through: a) supporting actions research (enhancement of educational skills, identification of opportunities, selection and experimentation of reference research ideas, protection and management of intellectual property, creation and support of spin-offs); b) actions to support economic and social development (creation of a distinctive image for university-brand activities, development of innovative communication, website, presentations, social networks; c) development of structured relationships with local civil society of reference, associations, communities of citizens, local realities; d) development of relations with national and international institutions; e) opening of the innovation counter; f) support for the development of national and international partnerships; g) actions to incentivize funding to the University by public and private structures [23]. The traditional areas of the Third Mission are therefore focused on the enhancement of research and articulated as follows: a) management of industrial property, companies, spin-offs, activities for third parties (activities on behalf of third parties no longer fall within the of Third Mission, they could be included in public engagement activities; in fact the commissioned research, even if it plays a role of strategic importance for the universities from the point of view of external relations, is not considered a field of action of TM, but a financing instrument of the latter); b) intermediation structures; c) with regard to the production of public goods: the management of cultural heritage and activities, activities for public health, continuous training and public engagement (organization of cultural and public utility activities, concerts, shows, reviews, exhibitions, exhibitions and other events open to the community, scientific dissemination, non-academic publications, radio and TV programs, publication and management of websites and other social media channels for scientific dissemination, initiatives to involve citizens in research, debates, festivals, scientific cafes, online consultations, contamination labs, activities of involvement and interaction with the school world, laboratory activities, simulations and hands-on experiments).

For a long time, the University was perceived as an environment of excellence capable, on the one hand of forming the minds of students wishing to pour their know-how into the world of work, on the other hand, it went through a phase in which it was unable to completely shorten the distance from the realities present in the territory. Today this gap has fortunately been overcome and there are many testimonies that go in this direction and an increasing attention is paid to the fact that the managers and professionals of the future will require specific knowledge to understand the interconnection of economic, legal, environmental and social skills, as well as the skills to manage and contribute to change towards a more sustainable world.

4 Place-Based Approach as an Empowerment of the University-Community Relationship

The university, in its formative function, is called to "act locally and think globally" and therefore to connect and connect students with their community in order to reconstruct learning experiences based on the Deweyan idea of direct interaction with the environment [14]. The design of innovative teaching and training interventions must take into account the individual and social needs expressed by citizens which must be related to the development needs of the territory, in order to promote active and conscious participation and a responsibility towards other and the future of the earth [24–26]. The promotion of the culture of sustainability in the university system requires the adoption of a model, an approach capable of developing an integrated training project, as Silvia Mongili suggests, "in the sense of establishing a systemic and complex relationship with the territory" [27]. In fact, there are several aspects of the community to consider (experiences, needs, knowledge) for the design of training interventions that, starting from the local, can then arrive at an interpretation of the global, an aspect also supported by the 2030 Agenda itself. These aspects can be favored by a place-based approach that allows to understand and explore a territory in depth, to know its history and problems, activating an active involvement of the community, institutions and citizens in the resolution of real problems and improvement of places [28].

Attention to the territory, to the dialogic development of the person in his habitat, can become a good starting point which, if well managed, leads to forms of awareness of the footprint of men and organizations and therefore, indirectly, puts under observation the habits and lifestyles of the moment as they are configured in the social group to which they belong. The place-based approach was born in the economic field following the acknowledgment of the unproductiveness of universal and neutral development measures with respect to the place of application. Otherwise, this orientation makes it possible to respond to complex, interconnected or demanding issues, such as social and economic disadvantages, natural disasters or environmental problems, anchoring the solutions to the real needs of the context [29]. The place-based approach makes it possible to design intervention strategies related to the needs of the community by exploiting local resources in a collaborative and shared way between the various parties involved to create systemic change. A fundamental aspect of place-based is precisely the union of interested parties and members of the community which facilitates the definition of a collective vision for the future and therefore the commitment and investment by them to achieve the goal. This implies the ability to work together, to combine the skills and resources of community members (industry, university, government) for the creation of a territorial network and greater effectiveness of interventions and overcoming the fragmentation and isolation of systems [30]. "Collaboration with community members fosters social inclusion, self-efficacy and civic empowerment" [30].

For these aspects, the place-based approach can also be considered a tool for the design of pedagogical-didactic interventions for education to local and global citizenship. The objective of citizenship education, as Ben Kisby writes, is "to improve the levels of political knowledge and understanding of citizens and to educate citizens as actors of civil society in order to promote a critical and active citizenship, with citizens able to develop their own ability to engage in civic and political activities to bring about the social changes they wish to see" [31]. The place-based approach allows to know and deepen the history of places, to understand the social and ethical aspects, it allows to reflect critically and self-critically on what has been and on possible future developments through processes of sharing and comparison with the 'other that leads to the recognition of a shared responsibility for the common good, the development of valuable attitudes and the acceptance of otherness [32].

5   The WE Project: A Place-Based Practice and Third UniBa Mission

An example of intervention planning based on the place-based approach in the context of Third Mission actions is represented by the "WE. Sustainable innovations - Bari". This is an action-research project developed by the Sustainable Development School for education for global citizenship which has seen the department of education sciences, psychology and communication of the University of Bari involved in the management of local working groups (Community hub) and in sharing intervention strategies. The goal is to promote education for sustainable development through a participatory process of the educating community and local stakeholders and to make cities more sustainable. The project, already implemented in two other cities, involved 10 first and second grade schools in the city of Bari located mainly in the peripheral areas of the city, the municipality of Bari and associations, businesses and the third sector of the city itself and the University of Bari.

A first phase of the project saw the teachers of the schools involved, engaged in a training course aimed at providing knowledge of the territorial reality examined on the basis of the sustainable development goals of the 2030 Agenda. In this phase the teachers were able to become aware of the problems present in the area by identifying possible systemic and social causes, developing a shared vision of the problems and realizing the need for change. Subsequently, a mapping of the community, the stakeholders (companies, associations) present in the Bari area who work for sustainability and their involvement within the project was carried out. The creation of a network between stakeholders, schools and community members is fundamental both to establish a collective vision for the future and to engage in change for the achievement of common objectives but also because through the involvement of the public it is possible to implement initiatives or programs that are truly responsive to the needs of the community and therefore be more effective [30].

The community hubs represented the operational moment of the project, in which it was possible to activate moments of reflection, sharing, collaborative work and participatory planning that favored the construction of a plan for change through the design of sustainable development paths and social innovation initiatives for the whole community. The community hubs were held in three meetings and divided into three working groups: a) community, b) circular economy, c) school and innovation. During the meetings, after a first phase of presentation and sharing of best practices with respect to the topic of the worktable, we thought about the key words of each table and worked on the design of a social innovation project. The project presented shows the need for a recovery of social and educational spaces both inside the school and outside the school, in the neighborhood, for example, for a re-evaluation of the same. This path, focused at the local level, has offered the community the possibility of:

– develop a shared vision,
– foster motivation for change,
– generate innovative solutions and different ways of working and thinking in a collaborative and participatory way,
– facilitate evidence-based approaches, build partnerships between schools and between them and local stakeholders, involve companies capable of influencing the social impacts of projects.

6 Conclusion

The universities with the many initiatives have welcomed the challenge of sustainability with interest and commitment by working tenaciously for the construction of solid and lasting alliances with the territory, to promote its development by activating its unexpressed powers through a smart, sustainable and inclusive strategy in line with the 2030 agenda. It is only by promoting forms of collaboration on the part "of all the components of society down to individual citizens that sustainable development will go from being an auspice to an effective path" [33], working on a vision of the world which, by placing itself in the protection of an eco-sustainable model also brings with it the right to peace to positive forms of coexistence in diversity and to social justice that we can imagine translating the aforementioned objectives of the 2030 Agenda into a reality plan. it has been said, it is necessary to focus on the educational aspect. As argued by L. Santelli Beccegato "in the current pedagogical setting, despite the multiplicity of developments, a common approach is recognizable: assuming the awareness of the complexity and fragility of existence is configured as a basis for recognizing the need to acquire the appropriate cognitive tools, affective, social, ethical, religious" [33] to face and govern reality in the awareness of the importance and centrality of education. Educating to empower the citizens of tomorrow to actions that go towards the other, that take care of the other, overcoming self-centeredness and to "prepare to understand the needs of others" [26]. It is therefore necessary to invest in universities to promote a responsible society that respects differences, capable of recognizing the consequences of one's actions and of assuming "attitudes of moral, social and cultural co-responsibility towards sustainable development" [19]. This requires a rethinking of the models by focusing on systemic and integrated approaches that make it possible to establish a relationship with the territory, with the needs and urgencies expressed by the community, such as the place-based approach.

The focus on the place, as emerged from the experience of the "We" project, offers the university the opportunity for a more effective dissemination of the culture of sustainability as it provides a lens that helps to see, examine and analyze the systemic problems present on the territory and find the most suitable and guided pedagogical-educational, organizational solutions at the local level. It is only by looking "at reality through the knowledge of the various aspects of the society in which it is inserted" that the university can develop "the ability to effectively balance both the economic impact and that of social and environmental sustainability" [19].

References

1. Mattioli, G., Scalia, M.: Introduzione. In: Angelini, A., Pizzuto, P. (ed.) Manuale di ecologia, sostenibilità ed educazione ambientale, FrancoAngeli, Milano (2007)
2. Strongoli, R.C.: Ecodidattica. Una proposta di educazione ecologica. Ricerche di Pedagogia e Didattica. J. Theories Res. Educ. **14**, 3 (2019)
3. Goleman D.: Intelligenza ecologica. La salvezza del pianeta comincia dalla nostra mente, Milano, trad. it. Rizzoli (2009)
4. Farnè, R.: L'insostenibile pesantezza dell'educazione. Pedagogia Oggi/Rivista SIPED/anno XVI, vol. 1 (2018)
5. Riva, M.G.: Sostenibilità e partecipazione: una sfida educativa. Pedagogia Oggi/Rivista SIPED/anno XVI, vol. 1 (2018)
6. Kocher, U.: (a cura di): Educare allo sviluppo sostenibile. Pensare il futuro, agire oggi, Erickson, Trento (2017)
7. Francesco, P.: Laudato si', Enciclica di sulla cura della casa comune, CdV (2018)
8. Mortari, L.: Filosofia della cura. Raffaello Cortina, Milano (2015)
9. Malavasi, P., Iavarone, M.L., Mortari, L.: Editoriale, Pedagogia Oggi/Rivista SIPED/anno XVI, vol. 1 (2018)
10. Safran Foer, J.: Possiamo salvare il mondo, prima di cena. Perchè il clima siamo noi, Guanda, Milano (2019)
11. Grange, T.: Qualità dell'educazione e sviluppo sostenibile: un'alleanza necessaria, una missione pedagogica. Pedagogia Oggi/Rivista SIPED/anno XVI, vol. 1 (2018)
12. Malavasi, P.: Pedagogia verde, La Scuola, Brescia (2008)
13. Loiodice, I.: Investire pedagogicamente nel paradigma della sostenibilità. In: Pedagogia Oggi - Rivista SIPED XVI, p. 1. Pensa MultiMedia, Lecce-Brescia (2018)
14. Thornton, S., Graham, M., Burgh, G.: Place-based philosophical education: re-constructing 'place', reconstructing ethics. Childhood & Philosophy, Rio de Janeiro, vol. 17, abr. 2021, pp. 1–29 (2021)
15. UNESCO: Education for Sustainable Development. A Road Map (2020)
16. Santerini, M.: Educazione sostenibile e giustizia in educazione. Pedagogia Oggi/Rivista SIPED/anno XVI, vol. 1 (2018)
17. Broccoli, A.: Dall'informazione alla formazione. Educare alla sostenibilità per un nuovo modello di sviluppo. In: Sannella, A., Finocchi, R. (ed.) Connessioni per lo sviluppo sostenibile, Ediz. Unicassino, Cassino (FR) (2019)
18. MIUR: Piano per l'educazione alla sostenibilità (2017)
19. Cajola, L.C.: Scuola-Università: fare sistema e creare sinergie per il Piano di educazione alla sostenibilità. Pedagogia Oggi/Rivista SIPED/anno XVI, vol. 1 (2018)

20. Parricchi, M.A.: "Green" University: un orientamento pedagogico alla sostenibilità. In: Pedagogia Oggi/Rivista SIPED/anno XVI, vol. 1 (2018)
21. Agenzia nazionale di valutazione del sistema universitario e della ricerca: La terza missione nelle università e negli enti di ricerca italiani. In: Documento di lavoro del Workshop del 12 Aprile 2013, p. 3 (2013)
22. Dato, D., Cardone, S., Mansolillo, F.: Pedagogia per l'impresa. Università e territorio in dialogo. Progedit, Bari (2016)
23. De Bortoli, A., Predazzi, E., Susa, I.: La Terza Missione dell'Università, Analysis, Rivista di Cultura e Politica Scientifica, vol. 2/3 (2011)
24. Bertolino, F., Perazzone, A.: Il valore educativo del mondo rurale: la fattoria come contesto ponte tra bosco e città. In: Salomone, M. (ed.) (a cura di), Prepararsi al futuro: ambiente, educazione, sostenibility, Istituto per l'Ambiente e l'Educazione Scholè Futuro Onlus, Torino (2015)
25. Zakri, A.: Promuovere il decennio di educazione ambientale per lo sviluppo so-stenibile. In: Salomone, M. (ed.) (a cura di), Educational Paths towards Sustainability, atti del 3° Congresso mondiale di educazione ambientale, Istituto per l'Ambiente e l'Educazione Schole Futuro Onlus, Torino (2006)
26. Perla, L.: L'insegnamento dell'educazione civica: prodromi educativo-didattici e "prove tecniche" di curriculo. Nuova Secondaria, vol. 10, giugno 2020 - Anno XXXVII, pp. 222–238 (2020)
27. Mongili, S.: Il ruolo educativo del territorio per la sostenibilità e il benessere, Lifelong Lifiwilde Learning, vol. 7(17/18) (2011)
28. Smith, G.A.: Place-based education. Practice and impact. In: Stevenson, R.B., Brody, M., Dillon, J., Wals, A.E.J. (eds.) International Handbook of Research on Environmental Education. AERA, Routledge Publishers (2013)
29. Scarinci, A.: Il bosco nell'aula. Progettare l'educazione ambientale, Progedit, Bari (2021). ISBN 978-88-6194-522-7
30. Queensland Council of Social Service: Place-based approaches for community change: QCOSS' guide and toolkit. West End, Queensland (2019)
31. Kisby, B.: Citizenship education and civil society. Societies 11, 11 (2021). https://doi.org/10.3390/soc11010011
32. Majchrzak, K.: Places of remembrance in citizenship education. J. Educ. Cult. Soc. 3(1), 7–14 (2012)
33. Santelli, L.: Educazione allo sviluppo sostenibile. Un importante impegno da condividere, Guerini, Milano (2018)

Educational Interventions for Civil and Democratic Society: A Research-Training Project on Bullying and Cyberbullying at Apulian Schools

Loredana Perla^(⊠) and Ilenia Amati

University of Bari Aldo Moro, Bari, Italy
{loredana.perla,ilenia.amati}@uniba.it

Abstract. Social inclusion, closely linked to the concept of civic education and the removal of obstacles to participation and forms of discrimination, requires specific and widespread training on tackling bullying and cyberbullying, understood as complex social phenomena. A collaborative research-training with a network of I and II grade secondary schools in Puglia is described, aimed at designing a device for education for civil and democratic coexistence (the SEP device: sensitize, educate, protect). The phases of the experimentation included: a mapping of the dynamics between peers inside and outside the school context; the emergence of representations, dynamics and interactions between adolescents; the identification and training of opinion leaders, who have experimented with workshops with the tutorial support of the trainers. The following were used as data collection tools: a questionnaire administered before and after the training intervention, video recordings of the laboratory sessions, trainers' reports, tutor teacher logbook, researcher monitoring sheets. The results of the experiment made it possible to learn about the phenomena of bullying and cyber-bullying among the adolescents of the schools participating in the network; sensitize all actors in the school context to non-violent management approaches to conflict and forms of participation that counter homologation and discrimination of adolescents; empower young people to make informed use of technological media and social networks; educate to civil and democratic coexistence, promoting the culture of legality, intercultural communication and the enhancement of diversity.

Keywords: Bullying · Cyberbullying · Inclusive didactics

1 Theoretical Framework

The concept of inclusion is closely linked to that of education for "active citizenship", understood as the responsible participation of all citizens in the construction of a civil society and in the political life of the community they belong [1–3]. Civic education, introduced in the Italian school curriculum in 2019 with Law 92, was created to stimulate the processes of students' active participation in the construction of a democratic society where the fight to combat all forms of discrimination translates into intentional actions

F. Calabrò et al. (Eds.): NMP 2022, LNNS 482, pp. 878–886, 2022.
https://doi.org/10.1007/978-3-031-06825-6_83

to promote respect for gender differences and for all forms of relational fragility. The art. 7 of the law affirms the need for schools to strengthen and consolidate collaboration with families in order to promote behavior based on a citizenship aware of the rules of coexistence. It places at the foundation of civic education the knowledge of rights, duties, tasks, personal as well as institutional behaviors aimed at promoting the full development of the person.

In order for civic education to be fully promoted at school level, specific and widespread training is needed on the practices of managing the fight against bullying in the group-class and on communication techniques with parents who sometimes collude - unknowingly - with the antisocial behavior of their sons or daughters. The fight against bullying and cyberbullying therefore becomes a fundamental object for educational research and for the promotion of an inclusive education capable of guaranteeing the principles of civic and democratic education [4].

1.1 The Pedagogical Approach to Bullying and Cyberbullying: Characteristics of the Phenomenon and Prospects for Intervention

The theoretical hypothesis underlying the experiment presented here considers the bully-victim relationship as the result of a complex intertwining of personal and social factors: the former refer to the individual characteristics of the involved subjects, to the family and educational context and to their previous experiences, the latter to the rules that are established within a group, to the roles and expectations of children. Bullying and cyberbullying are, in fact, systemic phenomena and, in order to be addressed, they require system actions, such as the inclusion of prevention and contrast objectives in the vision and mission of a school [5].

Heinemann and Olweus (1996) defined bullying by observing the first signs of aggression inflicted on peers. It is a phenomenon defined as the recurrence of direct or indirect behaviors and attitudes aimed at bullying another with the intention of harming, with the use of physical force or psychological abuse [6]. A student is subjected to bullying, that is, he is prevaricated or victimized, when he is repeatedly exposed over time to the offensive actions carried out by one or more classmates. Bullying includes aggressive actions or behaviors of social exclusion perpetrated intentionally and systematically by one or more people to the detriment of a victim who is often upset and does not know how to react [7]. Bullying behavior is thus a behavioral interaction in which one of the subjects, the victim, is progressively relegated to a marginal position within a group and this marginality increases inversely proportional to the power acquired by another abusive subject, the bully or the bully-girl or the group as a whole (baby-gang). The imbalance of power can derive from multiple causes: from a difference in physical strength between a group that decides to join forces to attack the victim [4], but above all from prejudices and stereotypes: homophobicals (homophobic bullying), ethnicals (cultural prejudices), from psycho-physical discards perceived towards the intended victim (towards a disabled person or towards gifted children, more gifted on a cognitive level). Cyberbullying [8] represents the last frontier of this behavioral phenomenology which is expressed through social networks and which takes on more obscure and pervasive traits because the bully can attack his victim through slander, dissemination of images, videos or sexually explicit contents without the consent of the interested part (revenge porn).

The victims in these cases are mostly women and the bullying tends to humiliate and harm them in their own image and dignity also to conditioning their school or working life [4].

Olweus [9] identified aggressiveness and insensitivity as motivational patterns of bullying behavior. Low levels of empathy combined with aggressive intentionality are always found in the analysis of the phenomenologies of bullying behaviors. This lack has been highlighted by numerous psychological studies [7, 10], from which it is possible to deduce the total lack of empathy in both conditions of victim or bully caused by a difficulty in managing emotions. Bullying also has specific characteristics and in order for this phenomenon to be talked about, the following aspects must emerge: intentionality, asymmetry of power between the parties, notoriety of the act, repetitiveness, vulnerability. Based on the attitude assumed by the bully [11], that is, the one who effectively bullies others, different categories are outlined: the dominant bully, gregarious, victim, proactive, reactive one. In turn, the victim can be: passive or provocative.

Bullying is therefore characterized as a dynamic and relational phenomenon which, in addition to stiffening the link between prevaricator and prevaricated, respectively imprisoned in the role of bully and victim, systematically involves the entire group, which very often coincides with the class, with students of the same school, with teachers, but also with other protagonists of the school, educational (Managers, administrative and technical staff, parents) and territorial network. It appears as an articulated, complex and stratified phenomenon on several levels: the social, political and cultural level; the institutional one; the family one; the group one; the individual one [12].

The role of the context is crucial in curbing bullying dynamics and introducing transformative variables. The school and/or educational institution therefore plays a key role in preventing the phenomenon, especially if the educational intervention starts early and not in the older age groups where the phenomenon is already consolidated. Therefore, it is necessary to think about preventive approaches that start from early years of schooling [4].

In Italy, with the approval of the Guidelines adopted in application of the law of 20th August 2019, n. 92, the fight against bullying and cyberbullying events has fully entered the school educational action, also thanks to the updating of school curricula and the integration of the educational co-responsibility pact, extended to primary school courses. The Law of 29th May 2017 n. 71 containing "Provisions for the protection of minors for the prevention and contrast of the phenomenon of cyberbullying" filled a pre-existing regulatory void on the instruments of prevention and contrast to the phenomenon, often acted by minors unaware of the criminal relevance of their actions and, for this reason, to be educationally contrasted as early as possible.

Pedagogically coping with bullying and cyberbullying means choosing to accept the realities of these phenomena with an above all observational and receptive posture [4, 13–15]. To face and solve these challenges, adaptive, complex, inclusive, context-based and group-oriented approaches must be adopted rather than the bully-victim dyad. That is, it is necessary to assume the cornerstones of an inclusive action that involves the entire school and / or educational institution and the students' families [16–20].

2 Method

We present the results of a collaborative research-training conducted according to a self-study phenomenological methodological protocol [15, 21] (based on the analysis of practices and on the theoretical framework of professional didactics [15, 22–25]. The research, promoted by the University of Bari (Italy), began with a joint reflection between school leaders of the network of schools involved, teachers and university researchers. The main purpose was to experiment professional development itineraries [26] for the construction of skills and tools that could lead to the improvement of the observation and intervention skills of the teachers involved [25].

Among other objectives pursued by this project there were the education to civil and democratic coexistence, understood as an instrument for the protection of one's own rights, for the recognition, protection and enhancement of the rights of others as well as for the recognition of the equal dignity of everyone and the creation of a common relational place in which the aspirations and divergences of each one could find "home", the promotion of inclusion in a horizontal way, the support of attitudes of mutual support among the students involved and, more generally, between the student population, and education to intercultural communication and to enhancement of diversity. Finally, the research aimed to promote the culture of legality also through the knowledge of crimes related to the network and the careless use of social networks.

The construction of the survey protocol was based on three axes: legislation, actions, monitoring and evaluation. Initially teachers and students were invited to read and learn about the relevant legislation and in particular the Law of 29th May 2017 n. 71 containing "Provisions for the protection of minors for the prevention and contrast of the phenomenon of cyberbullying", the Guidelines for the prevention and contrast of the phenomenon of cyberbullying "(2017, 2021), the Guidelines for the positive use of digital technologies and risk prevention in schools (2019)".

After the in-depth study of the legislation, the protocol provided for five actions of an ideational-creative nature: a training action for teachers identified by the Managers and identification of opinion leaders by teachers, a training action for opinion leaders, an action to activate workshops conducted by opinion leaders with the support of trainers, an action to create a commercial.

The whole process was subjected to monitoring-evaluation actions aimed at verifying the effectiveness of the educational action and recalibrating the intervention based on the results emerging from the research. The monitoring tools used were: questionnaires for pupils and teachers, video recordings of laboratory sessions, trainers' reports, teacher-tutor logbook, researcher monitoring sheets.

The phases of the research are summarized below (Table 1):

Table 1. Research phases

I	Exploratory research by the Department of Education, Psychology and Communication at University of Bari Aldo Moro: mapping of the dynamics between peers inside and outside the school context; emergence "from below" of representations, dynamics and interactions between adolescents. Administration of the entry questionnaire
II	Selection of opinion leaders identified within the student population of classes I (for second grade educational institutions) and classes III (for first grade educational institutions)
III	Training of opinion leaders by raising awareness of the issue and building tools to act within their school and/or class
IV	Experimentation of workshops managed by opinion leaders with the tutorial support of trainers, administration of exit questionnaire and evaluation forms
V	Data analysis and synthesis

Some partial results of the experimentation are described, relating to the answers of the pre and post intervention questionnaires.

3   Data Analysis

The pre and post intervention questionnaires were administered to students and teachers of the six schools involved. 830 students (86.5% of the total) and 130 teachers (53%) answered the questionnaire. The percentage of teachers who participated overall is lower than that of students. It should be noted that the distribution by age and gender of the sample of students is fair, with no appreciable differences in age; 420 male pupils (51%) and 410 female pupils (49%).

The participation of teachers was 16% for secondary school in grade I and 37% for secondary school teachers in grade II. It should be noted that the distribution by gender of the sample of teachers is fair.

The first questions of the questionnaire intended to probe the perceived and declared relationship with regard to "classroom well-being". The answers show that the well-being within the class group is very high. 49.5% indicated that they were "very well" with their mates, 44.6% "generally well" (and only 5.9% answered "generally badly/very badly"). Looking at the relationship with other students of the same school, there is a difference to pay attention as researchers. In fact, those same students who get on very well with classmates say they are generally happy, 74.7%, with boys from the same school. Although this data is positive, it reveals a veiled note of problematicity that is explicitly declared only by 21.1% of the students. The second block of questions was intended to probe the perception that students have towards bullying, those who exercise it and their attitude in responding (Fig. 1).

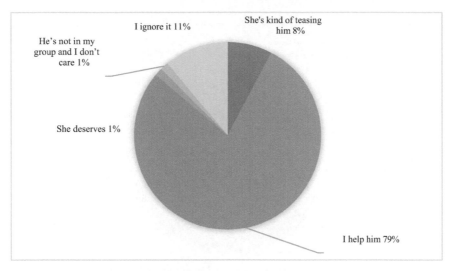

Fig. 1. Students' perception in relation to bullying.

It can be noted that compared to 59.4% who declare that they are not bullied, 11.8% declare that they have been ill, 9.7% feel helpless and without someone who could help them, 2.4% worried about what others were saying about them.

As for the analysis of the behavior of students who assist bullying, 52.2% declared that their friends take their side; 21.1% feel sorry but are afraid to intervene, 14% show disinterest in him and his state of mind. 48.4% declare that the adult intervenes to defend the victim, 27.5% that the adult is never present, 16.6%, pretend not to see (6.5%), laughs and has fun with the boys (1%). The answers highlight an adult attitude that is not always welcoming to the needs of all students. Looking at the extra-school as well, 39.5% of students confide in their parents, 34% with a friend, 17.8% with friends who attend outside the school; few respond with other adults (2%) or with other family members (6.7%).

Therefore, another sign of difficulty in the relationship between teachers and pupils concerns the attitudes of the young victims with respect to the different strategies for solving the problem. 57.4% of pupils admit that in their school there is someone who bullies. 32.4% say they don't know. Only 10.2% of students said there was no bully. From the answers to the question What kind of bullying is done? It is well understood how the students are aware that even "teasing" - if systematic and done with the aim of hurting the other - is an arrogance, not surprisingly it was reported by 40.4% of the participants. The other interesting data are the offenses and insults, chosen by 25.1% of the sample and the exclusion from the group chosen by 11.9%. Investigating the fear of being bullied at school, students who are not afraid are 39.4% compared to 43.8% who say they are a little afraid, 9.9% who say they have enough fear and 6.9% who are very afraid. This fear makes us understand that the problem in schools is real. Analyzing the reasons that according to teachers push a student to use a violent mode of interaction we observe (Fig. 2):

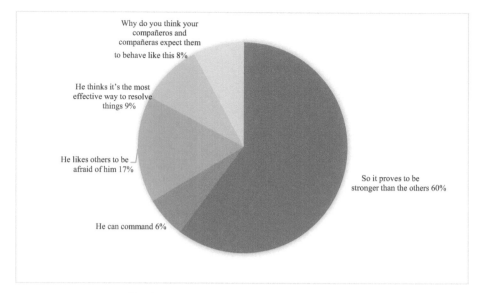

Fig. 2. Motivations mode of violent interaction.

Surely teachers are aware that the phenomenon exists in their classrooms as to the question Have there been bullying incidents among the pupils in your class in the last three months? 56% answered no, 34.3% answered yes and 9.7% I don't know: these percentages are quite high if we consider that they refer to a limited observation period, that is three months. When asked What are causes of bullying in their opinion, 51.3% attribute it to social causes; there follow causes related to the character of the protagonists for 30.8% and familiar causes for 14.5% follow; only 3.4% attribute a responsibility even to the school. The latter data underlines that the quality of the relationship with teachers is considered by themselves as one of the components that most contribute to a student's academic adaptation. 47.6% of the interviewed teachers believe that students ask for help to the teachers, for 31.7% they call on their parents and 20.6% believe that they call on their classmates.

4   Conclusions

From the monitoring and summary sheets of experts and researchers, it is clear that the approach proposed for the prevention and contrast of bullying and cyberbullying was indirect. An attempt was made to accompany girls and boys towards the issues covered by this path, through reflection on their own identity, by formalizing a device that makes it usable and clear all the constituent variables of the learning environment, the various steps and stages aimed at achieving certain goals, provides for the assignment of precise roles and functions to all the actors involved in the project action, establishes content, tasks and responsibilities, the actions that are carried out, the relational networks that are established in the situation, documents the products of the path. It offers a guide for co-constructing didactic and training planning, in the awareness, however, that its character is situated and not universal, therefore it is impossible to indicate activities and contents valid for all school institutions.

The project has allowed the achievement of some important results: the knowledge of the consistency of the bullying and cyber-bullying phenomena among adolescents of the schools participating in the network; raising the awareness of all actors in the school context - students, teachers, managers, families, administrative and technical staff - to non-violent conflict management approaches and methods of participation that counter homologation and discrimination of adolescents; the empowerment of young people towards the conscious use of technological media and social networks, deepening the possible risks and promoting their unexpressed potential also enhancing useful, constructive, funny and educational actions and navigation methods; education for civil and democratic coexistence: understood as an instrument for the protection of one's rights, for the recognition, protection and enhancement of the rights of others as well as for the recognition of everyone's equal dignity and for the creation of a common relational place in which one's aspirations and diversity find "home"; the promotion of the culture of legality through the knowledge of crimes related to the web and the social networks; promoting inclusion: horizontally, supporting mentoring attitudes among pupils and educating to intercultural communication and enhancement of diversity [27].

The experimentation suggests the opportunity to create itineraries for the construction of educational and didactic conditions to combat bullying and cyberbullying and to promote specific relational and clinical trainings for teachers and educators.

References

1. Perla, L.: Appunti per una teoria mediale dell'insegnamento di educazione civica. Tratteggio didattico. In: Rivista Lasalliana, vol. 88, no. 3, pp. 351–356. Provincia Italia, Torino (2021)
2. Scarinci, A., Fornasari, A., Massaro, S., Perla, L.: The place-based approach for citizenship education: new didactic scenarios. In: Proceedings of the 2nd International Conference of the Journal Scuola Democratica Reinventing Education, Citizenship, Work and The Global Age, vol. I, pp. 407–416. Il Mulino, Bologna (2021)
3. Perla, L., Agrati, L.S., Scarinci, A., Amati, I., Santacroce, M.: Riscrivere il Patto di corresponsabilità per l'insegnamento di educazione Civica: prime risultanze di una ricerca-formazione. In: Ricerca e Didattica per promuovere intelligenza comprensione e partecipazione, Proceeding X Congresso scientifico SIRD, II tomo, pp. 250–266. Pensa Multimedia, Lecce (2021)
4. Perla, L.: La forza mite dell'educazione. Contro bullismo e cyberbullismo. In: La forza mite dell'educazione. Un dispositivo pedagogico di contrasto al bullismo e cyberbullismo. FrancoAngeli, Milano (2021, in press)
5. Fedeli, D., Munaro, C.: Bullismo e cyberbullismo. Come intervenire nei contesti scolastici. FrancoAngeli, Milano (2010)
6. Farrington, D.P.: Understanding and preventing bullying. In: Tonny, M., Morris, N. (eds.) Crime and Justice, vol. 17. University of Chicago Press, Chicago (1993)
7. Olweus, D.: Bullismo a scuola. Ragazzi oppressi, ragazzi che opprimono. Giunti, Firenze (1996)
8. Genta, M.L.: Bullismo e cyberbullismo. FrancoAngeli, Milano (2017)
9. Olweus, D.: Aggression in the Schools: Bullies and Whipping Boys. Hemisphere, New York (1978)

10. Menesini, E., Nocentini, A., Palladino, B.E.: Prevenire e contrastare il bullismo e il cyberbullismo. Il Mulino, Bologna (2017)
11. Daffi, G., Prandolini, C.: Mio figlio è un bullo? Soluzioni per genitori e insegnanti. Erikson, Trento (2012)
12. Pennetta, A.L.: La responsabilità giuridica per atti di bullismo. Giappichelli-Linea Professionale, Torino (2014)
13. Massa, R., Cerioli, L.: Sottobanco. Le dimensioni nascoste della vita scolastica. FrancoAngeli, Milano (1999)
14. Riva, M.G.: Il lavoro pedagogico come ricerca dei significati e ascolto delle emozioni. Guerini Scientifica, Milano (2004)
15. Perla, L.: Didattica dell'implicito. Ciò che l'insegnante non sa. La Scuola, Brescia (2010)
16. Perla, L.: Per una didattica dell'inclusione. Orientamenti per l'azione. Pensa Multimedia, Lecce-Bergamo (2013)
17. Cottini, L.: Didattica speciale e inclusione scolastica. Carocci, Roma (2017)
18. D'Alonzo, L., Bocci, F., Pinnelli, S.: Didattica speciale per l'inclusione. La Scuola, Brescia (2015)
19. Medeghini, R.: Disability Studies. Emancipazione, inclusione scolastica e sociale, cittadinanza. Erickson, Trento (2013)
20. Medeghini, R.: Norma e normalità nei Disability Studies. Riflessioni e analisi critica per ripensare la disabilità. Erickson, Trento (2015)
21. Loughran, J.: Researching teacher education practices: responding to the challenges, demands, and expectations of self-study. J. Teacher Educ. **58**, 12–20 (2007)
22. Altet, M.: La ricerca sulle pratiche d'insegnamento in Francia. La Scuola, Brescia (2003)
23. Vinatier, I., Altet, M.: Analyser et comprendre la pratique enseignante. Presses universitaires de Rennes, Rennes (2008)
24. Laneve, C.: La didattica fra teoria e pratica. La Scuola, Brescia (2005)
25. Maubant, P., Martineau, S.: Fondements des pratiques professionnelles des enseignants. Les Presses de l'Université d'Ottawa, Ottawa (2011)
26. Perla, L.: Scritture professionali. Metodi per la formazione. Progedit, Bari (2012)
27. Amati, I., Santacroce, M.T.: Bullismo e cyberbullismo a scuola. In: La forza mite dell'educazione. Un dispositivo pedagogico di contrasto al bullismo e cyberbullismo. FrancoAngeli, Milano (2021, in press)

Digital Competence for Citizenship: Distance Learning Before and During the Covid-19 Emergency

Maria Sammarro[(⊠)] [iD]

Mediterranea University of Reggio Calabria, Reggio Calabria, Italy
maria.sammarro@unirc.it

Abstract. The Covid-19 emergency is undoubtedly an exceptional situation that has forced us all to reshape and transform spaces, times and activities of all kinds. Even the school, driven by need and emergency, has re-thought its teaching practices through a re-evaluation of digital technology and organizational experimentation. An opportunity to keep relationships alive, beyond distance. The presented study analyzes the perceptions of a sample of secondary school teachers about online learning and the use of new technologies, comparing the teaching practices adopted in the pre-pandemic Covid-19 period, during the lockdown and during the return to presence. The results highlight both the critical issues that emerged during distance learning – digital divide, accessibility, workload, support for students with Special Educational Needs, the teacher-learner relationship, digital skills – and the positive aspects – flexibility, availability of tools, adaptability, innovation, student self-determination and self-regulation. This is all in line with a media education that allows the potential of digital devices to be exploited in education and to support the digital skills of teachers and pupils.

Keywords: Distance learning · Digital competence · Citizenship

1 Theoretical Framework

The emergency from Covid-19 has caused an unprecedented shock wave, investing contemporary society and determining a change of course, a metamorphosis at all levels. The fragility that has emerged has not spared the school world [1, 2]; the sudden closure of the school in presence has represented a critical moment for every teacher and for own idea of teaching [3]. A new and unexpected event that has generated great turmoil in the school world because the assumptions within which the educational experience is realized, namely presence, contact, relationship, motivation of students and teachers, have disappeared. Everything that was taken for granted in the face-to-face meeting suddenly became something precious to seek and preserve with all available means. Thus, it became necessary to rethink the teaching strategies to be adopted [4], a reshaping of the educational program, a re-evaluation of digital [5] – from multimedia language in the classroom to electronic devices, such as Interactive whiteboards and tablets, from 2.0 classrooms to the culture of eLearning, to the use of technology in the service of

F. Calabrò et al. (Eds.): NMP 2022, LNNS 482, pp. 887–895, 2022.
https://doi.org/10.1007/978-3-031-06825-6_84

learning [6] – to transform what had been proposed as distance education into "prox-imity education" [7]; all this to create a new educational alliance, paradoxically even stronger because it is united by the particular situation of reduced social contact and the emotional experience of the moment. Even in the distance, it was necessary to keep alive the motivation of the students in order to cope with frustrations, distractions, external stresses: the students were asked to resist, to be resilient, to get up in front of life's events.

In this hectic phase, the new technologies have played a crucial and decisive role as they have prevented us from being deprived of everything that school represents, keeping alive, albeit at a distance, relationship between teacher and learner, in line with what Moore [8] states through the *phenomenology of distance*: there is a "distance in presence" – one can be close physically, but distant affectively, cognitively, relationally – but also a "proximity in spatial distance" – being empathically and intellectually close to people who are physically distant and connected virtually through technologies. As Rivoltella states, in fact, «the relationship does not depend on distance or presence, but on the intentionality of the speaker and the ability to translate it into gestures. One can be relational at a distance and be surprised to be able, despite the distance, to develop a relationality that does not always happen in presence. It is the educational intentionality that translates it into practice, that makes the relationship» [5].

Therefore, precisely because proximity is not a guarantee of presence, in an educa-tional context one cannot rely on the idea of presence by disconnecting it from participa-tion and involvement. Among the main dimensions of teaching redefined by technology we find participation: if conventionally "participate" means to be present in a certain place and time with other people, the means of communication have led to a rethinking of time and space, as it is possible to attend a lesson at a distance and have new partici-patory experiences. Among the strengths of the presence we can include bodily contact, construction and manipulation of objects, sharing of physical and sensory sensations; as regards the potential of distance, however, technologies allow to expand the time and space of learning, give the opportunity to participate with people far away or unable to move, allow synchronous and asynchronous communication, redefine teaching prac-tices, for example through the flipped classroom, which provides a reversal of the classic phases of work and distinction between autonomous activities and in presence [9].

The role of the teacher is redefined as a designer of learning [10] within an integrated multimedia environment for teaching: if properly trained, he becomes a point of refer-ence to promote students' digital skills and make them aware of the risks and potential of new digital media, through a responsible and reflective use of the same, accepting the new information coming from the web with a critical spirit. Rivoltella [11] introduces the figure of the technological teacher, identifying four different profiles: the first is the teacher who uses technologies in ordinary teaching, adopting a reflective attitude [12] and searching for the most suitable solution for each situation; the second is the teacher who uses Network technologies for communication with his students, their parents and colleagues, through different possibilities of interaction made available by ICT; the third is the teacher who builds a learning environment with a strong technological character-ization, organized around Network communication; the fourth profile is related to the teacher who conducts courses or workshops on technologies, maintaining a critical look at the discipline.

As supported by recent theories on Evidence-based education [13], a good teacher, in addition to taking care of students, must be able to master planning, teaching, organizational, evaluative and psycho-relational skills, resulting from integration between theoretical knowledge and operational skills. Evidence [14], in fact, shows that the mere use of new technologies does not imply a positive impact on students' learning, but it is necessary that there is a focused and competent planning of the digital world. Crucial, therefore, is the training of teachers since «technologies alone do not work if teachers do not know how to define well the objectives and implement them, so it is necessary that the usual frameworks of digital competence are integrated with references of a didactic and pedagogical nature» [9]. In this perspective stands the TPACK Model of Education, Technology, Pedagogy and Content Knowledge, proposed by Mishra and Koehler [15] starting from Shulman's studies, according to which teachers should be able to master both disciplinary and pedagogical content through the integration of three components: Content Knowledge (CK), Pedagogical Knowledge (PK), and Technology Knowledge (TK).

As mentioned above, it is necessary that there is continuous training on digital issues and, above all, that there are teachers, educators, trainers who, embodying the role of mediators, are able to filter or otherwise provide tools to be able to untangle within this deluge of information [16]. In a society in which «the diffusion of intelligent automata, the availability of any kind of data, the processes of disintermediation of communication favor the development of forms of self-learning and self-literacy», in a society in which «it is difficult to distinguish what is information and what is not, what information is reliable, to which authors this reliable information can be attributed» [17], didactic and educational mediation plays a crucial role.

2   Method

On the basis of these premises, we present the partial results of a survey conducted at the Liceo Scientifico Guerrisi of Cittanova (RC), at the end of a training experience aimed at in-service teachers on the promotion of digital skills and the use of new technologies.

The purpose of this research is to analyze teachers' experiences, perceptions, and perspectives on distance education during the COVID-19 period [3], the methodologies and tools used, and to compare teaching practices used in the pre-Covid-19 period, during the lockdown, and upon return to presence. The questions were worded to elicit completely anonymous responses, and the information provided in the responses was processed in an aggregate manner. The areas investigated concern the possible reshaping of the teaching program, the technological tools used, the ways in which teaching is carried out, the teaching strategies used, the critical issues encountered in DAD. The questionnaire consists of 20 multiple choice questions administered to a reference sample of 37 teachers.

3   Data Analysis

With regard to the reference sample, 78.4% of those interviewed were female, 21.6% male; 18.9% of the sample consisted of teachers in the 30–40 age bracket, 27% of teachers in the 41–49 age bracket, 40.5% of teachers in the 50–60 age bracket and 13.5% of teachers over 60.

Among the Areas of Expertise (DigComp 2.1) in relation to the use of ICT in teaching, particularly relevant are those related to Communication and Collaboration (66.7%), Information and Data Literacy (50%), Digital Content Creation (33.3%) and Security (33.3%).

As far as the use of new technologies in the pre-pandemic period of Covid-19 is concerned, 81.1% used them mainly for educational purposes, 56.8% for social relations, 29.7% for entertainment, 13.5% for recreational activities. Going into more detail, 19.4% have used them in class every day, 47.2% often, 30.6% rarely, while 3% have never used them in a teaching context; in particular, 33.3% have exploited the potential of the Net in a teaching context to view multimedia content, 27.8% to share teaching materials, 25% to search for information, 13.9% to manage virtual learning environments.

The devices most frequently used in the educational sphere, again before the Covid-19 pandemic, were the PC (38.9%), the interactive whiteboard (33.3%), the smartphone (13.9%) and the tablet (11.1%).

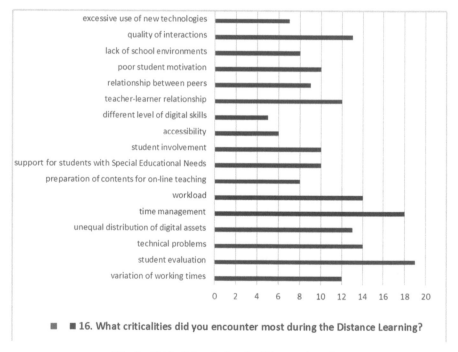

Fig. 1. Criticalities during the Distance learning.

In full lockdown from Covid-19, a remodeling of the didactic programming was necessary for 51.4%, while 40.5% felt it was necessary to partially remodel what had already been programmed, and finally 8.1% responded negatively. In particular, during Distance Learning, among the teaching strategies adopted we find the participatory lesson (70.3%), the transmissive frontal lesson (64.9%), debate (43.2%), problem solving

(35.1%), brainstorming (32.4%), cooperative learning (24.3%) and flipped classroom (18.9%). In any case, during the emergency period, distance learning proved particularly effective for sharing study materials (89.2%), creating work groups (45%), creating and sharing structured or semi-structured tests (48.6%), carrying out school-family communications (40.5%), communicating via chat (37.8%), creating and sharing videos (35.1%), structuring discussion groups (5.4%).

Among the critical issues encountered during the implementation of distance learning, due to the closure of schools, we can include the digital divide, accessibility, workload, support for students with Special Educational Needs, the teacher-student relationship, the different level of digital skills, the variation in work time, student assessment, technical problems, preparation of content for online teaching, student engagement, peer relationship, low student motivation, lack of school environments, quality of interactions, excessive use of new technologies (Fig. 1).

Among the positive aspects, on the other hand, that have pleasantly impressed teachers we find flexibility, the possibility of maintaining contact with students, albeit at a distance, the wide range of tools available, adaptability (i.e., the ability to customize learning for students), innovation (i.e., the freedom to experiment with teaching), the ability to easily access platforms, materials and multimedia resources, participation, autonomy and motivation of students, as well as their self-determination and self-regulation (Fig. 2).

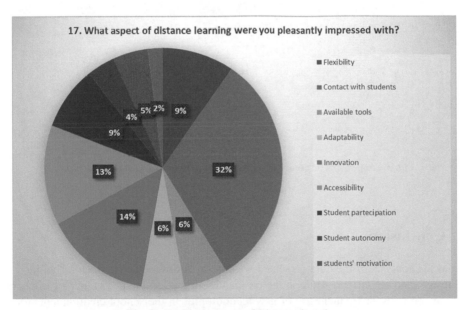

Fig. 2. Positive aspects of Distance learning.

With regard to online teaching, 43.2% believe that it could be regularly integrated into teaching even in normal conditions, 37.8% think that in normal conditions it can be useful only for special needs, while for 18.9% it should be used only in emergency conditions. These data are in line with what emerges in question no. 18, in which it is asked to indicate the exact definition of the didactics carried out during the pandemic; the answers provided highlight discordant opinions and diversified views: "Distance learning" (32.4%), "Distant learning" (8.1%), "Proximity learning" (13.5%), "Emergency learning" (32.4%), "Continuity of learning" (5.4%), "We stay together not-so-all" (2.7%), "Integrated digital learning" (2.7%), "Learning for the few" (2.7%).

With regard to the return to presence, after the experience of Distance learning, it is necessary to focus more on re-establishing the teacher-learner relationship, to enhance and improve the digital skills of teachers and pupils, to integrate the experience of online teaching with in-presence teaching, to manage the emotional climate in the classroom, to avert the "fear of others", to promote resilience (Fig. 3).

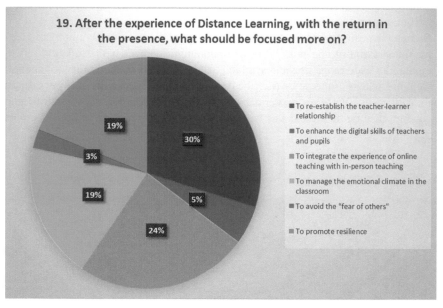

Fig. 3. Aspects to be considered after the experience of Distance learning.

4 Conclusion

From the analysis of the data presented we can see a difference between a "before", a "during", and an "after": while in the pre-pandemic period the percentage of teachers who made daily or frequent use of new technologies in the school environment stood at 65%, mainly to share multimedia resources or search for information, in the midst of the Covid-19 emergency all the teachers interviewed used digital devices to be able to ensure teaching and keep the teacher-learner relationship active [18]. Certainly, cooperative teaching strategies have been affected, in favor of frontal lessons and participatory lessons. The return to face-to-face teaching does not necessarily coincide with a mere return to the past, in fact, about half of the sample is convinced that it is possible to exploit, even in face-to-face teaching, the positive aspects experimented during distance teaching; while the remaining part believes that distance teaching should be used only for special needs or in emergency conditions.

The data, after all, reflect the contrasting opinions that emerge daily regarding the use of new technologies, the dichotomy between those who demonize [19] digital tools - because they destroy social relationships, promote the dangers of the network and generate addiction - and those, instead, consider them an educational resource, promoters of good practices that can be used at the educational level to create effective learning environments. Accepting the greek ideal of the measure [20], it is necessary to pursue a third way that manages to find a balance between the two positions, which makes us aware of the risks and potential of technological tools and social media, through a responsible and critical use of them.

As Rivoltella says, echoing the thought of Paulo Freire, "one of the saddest things for a human being is not to belong to his time", showing skepticism towards digital technology or eliminating new technologies from the educational environment: a school that stands on these assumptions fails to fully interpret and experience contemporaneity.

The pandemic emergency has taught us not to remain anchored to the past, to conventional practices, but has pushed us to go beyond, to dare, to accept and welcome change, to rethink teaching, to experiment even in the school environment, treasuring all that was positive during this time and trying to insert it into the new process. Moreover, it is precisely in conditions of normality that we need to reflect on the critical issues that have emerged in order to overcome them and create new opportunities: «to pose the problem of the gap and eliminate it, to realize that many people do not have connections, do not have tools, do not possess the alphabets, and to create the conditions for these impediments to be overcome» [21].

Therefore, there is a need for a media education [22] that, in line with European frameworks such as DigComp and DigCompEdu, allows to exploit the potential of digital devices in education and to support the digital skills of teachers and students, promoting digital and media literacy [23, 24] (Fig. 4).

The development of full digital citizenship depends on students' ability to take ownership of digital media, moving from passive consumers to critical consumers and responsible producers of content and new architectures. From the critical spirit and responsibility comes the ability to know how to make the most of the potential of technology, in terms of education, participation, creativity and sociality, minimizing the negative ones, such as social addiction [25, 26], cyberbullying [27], hate speech [28, 29], illegal online behaviors [30]. The same definition of digital citizenship, an expression increasingly used in recent years within the debate on new technologies and everyday life, indicates «the ability of an individual to consciously and responsibly use virtual means of communication, with a view to the development of critical thinking, awareness of the possible risks associated with the use of social media and surfing the Web, and contrasting hate speech»; together with "Constitution" and "Sustainable Development", represents one of the three axes of the macro-category of Civic Education, whose latest definition is contained in the Guidelines for the teaching of Civic Education, in application of Law 92 of 20/08/19.

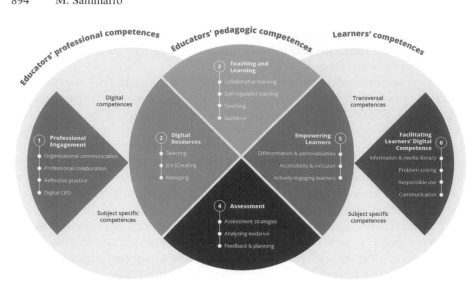

Fig. 4. DigCompEdu (Digital Competence Framework for Educators)

References

1. Huang, R.H., et al.: Handbook on Facilitating Flexible Learning During Educational Disruption. The Chinese Experience in Maintaining Undisrupted Learning in COVID-19 Outbreak. Smart Learning Institute of Beijing Normal University, Beijing (2020)
2. Arora, A.K., Srinivasan, R.: Impact of pandemic COVID-19 on the teaching–learning process: a study of higher education teachers. Prabandhan Ind. J. Manage. **13**(4), 43–56 (2020)
3. Hebebci, M.T., Bertiz, Y., Alan, S.: Investigation of views of students and teachers on distance education practices during the Coronavirus (COVID-19) pandemic. Int. J. Technol. Educ. Sci. (IJTES) **4**(4), 267–282 (2020)
4. Unger, S., Meiran, W.R.: Student attitudes towards online education during the COVID-19 viral outbreak of 2020: distance learning in a time of social distance. Int. J. Technol. Educ. Sci. (IJTES) **4**(4), 256–266 (2020)
5. Lee, M.J.W., McLoughlin, C.: Beyond distance and time constraints: applying social networking tools and web 2.0 approaches to distance learning. In: Veletsianos, G. (ed.) Emerging Technologies in Distance Education, pp. 61–87. Athabasca University Press, Edmonton, AB (2010)
6. Rivoltella, P.C.: COVID-19, tra didattica a distanza, eLearning e tecnologie di comunità. Avvenire (2020)
7. Iosa, R., https://www.erickson.it/it/mondo-erickson/articoli/settimana-slow-scuola-vicina nza/. Accessed 21 Dec 2021
8. Moore, M.G.: The death of distance. Am. J. Distance Educ. **9**(3), 1–4 (1995)
9. Bonaiuti, G., Calvani, A., Menichetti, L., Vivanet, G.: Le tecnologie educative. Carocci, Roma (2017)
10. Limone, P.P.: Media, tecnologie e scuola. Per una nuova cittadinanza digitale. Progedit, Bari (2012)
11. Ardizzone, P., Rivoltella, P.C.: Media e tecnologie per la didattica. Vita e Pensiero, Milano (2007)
12. Dewey, J.: Come pensiamo. Una riformulazione del rapporto fra il pensiero riflessivo e l'educazione. La Nuova Italia, Firenze (1961)

13. Hattie, J.: Apprendimento visibile, insegnamento efficace. Metodi e strategie di successo dalla ricerca evidence-based. Erickson, Trento (2016)
14. Calvani, A., Vivanet, G.: Tecnologie per apprendere: quale il ruolo dell'Evidence based education? J. Educ. Cult. Psychol. Stud. (ECPS J.) **10**, 83–112 (2014)
15. Mishra, P., Koehler, M.J.: Technological pedagogical content knowledge. A framework for teacher knowledge. Teachers College Rec. **108**(6), 1017–1054 (2006)
16. Levy, P.: Intelligenza collettiva. Per un'antropologia del cyberspazio. Feltrinelli, Milano (2002)
17. Rivoltella, P.C.: Nuovi alfabeti. Morcelliana, Brescia (2020)
18. Dhawan, S.: Online learning: a panacea in the time of COVID-19 crisis. J. Educ. Technol. Syst. **49**(1), 5–22 (2020)
19. Quèau, P.: Le virtuel. Vertus et vertiges. Champ Vallon, Ceyzérieu (1993)
20. Rivoltella, P.C.: Le virtù del digitale. Per un'etica dei media. Morcelliana, Brescia (2015)
21. Rivoltella, P.C.: COVID-19, tra didattica a distanza, eLearning e tecnologie di comunità. Avvenire (2020)
22. Cheung, C.K.: Web 2.0: challenges and opportunities for media education and beyond. E-learn. Digit. Media **7**(4), 328–337 (2010)
23. Ranieri, M., Bruni, I.: Digital and media literacy in teacher education: preparing undergraduate teachers through an academic program on digital storytelling. In: Cubbage, J. (ed.) Media Literacy in Higher Education Environments, pp. 90–111. IGI, Hershey, PA (2018)
24. Ranieri, M., Bruni, I., Orban de Xivry, A.C.: Teachers' professional development on digital and media literacy. Findings and recommendations from a European project. REM-Res. Educ. Media **10**, 10–19 (2017)
25. Zona, U.: Narcissus on the net. Narcisismo digitale e seduzione della merce. IAT J. **2** (2015)
26. Papacharissi, Z.: A networked self: identity, community, and culture on social network sites. Routledge, Londra (2010)
27. Smith, P.K., del Barrio, C., Tokunaga, R.S.: Definitions of bullying and cyberbullying: how useful are the terms? In: Bauman, S., Cross, D., Walker, J. (eds.) Routledge Monographs in Mental Health. Principles of Cyberbullying Research: Definitions, Measures, and Methodology, pp. 26–40. Taylor & Francis Group, Routledge (2013)
28. Pasta, S.: Razzismi 2.0. Analisi socio-educativa dell'odio online. Morcelliana, Brescia (2018)
29. Santerini, M.: La mente ostile. Forme dell'odio contemporaneo. Raffaello Cortina editore, Milano (2021)
30. Boyd, D.: It's Complicated: The Social Lives of Networked Teens. University Press, Yale (2014)

Juveniles and Mafias: The Project Free to Choose for *Growing up Differently*

Giuseppina Maria Patrizia Surace(✉) (iD)

University of Foggia, Foggia, Italy
patriziasuracegm@gmail.com

Abstract. The preeminent interest in the child represents the general informer principle of the entire system of the protection of minors. In the Italian context, the Constitution, in the articles 2, 30, 31, 34, outlines the "alternative" guarantees to set up in cases of parental inability, deciding on the necessity to exceptionally provide suitable steps that the judicial authority will arrange in order to insure the efficacy of the educative duties and assistance. The Juvenile Court of Reggio Calabria addresses the problem with a systemic intervention in a preventive prospective of the social marginality, with regard to minors coming from organized crime contexts. The project 'Free to choose' aims to guarantee the effective and targeted actions for the safeguarding of the regular psycho-physical growth of the minors coming from this family context. The aim is to interrupt the perverse spiral that feeds the use of the most precious human patrimony represented by children and teenagers, employed in carrying out criminal activities and for the capillary reproduction of the mafia power. The Juvenile Court of Reggio Calabria, in the last eight years, has ordered the forfeiture or limitation of parental responsibility, of those who belong to criminal or 'ndranghetist'-like organizations, when there is a concrete prejudice to the psychophysical integrity of the minors. This happens in cases of indoctrination and of emulation of parental criminal behavior, namely of a real involvement of the teenagers in illicit affairs by the reference adults, including the participation of young people in local feuds.

Keywords: Minors' integrity · Parental responsibility · Mafia power

1   Introduction

Mafia-type criminal organisations have been able to update and interact with evolved social models, preserving the status quo and acquiring some modern elements to their advantage [1]. At the same time, they have acquired space by filling the gaps left by the State in those places - such as the South of Italy - where the institutions do not offer employment and socio-cultural integration. The mafias [2], in such contexts, are the driving force behind work activities in the illegal area and fully meet the expectations of belonging that many adolescents and children have.

Young people can be affiliated and employed in drug dealing or in extortion or damage, or they are fully involved in the dynamics of family associations through the commission of very serious crimes such as homicide. What does not change is the attraction that violence, immediate economic availability and leadership exerts on adolescents who, without the sacrifice of study or respect for rules, are introduced into a world of power and abuse where institutions are perceived as enemies.

At a cultural level, therefore, the mafias express a universe of values recognised by society and which has a contaminating effect, especially on minors. Recalling Habermas [3] thought on the concept of culture, as a set of implicit and explicit knowledge that affects the processes of understanding and learning, and specifically on the biographical situation as the "starting point for the reconstruction of the world of life", we focus on interventions that can be used in order to improve the understanding of the world of life, we focus on interventions that affect the symbolic apparatus of the mafias and weaken the sense of belonging that represents their vital energy. We are addressing in particular those children and young people who, from an early age, are in close contact, for family or environmental reasons, with organised crime contexts. However, a distinction must be made. In the case of minors who are gregarious due to their contiguity to the mafia context, we are faced with situations of socio-economic disadvantage and the frustrations deriving from a labelling social rigidity nurturing in these children the idea of negative myths and false values, allowing them to become convinced followers of the criminal circuits [4].

In Calabria [5], on the other hand, what is of utmost interest is the family context and the strong ties that distinguish it. As well as acting as a shield against the outside world, it is an instrument of alliance between the various families and of collaboration in important economic operations (arms and drug trafficking). Internally, it allows the transmission of the Mafioso sentiment, i.e. that set of values, beliefs and customs, typical of the culture and population of the territory, reinterpreted and recoded in a mystified, symbolic culture, with a constitutive value, acquired through internalisation and transformed into belonging, power, intimidation, loyalty and environmental recognition.

The familiar context of the 'ndrangheta', reproducing that territorial lordship typical of the environmental contexts in which criminal powerful families are rooted, assumes a paving role on the psychic structure of its components, especially on minors. In fact, there are no mediators or nuances, but everything revolves around dynamic dichotomous values that do not admit exceptions (life-death; we-they; friend-enemy...). The family, therefore, assumes a role of "matrix" qualitatively saturated, that cannot be refused, which organizes values such as loyalty, obedience, respect, friendship, honour, in such a way that these are lived for one's own Family or for the membership clan from which those mechanisms of moral release originate which underpin the criminological dimension of the mafiastyle organizations.

In other words, mafiosi also educate, transmit knowledge and values, guarantee the generational continuity of power, regenerating themselves through those who will come next. The judicial history of the Juvenile Court of Reggio Calabria (Southern Italy) confirms the direct involvement of minors in the criminal activities of the families [6].

In concrete terms, the mafias educate [7] and predetermine a future of thought, feeling and actions in the sense and sign of their tyrannical sovereignty [8]. The conviction that it is necessary to censure the mafia educational models, in the same way as one intervenes against other violent or abusive parents, has oriented the interventions of the Juvenile Judicial Office of Reggio Calabria.

What we have detected and continue to detect in the educational and judicial practice, in the history of every child, girl, boy or girl we know, is the anaesthetisation of emotions, an apparent coldness and anaffectivity that transforms them into non-persons, in order to allow them the exercise of violence as a form of indifference towards others and the only instrument of communication and relationship. In other words, these are children who are denied their childhood, the possibility of being other than the universe they come from, and whose destiny, already predetermined, oscillates between imprisonment or, even worse, death. It is a destiny they can hardly oppose, but against which it is necessary to ask whether these same children, having known other ways of growing up, would be willing to give up their freedom to choose who they want to be and become.

2 The Italian Legal Framework and the 'Free to Choose' Project

The best interest of the child is the general guiding principle, both international and Italian, of the whole system of protection and tutelage of minors. The interpretative turn in favour of a full protection of rights, summarised - in the last instance - in the "Guidelines of the Committee of Ministers of the Council of Europe on child-friendly justice"[1], highlights the recognition of a full protection of the child's rights [9]. In the Italian Constitution, the superiority of the child's right to a fair trial is recognised by the Court of Justice.

In the Italian Constitution, the superiority of the child's interest is a founding super-value, an expression of the constitutional centrality of human dignity and, therefore, of the same personalist principle, the latter understood as a synthesis and value of the entire constitutional structure.

When, in practice, the physical or psychic safety of minors is harmed, as can happen in the contexts we are talking about, the only choice is to intervene to protect these children[2]. In the current Italian legal system[3], the family is not merely an institution of social importance, but a place where the human personality is promoted, an ideal and protected space leading to full self-expression and the realisation of the interests of its family members, primarily children.

This means that parental responsibility is attributed in the interests of the child and their education cannot be dissociated from the general values of the community and the social structures of which the family is an integral part. At the same time, it must reflect the child's interest in being educated and socially instructed in order to become a citizen endowed with the necessary maturity to live in a democratic community[4], such as that which emerges from the Italian constitutional system [10].

---

[1] Adopted on 17 November 2010.

[2] This approach is unquestionably recognised in Italian case law, both substantive and of legitimacy.

[3] In particular, Articles 2 and 30 of the Constitution and Articles 147 and 315 bis of the Italian Civil Code.

[4] The same applies, in substance, to the Convention drawn up in New York in 1989, ratified by Italy with Law No 176 of 1991, which states that "parents must educate the child to become a useful member of society and develop his or her sense of moral and social responsibility".

In cases where parents systematically involve their minor child in delinquent activities, one is faced with a chosen pattern, an impediment that cannot be otherwise, i.e. that it must be as prescribed by the mafia family or community. However, there is no automatism based on the mere demonstration of a link between a family and the mafia order. Individualised interventions are modulated according to the specificities of each child and the seriousness of family situations. Sometimes it has not been necessary to order the removal of the child from the family context, as it is sufficient to place the child in a semi-boarding school in the relevant district or within the region. On the other hand, in the most serious situations, the temporary removal of the juvenile from the family context has been arranged, with placement in a community or in available foster families.

Even in extreme cases, at the same time as the removal order, the Court adopts restrictive or abrogative decrees of parental responsibility against one or both parents, if they do not significantly distance themselves from the Mafia cultural models. The prospect of contacts with the family and the re-establishment of relations with the parents remains firm, and these should be modulated on the basis of compliance with the prescriptions and their participation in the children's educational process.

The experience gained in recent years has highlighted the need to plan interventions that accompany mafia minors until they achieve existential and working autonomy. The ordinary public network has not always been able to guarantee the effectiveness of the interventions planned by the juvenile court. The peculiarity and complexity of the phenomenon require a more intensive educational approach through a targeted strategy that strengthens public resources or integrates them with those of the private social sector.

The first Italian governmental framework agreement, dating back to 2017, was aimed at developing an experimental programme to prevent social marginality through training and work opportunities to be implemented throughout the Calabrian territory. To strengthen the intervention tools for the protection of mafia minors, the protocol of 2.2.2018[5] was signed, between the Presidency of the Council of Ministers -Equal Opportunities Department-, the National Anti-Mafia Prosecutor's Office, the Court and the Prosecutor's Office for Minors of Reggio Calabria, the Public Prosecutor's Office at the Court of Reggio Calabria and the association Libera (associations, names and numbers against mafias), which gave a further push in the concrete direction of accompanying these young people towards change.

In 2019 (on 5 November) and subsequently with the updated version of 30 July 2020, the latest protocol was signed, to which the Ministry of Education, the Ministry of Scientific Research, the Ministry of Justice, the Ministry of the Interior, the Presidency of the Council, the Minister for Equal Opportunities and the Family and the C.E.I.

---

[5] The protocol was signed by the Presidency of the Council of Ministers -Department for Equal Opportunities-, the National Anti-Mafia Prosecutor's Office, the Court and the Juvenile Prosecutor's Office of Reggio Calabria, the Public Prosecutor's Office at the Court of Reggio Calabria and the association Libera (associations, names and numbers against mafias), which gave a further push in the concrete direction of accompanying these young people towards change. Although it did not subscribe to the 2018 protocol, the Italian Church, through the CEI (Italian Episcopal Conference), has actively joined the initiative, becoming a concrete supporter, with 8 per thousand funds, and a companion of the institutions in favour of the most fragile and courageous people.

were added, with the aim of ensuring a concrete alternative life for minors from families involved in organised crime or who are victims of mafia violence, and for family members who dissociate themselves from the criminal logic.

As we were saying, the protection network includes the presence of qualified volunteers through the association Libera and the Italian Committee for Unicef, the latter outside the national protocol, but already working alongside the Juvenile Court of Reggio Calabria since 2016.

At present, compared to the hundred or so cases dealt with over the course of eight to ten years, about fifty cases are still being followed by qualified volunteers and institutional agencies, including entire protected family units (especially mothers and minor children), very young people in the criminal circuit and children or young people in the community or with foster families. The involvement of these third-sector organisations refers to a pedagogy of solidarity in which volunteering, in a generative welfare system [11], is characterised by practical action imbued with humanity and ethical value. In our case, the systematic intervention of volunteers is an indispensable resource, allowing for an integration between institutional skills and resources and the availability of a reception network, structured throughout the country and widely available thanks to the capillary organisational capacities of the associations involved.

3 Educational Interventions for Growth Otherwise

What is of interest, beyond the judicial process and its judicial motivations, is the incidence of educational factors on criminal behaviour and, conversely, what the possibilities are for a counter-criminal educational intervention. We understand that there may be cultural resistance to the idea of a criminal pedagogy, so much so as to qualify the stylistic concept as a contradiction in terms, but the stories told, experienced and recognised tell us something else. They tell us of a harmful (in legal jargon we would say prejudicial) education, not oriented to guarantee growth. But what does growth mean? For some, growth coincides with the understanding of what surrounds us and the ability to relate to it; for others, it consists in the conquered ability of the self-control of impulses and respect for rules; for others, it means taking over the illicit affairs of the family or of the criminal group to which one belongs, or acquiring criminal skills, matured in delinquent conducts of greater and greater social alarm that allow one to climb the ranks of gangs or clans.

The importance of education and its function in the growth of the child from birth is a crucial issue that, in this case, is declined in terms of 'black pedagogy' [12]: the Mafia world is totalitarian, fundamentalist, has a closed horizon, does not allow any subjectivity, demands obedience, silence and ruthlessly condemns dissent [13].

The children we are talking about have different stories, not necessarily of criminal relevance, but they all share a common implication, namely that they are the expression of a totalising world, aimed at guaranteeing the Mafia identity profile in order to strengthen internal relations, beliefs and representations. Their diversity, therefore, is oriented by a specific mafia-type pedagogy [14] that is constantly engaged in intervening on growth paths. Mafia children and young people grow up, and therefore expect to grow up, without any external interference, without any polemical issue: the educational proposal is not dialogued or debatable, any educational dialogue is missing.

The alternative pedagogical proposal aims at an educational process of freedom [15], which is even more necessary if it is aimed at minors, through the full recognition of the dignity of the person, in the awareness that "(…) the formation of the individual must aim at the growth of a critical mentality that clears the field for the ethics of responsibility of man towards another man and man towards the living environment that surrounds him" [16]. In Freire's pedagogy" [17] we see some common aspects of our daily commitment: the anthropological-ethical vision of educational action and the contrast to an oppressive existential reality; the teleological horizon of a path of humanizing restitution that offers children a 'being more' starting from themselves and their infinite possibilities; the role of dialogue-listening as a comparison and horizon of consciousness of self and others with the implications that derive from it for the institutions deputed to the protection of these young people and, more generally, of all the educational agencies (formal and informal).

As the right to be heard of the child[6], the hearing of minors is a necessary requirement in the Italian legal system and in the judicial procedures concerning them, with the consequence that listening to a child of at least 12 years of age - and even younger if he or she is capable of discernment - is one of the most important ways of recognising his or her fundamental right to be informed and to express his or her opinions, as well as an element of primary importance in assessing his or her interests.

Beyond the legal aspects of the hearing, it is certain that listening to one's neighbour is the indispensable basis of any authentic presence of oneself as proximity to the other [18]. The minor is not asked for an automatic readjustment to the world of counter-mafia values, i.e. beyond his/her own individual existential horizon. If anything, it is the latter that must be re-founded, becoming the starting point for a transformation, as Bertolini [19] would say, of the child's worldview. The objective is to provoke a modification of that deep system of meanings (implicit and unconscious), favouring a new vision of oneself and of reality that allows the construction of a different (and critical) model of behaviour in reality [20].

In view of the reluctance to communicate verbally, especially at an early stage, it is important to give priority to offering concrete experiences that serve as a significant and concrete example of the alternative lifestyle that one wishes to propose. These experiential initiatives, which are part of a broader educational project, respond to the motivational urge of wanting to use a solid 'philosophy of experience', the same one that Dewey [21] (in the field of learning) considers necessary so that what we do merits the name of education, which is even more delicate and difficult for these minors.

In fact, meeting people and environments that embody an alternative lifestyle to the criminal one has become the most effective way, beyond words, to make unknown emotions or feelings come alive. Thanks to the support of Libera, some of our young people (anonymously) have experienced the confrontation with the victims of the mafias, the story of the suffering they have experienced, of the bereavements they have suffered, of the economic difficulties linked to criminal extortion; just as they have experienced, again with a view to transformative experiential pedagogy, a new way of relating to

---

[6] To be understood as an obligatory fulfilment for the judge, by virtue of Articles 3 and 6 of the Strasbourg Convention of 25 January 1996, as well as Article 12 of the New York Convention on the Rights of the Child (ratified in Italy by Law 1991/176).

reality, rediscovering the world they have always perceived as an enemy, wishing to experience it with other eyes, learning to trust those who accompany them on this journey.

But discovery alone is not enough, they need to re-elaborate the experiences and emotions they have lived through to allow them to be internalized [22]. This passage is essential and must take into account the aftermath: the aftermath of being "put to the test" or the juvenile justice aftermath is substantially uncovered at an institutional level.

The positive outcome of the hearing and the extinction of the offence does not exclude the danger of recidivism, which, on the other hand, occurs when the juvenile returns to highly criminal environments. The reintegration into the territory through a prolonged accompaniment, beyond the judicial experience, is a challenge aimed at preventing deviant and delinquent stumbling. It is a choice of serious support for these young people, which involves listening to them and building a strong motivation for legality [23]. In other words, in the balance between the legal system - sometimes inattentive, untimely and fragmentary in responding to the needs of growth - and the mafia system, the latter risks prevailing because it is already known and attractive due to family or clan power.

This is why it is necessary to pursue a socially indispensable ethical ideal [24] which, translated into educational terms, coincides with that living otherwise through which young people are given a space to experiment and learn to be different, to give existential possibilities unknown to them and to kindle their desire for change [25].

Reassessing the value of democratic culture means thinking of an education that deals with growing up 'with others', respecting the inviolable principle of the otherness of each person. This cultural, existential and social proposal for change is primarily a reciprocal intentionality along a continuum between subjectivity and collectivity, where the 'I' and the 'we' combine in the function of the growth and emancipation of individuals and the community.

References

1. Fiandaca, G., Visconti C.: Scenari di mafia, Giappichelli, Torino (2010)
2. Santino, U.: Dalla mafia alle mafie. Scienze sociali e crimine organizzato, Rubbettino, Soveria Mannelli (2006)
3. Habermas, J.: La problematica della comprensione del senso nelle scienze dell'azione empirico-analitiche, in Id., Logica delle scienze sociali (LWS), il Mulino, Bologna (1970)
4. Carta, M., Chirico, D.: Under. Giovani, mafie, periferie, Giulio Perrone, Roma (2017)
5. Ciconte, E.: Ndrangheta, Rubbettino, Soveria Mannelli (2011)
6. Di Bella, R., Surace, G.M.P.: Il progetto 'Liberi di scegliere. L'esperienza giudiziaria del Tribunale per i minorenni di Reggio Calabria', Rubbettino, Soveria Mannelli (2019)
7. Schermi, M.: Crescere alle mafie. Per una decostruzione della pedagogia mafiosa, FrancoAngeli, Milano (2010)
8. Lo Verso, G.: La mafia dentro. Psicologia e psicopatologia di un fondamentalismo, FrancoAngeli, Milano (2002)
9. Lamarque, E.: Art. 30 Cost., in Commentario alla Costituzione, a cura di R. Bifulco, A. Celotto, M. Olivetti, vol. I, Utet, Torino (2006)
10. Casabona, S.: Limiti alla funzione educativa dei genitori tra strumenti di controllo giudiziari e automatismi legislativi, in Minori giustizia, n. 3 (2016)

11. Balzano, V.: Il volontariato come pratica di dono e relazione d'aiuto nella costruzione del progetto di vita, in Attualità pedagogiche, vol. 1, n. 1 (2019)
12. Miller, A.: La persecuzione del bambino. Le radici della violenza, Bollati Boringhieri, Torino (1980)
13. Lo Verso, G., Lo Coco, G., Mistretta, S., Zizzo, G.: Come cambia la mafia. Esperienze giudizia-rie e psicoterapeutiche, Franco Angeli, Milano (1999)
14. Barone, P.: Pedagogia della marginalità e della devianza, Guerini Scientifica, Milano, 2011, 2nd edn. (2019)
15. Cambi, F.: La "questione del soggetto" come problema pedagogico, in Studi sulla formazione, University Press, Firenze, 2 (2008)
16. Freire, P.: Pedagogia degli oppressi (1971), EGA, Torino, p. 32 (2002)
17. Freire, P.: L'educazione come pratica della libertà. I fondamenti sperimentali della "pedagogia degli oppressi" (1973). Arnoldo Mondadori, Milano (1974)
18. Mannese, E., Ricciardi, M.: Per una pedagogia dell'Ascolto: raccontare per conoscersi, in "MeTis. Mondi educativi. Temi indagini suggestioni", anno VI, n. 1, giugno (2016)
19. Bertolini, P., Caronia, L.: Ragazzi difficili. Pedagogia interpretativa e linee di intervento, FrancoAngeli, Milano (2015)
20. Bertolini, P.: Esistere pedagogico. Ragioni e limiti di una pedagogia come scienza fenomeno-logicamente fondata, La Nuova Italia, Firenze (1999)
21. Dewey, J.: Esperienza e educazione (1938). Raffaello Cortina, Milano (2014)
22. Demetrio, D.: Raccontarsi, Raffaello Cortina Editore, Milano (1995)
23. Regolosi, L.: Per un intervento socioeducativo nei confronti di minori coinvolti nel contesto mafioso, in Progetto "Mafia minors" – Programma AGIS 2004 – JAI/2004/AGIS/135 – Dossier ITALIA, Dipartimento Giustizia minorile (2004)
24. Bauman, Z.: Le sfide dell'etica, Feltrinelli, Milano (2010)
25. Dewey, J.: Democrazia e educazione. Una introduzione alla filosofia dell'educazione (1916). Spatafora, G., (a cura di), Anicia, Roma (2018)

Is City Love a Success Factor for Neighbourhood Resilience? Results from a Microcosmic Analysis of Rotterdam

Karima Kourtit[1,2,3,4] (iD), Peter Nijkamp[1,2,3(✉)] (iD), Umut Türk[5] (iD), and Mia Wahlstrom[6] (iD)

[1] Open University, Heerlen, The Netherlands
pnijkamp@hotmail.com
[2] Alexandru Ioan Cuza University, Iasi, Romania
[3] Mohammed VI Polytechnic University, Benguerir, Morocco
[4] Uppsala University, Uppsala, Sweden
[5] Abdullah Gül University, Kayseri, Turkey
umut.turk@agu.edu.tr
[6] Tyréns, Stockholm, Sweden
Mia.Wahlstrom@tyrens.se

Abstract. This study examines and tests the concept of 'city love' in the context of social resilience for urban neighbourhoods. It introduces the notion of 'city body' and 'city soul' so as to create an operational framework for measuring the citizens' appreciation and attachment for the local neighbourhood. Particular attention is given to the social bonds in urban community networks and language groups. A quantitative statistical analysis is carried out to test the relationships and determinants of city (or neighbourhood) love, based on extensive statistical, survey and social media data on the city of Rotterdam.

Keywords: City love · Body · Soul · Neighbourhood · Microcosmic · Bonds · Urbanometric

1 Urban Geography and Happiness

Research in economic and social geography, in regional and urban sciences, and in spatial planning has over the past years increasingly focused the attention on 'place-based' approaches. This new orientation in research and policy aims to address spatial issues in society from a place-specific angle, in which geographic uniqueness and specificity are emphasised rather than general or nomothetic principles (see e.g. Kourtit 2021).

Such a place-based perspective has recently also led to novel analytical departures for urban research, where heterogeneity and diversity in urban systems and in their actors' operations are receiving a prominent position. This is exemplified in the shift from urban macro-indicators (e.g., average unemployment, education or health condition) to place-specific or group-specific meso- or micro-indicators (e.g., citizens' well-being).

This change in scientific interest has clearly prompted much attention for specific urban neighbourhood issues or urban communities (e.g. migrant enclaves, district liveability).

The new systemic interest in small-scale constituents of cities is called a *microcosmic* approach: cities are examined from a disaggregate decomposition paradigm (Kourtit 2021), a vision which has already been advocated by Berry (1964), who regarded the urban landscape of a country as 'cities as systems within systems of cities'. Such a multi-scalar vision on spatial and urban systems does not only relate to geographical differentiation (e.g., the built environment, urban amenities, accessibility to urban infrastructure), but also to heterogeneity among people in urban agglomerations (e.g. income, education, age, cultural or ethnic background). Consequently, people and places exhibit a high degree of pluriformity and heterogeneity in urban agglomerations. Despite common background factors and common driving forces for human behaviour, cities – and their districts, neighbourhoods and communities – are diverse in nature. The '*New Urban World*' (Kourtit 2019) is pluriform.

Urban density and place quality have also great implications for the level of satisfaction or well-being of citizens. Research in this field has witnessed interesting shifts in the past decades, which has culminated in intensive recent research interest in urban happiness from a human perspective and left behind districts from a planning perspective. Traditional socioeconomic focal points on e.g. income or employment have increasingly been substituted for happiness or well-being studies of residents. Also related new concepts have over the past years been introduced in different disciplines (e.g. sociology, psychology, anthropology, human geography), such as happiness, contentment, life satisfaction, liveability, eudaimonia etc. (see e.g. Ballas 2018; Frey 2018).

With the introduction of the 'quantitative revolution' in the social sciences, a most challenging task has become the measurement of well-being or happiness in cities and their neighbourhoods, as part of a new urban modelling field called '*urbanometrics*'. The quantification of urban well-being or urban happiness is not only interesting from the perspective of socioeconomic disparities in cities, but also from the perspective of comparative monitoring studies (e.g. Martin et al. 2021). An illustrative example can be found in a study by Marlet and van Woerkens (2017) who created an extensive atlas (both statistical and cartographic) on the multidimensional socioeconomic profiles of municipalities in the Netherlands.

Against the background of place-based well-being approaches, the present study examines the micro-constituents of place-based well-being in a particular city, viz. Rotterdam, with a focus on neighbourhoods. Based on an extensive database at a neighbourhood scale and complemented with individual survey data of residents on neighbourhood satisfaction or liveability as well as with TripAdvisor platform data, a monitoring exercise and quantitative modelling study is carried out, with a view to a better understanding of the drivers of the residents' neighbourhood satisfaction, and the disparities therein.

The paper is organised as follows. After this Introduction, Sect. 2 will provide a concise research overview. Then Sect. 3 will offer a new perspective on the measurement of place-specific well-being using the concept of '*city love*'. In Sect. 4 the database used is presented, while the statistical-econometric modelling exercise serving to measure neighbourhood love is described in Sect. 5; this section offers also empirical results and an interpretation of the empirical findings. Finally, Sect. 6 provides conclusions and prospects.

2   Urban Well-Being: An Overview

Research on well-being – and related concepts like happiness or liveability – has led to an abundance of literature in many disciplines. We have, for instance, seen the rise of the economics of happiness (see Frey 2018), the sociology of happiness (see Veenhoven 2008), or the geography of happiness (Ballas 2018). An interesting collection of empirical studies from different backgrounds can be found in a recent special issue of the journal *Sustainability* guest-edited by Toger et al. (2021). It appears that, in general, modern quantitative applied well-being and happiness research is characterised by a multiplicity of approaches, databases and focal points of research. Some distinctions and research orientation are:

- individual well-being research vs. group well-being research (e.g. local communities).
- differences in explanatory frameworks, e.g. contextual analysis, neuro-psychological approaches or social network analysis.
- place-specific and geographical moderator analysis at different spatial scales ranging from countries as a whole to urban neighbourhoods.
- socioeconomic and socio-demographic analysis by distinguishing background variables like education, age or health conditions.
- different methodological approaches, like objectively measured proxy variables for well-being or self-reported well-being statements.
- different disciplinary orientations, e.g. related to economics, sociology, psychology or geography.
- different research scopes, ranging e.g. from the relationship between well-being and prosperity or income to the role of intangible background variables like political stability or community safety.

In almost all well-being studies the main interest is focused on disparities in well-being, either individually or collectively. In our applied study on the city of Rotterdam, we zoom in on a mix of meso-well-being drivers (at neighbourhood or community level) and micro-perceptions (derived from extensive bi-annual residents' surveys), which clearly characterise our approach as a microcosmic approach. In the subsequent section we will outline the novel methodological peculiarities adopted in our research. This focus on sub-local (i.e., neighbourhood) well-being factors is also important from the perspective of social resilience. The latter concept refers to the capacity of an organisation (in this case, a city) to survive, adapt and grow, or transform itself after shocks or crises, in particular from the perspective of social capital. Neighbourhood well-being is then seen and tested as one of the critical success factors for social resilience.

3   Measurement of Urban Well-Being

In recent years, much scientific effort has been devoted to the measurement of well-being or happiness (see also Östh et al. 2020 for an overview). A seminal contribution has been made by Veenhoven (2008) who as one of the first sociologists managed to develop a measurement framework for human happiness that could be applied to different population groups and areas. Blanchflower (2021) made a successful attempt in economics to measure well-being, taking into account distributional effects (e.g. based on age), which appeared to generate an inverse u-shaped curve.

Several attempts seeking to create systematic measurements of well-being or 'broad welfare' (or 'beyond-GDP') indicators can be found in the recent empirical literature showing a wide range of well-being indicators, such as the OECD Better Life Index, the UNDP Human Development Index (HDI), the UN World Happiness Index, the Happy Planet Index, the Global Liveability Index or the Beyond-GDP Index (Kalimeris et al. 2019). All such indicators are empirically very useful and also appropriate for comparative quantitative research. A limitation of the above indicators is however, that the place-specific elements in these measurement schemes are weakly developed. In other words, the micro-geography is largely missing in these well-being calculation frameworks. Given the rising interest in sub-local quality of city life (the resident as a villager in a large city), it is desirable to develop a measurement framework for place quality that takes into consideration the human living conditions in cities at the level of districts or neighbourhoods that are in accordance with welfare, inclusiveness and sustainability factors.

The above considerations have recently led to the development of a new operational measure for well-being of citizens, called '*city love*'. This concept refers to the operational appreciation of citizens for their urban environment; it comprises both satisfaction and attachment to urban life. The 'city love' indicator is the result of two underlying constituents of city life. In the first place, the attractiveness of cities is shaped by its physical and material appearance, in particular, built environment, infrastructure, amenities, urban green etc. This is called the '*body*' of the city. The quantitative representation of the city's body is called here '*human habitat*' index. In the second place, a city derives its attractiveness by its intangible assets, such as atmosphere, historical places, local pride, and symbolic structures. This is called the city's '*soul*'. The relevant quantitative index here is called '*feelgood*' index. To assess the city love at neighbourhood level, we have to collect multivariate measurable indicators on both the body and the soul in a microcosmic context. We refer to Kourtit et al. (2021; 2022 a,b) for more details on the methodological backgrounds and operational tests models of the city love concept.

In the present paper, we introduce two additional factors to the original city love idea, viz. *neighbourhood* love and *social cohesion*. Neighbourhood love refers to the love of residents for urban neighbourhoods or localised communities in cities (the microcosmic city). Social cohesion is related to bonds and social support or community systems that create a sense of social linkage and place-based cultural identity. In the latter case, indicators like mutual interaction, community communication, shared language or joint cultural bonds may be used. These will be summarised in a '*social bonds*' index.

As mentioned before, an interactive neighbourhood love mechanism will provide a solid cornerstone for a high social resilience capacity in urban neighbourhoods and cities. The conceptual 'neighbourhood love' model – to be applied in the second part of the present study – can be represented as follows (see Fig. 1).

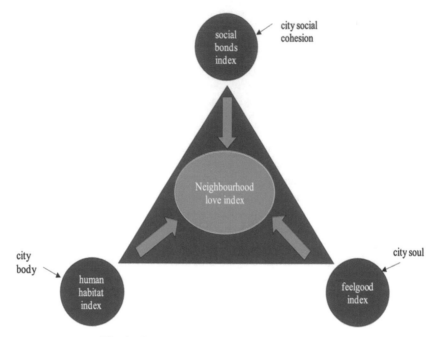

Fig. 1. Conceptual model of neighbourhood love.

The above conceptual model will now be applied and empirically tested for the case of Rotterdam. We will first describe the database used (Sect. 4) and then present the statistical-econometric model (the 'urbanometric' model) used, followed by a presentation of results (Sect. 5).

4   Database on Rotterdam

Rotterdam is a dynamic, internationally oriented trade city, which has since the beginning of WWII a turbulent history and shows the sign of frequent change in its physical appearance ('body'), of shifts in its identity ('soul'), and of population heterogeneity due to a wealth of cultural minorities ('social segmentation') or 'weak social cohesion'. The city has a rich multi-annual database at neighbourhood level, which comprises not only detailed statistical data on socio-economic, geographic and demographic items in each of the urban neighbourhoods (63 in total) up to the year 2020, but also detailed individual survey data on residents regarding their wellbeing and their appreciation of city life. In addition, we have a set of moderator variables originating from TripAdvisor information. Thus, we have a broad database that is in principle relevant for an application of our neighbourhood love model. Besides, for comparative purposes of the Rotterdam data with data from other Dutch municipalities, there is also a cross-sectional database collected by Marlet and van Woerkens (2017).

For the empirical analysis in this paper, we make use of the Rotterdam Neighbourhood data system. This database withholds both residents' survey and statistical register data up to 2020. Detailed statistics on place-based quality-of-life (liveability) indicators and individual survey information on the experiences and perceptions of the habitants are available for all 63 neighbourhoods in Rotterdam. The data is categorized into three indices: physical, social and community safety. The physical index contains information on housing, public space, services, environment, and living experiences. The social index covers information on people: capacities/skills, participation, living environment, bonding, and quality of life experiences. Finally, the safety index is based on information about crime, vandalism, nuisance, and safety experiences. The data are presented in a standardised form as percentages and index scores, where the average score for the whole city of Rotterdam has been set at a value of 100 as the baseline measurement (2014). The neighbourhood scores are calculated in relation to this city average. This rich dataset is used as the source of information for the various elements included in the city love triangular system in Fig. 1.

5   Empirical Results of Multivariate Analysis

In some previous studies a modelling exercise has been carried out to test the proposition that 'city love' is an important well-being index for cities and that a downscaling to neighbourhood levels ads new explanatory power (see Kourtit et al. 2021; 2022 a,b). According to earlier findings based on the conceptual model presented in Fig. 1, the 'feelgood' index has a strong and positive influence on the 'neighbourhood love', i.e. the appreciation residents feel for their neighbourhood. Recent results presented by Kourtit et al. (2022 a,b) indicated that social capital is strongly related to the feelgood factor, alongside with feelings of safety and the neighbourhood wealth. The results in Kourtit et al. (2022 a,b) also indicated that a neighbourhood with strong social capital will have a higher feelgood index and a stronger individual involvement in the neighbourhood and city. In this study, we dig deeper into the influence of social capital on neighbourhood love, also by taking the education and language skills of residents into account.

In our paper the human habitat attractiveness dimension will be proxied in terms of proximity to tourism attractions, which will be further explained later on. Clearly, a microcosmic perspective on the systemic complexity of Rotterdam prompts many research challenges.

In the light of previous research findings and with a view to an examination of the significance of social-cultural factors in neighbourhood well-being a few new statistical experiments are described here. We will conduct a multivariate analysis at neighbourhood level, in which neighbourhood love is defined as a function of bonding and bridging types of social capital and of additional socioeconomic factors defined at a neighbourhood level. While bonding social capital refers to inner group connections among similar socioeconomic groups and hence provides fewer opportunities for rich social interaction, bridging social capital describes interactions among heterogeneous groups and provides access to new knowledge pools and social relations (Granovetter 1973). We measure bonding social capital by the percentage of residents who know each other and spend much time together and who say that they have frequent interesting contacts and interactions with their families and friends. Meanwhile, bridging social capital is the percentage of residents who interact with each other, irrespective of age and ethnicity.

As a proxy for socioeconomic characteristics, we use the *education index* and the *language index*. Here the education index is defined as the weighted percentage of people without basic qualifications (age 18–64), and the language index is the weighted percentage of people with difficulties in reading, speaking and/or writing Dutch and/or in need of language help.

We also need an indicator for the physical attractiveness of local neighbourhoods. It is plausible to assume that tourism activity and tourist mobility will influence the residents' perceptions of their neighbourhoods (see Dirksmeier and Helbrecht 2015 for a review on residents' perception of new urban tourism). We scraped the information about 534 tourist attraction points (museums, historical sites, etc.) located in 44 neighbourhoods in Rotterdam using information from the TripAdvisor platform to approximate this effect. Scraped material includes the average review score per attraction. Using Geographical Information Systems (GIS), we identified the ten nearest attractions from each neighbourhood and next calculated the distance to reach these attractions from the centroid of neighbourhoods. Using the average review score of tourists and the distance to reach the attractions, we are able to construct a measure of access to tourism attractions from each neighbourhood. In constructing the index, our first assumption is that reviews will convey information about the tourists' appreciation of attraction points. The second assumption is that from a residents' perspective, as the distance from a neighbourhood to the attraction points increases, the residents' benefit of – or their exposure to – both the touristic attractions and the opinions of tourists decreases. The approach is in agreement with Hansen's (1959) standard gravity equation comprising both distance friction and attraction points. We use next the newly constructed variable Tourism Attraction for analysing the interplay between the residents' appreciation of their neighbourhoods and tourism activity in and around their neighbourhoods. Based on these propositions and related data, we estimate the neighbourhood love model by an OLS regression analysis. Thus, this urbanometric model tests the proposition that neighbourhood love is a function of social capital (bonding, bridging), cultural linkage (language), education, and physical/environmental attractiveness.

Table 1 summarises the regression outputs, where the first column shows the main model, the second and third columns include models with interaction terms between the variables incorporating language difficulty, and bonding and bridging social capitals, respectively. Finally, the last two distinct columns incorporate interaction terms between education difficulty and the two types of social capital.

From the main model, we can infer that neighbourhood love increases with both types of social capital. This means that when residents have good relationships with their inner circles and with other groups, they appreciate their neighbourhood higher. The main model also shows that neighbourhood appreciation considerably decreases in neighbourhoods with a high share of low educated population and those who experience the Dutch language as a barrier. Our measure of access to tourism attraction shows a strong and positive association with neighbourhood love. The latter finding implies that residents appreciate nearby tourist attractions. It also means a positive correlation between tourists' and residents' appreciation of points of interest in the neighbourhoods of a city.

Table 1. Multivariate regression analysis results

Variables	(1)	(2)	(4)	(5)	(6)
	Main Model	BondingXLanguage	BridgingXLanguage	BondingXEducation	BridgingXEducation
Bonding	4.330***	5.835***	6.257***	−1.896	4.524***
	(0.939)	(1.505)	(0.958)	(1.532)	(0.924)
Bridging	5.444***	−4.938***	−0.211	5.520***	8.865***
	(1.286)	(1.340)	(1.797)	(1.095)	(2.186)
Language_index	−47.540***	−55.540***	−30.325**	−33.230***	−45.681***
	(13.926)	(15.205)	(13.047)	(12.225)	(13.647)
Education_index	−23.455*	−23.776*	−38.406***	−22.968*	−24.609*
	(13.143)	(14.178)	(13.156)	(12.130)	(13.945)
TourismAttraction	26.029***	28.552***	26.210***	22.980***	26.073***
	(8.027)	(8.225)	(7.114)	(6.860)	(7.846)
BondingXLanguage_index		−11.859			
		(3.301)			
BridgingXLanguage_index			34.513***		
			(8.479)		
BondingXEducation_index				25.999***	
				(5.455)	
BridgingXEducation_index					−10.53
					(5.505)
Constant	75.762***	75.843***	79.236***	73.741***	74.943***
	(5.285)	(5.257)	(4.761)	(4.518)	(5.184)
Observations	63	63	63	63	63
R-squared	0.834	0.838	0.872	0.882	0.844

Standard errors in parentheses.
*** p<0.01, ** p<0.05, * p<0.1.

In our model experiments, we interact the two types of social capital with language (difficulty) and education variables to examine if social relations are a substitute for human capital. In this respect, the second and third models provide interesting results. While the interaction of the language index with a bonding type does not produce significant coefficients (Model 2), the interaction with bridging social capital does reverse the negative effect (Model 3). Comparing the outputs from Models 2 and 3, we may conclude that when interpersonal trust and reciprocity is high among different groups of individuals, neighbourhood appreciation increases, even in neighbourhoods with low language skills. Note that the language index can also be thought as a proxy for minority population. Finally, we find opposite results for the interaction of the education index with the two types of social capital (Model 5 and 6). Residents of neighbourhoods with a high concentration of lower-educated population express higher levels of neighbourhood love, if the inner-group interaction (among family members and close friends) is strong.

It is thus clear from the empirical results that social bonds variables (language, group and individual interaction, community sense etc.) play a decisive role in shaping neighbourhood love.

6   Conclusions and Outlook

This study has addressed neighbourhood well-being factors in the city of Rotterdam. It is part of a more extensive series of studies on urbanometric modelling of the new concept of city love. The present paper has addressed in particular the significance of neighbourhood love.

From a research perspective, we were able to demonstrate that the translation of the 'city love' concept into a 'neighbourhood love' notion is an important methodological step forward in understanding well-being of citizens from a microcosmic perspective.

It is noteworthy that, next to the core constituents of 'city body' and 'city soul' – quantitatively translated into a 'human habitat index' and a 'feelgood index', another factor appears to play a critical role in shaping citizens' well-being feelings, viz. 'social cohesion' in relation to social capital and networking. This third cornerstone of citizens' love is measured through a 'social bonds' index.

It goes without saying that the urbanometric approach to assessing the determinants of 'city love' is not only a promising approach, but prompts also the need for extensive data collection on urban life and its residents: disaggregate spatial data, survey data, digital data for ICT provisions/use, and social media data. The present study has only made a modest use of the rich data needs in urbanometric research.

And finally, 'city love' research is not a stand-alone research activity. It ought to be incorporated in a broader 'beyond-GDP' planning context as well as in foresight and trend analysis on social resilience in cities and their neighbourhoods.

References

Ballas, D.: The Economic Geography of Happiness, Inaugural Address, University of Groningen, Groningen (2018)

Berry, B.J.L.: Cities as systems within systems of cities. Pap. Reg. Sci. Assoc. **13**, 146–163 (1964)

Blanchflower, D.G.: The U shape of happiness. J. Popul. Econ. (2021). https://doi.org/10.1177/1745691620984393

Dirksmeier, P., Helbrecht, I.: Resident perceptions of new urban tourism: a neglected geography of prejudice. Geogr. Compass **9**(5), 276–285 (2015)

Frey, B.S.: Economics of Happiness. Springer, Cham (2018). https://doi.org/10.1007/978-3-319-75807-7

Granovetter, M.S.: The strength of weak ties. Am. J. Sociol. **78**(6), 1360–1380 (1973)

Kalimeris, P., Bithas, K., Richardson, C., Nijkamp, P.: Hidden linkages between resources and economy: a 'beyond GDP' approach using alternative welfare indicators. Ecol. Econ. **169**, 106508 (2019). https://doi.org/10.1016/j.ecolecon.2019.106508

Kourtit, K.: The New Urban World. Shaker, Aachen (2019)

Kourtit, K.: City intelligence for enhancing urban performance value: a conceptual study on data decomposition in smart cities. Asia-Pac. J. Reg. Sci. **5**, 191–222 (2021)

Kourtit, K., Nijkamp, P., Turk, U., Wahlstrom, M.H.: City love and place quality – assessment of liveable and loveable neighbourhoods in Rotterdam. Land Use Policy **119**, 106109 (2022)

Kourtit, K., Nijkamp, P., Turk, U., Wahlstrom, M.H.: City love and neighbourhood resilience in the urban fabric: a microcosmic urbanometric analysis of Rotterdam. J. Urban Manage. (2022b, in press). Conditionally accepted with minor revision

Marlet, G., van Woerkens, C.: Atlas 2017 voor Gemeenten. VOC, Nijmegen (2017)

Martin, R., Gardiner, B., Pike, A., Sunley, P., Tyler, P.: Levelling up Left Behind Places. Taylor & Francis, Milton Park (2021)

Toger, M., Kourtit, K., Nijkamp, P.: Special issue on: happy and healthy cities. Sustainability (2021). https://doi.org/10.3390/su132212817

Osth, J., Kourtit, K., Nijkamp, P.: My home is my castle: assessment of city love in Sweden. Int. J. Inf. Manage. **58**, 102213 (2020)

Veenhoven, R.: Sociological theories of subjective well-being. Sci. Subj. Well-being **9**, 44–61 (2008)

Hansen, W.G.: How accessibility shapes land use. J. American Plann. Assoc. **25**(2), 73–76 (1959)

Agents of Change and Window of Locational Opportunity (WLO) in Crypto Valley in Zug, Switzerland

Arnault Morisson[1](✉) [iD] and Clara Turner[1,2] [iD]

[1] Institute of Geography and Centre for Regional Economic Development (CRED),
University of Bern, Bern, Switzerland
arnault.morisson@giub.unibe.ch
[2] Department of City and Regional Planning, University of California, Berkeley, USA

Abstract. The article explores the role of human agency in the construction of the opportunity space during the emergence of a new technology—blockchain—in the town of Zug, Switzerland. Trinity of change agency (TCA)—Schumpeterian innovative entrepreneurs, institutional entrepreneurs, and place-based leadership—were able to seize a brief Window of Locational Opportunity (WLO) and construct the opportunity space to promote path creation in Zug. Branded as Crypto Valley, Zug is a medium-sized town that is home to a thriving blockchain and crypto start-up ecosystem. The case of Crypto Valley contributes to the debate on the interplay between the broader institutional context and with time-specific, region-specific, and agent-specific opportunity spaces during the emergence of a new technology—blockchain. The article also highlights the role of institutional relatedness in the emergence of Crypto Valley in Zug.

Keywords: Path creation · Institutional relatedness · Agents of change · Window of locational opportunity · Blockchain

1 Introduction

The literature in economic geography is increasingly interested in the role of agents of change in fostering new industrial path development [1, 2]. With a focus on change agency, trinity of change agency (TCA) involving Schumpeterian innovative entrepreneurs, institutional entrepreneurs, and place-based leadership have an important role to play during path creation [1]. Those agents of change do not act in an institutional void and are embedded into the broader institutional context that determines the agents' behaviours and expectations linked to the agency-structure debate [3]. As a result, changes are mediated by pre-existing and emerging institutions embedded in a structure that is time- and space-specific [4].

Agents of change play a fundamental role in successfully identifying and seizing Windows of Locational Opportunity (WLO) and thus fostering path creation [2]. The WLO-concept states that the emergence of new technologies offers greater opportunities for chance in new industrial path development because it represents a fundamental break with the past [5, 6]. The emergence of a new technology thus offers the possibility for agents of change to construct an opportunity space that makes path creation more likely to happen. The opportunity space, which is the agents' perception that change is possible at a certain time and space, offers an understanding of the interplay between agency and structure [1].

Blockchain, the technology behind cryptocurrencies that emerged with the creation of Bitcoin in 2008, can integrate into multiple applications in a decentralised manner. One example is decentralized finance or DeFi, which offers a new decentralised financial infrastructure [7]. Blockchain technology, a disruptive computing technology, thus represents a clear technological break from the past and presented a WLO. Switzerland emerged as a global leader in the development of blockchain technologies and crypto assets [7]. Agents of change in the town of Zug, Switzerland, were pioneers in identifying and successfully seizing the WLO related to blockchain technologies. TCA in particular were central to the construction of the opportunity space and path creation.

The article explores the role of human agency in the construction of the opportunity space during the emergence of a new technology, blockchain, in the town of Zug, Switzerland. The authors ask how the institutional context supported agents of change in constructing the opportunity space to seize a Window of Locational Opportunity (WLO)? The case of Crypto Valley makes three contributions. First, it sheds light on the role of institutional relatedness for path creation during a WLO. Second, it captures how agents of change constructed the opportunity space to seize a brief WLO and foster path creation. Third, it provides new evidence on the interplay between structure and agency in path creation. The article finds that institutional relatedness and the construction of the opportunity space were central for the emergence of Crypto Valley in Zug.

2 Theoretical Framework

The paper intends to show that, contingent upon generic resources relevant to the development of a disruptive technology, agents of change can construct a regional opportunity space to successfully seize a Window of Locational Opportunity (WLO) and thus foster path creation. Path creation refers to the emergence and growth of new industries and economic activities in a region [8]. Path creation follows a branching logic of related diversification based on pre-existing knowledge assets [9]. However, the role of agents of change, system-level agency, and the institutional context are increasingly recognised as important conditions for path creation [1]. Moreover, institutional relatedness, which refers to the degree institutions supporting an industry can be used in another industry, can explain path creation beyond technological knowledge [10].

The WLO-concept provides an analytical framework to explain path creation in unlikely places during the emergence of an innovative technology [5, 6]. Contingent upon generic resources relevant to the development of that innovative technology, the role of chance and agency make path creation more likely in unlikely places before increasing returns and supporting institutions curb the WLO. The WLO-concept suggests a greater unpredictability in path creation and unrelated diversification. Agents of change can seize a WLO to foster path creation in unlikely places, such as with the case of the cryptocurrency hardware wallet start-up Ledger in Vierzon, France [2]. The example highlights the role of the institutional context to increase the likelihood for Schumpeterian innovative entrepreneurs to seize a WLO. As a result, national institutional frameworks and institutional relatedness can enable or hinder the capacity of agents of change to seize a WLO and further construct the opportunity space.

The concept of the trinity of change agency (TCA), which consists of Schumpeterian innovative entrepreneurs, institutional entrepreneurs, and place-based leadership, emphasizes the role of human agencies in the creation of new industrial path development [1]. A Schumpeterian innovative entrepreneur is a person or group of persons who has the capacity, motivation, and willingness to transform an idea into an innovation and can identify emerging technologies and novel market opportunities allowing them to seize a WLO and expand opportunity spaces [2]. Institutional entrepreneurs are actors who mobilise resources, power, and competences to transform existing institutions by introducing new institutions or institutional arrangements [3]. These entrepreneurs usually have a strong interest in shaping institutional arrangements and institutional changes. They tend to act when Schumpeterian innovative entrepreneurs identify WLO to construct the opportunity space [3]. Place-based leadership is a form of collective strategic agency to coordinate and rally diverse stakeholders around a common vision of the future to drive regional changes [11].

The opportunity space, which is the agents' perception that change is possible at a certain time and space, offers an understanding of the interplay between agency and structure. There are time-specific, region-specific, and agent-specific opportunity spaces [1]. Agents of change can construct or exploit opportunity spaces, which mediate the interplay between structure and agency, to foster changes. They reflect in a strategic manner considering how structures may evolve in the future and considering how their actions might affect this evolution [1]. Yet, it is unclear how and at what point in time the opportunity space is constructed in the first place. The WLO-concept fills this gap by demonstrating that Schumpeterian innovative entrepreneurs can also construct the opportunity space in regions that are not yet structured to optimally support emerging technologies.

3   Methodology

This article's research methodology employs a single in-depth case study of Zug, Switzerland's "Crypto Valley" to explore path creation in blockchain technology and crypto assets. A case study approach was selected 'out of the desire to understand complex social phenomena' in which the researcher has no control [12]. Research on agents of change implies asking the question who did what, when, where, why, how, with whom and to what consequence making qualitative research a particularly suitable methodology [13]. Using multiple sources of information, the research strategy aimed to generate a timeline of events and understand the actions directed towards achieving change [13].

Researchers conducted four semi-structured interviews in November 2021—lasting from 30 to 120 min—to gather extensive data on aspects of the research question with questions on who did what, when, where, why, how, with whom and to what consequence in the emergence and development of Crypto Valley. The interviewees were selected as key informants. The documents collected for this research aimed to generate a timeline of events and actions related to the emergence of Crypto Valley in Zug. The data analysis consisted of triangulating the gathered information to 'produce empirically based findings' [12], with the objective of exploring path creation in blockchain technologies in Zug. Validation is achieved through triangulation of multiple sources of evidence to ensure that 'the right information and interpretations have been obtained' [14].

4 The Case

4.1 The Town of Zug

Zug is a medium-sized town located 30 min by car from the city of Zurich in the Canton of Zug in Switzerland. Historically, the Canton of Zug had a strong agricultural and industrial sector that went into a deep structural crisis in the 1930s. Influenced by Zurich's financial interests, the Canton of Zug revised its tax system in 1946 to create tax conditions favourable to holding companies. The Canton and the town of Zug attracted numerous holding companies—from 251 limited companies in 1955 to 3,768 in 1970—due to the combination of its status as a tax haven and proximity to Zurich [15]. From the 1980s to 2000s, with the financialisation of the world economy, the Canton of Zug became a popular location for international headquarters, with firms like Amgen, Johnson & Johnson, Siemens, and Shell, and commodity trading company Glencore all establishing headquarters in Zug. This was due to its "business-friendly" reputation, efficient bureaucracy, low tax rates, secrecy regulations, and Swiss double taxation treaties [16].

The Canton of Zug does not offer monetary incentives or subsidies to international companies, but rather provides favourable conditions. As per a representative of the Canton, "we understand international companies' needs, our tax authorities know what they want, we speak the same language" (personal communication, 23 November 2021). The Canton of Zug has a corporate profit tax of 12% and an individual income tax rate of 22%, which are the lowest in Switzerland and some of the lowest tax rates among developed countries. It hosts 115,800 workplaces (39,200 of these positions filled by workers commuting in) in 35,300 companies [17]. Both the Canton's policies and reputation have generated numerous criticisms for being too accommodating with "letterbox" companies, and the Canton and town of Zug are frequently at the centre of investigative stories written by the International Consortium of Investigative Journalists (ICIJ) [18].

4.2 The Emergence of Crypto Valley (2013–2014)

The emergence of Zug as Crypto Valley was the outcome of Schumpeterian innovative entrepreneurs and institutional entrepreneurs who successfully seized a WLO that opened with the emergence of blockchain technologies and cryptocurrencies. Two key Schumpeterian innovative entrepreneurs were Niklas Nikolajsen, who founded Bitcoin Suisse in August 2013, and the founders of Ethereum, Vitalik Buterin, Mihai Alisie, Joseph Lubin, and Gavin Wood, who spearheaded the successful Initial Coin Offering (ICO) of the Ethereum Foundation in 2014. Two key institutional entrepreneurs were

Johann Gevers, who co-founded Bitcoin Association Switzerland, the Digital Finance Compliance Association (DFCA), and relocated his crypto-based platform company Monetas from Vancouver to Zug in July 2013, and Lukas Müller, one of founding partners at the law firm MME, who incorporated Ethereum as a foundation. With tacit support from Zug's economic promotion office, these agents of change quickly developed a shared vision to position and brand Zug as Crypto Valley [19].

The Swiss institutional context was a key factor in these entrepreneurs' decisions to locate their crypto companies in Zug. The Schumpeterian innovative entrepreneurs were foreigners (American, British, Canadian, Danish, Romanian, and South African) looking for a favourable country to locate to. "I shortlisted three countries to relocate my company: I looked at Liechtenstein, Singapore, and Switzerland," reported one entrepreneur. "Liechtenstein was too small and difficult to immigrate to. Singapore has limited freedom, is too top-down and could easily be pressured by the United States to stop crypto. Switzerland, with its decentralised bottom-up political system and culture, financial sector, data privacy, and stability, was the obvious choice. Switzerland has resisted international pressure with banking secrecy for instance, it took many years before it was taken away and with Cryptos, we only needed a few years" (personal communication, 19 November 2021).

Bitcoin Association Switzerland organised meetups and helped establish the crypto network around Zurich and Zug. Bitcoin Association Switzerland was a key communicator during international crypto events, promoting Switzerland and Zug as favourable locations for Crypto companies and successfully convincing firms (one example being Ethereum). Moreover, the Association inspired the Federal Council to write a report in 2014 on virtual currencies and to assess the regulatory situation of Bitcoin and cryptocurrencies under Swiss law. This provided some legal certainty and regulatory visibility to crypto companies [20]. The Digital Finance Compliance Association (DFCA)—involving Bitcoin Suisse, Ethereum, Monetas, MME—was also created in 2014 to advocate for technology neutrality and to promote regulatory and legal certainties for crypto companies.

In 2014, the Ethereum founding team consulted with Lukas Müller from MME on legal structures for their firm. Ethereum's technology works as a decentralised open-source public protocol. The idea was floated to incorporate it as a foundation that operates similarly to a smart contract, as it has a predefined structure, a pre-defined functionality and where functionality will be matched with assets and the assets will be used according to the predefined purpose [21]. The Foundation structure consists of a neutral authority devoted to advancing the development of and non-profit technology development open-source protocols, rather than a for-profit company. The Foundation structure allowed to finance itself through token sales which permitted Ethereum to raise USD $17.3M and thus became the largest cryptographic token sale to date as of 2014 [22].

Table 1. Key indicators in Zug, the Canton of Zug, and Switzerland. Source: [23, 24].

	Zug	Canton of Zug	Switzerland
Population 2019	30,618	127,642	8,606,033
Population change 2010–2019	16.3	12.9	9.4
Foreigners in % 2019	34.4	28.3	25.3
GDP per capita in CHF in 2018	n/a	160,884	84,518
Employment per sector in 2018 (in %)			
Primary	146 (0.3)	1,866 (1.6)	161,497 (3.1)
Secondary	6,293 (15.2)	22,815 (19.7)	1,091,626 (20.8)
Tertiary	34,936 (84.4)	91,111 (78.7)	3,996,835 (76.1)
Crypto sector (number of companies)			
Blockchain/crypto companies	–	433	877
Top 50 blockchain/crypto companies	27	35	53
Blockchain focused VC firms	4	4	13
Blockchain and Crypto Savvy law firms	6	7	16
Banks active in crypto	1	1	15

4.3 Constructing the Opportunity Space for Crypto Valley (2015–2018)

With the highly successful Ethereum token sales, the promotion of Crypto Valley by local entrepreneurs and Zug's economic promotion office, and the increasing market valuation of cryptos, many blockchain and crypto companies located themselves in Zug leading to many ICOs in 2016–2017. In 2017, Crypto Valley Venture Capital (CV VC), an early-stage venture capital investor established itself in Zug. CV VC also founded CV Labs, a private support organisation that offers co-working spaces and a three-month incubation program for crypto projects. These contributions helped structure the emerging crypto ecosystem. As of 2021, the CV Labs coworking space accommodates more than 130 blockchain projects [24]. In 2017, the Crypto Valley Association was founded to support the development and dissemination of crypto technologies, blockchain, and other distributed ledger technologies through conferences and meetups and by supporting startups and other companies in Zug.

In 2016, the town of Zug collaborated with Bitcoin Suisse to allow residents to pay taxes up to CHF 200 in bitcoin, becoming the first public institution in the world to do so [25]. Other institutions, such as the Commercial Register Office of the Canton of Zug and the University of Lucerne, followed the town's example. In 2017, the town of Zug offered its citizens the opportunity to get a blockchain-based digital identity. The public initiatives promoted by Mayor Dolfi Müller in collaboration with local crypto companies, namely Bitcoin Suisse, created media buzz and coverage on Zug and its Crypto Valley, raising the town's profile as crypto-friendly.

The rapidly growing Crypto Valley ecosystem was accompanied by increasing regulatory and legal clarifications at the national level regarding cryptos. In 2015, the Swiss financial market regulator (FINMA) issued several guidance documents intended to clarify when entrepreneurs need to comply with anti-money laundering and securities laws. In 2015, FINMA announced that bitcoin would be treated as a foreign currency, meaning that no new regulations were needed and that bitcoin transactions would be exempt from sales taxes. In 2017, the Blockchain Taskforce, a group of regulators and crypto entrepreneurs, was created to discuss regulations for ICO/tokens, banking, and cybersecurity. In 2018, FINMA issued the ICOs guidelines distinguishing between three categories of ICOs, namely, payment, utility, and assets ICOs. In 2018, the Swiss economics minister, Johann Schneider-Ammann, declared at the Crypto Finance Conference in St. Moritz that he wanted Switzerland "to be the crypto-nation in five or ten years" [26].

4.4 Crypto Valley (2019 -)

Crypto Valley is one of the most vibrant crypto ecosystems in the world. At the end of 2020, 433 of the 877 crypto companies in Switzerland (49.4%) were in Zug (see Table 1). Moreover, of the 141 crypto actors—banks, law firms, venture capital firms, blockchain service providers, and crypto companies—identified in the CV report, 65 are in Zug. Finally, 8 of the 11 unicorns—crypto startups with a market valuation at more than USD 1 billion—are in Zug [24]. As of 2019, the Information and Communication Technology (ICT) cluster is an important sector in the Canton of Zug–encompassing more than 1,400 companies and over 8% of the total workforce [27]. In 2021, the Canton of Zug allowed for taxes for individuals and companies up to CHF 100,000 to be paid using cryptocurrencies Bitcoin and Ether. Zug thus becomes the first Swiss canton in which taxes can be paid with cryptocurrencies.

The Crypto Valley ecosystem is further enhanced with higher education institutions providing human capital, education and training, and research. The Institute of Financial Services Zug (IFZ) competence centre for finance and the Lucerne School of Computer Science and Information Technology (HSLU) at the Zug-Rotkreuz campus offer bachelor's and master's degree programs, applied research and development, and continuing education programs in Computer Science, Information Technology and Business Information Technology.

In December 2017, following the Blockchain Taskforce, public and private actors—Federal government, the cantons, private sector, universities, and civil society—formed the Swiss Blockchain Federation (SBF) to lobby for political and regulatory reforms attractive to the crypto ecosystem [28]. The SBF is led by Heinz Tännler, Financial Director of the Canton of Zug. In December 2018, the Federal government presented the blockchain report, which led to the passage of the Blockchain Act. In September 2020, the Swiss parliament passed a law to update existing corporate and financial regulations for Distributed Ledger Technology (DLT). FINMA also issued two banking licenses to SEBA and Sygnum in August 2019, followed by the first license for an independent regulated crypto exchange in 2021 [29]. Founded in 2019, FiCAS, a crypto investment management boutique in Zug, began offering the world's first actively managed Exchange Traded Product (ETP) with the top 15 cryptocurrencies as an underlying asset class.

Although many crypto companies are incorporated as foundations in Switzerland, most have few employees. Bitcoin Suisse is the largest company in terms of employment, with 236 employees in Zug. The leading blockchain companies, Tezos or Cardano have around 10 employees in Zug while most of their employees are in other countries—namely United States or United Kingdom [24]. Moreover, the ICOs craze of 2017–2018 led to fraud and scandals such as the Zug-based cryptocurrency exchange platform Bitfinex in an $850 million fraud scandal [30].

5 Discussion and Conclusions

Trinity of Change Agency (TCA) played a decisive role in the emergence of Crypto Valley in Zug. Schumpeterian innovative entrepreneurs like Bitcoin Suisse and Ethereum created successful crypto companies in 2013 and 2014, and institutional entrepreneurs who created associations to promote blockchain/cryptos and designed the foundation model were decisive in the emergence of Crypto Valley in Zug. These agents of change were supported by a proactive, business-friendly, and open Cantonal administration to new business opportunities that, for instance, helped Ethereum founder, Vitalik Buterin, to have a Swiss residency permit in 2014. The actions taken by agents of change led to an informational cascade that attracted many crypto companies in Zug, thus cementing its reputation as crypto hub for crypto entrepreneurs. This also led to place-based leadership which formed a vision to transform Zug into Crypto Valley. They used bold public initiatives to actively promote crypto at the institutional level in Zug with Zug's municipality, canton and Zug's economic promotion office, the creation of multiple associations, and the creation of public-private institutional arrangements to lobby for regulatory and legal reform to promote crypto companies.

TCA were able to seize a window of locational opportunity (WLO) in crypto and blockchain technologies. The emergence of the disruptive blockchain and crypto technology was successfully identified by Schumpeterian innovative entrepreneurs and institutional entrepreneurs in the early stages—2013 to 2016—allowing to take critical actions to signal openness to the technology but also to provide legal and regulatory certainties for crypto companies. TCA allowed the further construction of the opportunity space, thus expanding expectations about the future. TCA promoted the actions to construct time-specific, region-specific, and agent-specific opportunity spaces. Zug's economic promotion office actively connects newly arrived crypto companies to the ecosystem. The result was the creation of a tightly networked cluster; in the words of one stakeholder, "Zug is a dense and small community where you get to know everyone extremely rapidly" (personal communication, 23 November 2021).

Institutional relatedness played a central role in the emergence and development of Crypto Valley in Zug. The Swiss institutional context, with its unique features of decentralisation, legal and regulatory stability, predictability, data and banking secrecy, financial centre, and a neutral political system, gave the stability for crypto and blockchain technology to thrive. The institutional context in Zug, with its known low-tax rates, business-friendly attitudes to offshore headquarters, and related knowledge-intensive support services—lawyers, bankers, accountants—were critical in attracting crypto companies. Moreover, the adoption of bilateral treaties, namely the automatic exchange of information on financial accounts (AEOI) according to the Multilateral Competent Authority Agreement on the Automatic Exchange of Financial Account Information (MCAA) leading to the end of banking secrecy that entered into force on 1 January 2017 motivated the Swiss authorities to promote financial diversification in Crypto technologies [31]. The Swiss regulatory and legal institutions were very proactive in providing a legal framework for crypto companies to operate in.

The article provides two main policy recommendations for national and regional policymakers. First, national policymakers should identify the national institutional relatedness during the emergence of a new technology to design early on policy actions to seize the WLO. Second, regional policymakers should prioritize the formation of a shared vision and place-based leadership in the early stage of emerging technology, and put in place policy actions that will attract media coverage and position early on the region as a centre of the new technology with the objective to seize a WLO. Future research could investigate the factors leading to divergence within the same national institutional during a WLO.

References

1. Grillitsch, M., Sotarauta, M.: Trinity of change agency, regional development paths and opportunity spaces. Prog. Hum. Geogr. **44**(4), 704–723 (2020)
2. Morisson, A., Mayer, H.: An agent of change against all odds? The case of Ledger in Vierzon, France. Local Econ. J. Local Econ. Policy Unit **36**(5), 430–447 (2021)
3. Battilana, J., Leca, B., Boxenbaum, E.: How actors change institutions: towards a theory of institutional entrepreneurship. Acad. Manag. Ann. **3**(1), 65–107 (2009)
4. Moulaert, F., Jessop, B., Mehmood, A.: Agency, structure, institutions, discourse (ASID) in urban and regional development. Int. J. Urban Sci. **20**(2), 167–187 (2016)
5. Boschma, R.A.: New industries and windows of locational opportunity. A long-term analysis of Belgium. Erdkunde **51**, 12–22 (1997)
6. Scott, A.J., Storper, M.: High technology industry and regional development. A theoretical critique and reconstruction. Int. Soc. Sci. J. **112**, 215–232 (1987)
7. European Blockchain Observatory and Forum: EU Blockchain Ecosystem Developments. European Commission, Brussels (2020)
8. Hassink, R., Isaksen, A., Trippl, M.: Towards a comprehensive understanding of new regional industrial path development. Reg. Stud. **53**(11), 1636–1645 (2019)
9. Boschma, R.A.: Relatedness as driver of regional diversification: a research agenda. Reg. Stud. **51**(3), 351–364 (2017)
10. Content, J., Frenken, K.: Related variety and economic development: a literature review. Eur. Plan. Stud. **24**(12), 2097–2112 (2016)
11. Beer, A., Clower, T.: Mobilizing leadership in cities and regions. Reg. Stud. Reg. Sci. **1**(1), 5–20 (2014)
12. Yin, R.K.: Case Study Research: Design and Methods. SAGE Publications, Thousand Oaks (2013)
13. Grillitsch, M., Rekers, J.V., Sotarauta, M.: Investigating agency: methodological and empirical challenges. In: Handbook on City and Regional Leadership. Edward Elgar Publishing, Cheltenham, UK (2021)
14. Stake, R.E.: Multiple Case Study Analysis. Guilford Press, New York (2013)
15. Hoppe, P.: Zug (Canton) (2019)

16. van Dorp, M., Rácz, K.: The Swiss Connection the Role of Switzerland in Shell's Corporate Structure and Tax Planning. Stichting Onderzoek Multinationale Ondernemingen (SOMO), Amsterdam (2014)
17. Canton of Zug: Welcome Zug: small world – big business. Department of Economic Affairs of the Canton of Zug, Zug (2021)
18. Regenass, R., Budry Carbó, A.: Zug – an offshore paradise for shell companies. Public Eye (2021)
19. Brunner, A.: Crypto Nation. Stämpfli Verlag AG, Bern (2019)
20. Federal Council: Bericht des Bundesrates zu virtuellen Währungen in Beantwortung der Postulate Schwaab (13.3687) und Weibel (13.4070) (2014)
21. Müller, L.: MME – The Law Firm Behind the Rise of Switzerland's CryptoValley. Epicenter.tv (2019)
22. Buterin, V.: Ether Sale: A Statistical Overview. Ethereum Foundation Blog (2014)
23. BFS: Portraits of the communes (2021)
24. CV VC: Top 50 Report H2/2020 The blockchain industry in Crypto Valley - Switzerland and Liechtenstein - analyzed and visualized. CV VC, Zug (2021)
25. Miller, H.: Welcome to Crypto Valley. Bloomberg Businessweek (2017)
26. Seele, P.: Let us not forget: crypto means secret. Cryptocurrencies as enabler of unethical and illegal business and the question of regulation. Humanist. Manage. J. 3(1), 133–139 (2018)
27. Canton of Zug: Zug: ICT cluster. Department of Economic Affairs of the Canton of Zug, Zug (2019)
28. Brunner, A.: Swiss Digital Asset and Wealth Management Report 2021. CV VC AG, Zug (2021)
29. Federal Department of Finance. Blockchain (2021)
30. Emmel, C., Pilet, F.: The dark side of Zug's Crypto Valley. Swissinfo (2019)
31. State Secretariat for International Finance SIF: Financial accounts (2017)

Using Digitalization to Boost Lucanian Agriculture

Maria Assunta D'Oronzio[1]([⊠]) [iD] and Carmela Sica[2] [iD]

[1] CREA, Research Centre for Agricultural Policies and Bioeconomy, Potenza, Italy
massuntadoronzio@crea.gov.it
[2] Freelance, 85100 Potenza, Italy

Abstract. The United Nations have entitled 2020–2030 as "the Decade of Action", a decade in which sustainable development objectives must be achieved. Digitalization is one of the 2030 Agenda goals for the primary sector, however, it is not as widespread as it should be in some Italian territories. Due to its geography, parts of inland rural Basilicata lack digital networks, blocking farmers, particularly more mature farmers, from using technology to manage agricultural activities.

Much remains to be done to bridge the gaps in connectivity, accessibility and digital skills, nonetheless, rural development regional policies and the national recovery and resilience plan appear to be moving in this direction. Digital agricultural strategies have been launched in the 2014–2020 programming period, contributing to the rationalization of productive resources, exploiting opportunities from the use of Innovative Technologies (including Information and Communication, ICT) generating new incomes and improving quality of life in rural areas.

This study is an update on the diffusion of digital technology in the Lucanian agri-food and forestry chains which was carried out by the European Partnerships (EIPs) Operational Groups (OGs) who, due to their multi-company characteristics and the adoption of "mature" innovative solutions suitable for the digitalization of the Lucanian agri-food industry 4.0.

Keywords: Digital tools · Sustainable development · Operational Group

1 Introduction

Europe is a leader in the development of digital agriculture [1] and Italy as a whole is evolving at great strides [2], so much so that in the four-year period, 2017–2020, the Agriculture 4.0 market went from 100 million to 540 million €, increasing by 270% in just one year (2017–2018) [3].

According to 2016 ISTAT data, only 19% of Italian agricultural enterprises used electronic tools with only 10% in Basilicata, characteristic of central-southern statistics [4]. Some recent studies have confirmed a growing interest by farmers in Precision Agriculture (PA). A 2021 study by the Milan Polytechnic University, Smart Agrifood Observatory found a 20% increase in expenditure by farmers for Agriculture 4.0 services,

F. Calabrò et al. (Eds.): NMP 2022, LNNS 482, pp. 924–932, 2022.
https://doi.org/10.1007/978-3-031-06825-6_88

despite the pandemic which contracted the first mid-year 2020 statistics, this is essentially due to the travel restrictions, while the second half showed a clear recovery made possible by the Covid19 tax incentives.

In 2020, the main digital solutions used in Italian agriculture were: farm management software (37%), monitoring and control systems for agricultural machinery and equipment (33%), crop and land mapping services and precision irrigation systems (both 27%), crop and land monitoring and control systems (17%), decision support systems and remote monitoring systems for farm infrastructure (both 15%) and variable rate distribution systems (10%) [5]. However, there are still parts of inland rural areas in Italy and Basilicata that suffer from under-funding in digital technologies, a weakness in the primary sector that must be addressed whilst also considering sustainability which is one of the strategies to achieve 2030 Agenda goals [6].

The recent Covid19 pandemic, turned out to be an opportunity for the digitalization of small local Lucanian producers who have managed to increase their visibility and experiment with new forms of digital marketing that have been favourable with consumers [7]. In fact, digital agriculture has the potential to deliver economic benefits through increased agricultural productivity, cost efficiency and market opportunities and social and cultural benefits through increased communication and inclusivity and environmental benefits through optimized resource use as well as adaptation to climate change [8].

It is equally true, however, that farmers must acquire new and improved knowledge on digital innovations that work in proximity, such as Internet of Things (IoT), or remotely (remote sensing, decision support system-DSS, ...) to face more complex decisions whilst conducting their activities (whether it is crops or livestock, markets choices or means of subsistence and the well-being of society) as well as to fill information and knowledge gaps in the sector. Farmers therefore need to change the way they access and use information.

Digital technology favours a production system known as "precision agriculture" which is supported by the increasing use of Innovative Information and Communication Technologies (ICT) required to manage the large amount of information available. As a result, ICT can be considered the key to innovation and change and essential for the economic growth of the sector [1].

Currently, digital agriculture solutions have mostly captured the attention of large agricultural entrepreneurs, producer associations and large distributors, however, they must also be advocated to smaller farms to bridge the information and production capacity gap between large and small businesses. This will help solve production and economic problems and support them in sustainable practices and specific and complex agricultural systems (*e.g.* organic farming). Experience and knowledge strongly influence farmer behaviour with some organic farms adopting technological innovations already in use in conventional agriculture, paving the way for profitability and efficiency [9].

Research institutions, advisors and farmers have developed various digital innovations in Italy. CREA-PB carried out a recent survey on 343 OGs providing agriculture digital information [10].

Partnerships who recognize the need for small farms to follow these strategies have been activated in Basilicata [11] where, until a few years ago, digitalization was only present in a few places characterized by a higher entrepreneurial capacity than some operating producers, associations in the olive growing, viticulture and fruit and vegetable sectors.

2 Materials

Lucanian farmers have used various funding sources to introduce technological innovations in their activities, including the European partnership for agricultural innovation (EIP-AGRI), as part of the Rural Development Program 2014–2020, the S3 strategy of the 2014–2020 Regional Operational Program and the "research and innovation - Horizon 2020 (H2020) program.

Many EIP-AGRI professional events and publications focus on digitalization and the Agricultural Knowledge and Innovation System.

The OGs, set up through the voluntary aggregation of public and private sector, researchers, scientists, advisors and agricultural-forestry operators (see Fig. 1), have a common goal: to increase productivity, using already mature innovations that involve a more rational use of production inputs. The OGs allowed for an increase in the sustainability of production processes from a technical, economic and environmental point of view.

Of the eleven OGs in Basilicata, no less than seven were precursors of digital innovations, aimed at contributing to the growth and development of the Lucanian agricultural-food and forestry sector. The remaining OGs can also be classified as digital as they present robotic-automation digitization processes regarding information collection systems, software and data analysis [3]. The RDP measures are sub-measures 16.1 - Support for the establishment and management of the Operational Groups of the EIP and 16.2 - Support for pilot projects and the development of new products, practices, processes and technologies - under Measure 16 - Cooperation of the 2014–2020 Basilicata RDP.

This study is the advanced stage of a study carried out between September 2019 and July 2020 [3] which analysed all the partnerships and the innovations transferred (16.1) and tested (16.2) to obtain an overall picture of the use of digital technology in Lucanian agriculture. The main objectives were to verify progress relating to the diffusion of digital technology in the Lucanian agri-food and forestry supply chains and the response of agricultural, agri-food and forestry companies. In fact, the progress of the OGs activities has varied over time and there have been changes to implementation timelines and methodologies due to the pandemic.

In-person and/or telephone interviews, participation in informative webinars and the administration of a questionnaire for the scientific project managers (and/or representatives of the main partners) were carried out to understand the degree of application of innovation in the farms involved in the partnership. Specifically, the questionnaire was based on twelve closed-ended questions, some of which investigated the degree of satisfaction by the farmers, their availability and interest in continuing to use digital and introduce further innovations even at the end of this project phase, the impact of innovation on production and in terms of environmental and economic sustainability.

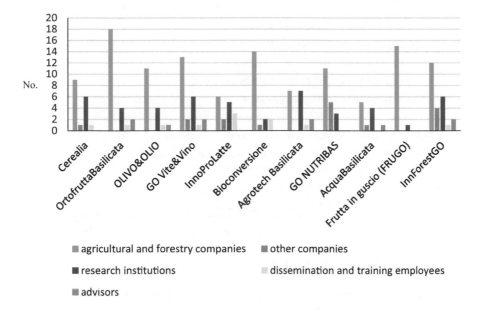

No.

■ agricultural and forestry companies   ■ other companies

■ research institutions   ▨ dissemination and training employees

■ advisors

Fig. 1. Partnership composition of each OG according to the number of partners by typology

3 Results and Discussion

Thanks to Basilicata OGs planning of innovation, agriculture and forest management is changing, moving towards more modern and technological standards whilst preserving the traditional nature of the locations and the production of typical products that characterize them.

The OGs managers declared that the farmers (single farm -family business or SME) or farm associations (consortia and cooperatives) (see Fig. 2) have actively taken part in the project phases. 20% of farmers are satisfied and 80% are sufficiently satisfied with the digital innovations.

The analysis of the partnerships revealed that many farms participated in the EIP partnership, in testing (sub-measure 16.2) and in other relevant European experiences. For example, the AcquaBasilicata OG is still testing the development of RDPs in other Regions as well as in national and European partnerships. The AcquaBasilicata OG maintains relations with the European Water Network (EIP-WATER, EIP-AGRI, ERIAFF, etc.) and participates in the LIFE AgroClimaWater (LIFE14 CCA/GR/000389) promoting water efficiency and supporting the shift towards a climate resilient agriculture in Mediterranean countries.

80% of the OGs interviewed said they had tested and introduced two digital innovations while 20% introduced more than two.

The digital innovations used in the agri-food supply chains are mostly related to Precision Agriculture (PA) suitable for guaranteeing agricultural businesses a reduction in production costs and improved environmental performance.

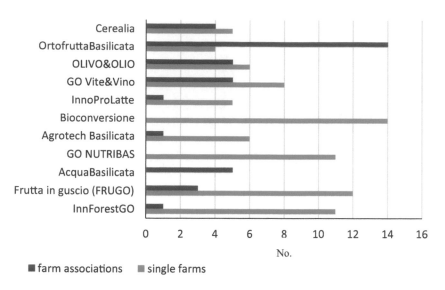

Fig. 2. Number of single farms and of farms associations with each OG

The three OGs using PA to be followed were Cerealia from the cereal sector, and the remaining OGs, AgrotechBasilicata and AcquaBasilicata, operating across several sectors, including fruit and vegetables, olive growing and viticulture.

Some farms, who are members of consortiums and cooperative partners from the Cerealia OG, have tested PA techniques and methodologies on wheat crops. Specifically, technologies have been put into place for the variable rate application of nitrogen fertilizers in experimental fields. The farmers, under the guidance of advisors and researchers, used drones to produce prescription maps, automatic GPS guidance systems and proximity sensors for the continuous identification of the nutritional potential of the soil.

The two transversal OGs operated PA practices for rational management at variable rates of fertilization and irrigation (AgrotechBasilicata) and irrigation (AcquaBasilicata) of the crops. Innovative sensors and systems are being used for the irrigation of the experimental fields, to control and monitor the soil-plant-atmosphere system which are allowing farmers to access soil moisture readings in real time. Sensors, put in several points, allow the activation of the irrigation system only where necessary, giving rise to more efficient usage, a reduction of the drained water and a lower consumption of water. Furthermore, if controlled remotely, the system allows for quicker responses compared to field inspections. All this translates into greater flexibility of field management and greater efficiency of time and water.

The InnoproLatte OG for the dairy livestock sector, has developed an innovative plant for the production of stuffed cheeses [12]. SMEs in this group were the most predisposed to the introduction of digital technology, using data sharing systems and sensors for process controls. Family farms and units belonging to consortia have shown an interest in digitalization, using systems, software and precision sensors to collect field data using automation and decision support systems to support the farmer/breeder.

Digital technological innovations using the olive ripening index of have been extremely useful in determining the best harvest times, ensuring the best compromise between potential yield and quality (high content of polyphenols and high organoleptic profiles) in the production of extra virgin olive oils of the same quality. Data is also being collected on different varieties of olives and best harvesting times which farmers can access via smartphones to compare the veraison state of the olives in his field with the digitized one (similar in variety and pedoclimatic area).

Forest-wood supply chain innovations include the continual updating of the KBS (Knowledge-Based System) Platform, by users (information bodies, forestry entrepreneurs and consultants) and scientific representatives (research bodies). The platform aims to implement an information database and an interactive desk to disseminate knowledge that can be of support to the stakeholders involved in forest management and the resulting supply chains, as well as to unite/economic subjects that are difficult to reach due to the lack of infrastructure.

All the interviewees stated that specific difficulties were not encountered in introducing the innovations, and initial difficulties were easily overcome thanks to practical and theoretical informative meetings or, as in the case of the wood supply chain, using the platform, representing the core elements of the OG methodology. Thanks to the platform, small regulatory problems posed by some chestnut growers have been resolved.

Satisfactory results have been achieved regarding the impact of digital innovation on the quality and quantity of the product as well as sustainability and profitability, verifying the impact on the socio-economic aspect (see Fig. 3).

Positive impacts in terms of environmental sustainability: the innovations introduced using PA techniques reduce surplus (including economic ones) as they allow for the dosage of production inputs in a timely manner, if and where necessary. The benefits of chemical fertilizers (such as pesticides) in agriculture is well known, however their harmful effects are also becoming apparent if applied incorrectly or used in excess (damage to the plant to which they are administered and pollution of groundwater and rivers through excess washout). Water, on the other hand, is mistakenly considered an abundant or even unlimited resource has unfortunately been used improperly, often excessively with huge waste, particularly in the agricultural sector. An increase in water pollution and the ongoing climate change (increased in rainy seasons, increased periods of drought, desertification processes, etc.) are all making water an increasingly scarce resource. The proposed innovations allow for interventions on the control and management of water consumption by limiting quantities to the plant needs. Moreover, the reduction of these production inputs translates into the reduction of both direct costs, for their purchase, and indirect, such as the costs incurred to repair the damage caused to the ecosystem.

The farmers, either individually or as part of an association, are an active part of the OGs and are testing and using the online communication channels and are contributing to economic sustainability. The hope is that these channels will be increasingly used both to ensure timely interventions and to support the direct sale and delivery of agricultural products to consumers, these actions are unfortunately still only the prerogative of a few Lucanian entrepreneurs (olive oil sectors and wine).

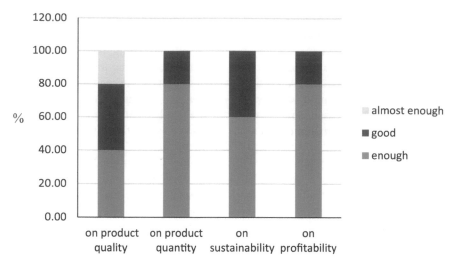

Fig. 3. Satisfaction survey: farmers' response to the use of digital technology

To date, no innovation has been shared with companies outside of the various partnerships, however, the interviewees believe that the companies are interested in continuing to use digital and introduce further innovations even at the end of this project phase.

The OGs have organized numerous webinars for the dissemination of partial results, recording an increase in the use of digitalized technology during the pandemic similar to other sectors such as commerce and education. An optimistic way forward which allows for more farmers to participate in training courses not only related to agroforestry issues but also to the daily use of digital technology.

In summary, precision agriculture is the expression of digitalized production, using satellites, drones and proximity sensors, etc. All Lucanian OGs can be classified as digital as they use information systems and software to analyse data.

4   Conclusion

The push towards digitalization [13] is highlighted in the main European strategies for the sector, such as the Green Deal [14], Farm to Fork [15] and by other various policies, in particular the "National Recovery and Resilience Plan" [16, 17] and the CAP National Strategic Plan 2023/2027 [18], which must work jointly, combining the various opportunities from the range of tools and resources available.

It is therefore quite clear that the new model of economic development is also linked to the diffusion of technologies [19] that have an impact on the production process and a high potential to help address the important and urgent economic problem which has also been impacted by Covid19.

The Lucanian agricultural sector continues to play a significant role within the regional economy, contributing 4.9%, as of 2016, to the formation of the total added value [20], however, it will still face many challenges and will need to establish a prominent position at national and/or international level. Digitalized technologies such as artificial intelligence, robotics, blockchain, High Performance Computing, Internet of Things and 5G have the potential to increase farm efficiency while improving the economy and environmental sustainability [21]. In fact, new technologies can, indeed, allow not only the monitoring of crops, but above all, their management according to some objectives defined by the 2030 Agenda. The greater use of digitalized technologies could finally attract a generation of younger farmers to commence agricultural activities and specialized rural businesses capable of producing both quality products and improve the quality of life in extra-urban areas.

References

1. VV.AA. Status of Digital Agriculture in 18 Countries of Europe and Central Asia. Published by ITU & FAO 2020. Geneva, Switzerland
2. Corbo, C., Pavesi, M.: Agricoltura 4.0: lo stato di adozione delle aziende agricole italiane, PianetaPSR numero 89 marzo (2020)
3. D'Oronzio, M.A., Sica, C.: Innovation in Basilicata agriculture: from tradition to digital Economia agro-alimentare. Food Econ. FrancoAngeli J. **23**(2), 1–18 (2021). https://doi.org/10.3280/ecag2-2021oa12210
4. Italian National Institute of Statistics (ISTAT). Updating of the 6° Agricultural Census (2010)
5. Valmori, I.: Smart agrifood: condivisione e informazione, gli ingredienti per l'innovazione. Live streaming event on the website www.osservatori.net. Accessed 5 Mar 2021
6. ONU Italia La nuova Agenda 2030 per lo Sviluppo Sostenibile (unric.org)
7. D'Oronzio, M.A., De Vivo, C.: COVID-19 and agri-food: the effects on the Lucanian agri-food sector and future challenges (in Italian). CREA (2021)
8. Trendov, N.M., Varas, S., Zeng, M.: Digital Technologies in Agriculture and Rural Areas. Briefing Paper of the Food and Agriculture Organization of the United Nations Rome (2019)
9. D'Oronzio, M.A., De Vivo, C.: Organic and conventional farms in the Basilicata region: a comparison of structural and economic variables using FADN data Economia agro-alimentare. Food Econ. FrancoAngeli J. (awaiting publication)
10. Bonfiglio, A., Carta, V.: Digitalizzazione in agricoltura: la trasformazione digitale passa attraverso i Gruppi Operativi, PianetaPSR n. 92 giugno (2020)
11. D'Oronzio, M.A., Costantini, G.: Knowledge agriculture systems in Basilicata, Southern Italy. In: New Metropolitan Perspectives 2020 - Knowledge Dynamics and Innovation-Driven Policies Towards Urban and Regional Transition, vol. 2, pp. 1552–1561 (2021). https://doi.org/10.1007/978-3-030-48279-4_145
12. D'Oronzio, M.A., Sica, C.: Basilicata, innovare per valorizzare le tipicità casearie lucane. PianetaPSR numero 94 settembre (2020)
13. Bacchetti, A., Corbo, C., Pavesi, M., Rizzi, F.M.: La corsa dell'innovazione digitale in agricoltura non si ferma. Pianeta PSR numero 101 (2021)
14. Deal, T.E.G.: Communication from the Commission 2019, 640 Final. EU Publications Office, Luxembourg (2019)
15. A Farm to Fork Strategy for a Fair, Healthy and Environmentally-Friendly Food System; Communication from the Commission 381 final. EU Publications Office, Luxembourg (2020)
16. Pierangeli, F.: Il Piano Nazionale di Ripresa e Resilienza. Quali opportunità per il settore primario? PianetaPSR numero 101 aprile (2021)
17. Pergamo, R.: Il piano della ripresa per l'agricoltura (2021). www.meridianoitalia.tv

18. Council of the European Union. Council General Approach on the Proposal for a Regulation of the European Parliament and of the Council Establishing Rules on Support for Strategic Plans to be Drawn up by Member States under the Common Agricultural Policy (CAP Strategic Plans), 12148/1/20 REV 1, Bruxelles, 14 December (2020). https://data.consilium.europa.eu/doc/document/ST-12148-2020-REV-1/en/pdf

19. Parviainen, P., Tihinen, M., Kääriäinen, J., Teppola, S.: Tackling the digitalization challenge: how to benefit from digitalization in practice. Int. J. Inf. Syst. Proj. Manag. 5(1), 63–77 (2017)

20. Documento di Economia e Finanza Regionale 2019–2021 - Nucleo Regionale Valutazione e Verifica Investimenti Pubblici (NRVVIP). https://www.regione.basilicata.it/giunta/files/docs/DOCUMENT_FILE_3055237.pdf

21. Commissione Europea. Shaping Europe's Digital Future. Publications Office of the European Union, Bruxelles (2020)

Printed by Printforce, the Netherlands